Bruce A. Bohm
Tod F. Stuessy

Flavonoids of the Sunflower Family (Asteraceae)

SpringerWienNewYork

Prof. Dr. Bruce A. Bohm
Botany Department, University of British Columbia, Vancouver, Canada

Univ.-Prof. Dr. Tod F. Stuessy
Department of Higher Plant Systematics and Evolution, Botanical Garden, Institute of Botany, University of Vienna, Vienna, Austria

Printing was supported by the Fonds zur Förderung der wissenschaftlichen Forschung (FWF)/
Austrian Science Fund

Printed in Austria

Typesetting: Composition & Design Services, Minsk, Belarus
Printing and Binding: Druckerei Theiss GmbH, A-9400 Wolfsberg

Printed on acid-free and chlorine-free bleached paper
SPIN: 10758972

With 56 Figures

CIP-data applied for

ISBN 3-211-83479-6 Springer-Verlag Wien New York

This book is dedicated to:

Arthur Cronquist (1919–1992)
Outstanding plant systematist and
major contributor to understanding
taxonomic relationships within
Asteraceae

and

Ralph E. Alston (1925–1967)
Visionary natural products chemist
and geneticist who pioneered the use
of chemical data in studies of plant
systematics and evolution

Preface

Certainly many of the readers of this book will remember the early developments of chemosystematics in the late 1950s. We ourselves remember the excitement of these new data bearing on plant interrelationships. The hopes were high, the techniques were rapidly evolving, the quantities of data were large, and young workers were eagerly enthusiastic. Some even went so far as to predict that chemical data would soon replace morphological information as the basis for plant classification.

These overly zealous predictions did not materialize, of course, as the history of plant systematics has amply shown during the past two centuries. If there is any distinct characteristic of systematics it is synthesis, as Lincoln Constance (1964) called it, the "unending synthesis." New tools generate new data and provide insights on additional dimensions of plant relationships. This will never change—we will continue to discover new tools and new data in the years ahead, and they will continue to be incorporated into the predictive general reference system of classification.

The earlier efforts in chemosystematics focused on secondary plant products: alkaloids, betacyanins, carbohydrates, cyanogenic glycosides, glucosinolates, lipids, terpenoids, and especially flavonoids. The latter were particularly well suited for chemosystematic investigation for several reasons: ease of isolation and characterization, small amounts of plant material needed for analysis, stability of compounds especially through routine preparation of herbarium specimens, and low cost to obtain useful information. As a result of these considerable advantages, literally thousands of studies on use of flavonoids in plant systematics have been published. Although now with the present zest for macromolecular data from DNA restriction sites and sequences, there are fewer workers and laboratories dedicated to flavonoid chemosystematic studies; twenty years ago they were the new currency of exciting data in plant systematics.

Because of many workers historically interested in the sunflower family, Asteraceae (or Compositae), in part due to its large size (approximately 23,000 species; Bremer, 1994), and because hundreds of flavonoid compounds were discovered in this family, many chemosystematic investigations have been completed on various taxa. Many studies came from the laboratories of R. A. Alston, B. L. Turner, and T. J. Mabry in the Department of Botany of the University of Texas, which had a long tradition of interest in systematics of the family. Other active laboratories were those of Dan Crawford, Jeffrey Harborne, Werner Herz, and Eckhard Wollenweber, among others.

Through all these flavonoid chemosystematic investigations on Asteraceae, considerable was learned about patterns of variation of these secondary products at different taxonomic levels. Most workers focused on the specific and infraspecific levels, with fewer studies addressing correlations with other data at intergeneric levels and above. In fact, literally thousands of papers dealt with the lower levels of the hierarchy in the family, and very few considered higher level relationships. Turner (1977b, p. 1116) himself remarked: "Too little is known at present about the distribution of secondary compounds across the whole of the Compositae to say with certainty that the absence or presence of a given compound, or a group of compounds, indicates any definite relationship."

Recent macromolecular studies by Robert Jansen and collaborators (e.g., Jansen and Palmer a, b, 1987; Jansen et al., 1990, 1991a, 1992; Jansen and Kim, 1996) and many others on higher level systematics of the family, have provided new and meaningful insights on relationships among subfamilies, tribes, and subtribes. For the first time, we have glimpsed that Barnadesioideae, an enigmatic group of nine genera long nestled within tribe Mutisieae, are evolutionarily basal for the family (Jansen and Palmer, 1987a, b). We have also confirmed that the rest of the tribes group reasonably well into two additional subfamilies, Asteroideae and Cichorioideae (Bremer, 1994; Jansen and Kim, 1996).

In view of these new perspectives on higher level classification of Asteraceae, it seems appropriate to review the extensive data on flavonoids for purposes of seeking correlations with these new insights. That is, now armed with new phylogenies of higher level relationships within the family, how well and in what ways do flavonoids correlate with the macromolecular data, and what new insights can they offer? Further, there are now new opportunities to address the question of evolution of flavonoid biosynthesis within the family using the newly generated macromolecular phylogenies as guides.

Basically, then, this book brings together for the first time the majority of the information available on the occurrence of flavonoids in Asteraceae in a comprehensive synthesis that seeks correlations with existing taxonomic structure at all levels of the hierarchy. The new macromolecular studies provide a stimulating framework for considering phylogenetic questions, and the new useful taxonomic synthesis by Bremer (1994) offers the chance to seek correlations with the most recent morphological conspectus of the family. The two new volumes on Compositae emanating from the 1994 Kew conference (Hind, 1996) also provide a wealth of new ideas on relationships within selected groups.

A note on the nomenclature used in this book is in order. For the most part, we have employed names of subfamilies, tribes, and subtribes following literature and dates in Solbrig (1963). Hence, Cardueae Cass. (1819) has priority over Cynareae Less. (1830; also *nom. illegit.*), Lactuceae Cass. (1819) has priority over Cichorieae Dumortier (1829), and Cichorioideae W. D. J. Koch [1837; ICBN, Greuter et al. (2000)] has priority over Lactucoideae Solbrig (1963). DeWolf's (1995) criticism that Cichorieae instead of Lactuceae should be used within subfamily Cichorioideae is incorrect; there is no such requirement in subfamilies that do not contain the type genus of the family [ICBN, Article 19.4 Note 1 and Ex. 4; (Greuter et al., 2000)]; it does apply, however, in Asteroideae (Asteraceae, Asteroideae, Astereae, Asterinae, *Aster*). This same nomenclature we have used has also been employed by Bremer (1994), against which classification the flavonoid data in this present book are compared.

Recently, however, Reveal (1997) has found earlier places of publication for many suprageneric names in Asteraceae, which would alter the names adopted herein and elsewhere. In particular, Cynareae Lam. & DC. (1806) would have prioritiy over Cardueae Cass. (1819), Cichorieae Lam. & DC. (1806) over Lactuceae Cass. (1819). We do not take nomenclatural action on these new dates in this book, however, for several reasons: (1) it remains to be seen if other references will be located that might alter still further the use of these tribal names (as they are not indexed in any one convenient place, it is somewhat a matter of chance for them to be located); (2) we embrace the *general spirit* of caution in not changing names unnecessarily as exemplified by the Names in Current Use initiative (e.g., Hawksworth 1991, 1994; Greuter et al., 1993), even though we have *serious reservations* about applying these concepts in practice, especially at the specific level; and (3) most importantly, we wish to maintain name correspondence with current literature on Asteraceae both in the many separate published papers (e.g., Jansen and Kim, 1996) as well as in Bremer's (1994) book.

We believe strongly that the most efficaceous use of flavonoids, indeed of all classes of secondary compounds, will be greatly enhanced when we learn better what they actually are doing in the plants they inhabit so consistently. This enlarged ecological understanding will do much to help interpret evolutionary homologies and also to place the diversity of compounds in adaptive context. Much has been written about secondary products and their roles as insect and herbivore feeding deterrants; physiological roles also have been suggested (see Chapter 6). It will take decades before we have a full understanding of the importance of flavonoids in the ecosystems in which they occur (virtually everywhere and in nearly all plant groups). Renewed interest in biodiversity prospecting, often focusing on natural products from tropical forests (e.g., Seidl et al., 1995; Miller, 1996) will stimulate these discoveries.

In the meantime, this book brings together a large percentage of the abundant and scattered literature on flavonoids in one of the largest families of flowering plants. The results show surprising relationships in some cases, offering new hypotheses in certain groups, and they also correlate well in other instances with new morphological and macromolecular data.

We applaud the new sources of DNA data in plant systematics and their use in the sunflower family. We look forward to new insights as more and more taxa are investigated at different levels. At the same time we demonstrate clearly in this book that flavonoids are still an exciting source of systematic information in Asteraceae, and we sketch numerous areas for further investigation. Systematics will be best served by continued aggressive investigations of all types of data at all levels of the taxonomic hierarchy. This book will hopefully remind us of the power and importance of flavonoids in Asteraceae and encourage continued studies into the future.

Bruce A. Bohm
Tod F. Stuessy

Contents

Acknowledgements

Many persons have helped bring this large work to successful completion. We especially thank the staff of Springer-Verlag (Vienna), in particular Dr. Sigrid Ofner and Mag. Elisabeth Bohunovsky, who have been extremely helpful in all phases of development and production of this book.

Continuous support from granting agencies to the laboratories of both authors has permitted orderly progress in the preparation of this book, although grants were not received specifically for writing it. Such helpful background funding to BAB came from the Natural Sciences and Engineering Research Council of Canada and to TFS from the National Science Foundation (U.S.A.) and National Geographic Society. A subvention to assure publication of the book has been generously provided by the Fonds zur Förderung der wissenschaftlichen Forschung (FWF; D3073-BIO).

We thank our wives: Lesley Bohm, for a thousand little reasons; and Patricia Stuessy, for encouraging support of this academic passion.

We appreciate the help of other individuals who have contributed in different ways: Jesse Cantley for drawing Figs. 1–1 through 1–3; Heilman and Fowler, botanical photographers, for the cover photo; Peter Stevens for sending a crisp reproduction of the original plate from Cassini that serves as Fig. 2–1; Kåre Bremer for permission to republish cladograms presented in Figs. 2–3 and 2–4; Robert Jansen for permission to use his cladograms illustrated in Figs. 2–5 through 2–8; the Board of Trustees of the Royal Botanic Gardens, Kew for permission to republish Figs. 2–4 through 2–8 from the 1996 Kew Compositae Symposium, Systematics volume; J. Muona, current editor of the journal Cladistics, for permission to republish Fig. 2–3; Beatrice Beck, Librarian of the Rancho Santa Ana Botanic Garden, for help in locating and copying needed taxonomic literature; additional taxonomic and chemical information for members of Mutisieae, *Tithonia*, and *Viguiera*, supplied by Dan Crawford, John La Duke, and Edward Schilling, respectively; Lara Menon and Maria Morales for library work and xeroxing; Hanna Weiss for checking all text citations for consistency against the references; Karin Vetschera for carefully reading Chapter 7; and our many graduate students and post-doctoral fellows who, over the years, have contributed to the chemical database presented in this book.

Section I

Introduction to the Sunflower Family

Daisies, chrysanthemums, dahlias, ageratum, lettuce, coreopsis, chicory, endive, sunflowers, ragweed, and thistles all belong to the sunflower family (Asteraceae). The more than 23,000 species that comprise the family reveal a remarkable morphological diversity, including trees to 30 m tall and small herbs barely 1 cm high. Flowering heads, or capitula, occur in a bewildering array of sizes, shapes, and colors, from minute structures to the giant heads of the cultivated sunflower. Sexual conditions vary with flowers (or florets) being hermaphroditic, female, functionally male, or neuter (with no sexual parts; contributing to reproductive success only through attraction of pollinators).

Distributionally, Asteraceae are found on every continent except Antarctica. They abound in temperate regions of the world, or in tropical regions in cooler montane habitats. They are not common dwellers of hot lowland tropical rain forests. Sometimes they occur solitarily, occupying small patches of forests or rocky cliffs. At other times they fill vast areas with color such as demonstrated by *Lasthenia*, the "goldfields" of central California in the spring or by the common goldenrods (*Solidago*) in the autumn in many parts of North America.

The function of Part One of the book is to introduce the reader to the family in a general way so that the material on flavonoids can be interpreted within an understandable biological and systematic context. This section is especially appropriate for flavonoid chemists who seek a basic overview of the family. Chapter 1 deals with the biology, distribution, and classification of Asteraceae, giving details of morphology, cytology, and reproductive biology as well as distributional comments and outlines of classification for subfamilies and tribes. Chapter 2 presents different evolutionary and biogeographic viewpoints that have been proposed within the family, focusing on the higher levels from subtribe through subfamily, and to related families. A strong historical emphasis is given here in the belief that this will enrich understanding of current hypotheses of classification and evolution.

CHAPTER 1

Biology and Distribution

This chapter is designed to teach about biology and distribution of Asteraceae (or Compositae*) in a broad sense so that correlations drawn with flavonoid chemistry in other chapters of the book will be more meaningful. This is a difficult task; so much material on the family exists that it is impossible to present successfully (and succinctly!) all of it on structural, biological, and spatial dimensions in just one short chapter. We have selected information concerning the family that seems pertinent, interesting, and helpful for understanding relationships at different levels of the taxonomic hierarchy.

Biology

The most commonly used data in systematics of Asteraceae are morphology, reproductive biology, and cytology. Other significant data also have been used, such as anatomy, cytogenetics, genetics, macromolecular information, or other classes of secondary plant products including polyacetylenes, diterpenes, or sesquiterpene lactones. We have chosen to discuss, therefore, those sets of data that we judge most appropriate in the context of this book. We also comment briefly on modes of speciation.

---

* The International Code of Botanical Nomenclature (Greuter et al., 2000) allows two acceptable names for eight plant families, including Asteraceae. The name Compositae is a very old descriptive term referring to the head inflorescence. Although it is the oldest name, it is not formed correctly with -aceae ending. The former *is* constructed properly from a genus in the family and bears the acceptable ending, but it is not the oldest name. Hence, Asteraceae has been conserved, whereas the older name, Compositae, is sanctioned as an alternative (*nom. alt.*). Both are regarded as acceptable. We have elected to follow the modern trend in this book and have used Asteraceae throughout.

Morphology

Morphology is obviously the basis for classification of Asteracaee, as well as for that of most other higher organisms. The external form and aspect of a member of the sunflower family is the way we interact with it, and the way in which other life forms also interact with it, such as insects, mammalian herbivores, etc. It is also the shape and form of the plant that interacts with the physical environment with regard to temperature, wind, water relations, architectural support, etc. To understand Asteraceae, we must understand their morphology.

Vegetative features – Vegetative features in the family vary considerably. Some members are large trees, such as the monstrous *Vernonia arborea* in Malaysia or *Dasyphyllum excelsum* from Chile that reach to more than 30 m tall and have trunks to more than 1 m in diameter. There are a variety of shrubs, too, occurring on nearly every continent, such as the well-known rabbit brush (*Chrysothamnus*) of the western United States. Some are well armed, such as seen in many taxa of subfamily Barnadesioideae, as in *Chuquiraga* or *Barnadesia*. Some are rosette-trees, standing like soldiers against the stark landscape of African mountains (*Dendrosenecio*) or enswirled with mist in Colombian páramos (*Espeletia*). Many others form bizarre candelabra-branched shapes on oceanic islands as in the Juan Fernandez archipelago (*Dendroseris*) or Hawaii (*Argyroxiphium* and *Wilkesia*). But, most of the members of the family are perennial, or less often annual, herbs inhabiting numerous ecological zones, although with a preference toward temperate regions. Some are typical upright sunflowers, standing 1–3 m tall, whereas others are of reduced size down to being flat on the ground as in *Hypochaeris sessiliflora* of the Ecuadorean alpine zone. One of the smallest genera is *Mnioides* of the Peruvian Andes, which measures only several cm tall. There are even a few aquatic Asteraceae (Hutchinson, 1916).

As with habit, leaves vary widely within the family. Some leaves are very large, e.g., nearly 1 m long in *Clibadium grandifolium* (Heliantheae) from Costa Rica and Panama. Others are small, being almost nonexistant as in the case of *Baccharis sagittata* from Chile, in which the stems are green, photosynthetic, and conspicuously winged, and have taken over the function of leaves. Some leaves are very spiny, as anyone knows who has attempted to weed out members of the thistle tribe from their gardens (e.g., *Carduus* spp.). Other leaves are covered with varying degrees of indumentum, hairs of all densities, lengths, and colors. Many species of the well-known genus *Artemisia* are attractive for their interesting hairy leaf surfaces, making desirable horticultural plantings.

Reproductive features – Despite the considerable morphological variation encountered in vegetative features in Asteraceae, the variety of form of

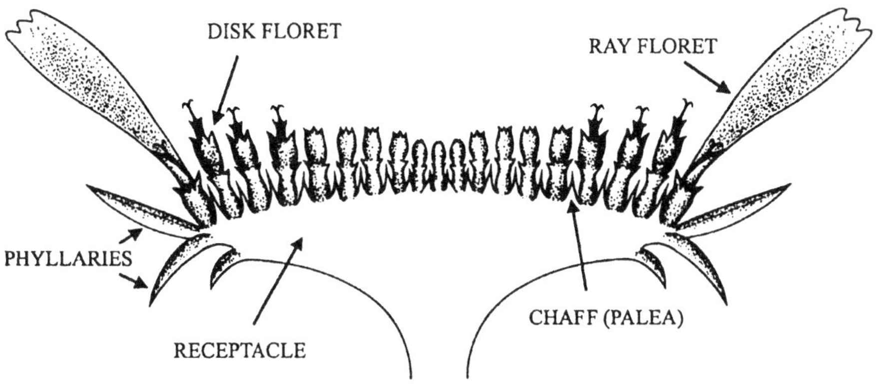

Fig. 1–1. Longitudinal-section of a head inflorescence (capitulum) of a typical member of Asteraceae showing basic morphological structures

reproductive aspects is even greater. This follows the general rule within flowering plants that speciation deals with closely related taxa that become reproductively isolated. Such isolation often becomes effective through modification of reproductive structures, or at the very least, these changes occur over time owing to differential selection after initial isolation (which may have been purely geographical). To unravel evolutionary relationships among members of the family, therefore, usually involves a close scrutiny of comparative reproductive features.

The basic reproductive unit of Asteraceae is the head or capitulum (Fig. 1–1). This is a cluster of flowers, or an inflorescence. The term "Compositae" refers to a composite inflorescence in which the flowers (or florets, a term simply referring to small flowers) are arranged sessile (i.e., unstalked) on the point of attachment (to the receptacle). Heads of Asteraceae are remarkable in variation of their stuctures, and these modifications often have taxonomic importance at different levels in the hierarchy.

One aspect of variation in heads is their arrangement into higher levels of aggregation, or capitulescences (= synflorescences of the polytelic type; Weberling, 1989). Sometimes heads occur solitary or with just a few heads, as may be seen in *Dahlia* or even *Helianthus*. More often, however, the heads occur in groups that themselves resemble inflorescences of single flowers. Heads may be aggregated in spikes (*Liatris*), racemes, corymbs, and even secondarily clustered into higher-level heads (as in *Lagascea*; Stuessy, 1978; Harris, 1994), in a sense repeating the cycle of aggregation of flowers into heads, perhaps in response to adaptation for increased targets for pollinators. The number of heads arranged into capitulescences in one individual can vary tremendously from just one to literally more than 250,000 in some of the common goldenrods (*Solidago*; Stuessy et al., 1986).

Surrounding the head is a series of involucral bracts, or phyllaries. If one hypothesizes that the head represents a condensation of a more open inflorescence, such as a raceme at the base of which are often a series of involucral bracts, then it is easy to comprehend the origin of the phyllaries in the head. These may occur in a single series or in many series, may be separate or fused (such as the single fused row in *Senecio* and relatives, which is special enough to be provided a separate term, a calyculus), may have hairs or glands, be fringed or spiny, and so on. There is almost no end to the patterns of variation in phyllaries, which provide much useful taxonomic information.

Another type of bract is found in some, but not all, members of the family. This is the chaff (also called receptacular bract, or palea). If we return to the hypothesis of origin of head from racemose ancestry, in addition to collective bracts found at the base of the inflorescence, each individual flower also is subtended by a small bract. With evolutionary condensation, therefore, each floret in the Asteraceae head would then be subtended by its own bract or palea (pl. paleae). When present, these paleae also vary in size, shape, and ornamentation, which can have strong taxonomic value, as well as protective functions (Stuessy and Spooner, 1988).

The point of attachment of florets in the head is the receptacle. This can be smooth with scarcely any notice of attachment of florets after they have matured and fallen, or, in other instances, it can be honeycombed, with the florets nestled into each little cubicle. The receptacle can also have hairs, or enations, that encircle the florets, a trait often seen in the thistle tribe (Cardueae).

Even more important diagnostic features of the Asteraceae head relate to flowers and their specific reproductive structures, namely, corolla, stamens, style, ovary, and pappus (Fig. 1–2). The most conspicuous part of the floret is the corolla (Fig. 1–3). The basic corolla type in the family is tubular and five-lobed (Fig. 1–3A), from which the others presumably have been derived evolutionarily (e.g., Koch, 1930a,b; Jeffrey, 1977). The bilabiate corolla (Fig. 1–3B), so prevalent in Mutisieae, is essentially a modified 5-lobed tubular corolla with two of the lobes significantly reduced, and sometimes coiled outward. The radiate corolla (Fig. 1–3C) is different from the tubular types in that three of the lobes have been expanded outward, and two on the inside have been suppressed. This type of corolla is found only on the outside of radiate heads (i.e., those with tubular corollas in the center and radiate corollas on the perimeter). Florets that have radiate corollas may be either only carpellate (with style and bifid stigmas but without stamens) or neuter (with no male or female parts). The final type of corolla is the ligulate one (Fig. 1–3D) in which all five corolla lobes have been modified and flattened on one side of the corolla. In contrast to radiate

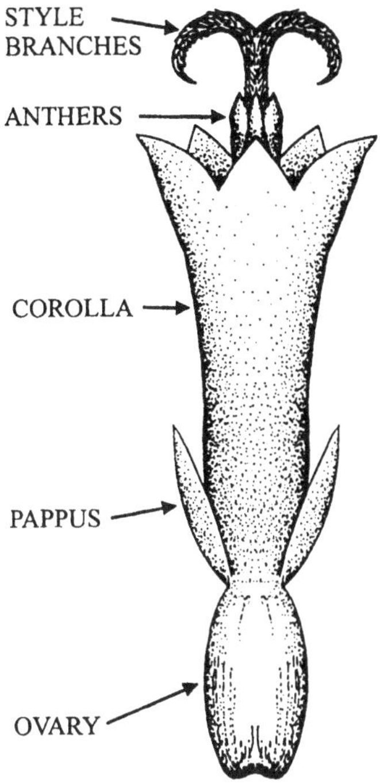

Fig. 1–2. Representative flower (floret) of Asteraceae showing reproductive (and associated) structures

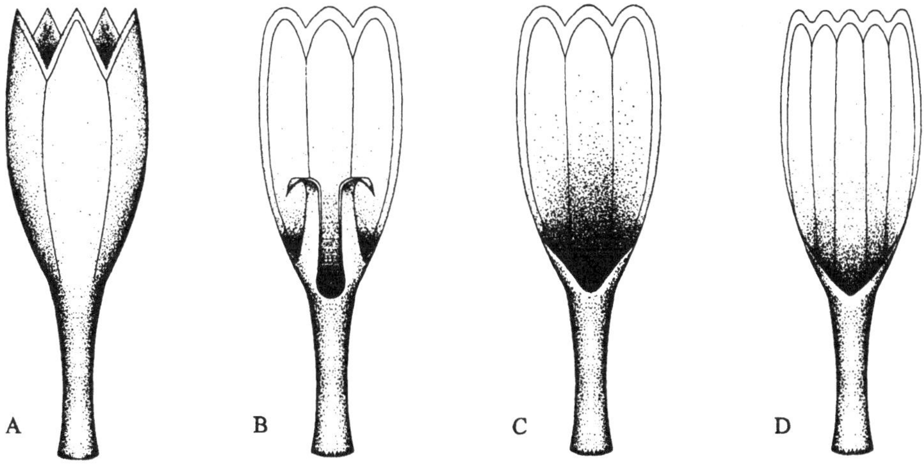

Fig. 1–3. Basic corolla types in Asteraceae: A, tubular; B, bilabiate; C, radiate; D, ligulate

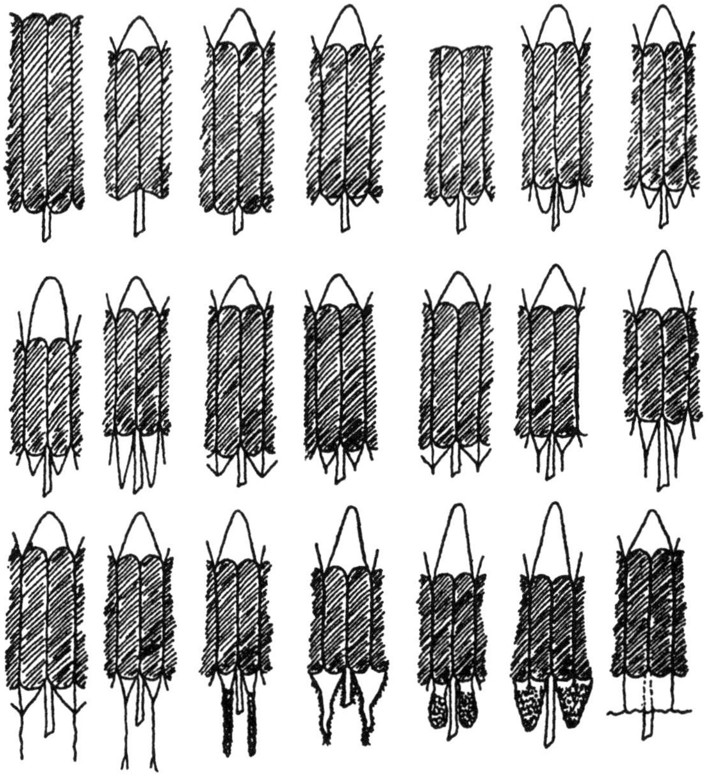

Fig. 1–4. Representative types of stamens in Asteraceae. From Small (1919)

corollas, ligulate corollas always have both male and female parts associated (i.e., the florets are hermaphroditic), and taxonomically they are mostly restricted to tribe Lactuceae. Numerous modifications of these four basic corolla types occur in the family.

Presence of one or two types of corollas in the same head leads to different structural types of heads. If all florets have tubular corollas, the heads are said to be discoid. If they are all ligulate corollas, then they are likewise called ligulate heads. When heads have exterior florets with radiate corollas and inner florets with tubular (discoid) corollas, the heads are termed radiate. In a very broad sense, these distinctions correlate with the two major subfamilies: Asteroideae, which are largely radiate; and Cichorioideae, which are often discoid (and ligulate).

Stamens offer a bewildering array of morphological variations involving apical appendage, basal appendage, and nature of cells of the thecae (Fig. 1–4). These variations have been extensively used taxonomically throughout the history of classification of the family. The apical appendage

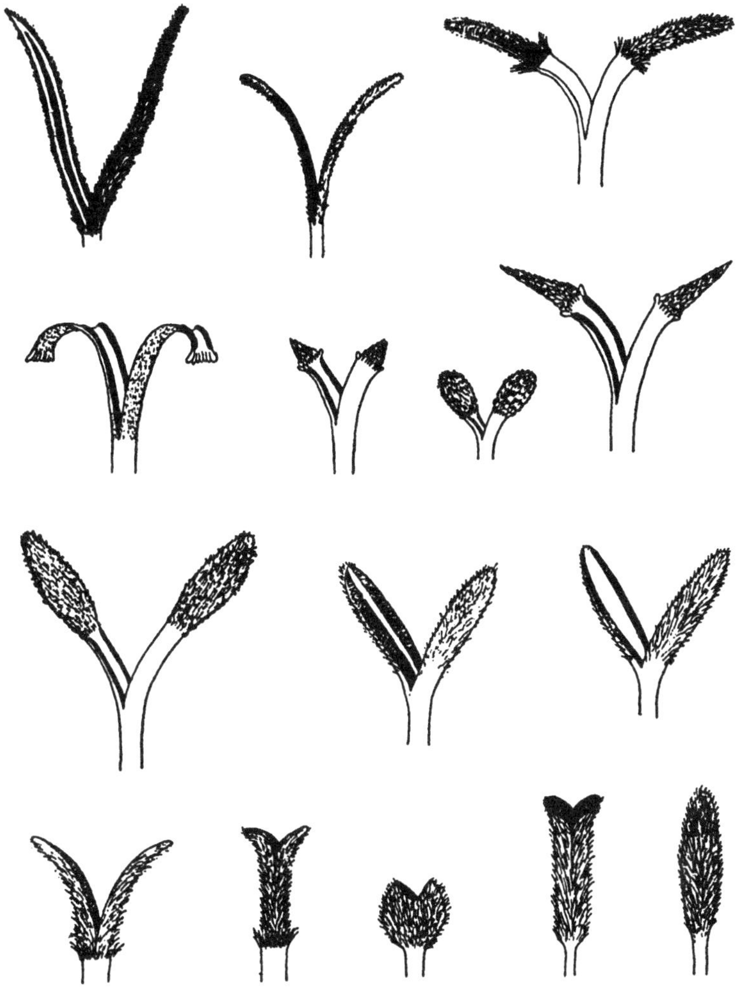

Fig. 1–5. Representative stylar variations in Asteraceae. From Small (1919)

varies from being absent (as in Piqueriinae of Eupatorieae) to greatly expanded in many genera. Likewise, the basal portions of the thecae may be essentially unmodified, whereas in other cases they may be tailed with all manner of fantastic variations. What these features actually do functionally has yet to be determined, although they may relate to protection of pollen from unwanted moisture and insects. The microscopic differences in thecal cells may relate to patterns of dehiscence; this likewise has been little studied from a functional standpoint (see Dormer, 1962, for some suggestions). It is well known that stamens can show irritability, and recent studies have shown that these rapid movements relate to elasticity of the stretched cuticle of the filaments and rapid changes in turgor pressure (Pesacreta et al.,

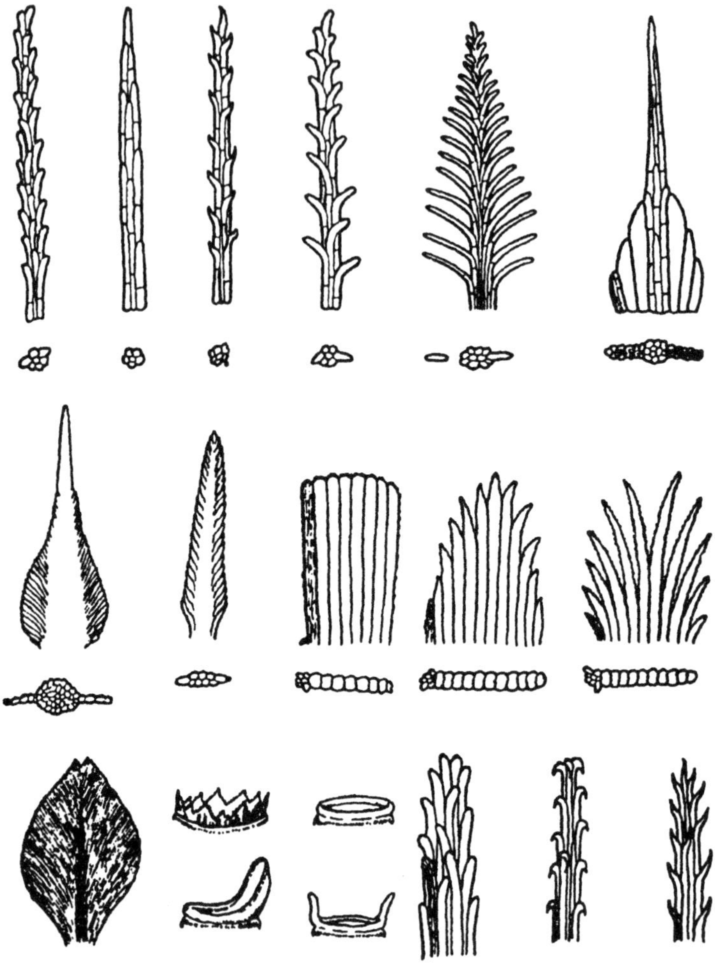

Fig. 1–6. Variation in forms of pappus in Asteraceae. From Small (1919)

1991; Hasenstein et al., 1993). Another feature of anthers that bears mention is the basal portion of the connective, which extends downward from between the two thecae. This "article anthérique" (Cassini, 1814), "anthéropode" (Beauverd, 1910; Briquet, 1917, 1918), "collar" (King and Robinson, 1977), or "connective base" (Pesacreta et al., 1993; Pesacreta and Stuessy, 1996), does show taxonomically important variations. It also may function in facilitating movement of anthers as part of dehiscence (Thiele, 1988).

As with male reproductive structures of florets, the female features also reveal considerable variation. The basic female reproductive unit of the sunflower family consists of two fused carpels in one inferior ovary, but with only a single ovary (and resultant seed). Attached to the ovary is a single style with usually two bifid style branches. These occur in a wide pattern of variation (Fig. 1–5). As with stamens, features of the styles including length, and shape, plus variations in placement and length of papillae of stigmatic surfaces, have import at tribal, subtribal, and lower levels in the hierarchy. These differences also have obvious functional significance in some instances (as suggested by Robinson, 1984), but we are still largely ignorant of the biological role of these numerous variations.

Young ovaries are usually not encountered in routine taxonomic work for the simple reason that we tend to study mature flowering heads and florets that have the male and female parts well developed. Hence, little has been done in comparing features of ovaries. Rather, most attention has been placed on the mature achenes (or cypselas), that ripen from the fertilized ovaries. The term cypsela is used preferentially by some workers (e.g., Spjut, 1994) because, strictly speaking, an achene is a single-seeded, dry fruit that comes from a single superior ovary. In Asteraceae, the fruit is one-seeded, but presumably by reduction because of the obvious two-carpellate nature of the styles, and the ovary is inferior. This is quite different from a more typical achene in the angiosperms, such as found in *Ranunculus* (Ranunculaceae). Elaborations of the achenial wall in Asteraceae are numerous involving sculpturing, coloration, and with the presence of wings, hairs, spines, etc.

Occurring on top of the inferior ovary is a structure that has been interpreted as the calyx, but here in Asteraceae it is so significantly different in appearance that it is called by a special term: the pappus (Fig. 1–6). The pappus has been hypothesized to have a defensive function in the young head (Stuessy and Garver, 1996), subsequently assuming in many cases a dispersal function when the achenes are mature (as seen in the common dandelion, *Taraxacum officinale*). The variations in this part of the floret are equally striking, ranging from being absent to existing as crowns, scales, awns, or long fluffy bristles.

Reproductive Biology

The many reproductive morphological variations sketched and shown in the figures above all combine to yield a complex reproductive biology for the family. In fact, it is this complexity that may well be the reason for the astounding evolutionary success of the family yielding more than 23,000 species! The possible combinations of morphological variations are virtually limitless.

Different combinations of male and female parts in florets and different arrangements of florets in heads yield varying gender strategies for members of the family. The basic condition for florets is hermaphroditic, i.e., with both stamens and carpels. Typical discoid heads are often hermaphroditic, with all tubular florets possessing both functional male and female parts. Heads with all ligulate florets, which also have both functioning stamens and carpels, are also hermaphroditic in gender condition, even though the morphology of the individual flowers is quite different. A combination of carpellate ray florets with hermaphroditic disc florets results in a gynomonecious condition (i.e., with female and hermaphroditic flowers on the same plant). If the rays are neuter, as happens in the common sunflower, *Helianthus*, the central florets are hermaphroditic but the head is radiate rather than discoid. Sometimes in a radiate head with carpellate florets, the disc florets lose the female structures, yielding functionally male florets. This is common in subtribe Melampodiinae of Heliantheae. This is a simple monecious condition: female flowers on the outside of the head, and male flowers on the inside. And, completely unisexual heads occur too, often distinguishable morphologically, such as in *Robinsonia* of the Juan Fernandez Islands, in which the male heads are larger than the carpellate ones, and each is found on separate plants (true dioecy). The pussytoes, *Antennaria*, provide another example of this separation of sexes. The pernicious ragweed, *Ambrosia*, provides yet another example of unisexual heads, this time on the same plant (monoecy). The evolutionary advantage that this variation in gender condition confers is for increased outcrossing, and increased genetic recombination, perhaps yielding morphological novelties and evolutionary divergence.

Coupled with gender strategies are breeding systems. Here genetic, structural, and developmental constraints come together to determine what pollen can successfully germinate on which florets. Genetic systems exist that confer incompatibility of self pollen on the stigma of the same floret, in the same head, or even on the same plant. The pollen presentation system in Asteraceae is designed to promote outcrossing, with the receptive inner surfaces of the styles remaining closed and inaccessible until after they push upward, plunger-like, through the stamen column, sweeping the pollen outward to facilitate dispersal. Only then do they open and become receptive to pollen from other florets, heads, or plants.

Obviously, pollen cannot germinate on a receptive stigma unless it can make the journey from a dehisced stamen to this receptive site. This is pollination. Pollen grains in the family tend to be somewhat uniform, with numerous spines. There are some variations, however, with lophate (i.e., ridged and honeycombed) grains in many Lactuceae and nearly smooth grains in Barnadesioideae, etc. (Skvarla et al., 1977). In some families, there are very specific pollination vectors involving, in some cases, individual species of insect as seen in Orchidaceae. In Asteraceae, however, the head is best suited for visits by many insects, and pollination is largely promiscuous (e.g., Robertson, 1928). This biological flexibility may have been another reason for the evolutionary diversification of the family. For a good recent review of pollination in the family, see Lane (1996) and Mani and Saravanan (1999).

Because sunflower heads represent a substantial reservoir of food resources concentrated in one place on the plant, predation pressure by insects is high. Hence, members of the family have elaborated a series of defensive measures to survive this vulnerability. Structural defenses such as laterally positioned phyllaries, apically directed pappus bristles, and tightly packed florets sometimes with intervening bracts (paleae), all help provide physical barriers to insect ravages. Further, many secondary compounds are synthesized by members of the family, such as the bitter sesquiterpene lactones, that have been shown in laboratory tests to deter feeding on leaves of the family (e.g., Burnett et al., 1974).

It does little good for successful pollination, fertilization, and defense of the maturing achenes if they cannot be effectively dispersed from the parent plant. Numerous mechanisms, therefore, have been adapted for moving these propagules by wind, water, and attachment to various animal hosts, especially birds. In some cases, the entire head remains intact with outer phyllaries developing hooks that can attach to unsuspecting passing animals (e.g., the common burdock, *Arctium lappa*; this can be extremely effective as anyone knows who has had several heads attached to a loose-knit sweater!). Several studies have been described that measured the capacity for wind dispersal of Asteraceae achenes equipped with capillary pappus (e.g., Burrows, 1973; Sheldon and Burrows, 1973; Anderson, 1992, 1993).

Despite numerous mechanisms that ordinarily limit hybridization between species within the family, it should not be surprising to learn that occasionally hybrids are formed. Sometimes there is a breakdown of separation of flowering time, or perhaps species once geographically separated come into contact again through habitat disturbance caused by human or natural activities. Such interspecific crosses are reasonably common in Asteraceae, but intergeneric crosses less so (a few examples are known, however, e.g., Anderson and Reveal, 1966; Kyhos, 1967; Yeo, 1971; Hartman and Lane, 1991; Carr, 1995; for an earlier review of documented cases, see Knobloch, 1972). Chapter 25 discusses many of these examples.

Cytology

Although morphology offers the most important source of taxonomic characters for Asteraceae, additional useful information has come from cytological data, especially from chromosomes (e.g., Jones, 1985, on Astereae). The most frequently used information has been chromosome number and the size and shape of the chromosomes (the karyotype). The principal reason that these data have been so useful is that they contain the hereditary material, the DNA, and thus reveal how the genetic material is packaged. Crossing between two individuals, therefore, is conditioned by the behavior of the chromosomes during gamete production in the parents (the process of meiosis) and by their pairing in any successfully produced hybrids. Degree of pairing of chromosomes in meiosis in hybrids can be a helpful index to the closeness of relationship between two taxa; the more regular the pairing (as normal bivalents) the more closely related the parents are likely to be. The size and shape of the chromosomes can best be assayed in mitosis through root tip squashes and use of an appropriate stain such as acetocarmine.

A considerable amount of work has been done on chromosome numbers in the family. These are now conveniently surveyed in several chromosomal indexes: Darlington and Wylie (1955); Cave (1958–1965); Ornduff (1967–1969); Fedorov (1969); Moore (1970–1977); Goldblatt (1981, 1984, 1985, 1988); Goldblatt and Johnson (1991, 1994, 1996). For New World taxa, many counts were obtained through two impressive research surveys conducted by B. L. Turner and P. H. Raven, and their associates, in the 1960s and 1970s (these, in fact, are continuing; Carr et al., 1999), but many other investigators throughout the world have also contributed new observations. The results have given us information of great value for classification and interpretation of evolutionary relationships as well as suggesting possible modes of speciation.

The range of chromosome numbers in Asteraceae is from $n = 2$ in *Brachyscome lineariloba* (from Australia; Smith-White, 1968; Watanabe and Smith-White, 1987) to $n = 110–120$ in *Montanoa guatemalensis* (Funk and Raven, 1980). Between these extremes lie every conceivable number, but most haploid levels tend to be between $n = 5$ and $n = 20$. Solbrig (1977) has shown modes of $x = 8$, 9, and 12. It has been hypothesized recently that $x = 8$ might be originally primitive for the family as it occurs in Calyceraceae, Goodeniaceae, and primitive Barnadesioideae (Stuessy et al., 1996). Variations up or down from a basic number evolutionarily are regarded as dysploids (sometimes called aneuploids, but this often refers to chromosomal variation within a species), and strict multiples are called euploids or polyploids.

Differences in chromosome number within particular genera can be very helpful in suggesting evolutionary lines, such as was seen in *Melampodium* (tribe Heliantheae; Stuessy, 1971). Here, morphological groupings of species correlated very well with different base numbers of $x = 9$, 10, 11, and 12. Cladistic analyses of relationships within the genus also confirmed these morpho-cytological evolutionary units (Stuessy, 1979), which have been recognized formally as taxonomic sections.

Variation in chromosome number within and between species is important in developing concepts of species and infraspecific taxa. Differences in chromosome number between closely related species usually reflect reproductive barriers between them. As a result, such taxa may be at least partially sympatric with no viable hybrids observed (see Sundberg and Stuessy, 1990, for examples in Heliantheae). If closely related taxa occur at the same chromosomal level, however, the potential for hybridization is greater, although by no means certain because of all the other pre- or postzygotic isolating mechanisms that may be operative. In the discussion of chemical variation within species in this book, one will note frequent mention of chromosomal information, because it bears so directly on the ideas of reproductive isolation and concepts of taxa.

Speciation

As a correlate to brief comments regarding chromosomal variation in taxa within Asteraceae, a few comments may be appropriate regarding modes of speciation. Once again, this information may be helpful when considering variation in chemical features within particular taxa. Three basic modes of speciation may be mentioned: (1) geographic (allopatric); (2) catastrophic (peripheral isolates); and (3) allopolyploidy.

Geographic speciation is surely one of the most common modes within Asteraceae. This has been suggested as being prevalent in tribe Heliantheae (Sundberg and Stuessy, 1990), and it has been shown to be so in other tribes as well (e.g., *Crepis*, Lactuceae; Babcock, 1947). The basic idea is that populational systems become partitioned slowly through time, perhaps in response to geographic barriers such as mountain building, volcanic activity, stream and canyon development, etc. These spatial isolating mechanisms lead to divergent gene pools that eventually accumulate enough differences to be recognized as distinct species. When this stage is reached, sometimes the species may come back together and occur sympatrically, especially if the original spatial barrier ceases to exist (e.g., erosion of previous mountains). Recently, Levin (1993) has argued that such broad and gradual geographic speciation is unlikely, and that local populational differentiation is the preferred concept. We believe that these are totally compatible processes, with speciation initiating at the local populational level through gradual isolation that becomes more broadly divergent as new derivatives proliferate and migrate into new adjacent habitats.

Another common mode of speciation is by more rapid processes, called catastrophic speciation, quantum speciation, or speciation through peripheral isolates (Grant, 1981). The original view of this mode was for chromosomal variants to become stabilized rapidly in ecotones somewhat different from that of the parents (e.g., Lewis, 1962). This provides opportunity for new derivatives to proliferate and survive without direct competition from parental types. This type of origin of species has also been referred to as peripatric speciation because the derivative taxon originated on the periphery of the parental species range.

A final mode of speciation in Asteraceae would be by polyploidy, i.e., by doubling of chromosome number. It is suspected that at least half of the angiosperms originated in this fashion (Grant, 1981). Two specific modes exist: allopolyploidy and autopolyploidy. In the latter, a new taxon is created relatively rapidly by simple chromosomal doubling via fertilization through unreduced gametes or apical meristem somatic doubling. In the former, interspecific hybridization occurs and chromosomal doubling then takes place resulting in offspring that have double the chromosome number of either parent and with features of both of them. The hybrid is also reproductively isolated from the parents. This has been a common mode of speciation in Asteraceae as documented by numerous examples: Ownbey (1950); Solbrig (1977); Grant (1981); Soltis and Soltis (1989); Sundberg and Stuessy (1990). Until recently, it was believed that the evolutionary import of allopolyploidy was vastly greater than autopolyploidy (e.g., Stebbins, 1971). The reason for this view was that the autoploid has the same genetic material as the one parent—only twice the amount of it. Nonetheless, recent studies (see review by Soltis and Soltis, 1993) have shown that over time such polyploids can become diploidized. Use of macromolecular markers has allowed workers to assess the relative importance of autopolyploidy, the outcome of which is the appreciation that it is more important than previously believed. Another important finding is that polyploids sometimes have originated several times via similar hybridization events (e.g., Soltis and Soltis, 1991; see also review by Soltis and Soltis, 1993).

Distribution

It is relatively easy to summarize the geographic distribution of Asteraceae: the family occurs on every continent except Antarctica (where there are only two angiosperms, *Colobanthus quitensis*, Caryophyllaceae, and *Deschampsia antarctica*, Poaceae; Moore, 1983). This means that the more than 23,000 species of the family have been extraordinarily successful in dispersing to new habitats and becoming established. Once established, they have speciated and radiated into the numerous forms we now see.

Within this general occurrence on the globe, tribes of Asteraceae do tend to have at least continental concentrations (after Bentham, 1873). The typically New World tribes would include Eupatorieae, Heliantheae, Helenieae, and Mutisieae. The first three tribes together have greater concentrations of taxa in North and Central America, although all have considerable numbers of species in South America as well. Mutisieae, however, are clearly centered in southern South America and on into the Andean zone. Tribes that are clearly centered in the Old World would include Anthemideae, Arctoteae, Calenduleae, Cynareae, and Inuleae. Inuleae are particularly diverse in South Africa and Australia. Anthemideae occur preferentially in the Mediterranean region and in Europe and Asia in general. Calenduleae and Arctoteae are largely found in South Africa. Cynareae are most concentrated in the Mediterranean area followed by Europe and Asia. Vernonieae are more abundant in the New World, but numerous taxa also occur in the Old World, in tropical and South Africa and tropical Asia. Astereae are nearly twice as abundant in the New World especially in Mexico and the United States, but they also have concentrations in South Africa and Australia. Senecioneae are nearly evenly distributed on both major areas of the earth. In the Old World, the tribe is most common in the Mediterranean region, Europe-Asia, and South Africa. In the New World, the tribe occurs abundantly in Mexico and the Andean region down into Chile (where there are numerous species; e.g., Cabrera, 1949). Lactuceae are concentrated in the northern hemisphere, especially in the Old World.

Within particular continents, Asteraceae occur nearly everywhere in just about every conceivable habitat. As a rule, however, the family is more abundant in temperate areas or in cooler, higher elevation regions of the tropics. Few members of the family are fond of hot, lowland rainforests, although some are found there (e.g., *Ichthyothere scandens*, *Clibadium asperum*, Heliantheae). Only a few species are aquatic (e.g., *Megalodonta*, a segregate from *Bidens*, Heliantheae), but many taxa occur in deserts, mountains up to timberline, valleys, river borders, disturbed rocky outcrops, in forests, and just about everywhere in between.

CHAPTER 2

Classification, Phylogeny, and Biogeography

The information in Chapter 1 offered an overview of the general biology of Asteraceae and their distribution. What has not been discussed are changing concepts of tribes and subfamilies, phylogeny, and biogeography of the family. These are the syntheses of information about groups and subgroups in space and evolutionary time. This chapter begins with general concepts of classification, phylogeny, and biogeography, designed for flavonoid chemists who might lack this background. The chapter then turns to details of these same issues.

General concepts of classification, phylogeny, and biogeography

Classification and phylogeny

Classification is the grouping and ranking of taxa based on similarities and/or differences (Stuessy, 1990). Very early (i.e., pre-1780s) classifications were based on features selected for specific reasons, such as ease of observation, or reflection of the intent of the Creator. In pre-Darwinian times (1780–1859), natural classifications were constructed by assessing the presumptive stability and correlative importance of characters and their states. Post-Darwinian classifications focused conceptually more on features of likely evolutionary value, and closely related groups were viewed as being similar owing to descent from a common ancestor. Although the perspectives on classification changed dramatically after Darwin's (1859) theory of evolution by means of natural selection was proposed, the actual procedures of classification changed little (Stuessy, 1990; Stevens, 1994). All of these attempts at classifications were accomplished intuitively, i.e., they were produced without clear rules or procedures. Workers were encouraged to gain experience with groups by studying under established experts, almost in an apprenticeship mode. In the last half of the 20th century, the intuitive approach to classification came to be viewed as not sufficiently objective nor repeatable for modern scientific tastes.

This viewpoint spawned several explicit approaches to biological classification.

The earliest explicit method for classification was phenetics. This approach specifies that the most predictive classification will be achieved by taking the largest number of characters possible and giving them equal weight. There should be no *a priori* judging of features; all of them potentially contribute in a meaningful way to the final classification. The more characters that are used, the more robust should be the final classification. Although the early proponents of this approach eschewed evolutionary interpretations as being too speculative (e.g., Sokal and Sneath, 1963), in later phases of development a greater interest in providing explicit methods for interpreting evolutionary relationships appeared (e.g., Camin and Sokal, 1965; Sneath and Sokal, 1973). The phenetic approach to classification is still valid as a means of sorting out complex character variation, especially at lower levels of the hierarchy (e.g., infraspecific levels, or dealing with cultivars), for helping to understand hybridization and introgression, and for dealing statistically with genetic frequencies within and between populations. However, it is no longer used as a general purpose approach to classification, because it does not reflect evolutionary relationships. It is possible to prepare a phenogram (based on phenetic concepts and methods) and to interpret this as if it were an evolutionary branching diagram, but few workers have opted for this approach because the initial assumptions do not deal directly with evolutionary concepts.

A second explicit approach to classification is based on phylogeny, that is, the pattern of evolution of a taxon in times past. Understanding the phylogenetic history of a group of organisms is judged by most systematists as the most important element in classification. The reconstruction of phylogeny is based primarily on study of extant forms, simply because in most cases few fossils exist—as is the case with Asteraceae. [The macrofossil record of the family is essentially nonexistent except for Pleistocene remains (Crepet and Stuessy, 1978). The microfossil record (pollen) is more informative; a recent summary shows the earliest reports in Upper Oligocene (Graham, 1996). However, the number of fossil reports are few in comparison with the size of the family.] Phylogeny can be estimated intuitively, that is, by drawing branching diagrams (phylograms) that reflect evolutionary relationships as viewed best by the worker. This was, in fact, the approach taken in the period between 1859 and the late 1960s. During the 1970s and 1980s, however, numerous efforts developed to make such evolutionary reconstructions more explicit and repeatable. Thus was born cladistics: the concepts and methods for determining branching sequences of evolution (phylogeny). Many variations to this approach exist; the following sources provide relevant information: Wiley (1981), Joysey and Friday

(1982), Schoch (1986), Stuessy (1990), Harvey and Pagel (1991), Forey et al., (1992), Quicke (1993), Scotland et al. (1994), Smith (1994), and Kitching et al. (1998). In general, branching patterns are constructed based on shared derived character states (synapomorphies) of taxa, and maximum numbers of these states are used to form common branches between taxa. Parsimony is invoked at least at some stage to provide a means of doing this effectively (by determining character state evolution as well as branching patterns). The guiding principle is that the simplest explanation is most likely the correct one, which is actually a complex philosophical concept (e.g., Sober, 1983, 1988). Although evidence exists that evolution does not always occur parsimoniously (e.g., Gastony, 1986; Lamboy, 1994; also consider the known cases of reticulate evolution via allopolyploidy discussed in Chapter 1), if there is no evidence to the contrary, at least this does provide a useful operational guideline (Crisci and Stuessy, 1980). The importance of cladistics in Asteraceae is underscored by mentioning that the most recent classifications (Bremer, 1994, 1996; Jansen and Kim, 1996) have been constructed, in large measure, through this approach.

However useful cladistics has been, some workers have attempted still another explicit approach to classification, labelled explicit evolutionary classification or explicit phyletics. The goal of this approach is to go beyond just branching patterns of phylogeny to incorporate more information in classifications, specifically evolutionary divergence within lineages [the patristic distance; e.g., Estabrook (1986), Stuessy (1987, 1990, 1997), Hall (1988, 1991, 1995), Ashlock (1991), Carpenter (1993), and Mayr (1995)]. In general, this represents an attempt to provide explicit methods for the intuitive approach to evolutionary classification that has been in use successfully for more than 100 years. This is most challenging, and so far no single method has emerged as best. Further, there does not yet exist a user-friendly computer program that allows one to present these deeper dimensions of phylogeny and base classifications directly upon them. Future work, however, will hopefully yield important new avenues for explicit phyletic classification.

Biogeography

Biogeography attempts to explain present distributions of organisms with reference to phylogeny, major earth events, climatic regimes, and biological aspects of taxa (Brown and Gibson, 1983; Brown and Lomolino, 1998). In the 1930s through the 1950s, explanations of present distributional patterns of taxa were commonplace within taxonomic monographs. Since that time, the appearance declined substantially (for a good exception, however, see Stuckey, 1972), owing to negative attitudes regarding too much speculation. With the advent of plate tectonic theory and the realization that continents had changed positions over time, older biogeographic hypotheses also were in need of reevaluation. Further, significant biogeographic syntheses demanded a thorough understanding of pertinent geological literature, often filled with uncomfortable concepts and new vocabulary. As a result, biogeography began to fade from currency within taxonomic works. During the 1970s and 1980s, however, a renewed interest developed mainly from cladists (e.g., Nelson and Platnick, 1981) who were also seeking testable hypotheses in a fashion similar to what they had achieved with the reconstruction of phylogenies. This has led to many new biogeographic concepts and methods of analysis, with the result that the field is currently in the midst of great change (for a good review of available methods, see Morrone and Crisci, 1995). Few detailed biogeographic studies are being carried out in Asteraceae, however, but this situation will soon change, especially with new DNA data being generated in the context of reconstructing phylogenies.

Biogeography of the 19th century and early part of the 20th century sought explanations for present distributions of plants largely on the basis of dispersal of diaspores (propagules) throughout the world by biotic and abiotic vectors. Physical connections between present land areas were often postulated via ancient land bridges, or submerged island stepping-stones. Fundamentally, however, the world was viewed as consisting of stationary continents with the organisms providing dynamic movements, although more minor geological occurrences, particularly vulcanism and subsidence and erosion, were recognized as possible contributing factors.

Plate tectonic theory changed all that. The idea that major land masses have moved had a profound impact on interpretations of distributions of plants and animals on the planet. For the first time, it was acceptable, indeed recommended, to consider that it was not the organisms that moved to their present patterns of distribution, but rather that the land masses themselves moved, with the organisms rafting upon them. Continental drift provided new and powerful concepts for approaching biogeographic questions. An excellent early review of the possible impact of these new geological concepts on the origin and distribution of angiosperms was provided by Raven and Axelrod (1974). The awareness of the importance of continental drift in biogeography suggested the need to integrate thinking in general about the impact of major earth events on distributional patterns.

At the same time, workers began to wonder if biogeographic hypotheses could be made more explicit and more testable. Most of the earlier hypotheses were advanced in a narrative form that attempted to explain the origin of a group and its colonization and diversification over the globe.

The influences of phenetics and cladistics were soon felt, however, with the development of more quantitative methods for biogeographic research (e.g., Crovello, 1981).

The coming together of acceptance of continental drift, or other major earth events, as a means of explaining distributions, and cladistics as an explicit means for revealing phylogeny, led to vicariance and cladistic biogeography. Vicariance biogeography (Nelson and Platnick, 1981; Nelson and Rosen, 1981) was simply the interpretation of present and past distributions in the face of data regarding major earth events, such as drifting continents, vulcanism, sea level changes, etc., that could have markedly affected distributions of organisms. By looking at distributions of different groups of organisms, repeating patterns might be found that could suggest a common major geological cause. Careful geological study (or synthesis) might reveal the factors involved. Cladistic biogeography emphasizes use of cladograms reflecting branching patterns of phylogeny that could be compared one to another (e.g., Humphries and Parenti, 1986). Methodologically, areas are substituted for taxa and these then became hypotheses of relationships among areas. Congruence among area cladograms suggests common major earth events. Dispersal is ignored as being impossible to deal with directly, because it is so chance an occurrence (although still very significant, Carlquist, 1981).

Additional quantitative biogeographical concepts and methods were developed. A major alternative was panbiogeography, coming initially from Croizat (1958) and recently discussed in detail by Craw et al. (1999). This method stresses looking at evolutionary relationships among taxa, and drawing lines connecting distributions of these relatives on maps. When group after group trace the same lines on a map, these are regarded as "tracks" and imparted biogeographical significance. Some common explanation, especially major earth events, should be involved or so many organisms would not have been affected simultaneously. Quantitative methods have been introduced recently for these concepts (e.g., Craw, 1989; Alroy, 1995; Craw et al., 1999).

Most of these newer methods have not yet been applied to biogeographic problems in Asteraceae (however, see Bremer, 1992, and Swenson and Bremer 1997). Because successful biogeographic studies depend upon having clear understanding of evolutionary relationships of the taxa in question (Thorne, 1989, 1996), new molecular phylogenies that are being generated at a rapid rate for different parts of the family will doubtless stimulate quantitative biogeographic analyses in the near future.

Classification and phylogeny within Asteraceae

As an aid to appreciating classification and phylogeny of tribes and subfamilies of Asteraceae used in this book, we believe that it is appropriate to give an overview of past and present efforts. The focus is on higher levels of the family, although much could be said about subtribal and generic limits as well. To some extent, these lower levels of the hierarchy are touched upon within each of the chapters of this book dealing with individual tribes and different categories. For convenience, we will divide the history of classification and phylogeny reconstruction of the family into four phases: (1) early circumscriptions and subdivisions (through the 18th century); (2) initial tribal concepts (1800–1872); (3) stabilization of tribes and subtribes (1873–1977); and (4) recent developments (1978-present). Much of our discussion of the earlier classifications has been drawn from Small (1919).

Early circumscriptions and subdivisions (through the 18th century)

For a family as natural as Asteraceae, it should not be surprising that the history of classification of the group goes back quite some time. It goes back, in fact, to ca. 300 B.C. with Theophrastus, who is often referred to as the Father of Botany. In his *Enquiry into Plants* (Hort, 1916), Theophrastus recognized, in a very general way, the two tribes Lactuceae and Cynareae. For the former he commented (vol. 2, p. 105) on the "chicory-like" plants having similarity of leaves, and for the latter (vol. 1, p. 93) on the "thistle-like" plants with the flower attached "on top of the actual seeds." These are certainly two of the most distinctive and cohesive tribes of the family.

During the age of the early classifiers after the Dark Ages, several treatments of Asteraceae were provided. Andraea Caesalpino (1583) was the first botanist to suggest and group the family as a whole as "Herbaceae pluribus seminibus." This was divided into Anthemideae, Cichorieae (= Lactuceae), and Cynareae (= Cardueae). The first tribe, however, contained essentially all of the other members of the family exclusive of the dandelion and thistle tribes. John Ray, an Anglican minister, also recognized (1682) Asteraceae as a distinct family (as "flore composito"), but his view of the family also encompassed taxa from other families (e.g., *Eryngium*, Apiaceae) that also had head inflorescences. He highlighted in *tabulae* (diagnostic tabular graphics) groups that we would now regard as tribes, especially Lactuceae and Cardueae.

Detailed studies on Asteraceae continued into the early 18th century. The first separate monograph of the family, which was done by Vaillant (1718–1743), appeared in four papers in the *Histoire de L'Academie Royale des Sciences*. In this work, he recognized four sections: (1) Cynarocephales (the thistles); (2) Corymbiferes (miscellaneous taxa); (3) Cichoracées (Lactuceae); and (4) Dipsacées (*Viburnum*, *Scabiosa*, etc.). This system, however, was much amplified by Pontedera (1720) in another treatment of the entire family. Here he made additional advances and suggested tribal groupings similar to Senecioneae, Calenduleae, Astereae, and Helenieae, but without named recognition. His major classes (from Adanson, 1763, p. lxxxiii) were: (1) demi-fleuronées; (2) les fleuronées; and (3) les radiées.

No discussion of classification of any plant family is complete without comments on the contributions of Carolus Linnaeus. Linnaeus is perhaps the most influential botanist who ever lived, and he has been rightly regarded as the Father of Taxonomic Botany (and Zoology; Lawrence, 1951). He dealt with all flowering plant families known to him during the mid-1700s (e.g., *Species Plantarum*; Linnaeus, 1753). Despite his overwhelming success in providing order to many plant groups, he did not have much luck with Asteraceae. Most of the members of the family were contained in "Syngenesia", referring to the fused anthers. Many other genera were included here that also possess fused anthers, such as the violets (*Viola*). Nonetheless, Linnaeus did stress different sexual conditions of flowers in the head, following the general emphasis of his entire sexual classification system, and in some ways prefigured the different gender strategies that we now recognize (see Chapter 1). His classes were: (1) Polygamia aequalis, consisting of heads with florets all hermaphroditic; (2) Polygamia superflua, with hermaphroditic disc florets and carpellate ray florets; (3) Polygamia frustranea, with neuter ray florets; (4) Polygamia necessaria, with disk florets hermaphroditic, but with ovaries sterile (i.e., functionally male only) and with ray florets carpellate; and (5) Monogamia, which contained taxa without heads but with flowers with syngenesious anthers, such as *Lobelia*, *Viola*, etc. Often included within these classes, however, were disparate elements. For example, both Lactuceae and Cardueae, as well as the discoid-headed taxa, were included within Polygamia aequalis. Linnaeus did, however, publish in different places his sketch of a more natural system. In the *Philosophia Botanica* (1751), he gave the outline of the "Methodi Naturalis" (pp. 27–36) and here he included "Compositi." Four subdivisions were included (but not defined): (1) Semiflosculosi; (2) Capitati; (3) Corymbiferi; and (4) Oppositifolii. The first group contained Lactuceae, the second Cardueae. The third was a broad mixture of discoid and radiate groups, apparently so constructed for the capitulescence. The last grouping contained mostly radiate taxa and largely Heliantheae. This natural classification contained many elements from Caesalpino as well as from Ray (Cain, 1995).

Berkhey (1760) presented an elaborate morphological discussion (with excellent drawings) of the family in which he presented and discussed the Linnaean sexual classification. However, he added a class that he called Polygamia spuria, in which the heads are still discoid and largely hermaphroditic, but have marginal tubular capellate florets. He also introduced a new system of "ordines" that was based on the natural orders of Linnaeus in which he recognized: (1) Semiflosculosae, containing Lactuceae (including mention of milky juice); (2) Capitatae, the Cardueae; (3) Flores Nudes, of discoid headed taxa; and (4) Radiatae, those taxa with radiate heads. In a sense, then, he treated radiate, ligulate, and discoid heads as fundamental in the classification, with the thistles separated out as well (following long tradition in previous works). This treatment was very similar to Linnaeus' natural approach, but Berkhey focused better on the concepts of the last two orders.

Because workers in Paris never followed the Linnaean sexual system of classification, and early struggled with concepts of more natural approaches (Stevens, 1994), it is useful to examine how they dealt with the diversity that characterizes Asteraceae. Adanson was a most remarkable worker who, despite his originality of botanical contributions, never received the attention that he may have deserved (Stafleu, 1963, 1966). He attempted to provide a natural system that markedly departed from previous efforts. He recognized "famille" Compositae, that had elements of previous workers (such as Pontedera). The major subdivisions (sections) in Adanson's system were (1) Lactucae (dandelion tribe); (2) Echinopi; (3) Cardui (thistles); (4) Xeranthema; (5) Ambrosiae; (6) Tanaceta; (7) Conyzae; (8) Jacobeae; (9) Calthae; and (10) Bidentes. This was definitely a major departure from all previous workers, except that some elements here were drawn from the past (e.g., Ambrosieae).

The natural system of Antoine Lauren de Jussieu, as presented in his *Genera Plantarum Secundum Ordines Naturales Disposita* (1789), represents another departure from the Linnaean system, but less so than that of Adanson. Jussieu placed Asteraceae in Classis X, dicots with epigynous flowers and with connate anthers. Within this class were contained the following subdivisions: Ordo I was Cichoraceae, the dandelion tribe; Ordo II, Cinarocephalae, the thistles; and Ordo III, Corymbiferae, which included the rest of the family. Within Ordo III there were subdivisions based on paleae, types of head (i.e., discoid or radiate), and pappus. Parts of Heliantheae were clearly suggested (subunit VI). This classification, in fact, is quite similar to the earlier one described by Vaillant (1718–1743).

Initial tribal concepts (1800–1872)

The most significant initial contributions to the classification of Asteraceae at the tribal level came from Henri Cassini. A lawyer in Paris (King and Dawson, 1975), Cassini devoted much of his free time to the study of

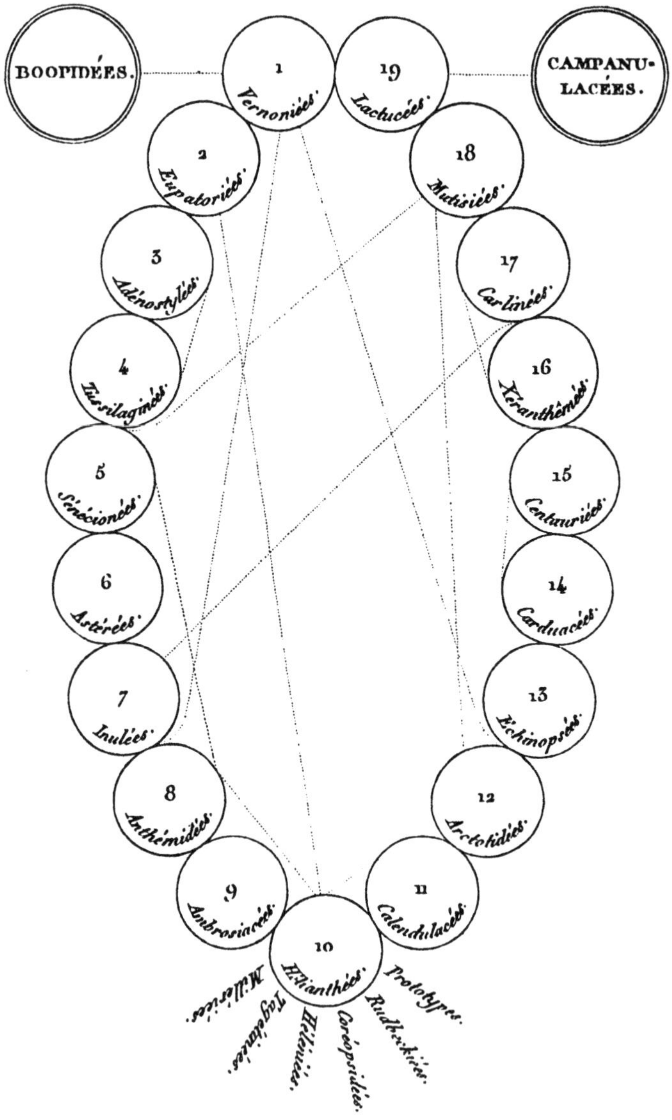

Fig. 2–1. Relationships of tribes of Asteraceae in Cassini's (1834) classification system. Spatial proximity of circles to each other presumably indicates closeness of resemblance; dotted lines apparently show secondary affinities

Compositae.These studies resulted in many publications during the period 1812–1831, many of which appeared in Cuvier's *Dictionnaire des Sciences Naturelles*. As an aid to workers in the family, these contributions have been assembled by King and Dawson (1975) and King et al. (1995a,b).

Cassini recognized 19 tribes in the family (Fig. 2–1). Although he did maintain some of the units treated earlier, such as the thistles (Carduacées) and dandelion group (Lactucées), he also made new groupings that greatly resemble present tribes. Cassini was also a very careful observer of morphological features. Sometimes, however, he carried this to extremes which resulted in the creation of many genera that have since been relegated to synonymy (e.g., creating *Zarabella* Cass., which clearly belongs in *Melampodium*; Stuessy, 1972). Nonetheless, there can be no doubt that Cassini's ideas have had a significant effect upon modern tribal perspectives within Asteraceae.

Another most important contributor to the tribal classification of Asteraceae was Lessing (1832). He modified Cassini's system by placing more emphasis on stylar features and reducing the number of tribes from 19 to eight. Many of Cassini's tribes, however, appeared as subtribes in Lessing's system. The tribes were: Cynareae, Mutisiaceae, Cichoraceae, Vernoniaceae, Eupatoriaceae, Asteroideae, Senecionideae, and Naussauviaceae.

Agustin Pyramus de Candolle treated Asteraceae in the monumental *Prodromus* (1824–1838 for the first seven volumes; the remaining ten volumes were published by his son, Alphonse, 1844–1873), the last complete flora of the world at the specific level. In this floristic work, and in separate publications (e.g., Candolle, 1838), he outlined the general classification of Asteraceae. He recognized three tribes: Chicoracées (or Semi-flosculeuses); Labiatiflores; and Tubuleuses. Three sections were included in the latter: Cinarocéphales; Corimbifères; and Helianthées. An odd aspect, perhaps, of this treatment of the family is that it is based largely upon pre-Cassinian concepts. Candolle was aware of the work of Cassini as evidenced by his discussion (Candolle, 1838) of Cassini's new system that contained ten new tribes. Candolle (1838, p. 6) was not impressed by the numerous genera created by Cassini, which were often based upon small morphological differences. This may have led him to follow the system of Lessing more closely in his infrafamilial classification.

Stabilization of tribes and subtribes (1873–1977)

After Lessing, Cassini, and Candolle, the greatest contributor to the family was George Bentham, an independently wealthy investigator stationed at

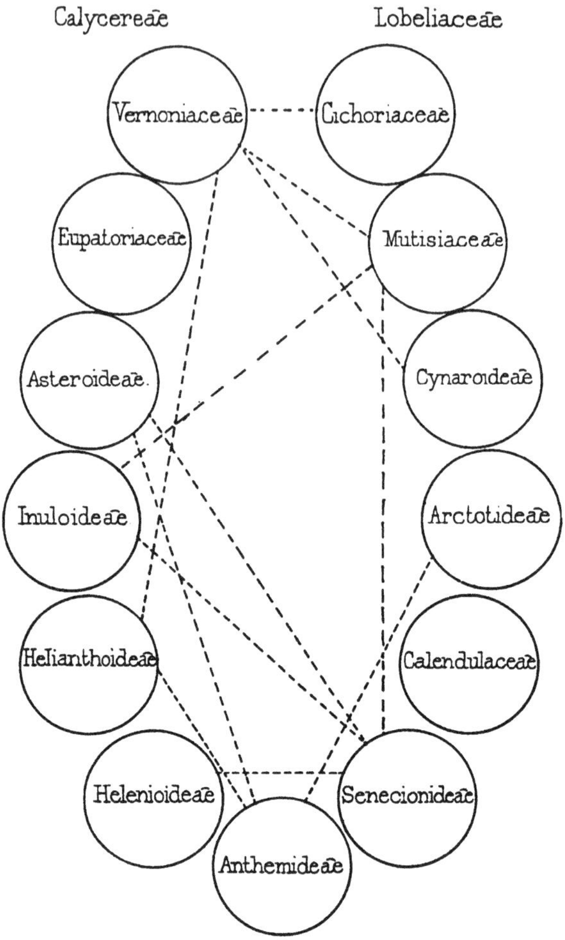

Fig. 2–2. Bentham's (1873) scheme of relationships of tribes of Asteraceae. Mode of data presentation similar to that of Cassini in Fig. 2–1

Kew Gardens. As part of the effort in producing the monumental *Genera Plantarum* (1862–1883) with J. D. Hooker, Bentham took responsibility for Asteraceae. In addition to the treatment for the joint venture, he published (1873) a separate, lengthy discussion of the family, which truly represents a masterpiece of taxonomic balance in classification. Bentham produced a system of 13 tribes (Fig. 2–2) that is largely still in use today. His treatment consisted of: Vernoniaceae, Eupatoriaceae, Asteroideae, Inuloideae, Helianthoideae, Helenioideae, Anthemideae, Senecionideae, Calendulaceae, Arctotideae, Cynaroideae, Mutisiaceae, and Cichoriaceae. Bentham tended to use a wider spectrum of different morphological characters than did Cassini to build his system, and it has been more widely followed, no doubt in part owing to its use in the extremely popular *Genera Plantarum*.

The first worker to address evolutionary issues within Asteraceae was also Bentham. He was slower to accept the theory of evolution by means of natural selection, however, than his colleague Joseph Hooker who was a close friend of Darwin. The *Genera Plantarum* project (1862–1883) was already well advanced before Darwin's book appeared, and therefore its content is not evolutionary. Bentham, however, could not have escaped the influence of Hooker, nor for that matter of Darwin, and therefore he did submit to evolutionary speculation regarding the family in his separately published monograph (1873, pp. 481–484). The principal points of Bentham's views were that Heliantheae (Helianthoideae) represented the most primitive elements and Lactuceae (Cichoriaceae) the most derived. His arguments were quite well reasoned: Heliantheae contained much morphological diversity, and taxa had often foliaceous phyllaries, palae on the receptacle, and more calyx-like pappus members. On the other extreme, he stressed the advanced nature of Lactuceae (Cichoriaceae) with the more uniform morphology throughout the group, the pappus highly modified into bristles, and usually without receptacular paleae (but when present, thin and less bract-like).

During the period from Bentham's time up to the early part of the 20th century, a number of important taxonomic and evolutionary contributions appeared. Hoffmann (1890) authored the section on Asteraceae for Engler and Prantl's *Die natürlichen Pflanzenfamilien*, but the treatment was basically the same as that contained in Bentham's 1873 work with the addition of many new genera. Britton and Brown (1896–1898) departed from past perspectives and elevated all tribes of Asteraceae to independent familial status. This treatment was followed by other American authors, such as Bessey (1915) and Small (1903), but the viewpoint has never been well accepted owing to the overwhelmingly natural aspect of the entire assemblage in comparison with other related families. Bessey (1914, p. 71) also believed that Heliantheae (as "Helianthaceae") were primitive for the family. He stressed that, "In the following arrangement the *Helianthaceae* are regarded as the lowest, from which the two principal phyletic lines have arisen, culminating on the one hand in the *Eupatoriaceae*, and on the other in the *Lactucaceae*."

Small (1919) departed significantly from Bentham's evolutionary views of Asteraceae and believed Senecioneae as most primitive. His reasoning was based largely on the concept of age and area of Willis (1918, 1922), which stated that those groups with large numbers of taxa and broad geographical distributions were primitive within a particular taxon. The reason for this viewpoint was that they had been around the longest and hence diversified more, both morphologically and spatially. Because *Senecio* is the largest genus of the family, with upwards of 1250 species (Bremer, 1994), and is widely distributed throughout the world, this was selected as the ancestral type.

Other workers have suggested at various times that other tribes might be primitive. Leonhardt (1949) regarded Cardueae (thistles) as likely to be ancestral for the family, based on presumptive close relationships with Campanulaceae. He viewed Asteraceae as having had two separate origins, however, one that led into Lactuceae and the other into Cardueae (summarized by Bremer, 1987). Augier and DuMerac (1951) suggested that Vernonieae were the basal tribe of Asteraceae, having been derived from Campanulaceae, subfamily Lobelioideae.

One of the major contributors to classification and phylogeny of the family in the second half of the 20th century was Arthur Cronquist. Although agreeing in large measure with the classification of Bentham, Cronquist believed that tribe Helenieae, separated from Heliantheae by absence of receptacular paleae, should be merged into that tribe. He also believed Heliantheae primitive for the family. Cronquist (1955, 1977) summarized the arguments for regarding Heliantheae as basal (founded in part on criteria articulated earlier by Bentham, 1873) including their woody nature (woodiness was largely believed as primitive and herbaceousness as derived, in part owing to gymnosperms and early angiosperm fossils being woody), their large and often leaf-like receptacular bracts (= paleae), and their often scaly pappus (i.e., more calyx-like). However, there were some who questioned this perspective (e.g., Stuessy, 1977; Stuessy and Spooner, 1988) on several grounds, such as that large paleae might simply be an adaptation for protection of larger achenes often found in the tribe, rather than representing ancestral modified bracts.

Participants at the 1975 International Compositae conference in Reading, England, added many perspectives toward modifying the classification and phylogeny of the family. Because the conference was organized along existing tribal boundaries (i.e., chapters were devoted to morphology and chemistry of each of the tribes), no new framework of tribes is found in the published volumes (Heywood et al., 1977). Dittrich (1977) did attempt to divide Cardueae into three separate tribes, however, but this approach has so far received little support.

Jeffrey (1977) provided an alternative view of the primitive tribe of the family. He reasoned that since Asteraceae were derived from some other family (or ancestor of another family) in Asterales, and since many of these other families (e.g., Campanulaceae) have racemes or cymes in which the individual flowers are bilabiate, then the early Asteraceae would possess a condensed head inflorerscence composed of bilabiate flowers. The tribe that contains most of these types of florets is Mutisieae, concentrated in

South America. This was an interesting idea, but it had little impact when it was published.

The great interest in plant secondary chemistry during the 1960s and 1970s also led to hypotheses regarding which tribe of the Asteraceae was most primitive. Mabry and Bohlmann (1977) summarized many different classes of compounds within the family and cautiously suggested (p. 1101) that the "...Heliantheae may be the most primitive group, even though some species or genera are advanced relative to other members of the tribe."

Recent developments (1978-present)

Gerhard Wagenitz (1976) and Sherwin Carlquist (1976) independently provided a careful morphological examination of the tribes of the family and concluded that two subfamilies existed. Previous workers (e.g., Cronquist, 1955) had also assumed there to be two subfamilies, but configured in the following way: Lactuceae constituted one subfamily (Cichorioideae), with all other tribes in the second subfamily (Asteroideae). Wagenitz (1976) followed this initial subdivision but then broke up subfamily Asteroideae into two unranked groups: (1) Arctoteae, Vernonieae, Cardueae, and Mutisieae; and (2) Eupatorieae, Astereae, Inuleae, Heliantheae, Helenieae, Anthemideae, Senecioneae, and Calenduleae. Carlquist (1976) placed Lactuceae in with other tribes and recognized two broad subfamilies, Asteroideae and Cichorioideae. Asteroideae contained Astereae, Inuleae, Heliantheae (including Helenieae), Anthemideae, Senecioneae, and Calenduleae. In Cichorioideae he placed Arctoteae, Vernonieae, Cardueae, Mutisieae, Cichorieae (= Lactuceae), and Eupatorieae. Jeffrey (1978) had basically the same tribal and subfamilial organization, except that he recognized in addition several segregate tribes (Eremothamneae, Liabeae, Ursinieae, and Cotuleae). Jansen and Stuessy (1980) followed Wagenitz but elevated within subfamily Asteroideae his two informal subgroups to Asteroideae and Vernonioideae, thus making three subfamilies within the family. Robinson (1981, 1983b) followed with the same classification as Carlquist, except that he moved Eupatorieae into Asteroideae following Wagenitz. Thorne (1983) also adopted this approach.

In addition to evolutionary investigations in Asteraceae using morphology, a number of chemical studies have also addressed broad issues of phylogeny in the family. Otto Gottlieb and collaborators (e.g., Emerenciano et al., 1987) have examined sesquiterpene lactone and flavonoid data for Asteraceae with regard to relative biosynthetic complexity among tribes. The general approach is to assess the degree of skeletal specialization, oxidation, *O*-glycosylation, and *O*-methylation in each of the tribes and determine their relative primitive vs. advanced conditions. In general, Arctoteae, Astereae, Inuleae, and Liabeae are shown as more basal for sesquiterpene lactone data, but Arniceae, Calenduleae, Mutisieae, Tageteae, and Senecioneae as more structurally primitive for flavonoids. Despite the data not being completely congruent, they do suggest, in a general way, that subfamily Asteroideae is more derived than subfamily Cichorioideae.

Maffei (1996) has recently assayed leaf wax alkanes in more than 85 species of Asteraceae and subjected the data to phenetic analysis. His sampling involved representatives of ten tribes, with Lactuceae and Cardueae the only tribes included for Cichorioideae. The latter two tribes appear very similar to each other in leaf wax alkane composition. Relationships among other tribes in subfamily Asteroideae, based on clustering and principal component analysis (PCA) techniques, reveal two general groups. The first group has Heliantheae and Helenieae placed closely together, followed by Eupatorieae and Anthemideae. The second major group reveals Astereae and Inuleae as close, followed by Gnaphalieae and Senecioneae.

Proksch and Kunze (1996) reveal that prenylated acetophenones are found in many species of Asteroideae, and hence help unify that subfamily, but are nearly absent from Cichorioideae. This same general pattern has been documented for other classes of secondary compounds in the family (Jansen et al., 1991a). Proksch and Kunze (1996) also show tribes Eupatorieae and Heliantheae as having "close affinity," a relationship that was recently re-emphasized by Robinson (1996b).

Also, in recent years, explicit morphological cladistic studies have been completed on numerous genera and on higher-level relationships within the family. Bremer (1987) was the first to produce a comprehensive cladistic analysis for the entire family (Fig. 2–3). This, and additional efforts, have been summarized by Bremer et al. (1992) and Bremer (1994). In Bremer's most recent (1994) treatment, Barnadesioideae are regarded as a single monotribal subfamily, and then with two other subfamilies, Cichorioideae (with six tribes: Mutisieae, Cardueae, Lactuceae, Vernonieae, Liabeae, and Arctoteae), and Asteroideae (with 10 tribes: Inuleae, Plucheeae, Gnaphalieae, Calenduleae, Astereae, Anthemideae, Senecioneae, Helenieae, Heliantheae, and Eupatorieae). The major changes are recognition of subfamily Barnadesioideae, acceptance of Liabeae as a tribe (earlier stressed by Robinson and Brettell, 1973a and Nordenstam, 1977), and splitting of Inuleae into three independent tribes. The tribe Helenieae is resurrected as well. Bremer (1996) has recently made additional modifications to this scheme (Fig. 2–4) in which he treats Cichorioideae as containing only Lactuceae, Vernonieae, Liabeae, and Arctoteae, with Mutisieae in an uncertain position near basal Barnadesioideae. Cardueae are placed in a fourth subfamily: Carduoideae Cass. ex Sweet. The tribal composition of Asteroideae

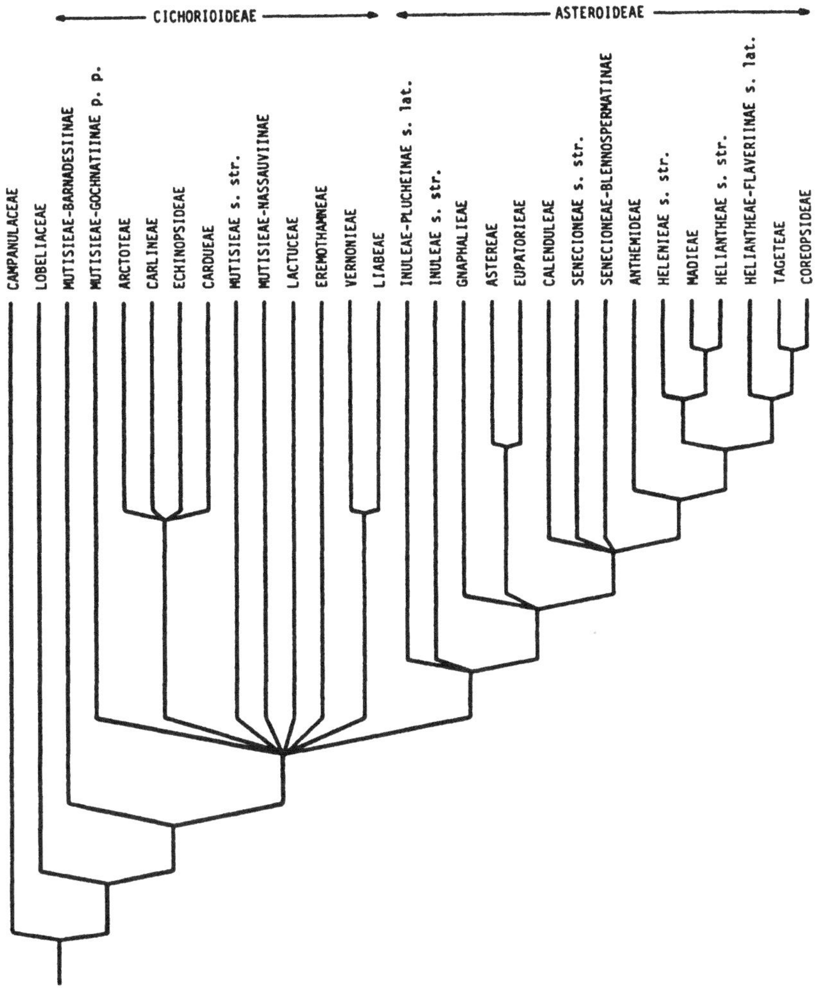

Fig. 2–3. Strict consensus tree for all equally parsimonious trees in a morphological cladistic analysis of tribes of Asteraceae. From Bremer (1987)

remains unchanged. Bremer (1987, 1994) also concluded that subtribe Barnadesiinae of tribe Mutisieae was basal for the family, but this was influenced at least in part by the DNA work of Jansen (see below). He also believed that other parts of this same tribe might be quite primitive. The arguments for these conclusions, however, were not entirely coincident with those of Jeffrey (1977). Barnadesioideae, the hypothesized most primitive subfamily, mostly do not have bilabiate corollas; rather, they have modified tubular 5-lobed corollas.

Further significant contributions to classification and phylogeny of Asteraceae came with macromolecular DNA studies by Robert Jansen and

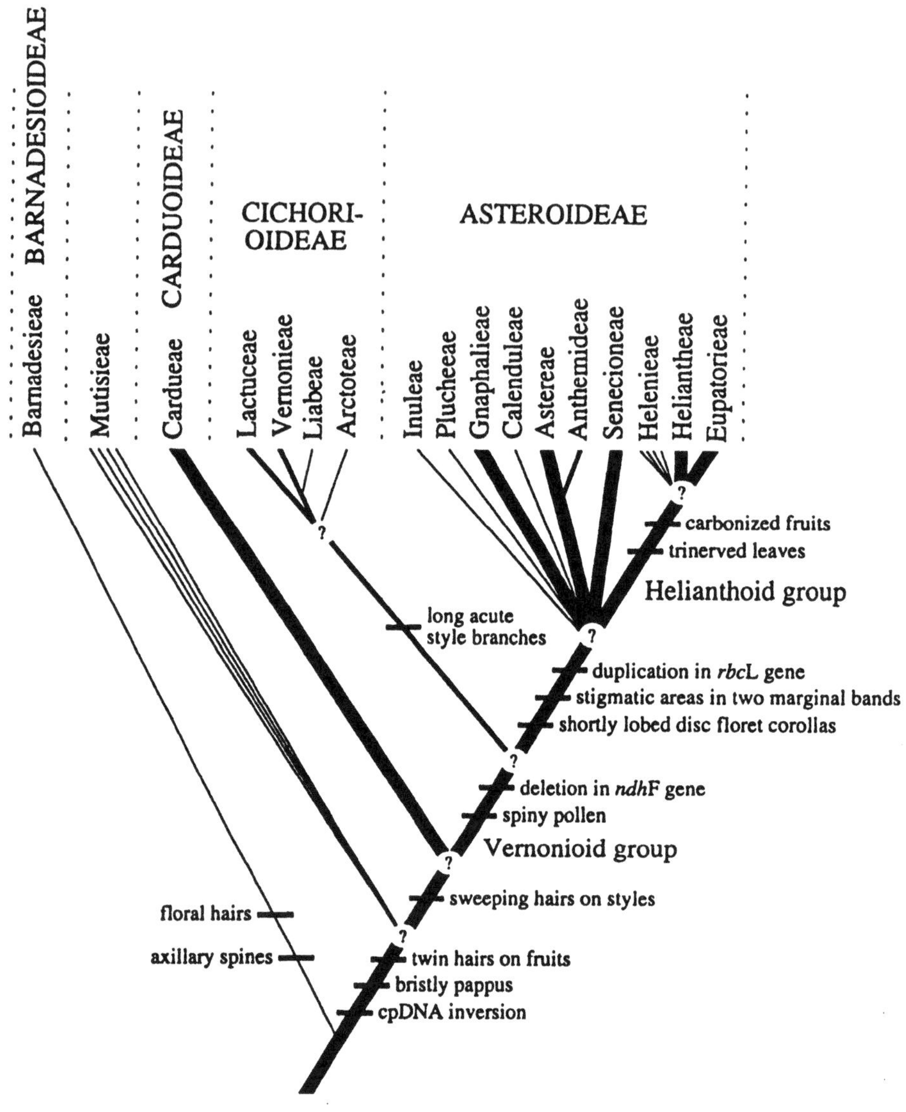

Fig. 2–4. Diagram of cladistic relationships among tribes and subtribes of Asteraceae. Width of lines is proportional to numbers of species in each group. From Bremer (1996)

colleagues (Figs. 2–5 to 2–8). The initial surprising discovery (Jansen and Palmer, 1987a,b) was that all sampled genera of the family (57 at that time; now more than 250, Jansen et al., 1992) possessed a 22kb inversion in the chloroplast genome, except for three genera (*Barnadesia*, *Chuquiraga*, *Dasyphyllum*) all belonging to subtribe Barnadesiinae of tribe Mutisieae. This was the first clear indication that Mutisieae were, indeed, part of the basal complex of the family. Barnadesiinae were themselves somewhat un-

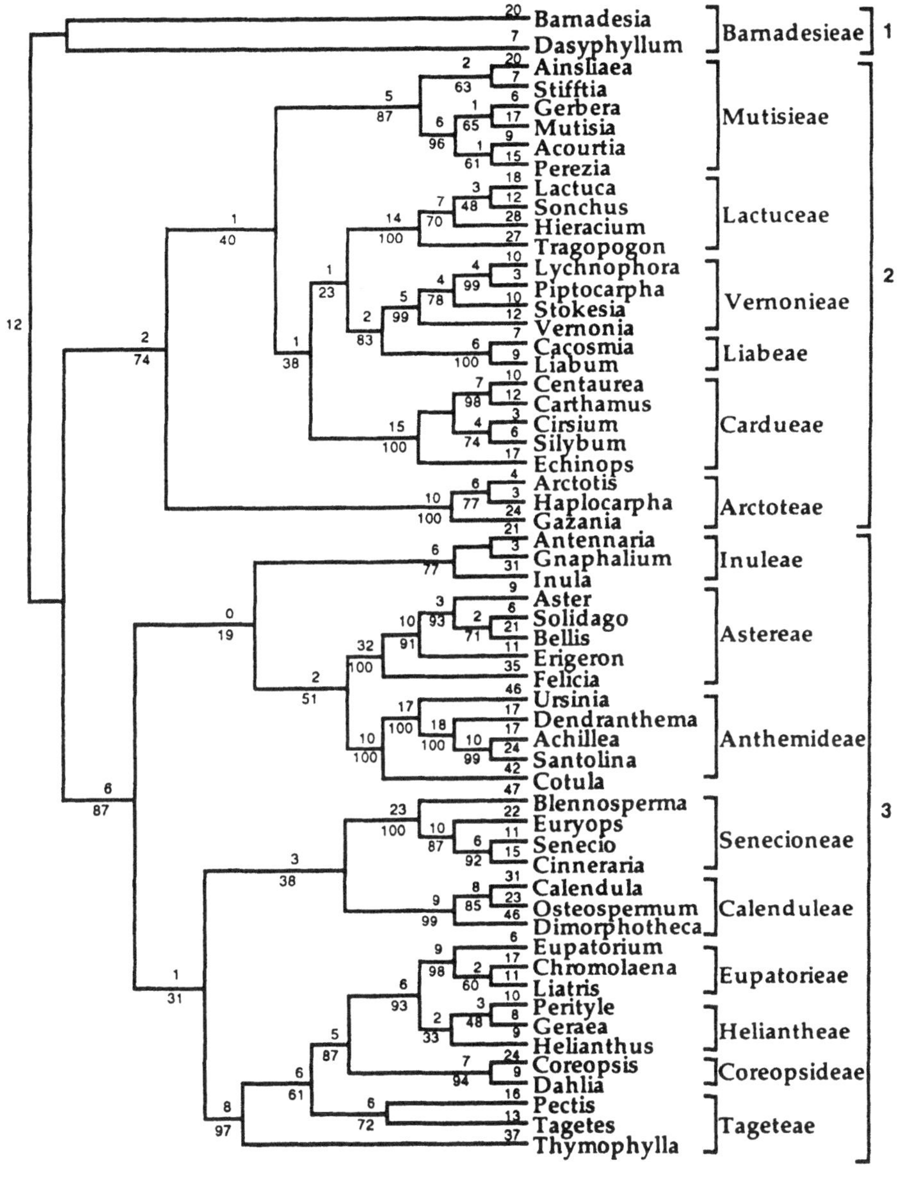

Fig. 2–5. Majority-rule consensus tree of Asteraceae based on cpDNA restriction site data and Dollo bootstrap analysis. The numbers above and below nodes indicate the number of restriction site changes and bootstrap value (100 replicates), respectively. Brackets indicate tribal and subtribal circumscriptions: 1 = Barnadesioideae; 2 = Cichorioideae; 3 = Asteroideae. From Jansen and Kim (1996; modified from Jansen et al., 1991a)

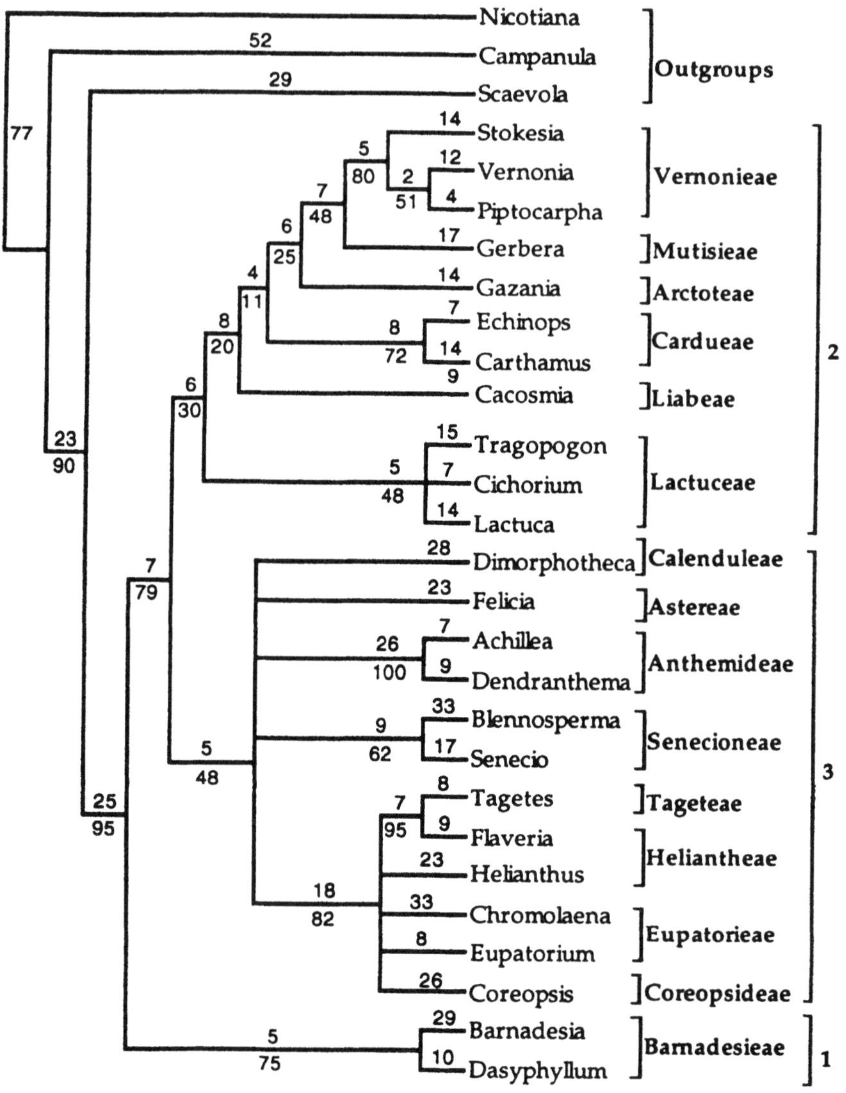

Fig. 2–6. Strict consensus tree of eight equally parsimonious trees for Asteraceae based on *rbc*L sequence data with bootstrap analysis. Above nodes are numbers of nucleotide changes; the rest as in Fig. 2–5. From Jansen and Kim (1996; modified from Kim et al., 1992)

usual morphologically, being mostly shrubby, discoid, and endowed with many spines on leaves and stems (Cabrera, 1977; Hansen, 1991). *Barnadesia* is even hummingbird-pollinated which is an uncommon pollination syndrome within the family (usually visited by insects). Nonetheless, the data seem incontrovertible; either Barnadesiinae had the inversion as did the rest of

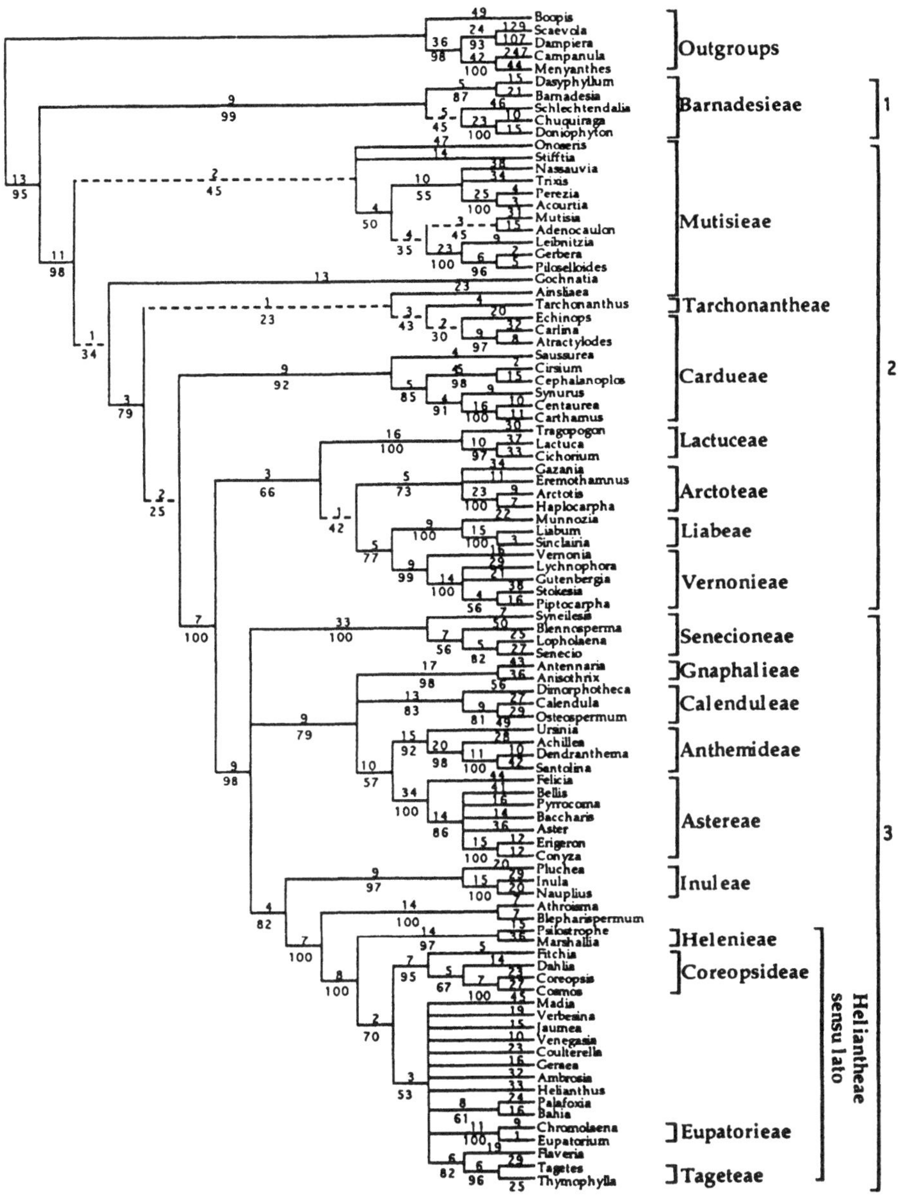

Fig. 2–7. Strict consensus tree of 5,235 equally parsimonious trees for Asteraceae based on *ndh*F sequence data and bootstrap analysis (300 replicates). Numerical and graphic format as in Fig. 2–6. From Jansen and Kim (1996; modified from Kim and Jansen, 1995b)

the family and then lost it, or it never initially had it (as is the case with related families in Asterales). The latter explanation is more parsimonious

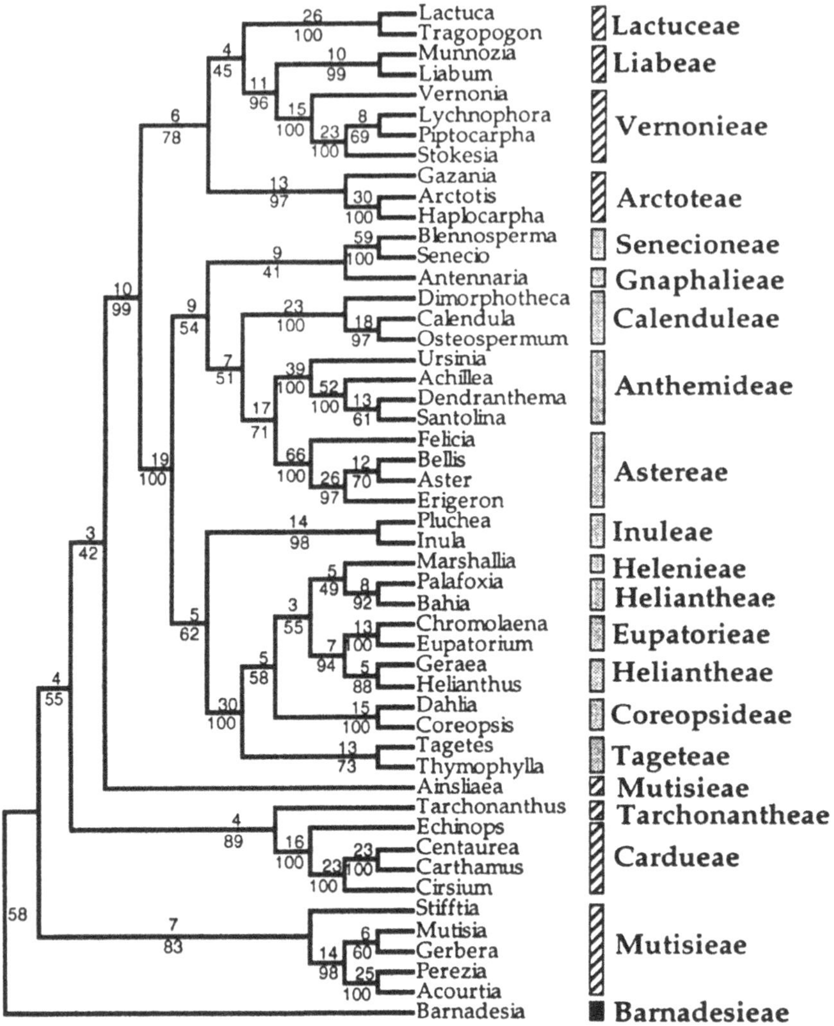

Fig. 2–8. Single most parsimonious tree of Asteraceae based upon combined *ndh*F and cpDNA restriction site data. Numerical format as in Fig. 2–5. Rectangles show subfamilial affinities: black = Barnadesioideae; gray = Asteroideae; hatched = Cichorioideae. From Jansen and Kim (1996)

as well as being confirmed by sequence data of the *rbc*L gene (Jansen et al., 1992). In addition to suggesting a basal complex for the family, later elevated to subfamilial status (as Barnadesioideae; Bremer and Jansen, 1992), Jansen's DNA data revealed two large clusters of tribes, with Cichorioideae (or parts of it) always being seen as more primitive than Asteroideae (Figs. 2–5 to 2–8).

Recent morphological cladistic studies also evaluate intertribal relationships within Asteroideae. Bremer (1996; Fig. 2–4) sees a basal heptachotomy of unresolved relationships including Inuleae, Plucheeae, Gnaphalieae (these three comprise Inuleae s.l.), Calenduleae, Astereae (with Anthemideae as sister group), and Senecioneae, followed by a derived complex of Helenieae, Heliantheae, and Eupatorieae (the "Helianthoid group"). Karis (1993a, 1996), using different tribes of Cichorioideae as outgroup, regards Inuleae as basal followed by Calenduleae, next Astereae with its sister taxon Anthemideae, then Senecioneae, and finally Heliantheae s.l. with Eupatorieae.

Additional independent molecular phylogenies have been completed by Jansen and colleagues for taxa of the family using *rbc*L nucleotide sequences (Fig. 2–6) and *ndh*F sequences (Fig. 2–7). Figure 2–8 is a combined cpDNA and *ndh*F molecular phylogeny. These trees all have several points in common. First, they all point to a well circumscribed subfamily Asteroideae consisting of Anthemideae, Astereae, Calenduleae, Eupatorieae, Heliantheae (s.l.), Inuleae (s.l.), and Senecioneae. These results are the same as those based on morphology of Wagenitz (1976), Robinson (1981, 1983b), Thorne (1983), and Bremer (1987, 1994, 1996). The uncertainty lies not with subfamily Barnadesioideae nor subfamily Asteroideae, but rather with Cichorioideae. Second, there is much branch shifting among tribes of Cichorioideae in the cladograms derived from the independent molecular data. Tribes Arctoteae, Lactuceae, Liabeae, and Vernonieae form a cohesive unit. Hence, from cpDNA restriction site and *rbc*L data, the subfamily is holophyletic (i.e., the group contains all descendants from the common ancestor; Ashlock, 1971, 1984), but with *ndh*F it is paraphyletic (i.e., the group contains some, but not all, descendants of the common ancestor; Henning, 1966). The basal tribe of the subfamily also changes from Arctoteae (cpDNA) to Lactuceae (*rbc*L) or Mutisieae (*ndh*F, but weakly supported). A cladogram (Fig. 2–8) based on combined cpDNA and *ndh*F data reveals a more complex picture with most of Mutisieae basal followed by Cardueae and then the rest of Cichorioideae and all of Asteroideae as holophyletic groups. Mutisieae, Cardueae, and Tarchonantheae (problematic, earlier placed in Anthemideae, Astereae, Inuleae, Mutisieae, or Vernonieae; see Keeley and Jansen, 1991, for a review) are listed as "uncertain" as to subfamilial placement. The more complex and weaker structure within subfamily Cichorioideae has also been obtained in morphological cladistic studies by Bremer (1996; Fig. 2–4). Further, reanalysis by Mishler et al. (1996) of the chloroplast restriction site data of Jansen et al. (1990, 1991a), with special focus on the suitability of Dollo parsimony for tree construction with this type of data, tentatively also supports the hypothesis of a paraphyletic Cichorioideae. More recent phylogenetic studies on the entire family by

Bayer and Starr (1998) using *trn*L intron and *trn*L/*trn*F intergenic spacer regions of cpDNA support these same general topological relationships.

These same macromolecular (DNA) cladistic studies (summarized in Jansen and Kim, 1996; Figs. 2–5 to 2–8) also provide representations of tribal interrelationships within Asteroideae. The *rbc*L data reveal only that Eupatorieae and Heliantheae are closely related, a conclusion also seen in all other morphological and macromolecular analyses. The cpDNA restriction site and *ndh*F analyses also show clearly that Astereae are sister to Anthemideae, again supported by morphology (Karis, 1993a, 1996). The position of Inuleae is not as clear, however, nor are the positions of Senecioneae and Calenduleae.

Even newer insights on classification of Asteraceae came from the second international conference on Compositae held in July of 1994 at the Royal Botanic Gardens, Kew. One volume of pre-symposium papers appeared just after the conference (Hind et al., 1995), and two volumes of symposium papers have recently been published (Hind, 1996). This conference, in contrast to that of 1977, was not organized around tribal concepts, perhaps because so much of the high-level classification is now being actively investigated. Many useful contributions exist, and these are discussed in this book when appropriate in relation to flavonoid data and classification in the family.

The order of tribes and subtribes in this book follows the recent ideas emanating fom Wagenitz, Carlquist, Thorne, Bremer, and Jansen. Although the flavonoid data are presented within this basic framework, we assess the implications that flavonoid data have for this new classification. We are, in effect, testing the morphological and DNA classifications by comparison with available flavonoid data. The newer phylogenies are also used to suggest trends in flavonoid evolution within the family. We also offer numerous comments on lower-level taxonomic problems, and point out in many instances where flavonoid data have proven critical for helping resolve taxonomic difficulties.

Phylogenetic relationships between Asteraceae and other families

A number of workers prior to Darwin opined on the relationships of Asteraceae to other families of angiosperms. Cassini (1816) in describing Calyceraceae as new (as "Boopideae"; rejected against Calyceraceae, Richard, 1820; ICBN, Greuter et al., 2000) regarded this new family as (p. 161) "une transition très-naturelle et très-satisfaisante de la famille des Synanthérées à celle des Dipsacées,..." Later (1834) he presented a diagram of relationships that showed "Boopidées" and "Campanulacées" as graphically close to Asteraceae (Fig. 2–1).

Other workers followed with this view of a close tie with Calyceraceae. Bentham (1873) revealed close ties with Calyceraceae and Lobeliaceae in a diagram at the end of his treatise (Fig. 2–2) that was similar in graphic format to that of Cassini (1834). Bessey (1897, p. 34) listed the same basic idea as contained in Cassini (1816) in his outline of a new phylogenetic system of classification of the angiosperms: "Valerianaceae; > Dipsacaceae; > Calyceraceae; > Compositae", i.e., Calyceraceae as being transitional between Dipsacaceae and Asteraceae.

Small (1919) emphasized the relationship of Campanulaceae subfamily Lobelioideae as closest to Asteraceae. This was previously mentioned by several authors such as Delpino (1871), who was impressed with similarities in pollen presentation mechanisms between the two groups. The closest tie has always been suggested as between Lobelioideae and tribe Lactuceae, which has ligulate florets and milky sap, features found in both groups. More modern thinking (e.g., Cronquist, 1955) interprets these similarities as parallelisms within both families rather than as characteristics inherited directly from a common ancestor.

Hutchinson (1926) placed Asteraceae within his order Asterales along with Adoxaceae, Calyceraceae, Dipsacaceae, and Valerianaceae, but with comment (p. 26) that this is "A polyphyletic group, probably derived from several lower groups, including the *Umbelliflorae*, *Rubiales*, *Campanales*."

Rendle (1938) summarized the features that Bentham (1873) regarded as primitive for Asteraceae, but he suggested (p. 610) that "Such a flower-form resembles that of present-day Rubiaceae, but with a reduction in the gynoecium, and might well be an offshoot of the same group or plexus to which Rubiaceae and allied families owe their origin."

Cronquist (1968) continued this theme and regarded the ancestors of Asteraceae as woody, following suggestions based on wood anatomy by Carlquist (1966), which led him to hypothesize (p. 311) "the Rubiales-Dipsacales complex as more likely ancestors than the Campanulales." He narrowed this view to focus on Rubiaceae in latter publications (Cronquist, 1981), and he regarded Calyceraceae as having a parallel evolution also from out of Dipsacales (Cronquist, 1981, 1988).

Takhtajan (1969) perceptively argued that the family closest to Asteraceae was Calyceraceae, alone in its own order, Calycerales, but (p. 233) standing in the other direction "...nearest to the Campanulales, especially to the Goodeniaceae, which it definitely approaches in the structure of the pollen grain wall. Both these orders evidently had a common origin."

Thorne, in a series of papers (e.g., 1968, 1976, 1983, 1992), presented a novel system of classification of the angiosperms, but few phyletic insights are given for the close relatives of Asteraceae. Never satisfied with his own previous attempts at classification as portrayed in a two-dimensional "shrub" of relationships, he continually modified them in subsequent publications. In 1983 he portrayed (p. 99) Asteraceae (as "Asteriflorae") as closer in point of origin to Dipsacales, but in 1992 (p. 249) he attaches Campanulales to Asterales in a joint "Asteranae" more distant from Dipsacales in Cornanae.

Before the unfortunate accidental death of Rolf Dahlgren, he sketched outlines of a new system of classification of angiosperms (1975, 1980, 1983) as well as completing the first volume on monocotyledons (Dahlgren et al., 1985). Most recently, he (Dahlgren, 1983) placed Asteraceae alone in Asterales but joined with Campanulales, which included Pentaphragmataceae, Campanulaceae, and Lobeliaceae, in "Asteriflorae." This position was retained in his two-dimensional "bubble diagrams" (e.g., Dahlgren et al., 1981; Dahlgren, 1989) showing relationships (similar to those presented by Thorne) whereby Asterales and Campanulales are joined graphically. The same relationship has been emphasized based on secondary pollen presentation morphological data by Erbar and Leins (1995).

Because of the plethora of secondary products in Asteraceae, and interest in the use of these compounds as taxonomic characters during the 1960s and 1970s, a number of studies synthesized information about relationships between Asteraceae and other families. One of the most useful summaries came from the 1975 Reading Conference on the family, in which Mabry and Bohlmann (1977) discussed the chemical relationships with other families. They concluded (p. 1097) that the "available chemical data suggest that the Umbelliferae and the Campanulaceae are the extant families phyletically closest to the ancestors of the Compositae." The strong chemical tie to the former family was especially stressed. This was based on shared possession of sesquiterpene lactones, which are relatively uncommon in angiosperms, biogenetically related acetylenes, and certain coumarins. A strong secondary product relationship with Calyceraceae was rejected. Zdero and Bohlmann (1990d) more recently continued to point out the close ties with Umbelliferae, with comment (p. 13) that they may have arisen "independently from an ancestral stock..."

In a morphological cladistic analysis of Asteraceae, Bremer (1987) suggested Lobeliaceae as the sister-group (i.e., closest immediate close relative) and used features of this family for polarizing character states within Asteraceae. [Based on newer insights, however, he has modified that view and now regards Calyceraceae as closest relative (Bremer, 1994; Gustafsson and Bremer, 1995).]

That Calyceraceae are the sister-group (i.e., closest relative) to Asteraceae was speculated upon by Turner (1977b) based particularly on the pollen data presented by Skvarla et al. (1977). Hansen (1992) directly addressed the relationships of Calyceraceae to Asteraceae, and initially concluded

(p. 73) "to me, the Calyceraceae therefore appears to be the best sister-group choice of Compositae" based on having more synapomorphies (shared derived character states) in common between the two families. He equivocated, however, finding (p. 73) "no strong evidence against" Lobeliaceae. His final conclusion was that (p. 74) "...the difficulty of deciding whether the Calyceraceae or the Lobeliaceae is the sister-group of Compositae might be most parsimoniously explained by considering all three families members of a monophyletic taxon."

These numerous difficulties in determining the closest relative of Asteraceae are due to the family having no clear multiple character state associations with only one relative. All characters that we use to diagnose the family, such as head inflorescence, uniovulate carpel, inferior ovary, pappus, etc., are found scattered in other families as well. It is the mosaic of features in combination that so well characterizes Asteraceae, but at the same time also makes it so difficult to trace its origin.

The real breakthrough in understanding close relatives of Asteraceae came with the availability of macromolecular data. Just as cpDNA data helped to reveal subfamilial structure more clearly (e.g., Jansen et al., 1992), sequences of the *rbc*L chloroplast gene (Olmstead et al., 1990, 1992; Gustafsson et al., 1996) and restriction site mapping studies of the cpDNA inverted repeat (Downie and Palmer, 1992) have shown clearly that Asteraceae relate closest to Calyceraceae and Goodeniaceae. Within that new context, therefore, DeVore and Stuessy (1995) provided a careful review of the morphology of these three families and concluded that Calyceraceae are considerably closer to Asteraceae than either is to Goodeniaceae. This has been independently confirmed in morphological cladistic analysis by Gustafsson and Bremer (1995). General referral of Calyceraceae to the Campanulales-Asterales complex has also been confirmed by Erbar (1993) based on floral development and pollen-presentation data. See additional discussion in Chapter 26.

Through all these many years, then, we return to the simple intuitive judgements of Cassini (1816) and Bessey (1897) that Calyceraceae are the closest living relatives of Asteraceae. This small family of approximately six genera and 50 species is confined to southern South America with concentrations in Argentina and Chile. The significance of this geographical pattern will be explored in the following section.

Biogeography of Asteraceae

Within Asteraceae, little detailed biogeographic work has been done in a meaningful way. The most recent book by Bremer (1994) summarizes nicely the present distributions of genera and tribes, but interpretation of distributions across continents had been hindered by the lack of meaningful phylogenies at the higher levels of the hierarchy. Now that more insightful phylogenies are being generated, at least for restriction site chloroplast DNA data and for some selected nuclear and chloroplast gene sequences, the broader relationships are beginning to clarify. This will greatly stimulate biogeographic syntheses in the near future within the family. Further, it is not a question of whether vicariance or dispersal can explain all the patterns of distribution seen—both have undoubtedly been involved. The particularly good dispersal capability of achenes of the family, with their attached pappus, argues for dispersal over long distances in many cases, and helps explain the very rapid distribution of the family over many continents from an hypothesized Oligocene origin in southern South America (DeVore and Stuessy, 1995). The principal biogeographic points of interest deal with where and when the family originated and its early patterns of dispersal and migration.

Post-Darwin workers hypothesized on the time and place of origin of the family. Bentham (1873) suggested Africa, Western America, and possibly Australia, as the sites for ancestors of the family. Among extant primitive taxa, he pointed to their distributions in the high Andes and Pacific Islands. Small (1919) suggested a late Cretaceous or early Eocene time of origin and pointed to the Amazon or Andes as possible early locations. Cronquist (1955) focused on Heliantheae as most primitive within the family, and because this tribe is most diverse in Mexico, this region was indirectly suggested as an area of possible origin. Raven and Axelrod (1974) presented a mid-Oligocene time and a South American geographic place of origin. Turner (1977a) attempted to push back the time of origin of the family by relating extant distributions of tribes to continental drift. Bremer (1992) offered a useful technique of determining "ancestral areas" derived from cladistic analysis and concluded that the place of origin for the family would be somewhere in South America and the Pacific Basin.

Based on renewed confidence from DNA data that Calyceraceae and Goodeniaceae are closest relatives of Asteraceae, and owing to findings that Barnadesioideae represent the basal complex, it became possible to address biogeography of the family more comprehensively than ever before (Stuessy et al., 1996). DeVore and Stuessy (1995) examined distributions of the three families and noted that Goodeniaceae are largely found in Australia, whereas Barnadesioideae and Calyceraceae are confined mostly to southern South America. This suggested that the three may have diverged from an ancestral complex in Gondwanaland, particularly in Antarctica, during Paleocene time (ca. 60 mya), in a period when mixed temperate-tropical forests covered the continent (Torres, 1985; Drinnen and

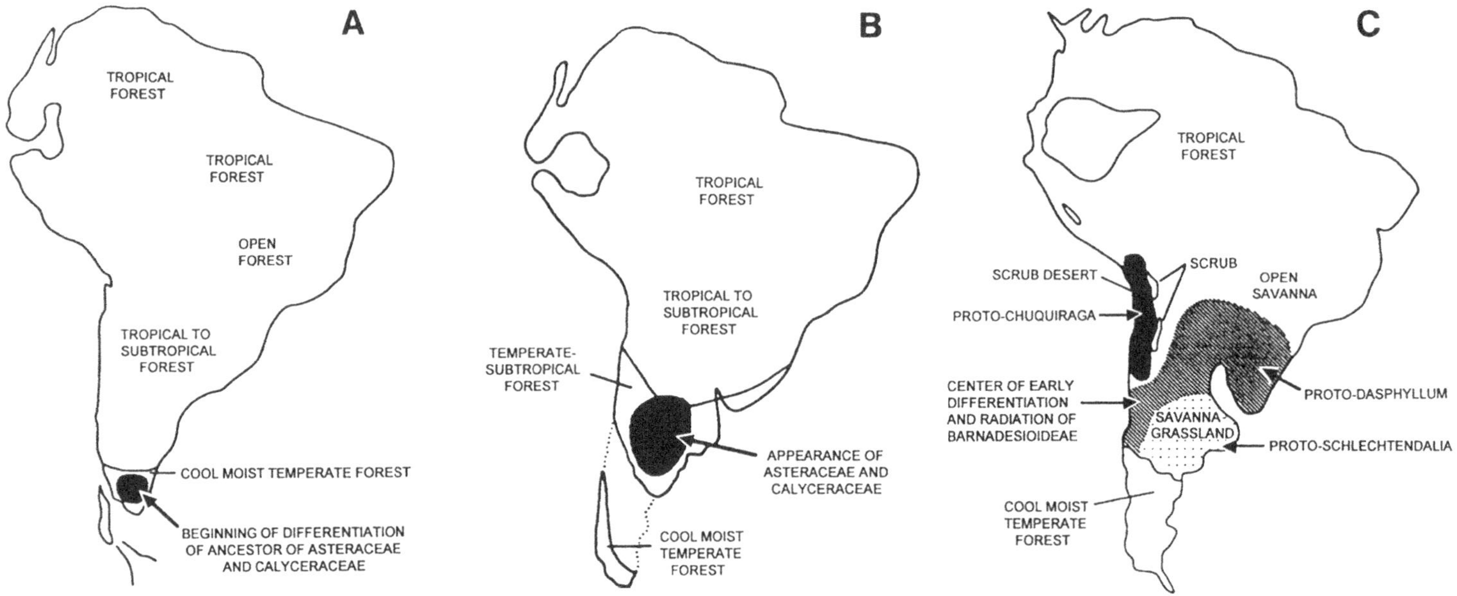

Fig. 2–9. Early biogeography of Asteraceae. A, Eocene (55–38 mya), showing suggested location of ancestor of Asteraceae and Calyceraceae; B, Oligocene (38–25 mya), showing appearance of Calyceraceae and Asteraceae as distinct families; C, Early Miocene (ca. 25 mya) showing center of differentiation and radiation of Barnadesioideae (in early Miocene). From Stuessy et al. (1996); base map modified after Simpson (1983)

Crane, 1990). Goodeniaceae may have diverged on the eastern edge of Gondwanaland as Australia separated (see Bremer and Gustafsson, 1997, for agreement), and likewise, the common ancestor of Asteraceae and Calyceraceae did so on the western side as South America continued its drift (Fig. 2–9A). It is hypothesized that Asteraceae and its sister family, Calyceraceae, diverged as distinct entities in late Eocene or early Oligocene (ca. 38 mya; Fig. 2–9B). Subsequent early differentiation of what we now call subfamily Barnadesioideae (Fig. 2–9C) resulted in a geographic pattern covering the lower part of South America by the early Miocene (ca. 25 mya). The stimulus for development of Asteraceae from out of a common ancestor with Calyceraceae is buried in history, but it may have had something to do with increasingly temperate and more arid environments developing in southern South America during this time. Perhaps this led to new adaptive strategies to deal with water loss, a shortage of pollinators, and changing dispersal agents.

From this center of origin in southern South America, Asteraceae continued evolutionary development and distributional expansion in South America to yield the many Mutisieae still localized there. Cladistic analysis of macromolecular data (Fig. 2–8) provides some support for this possibility, although there is still uncertainty as to exactly how Mutisieae relate to other tribes (Jansen and Kim, 1996).

From these early origins in South America, there occurred dispersal and vicariance and eventual divergence that yielded the present subfamilies and tribes (with no doubt intervening extinctions). That this process occurred rapidly is evidenced by fossil pollen records showing palynomorphs representing different tribes on many continents in later Miocene (Graham, 1996). Distributions of the two subfamilies today obviously contain both tribes with Old and New World concentrations. This means that the present distributional patterns of tribes have resulted from complex dispersal events criss-crossing the major continents, probably several times. Certainly the small diaspores, sometimes equipped with means of obvious dispersal such as hooks, awns, and sticky substances, would have been especially amenable for bird transport. Perhaps this should not be so surprising, for members of the family are well known for their capacity to arrive on oceanic islands and diversify and become distributed rapidly as is seen in such examples as the silverswords in Hawaii, *Scalesia* in Galapagos, and *Dendroseris* and *Robinsonia* in Juan Fernandez.

Section II

Introduction to the Flavonoids

In Section Two we present information on flavonoids designed primarily for the botanical reader. Chapter 3 emphasizes the systematic importance of flavonoids through a presentation of selected examples in different angiosperm and gymnosperm groups. This chapter makes the point that flavonoids have already proven very useful taxonomically at different hierarchical levels in diverse vascular plant taxa and sets the stage for seeking pertinent insights within Asteraceae. Chapter 4 presents a moderately detailed view of flavonoid structural variation in Asteraceae. The primary goal of Chapter 5, wherein we review flavonoid biosynthesis and its genetic control, is to demonstrate that information gleaned from a variety of plant species applies equally well to understanding how these compounds are made in Asteraceae. Chapter 6 rounds out this section with a look at some of the functions played by flavonoids in Nature.

Chapter 3

The Use of Flavonoids as Taxonomic Markers

E. C. Bate-Smith, one of the great pioneers in the application of plant chemicals to taxonomic problems, once described flavonoids as "privileged compounds," (pers. comm.). Flavonoids, more than any other secondary metabolite, best meet the requirements for use of chemicals in taxonomic studies: (1) structural diversity; (2) widespread occurrence; (3) stability; and (4) ease of identification (Harborne, 1967b). Let's take a closer look at each of these.

A detailed treatment of flavonoid structural diversity within Asteraceae can be found in the next chapter. At this point, however, it is our objective to present a very general introduction to flavonoid structure, give examples of a few common compounds, and introduce the reader to the numbering system by which substituted flavonoids can be precisely identified. Members of the family produce nearly all of the classes of flavonoids known (Fig. 3–1). These include the familiar anthocyanin floral pigments, as represented here by pelargonidin 3-*O*-glucoside **[3–1]***. This is a comparatively simple member of this class of pigments; much more complicated ones, such as those obtained from the straw flower, *Centaurea cyanus*, will be encountered in the next chapter. Chalcones and aurones, often collectively referred to as anthochlors, are also found as floral pigments, particularly in members of *Coreopsis* and related genera. The common and co-occurring pair coreopsin **[3–2]** and sulfurein **[3–3]** exemplify these important classes of pigments. Chalcones are characterized by the lack of a heterocyclic ring, whereas aurones are characterized by having a five-membered heterocyclic ring (called the C-ring). All other flavonoid types, the anthocyanins as shown in **[3–1]** and others shown below, have six-membered C-rings. The most common flavonoids that feature six-membered C-rings are the flavanones, represented by naringenin **[3–4]**, flavones, represented by apigenin **[3–5]**, and the flavonols, here represented by kaempferol **[3–6]**. These classes of flavonoids differ in the presence or absence of a double

* These numbers within brackets refer to specific chemical structures that are numbered consecutively within each chapter. Consult figures and legends for details.

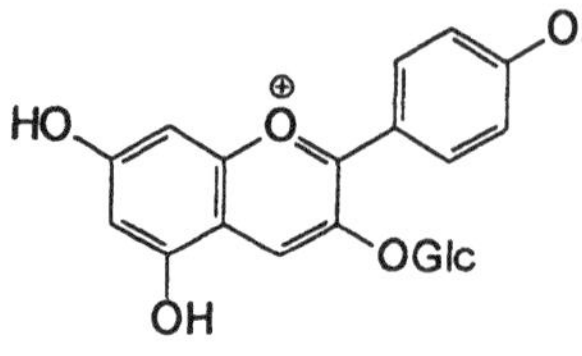

[3-1]

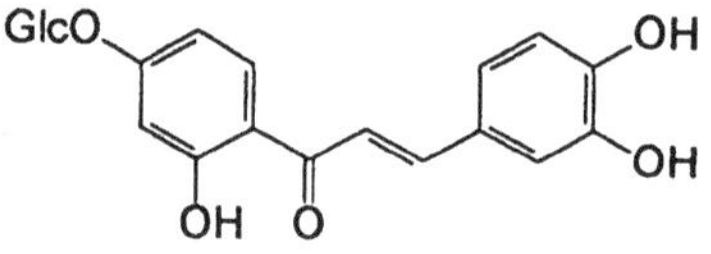

[3-2]

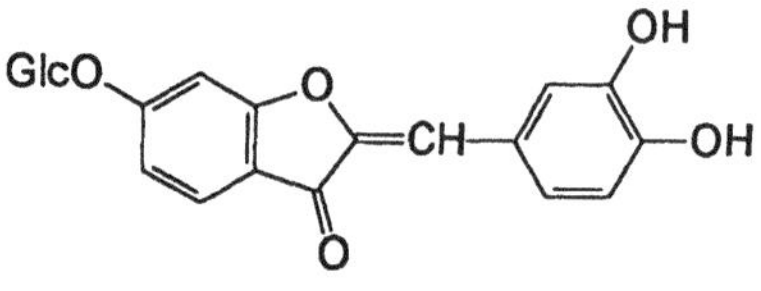

[3-3]

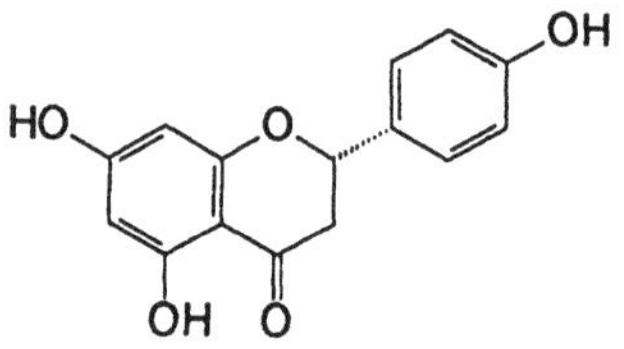

[3-4]

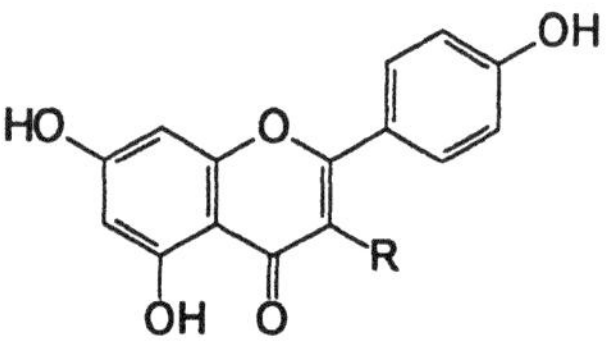

[3-5] R = H

[3-6] R = OH

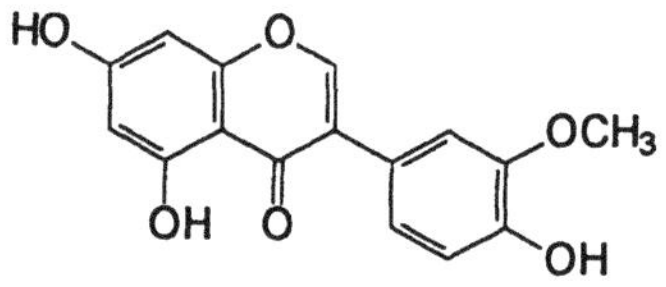

[3-7]

Fig. 3–1. Basic types of flavonoids. Anthocyanin **[3–1]**, chalcone **[3–2]**, aurone **[3–3]**, flavanone **[3–4]**, flavone **[3–5]**, flavonol **[3–6]**, and isoflavone **[3–7]**. See text for detailed nomenclature, description, and discussion. See Fig. 4–1 for lettering of basic rings and numbering of positions in flavonoid molecule

bond between C-2 and C-3 and in the presence of an hydroxyl group at the C-3 position in flavonols. Another variation involves the position at which the B-ring is attached. In flavanones, flavones, and flavonols the B-ring, the one that conventionally appears on the right in the drawings, is attached at C-2. In isoflavones the B-ring is attached at C-3 as seen in 3'-methylorobol **[3–7]**, which happens to be one of the few isoflavones known from the family. As the reader can plainly see, 3'-methylorobol has a level of complexity, that is, the presence of a methoxy group, slightly higher than the other compounds seen so far. It is also obvious that the name "orobol" is likely to be meaningless to most readers! A formal numbering system is available that resolves this difficulty. Using the conventional numbering system, orobol becomes 5,7,3',4'-tetrahydroxyisoflavone and its 3'-methyl ether becomes 5,7,4'-trihydroxy-3'-methoxyisoflavone. This system, of course, requires knowledge of what the base structure of an isoflavone is and what the numbering system is that is used to locate substituents. The system is straightforward and starts by assigning the heterocyclic atom, the ring oxygen in this case, the lowest possible number (for a graphic view of the numbering system, see Fig. 4–1 **[4–3]**). The carbon to which the B-ring is attached in flavanones, flavones, and flavonols is position-2, routinely referred to as C-2. We have already seen C-3 identified in flavonols as the position bearing the characteristic C-ring hydroxyl, in pelargonidin as the point of glucosylation, and in isoflavones as the point of attachment of the B-ring. The carbonyl group characterizes C-4 in most flavonoids with the obvious exception being the anthocyanidins where oxygen has been lost from that position altogether. Continuing in a clockwise direction, the A-ring carbons become C-5, C-6, C-7, and C-8. Positions on the B-ring are identified by primed numbers. Returning to apigenin **[3–5]** as our example, we see there are hydroxyl groups at C-5, C-7, and C-4'. The reader should be aware that the numbering of chalcones, their derivatives, and aurones deviates from this system. The differences, which will be described in Chapter 4, need not concern us at this point.

Other flavonoid classes have been identified from members of Asteraceae including dihydrochalcones, dihydroflavonols (3-hydroxyflavanones), flavans, *C*-glycosylflavones, and even a biflavonoid. Details of the structures of these compounds can be found in Chapter 4. In addition, extensive structural modifications are known: *O*-glycosylation, *O*-methylation, examples of both of which were given above, *C*-alkylation, sulphation, and various combinations of these. It suffices to say that the variation in flavonoids obtained from Asteraceae reflects variation seen in these compounds across the plant kingdom. Very few structural modifications of flavonoids have been discovered in Nature that are not found at least somewhere in Asteraceae.

It is generally accepted that flavonoids occur ubiquitously in angiosperms, gymnosperms, and pteridophytes. In the few cases where they have not been found, the absence almost certainly can be attributed to loss. This breadth of occurrence allows use of flavonoids to assess evolutionary trends across the entire spectrum of vascular plants. One of our goals in this book will be to assess the usefulness of applying flavonoid occurrences in Asteraceae to test relationships and evolutionary trends at all taxonomic levels within the family, from subfamily to assessing the significance of patterns of variation at the population level. We also examine the usefulness of flavonoids in assessing relationships of Asteraceae with other families that comprise Asteridae.

Chemical stability of flavonoids is one of the main advantages of their use. Ubiquitous occurrence would be of little value if a compound were so unstable that isolation destroyed it, or if very special storage requirements had to be met. Flavonoids are very forgiving compounds. With normal laboratory care, flavonoids can be stored and examined at leisure. In most cases, reliable results can be obtained equally well with fresh or dried plant material. It is even possible to obtain highly reliable results from analysis of herbarium specimens, providing they have been properly dried and cared for. In most cases, degradation of flavonoid glycosides to aglycones is minor. Flavonoids have, in fact, passed the supreme test for stability! In several instances, flavonoids have been recovered from fossil plant specimens several million years old (Giannasi and Niklas, 1977).

Other concerns are the physiological, developmental, and ecological stability of a given flavonoid profile. Examples of variation arising from these factors have been discussed in a general review of infraspecific flavonoid variation (Bohm, 1987). Although flavonoid profiles are remarkably stable, qualitatively at least, most investigators are aware that growth and development as well as ecological factors may influence the pigment profile, and have generally worked with the same tissues and have made efforts to compare plants of corresponding age.

Flavonoids are among the most easily accessed secondary compounds known. They are solid, easily soluble in common solvents, lack pesky and complicating chiral properties, and, because of their phenolic nature, have a very approachable chemistry, which enables ease of detection by means of various chromogenic reagents. They absorb ultraviolet radiation conveniently, which provides highly diagnostic methods that have been standardized (Mabry et al., 1970; Markham, 1982). Paper chromatography provided a means of separation and purification of flavonoids that was readily and inexpensively available for most laboratories. The combination of paper chromatography, spray reagents, and ultraviolet spectroscopy provided excellent information that was easily available in botany laboratories and for people with only minimal experience (and interest) in chemistry. Ultraviolet spectroscopic methods, well established as indispensable, have been fine-tuned, paper chromatographic methods have been largely replaced by thin layer or column chromatographic techniques, mass spectroscopy has entered the arena (Markham, 1982), and a variety of nuclear magnetic resonance techniques (NMR) is now available. The outcome of these improvements is the realization that some very complex flavonoid structures exist in Nature. It would have been discouragingly difficult to arrive at the correct structure of some of the more complex acylated anthocyanin derivatives, for example, without access to currently available analytical instrumentation.

Despite the level of analytical sophistication currently available, one only needs access to chromatographic gear and an ultraviolet spectrophotometer for realizing the potential usefulness of flavonoids. The approach outlined in the paragraph above presupposes an interest in determining what the structures of the observed flavonoids are. This is not strictly required. Much early work was done using only chromatographic spot patterns. Although this approach has its limitations, it can still be of value in some situations. Determination of structures is the recommended approach, of course, because with structures in hand, and a knowledge of flavonoid biosynthesis, it becomes possible to assess degrees of similarity or difference between flavonoid profiles. With increased knowledge of both the biosynthesis and genetics of flavonoids, it is possible to discuss differences between taxa in terms of genetic differences.

In the following section we will present a selection of cases where flavonoids have been used to good purpose in taxonomic studies. We have chosen examples that show applications at various levels of the taxonomic hierarchy. We will start with an example of flavonoid composition significant at the ordinal level, proceed with examples involving families, subfamilies, generic relationships, species relationships, and conclude with an example or two involving horticultural varieties.

The literature on the general subject of flavonoid systematics is vast and what follows is little more than a sampler. For more extensive treatments of the subject, the following treatments can be consulted: Alston and Turner (1963a), Harborne (1967b), Harborne and Turner (1984), *The Flavonoids* series (Geissman 1962; Harborne et al., 1975; Harborne and Mabry, 1982; Harborne, 1987, 1993; Bohm, 1998b). So much has been written on the subject, in fact, that any additional treatment runs a serious risk of sounding repetitious. So, with apologies to those readers who have heard it all before, the following examples are offered.

Centrospermae/Caryophyllales

One of the most striking examples from the flavonoid literature actually involves the absence of an otherwise widely distributed compound type from a well defined taxon. In the early 1960s, work appeared describing a new class of pigments obtained from members of Centrospermae (Mabry, 1964, 1966). Originally called nitrogenous anthocyanins because of their elemental composition and their resemblance to anthocyanins, these pigments were formally renamed "betalains" (Fig. 3–2). The structure of betanin, the constituent pigment of beets (*Beta vulgaris*), is shown as structure **[3–8]**. The structure of cyanidin 3-*O*-glucoside **[3–9]** is given for comparison. Betacyanins have been identified from a wide sampling of Caryophyllales including members from Aizoaceae (e.g., *Mesembryanthemum*, the ice-plants), Amaranthaceae (e.g., *Celosia*), Cactaceae, Chenopodiaceae (e.g., beets),

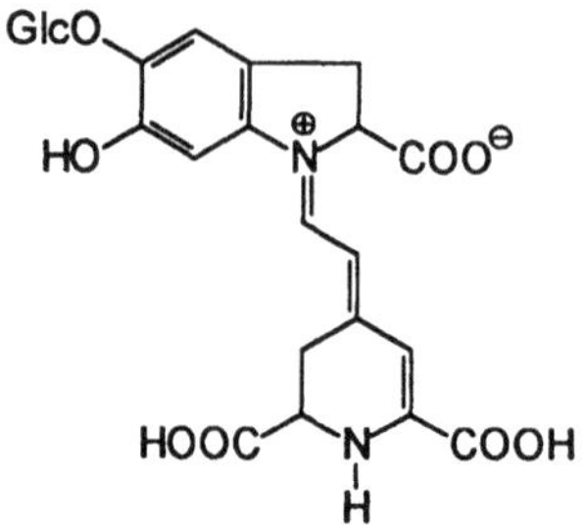

[3-8]

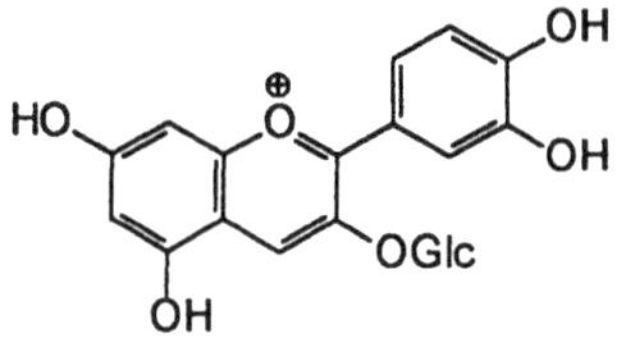

[3-9]

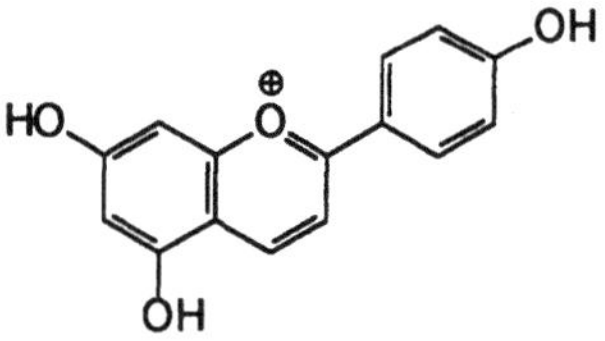

[3-10]

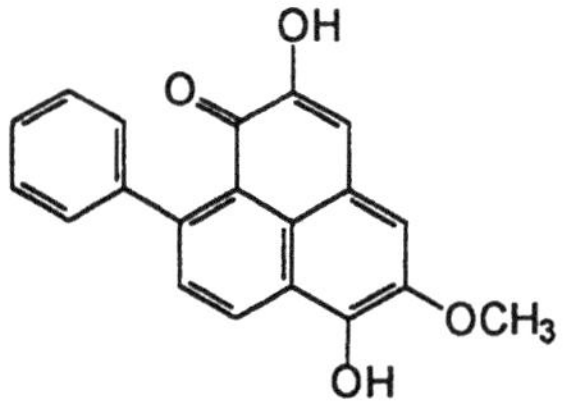

[3-11]

Fig. 3–2. Comparison between betalains **[3–8]** and flavonoids **[3–9]**, and representative flavonoids in Gesneriaceae **[3–10]** and major pigments of Haemodoraceae **[3–11]**

Nyctaginaceae (e.g., *Mirabilis*, the 4-o'clocks; *Bougainvillea*), Portulacaceae (e.g., *Lewisia*) and Phytolaccaceae (poke-weed). Caryophyllaceae and Molluginaceae are unusual in Caryophyllales insofar as their floral pigments are true anthocyanins. In view of the strikingly different structures of these two classes of pigments it is should come as no surprise that their biosynthetic origins are also quite different. Anthocyanins are products of the normal flavonoid pathway deriving their carbon from *p*-coumaroyl coenzyme A and malonyl coenzyme A, whereas betacyanins are formed from two tyrosine units, one of which has undergone a drastic rearrangement to yield an intermediate that comprises the bottom part of compound **[3–8]**. The replacement of anthocyanins by betacyanins as flower pigments in members of Caryophyllales represents one of the best documented, and certainly most evident, cases of convergent evolution seen in plants.

Ulmaceae, one family or two?

Although Ulmaceae have generally been considered a natural assemblage of genera, there has been discussion about subdivisions within the family. The most generally accepted view of the family is that it consists of two subfamilies, Ulmoideae and Celtoideae (see Sweitzer, 1971, for a review of the taxonomy). However, an older view that the two groups of genera should best be considered as comprising two families, Ulmaceae and Celtidaceae, has been resurrected by Grudzinskaya (1965). Flavonoids entered the debate when David Giannasi and Karl Niklas, then at the New York Botanical Garden, became interested in chemicals contained in fossil members of the family (Giannasi and Niklas, 1977; Niklas and Giannasi, 1977a,b). They observed that fossil *Celtis* afforded only *C*-glycosylflavones identical to those from extant species, while fossil *Ulmus* similarly gave only flavonols identical to those of living species.

In order to assess the significance of this difference, Giannasi (1978) undertook a study of representatives of 19 genera of Ulmaceae s.l. Flavonols were reported from *Ampelocera, Aphananthe, Barbeya, Chaetoptelea, Hemiptelea, Holoptelea, Mirandaceltis* (data from the literature), *Phyllostylon, Planera, Ulmus, Zelkova*, and *Gironniera* subgen. *Galumpita*, whereas *C*-glycoflavones were reported from *Celtis, Chaetachme, Lozanella, Parasponia, Plagioceltis, Pteroceltis, Trema*, and *Gironniera* subgen. *Gironniera*. Other than *Ampelocera* and *Aphananthe*, whose affinities within the family are not clear, the flavonol-bearing genera are of Ulmoid affinity. The split in chemistries between the two subgenera of *Gironniera* is interesting since the relationships of these sets of species are also moot. It has been suggested that the flavonol-positive subgen. *Galumpita* may in fact be more closely related to *Aphananthe*, which is a flavonol producer. While flavonoids cannot, on their own, be used to define family, as opposed to subfamily, status, the pigment differences between the two sets of genera give additional clear evidence that differentiation between them has occurred.

Subfamilies of Gesneriaceae

Gesneriaceae are a moderately large family of some 2400 species in 146 genera (Mabberley, 1987). Two subfamilies are thought to exist, Gesnerioideae and Cyrtandroideae, which are distinguished by, among other things, the developmental fate of the cotyledons; they remain equal after germination in members of Gesnerioideae, whereas in Cyrtandroideae they become unequal with one of them sometimes developing into the plant's only photosynthetic organ (Mabberley, 1987). They are also phytogeographically distinct with members of Gesnerioideae native to the New World and Cyrtandroideae native to the Old World. Harborne (1967a) found an interesting biochemical difference between the two groups as well.

A 3-deoxyanthocyanidin **[3–10]** was first reported from members of the family by Robinson et al. (1934) and identified as the 3-deoxy equivalent of pelargonidin **[3–1]**. The new compound was named apigeninidin (its glucoside was called gesnerin) to recognize its relation to apigenin **[3–5]**. (3-Deoxyanthocyanidins bear the same relationship to flavones as anthocyanidins do to flavonols.) The luteolin analogue (two hydroxyl groups on the B-ring) was reported from the family more recently (Harborne, 1960). In his 1967a paper, Harborne reported the presence of 3-deoxyanthocyanins in 18 of 21 species from Gesnerioideae and their absence from 25 species of Cyrtandroideae. A further distinction was noted in that work, namely, that certain chalcones and aurones occurred in species of Cyrtandroideae but were not seen in any members of Gesnerioideae.

Family affiliation of *Lanaria*

Flavonoids need not confirm relationships. In the case of the monotypic South African genus *Lanaria*, flavonoid data argue for removal of the genus from Haemodoraceae, the family with which it had been traditionally associated. Information from embryological and palynological studies (cited by Dora and Edwards, 1991) suggested that the affiliations of *Lanaria* lay elsewhere, possibly with Tecophilaeaceae. Haemodoraceae are characterized by the presence of 9-phenyl phenalenones, whose fundamental ring system is shown as **[3–11]**, and related compounds; no flavonoids have been reported from the family. Dora and Edwards (1991) undertook a chemical study of *L. lanata* with the interesting outcome that none of the typical Haemodoraceae pigments could be detected in any parts of the plant. Instead, they isolated two biflavones, one, the well known amentoflavone, the other a novel compound identified as 4''',5,5'',7,7''-pentahydroxy-4',8''-biflavonyl ether, which they called "lanaroflavone." Thus, chemical data can be added to the list of features by which *Lanaria* can be distinguished from Haemodoraceae. It is also of interest to note that a biflavone has been isolated from *Lophiola aurea* (Zavada et al., 1983), which had been affiliated with Haemodoraceae at one time, but is now considered a more likely candidate for inclusion in Tecophilaeaceae.

Family affinity of *Itea*

Itea is a genus native to eastern Asia with a single species known from eastern North America; in all, about 10 species are recognized (Mabberley, 1987). Engler (1891) first placed *Itea* in Escallonioideae in his broadly-based Saxifragaceae. He later reassigned *Itea* to its own subfamily, Iteoideae (Engler, 1928). A detailed study of the flavonoids of three species of *Itea* along with a chromatographic analysis of several others (Bohm et al., 1988) showed that the genus is characterized by the presence of derivatives of the

C-glycosylflavones orientin, isoörientin, vitexin, and isovitexin. Detailed study of the flavonoids of some 30 genera from Saxifragaceae (Nicholls et al., 1986, and citations therein) showed the profiles to consist principally of flavonol 3-*O*-mono-, 3-*O*-di- and 3-*O*-triglycosides. *C*-Glycosylflavones were not detected in any of the species from Saxifragaceae.

Unfortunately, flavonoid data alone do not allow us to find a home for *Itea*; all we can state with certainty is that, with reference to flavonoids, it does not appear at home with the saxifrages. This is in contrast to the observations of Morgan and D. E. Soltis (1993), however, who, using *rbc*L sequence data, found *Itea* to occupy a clade along with *Astilbe*, *Boykinia*, *Darmera*, *Heuchera*, *Leptarrhena*, *Pterostemon*, and *Saxifraga*. The sequence data suggest that the closest relative of *Itea* is a member of the Mexican genus *Pterostemon*.

Recent studies (Bohm et al., 1998) of both species of *Pterostemon* have shown the presence of an array of *C*-glycosylflavones closely matching the array observed in *Itea*. The *Pterostemon* flavonoid profile also contained flavonol 3-*O*-glycosides, the flavonoids normally expected in species of Saxifragaceae. It is of interest to speculate on the origin of such profound biosynthetic differences in groups of taxa that are clearly related, i.e., *Itea* and *Pterostemon* and the rest of the saxifrage group. A possible scenario would involve the sequential loss of the capacity to biosynthesize flavonols in association with the gain of the capacity to make the *C*-glycosylflavones. *Pterostemon* would occupy a position in the middle, having acquired the capacity to make *C*-glycosylflavones while still maintaining the capacity to make flavonols. A detailed study of the molecular genetics of these systems with regard to flavonoid biosynthesis is clearly needed.

Datiscaceae

Datiscaceae Lindl. consists of three genera with a total of four species. *Datisca cannabina* occurs from northwestern India to Crete, while *D. glomerata* can be found from northern Baja California to northern California. *Octomeles sumatrana* and *Tetrameles nudiflora*, both monotypic tree genera, are native to southeast Asia, New Guinea, and parts of Australia. An alternative view, that Datiscaceae consist solely of the two *Datisca* species and that *Octomeles* and *Tetrameles* comprise Tetramelaceae, has been advanced by Airy-Shaw (1964).

A comparison of the flavonoid chemistry of all four taxa showed interesting differences between *Datisca* and the tree species (Bohm, 1988). That study confirmed the literature reports of B-ring deoxyflavonols and 2'-hydroxyflavonols from *D. cannabina* (Grisebach and Grambow, 1968, and citations therein) and showed that the same compounds occur in *D. glomerata*. Common glycosides of kaempferol and quercetin were also observed in both *Datisca* species, but were the sole flavonoids present in *Octomeles* and *Tetrameles* (Bohm, 1988). The flavonoid data, therefore, provide additional support for distinguishing between the two sets of taxa, although they offer no unequivocal support for suggesting correct family relationships.

A recent study of Datiscaceae s.l. using restriction site mapping of PCR-amplified chloroplast DNA fragments (Rieseberg et al., 1992) also showed significant differences between *Datisca*, and the two tree genera. Nucleotide sequences from the large subunit of *rbc*L suggested that *Datisca* is more closely related to members of Cucurbitaceae or Begoniaceae than it is to *Octomeles* and *Tetrameles* (Swensen et al., 1994).

Heartwood flavonoids of pine

Erdtman (summarized in 1963) examined more than half the known species of pines (*Pinus*) for their heartwood components. The flavonoids identified belonged to three structural classes, flavones, flavanones, and dihydroflavonols. Eight compounds were identified, all of which lacked B-ring oxygenation. With the exception of subsect. *Gerardiane*, subgen. *Diploxylon* is characterized by the absence of flavones. 5,7-Dihydroxyflavanone (pinocembrin) appears to be universally present in both subgenera *Diploxylon* and *Haploxylon*, as is pinobanksin, the corresponding dihydroflavonol. The two subgenera are clearly distinguishable, however, on the basis of the uniform occurrence of pinostrobin (5-hydroxy-7-methoxyflavanone) in all subsections of *Haploxylon* and its absence from all members of *Diploxylon*. Two subsections of subgenus *Diploxylon*, *Gerardiane* and *Strobi*, are further distinguished from all others in the genus by their capacity to make *C*-methylated flavonoids.

The podocarps

Podocarpaceae, taken in the broadest sense, are a large and heterogeneous assemblage of genera distributed in Japan and Central America and widely in the Southern Hemisphere. Re-examination of the New Zealand members resulted in the recognition of three genera (de Laubenfels, 1969, 1978): *Podocarpus*, with four species, *Dacrycarpus* with one, and *Prumnopitys* with two. All seven species had been included in *Podocarpus* by Allan (1961). Markham et al. (1985) showed that the flavonoid profiles of all seven species, plus four non-New Zealand species, corresponded closely with the revised taxonomy. Noteworthy differences include the restriction of flavanones and dihydroflavonols to *Podocarpus* and flavonol 3-*O*-diglycosides to *Prumnopitys*.

C-Glycosylflavones, seen as dominant compounds in *Dacrycarpus* and *Podocarpus*, were not seen in the New Zealand members of *Prumnopitys* (a trace was observed in a non-New Zealand species). Flavonol 3-methyl ethers and compounds with B-ring trihydroxylation occurred as major components of *Dacrycarpus* (these occurred as minor components in two species of *Podocarpus*).

North American lupines

A detailed study of North American lupines provide us with several cases where flavonoids figured significantly in assessing relationships. Two cases will be described below, one that featured qualitative differences and one wherein quantitative differences were of interest. The first example involves annual lupines with connate cotyledons, recognized by Watson (1873) as comprising sections *Platycarpos* (10 species with terminal, racemose inflorescences) and *Lupinellus* (1 species with solitary, axillary flowers). Of some 80 species of lupines examined in the main survey (Nicholls and Bohm, 1987), only the members of these two sections, and *L. arboreus*, were shown to have the *C*-glycosylflavone cytisoside (isovitexin 4'-methyl ether). A further distinction within sect. *Platycarpos* was made by Smith (1944) who described the informal groups Pusilli, characterized by scattered flowers with glabrous keels, and Microcarpi, characterized by whorled flowers with ciliate keels. The major flavonoids of Smith's Pusilli were the 7-*O*-glucosides of apigenin, luteolin, and chrysoeriol, while the principal compounds present in members of his Microcarpi were the *C*-glycosylflavones vitexin, orientin, and cytisoside (Nicholls and Bohm, 1987).

The second example involving lupines concerns *L. sericeus*. This is a highly variable taxon [23 synonyms in the Flora of the Pacific Northwest (Hitchcock et al., 1955)] that occurs from northern Arizona to southwestern Alberta and southeastern British Columbia. Representatives of 32 populations representing the entire range were compared by thin layer chromatography and found to possess the same qualitative array of compounds, but much larger concentrations (estimated by spot size) of what was determined to be orientin were observed in some individuals (Nicholls and Bohm, 1982). Differences in orientin concentration of 181 individuals were then quantified using high performance liquid chromatography (HPLC). Statistical examination of the data showed that significant quantitative differences in orientin content distinguished the southernmost populations (Kaibab Plateau of Arizona) from all others. Plants in the southern part of the species range (Kaibab Plateau to north-central Utah) are characterized by the presence of heavy anthocyanic pigmentation in their lower stems and have been accorded separate recognition as *L. sericeus* subsp. *huffmanii* (Fleak and Dunn, 1971). Left at this stage, our flavonoid data correlated comfortably with observations of others. However, we did not leave well enough alone! Seeds of *L. sericeus* collected from all the previously sampled populations were germinated and maintained in uniform conditions in a growth chamber. Re-analysis of the non-cyanic flavonoids by HPLC showed neither qualitative nor quantitative differences of pigment profiles among individuals representing the entire range of the species. Anthocyanic pigmentation of plants from the southern part of the species' range was maintained, however. The apparent environmental influence on the pigment differences seen in the southern populations has not been examined further.

The *Commelina* blues

The flower pigments of members of Commelinaceae, *Commelina* and *Tradescantia* in particular, have attracted a good deal of interest owing to the unusual nature of blue pigments in Nature. Acylated anthocyanins along with flavone co-pigments have generally been found to be responsible. In a comparative study of anthocyanin pigment patterns in Commelinaceae involving representatives of 10 genera, Stirton and Harborne (1980) found two distinct arrays of compounds (Fig. 3–3). In all members of the family tested, excluding *Commelina* itself, these workers identified the principal anthocyanin component as cyanidin 3,7,3'-tri-*O*-glucoside acylated with caffeic acid on each sugar **[3–12]**. In the case of *Commelina*, however, the pigment consists of delphinidin 3,5-di-*O*-glucoside acylated with *p*-coumaric acid residues **[3–13]** and co-pigmented with a *C*-glycosylflavone. The presence of the 3,7,3'-tri-*O*-glycosidic system also appears to distinguish Commelinaceae from other monocot families, which have various mixtures of anthocyanidin 3-*O*-mono and 3,5-di-*O*-glycosides.

Peperomia in the islands

Peperomia (Piperaceae) is a large genus consisting of perhaps 1000 species (Mabberley, 1987) four of which occur on the Juan Fernandez Islands, Chile. Three of these are endemic; the fourth one is also known from the South American mainland. Studies by Valdebenito et al. (1990a,b; 1992b) were

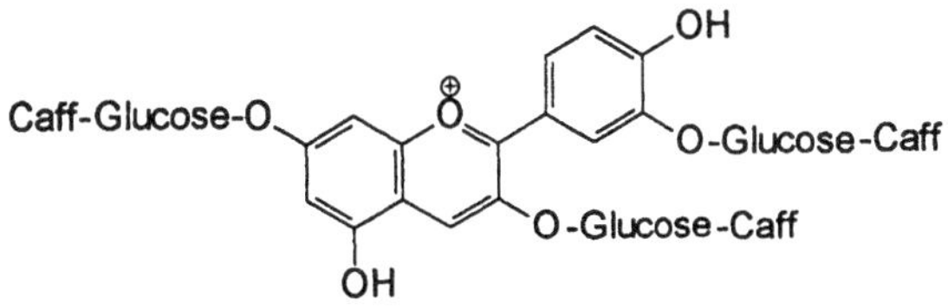

[3-12] Caff = caffeoyl =

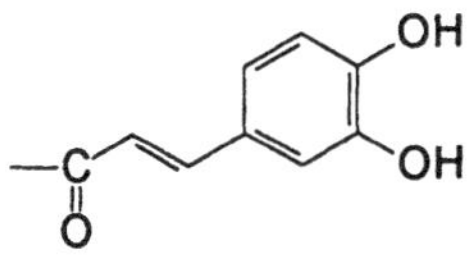

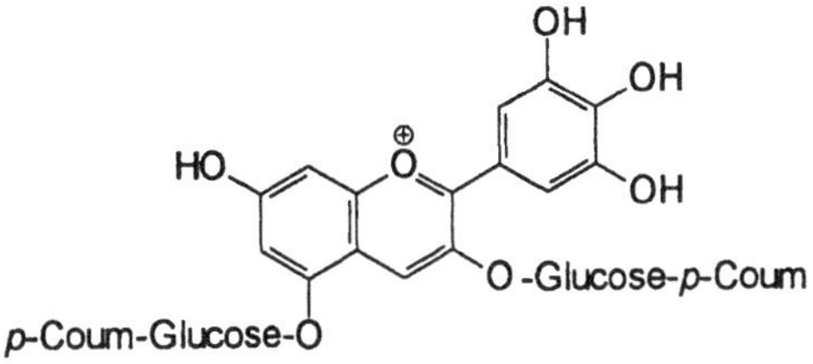

[3-13] p-Coum = p-coumaroyl =

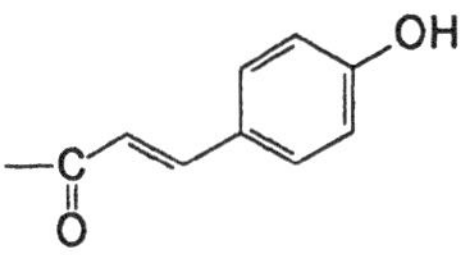

Fig. 3–3. Anthocyanins of Commelinaceae

directed toward the origin of the endemic species. An unexpected result was the observation that *P. tristanensis*, endemic to Inaccessible Island of the Tristan da Cunha Islands group in the southern Atlantic Ocean, is closely related to *P. berteroana* of the Juan Fernandez Islands, and that they ought to be considered subspecies of *P. berteroana*: *P. berteroana* subsp. *berteroana* of the Juan Fernandez Islands and *P. berteroana* subsp. *tristanensis* of Inaccessible Island. It is important to appreciate that the flavonoid profiles of *P. berteroana* subsp. *berteroana* from the two main Juan Fernandez Islands, Masatierra and Masafuera, are different. The finding, then, that the flavonoid profile of subsp. *tristanensis* is identical to that of subsp. *berteroana* from Masafuera not only adds additional support to the view that these two taxa are closely related, but also allows one to state, with some certainty, that the propagule(s) from which the south Atlantic population originated came from plants on Masafuera.

Subgeneric differences in *Galeopsis* (Lamiaceae)

Galeopsis is a genus of about 10 European species (Mabberley, 1987). A study of its flavonoids showed a remarkable division between members of the two subgenera, *Ladunum* and *Galeopsis*. Tomas-Barberan et al. (1991) reported that 8-hydroxyflavones occurred exclusively in members of subgenus *Ladanum* (six species examined). Species of subgen. *Galeopsis* (four species examined) had, in a few species, 6-hydroxyflavonoids. Members of both subgenera had luteolin 7-*O*-glucuronide, apigenin 7-*O*-glucuronide and either apigenin or chrysoeriol 7-*O*-glucoside in various combinations.

Some species of *Vitis*

Despite numerous studies on the genus *Vitis* in North America, there has apparently been little agreement on the delimitation of taxa. Moore and Giannasi (1987) examined the flavonoid composition of five closely related taxa from the southeastern U. S.: *V. aestivalis, V. cinerea* var. *cinerea, V. cinerea* var. *floridana, V. riparia*, and *V. vulpina*. Two noteworthy conclusions emerged from the flavonoid data. First, very similar flavonoid profiles for *V. riparia* and *V. vulpina* suggest that these may not be distinct species, at least in the area involved in this study. Morphological studies confirm this conclusion (Moore, 1990). Perhaps the more interesting finding was that the two varieties of *V. cinerea* had much less similarity to each other than *V. cinerea* var. *cinerea* had to *V. aestivalis*. The latter two shared a suite of *C*-glycosylflavones not otherwise seen in this study. This is not an isolated observation: *V. cinerea* var. *floridana* was originally described as *V. simpsonii*.

Hybrids within *Sideritis* (Lamiaceae)

Much of the taxonomic difficulty in *Sideritis* in southwestern Europe is thought to be due to hybridization (Heywood, 1972). Morphological similarities among species adds to the problem. In a study of hybrids between *S. serrata* and *S. bourgaeana* in Spain, Ferreres et al. (1989) analyzed morphological data and both vacuolar and exudate flavonoids, using Anderson's hybrid index. In most cases, hybrid individuals exhibited flavonoid profiles reflecting contributions from both parents. Supposed backcross individuals had proportionately more of the compound(s) characteristic of the recurrent parent. A factor that distinguishes the two species is the nature of the B-ring methyl ether derivative of sideritoflavone (5,3',4'-trihydroxy-

6,7,8-trimethoxyflavone) present in each. *Sideritis serrata* produces only the 4'-methyl ether; *S. bourgaeana* only the 3'-isomer. Hybrid individuals could be identified by the presence of both isomers. It is interesting to note that the 3',4'-dimethyl ether, which might have been anticipated as a "hybrid" compound, was not detected in any of the plants studied. This is in agreement with the general idea that neither of the monomethyl ethers would be accepted as substrate by either *O*-methyltransferase (De Luca and Ibrahim, 1985b).

Variety is the spice of horticultural life

The value of quantitative data for the resolution of taxonomic problems was well demonstrated with the *Sideritis* example above. An equally impressive application of HPLC analysis of phenolic compounds comes from the work of Biolley et al. (1992) on varieties of *Rosa hybrida*. These workers established the flavonol glycoside profiles of 18 color varieties of rose and showed, using principal components analysis (PCA), that each variety was unique. The method was sufficiently robust that analysis of new samples of cultivars, while increasing the levels of infracultivar variation, did not interfere with recognition of the cultivars. Included in these authors' presentation was a list of other ornamental plants that have been successfully characterized by HPLC methods: *Azalea*, *Petunia*, *Pelargonium*, *Poinsettia*, *Gerbera*, and carnations (*Dianthus*).

Gerbera was the subject of an earlier, qualitative, study. Eighteen varieties of *G. jamesonii* were examined in detail by Asen (1984). Involved in the production of the myriad of colors that make this an important commercial crop are pelargonidin and cyanidin 3-*O*-malonylglucosides, which comprise the major contributors to color, and pelargonidin and cyanidin 3-*O*-glucosides, present in lesser amounts. Accompanying the anthocyanins, and in all likelihood playing a contributing role as co-pigments, are apigenin and luteolin 7-*O*- and 4'-*O*-glucosides, kaempferol and quercetin 3-*O*- and 4'-*O*-glucosides, and their respective malonyl derivatives.

CHAPTER 4

Structural Variation of the Flavonoids of Asteraceae

We concern ourselves in this chapter with a moderately detailed view of the flavonoid chemistry of Asteraceae. Attention will be given to the individual flavonoid types that have been reported, flavones, flavonols, flavanones, chalcones, and the like, and to the diversity of substitution patterns that underlie the richness of the family's pigment profile. Adding to this richness are the various types of derivatives of each of the main structural types: *O*-methyl ethers, *C*-alkyl groups, both *O*- and *C*-linked glycosides, acylated forms of the glycosides, sulphates, etc. Combinations of these features add further to the variety of structures known. Owing to the shear mass of data available much of the information will be presented in tabulated form whenever practicable. The order of presentation will be as follows:

Flavones, flavonols, and flavonoid oxygenation patterns
Flavanones
Dihydroflavonols
Chalcones, dihydrochalcones, and aurones
Isoflavones
Anthocyanins
Glycosides and acylated glycosides
Sulphates
Flavans and C-ring hydroxyflavans
Flavonolignans
Biflavonoids

The nomenclature used in the preceding outline, and in the rest of this book, should be familiar to anyone acquainted with the chemosystematic or phytochemical literature. It is not, however, the formal nomenclature used by *Chemical Abstracts*, and if one needs to search for a specific compound they should be aware that looking for "apigenin 7-*O*-glucoside" may not yield a very fruitful harvest. A general entry under the key word "flavone" would likely exist, but one would have to look at all entries under that lead in order to locate specific references to apigenin 7-*O*-glucoside. A more direct approach exists.

A brief note on the formal conventions used by *Chemical Abstracts* will be useful for anyone not familiar with the official chemical rules of nomenclature. Chalcones are named as derivatives of 2-propen-1-one. Chalcone **[4–1]** itself would be properly known, as 1,3-diphenyl-2-propen-1-one, while isoliquiritigenin, more familiarly known as 2',4',4-trihydroxychalcone, becomes 1-(2,4-dihydroxyphenyl-3-(4-hydroxyphenyl)-2-propen-1-one. Note that the phenyl group attached to the number-1, or carbonyl carbon, is the A-ring (biosynthetically speaking). Dihydrochalcones are based on 1,3-diphenylpropan-1-one, with substituents located as with chalcones. By convention, the A-ring is written on the left and the carbonyl group at C-4 is written on the lower side.

Aurones are found in *Chemical Abstracts* by looking under the general entry 3(2*H*)-benzofuranone. Aurone **[4–2]** itself would be 2-phenylmethylene-3(2*H*)-benzofuranone. Sulfuretin, commonly referred to as 6,3',4'-trihydroxyaurone, would be found under 2-(3,4-dihydroxyphenyl)methylene-6-hydroxy-3(2*H*)benzofuranone. The reader should be aware that at times aurones have also been referred to as benzalcoumaranones in the literature.

The basic structural unit in flavonoids that possess a six-membered heterocyclic ring (the C-ring) consists of two rings, identified as "A" and "C" in structure **[4–3]**. Rings A and C, without further substitution, would be called 4*H*-1-benzopyran-4-one. The flavonoids that we will deal with in this book have a (usually) substituted phenyl group at either C-2 (flavones, etc.) or C-3 (isoflavones, etc.). Substitution of a phenyl group at C-2, as in **[4–3]**, would be represented by the name 2-phenyl-4*H*-1-benzopyran-4-one. Information on apigenin **[4–4]**, commonly known as 5,7,4'-trihydroxyflavone, can be found under 2-(4-hydroxyphenyl)-5,7-dihydroxy-4*H*-1-benzopyran-4-one. The equivalent isoflavone (genistein) **[4–5]** would be 3-(4-hydroxyphenyl)-5,7-dihydroxy-4*H*-1-benzopyran-4-one. Flavonols would add the additional hydroxyl group to the fundamental name so that kaempferol derivatives, for example, would be found under 2-(4-hydroxyphenyl)-3,5,7-trihydroxy-4*H*-1-benzopyran-4-one. Flavanones and dihydroflavonols would add the "2,3-dihydro" term to the appropriate name. As an example of the latter, naringenin, 5,7,4'-trihydroxyflavanone, derivatives would be found under 2-(4-hydroxyphenyl)-2,3-dihydro-5,7-dihydroxy-4*H*-1-benzopyran-4-one.

In the case of flavanones and dihydroflavonols, appropriate notation is also necessary to inform the reader what stereochemical forms are involved. The most frequently seen flavanone stereochemistry has the 2-phenyl group below the plane of the molecule, which makes it (2*S*) **[4–6]**. In the case of dihydroflavonols, the most commonly seen stereochemistry is (2*R*,3*R*) which also has the 2-phenyl group below the plane of the molecule but has the 3-hydroxyl group above it **[4–7]**. This is one of the *trans* or *E*-forms possible. As a further example, the structure shown as **[4–8]** has both substituents beneath the plane of the molecule; they exhibit one of the *cis*- or *Z*-forms; the molecule is identified as having (2*R*,3*S*) stereochemistry.

Another problem attendant upon any review of natural product chemistry involves the use of common or trivial names. We all know a few of these, mainly the ones we use regularly in our particular area of study. For example, the worker who deals with legumes on a regular basis is bound to recognize more common names of isoflavones than someone who spends most of their time working with Asteraceae where isoflavones are uncommon. With the exception of a few very widely used common names, such as naringenin and eriodictyol for the common flavanones, apigenin and luteolin for the common flavones, and kaempferol, quercetin, and myricetin for the common flavonols, we will provide chemical names when they are necessary.

From time to time we will also use a notation that avoids names altogether. This system involves writing the positions of hydroxyl groups and positions of methoxyl groups separated by a slash "/". Thus, quercetin 3,3'-dimethyl ether would be represented as 574'/33'-flavone using this notation. A flavonoid with only hydroxyl substitutions would have the slash at the end followed by a "-" sign so that quercetin would be written 3573'4'/-flavone. A compound with complete *O*-methylation would be written as -/3573'4'-flavone. Additional substituents are handled by providing a third division: 5,4'-dihydroxy-3,6,3'-trimethoxy-7,8-methylenedioxyflavone would be written as 54'/363'/78-MDO-flavone **[4–9]**. This notation will be used extensively in the tables accompanying the chapters.

One final note: the reader will have seen in the case of isoliquiritigenin above that we placed the primed numbers before the un-primed numbers, e.g., (2',4',4-...). In the case of chalcones and dihydrochalcones, this is strictly speaking, improper. However, this convention reflects our policy of *always* identifying A-ring substituents before B-ring substituents. In the case of flavones, and other flavonoids with a C-ring, this is not a problem. When dealing with flavonols, the substituent at the 3-position is always shown first.

Before we turn our attention fully to the flavonoid chemistry of the family, it is useful to remind the reader of a major problem in surveys of this sort. We are limited, obviously, to what has been published on the subject. In the case of workers dealing specifically with flavonoid occurrence from a systematic perspective, we can reasonably expect a given report to be, if not comprehensive, at least representative of the biosynthetic capacities of the taxon or taxa under consideration. In many instances, however, reports

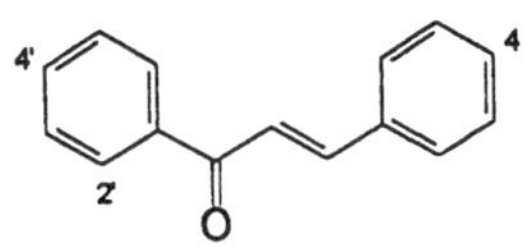

[4-1]

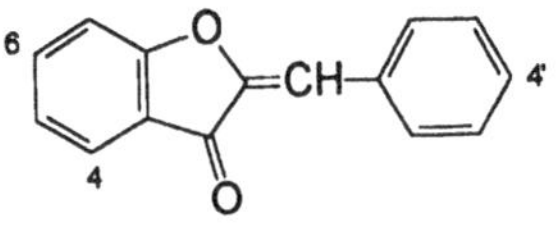

[4-2]

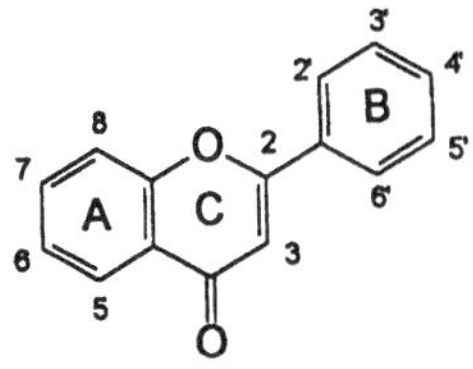

[4-3]

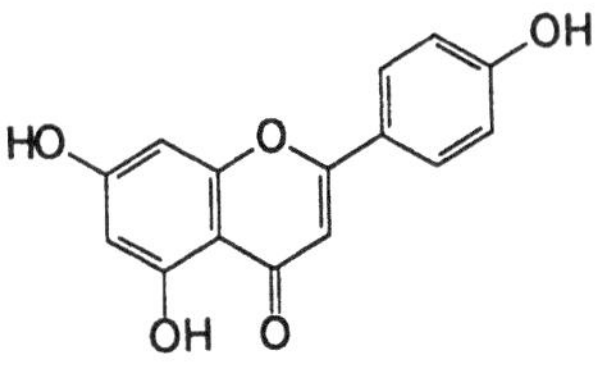

[4-4]

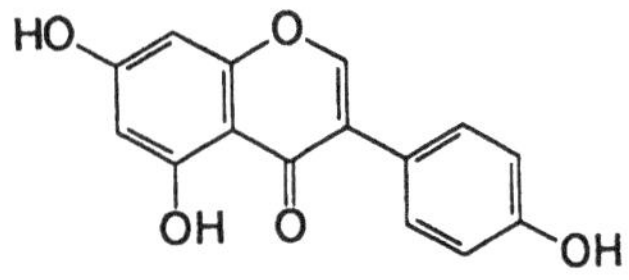

[4-5]

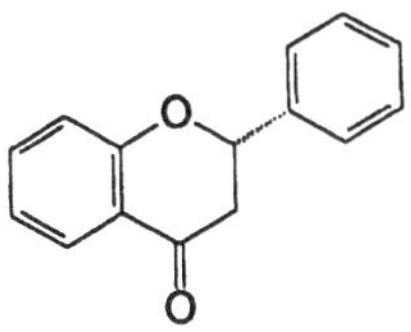

[4-6]

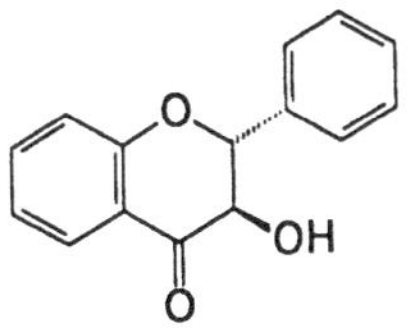

[4-7]

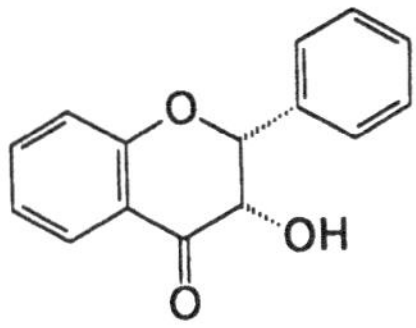

[4-8]

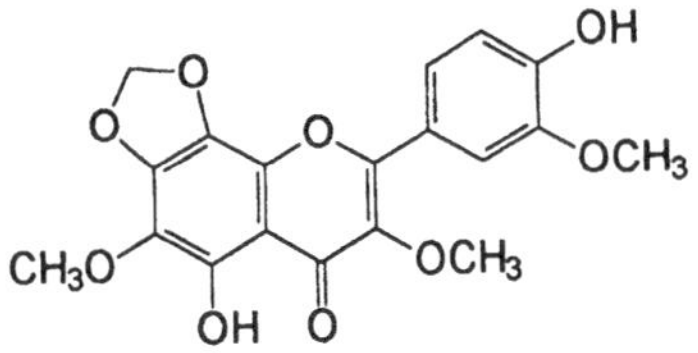

[4-9]

Fig. 4–1. Representative flavonoids showing basic rings, numbering system, variations in stereochemistry, and illustrating chemical nomenclature. Chalcone **[4–1]**; aurone **[4–2]**; letter and number system for flavonoids **[4–3]**; flavone (apigenin) **[4–4]**; isoflavone (genistein) **[4–5]**; flavanone with typical stereochemistry **[4–6]**; dihydroflavonol **[4–7]**; *cis*- (*Z*)-form **[4–8]**; complex methylenedioxyflavone **[4–9]**. See text for discussion

of flavonoids appear as by-products of the real purpose of the work. This is commonly seen in Asteraceae where workers, primarily interested in sesquiterpene lactones, have encountered one or more nonpolar flavonoids. Such information, nonetheless welcome, comes to us as a consequence of the compounds' solubility properties and not through specific interest on the part of the workers. In many, if not most, instances no attempts have been made by the workers involved to look at the total flavonoid profile of the taxa in question. As a case in point, we can cite our experience with *Marshallia*, a genus that has provided several very unusual nonpolar flavonoids (Bohlmann et al., 1979e, 1980c; Herz and Bruno, 1987b; Jakupovic et al., 1988e). Examination of the polar fraction clearly showed the presence of quercetin 3-*O*-glycosides (Bohm, unpubl.).

Another source of skewed results involves flavonoids that occur on plant surfaces. These glandular, or exudate, compounds have attracted considerable attention owing to their presumed ecological importance. They are also easy to isolate. In many instances we are left ignorant about the vacuolar components of these plants. Of this we are also guilty: in our studies of the *Balsamorhiza—Wyethia* complex and the Hawaiian silverswords, vacuolar flavonoids were examined in only one species of the former (Bohm and Choy, 1987) and two of the latter (Crins et al., 1988a), while most of our attention was paid to the non-polar fraction.

Occasionally, a flavonoid may be identified as the active component of a plant through biological-activity directed fractionation. We are fortunate when the investigators identify all or most of the components in the fraction, as seen in a recent study of anti-viral compounds from *Chrysanthemum morifolium* (Hu et al., 1994). We can hardly fault others who confine their work to the active compound(s) since they are motivated by different factors. We are, of course, happy to use any information we can find!

Sampling error will always plague a project of this sort. Gaps in the literature are huge and frequently involve key genera or even subtribes. There are also significant gaps at the tribal level, particularly Arctoteae, Calenduleae, and Liabeae. However, we feel that the number of taxa within most tribes and subtribes of Asteraceae that have been studied allow for useful family-wide comparisons to be made. So, on to the subject.

Flavones, flavonols, and flavonoid oxygenation patterns

Flavones and flavonols are the most commonly encountered flavonoid structural types in the family. The most widely occurring substitution patterns for flavones are 5,7,4'-trioxygenation (apigenin type) **[4–10]** and 5,7,3',4'-tetraoxygenation (luteolin type) **[4–11]**. For flavonols, 3,5,7,4'-tetraoxygenation (kaempferol type) **[4–12]** and 3,5,7,3',4'-pentaoxygenation (quercetin type) **[4–13]** are the most common. It should not be surprising that the most common flavones and flavonols reported from the family are, in fact, the four compounds just named. Apigenin, or one or more of its glycosidic derivatives, has been reported from at least one member of some 118 genera of over 430 for which flavonoid data are available. Similarly, luteolin or its glycosides occur in some 139 genera, kaempferol or its glycosides in 114, and quercetin or its glycosides in 212. The number of compounds having these oxygenation patterns jumps dramatically when all struc-

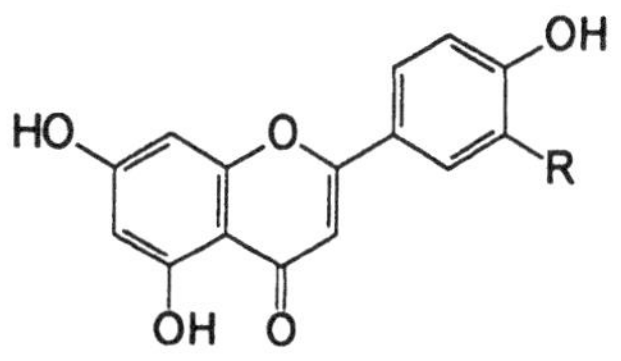

[4-10] R = H, Apigenin
[4-11] R = OH, Luteolin

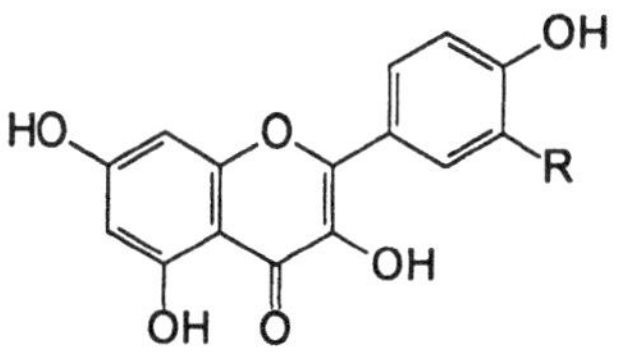

[4-12] R = H, Kaempferol
[4-13] R = OH, Quercetin

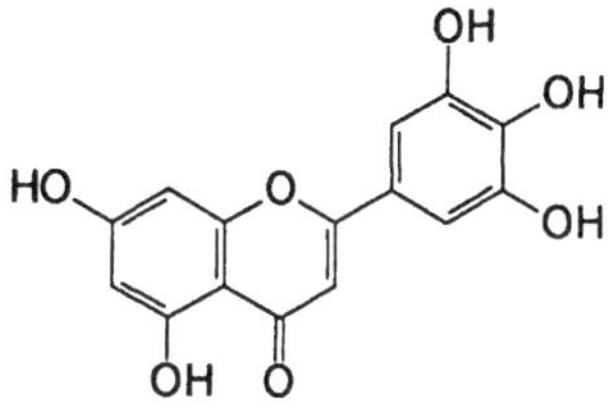

[4-14]

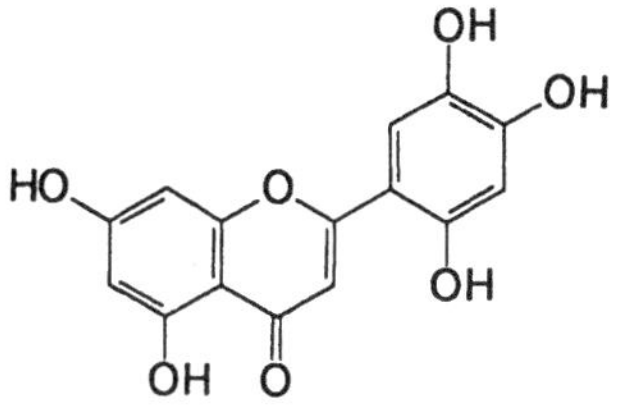

[4-15]

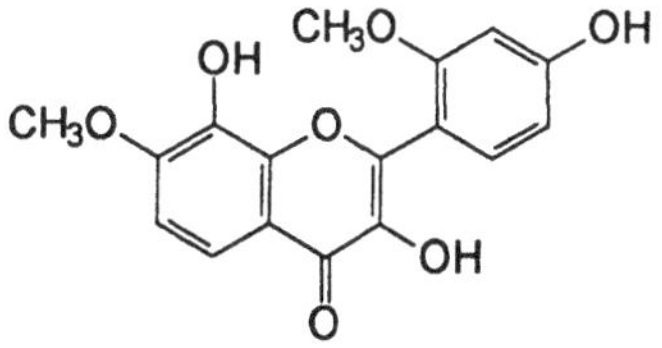

[4-16]

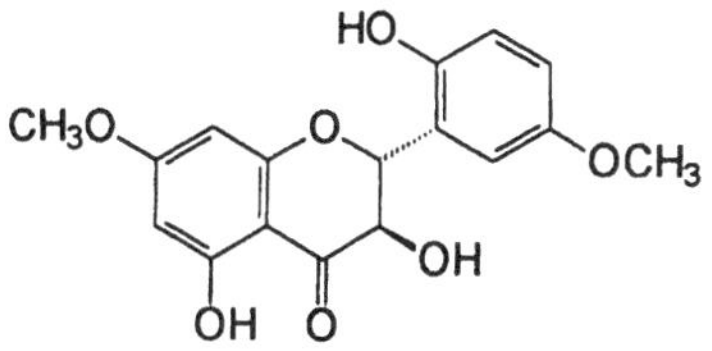

[4-17]

Fig. 4–2. Representative flavones and flavonols and B-ring substitution patterns for flavones **[4–10, 4–11]** and flavonols **[4–12, 4–13]**; trioxygenated B-rings **[4–14, 4–15]**; dioxygenated B-rings **[4–16, 4–17]**

tural classes are included in the census. Disregarding other structural features, flavonoids with 4'-oxygenation occur in over 270 genera, while 3',4'-oxygenated flavonoids occur in about 315.

An appreciation of the richness of the family's flavonoid chemistry can be gained by looking at the number of different A-ring and B-ring oxygenation patterns described, while keeping in mind that these are merely the foundations upon which all of the other structural features are built. For the purpose of assessing the variety of substitution patterns known, it is convenient to include flavanones and dihydroflavonols in this presentation since they also represent major structural types known from the family. It should be noted that not all types of substitution have been reported within each structural class, nor do all types occur in all taxa.

Table 4–1 lists oxygenation patterns for flavones, flavonols, flavanones, and dihydroflavonols. Reading down the columns gives oxygenation pat-

Table 4–1. Oxygenation patterns of flavones, flavonols, flavanones, and dihydroflavonols in Asteraceae

Flavones	Flavonols	Flavanones	Dihydroflavonols
57 (Chr)[1]	357 (Gal)	57 (Pc)	
567	3567	567	3567
578	3578	578	
5678	35678		
574'(Apig)	3574' (K)	574' (Nar)	3574' (Arom)
5674' (Scut)	35674'		35674'
5784'	35784' (Herb)	5784'	35784'
56784'	356784'		
573'4' (Lute)	3573'4' (Q)	573'4' (Er)	3573'4' (Tax)
5673'4'	35673'4' (Qt)	5673'4'	35673'4'
5783'4'	35783'4' (Gos)	5783'4'	35783'4'
56783'4'	356783'4'		
573'4'5' (Tri)	3573'4'5' (My)	573'4'5'	
5673'4'5'	35673'4'5'	5673'4'5'	
5783'4'5'	35783'4'5'		
56783'4'5'	356783'4'5'	56783'4'5'	
	3572'5'	572'5'	3572'5'
	35672'5'	5672'5'	
572'4'5'	3572'4'5'		
5672'4'5'	35672'4'5'		
	35782'4'5'		
56782'4'5'	356782'4'5'		
5672'3'4'5'			
56782'3'4'5'			
5672'3'4'6'			
73'4'	373'4' (5-dQ)	73'4'	373'4'
673'5'			
564'			
584'			

[1] Common names: Apig = apigenin; Arom = aromadendrin; Chr = chrysin; Er = eriodictyol; Gal = galangin; Gos = gossypetin; Herb = herbacetin; K = kaempferol; Lute = luteolin; My = myricetin; Nar = naringenin; Pc = pinocembrin; Q = quercetin; Qt = quercetagetin; Scut = scutellarein; Tax = taxifolin; Tri = tricetin; 5-dQ = 5-deoxyquercetin.

Table 4–1. (Continued)

Flavones	Flavonols	Flavanones	Dihydroflavonols
582'5'			
583'5'			
	356783'		
	3782'4'		
	3783'4'		
	3573'5'		
	35684'		
		7	
		257	
		74'	
		573'	

terns of increasing complexity with highly unusual, or unique, members at the bottom. These series start with B-ring deoxyflavonoids (57 = 5,7-dioxygenated; 357 = 3,5,7-trioxygenated), followed by 6-substituted B-ring deoxyflavonoids (567 or 3567), 8-substituted B-ring deoxyflavonoids (578 or 3578), and so forth, followed by compounds with oxygenation at position-4'. Reading across, one sees a family of compounds that shares A-ring and B-ring oxygenation patterns. For example, the familiar group consisting of flavone (574', apigenin), flavonol (3574', kaempferol), flavanone (574', naringenin) and dihydroflavonol (3574', aromadendrin) would be found in row five. Abbreviated names have been added to assist the reader in identifying some common flavonoids.

Twenty-eight different oxygenation patterns have been reported for flavones, 28 as well for flavonols, 18 for flavanones and nine for dihydroflavonols. Many of the oxygenation patterns are shared by flavones and flavonols and a few are shared by all four flavonoid types. As the oxygenation patterns become more complex, the frequency of occurrence drops, as might be expected. Only 34 genera have species that accumulate 3',4',5'-trioxygenated compounds, e.g. **[4–14]**. Fourteen genera have flavonoids with 2',4',5'-B-ring substitution, e.g., the isoëtin-type **[4–15]**. Nine of these are members of Lactuceae (see Chapter 13). Flavonoids with this B-ring pattern also occur in species of *Ageratum* (Quijano et al., 1992), *Artemisia* (Shilin et al., 1989), *Brickellia* (Iinuma et al., 1985), *Gutierrezia* (Fang et al., 1985a,b; Li et al., 1987), and *Gymnosperma* (Yu et al., 1988). In all cases in these genera, excepting *Artemisia hispanica*, from which isoëtin 5'-*O*-glucoside was obtained (Marco et al., 1988), flavonoids with the 2',4',5'-

pattern are *O*-methylated. In Lactuceae, isoëtin occurs in a variety of glycosidic forms, but never *O*-methylated.

Many members of Asteraceae make flavonoids that have oxygenation at C-6 or C-8 or at both positions. One hundred and seventy genera have been recorded as possessing at least one flavonoid with 6-oxygenation; 31 have at least one flavonoid with 8-oxygenation; forty-three have at least one with 6,8-dioxygenation. Breaking these numbers down a little, we see that over one hundred genera, 111 to be exact, contain species that make flavonoids with both a 4'-oxygenated B-ring and oxygenation at C-6; the combination of 3',4'- and 6-oxygenation occurs in compounds from about 136 genera. Flavonoids with (3),5,7,8,4'-oxygenation (the 8–4' combination) occur in 19 genera, while those with (3),5,7,8,3',4'-oxygenation (the 8–3',4' combination) have been reported from 17 genera. Flavonoids with the 6,8–4' combination occur in 25 genera, those with the 6,8–3',4' combination occur in 19. As complexity increases the number of genera exhibiting a given combination continues to decrease. The 6,8–3',4',5' combination has been observed in five genera, *Ageratum, Conoclinium, Eupatorium,* (Eupatorieae), *Hymenoxys* (*Tetraneuris*) (Helenieae), and *Psiadia* (Astereae). More complex combinations are commented on below.

The occurrences of oxygenation patterns determined for each tribe are given in Table 4–2 (these data will be encountered in a slightly expanded form in Table 18–2). It is important to point out here that, owing to ongoing problems of tribal affinities and nomenclatural changes, the number of genera representing each tribe and the frequencies of the various features are not necessarily exact. Running the risk of garnering criticism for being inexact in this compilation, we would point out that our primary interest lies in detecting trends in the occurrence data. It is abundantly clear, as will be discussed in more detail in Chapter 19, that absolute differences among tribes would be extremely unlikely.

Two unusual dioxygenated B-ring patterns warrant comment. The pattern is a familiar feature from some legumes and from members of Moraceae (mulberry family), but the appearance of flavonoids with 2',4'-dioxygenation in Asteraceae is rare. The report of 8,4'-dihydroxy-3,7,2'-trimethoxyflavone **[4–16]** from *Artemisia japonica*, Japanese wormwood, seems to be unique within Asteraceae (Gu and Tu, 1993). Four species, representing four genera, accumulate flavonoids with 2',5'-oxygenation: *Bidens parviflora* (De Tommasi et al., 1992), *Blumea balsamifera*, e.g., **[4–17]** (Barua and Sharma, 1992), *Inula grantioides* (Ahmad and Ismail, 1991a,b), and *Tanacetum sibiricum* (Stepanova et al., 1981).

Equally rare in Asteraceae are flavonoids with four oxygenations in the B-ring. The 2',3',4',5'-pattern forms the framework upon which elaborate arrays of *O*-methyl and methylenedioxy derivatives are based in three gen-

Table 4–2. Occurrences of flavonoid oxygenation patterns in Asteraceae by tribes

Tribe	Gen[3]	B-Ring[1]					A-Ring			Combinations[2]											
		Nil	K	Q	M	I	6	8	68	6K	6Q	6M	8K	8Q	8M	68K	68Q	68M	6I	8I	68I
Anthemideae	44	3	30	33	4	1	32	4	2	16	29	4	1	3	0	1	0	0	1	0	0
Astereae	48	5	30	35	12	2	19	4	9	11	13	4	4	4	2	4	7	4	2	1	2
Barnadesieae	7	0	7	7	0	0	0	0	0	0	0	0	0	0	0	0	0	0	0	0	0
Calenduleae	3	0	1	2	1	0	0	0	0	0	0	0	0	0	0	0	0	0	0	0	0
Cardueae	26	2	19	25	2	0	10	3	0	7	6	0	2	1	0	0	0	0	0	0	0
Eupatorieae	32	3	23	25	6	2	22	2	3	16	21	4	2	2	0	2	3	3	2	0	1
Helenieae	35	0	29	27	1	0	24	2	6	16	21	0	2	1	5	5	0	1	0	0	0
Heliantheae s.s.	75	3	48	53	0	0	27	3	13	17	24	0	3	4	0	9	8	0	0	0	0
Heliantheae s.l.	110	3	77	79	1	0	51	6	19	33	45	0	5	5	5	14	8	1	0	0	0
Inuleae s.s.	12	0	8	7	2	0	10	0	1	9	8	1	0	0	0	1	0	0	0	0	0
Gnaphalieae	25	9	18	18	1	0	8	5	8	5	0	0	1	1	1	3	0	0	0	0	0
Plucheeae	6	0	3	5	0	0	3	1	1	2	4	0	1	1	1	0	1	0	0	0	0
Inuleae s.l.	43	9	29	30	3	0	21	6	10	16	12	1	2	2	2	4	1	0	0	0	0
Lactuceae	39	0	21	27	1	9	5	0	0	5	0	0	0	0	0	0	0	0	0	0	0
Liabeae	4	0	2	4	0	0	1	0	0	0	1	0	0	0	0	0	0	0	0	0	0
Mutisieae	22	4	15	10	0	0	2	1	0	4	2	0	1	0	0	0	0	0	0	0	0
Senecioneae	21	0	11	18	2	0	0	2	0	0	0	0	2	1	0	0	0	0	0	0	0
Tageteae	7	0	5	7	1	0	7	0	0	3	7	0	0	0	0	0	0	0	0	0	0
Vernonieae	35	1	9	13	2	0	1	0	0	3	0	0	0	0	0	0	0	0	0	0	0
Totals	441	30	278	314	35	14	170	31	43	112	136	12	19	17	19	25	19	8	5	1	3

[1] Nil = B-ring deoxyflavonoids; K = kaempferol pattern (4'); Q = quercetin pattern (3'4'); M = myricetin pattern (3'4'5'); I = isoëtin pattern (2'4'5').

[2] 6K = flavonoids with both 6-oxygenation and a kaempferol B-ring pattern; 8M = flavonoids with both 8-oxygenation and a myricetin B-ring pattern; 68I = flavonoids with both 6,8-dioxygenation and an isoëtin B-ring pattern; etc.

[3] Number of genera for which appropriate flavonoid data are available.

[4] Totals calculated using Heliantheae s.l. and Inuleae s.l. data.

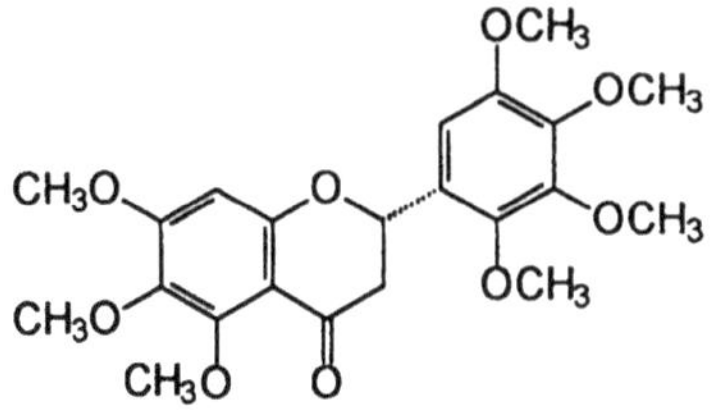

[4-18]

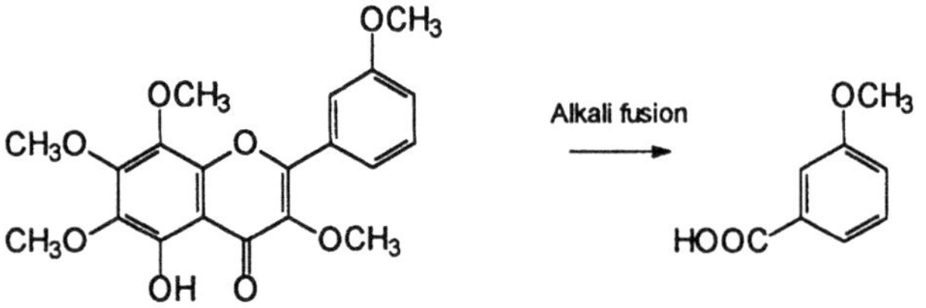

[4-19] [4-20]

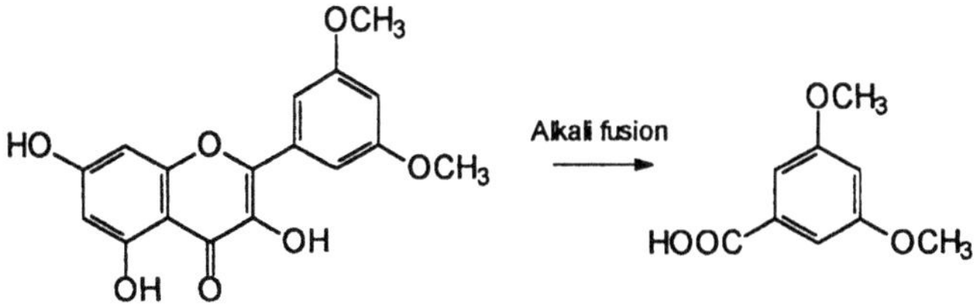

[4-21] [4-22]

Fig. 4–3. Flavonoids with unusual B-ring oxygenation patterns. Tetraoxygenated B-ring **[4–18]**; 3'-oxygenation **[4–19]** and alkali degradation product **[4–20]**; 3',5'-dioxygenation **[4–21]** and alkali degradation product **[4–22]**

era: *Ageratum*, e.g., **[4–18]**, (Quijano et al., 1982b, 1985, 1992), *Brickellia* (Goodwin et al., 1984), and *Psiadia* (Al-Yahya et al., 1987; El-Feraly et al., 1990; Mossa et al., 1992). A second tetra-oxygenated B-ring structure was proposed by Roberts et al. (1984) for a compound from *Brickellia chlorolepis* and *B. veronicaefolia*. It was concluded that "brickellin" was 5,6,6'-trihydroxy-7,2',3',4'-tetramethoxyflavone. A re-examination of spectral data, coupled with synthetic studies, led Iinuma et al. (1985) to suggest a revised structure for the compound, namely, 5,2'-dihydroxy-3,6,7,4',5'-pentamethoxyflavone.

A few flavonoids with other, highly unusual, B-ring oxygenation patterns have been isolated from members of the family (Fig. 4–3). A recent study of *Flourensia retinophylla* (Stuppner and Müller, 1994) afforded three flavonoids distinguished by the rarely seen 3'-oxygenation pattern. Two of the compounds, 5,7,3'-trihydroxyflavanone and its 3'-*O*-methyl ether, were previously known, while 5,7,3'-trihydroxyflavan-3-ol isobutyrate was identified as a new natural product. The only other report of a 3'-oxygenated flavonoid in the family comes from the work of Dominguez and Torre (1974) on *Gymnosperma glutinosum*. In addition to the known 5,7-dihydroxy-6,8,3',4',5'-pentamethoxyflavone (luiselizondin), these workers suggested that an accompanying compound was 5-hydroxy-3,6,7,8,3'-pentamethoxyflavone **[4–19]**. Consistent with this B-ring pattern was the production of *m*-methoxybenzoic acid **[4–20]** by fusion of the unknown with alkali.

Flavonoids with 3',5'-dioxygenation have been identified from three species, the earliest of which was 3,5,7-trihydroxy-3',5'-dimethoxyflavone (morelosin IV) **[4–21]** obtained from *Baileya multiradiata* by Dominguez et al. (1976). Fusion with alkali afforded 3',5'-dimethoxybenzoic acid **[4–22]** in this case. More recently, Ahmad and Ismail (1991b) isolated "grantionin" from *Inula grantioides* and showed it to be 7-hydroxy-6,3',5'-trimethoxyflavone. This unusual structure, with *meta*-substitution on the B-ring and a modified *ortho*-substitution on the A-ring, suggests the possibility that its biogenesis could have involved a retrochalcone, a small but well-known class of flavonoids that appear to have their A- and B-rings on backward (see Bohm, 1994, pp. 398–399, for details on retrochalcones). The third member of this small group is 5,8,3',5'-tetrahydroxyflavanone, which Shi et al. (1992) reported from *Artemisia subdigitata.*

The last compound has, in addition to an unusual B-ring, an A-ring feature that one sees only rarely in flavonoids, namely, the absence of oxygenation at C-7 (Fig. 4–4). It is of interest that another *Artemisia* species, *A. campestris* ssp. *maritima*, contains two such compounds, 5,8,4'-trihydroxyflavanone **[4–23]** and 5,6-dihydroxy-4'-methoxyflavanone **[4–24]** (Rauter et al., 1989). Gonzalez et al. (1991) reported 5,6,8,3',4',5'-hexamethoxyflavone from *Ageratum conyzoides*, but reinvestigation of the compound by Horie et al. (1993) showed that the correct structure is 5,6,7,3',4',5'-hexamethoxyflavone.

The vast majority of flavonoids from the family have oxygenation at C-5. The first report of a 5-deoxyflavonoid in the family appears to be that of Chumbalov and Fadeeva (1969) who identified 7,8,3',4'-tetrahydroxy-3-methoxyflavone **[4–25]** from *Artemisia transiliensis*. Demethylation gave the known 3,7,8,3',4'-pentahydroxyflavone (5-deoxygossypetin). Two years later, Correa and Cervera (1971) described the isolation 3'-hydroxy-7,4'-dimethoxyflavone **[4–26]** from *Tithonia tubaeformis*. Fusion with potassium hydroxide gave 2-hydroxy-4-methoxybenzoic acid as one of the fragments, which is consistent with the A-ring as written. The structure of "tithionine" was confirmed through synthesis (Correa and Cervera, 1971).

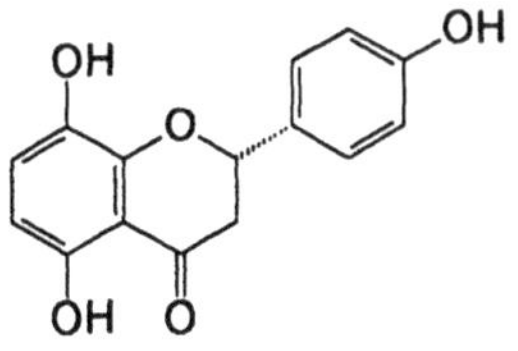

[4-23]

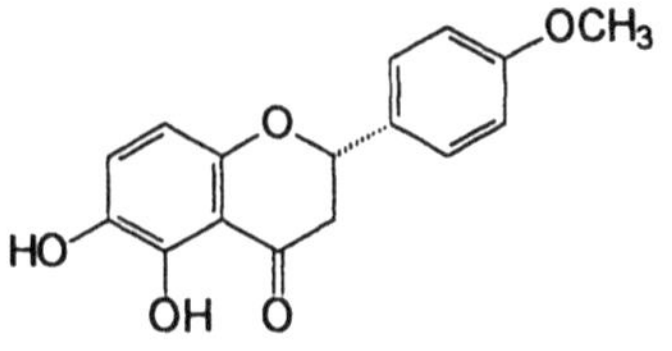

[4-24]

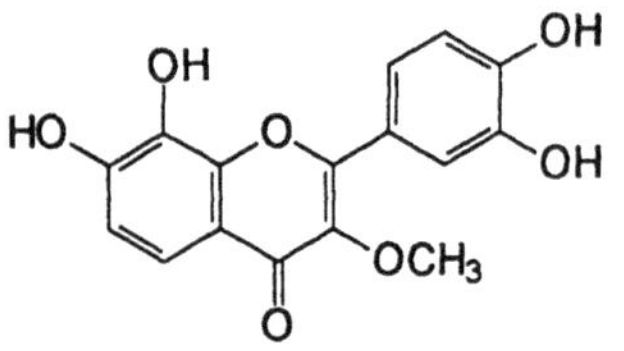

[4-25]

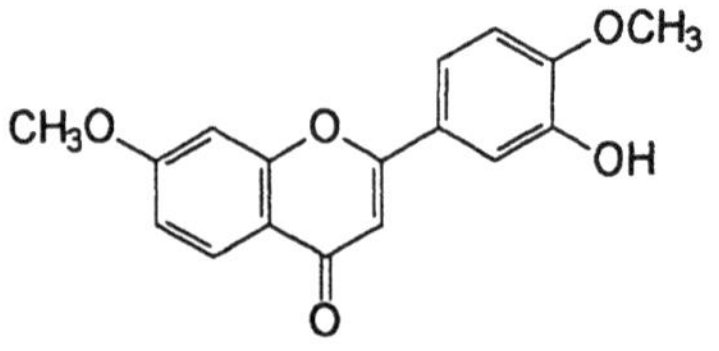

[4-26]

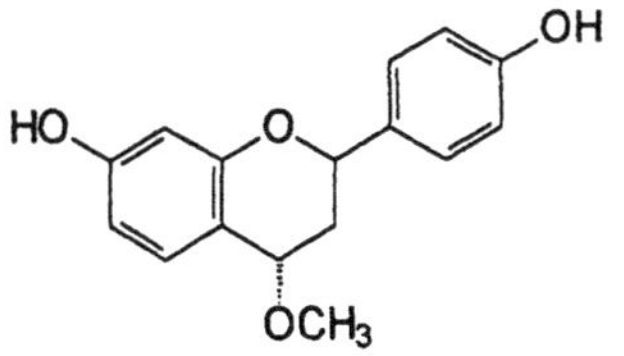

[4-27]

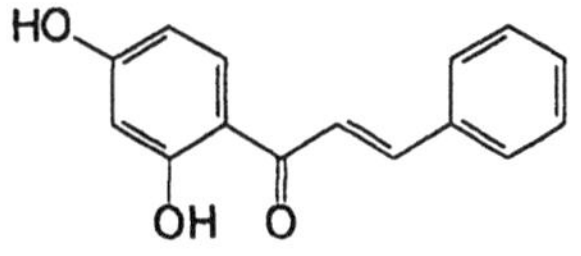

[4-28]

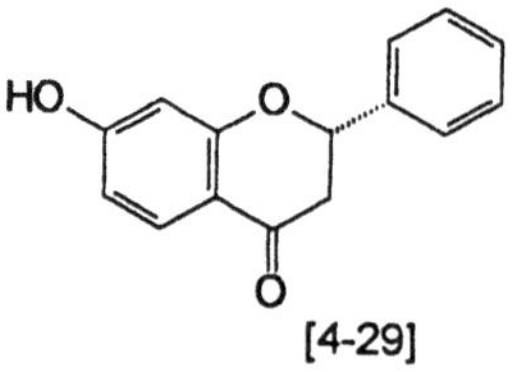

[4-29]

Fig. 4–4. Variations in A-ring oxygenation pattern. Absence of oxygenation at C-7 **[4–23, 4–24]**; absence of oxygenation at C-5 **[4–25 – 4–27]**; 2',4'-dioxygenation in a chalcone **[4–28]** and in the isomeric flavanone **[4–29]**

Grantionin, introduced above because of its unusual B-ring (3',5'-dimethoxy), also lacks oxygenation at C-5 (Ahmad and Ismail, 1991b). Grantionin occurs in *Inula grantioides* along with 5-hydroxy-3,6,7,2',5'-pentamethoxyflavone (grantionidin), normal with regard to the 5-OH but distinguished by the presence of the rare 2',5'-dioxygenation pattern. A recent report by El-Lakany et al. (1996) documented isolation of 3,7,3',4'-tetrahydroxy-6,5'-dimethoxyflavone from *Inula crithmoides* and named it "inucrithmin." 7,4'-Dihydroxyflavone occurs (as the 7-*O*-rhamnoside) in *Achillea fragrantissima* (Seida et al., 1990), while 7-Hydroxy-3',4'-dimethoxyflavone was reported from *Launaea asplenifolia* by Gupta et al. (1985a,b). Two 5-deoxyflavonols, identified as 6,7,4'-trihydroxy-3-methoxy- and 6,7,3',4'-tetrahydroxy-3-methoxyflavone, have been isolated from *Graziela mollissima* by Nakashima et al. (1994). A final example of this small group is 6,4'-dihydroxy-4-methoxyflavan **[4–27]**, which occurs in *Vernonia anthelmintica* as the 7-*O*-glucoside and the corresponding 6"-*p*-hydroxybenzoate (J. Wu et al., 1991).

It should be noted that the examples cited above involve only flavonoids with C-rings. Several chalcones and dihydrochalcones have been reported that lack oxygen at C-2'. Aurones lacking oxygen at C-4, which is biosynthetically equivalent to the C-2' position in chalcones, are also known. These 2'/4-deoxyanthochlors are quite common, in fact. Crawford and Stuessy (1981) pointed out the difference between phloroglucinol- and resorcinol-based chalcones in their anthochlor survey of Coreopsidinae.

The final consideration of B-ring oxygenation patterns involves flavonoids that lack substitution altogether on that ring. Species representing at least 30 genera of the Asteraceae accumulate B-ring deoxyflavonoids. This is a remarkably wide-spread structural phenomenon in plants; B-ring deoxyflavonoids have been reported from over 50 families of flowering plants (Bohm and Chan, 1992). This structural feature, involving chalcones, flavanones, dihydroflavonols, flavones, and flavonols, cuts a wide swath through Asteraceae as well, having been recorded in members of Anthemideae, Astereae, Cardueae, Eupatorieae, Heliantheae s.l., Inuleae s.l., Mutisieae, and Vernonieae. Within Inuleae s.l., *Helichrysum* is the clear winner with B-ring deoxyflavonoids having been identified from over 40 species.

Helichrysum is further distinguished by making compounds based on all seven possible combinations of A-ring substitution and oxygenation at C-3. That accounting assumes the presence of oxygens at positions-5 and 7. The simplest pattern is seen in *Flourensia* species, from which 2',4'-dihydroxychalcone **[4–28]** and the isomeric 7-hydroxyflavanone **[4–29]** (and 7-methoxyflavanone in one case) have been obtained (Bohlmann and Jakupovic, 1979; Guerreiro et al., 1979). Most unexpected, however, was the finding of chalcone itself as a constituent of *Centaurea calcitropa* (Dawidar et al., 1989). Details of the occurrence of B-ring deoxyflavonoids in the family can be found in Table 4–3.

O-Methylation is a major source of structural variation in flavonoids; this is no less true for Asteraceae. The number of combinations and permutations of methyl ethers possible based upon the flavonoid oxygenation patterns listed above is staggering. The level of complexity achieved, though not anywhere near the theoretical, is nonetheless substantial (Fig. 4–5). Some appreciation of the complexity may be gained by two examples. Four monomethyl ethers of quercetin have been reported, as have six dimethyl ethers, four trimethyl ethers, and a single tetramethyl ether. In the case of

Table 4–3. Occurrence of B-ring deoxyflavonoids in Asteraceae

Source	Substituents[1] (Class)	References
Achillea nobilis	35/678 (flavonol)	Adekenov et al. (1984)
Achyrocline alata	56/37 (flavonol)	Bohlmann et al. (1980b)
	35/78 (flavonol)	Bauer et al. (1985)
	57/38 (flavonol)	Bauer et al. (1987, 1989)
	5/78 (flavone), 5/378 (flavone), 35/678 (flavone)	Torrenegra et al. (1982), Broussalis et al. (1993)
	57/3 (flavonol)	Broussalis et al. (1993)
A. flaccida	57/ (flavanone), 57/3 (flavonol), 578/3 (flavonol), 57/38 (flavonol), 58/37 (flavonol)	Norbedo et al. (1984)
A. satureioides	357/ (flavonol), 57/3 (flavonol)	Ferraro et al. (1981)
	58/37 (flavonol)	Hänsel and Ohlendorf (1971), Hirschmann (1984)
	5/367 (flavonol), 7/358 (flavonol), /578 (flavone), /3578 (flavonol)	Mesquita et al. (1986)
A. tomentosa	35/78 (flavonol)	Ferraro et al. (1985)
Ambrosia camphorata	57/ (flavanone)	Wollenweber et al. (1987a)
Anaphalis araneosa	57/368 (flavonol)	Ali et al. (1979)
A. margaritacea	2'4'6'/3'-*C*-Pr (chal.), 57/3 (flavone), 357/6 (flavone), 57/36 (flavone), 57/38 (flavone), 357/68 (flavone, 35/678 (flavone), 57/368 (flavone)	Wollenweber et al. (1993b)
A. velutina	35/678 (flavonol), 5/3678 (flavonol)	Ikramov et al. (1986)
Artemisia campestris	57/ (flavanone), 5/7 (flavanone)	Hurabielle et al. (1982)
A. klotzchiana	35/678 (flavonol)	Dominguez and Cardenas (1975)
Aster squamatus	567/ (flavone)	Ross et al. (1984)
Baccharis bigelovii	2ß57/ (flavanone), 357/6 (flavanone), 57/ (flavone),	

Table 4–3. (Continued)

Source	Substituents[1] (Class)	References
	357/6 (flavonol)	Arriaga-Giner et al. (1986)
B. concinna	57/ (flavanone)	Bohlmann et al. (1981f)
B. glutinosa	57/ (flavanone)	Miyakado et al. (1976), Dominguez et al. (1986c)
	357/ (flavanone)	Dominguez et al. (1986c)
B. oxydonta	57/ (flavanone), 357/ (flavanone)	Bohlmann et al. (1981f)
B. viminea	57/ (flavanone), 5/7 (flavanone)	Bohlmann et al. (1984a)
Brickellia laciniata	57/ (flavone)	Jakupovic et al. (1989b)
Carthamus glaucus	57/6 (flavone)	Khafagy et al. (1979c, 1980), El-Din (1987)
	5/67 (flavone)	Khafagy et al. (1979c, 1980)
Cassinia arcutata	357/ (flavanone)	Zdero et al. (1991e)
C. quinquefaria	57/ (flavanone), 357/ (flavanone), 357/ (flavonol), 3567/ (flavonol), 357/6 (flavonol)	Wollenweber et al. (1993c)
C. amoena	2'6'/4' (chalcone), 2'6'/4' (diHchalcone), 5/7 (flavanone)	Reid and Bohm (1994)
C. fulvida s.l.	2'4'6'/ (chalcone), 2'6'/4' (chalcone), 2'6'/4' (diHchalcone), 5/7 (flavanone)	Reid and Bohm (1994)
C. leptophylla	57/ (flavanone), 357/ (flavone)	Reid and Bohm (1994)
C. vauvilliersii s.l.	2'/4' (chalcone), 2'4'6'/ (chalcone), 2'6'/4' (chalcone), 57/ (flavanone)	Reid and Bohm (1994)
Centaurea calcitropa	Chalcone	Dawidar et al. (1989)
C. clementei	56/7 (flavone)	Collado et al. (1985)
C. ragusina	57/ (flavanone)	Mahmoud et al. (1986)
Chromolaena chaseae	2'4'6'/ (chalcone), 57/ (flavanone), 357/ (flavanone), 357/6 (flavanone), 357/ (flavonol), 357/6 (flavonol)	Bohlmann et al. (1982e)
Chrysanthemum morifolium	567/ (flavone)	Hu et al. (1994)

Table 4–3. (Continued)

Source	Substituents[1] (Class)	References
Chrysothamnus nauseosus	57/ (flavanone), 5/7 (flavanone), 5/67 (flavanone)	Bohlmann et al. (1979a)
Dolichasium lagascae	5/7 (flavanone)	Zdero et al. (1986b)
Encelia stenophylla	57/8-*C*-Pr (flavanone)	Proksch et al. (1988)
Ericameria diffusa	57/3 (flavonol)	Urbatsch et al. (1976)
Flourensia heterolepis	2'4'/ (chalcone), 7/ (flavanone), /7 (flavanone)	Bohlmann and Jakupovic (1979)
F. ilicifolia	57/ (flavanone)	Dillon and Mabry (1977)
F. laurifolia	57/ (flavone), 57/6 (flavone), 57/8-*C*-Pr(cy)[4]	Guerrero and de Lourdes Flores (1990)
F. oolepis	2'4'/ (chalcone), 7/ (flavanone)	Guerreiro et al. (1979)
F. resinosa	57/ (flavone), 5/7 (flavone), 357/ (flavonol), 57/3 (flavonol), 35/7 (flavonol), 357/ (flavanone)	Wollenweber and Yatskievych (1985)
F. retinophylla	57/ (flavanone)	Dillon and Mabry (1977)
Gnaphalium[2] *chilense*	35/678 (flavone)	Wollenweber et al. (1993b)
G. elegans	57/368 (flavone)	Torrenegra et al. (1980)
G. gaudichaudianum	58/67 (flavone), 58/367 (flavonol)	Guerreiro et al. (1982)
G. lanuginosum	57/38 (flavone)	Torrenegra et al. (1979)
G. luteo-album	57/38 (flavonol)	Mericli (1980)
G. microcephalum	357/ (flavone), 57/38 (flavone), 35/678 (flavone)	Wollenweber et al. (1993b)
G. obtusifolium	57/-8-alkyl (flavanone)	Narayanan et al. (1970)
	35/78 (flavonol), 5/378 (flavonol)	Hänsel and Ohlendorf (1969)
	357/68 (flavonol)	Ohlendorf et al. (1971)
G. pellitum	5/78 (flavone)	Escarria et al. (1977)
G. robustum	5/3678 (flavonol)	Bohlmann and Ziesche (1980)

Table 4–3. (Continued)

Source	Substituents[1] (Class)	References
G. wrightii	57/36 (flavonol)	Bohlmann and Ziesche (1980)
Gutenbergia cordifolia	57/ (flavone)	Mungarulire (1990)
Haplopappus remyanus	5/7 (flavanone), 35/7 (flavanone)	Zdero et al. (1991d)
Helichrysum achryroclinoides	2'4'6'/3'-*C*-Pr (chal.)	Randriaminahy et al. (1992)
H. acutatum	2'4'6'/ (chalcone), 57/ (flavanone)	Bohlmann and Abraham (1979c)
H. aphelexioides	2'4'/6' (chalcone), 2'4'/6'/3'-*C*-Pr (chal.), 2'4'6'/4'-*O*-Pr (chal.), 2'4'/6'/4'-*O*-Pr, Complex chalcones, Complex dihydrochalcone	Randriaminahy et al. (1992)
H. arenarium	357/ (flavonol)	Vrkoc et al. (1973, 1975)
	35/678 (flavonol)	Vrkoc et al. (1973), Çubukçu (1982)
H. argyrolepis	2'4'6'/3'-*C*-Pr (chal.), Complex dihydrochalcone	Bohlmann et al. (1984d)
H. armenium	57/3 (flavonol)	Çubukçu (1982)
H. athrixiifolium	2'4'6'/3'-*C*-Pr (chal.), 2'4'6'/4'-*O*-Pr (chal.), 57/7-*O*-Pr (flavanone)	Bohlmann and Ates (Gören) (1984)
H. bracteiferum	2'4'/6' (chalcone), 5/7 (flavanone), 357/ (flavonol), 57/36 (flavonol), 57/38 (flavonol), 357/68 (flavonol), 57/368 (flavonol)	Randriaminahy et al. (1992)
H. calliconum	57/ (flavanone)	Bohlmann and Abraham (1979b)
H. cephaloideum	5/3678 (flavonol)	Jakupovic et al. (1986)
H. chrysargyrum	56/37 (flavonol)	Bohlmann et al. (1979c)
H. cymosum	2'6'/4' (chalcone), 2'4'6'/3'-*C*-Pr (chalcone), 2'6'/4'/3'-*C*-Pr (chalcone),	

Table 4–3. (Continued)

Source	Substituents[1] (Class)	References
	2'6'/4'/3'-*C*-Pr[4] (chal.), Complex dihydrochalcone	Bohlmann et al. (1979f)
	57/ (flavanone), 57/8/-7-*O*-Pr (flavanone), 57/-8-*C*-Pr (flavanone), 57/8 (flavanone), 58/7 (flavanone)	Jakupovic et al. (1989c)
H. decumbens	35/67 (flavonol), 35/678 (flavonol), 57/368 (flavonol)	Tomas-Lorente et al. (1989)
H. forskahlii	2'4'6'/4'-*O*-Pr (chalcone), 2'4'/6'/4'-*O*-Pr (chal.), 57/7-*O*-Pr (flavanone), Complex dihydrochalcones	Jakupovic et al. (1990b)
H. glaciale	57/6 (flavanone), 57/8 (flavanone)	Bohlmann et al. (1979c)
H. glomerata[3]	2'/6'/3'4'-MDO (chal.)	Bohlmann and Suwita (1979c)
H. graveolens	35/678 (flavonol)	Hänsel and Çubukçu (1972), Çubukçu and Damatyan (1980, 1986), Çubukçu (1982)
	57/3 (flavonol)	Çubukçu and Damatyan (1980, 1986)
H. herbaceum	7/5 (flavanone), /57 (flavone), 58/67 (flavone), 8/567 (flavone), /5678 (flavone)	Bohlmann et al. (1979c)
H. heterolasium	57/36 (flavonol)	Bohlmann and Abraham (1979b)
H. hirtum	57/7-*O*-Pr (flavanone)	Randriaminahy et al. (1992)
H. hypocephalum	57/6-*C*-Pr (flavanone), 57/8-*C*-Pr (flavanone), 57/8-*C*-Pr[4] (flavanone), 57/8-*C*-Ger	Bohlmann and Abraham (1979d)
H. italicum	57/ (flavanone)	Freitag (1979)
	35/78 (flavonol)	Hänsel et al. (1980)
	57/38 (flavonol)	Opitz et al. (1971)
H. kraussii	35/678 (flavonol)	Candy et al. (1975)

Table 4–3. (Continued)

Source	Substituents[1] (Class)	References
H. lepidissimum	57/-6-*C*-Alk (flavanone), 57/-6-*C*-Alk-8-*C*-Pr (flavanone), 357/ (flavanone)	Jakupovic et al. (1989c)
H. mimetes	58/67 (flavone)	Jakupovic et al. (1986)
H. mundii	/578 (flavone), /5/78-MDO (flavone)	Bohlmann et al. (1978a)
	Complex flavones	Bohlmann et al. (1980a)
H. nitens	5/67 (flavone), 5/678 (flavone), /57 (flavone), /567 (flavone), /5678 (flavone), /357 (flavone), /3567 (flavone), /35678 (flavone),	Tomas-Barberan et al. (1988)
H. noeanum	35/678 (flavonol)	Çubukçu (1982), Bingöl and Çubukçu (1984)
H. oreophilum	2'4'6'/ (chalcone), 2'6'/4' (chalcone), 2'/4'6' (chalcone), 57/ (flavanone)	Jakupovic et al. (1986)
H. orientale	35/678 (flavonol)	Çubukçu (1976, 1982)
H. oxyphyllum	2'6'/4' (chalcone), 2'/4'6' (chalcone), 57/ (flavanone)	Bohlmann et al. (1980e)
H. pagophilum	35/678 (flavonol)	Bohlmann et al. (1980e)
H. pallasii	35/678 (flavonol)	Çubukçu (1982), Çubukçu and Bingöl (1984)
H. picardii	57/38 (flavonol)	De La Puerta et al. (1990), Tomas-Lorente et al. (1991)
	57/3 (flavonol)	Tomas-Lorente et al. (1991)
H. platypterum	357/ (flavanone), 57/3 (flavonol)	Jakupovic et al. (1986)
H. polycladum	2'4'6'/ (chalcone), 2'6'/4' (chalcone), 2'4'6'/3'-*C*-Pr (chal.), 5/7 (flavanone), 57/6-*C*-Pr (chalcone), Complex dihydrochalcones	Bohlmann et al. (1980e)
H. retrorsum	2'4'6'/3'-*C*-Pr (chal.), 2'4'6'/3'-*C*-Ner. (chal.), 2'4'6'/4'-*O*-Pr (chal.),	

Table 4–3. (Continued)

Source	Substituents[1] (Class)	References
	57/7-*O*-Ner (flavanone)	Randriaminahy et al. (1992)
H. rugulosum	2'4'6'/3'-*C*-Me (chal.), Complex Pr chalcones, Complex Pr flavanones	Bohlmann and Misra (1984)
H. sanguineum	57/38 (flavonol), 357/68 (flavonol)	Çubukçu (1982), Mericli et al. (1984)
H. schimperi	5/7 (flavone), 5/67 (flavone), /567 (flavone), 5/78 (flavone), 5/678 (flavone), 5/367 (flavonol)	Jakupovic et al. (1990b)
H. stirlingii[3]	57/ (flavanone), 357/ (flavanone)	Jakupovic et al. (1987b)
H. sutherlandii	2'/4'5'6' (chalcone), /2'6'/3'4'-MDO (chalcone) /2'6'/3'4'-MDO (diHchal.)	Bohlmann et al. (1978a, 1980e)
H. tenuiculum	2'6'/4' (chalcone), 2'4'6'/3'-*C*-Pr (chalcone), Complex dihydrochalcones	Bohlmann et al. (1979f)
H. tenuifolium	57/ (flavanone), 57/ (flavone), 357/ (flavanone), 357/ (flavonol), 57/3 (flavonol), 2'4'6'/ (dihydrochalcone)	Bohlmann and Abraham (1979a)
H. thapsus	57/-6-*C*-Pr (flavanone), 357/6-*C*-Pr (flavanone), 357/6-*C*-Pr[4] (flavanone), 357/6-*C*-Ger (flavanone)	Bohlmann and Zdero (1983)
H. triplinerve	2'4'/6' (chalcone)	Randriaminahy et al. (1992)
H. umbraculigerum	2'4'6'/3'-Ger. (chal.)	Bohlmann and Hoffmann (1979)
H. zeyheri	57/ (flavanone)	Jakupovic et al. (1986)
Heterothalamus psiadioides	5/7 (flavanone), 357/ (flavone)	Kerber et al. (1993)

Table 4–3. (Continued)

Source	Substituents[1] (Class)	References
Hymenoclea monogyra	57/ (flavanone) 357/ (flavanone)	Miyakado et al. (1976) Zdero et al. (1989a)
Lophopappus tarapacanus	57/ (flavanone)	Hoeneisen et al. (1993)
Lucilia nivea	5/3678 (flavone)	Bohlmann et al. (1990)
Lychnophora diamantinana	57/ (flavanone), 357/ (flavanone)	da Costa et al. (1993)
Metalasia cymbifolia	57/7-*O*-Pr (flavanone), Complex chalcones, Complex dihydrochalcone	Zdero and Bohlmann (1990c)
Pleiotaxis cf. *rugosa*	2'4'6'/4'-*O*-Pr (chal.), 2'4'6'/3'-*C*-Pr (chal.), 2'6'/4'/3'-*C*-Pr (chal.)	Bohlmann and Zdero (1982)
Ratibida columnifera	57/ (flavanone)	Ellmauerer et al. (1987)
Tarchonanthus trilobus	57/ (flavanone)	Bohlmann and Suwita (1979a)

[1] Substitution: Hydroxyl positions/Methoxyl positions/Other; Pr = prenyl; Ger = geranyl.
[2] *Gnaphalium* species listed in this table are now considered members of the genus *Pseudognaphalium*. See Chapter 12 for comments on taxonomy.
[3] Now considered to be a species of *Ozothamnus*. See chapter 12 for comments.
[4] Modified prenyl group, e.g., cyclized to adjacent phenol group.

quercetagetin, four monomethyl ethers, five dimethyl ethers, 10 trimethyl ethers, seven tetramethyl ethers, two pentamethyl ethers, as well as the totally *O*-methylated derivative (six methoxyl groups) have been reported. Although some of these have limited distributions, others, such as 5-hydroxy-3,6,7,3',4'-pentamethoxyflavone **[4–30]** (quercetagetin-3,6,7,3',4'-pentamethyl ether, artemetin), occur widely. Very highly *O*-methylated flavonoids are limited to a relatively small number of genera, e.g., *Ageratum, Artemisia, Conoclinium, Eupatorium, Gutierrezia*, and *Gymnosperma*. Typical examples of extreme cases include 5,3'-dihydroxy-3,6,7,8,4',5'-hexamethoxyflavone **[4–31]** from *Gymnosperma glutinosum* (Yu et al., 1988), 5,6,7,8,2',3',4',5'-octamethoxyflavone, among others, from *Ageratum houstonianum* (Quijano et al., 1982b, 1987) and 5,6,7,8,3'-pentamethoxy-4',5'-methylenedioxyflavone from *Conoclinium coelestinum* (Le-Van and Pham, 1979).

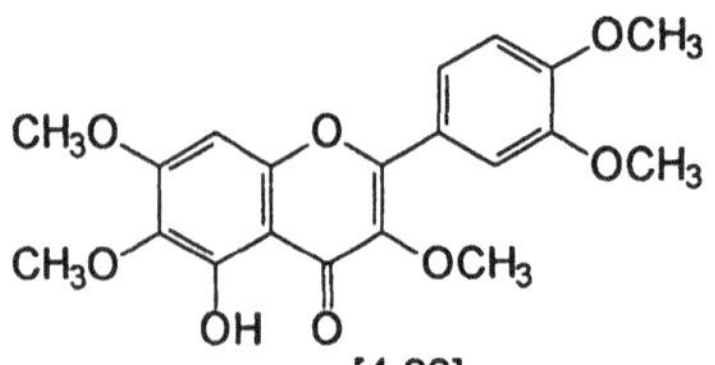

[4-30]

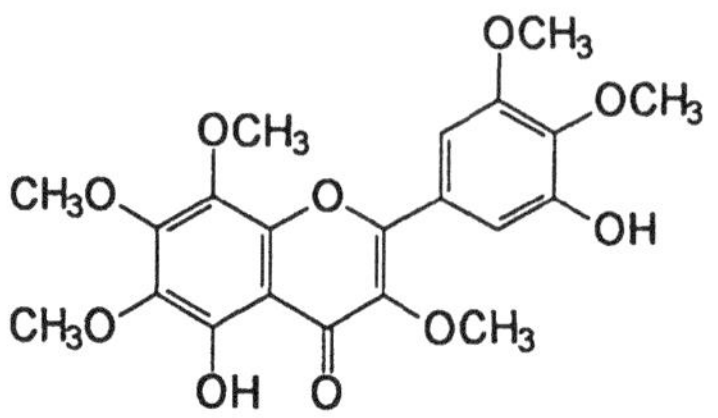

[4-31]

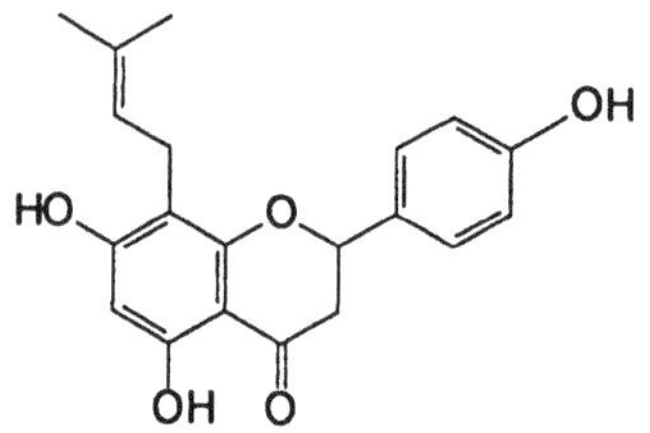

[4-32]

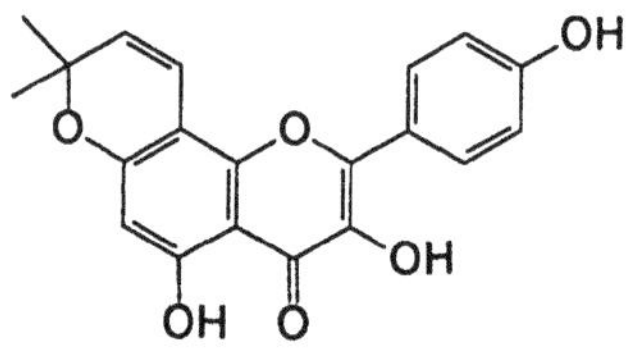

[4-33]

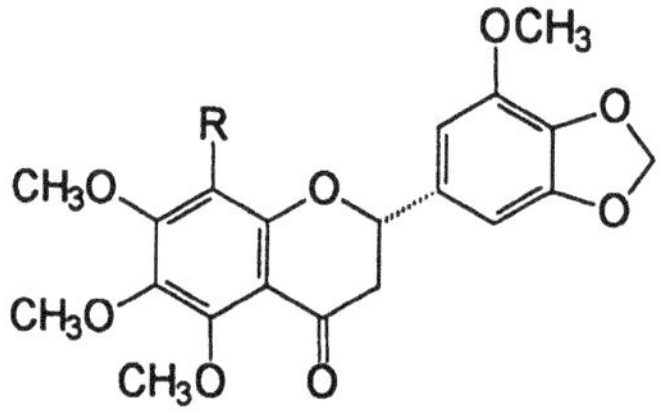

[4-34] R = H

[4-35] R = OCH_3

Fig. 4–5. Selected patterns of *O*-methylation and *C*-alkylation showing complex combinations of A-ring and B-ring patterns **[4–30 – 4–35]**

Several *C*-alkylated flavonoids have been isolated from members of Asteraceae. Most involve chalcones, flavanones, and dihydroflavonols, which will be discussed below, but a few *C*-alkylated flavones and flavonols have also been identified. The first of these came from the work of Pashchenko et al. (1966) who reported 8-*C*-prenylapigenin **[4–32]** from *Xanthium spinosum* (prenyl = 3-methylbut-2-enyl). This compound was encountered again as a constituent of *Marshallia grandiflora* where it was accompanied by the corresponding flavanone and a flavan-3,4-diol derivative (Bohlmann et al., 1979e). More recently, Gupta and Ahmed (1985) described 5,7,3',4',5'-trihydroxy-3-*C*-(3-methylbutyl) flavone and its 5-*O*-neohesperidoside from *Launaea asplenifolia*. (3-Methylbutyl is equivalent to dihydroprenyl.) *Artemisia austriaca* yielded 5-hydroxy-7,4'-dimethoxy-6-*C*-methylflavone in a study done by Adekenov et al. (1987a). Herz and Bruno (1987b) isolated a compound from *Marshallia tenuifolia* identified as kaempferol with a 6-*C*-prenyl group that had undergone cyclization with the adjacent 7-OH group to afford compound **[4–33]**. *C*-Prenylated isoflavones and flavan derivatives will be discussed below.

Flavanones

B-Ring deoxyflavanones, including dihydroflavonols, have been identified in members of 21 genera based upon the cited names (22 with the transfer of two species of *Helichrysum* to *Ozothamnus*). Details of these occurrences can be found in Table 4–2 where they are collectively identified as flavanones.

Naringenin, 5,7,4'-trihydroxyflavanone, is the most widely occurring flavanone in Asteraceae. It has been reported as such, in a variety of glycosidic forms, as mono and dimethyl ethers, and as both *C*- and *O*-prenylated derivatives. Occurrences of the simpler forms (prenyl derivatives will be commented upon below) have been recorded for over 50 genera. Eriodictyol, 5,7,3',4'-tetrahydroxyflavanone, and its simple derivatives have been reported from at least 45 genera.

Several flavanones with 3',4',5'-trioxygenation have been identified in members of the family, the simplest one of which is 5,7,3',4',5'-pentahydroxyflavanone, reported from *Helichrysum bracteatum* by Rimpler et al. (1963) and Forkmann (1983). [Current taxonomic opinion places *Helichrysum bracteatum* in the genus *Bracteantha* (as *B. bracteatum*); see discussion in Chapter 12]. *Artemisia fragrans* and *A. gurganica* yielded 5,7,3',4'-tetrahydroxy-6,5'-dimethoxyflavanone (Belenovskaya et al., 1992), while *A. subdigitata* was shown to accumulate 5,7,4'-trihydroxy-3',5'-dimethoxyflavanone (Shi et al., 1992). Boehrer (1968) also identified the latter compound in his work on *Silybum marianum*. 5,4',5'-Trihydroxy-7,3'-dimethoxyflavanone was identified as one of the many compounds present in the leaf resin of *Hazardia squarrosa* var. *grindelioides* (Clark and Wollenweber, 1985). *Ageratum corymbosum* accumulates two compounds that fit into this category, 5,6,7,5'-tetramethoxy-3',4'-methylenedioxy- and 5,6,7,8,5'-pentamethoxy-3',4'-methylenedioxy-flavanone (Quijano et al., 1980), **[4–34]** and **[4–35]**, respectively.

The simplest flavanones in the family, excluding the B-ring deoxy compounds, discussed above, are 7-hydroxyflavanone and its methyl ether from *Flourensia heterolepis* (Bohlmann and Jakupovic, 1979) and *F. oolepis* (Guerreiro et al., 1979). Other 5-deoxyflavanones identified include 7,3',4'-trihydroxyflavanone (butin) from species of *Bidens* (Serbin et al., 1972b, 1975; Borisov et al., 1980), and 7,8,3',4'-tetrahydroxyflavanone, also from *Bidens* (Serbin et al., 1973; Borisov et al., 1979). *Baccharis salicifolia* has also provided a member of this small group, 4'-hydroxy-7,3'-dimethoxyflavanone (Kavka et al., 1973).

5,6-Dihydroxy-4'-methoxyflavanone and 5,8,4'-trihydroxyflavanone, both from *Artemisia campestris* subsp. *maritima* (Rauter et al., 1989), were commented upon above as members of the unusual group of flavonoids that are characterized by the lack oxygenation at C-7.

Several flavanones with oxygenation at C-6 or C-8 and with various levels of *O*-methylation have been reported from members of the family. The structures and sources of these compounds are detailed in Table 4–4. Some of these compounds characterized by unusual oxygenation patterns were commented on above.

C-Alkylated flavanones, in some variety, have been identified from members of several genera, *Encelia, Flourensia, Gnaphalium, Helichrysum, Marshallia*, and *Wyethia*. The majority of alkyl groups encountered in this set involve the prenyl function or some derivative thereof. The derived forms may involve oxidation of one of the methyl groups, cyclization with an adjacent phenol group to form the dimethylchromeno ring system, or a modified form of the ring, or addition of a second prenyl group to yield a geranyl side chain (Fig. 4–6). An unusual *C*-alkyl function was encountered in *Gnaphalium obtusifolium* (Narayanan et al., 1970) and again in *Helichrysum lepidissimum* (Jakupovic et al., 1989c) and shown to consist of a highly substituted nine-carbon lactone whose structure was confirmed by X-ray crystallography **[4–36]** (Narayanan et al., 1970). Details of the occurrence of *C*-prenyl flavanones, and their derivatives, are given in Table

Table 4–4. Flavanones of Asteraceae with C-6 or C-8 oxygenation (other than compounds lacking B-ring oxygenation)

Taxon	Substitution[1]	Reference
Ageratum corymbosum	4'/5673'; -/567/3'4'-MDO	Quijano et al. (1992)
Ageratum strictum	6/573'4'; 6/57/3'4'-MDO	Quijano et al. (1982a)
Artemisia subdigitata	583'5'/-; 582'/5'	Shi et al. (1992)
Chromolaena heteroclinea	64'/573'; 6/573'4'; 4'/5673'; -/5673'4'	Boeker and Bohlmann (1986)
Chromolaena subscandens	57/84';54'/67	Amaro and Delgado-Mendez (1993)
Chromolaena tunariensis	4'/567; -/5674'	Bohlmann et al. (1979d)
Gutierrezia sphaerocarpa	573'4'/6; 574'/63'	Li et al. (1988)
Tanacetum sibiricum	572'5'/-; 572'5'/6; 574'/6	Stepanova et al. (1981)

[1] Substitution: Hydroxyl positions/methoxyl positions/other; MDO = methylenedioxy.

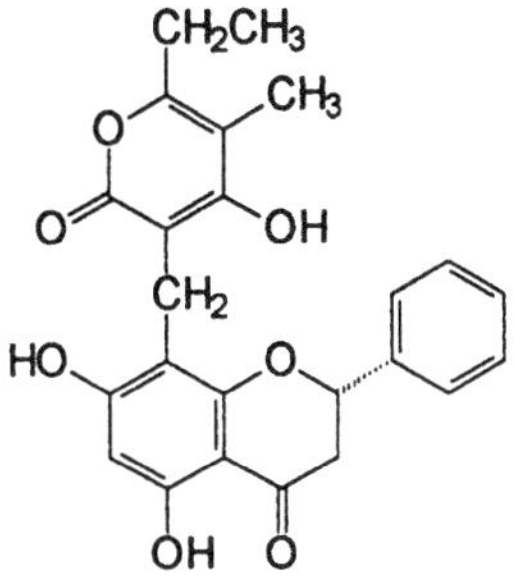

[4-36]

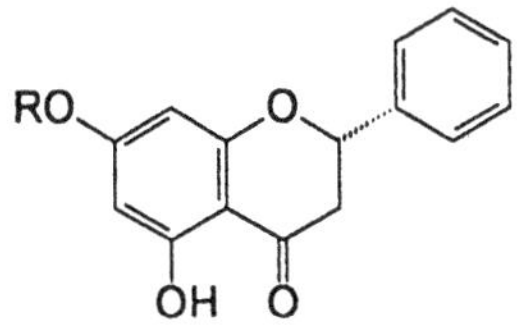

[4-37a] R = Prenyl

[4-37b] R = Neryl

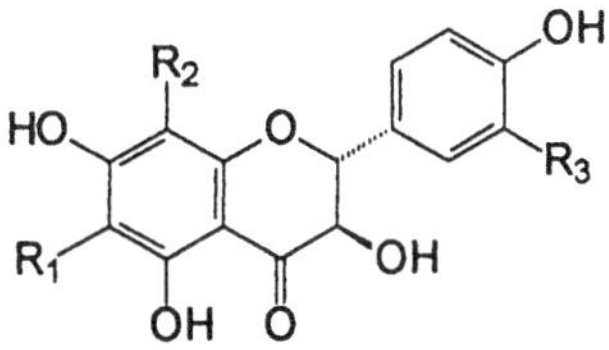

[4-38] $R_1 = R_2 = R_3 = H$

[4-39] $R_1 = R_2 = H, R_3 = OH$

[4-40] $R_1 = OCH_3, R_2 = R_3 = H$

[4-41] $R_1 = OCH_3, R_2 = H, R_3 = OH$

[4-42] $R_1 = R_3 = H, R_2 = OCH_3$

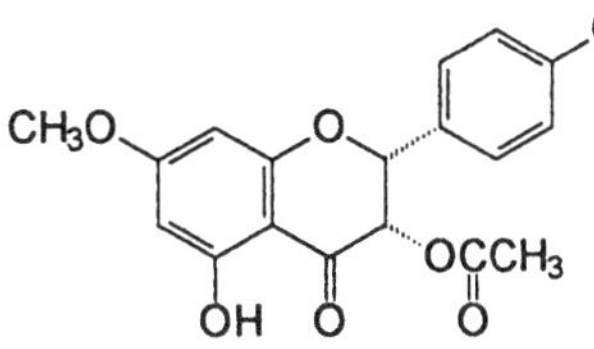

[4-43]

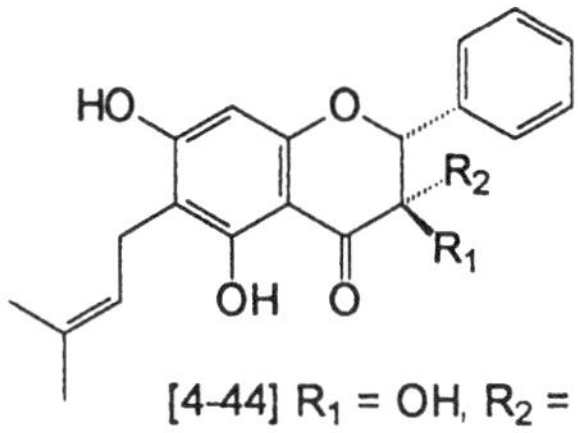

[4-44] $R_1 = OH, R_2 = H$

[4-45] $R_1 = H, R_2 = OH$

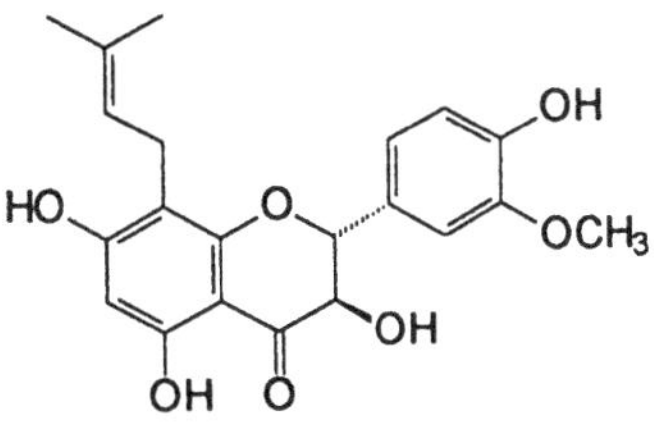

[4-46]

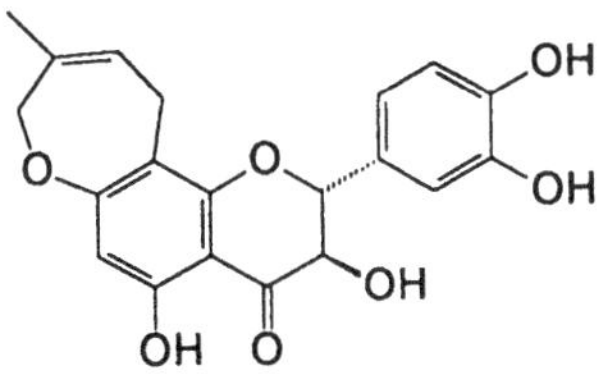

[4-47]

Fig. 4–6. Selected complex flavanones and dihydroflavonols. *C*-Alkylated **[4–36]** and *O*-alkylated flavanones **[4–37a,b]**; variations in dihydroflavonols **[4–38 – 4–42]**; dihydroflavonol stereochemical variation **[4–43]**; *C*-alkylated dihydroflavonols **[4–44 – 4–47]**

4–5. Accompanying *C*-prenylated flavanones in several species are flavonoid prenyl ethers. 7-Prenyl ethers of pinocembrin **[4–37a]** and/or naringenin (or *O*-methyl ether) have been reported from *Metalasia cymbifolia*

Table 4–5. *C*-Prenylated flavanones in Asteraceae

Compound[1]	Source	Reference
6-*C*-Pr-naringenin	*Wyethia angustifolia*	McCormick et al. (1986)
	Wyethia helenioides	McCormick et al. (1986)
	Wyethia invenusta	McCormick et al. (1987)
8-*C*-Pr-naringenin	*Marshallia grandiflora*	Bohlmann et al. (1979e), Jakupovic et al. (1988e)
	Helichrysum athrixiiflium	Bohlmann and Ates (Gören) (1984)
	Wyethia helenioides	Bohlmann et al. (1981j), McCormick et al. (1986)
	Wyethia mollis	Bohm and Choy (1987)
	Wyethia angustifolia	McCormick et al. (1986)
	Wyethia scabra	McCormick et al. (1987)
6-*C*-Pr-pinocembrin	*Anaphalis margaritacea*	Wollenweber et al. (1993b)
	Helichrysum thapsus	Bohlmann and Zdero (1983)
	Helichrysum hypocephalum	Bohlmann and Abraham (1979d)
	Helichrysum rugulosum	Bohlmann and Misra (1984)
6,8-Di-*C*-Pr-pinocembrin	*Helichrysum rugulosum*	Bohlmann and Misra (1984)
8-*C*-Pr-7-*O*-Pr-pinocembrin	*Helichrysum rugulosum*	Bohlmann and Misra (1984)
6-*C*-Lac[2]-pinocembrin	*Helichrysum lepidissimum*	Jakupovic et al. (1989c)
8-*C*-Pr-pinocembrin	*Encelia stenophylla*	Proksch et al. (1988)
	Helichrysum hypocephalum	Bohlmann and Abraham (1979d)
8-*C*-Pr[3]-pinocembrin	*Helichrysum hypocephalum*	Bohlmann and Abraham (1979d)
8-*C*-Ger-pinocembrin	*Helichrysum hypocephalum*	Bohlmann and Abraham (1979d)
6-*C*-Pr-eriodictyol	*Wyethia angustifolia*	McCormick et al. (1986)
	Wyethia helenioides	Bohlmann et al. (1981j), McCormick et al. (1986)
	Wyethia invenusta	McCormick et al. (1987)
8-*C*-Pr-eriodictyol	*Encelia stenophylla*	Proksch et al. (1988)
	Wyethia angustifolia	McCormick et al. (1986)
	Wyethia arizonica	Bohlmann et al. (1984h)
	Wyethia helenioides	Bohlmann et al. (1981j) McCormick et al. (1986)
	Wyethia mollis	Bohm and Choy (1987)
	Wyethia scabra	McCormick et al. (1987)
8-*C*-Pr[3]-pinocembrin	*Flourensia laurifolia*	Guerrero and de Lourdes Flores (1990)
8-*C*-Pr-7-*O*-methyleriodictyol	*Flourensia heterolepis*	Bohlmann and Jakupovic (1979)

Table 4–5. (Continued)

Compound[1]	Source	Reference
8-*C*-Pr-6-*C*-Lac[2]-pinocembrin	*Helichrysum lepidissimum*	Jakupovic et al. (1989c)
6-*C*-Pr-isoflavones	*Wyethia angustifolia*	McCormick et al. (1986)
	Wyethia mollis	Bohm et al. (1989a)

[1] Pr = prenyl; Ger = geranyl.
[2] Unusual *C*-lactone function.
[3] Prenyl group has cyclized with adjacent phenolic group.

(Zdero and Bohlmann, 1990c) and several species of *Helichrysum*: *H. athrixiifolium* [Bohlmann and Ates (Gören), 1984], *H. cymosum* (Jakupovic et al., 1989c), *H. forskahlii* (Jakupovic et al., 1990b; *H. hirtum* (Randriaminahy et al., 1992), *H. retrorsum* (Randriaminahy et al., 1992) and *H. rugulosum* (Bohlmann and Misra, 1984). *Helichrysum retrorsum* also accumulates the 7-*O*-neryl derivative of pinocembrin **[4–37b]** (Randriaminahy et al., 1992). These compounds are often accompanied by isomeric or closely related *O*-prenylated chalcones, which will be discussed below.

Acylation of sugar hydroxyls in glycosides is commonly encountered, direct acylation of a phenolic group much less so. It is interesting to note that direct phenolic acylation occurs, at least in Asteraceae, primarily with flavanones and dihydroflavonols, some examples of which have just been cited. A few cases involving acylated flavones and flavonols are also known. Acetic acid is the most commonly encountered acylating acid but several four- and five-carbon acids (e.g., butyric, isobutyric, 2-methylbutyric, 2-methyl-2-butenoic) have also been identified, and benzoic acid is seen in one or two instances. An interesting combination of biosynthetic events occurs in *Flaveria* where both acetylation (at the 3-OH) and sulphation (at 7-OH, 3'-OH and 4'-OH) have occurred (Varin et al., 1986a,b). Table 4–6 lists compounds that exhibit direct phenolic acylation. The source plants are listed alphabetically for convenience.

Dihydroflavonols

Dihydroflavonols (3-hydroxyflavanones) are of more limited occurrence in the family than are flavanones *per se*. Dihydroflavonols lacking B-ring oxygenation, for example, occur in eight of the 31 genera in which this feature is found. Dihydroflavonols based on the 3,5,7,4'-oxygenation pattern (dihydrokaempferol or aromadendrin) **[4–38]** occur in 20 genera in

Table 4–6. Flavonoids from Asteraceae that feature direct phenol acylation

Taxon	Compound[1]	Reference
Baccharis bigelovii	57/-flavone 7-benzoate	Arriaga-Giner et al. (1986)
B. glutinosa	357/-flavanone 3-acetate	Dominguez et al. (1986c)
B. varians	3574'/-flavone 3-acetate; 3573'4'/-flavone 3-acetate	Bohlmann et al. (1981h)
Cassinia arcutata	357/-flavanone 3-acetate	Zdero et al. (1991e)
Chromolaena heteroclinia	4'/5673'-flavone 4'-acetate	Boeker and Bohlmann (1986)
Flaveria bidentis	3573'4'/-flavone 3-acetate-7,3'4'-trisulfate	Cabrera and Juliani (1976), Varin et al. (1986a,b)
Flourensia retinophylla	3573'/-flavan 3-isobutyrate	Stuppner and Müller (1994)
Galeana pratensis	578/364'-flavone 8-tiglate; 584'/367-flavone 8-(2-methylbutyrate)	Maldonado et al. (1992b)
Geraea canescens	3574'/8-flavanone 3-acetate; 3574'/8-flavanone 3-angelate	Proksch et al. (1986)
Gnaphalium robustum	578/3-flavone 8-(2-methyl-butyrate; 578/3-flavone 8-angelate	Urzua and Cuadra (1990)
Gutierrezia dracunculoides	574'/33'-flavone 4'-(2-methylbutyrate); 574'/33'-flavone 4'-isovalerate	Bohlmann et al. (1981b)
G. texana	572'5'/34'-flavone 5'-acetate	Li et al. (1987)
Haplopappus remyanus	35/7-flavanone 3-acetate	Zdero et al. (1991d)
Helichrysum stirlingii	357/-flavanone 3-acetate	Jakupovic et al. (1987b)
Hymenoxys turneri	3573'4'-flavanone 3-acetate; 3574'/6-flavanone 3-acetate; 3573'4'/6-flavanone 3-acetate	Gao et al. (1990)

Table 4–6. (Continued)

Taxon	Compound[1]	Reference
Inula graveolens	3573'4'/-flavanone 3-acetate, 354'/7-flavanone 3-acetate, 353'4'/7-flavanone 3-acetate	Öksüz and Topcu (1992)
I. viscosa	3573'4'/-flavanone 3-acetate	Bohlmann et al. (1977), Grande et al. (1985), Ulubelen et al. (1987) Wollenweber et al. (1991)
	353'4'/7 3-acetate	Bohlmann et al. (1977), Grande et al. (1985), Ulubelen et al. (1987a) Wollenweber et al. (1991)
	3574'/-flavanone, 3573'/4'-flavanone 3-acetate	Grande et al. (1985), Wollenweber et al. (1991)
Lophopappus tarapacanus	57/-flavanone 7-benzoate	Hoeneisen et al. (1993)
Tanacetum microphyllum	3573'/4'-flavone 7-acetate	Abad et al. (1993)
Tessaria dodonaefolia	3573'4'/-flavanone 3-acetate	Kavka et al. (1977), Nanayakkara et al. (1988), Guerreiro et al. (1990)
Traversia baccharoides	3573'4'/-flavone 3-isobutyrate, 3573'4'/-flavone 4-isobutyrate, 3573'4'/-flavone 3'-isobutyrate	Kulanthaivel and Benn (1986)

[1] Substitution: Hydroxyl positions/Methoxyl positions.

the family, either as such or, frequently, as methyl ethers. The 7-prenyl ether was reported from *Pterocaulon virgatum* (Bohlmann et al., 1981a) where it occurred with the corresponding dihydroquercetin derivative. Dihydroquercetin (taxifolin) **[4–39]** and its simple derivatives occur in 17 genera. Only a few dihydroflavonols have been identified with oxygenation at C-6 or C-8, apparently none with both. 6-Methoxydihydrokaempferol **[4–40]** and 6-methoxydihydroquercetin (dihydropatuletin) **[4–41]**, along with dihydroquercetin itself, were identified as components of *Hymenoxys turneri* by Gao et al. (1990). 8-Methoxydihydrokaempferol (dihydroherbacetin 8-methyl ether) **[4–42]**, its 4'-methyl ether, and 5,7,3'-trihydroxy-8,4'-dimethoxydihydroflavonol (dihydrogossypetin-8,4'-dimethyl ether) were isolated from *Enceliopsis nudicalulis* where they occurred with several methyl ethers of the corresponding flavonols, herbacetin and gossypetin (Proksch et al., 1987). In the same paper, Proksch et al. (1987) reported 8-methoxydihydrokaempferol from *Geraea canescens* where it occurred as the 3-acetate and the 3-angelate.

The normal stereochemistry of dihydroflavonols is (2*R*,3*R*) as shown above. Other stereochemistries, although comparatively uncommon, are known, as is the case of certain compounds isolated from *Inula graveolens*. Öksüz and Topcu (1992) reported several dihydroflavonols from this species two of which, dihydroquercetin 7-methyl ether and dihydrokaempferol 7-methyl ether 3-acetate **[4–43]**, were shown to exist in the (2*R*,3*S*) conformation.

C-Alkylated dihydroflavonols have been identified from five species representing four genera. The first of these reports involved two compounds from *Marshallia obovata* which Bohlmann et al. (1980b) identified as 8-*C*-prenyldihydrokaempferol and its corresponding cyclization product involving the 7-OH (2,3-dihydroderivative of structure 4–30). The next report, also by the Berlin group (Bohlmann and Zdero, 1983), detailed the finding of two isomers of 6-*C*-prenylpinobanksin (3,5,7-trihydroxy-6-*C*-prenylflavanone), one with the normal (2*R*,3*R*) **[4–44]** and the other with (2*R*:3*S*) stereochemistry **[4–45]**. The third compound has a geranyl group at C-6 and (2*R*:3*S*) stereochemistry. 8-*C*-Prenyldihydroisorhamnetin (3,5,7,4'-tetrahydroxy-3'-methoxy-8-*C*-prenylflavanone) **[4–46]** was reported from two species of *Wyethia*, *W. angustifolia* by McCormick et al. (1986) and *W. mollis* by Bohm and Choy (1987). A novel cyclization product involving a *C*-prenyl group was identified as a constituent of *W. angustifolia* (McCormick et al., 1986). Normally, as we have seen in examples above, cyclization of a prenyl group with an adjacent phenolic group gives rise to a six-membered ring, e.g., **[4–33]**. An alternative cyclization to give a seven-membered ring would result in the formation of compounds such as **[4–47]**.

Returning to *Marshallia*, we note the identification, for the first time as a natural product, of 8-*C*-prenyldihydrokaempferol 5-methyl ether from *M. grandiflora* (Jakupovic et al., 1988e). Other *C*-alkylated flavonoids from *Marshallia* and *Wyethia* will be met below. The 7-*O*-prenyl ether of dihydrokaempferol was reported from *Pterocaulon virgatum*, where it occurred with the corresponding dihydroquercetin derivative (Bohlmann et al., 1981a), and from *Gochnatia paniculata* (Bohlmann et al., 1983a).

Chalcones, dihydrochalcones, and aurones

Chalcones

In the earlier literature aurones, a term coined by E. C. Bate-Smith and T. Geissman (1951), and chalcones were referred to collectively as anthochlors. These compounds figured significantly in early applications of chemosystematic methods to problems within the family, a topic that we will expand upon in discussions of Coreopsidinae, and other taxa, below. Much of what we know about anthochlor structure came from pioneering studies by Ted Geissman and his associates on the pigments of *Coreopsis* and *Cosmos* (Geissman, 1941a,b, 1942; Geissman and Heaton, 1944a,b; Geissman and Mojé, 1951; Geissman et al., 1956) and work by Shimokoriyama (1957), Shimokoriyama and Hattori (1953), and Shimokoriyama and Geissman (1960) on *Coreopsis*, *Lasthenia* (reported as *Baeria*) and *Viguiera*.

The majority of anthochlor reports in the literature involve chalcone-aurone pairs (Fig. 4–7). Butein (2',4',3,4-tetrahydroxychalcone) **[4–48]** and sulfuretin (6,3',4'-trihydroxyaurone) **[4–50]** comprise the most common pair of aglycones; butein 4'-*O*-glucoside (coreopsin) **[4–49]** and sulfuretin 6-*O*-glucoside (sulfurein) **[4–51]** are the corresponding pair of glucosides. Adding one hydroxyl group to the A-ring one gets the next pair of aglycones, 2',3',4',3,4-pentahydroxychalcone (okanin) **[4–52]** and 6,7,3',4'-tetrahydroxyaurone (maritimetin) **[4–54]**. Their respective glucosides are okanin 4'-*O*-glucoside (marein) **[4–53]** and maritimetin 6-*O*-glucoside (maritimein) **[4–55]**. *O*-Methylation at the 7-OH of okanin gives lanceolatin **[4–56]**; its 4'-*O*-glucoside is lanceolin **[4–57]**. *O*-Methylation of maritimetin affords leptosidin **[4–58]**; its 6-*O*-glucoside is leptosin **[4–59]**. Very often workers report the presence of these pairs of anthochlors on the basis of chromatographic behavior and color reactions rather than from detailed chemical analyses. Owing to the striking reaction that chalcones and aurones give with alkali (red to purple depending on degree of hydroxylation), such color reactions are quite reliable.

Several surveys of the family for anthochlors have been described. The first was that of Gertz (1938) who seems to have been the first person to use the color test in a systematic fashion. Shimokoriyama's (1962) contribution appeared in *The Chemistry of Flavonoid Compounds* (Geissman, 1962) and involved numerous first reports of positive anthochlor tests in the family. The most recent survey was that of Crawford and Stuessy (1981) who focused on Coreopsidinae (Heliantheae) but included representatives of other subtribes and other tribes. The latter survey has the added advantage of listing species that did not give positive anthochlor tests. The Crawford and Stuessy survey incorporated the results of Shimokoriyama's (1962) survey

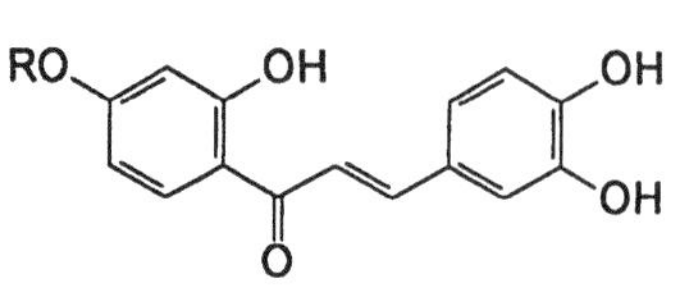

[4-48] R = H, Butein

[4-49] R = Glucose, Coreopsin

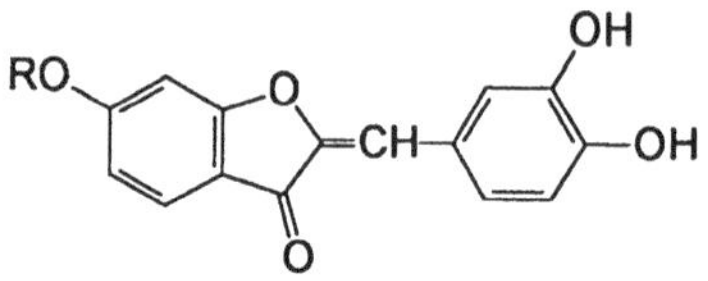

[4-50] R = H, Sulfuretin

[4-51] R = Glucose, Sulfurein

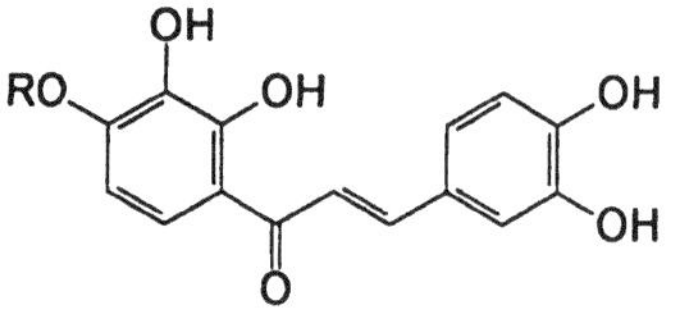

[4-52] R = H, Okanin

[4-53] R = Glucose, Marein

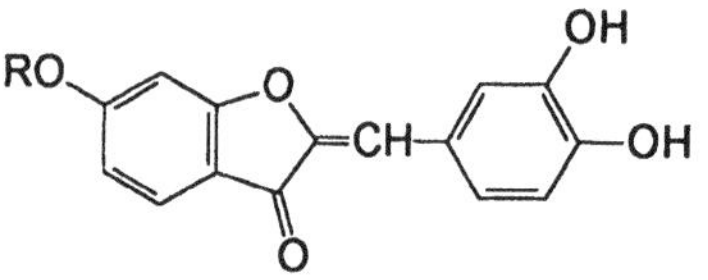

[4-54] R = H, Maritimetin

[4-55] R = Glucose, Maritimein

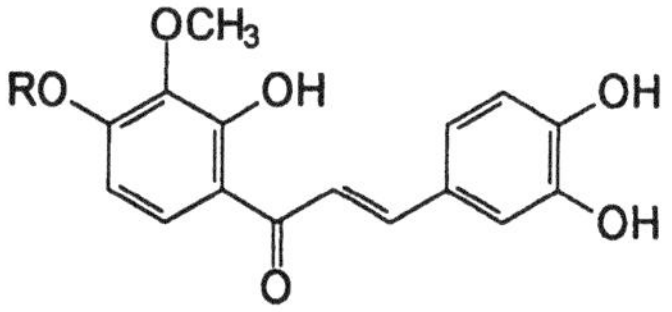

[4-56] R = H, Lanceoletin

[4-57] R = Glucose, Lanceolin

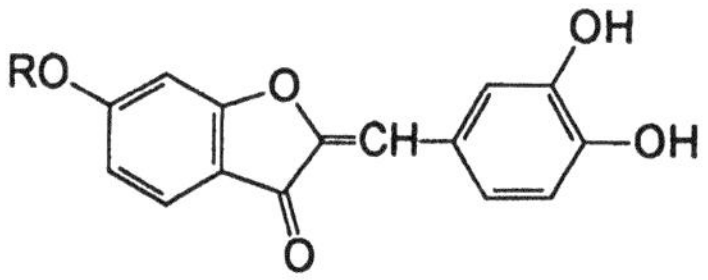

[4-58] R = H, Leptosidin

[4-59] R = Glucose, Leptosin

Fig. 4–7. Common chalcone-aurone pairs **[4–48 – 4–51]** with additional chalcone-aurone structural variations **[4–52 – 4–59]**

as well. Table 4–7 gives details of the surveys in which the common anthochlors have played a role.

Detailed studies of *Bidens* have yielded a number of glycosides of okanin or okanin methyl ethers. Okanin 4-methyl ether, 3,4-dimethyl ether, 3',3,4-trimethyl ether and 3',4',3,4-tetramethyl ether, along with glucosides of the first three, were isolated from *B. torta* (McCormick et al., 1984). Hoffmann and Hölzl (1988a,b,c, 1989a) obtained several glycosides, some linked through the 3'-OH, some through the 4'-OH and one involving both positions. Many of these compounds were also acylated. Acylated glucosides of okanin and okanin 4-methyl ether were also reported from *B. frondosa* by Karikome et al. (1992) and from *B. campylotheca* by Redl et al. (1993). Similar acylated okanin glycosides were reported by D'Agostino et al. (1991b) from *Senecio pseudotites*, which appears to be the sole report of this sort of compound from that genus. The final examples of okanin derivatives in-

Table 4–7. Sources of information on the occurrence of common anthochlors (butein/sulfuretin and okanin/maritimetin) and their glycosides in Asteraceae

Taxon	Reference
Bidens (Hawaii)	Ganders et al. (1990)
B. bipinnata	Wang et al. (1992)
B. cernua	Borisov et al. (1980)
B. cordylocarpa	Crawford (1971)
B. ferulaefolia complex	Hart (1979)
B. frondosa	Romussi and Pagani (1970)
B. sect. *Greenmania*	Roseman (1986)
B. parviflora	Ma et al. (1991)
B. pilosa complex	Ballard (1986)
B. tripartita	Serbin et al. (1972b)
Chrysanthellum indicum	Brasseur et al. (1987)
Coreopsis (North American)	Crawford and Smith (1983a)
Coreopsis sect. *Coreopsis*	Crawford and Smith (1985)
C. sect. *Palmatae*	Crawford et al. (1980)
C. grandiflora s.l.	Crawford and Smith (1980)
C. mutica s.l.	Crawford (1970)
C. nucensis	Harborne (1977b)
Cosmos bipinnatus	Saito (1974)
Dahlia	Giannasi (1975a,b)
D. variabilis	Price (1939), Bate-Smith and Swain (1953), Puri and Seshadri (1954), Bate-Smith et al. (1955), Nordström and Swain (1956), Saito et al. (1970)
Helianthus (three species)	Harborne and Smith (1978)
series *Angustifolii*	Schilling (1983)
series *Corona-solis*	Schilling and Mabry (1981)
series *Microcephali*	Schilling et al. (1987)
Iostephane	Schilling and Panero (1991)
Lasthenia	Bohm et al. (1974)
L. "chrysostoma"	Shimokoriyama and Geissman (1960)
Megalodonta beckii	Roberts (1980)
Pyrrhopappus	Harborne (1977b)
Simsia	Schilling and Spooner (1988)
Thelesperma megapotanicum	Ateya et al. (1982)
Viguiera series *Brevifoliae*	Schilling and Panero (1988)
sect. *Maculatae*	Schilling et al. (1988)
series *Viguiera*	Schilling (1989), Rieseberg and Schilling (1985)
Zinnia linearis	Harborne et al. (1983)

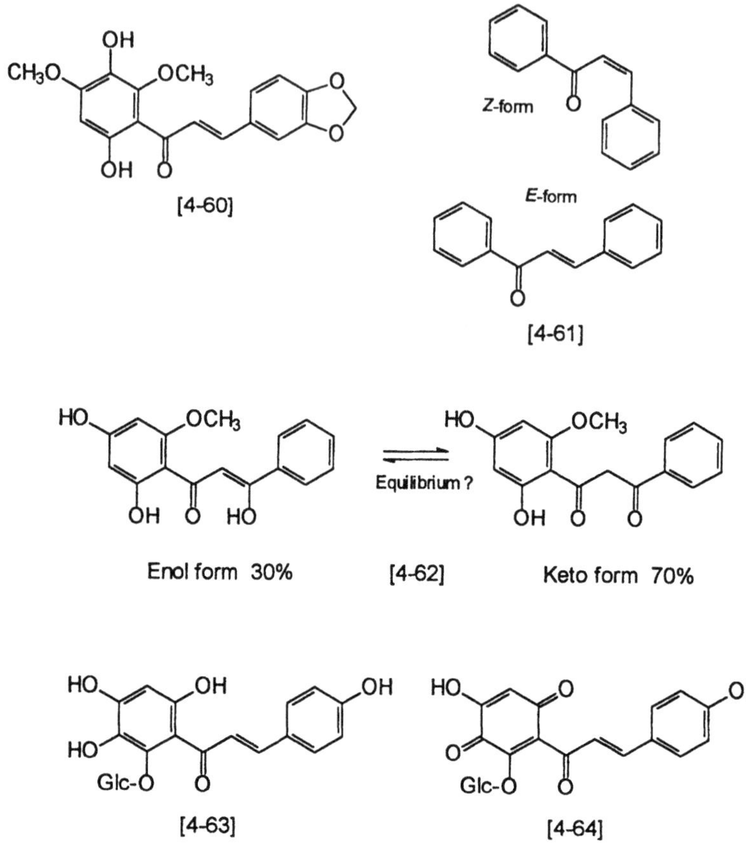

Fig. 4–8. Additional variations in chalcone structure showing a high level of substitution **[4–60]** and possible isomeric forms of the bridging carbons **[4–61]**. Keto-enol tautomerization is illustrated in **[4–62]**; hydroquinone **[4–63]** and quinone **[4–64]** forms of pigment from safflower

volve two compounds that feature additional oxygenation on the A-ring and a high level of *O*-alkylation (Fig. 4–8). *Ageratum strictum* afforded 2',5'-dihydroxy-4',6'-dimethoxy-3,4-methylenedioxychalcone **[4–60]** (Quijano et al., 1982a), while Barua et al. (1978) identified 2'-hydroxy-4',5',6',3,4-pentamethoxychalcone as a component of *Chromolaena odorata*. Two chalcones with only 4-oxygenation, but with extra oxygenation of the A-ring, were also found in *C. odorata* by those workers: 2',4-dihydroxy-4',5',6'-trimethoxy- and 2',3',4',6',4-pentamethoxychalcones (Barua et al., 1978). The levels of *O*-alkylation seen in these two examples from Eupatorieae are typical of their respective genera.

As we have just seen, chalcones with dioxygenated B-rings occur quite widely, particularly in Heliantheae. Chalcones with only a single oxygenation in the B-ring, derivatives of chalcononaringenin, are more restricted in their occurrence, having been reported from *Helichrysum* and *Gnaphalium* (Inuleae s.l.) and one species from Astereae (*Callistephus chinensis*). Chalcononaringenin 2'-*O*-glucoside, or isosalipurposide, has been reported from the following species of *Helichrysum*: *H. arenarium* (Vrkoc et al., 1959; Hänsel et al., 1960), *H. armenium* (Çubukçu and Yüksel, 1982), *H. bracteatum* (Kaufmann and El Baya, 1969), *H. graveolens* (Çubukçu and Damatyan, 1980, 1986), *H. noeanum* (Bingöl and Çubukçu, 1984), *H. orientale* (Çubukçu and Bingöl, 1981), *H. pallasii* (Çubukçu and Bingöl, 1984), and *H. plicatum* (Çubukçu and Mericli, 1977, 1979). The *Callistephus* report came from work of Kuhn et al. (1978). *O*-Methylation of chalcononaringenin at the 2'-OH yields a compound called dehydro-*p*-asebotin, which has been isolated as the aglycone from *Helichrysum heterolasium* (Bohlmann and Abraham, 1979b), *H. odoratissimum* (Van Puyvelde et al., 1989) and *Gnaphalium multiceps* (Ahluwalia and Rani 1976). The 4'-*O*-glucoside has been reported from *Helichrysum affine* and *Gnaphalium affine* (possibly synonyms) by Aritomi and Kawasaki (1974) and from the latter species by Itakura et al. (1975). Wright (1976) reported the glucoside from *H. cooperi*; Maruyama et al. (1974) found it in *G. multiceps*. 2',6',4-Trihydroxy-4'-methoxychalcone was reported as a component of the lipophilic phenols of *H. achyroclinoides* and *H. triplinerve* both of which are endemic to Madagascar (Randriaminahy et al., 1992).

Given that chalcones are obligate intermediates in the biosynthesis of flavonoids (see Chapter 5), and the abundance of 4'-oxygenated flavones and flavonols in the family, one might expect to see chalcononaringenin derivatives reported more often. Theoretically, this chalcone could be detected in all flavonoid-producing plants if sufficiently sensitive methods were employed. However, in only a few cases is a glycosylated or *O*-methylated derivative formed. We see this property manifested in the species discussed in the preceding paragraph. We do not know what factors are at play in the accumulation of these compounds in *Helichrysum* and *Gnaphalium*; in *Callistephus* we do. Kuhn et al. (1978) showed that the presence of isosalipurposide in flowers of *C. chinensis* was clearly correlated with deficiency of the allele for the isomerase; *chch* genotypes accumulate the chalcone whereas **Ch__** genotypes, which have colored flowers, have the allele required for the cyclization step.

A few chalcones based on the 2',4',4-trihydroxy oxygenation pattern have been described from members of the family. The base compound, called isoliquiritigenin, and its 4'-*O*-glucoside have been found in *Dahlia variabilis* by Saito et al. (1970). The work of Giannasi (1975a,b) and Lam and Wrang (1975) confirmed this sighting and reported several additional glycosides as well from species of *Dahlia*. Other sources of isoliquiritigenin derivatives, both glycosides and methyl ethers, include species of *Bidens* sect. *Greenmania* (Roseman, 1986), *Viguiera* (Schilling et al., 1988; Schilling and Panero, 1988; Schilling, 1989).

It may seen odd to begin a paragraph on structurally unusual chalcones by talking about chalcone itself, but totally unsubstituted flavonoids are rare as natural products. The unencumbered parent of this class of compounds was reported to occur in *Centaurea calcitropa* by Dawidar et al. (1989). As indicated by both infrared and nuclear magnetic resonance data, both *Z* (*cis*) and *E* (*trans*) isomers were present in the isolated product, but they could not be separated. After standing for one week in methanol, conversion of the *Z* isomer to the *E* isomer was complete **[4–61]**.

The remaining chalcones to be addressed have unsubstituted B-rings. Many of these also have *O*- or *C*-prenylation (occasionally both) and *O*-methylation is common. *Helichrysum* is highly represented in this group. *C*-Geranyl and *C*-neryl groups have been reported as well. These compounds are listed in Table 4–3.

From time to time, one encounters chalcones with additional oxygenation on one of the bridge carbons. A representative of this group was isolated from *Baccharis salicifolia* by Wollenweber et al. (1989a) and shown to be ß,2',4'-trihydroxy-6'-methoxychalcone. ß-Hydroxychalcones, e.g., **[4–62]**, can exist in two forms, however, the diketo (or dibenzoylmethane) form and the enol form. The *Baccharis* compound was shown to exist as a mixture consisting of keto:enol forms in the ratio of 7:3. It was not clear from the authors' description if this mixture was in equilibrium.

The flowers of safflower, *Carthamus tinctorius*, are an important source of coloring matter, and have been cultivated widely for many years. Isolation of a red, crystalline pigment was first described by Kametaka and Perkin (1910), which makes it, as Shimokoriyama (1962) noted, the first chalcone to be studied as a natural product. Early studies by Kuroda (1930a,b) and Seshadri's group (Seshadri, 1956; Seshadri and Thakur, 1960) resulted in recognition of a pair of compounds related as quinone and hydroquinone. Seshadri (1956) concluded that one of the compounds, called carthamin, was 2',3',4',6',4-pentahydroxychalcone 2'-*O*-glucoside **[4–63]** and that the other was the corresponding quinone **[4–64]** formed by oxidation of the 3' and 6' hydroxyl groups. Based upon studies of model compounds, Obara et al. (1974), however, concluded that the conventional structure of carthamin was incorrect. Subsequent studies demonstrated the existence of pigments with vastly more complex structures than originally thought (Obara and Onodera, 1979; Onodera et al., 1979, 1981a,b). Further work on these compounds, ultimately shown to be quinochalcone *C*-glycosides, was in substantial agreement (Takahashi et al., 1982, 1984; Meselhy et al., 1993a,b). The structure of safflor yellow B, as given in the Meselhy et al. papers, is

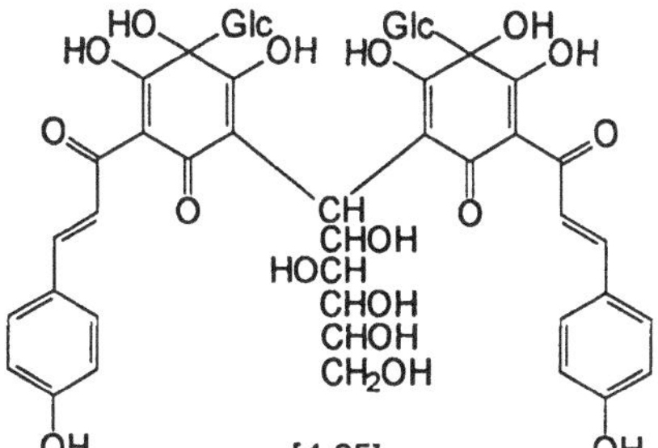

[4-65]

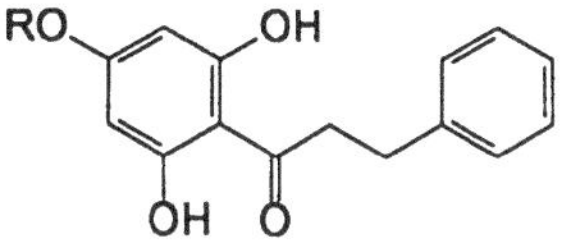

[4-66] R = H

[4-67] R = CH_3

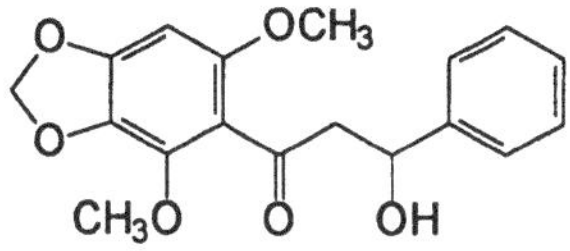

[4-68]

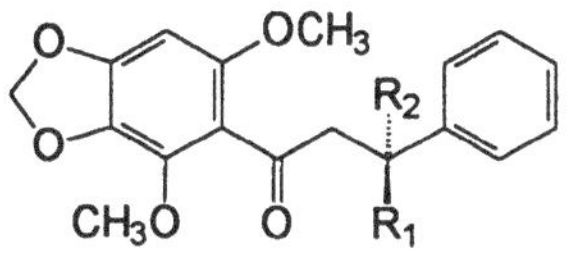

[4-69] OH at either R

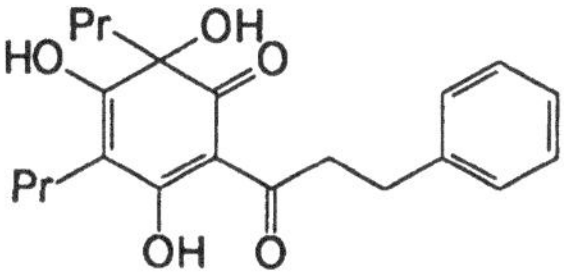

[4-70] Pr = Prenyl

Fig. 4–9. An unusually complex chalcone from safflower **[4–65]**; two simple dihydrochalcones **[4–66, 4–67]**; b-hydroxydihydrochalcones **[4–68, 4–69]**; an A-ring modified dihydrochalcone (a cyclohexadienone) **[4–70]**

shown in Fig. 4–9 **[4–65]**. More recent information on these and related molecules may be found in Kumazawa et al. (1994a,b), Kazuma et al. (1995), and Onodera et al. (1995).

So far as we are aware, these compounds are restricted to safflower so their systematic importance is limited. However, they serve as excellent examples of the level of structural complexity that can exist within a group of pigments (chalcones in this case) normally considered to be structurally straightforward.

Dihydrochalcones

Although dihydrochalcones comprise only a minor part of the flavonoid picture of Asteraceae, a remarkable variety of these compounds has been reported. Among the simpler ones are dihydroisoliquiritigenin and its 4-

and 4'-methyl ethers, which were obtained from *Artemisia palustris* by Huneck et al. (1985). Dihydrochalcononaringenin was identified as a component of *Helichrysum splendidum* (Bohlmann and Suwita, 1979b). All other dihydrochalcones so far reported from the family lack B-ring oxygenation, and have only been reported from members of Inuleae s.l.: 2',4',6'-trihydroxy-dihydrochalcone **[4–66]** from *H. tenuifolium* (Bohlmann and Abraham, 1979a), 2',6'-dihydroxy-4'-methoxydihydrochalcone **[4–67]** from species of *Cassinia* (Reid and Bohm, 1994) and 2',6'-dimethoxy-3',4'-methylenedioxydihydrochalcone from *H. sutherlandii*, where it was shown to occur with the corresponding chalcone (Bohlmann et al., 1978a). In the same paper these workers (Bohlmann et al., 1978a) reported isolating ß-hydroxy-2',6'-dimethoxy-3',4'-methylenedioxy-dihydrochalcone **[4–68]** from *H. mundii*. Accompanying the ß-hydroxy-dihydrochalcone in *H. mundii* were a set of isomeric propane-1,3-diols **[4–69]** that would be formed, presumably, by reduction of the carbonyl group.

C-Alkylated dihydrochalcones have been reported from species of *Helichrysum* and from *Metalasia cymbifolia*. *Metalasia* accumulates *O*- and *C*-prenylated compounds with the 2',4',6'-oxygenation pattern (Zdero and Bohlmann, 1990c), while *H. monticola* goes a step further in making 2',3',4',6'-tetrahydroxy-5'-*C*-geranyldihydrochalcone (Bohlmann and Zdero, 1980). Both the 2",3"-*Z* and 2",3"-*E* isomers of the latter compound were obtained from the plant.

A small but structurally complex set of dihydrochalcone derivatives has been reported from members of two genera, *Metalasia* and, not surprisingly, *Helichrysum*. These compounds have unsubstituted B-rings but are characterized by very complex A-rings. Alkylation has occurred to such an extent that the rings cannot assume the fully conjugated structure normally found in flavonoid A-rings. These compounds can be considered to be derivatives of cyclohexadienone; thus, the compound shown as [**4–70**] can be called 2-dihydrocinnamoyl-3,5,6-trihydroxy-4,6-di-*C*-prenylcyclohexa-2,4-dienone. Variations on this theme involve the number of prenyl groups and the presence or absence of a methoxy function (R = methyl). These compounds have been identified from *Metalasia* (Zdero and Bohlmann, 1990c), *H. aphelexioides* (Randriaminahy et al., 1992), *H. argyrolepis* (Bohlmann et al., 1984d), *H. forskahlii* (Jakupovic et al., 1990b), and from *H. cymosum* and *H. tenuiculum* (Bohlmann et al., 1979f).

Aurones

The majority of aurones encountered in the family are glycosides of sulfuretin (6,3',4'-trihydroxyaurone) **[4–50]** and leptosidin (6,7,3',4'-tetrahydro-

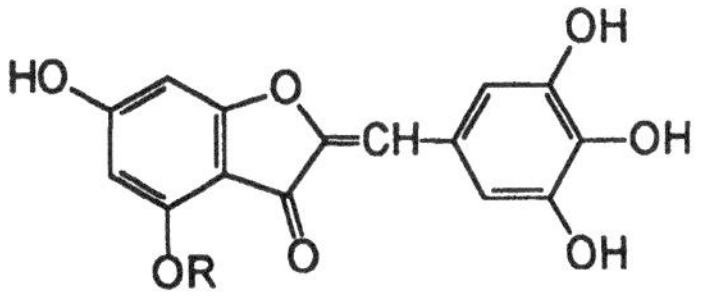

[4-71] R = H, Bracteatin

[4-72] R = Glucose, Bractein

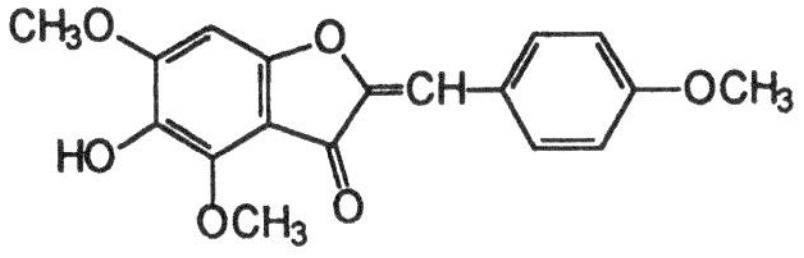

[4-73]

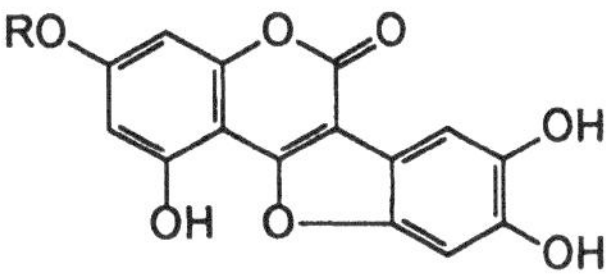

[4-74] R = CH_3

[4-75] R = H

[4-76] R = Glucose

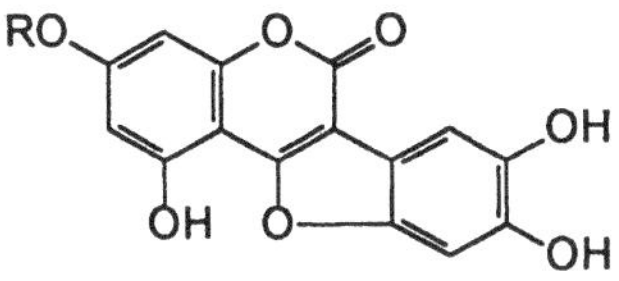

[4-77]

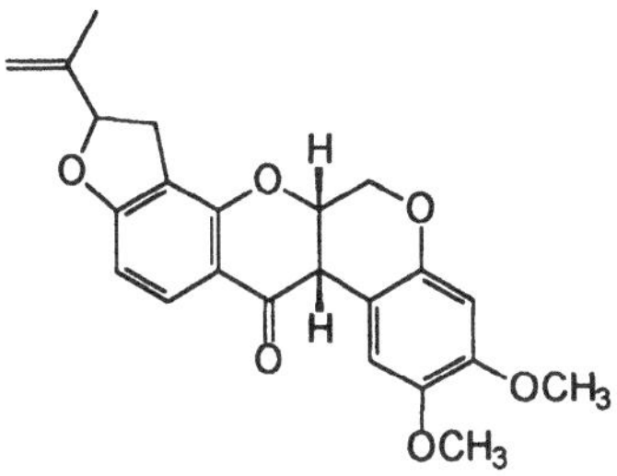

[4-78]

Fig. 4–10. Structural variations in aurones **[4–71 – 4–73]**; complex isoflavonoids **[4–74 – 4–78]**

xyaurone) **[4–58]**. Occasionally, methyl ethers of these aurones are encountered. These were presented in general terms in the introductory comments on chalcones above. The earliest work on naturally occurring aurones in the family happened to deal with compounds that lacked oxygenation at C-4 but had hydroxyl groups at C-3' and C-4' (see citations for Geissman and Shimokoriyama on p. 95). In the early 1960s, Hänsel et al. (1962) encountered a compound from *Helichrysum bracteatum* that had two new features, a 4-OH group and 3',4',5'-oxygenation on the B-ring. Synthetic studies soon confirmed the structure of bracteatin (Fig. 4–10), the new aglycone, as 4,6,3',4',5'-pentahydroxyaurone **[4–71]** (Hänsel and Langhammer, 1963; Hänsel et al., 1963). Rimpler et al. (1963) isolated a glycosidic derivative of bracteatin that had glucose linked through the 4-oxygen; the glucoside is called bractein **[4–72]**. In this and later papers (Rimpler et al., 1963; Rimpler and Hänsel, 1965; Krishnamoorthy and Seshadri, 1966; Kaufmann and El Baya, 1969) evidence was given for the existence of a chalcone derivative with the 2',4',6',3,4,5,-oxygenation pattern. There was also an indication of the equivalent flavanone, although it could have been an artifact since the chalcone just named cyclizes readily. The flavanone was again mentioned by Forkmann (1983). More recently, Nair et al. (1989) established the structures of two aurone glucosides from *Helichrysum buddleioides*, 4,6,3',4'-tetrahydroxyaurone 4-*O*-glucoside (cernuoside or aureusidin 4-*O*-glucoside) and 4,6,3',4',5'-pentahydroxyaurone 4-*O*-glucoside (bractein). In addition to the collection of okanin derivatives described from *Bidens pilosa*, Hoffmann and Hölzl (1989b) reported two acylated derivatives of maritimein, one with two acetate groups and one with three. An unusual aurone was obtained from an epicuticular extract of *Helianthus annuus* flowers and shown to be 5-hydroxy-4,6,4'-trimethoxyaurone **[4–73]** (Alfatafta and Mullin, 1992).

Two unexpected sightings of common aurones deserve comment. In a survey of *Clibadium*, we observed 4,6,3',4'-tetrahydroxyaurone-6-*O*-glucoside, and the equivalent chalcone, in one of eight populations of *C. triane* studied (Bohm et al., 1983). This was the only time that anthochlors have been encountered in the genus. More striking was the report of maritimein (6,7,3',4'-tetrahydroxy-aurone 7-*O*-glucoside) in *Cynara scolymus* (Hinou et al., 1989). This is the only report of an aurone in that tribe (Cardueae) as far as we are aware. The only other anthochlor reported from the tribe is chalcone itself, which was commented on above.

Milovanovic and Djermanovic (1994) reported isolation of 4,6,7,4'-tetrahydroxyaurone, which they called "helmon," along with 6-hydroxykaempferol and apigenin 7-*O*-glucoside, from "*Helminthia echoides*." *Helminthia* is not listed in the 1977 Compositae volumes, nor does it appear in Mabberely (1987). Bremer (1994) does not list *Helminthia* but does show *Helminthotheca*, one of whose species is *H. echioides* (L.) Holub. described as a widespread weedy species. *Helminthotheca* belongs to Lactuceae.

Isoflavones

Usually thought of as characteristic of Fabaceae (or parts thereof) or Iridaceae, isoflavone derivatives have also been reported from a few members of Asteraceae (Fig. 4–10). The coumestan wedelolactone **[4–74]** was isolated from *Wedelia calendulacea* and *Eclipta alba* some years ago (Govindachari et al., 1956a,b). Desmethylwedelolactone **[4–75]** and its 7-*O*-glucoside **[4–76]** were subsequently obtained from *Eclipta alba*

(Bhargava et al., 1970). More recently, Govindachari and Premila (1985) confirmed the isolation of wedelolactone from extracts of *W. calendulacea* but concluded that norwedelic acid may in fact be the naturally occurring form in the plant. An unusual coumestan was described by Daily and colleagues (1988) from *Mutisia acuminata* var. *hirsuta*. The compound was identified as 8,9-dihydroxy-1-methylcoumestan **[4–77]**. The absence of oxygenation on the A-ring and the presence of the *C*-methyl group strongly suggest that this compound is probably not formed via the flavonoid pathway.

Yet another unusual discovery concerning isoflavonoids in the family was reported by Lee et al. (1972). In their study of the factors responsible for the anti-tumor activity of *Balduina angustifolia* they isolated and identified several compounds including the flavone hispidulin and, most unexpectedly, rotenone **[4–78]**, a well known constituent of several genera in Fabaceae. (These were not the active components.) This remains the only report of rotenone in the family. While on the subject of unusual sightings, it is also of interest to note the identification of 7-hydroxy-4'-methoxyisoflavone (formononetin) 7-*O*-glucoside from roots of *Pluchea lanceolata* by Inderjit and Dakshini (1994). It would be of considerable interest to see how widely distributed in root tissue isoflavonoids might be in Asteraceae, especially in view of the known interactions among root exudate compounds and soil micro-organisms.

More or less normal isoflavones have been reported from above-ground parts of species representing three genera (Fig. 4–11). The largest array of isoflavones in the family comes from the genus *Wyethia* from which *O*-methylated derivatives of 5,6,7,4'-tetrahydroxyisoflavone (6-hydroxygenistein) **[4–79]**, 5,7,3',4'-tetrahydroxyisoflavone (orobol) **[4–80]** and 5,6,7,3',4'-pentahydroxyisoflavone (6-hydroxyorobol) **[4–81]** have been described (Bohlmann et al., 1984h; Waddell et al., 1982; McCormick et al., 1986; Bohm and Choy, 1987; Bohm et al., 1989a). In addition to methyl ethers, three *Wyethia* species afforded *C*-prenylated isoflavones (e.g., structure **4–80** with a prenyl group at C-8): *W. angustifolia* and *W. helenioides* (McCormick et al., 1986) and *W. mollis* (Bohm and Choy, 1987). *Gaillardia* was added to the list of isoflavone-producing genera with the report of 5,4'-dihydroxy-7,3'-dimethoxyisoflavone, 5,6,4'-trihydroxy-7-methoxyisoflavone and 5,6,4'-trihydroxy-7,3'-dimethoxyisoflavone from *G. suavis* (Herz et al., 1991). Most recently, Miles (1993) reported 2,7,4'-trihydroxy-5,3'-dimethoxyisoflavone **[4–82]** from *Wedelia biflora* where it occurred with a pair of quercetin dimethyl ether derivatives. The presence of the 2-OH function is unusual.

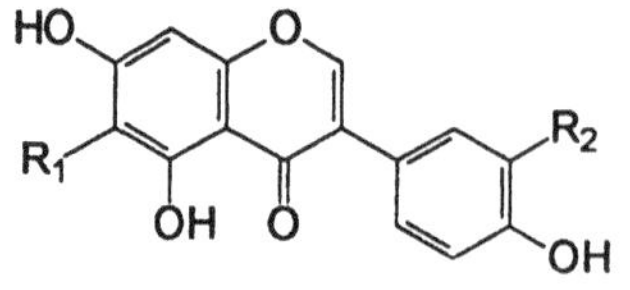

[4-79] R_1 = OH, R_2 = H
[4-80] R_1 = H, R_2 = OH
[4-81] R_1 = R_2 = OH

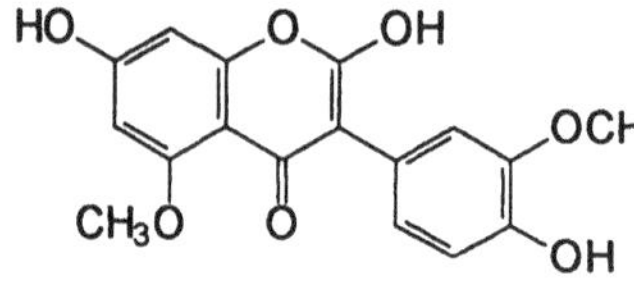

[4-82]

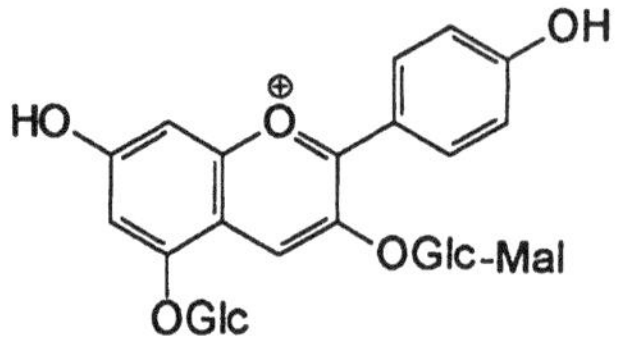

[4-83] Mal = Malonyl

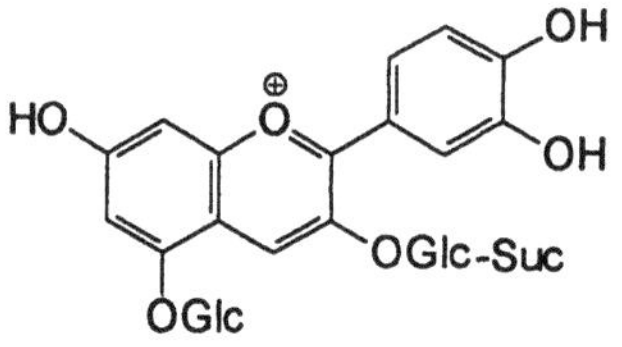

[4-84] Suc = Succinyl

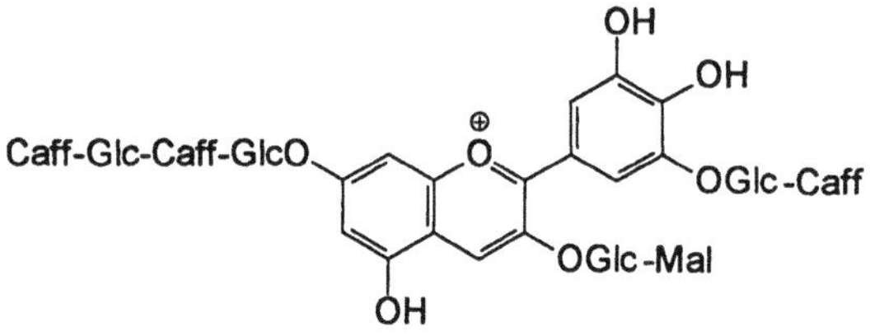

[4-85] Mal = Malonyl, Caff = Caffeoyl

Fig. 4–11. Representative isoflavones **[4–79 – 4–82]** and anthocyanins **[4–83 – 4–85]**

Anthocyanins

Anthocyanins are widespread in Asteraceae (Fig. 4–11) and have been the subject of investigation for many years. Among the earliest reports of flavonoid derivatives in the family, in fact, are reports of anthocyanidin derivatives in *Centaurea cyanus* by Willstätter and Everest (1913), and that of Willstätter and Bolton (1916) who identified cyanidin 3,5-diglucoside from *Helenium autumnale* and *Zinnia elegans*. There seems to be little of systematic value in the distribution of anthocyanidin derivatives in the family other than the nature of the acylating acid in acylated anthocyanins (see Chapter 19). Most interest has been directed towards species that have commercially valuable features, in particular, large flower heads and attractive colors. Many species of *Dahlia* meet these requirements as do species of *Zinnia* and *Gerbera*.

The anthocyanidins identified in members of Asteraceae are common ones, pelargonidin, cyanidin and peonidin, delphinidin and malvidin. Variations in structure involve sugars, how many and where they are attached, and, in some species, the nature and number of acylating groups. Giannasi (1975a,b) mentioned cyanidin glycosides from several species of *Dahlia* that had red to purple ray florets. He did not determine total structures of any of them, however. Detailed work by Harborne and Boardley (1985) with *D. variabilis* showed that pelargonidin 3-*O*-(6"-*O*-malonylglucoside)-5-*O*-glucoside **[4–83]** is a major component of the ray florets of this species. Other malonylated anthocyanins identified from the family are pelargonidin 3-*O*-(6"-malonylglucoside) from *Callistephus chinensis*, pelargonidin 3-*O*-(6"-*O*-malonylglucoside)-5-*O*-glucoside and pelargonidin 3-*O*-(6"-*O*-malonylglucoside)-5-*O*-malonylglucoside from *Dahlia variabilis* cv. Biddenham Strawberry, cyanidin 3-*O*-dimalonylglucoside from *Coleostephus myconis*, cyanidin 3-*O*-malonylglucuronosylglucoside from *Helenium* cv. Bruno, cyanidin 3-*O*-(6"-malonylglucoside)-5-*O*-glucoside and cyanidin 3-*O*-(6"-*O*-malonylglucoside)-5-*O*-malonylglucoside from *Dahlia variabilis* cv. Shandy and delphinidin 3-*O*-(6"-*O*-malonylglucoside)-5-*O*-malonylglucoside from *Cichorium intybus* (Takeda et al., 1986). *Cichorium intybus* cv. Rossa di Verona accumulates cyanidin 3-*O*-(6"-*O*-malonylglucoside) (Bridle et al., 1984). Asen (1984) identified pelargonidin and cyanidin 3-*O*-malonylglucosides, along with several flavone and flavonol glycosides, in 18 cultivars of *Gerbera jamesonii*.

Considerable attention has been focused on individual pigments of some well known species because of their unusual blue color and, subsequently, because of their complexity. The blue cornflower (*Centaurea cyanus*) is a case in point. Willstätter and Everest (1913) isolated "cyanin" from cornflowers and reported its structure to be cyanidin 3,5-diglucoside. During early stages of research on this pigment, it was thought that metallic complexation was involved, an aspect of the study that was examined by Bayer et al. (Bayer, 1958; Bayer et al., 1960). The apparent complexity of the system received additional support when Hayashi's group (Hayashi et al., 1961) hydrolyzed their "protocyanin" preparation and recovered cyanin, a flavone, a carbohydrate, a peptide, and ions of iron, magnesium, and potassium. The next contribution was that of Asen and Jurd (1967) who obtained a blue crystalline product from cornflower and concluded that it was an iron complex consisting of four molecules of cyanin and three molecules of a "bisflavone glucoside." Because they thought their compound was different from cyanin, they called it "cyanocentaurin." This work was followed by a study that showed the bisflavone glucoside to be apigenin 4'-*O*-ß-D-glucoside 7-*O*-ß-D-glucuronide (Asen and Horowitz, 1974). Although cyanin has been known as the major pigment of cornflower for many years, the anthocyanin actually involved in the complex is cyanidin 3-*O*-(6"-succinylglucoside) 5-*O*-glucoside **[4–84]** (Takeda and Tominaga, 1983; Tamura et al., 1983). More recently, Takeda et al. (1988) identified the equivalent pelargonidin derivative from cornflower. It is interesting to note that cell cultures of *Centaurea cyanus* retain the capacity to make cyanidin 3-*O*-(6"-malonylglucoside) (Kakegawa et al., 1987).

A study of the relationship between phenolic chemistry and flower color in *Cineraria cruenta* was undertaken by Ashtakala et al. (1971), but other than reporting apigenin, cyanidin, and delphinidin, no details were forthcoming. A complex pigment was isolated from the garden cineraria, *Senecio cruentus* (*Cineraria cruenta* cv. blue light), and shown to yield delphinidin, glucose and caffeic acid in the ratio 1:3:2 (Yoshitama and Hayashi, 1974). Some possible structures were mentioned but no conclusion was drawn. Degradation and other studies led to the suggestion that the pigment is a dicaffeoyl delphinidin 3,7,3'-triglucoside (Yoshitama et al., 1975). Goto and colleagues (1984) then demonstrated that the pigment is even more complicated; it is a malonyl tricaffeoyl delphinidin tetraglucoside and would be named 3-*O*-(6-*O*-malonyl-ß-D-glucopyranosyl)-7-*O*-(6-*O*-(4-*O*-(6-*O*-caffeoyl-ß-D-glucopyranosyl)caffeoyl)-ß-D-glucopyranosyl)-3'-*O*-(6-*O*-caffeoyl-ß-D-glucopyranosyl) delphinidin **[4–85]** (thus indicating the usefulness of common names). Cyanidin 3-*O*-(6-*O*-malonyl-ß-D-glucopyranosyl)-3'-*O*-(6-*O*-caffeoyl-ß-D-glucopyranoside was reported recently, also from *S. cruentus* (Terahara et al., 1993).

Recent surveys have shown that malonylated anthocyanins occur widely in the plant kingdom. A survey of 81 species representing 27 plant families was undertaken by Harborne and Boardley (1985). Of significance to us is the finding that nearly all of the species from Asteraceae tested were positive for zwitterionic anthocyanins, that is, anthocyanins that carry a positive charge on the flavonoid portion of the molecule and a negative charge on the acylating function, most often malonic acid. An extended study of Asteraceae confirmed their widespread existence in the family (Takeda et al., 1986). It was suggested that many other naturally occurring acylated compounds exist but they likely suffer loss of their acylating acids during isolation from the plant owing to their sensitivity to both alkaline and acidic conditions.

Anthocyanins are not, of course, restricted to floral tissue; seed coats may often be highly pigmented. Vaccari et al. (1982) identified peonidin, cyanidin, and malvidin as their 3-*O*-vanilloyldiglycosides from purple sunflower seed (*Helianthus annuus*). The acylating acid in this instance is vanillic acid (3-methoxy-4-hydroxybenzoic acid). From the same source, Mazza and Gao (1994) more recently reported cyanidin 3-*O*-malonylglucoside and 3-*O*-malonylxyloside, along with the nonacylated glycosides.

Glycosides and acylated glycosides

Many examples of flavonoids that occur as the free phenols have been seen above. These non-polar compounds are generally taken to be components of the plant surface, involving stems, leaves, or floral structures. Flavonoids can, of course, also exist bound to one or more sugar units to produce, theoretically at least, a vast array of phenolic glycosides. Two types of connections exist between flavonoids and their associated sugars, *O*-linkages and *C*-linkages. It is possible, of course, for a flavonoid to exhibit both types of linkages, and indeed, some of these are known in Asteraceae. *O*-Glycosides are the more common type in the family, but *C*-glycosides are nonetheless significant components of several genera. We will examine both types in this section, including a discussion of acylated derivatives of flavonoid glycosides. Flavonoids that enjoy direct acylation of one or more of their phenolic hydroxyl groups have been mentioned previously and are listed in Table 4–6. Sulphated flavonoids are discussed further on in this chapter.

O-*Glycosides*

The majority of naturally occurring flavone glycosides have their sugar attachment through the 7-oxygen. However, there have been reports of glycosylation of flavones at positions 5, 3', 4', and combinations of some of these. Luteolin 5-*O*-glucoside is known from a *Cotula* species (Glennie and Harborne, 1971); luteolin 3'-*O*-glucoside is known from *Centaurea regia* (Ulubelen et al., 1988), and luteolin 4'-*O*-glucoside has been identified from species of *Cynara* (El-Ansari et al., 1988; El-Negoumy et al., 1987). Luteolin 7,3'- and/or 7,4'-diglucosides have been found in *Antennaria dioica* (Mericli, 1983), species of *Cynara* (El-Ansari et al., 1988; El-Negoumy et al., 1987), *Launaea* (Mansour et al., 1983a), *Prenanthes* (Fusiak and Schilling, 1984), and in *Reichardia picroides* (Recio et al., 1992). Higher levels of glycosylation falling into this pattern are also known as evidenced by the finding of luteolin 7-*O*-rutinoside-4'-*O*-glucoside in *Cynara scolymus* (Dranik and Chernobai, 1966) and the 7-*O*-gentiobioside-4'-*O*-glucoside in species of *Launaea* (Mansour et al., 1983).

When more hydroxyl groups are available, more glycosides become possible. Such is the case with isoëtin (5,7,2',4',5'-pentahydroxyflavone) where glycosylation at position 2' becomes a possibility. *Hypochaeris maculata* makes the 2'-*O*-xyloside (and its 4"-acetate) and the 7-*O*-glucoside-2'-*O*-xyloside (and the corresponding acetate) (Gluchoff-Fiasson et al., 1991). *Heywoodiella* (= *Hypochaeris*) *oligocephala* focuses on the 7-position with the result that isoëtin 7-*O*-xylosylarabinosylglucoside is produced (Harborne, 1991). More triglycosides will be met with presently, but now let's look at two B-ring deoxyflavonoids.

Most B-ring deoxyflavonoids seem to occur in the nonpolar fraction of their respective plants. Some exceptions exist, however. In the case of *Centaurea ragusina* pinocembrin (5,7-dihydroxyflavanone) exists as the 7-*O*-diglucoside (Mahmoud et al., 1986). Baicalein 7-*O*-glucuronide (baicalein is 5,6,7-trihydroxyflavone) was reported from *Centaurea scabiosa* by Marsh (1957) while *Carthamus glaucus* makes oroxylin 7-*O*-glucoside (El-Din, 1987). Oroxylin is 5,7-dihydroxy-6-methoxyflavone.

In common with the flavones, glycosylated flavanones most frequently occur with the sugar attached through the 7-oxygen. A few exceptions are known. Naringenin 5-*O*-glucoside has been recorded from *Helichrysum arenarium* (Jerzmanowska and Grzybowska, 1958; Hänsel and Heise, 1959; Borkowski and Pasich, 1961) and from *H. polyphyllum* by Zapesochnaya et al. (1972). Naringenin 5-*O*-diglucoside has been reported as a component of *H. arenarium* by Vrkoc et al. (1959).

In flavonols the majority of glycosides have their sugar attachment through the 3-oxygen. 7-*O*-Glycosylation, however, is fairly common, and flavonol 5-*O*-glycosides are reported from time to time: *Cotula* and *Dendranthema* (Glennie and Harborne, 1971) and *Rhaponticum* (Varga et al., 1982). Of much lesser frequency are monoglycosides involving positions 3' or 4'. An example is the discovery of quercetin 3'-*O*-glucoside as a component of *Conyza bonariensis* where it is accompanied by the 3,3'-di-*O*-glucoside (El-Karemy et al., 1987). Diglycosides can exist with sugars at different positions, as just seen, or at one position. The latter would likely be the choice for more common arrangement when one considers the extremely common occurrence of quercetin 3-*O*-rutinoside (rutin), not only in the family, but in plants in general. Some of the combinations reported from Asteraceae are as follows: kaempferol-3-*O*-xyloside-7-*O*-glucoside from *Zinnia elegans* (Dembinska-Migas et al., 1983), isorhamnetin 3,7-di-*O*-galactoside from *Tagetes terniflora* (De Israilev and Seeligmann, 1983) and isorhamnetin 3-*O*-galactoside-4'-*O*-glucoside from *Senecio subdentatus* (Fadeeva and Nikischenko, 1979). We can also point to the existence of dihydrokaempferol-7-methyl ether 5,4'-di-*O*-glucoside, which has been reported from *Eupatorium micranthum* by Sagareishvili et al. (1981). A novel family of glycosides based on 6-hydroxykaempferol was reported from *Carthamus tinctorius* by Hattori et al. (1992). They identified the 3-*O*-glucoside, the 3,6-di-*O*-glucoside, the 3,6,7-tri-*O*-glucoside and the 3-*O*-rutinoside-6-*O*-glucoside from this species. The 3,6,7-triglucoside is one of only a few compounds known where sugars are attached at three different positions. Another is the anthocyanin from *Senecio* discussed above.

Triglycosides involving the 3,7-combination, which are seen somewhat more often, are exemplified by the existence of quercetin 3-*O*-gentiobioside-7-*O*-glucoside in *Grangea maderaspatana* (El-Karemy et al., 1987), quercetin 3-*O*-rhamnosylgalactoside-7-*O*-rhamnoside in *Senecio subdiscoides* (Reyes et al., 1986) and quercetin 3-*O*-xyloside-7-*O*-xylosylglucoside in *Lasthenia californica* (Gill et al., 1987).

Linear triglycosides are known from a few species detailed, structures of some of which have been published. *Calendula officinalis* has yielded mono-, di- and triglycosides of both quercetin and isorhamnetin (Vidal-Ollivier et al., 1989, 1991; Masterova et al., 1991). For example, in the 1989 paper by Vidal-Ollivier et al. isorhamnetin was found as the 3-*O*-glucoside, 3-*O*-rutinoside, 3-*O*-neohesperidoside and 3-*O*-(2^G-rhamnosyl)rutinoside. Liu et al. (1993) identified quercetin 3-*O*-(2^R-apiosyl) rutinoside from *Baccharis thesioides* along with simpler quercetin glycosides.

Naturally occurring chalcones most frequently have their sugars attached at either the 2'- or 4'-oxygen. (Remember that the 4'-oxygen of chalcones is the A-ring position biosynthetically equivalent to the 7-oxygen in flavanones.) A few chalcones have 4-, or 3'-monoglycosylation, or 3',4'-diglycosylation. Aurones lacking hydroxylation at C-4 occur naturally as the 7-*O*-glycosides. Bracteatin, 4,6,3',4',5'-pentahydroxyaurone, occurs in *Helichrysum bracteosum* as the 4-*O*-glucoside, bractein.

In the case of all flavonoid types, the most commonly encountered sugar is glucose, both in monoglycosides and as a component of di- and higher glycosides. Rhamnose, the next most frequently found sugar, is most often found as a component of rutinosides, which are arguably the most commonly encountered diglycosides, not just in Asteraceae. Other sugars are galactose, apiose, arabinose, xylose, glucuronic acid, and galacturonic acid. Occasionally a methyl or ethyl ester of the uronic acids is reported, but no significance is attributed to the occurrence of these esters owing to the ease of their formation as artifacts of the isolation procedure.

C-Glycosides

We have found reports of *C*-glycosylflavonoids from members of at least 30 genera of Asteraceae representing Anthemideae, Astereae, Cardueae, Eupatorieae, Heliantheae, Inuleae, Lactuceae, and Senecioneae. These compounds have not yet been reported from Barnadesieae, Calenduleae, Mutisieae, and Tageteae. The listing of some of the tribes as positive for *C*-glycosylflavonoids is misleading, however. In the case of Eupatorieae, for example, there are reports of only two species having these compounds, *Eupatorium serotinum* with 6,8-di-*C*-glucosylapigenin (vicenin-2) (Wagner

Table 4–8. *C*-Glycosylflavones in Asteraceae

Tribe and taxon	References
Anthemideae	
Achillea species (many)	Valant-Vetschera (1985b,c)
Ajania fastigiata	Chumbalov and Zhubaeva (1976)
Artemisia frigida	Liu et al. (1982)
A. judaica	Saleh et al. (1987)
A. sublessingianum[1]	Ryakhovskaya et al. (1987)
A. transilense[1]	Chumbalov and Fadeeva (1970)
Leucocyclus formosus s.l.	Valant-Vetschera (1982)
Astereae	
Erigeron species (8)	Valdebenito et al. (1992a)
Haplopappus scrobicularis	Ates et al. (1982)
H. sericeus	Ates et al. (1982)
Hazardia species (11)	Clark and Mabry (1978)
Psiadia punctata	Abou-Zaid et al. (1991)
Solidago graminifolia	Budzianowski (1990)
Stephanodoria tomentella	Fang et al. (1986a)
Cardueae	
Atractylis gummifera	Chaboud et al. (1988), Proliac et al. (1991)
Carlina acanthifolia	Proliac and Raynaud (1983)
C. acaulis	Raynaud and Rasolojaona (1979)
C. corymbosa	Besson et al. (1979), Raynaud et al. (1981)
Centaurea cyanus	Asen (1967)
C. macrophylla	Reynaud et al. (1992)
C. melitensis	Kamanzi et al. (1983b)
C. montana	Gonnet (1992)
C. ragusina	Mahmoud et al. (1986)
C. regia	Ulubelen et al. (1988)
C. thracia	Putun et al. (1989)
C. virgata	Öksüz et al. (1984)
Notobasis syriaca	Mericli and Dellamonica (1983)
Staehelina dubia	Raynaud et al. (1971)
Eupatorieae	
Ageratina calophylla	Fang et al. (1986e)
Eupatorium serotinum	Wagner et al. (1972e)
Heliantheae	
Balduina species (2)	Parker (1972)
Coreopsis species (5)	Crawford and Smith (1983a,b)
Dahlia species (several)	Giannasi (1975a,b)
Flourensia cernua	Dillon et al. (1976)
Gaillardia species (3)	Averett and Beaman (1975)

Table 4–8. (Continued)

Tribe and taxon	References
G. aristata	Zielinska-Stasiek and Gill (1977)
G. pulchella	Wagner et al. (1972d)
Helenium species (3)	Wagner et al. (1972c,g)
Inuleae s.l.	
Helichrysum bracteatum[2]	Rimpler et al. (1963)
Pluchea dioscoridis	Ahmed et al. (1987b)
Lactuceae	
Catananche caerulea	Proliac and Raynaud (1977)
Launaea asplenifolia	Gupta et al. (1985a,b)
Scolymus hispanicus	Romussi and Ciarallo (1978)
Tragopogon species (6)	Kroschewsky et al. (1969), Smolarz and Krzaczek (1988)
Senecioneae	
Senecio species (6)	Glennie et al. (1971)

[1] These species are currently recognized as belonging to the genus *Seriphidium*. See Chapter 7 for discussion of the taxonomy.

[2] *Helichrysum bracteatum* is currently recognized as *Bracteantha bracteatum*. See Chapter 12 for discussion of the taxonomy.

et al., 1972e) and *Ageratina calophylla* with 6-*C*-glucosylquercetin (Fang et al., 1986e). (Incidentally, this was the first report of a *C*-glycosylated flavonol from Nature.) The situation is identical within Inuleae s.l., where only two species have been found to accumulate *C*-glycosylflavones, 6,8-di-*C*-glucosylapigenin from *Pluchea dioscorides* (Ahmed et al., 1987b) and 8-*C*-glucosylluteolin from *Helichrysum bracteatum* (Rimpler et al., 1963). In the case of both these tribes, the study of flavonoids has been quite extensive, so the finding of only a few *C*-glycosylflavonoids must represent true biogenetic capacities of the organisms and is not likely the result of sampling error.

C-Glycosylflavones are by no means evenly distributed in the other tribes listed in the positive category. In Anthemideae, as a case in point, *C*-glycosylflavones have been detected in many, if not most, species of *Achillea*, in only four species of *Artemisia*, and in one species each of *Ajania* and *Leucocyclus*. In Astereae, *C*-glycosylflavones have been reported from eight species of *Erigeron*, 11 species of *Hazardia* and in one or two species from four other genera. This pattern repeats itself to a greater or lesser extent in the remaining tribes. Occurrences of *C*-glycosylflavones in the family are summarized in Table 4–8.

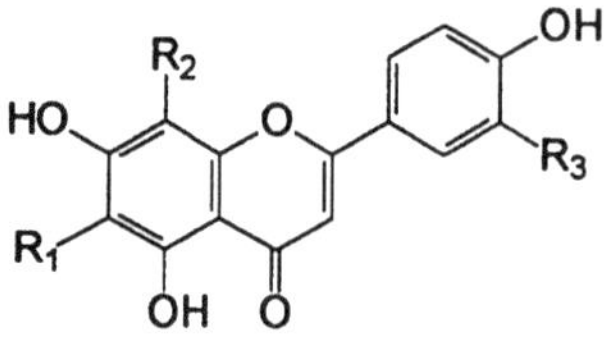

[4-86] $R_1 = R_3 = H$, R_2 = Glucose
[4-87] R_1 = Glucose, $R_2 = R_3 = H$
[4-88] $R_1 = H$, R_2 = Glucose, $R_3 = OH$
[4-89] R_1 = Glucose, $R_2 = H$, $R_3 = OH$
[4-90] $R_1 = R_2$ = Glucose, $R_3 = H$

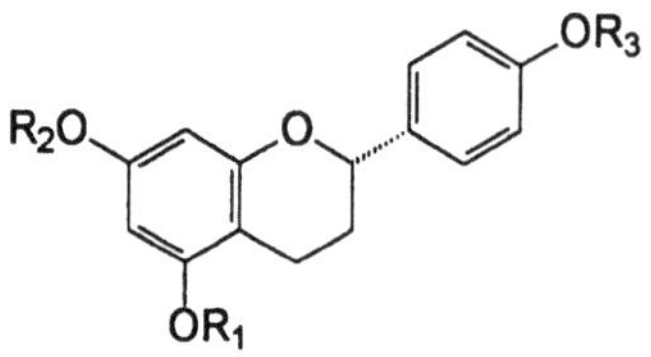

[4-91] $R_1 = H$, $R_2 = R_3 = CH_3$
[4-92] $R_1 = R_2 = CH_3$, $R_3 = H$

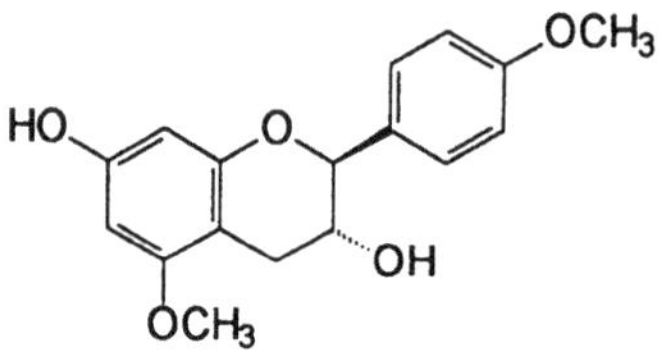

[4-93]

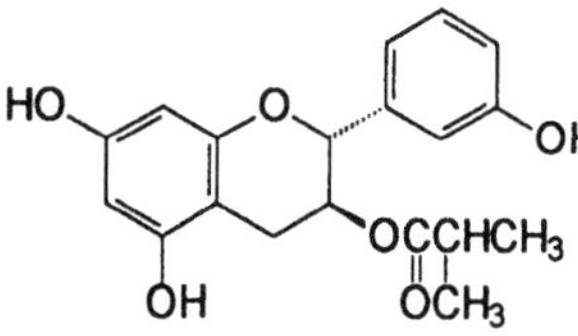

[4-94]

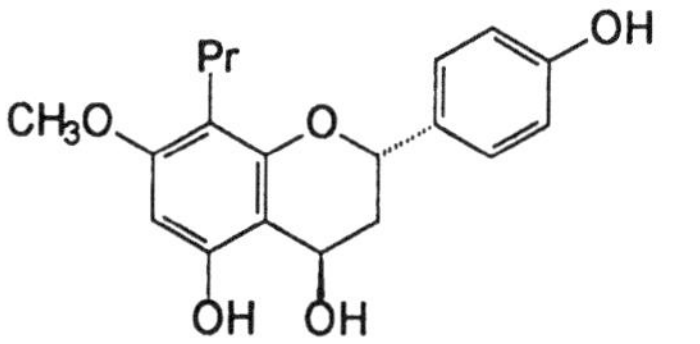

[4-95]

Fig. 4–12. Representative *C*-glycosylflavones **[4–86 – 4–90]**, flavans **[4–91, 4–92]**, flavan-3-ols **[4–93, 4–94]**, and a flavan-4-ol **[4–95]**

C-Glycosylflavones found in the family (Fig. 4–12) are based on well known compounds, vitexin **[4–86]**, isovitexin **[4–87]**, orientin **[4–88]**, isoörientin **[4–89]**, vicenins (e.g., **4–90**), their *O*-methyl derivatives, and some *O*-glycosides. Rich arrays of compounds have been found in some species, such as *Centaurea montana*, from which Gonnet (1992) isolated several vitexin, isovitexin, isoörientin, and isoscoparin derivatives, which he identified as 2"-*O*-glycosides of the parent compounds. The caffeic acid esters of vitexin and isoörientin 2"-*O*-glucosides were also reported, but the position of attachment of the caffeoyl function, presumably on the *O*-linked sugar, was apparently not determined. A similar situation exists in *Leucocyclus formosus* from which Valant-Vetschera (1982) isolated the caffeic acid ester of isoörientin 2"-*O*-arabinoside.

Sagareishvili et al. (1982) isolated "nivyaside" from *Leucanthemum vulgare*. The compound is an 8-*C*-alkylated flavone, based on apigenin, but the alkylating group is 5-deoxyquercitol (a 1,2,3,4-tetrahydroxycyclohexane) rather than the usual sugar. The compound exists as a glucoside with the sugar attached via an α-linkage to the C-1 hydroxyl of the quercetol moiety. To the best of our knowledge, this compound is unique, certainly within the family.

Acylated Glycosides

A prominent feature of many flavonoid glycosides is the presence of one or more acylating functions derived from either aliphatic or aromatic acids. Acylated anthocyanins from *Centaurea* and *Senecio* were mentioned above. Acetates are common, but other aliphatic acids are encountered, malonic and succinic for example. Singly acetylated flavone or flavonol glycosides have been reported from *Arnica* (Merfort, 1986, 1988a,b; Merfort and Wendisch, 1987), *Chrysactinia* (Harborne et al., 1991), *Chrysanthemum* (Hu et al., 1994), *Doronicum* (Reynaud et al., 1985; Reynaud and Raynaud, 1986), *Helenium* and *Plummera* (Wagner et al., 1971c), *Hymenoxys* (Wagner et al., 1972h), *Matricaria* (Kunde and Isaac, 1979; Redaelli et al., 1980; Pietta et al., 1987) and *Senecio* (Piegay et al., 1986). Chalcones and aurones have also been found to occur as acetylated glycosides in a few species. Thus, monoacetylated glucosides of okanin methyl ethers (okanin is 2',3',4',3,4-pentahydroxychalcone) have been reported from *Bidens torta* (McCormick et al., 1984) and from *Senecio pseudotites* (D'Agostino et al., 1991b). Multiple-acetylated glycosides have been identified from members of these genera as well. *Bidens pilosa* afforded okanin 4'-*O*-glucoside-2",4",6"-triacetate (Hoffmann and Hölzl, 1988a), *Senecio pseudotites* okanin 4'-*O*-glucoside-3",4",6"-triacetate (D'Agostino et al. 1991). The aurone maritimein occurs as the 4",6"-diacetate and 2",4",6"-triacetate in *Bidens pilosa* (Hoffmann and Hölzl, 1989b). Apigenin 7-*O*-glucoside 2",3"-diacetate and 3",4"-diacetate were identified in a study of *Matricaria chamomilla* by Redaelli et al. (1982). Quercetin 3-*O*-galactoside occurs as the 3",6"-diacetate and 2",3",4"-triacetate in *Tagetes elliptica* (D'Agostino et al., 1992).

Malonylated anthocyanin glycosides were encountered above, but malonic acid also occurs bound to non-anthocyanic glycosides: kaempferol 3-*O*-glucoside malonate was found in *Cichorium endivia* and the quercetin equivalent in *Lactuca sativa* (Woeldecke and Herrmann, 1974). Butein (2',4',3,4-tetrahydroxychalcone) 4'-*O*-glucoside and 4'-*O*-sophoroside exist as their respective malonate esters in both *Dahlia coccinea* and *D. variabilis* (Harborne et al., 1990). Eupafolin (6-methoxykaempferol) 7-*O*-glucoside 6"-2-methylbutyrate was isolated from a species of *Arnica* by Merfort (1988b), while kaempferol and quercetin 3-*O*-(6"-crotonyl) glucosides were shown to be constituents of *Saussurea japonica* by Shi et al. (1991).

Conjugation with benzoic or cinnamic acid derivatives is encountered fairly often, especially, it would seem, with anthocyanins. However, examples of phenolic acid conjugates with other flavonoid types are also known from Asteraceae. Two benzoic acid derivatives reported are isorhamnetin 7-*O*-galactose *p*-hydroxybenzoate from *Artemisia dracunculus* (Chumbalov et al., 1969b) and quercetin 3-*O*-glucoside 6"-gallate from *Tagetes elliptica* (D'Agostino et al., 1992). *p*-Coumaric acid esters of apigenin 7-*O*-glucoside have been reported from *Echinacea echinatus* (Chaudhuri, 1988) and *Echinops echinatus* (Chaudhuri and Thakur, 1986), while *p*-coumarates of kaempferol and/or quercetin 3-*O*-glycosides have been identified from *Helichrysum* species (Candy et al.,1975; Çubukçu, 1976; Gunasegaran et al., 1993; Sezik and Akdemir, 1986). Examples of caffeic acid as the acylating agent include the following flavonoid glycosides: sulfuretin 6-*O*-di- and triglucosides from *Coreopsis mutica* (Julian and Crawford, 1972); 6-methoxyluteolin 7-*O*-glucoside from *Gnaphalium uliginosum* (Konopleva et al., 1979a,b); quercetin 3-*O*-glucuronide from *Scolymus hispanicus* (Sanz et al., 1993), and vitexin and isoörientin-*O*-glucosides from *Centaurea montana* (Gonnet, 1992). One of the more unusual phenolic acids, 2,3-dihydroxycinnamic acid, was identified as the acylating acid in an apigenin 7-*O*-glucoside derivative from *Anthemis nobilis* (Hérisset et al., 1971).

The last set of acylated flavonoid derivatives to be covered are those that feature two different acylating acids within a single glycoside. We have seen these above with the more complex anthocyanins of *Senecio*. Examples of non-anthocyanic flavonoids with this added complication can be found in the work of Hoffmann and Hölzl (1988b) on *Bidens pilosa* from which they isolated okanin 4'-*O*-glucoside in the following acylated forms: 4"-acetate-6"-*p*-coumarate, 2",4"-diacetate-6"-*p*-coumarate, and 3",4"-diacetate-6"-*p*-coumarate. Mixed acetyl and caffeoyl conjugates were identified from *Bidens frondosa* by Karikome et al. (1992), who also reported the 2"-caffeoyl-6"-*p*-coumaroyl diester from the same plant.

Sulphates

Two sorts of flavonoid sulphate derivatives can exist, one with a sulphate group attached to a hydroxyl group of a sugar, and one with a sulphate group attached to a phenolic hydroxyl group. These negatively charged derivatives are routinely studied using high voltage electrophoresis. The counter ion associated with the sulphate entity is usually not described although potassium is a likely candidate.

Table 4–9. Sulphated flavonoids from members of Asteraceae

Taxa	Compound[1]	Reference
Amphiachyris amoena	"Sulphated flavonoids"	
A. dracunculoides	"Sulphated flavonoids"	Lane (1979)
Brickellia baccharidea	Pat 3-sulphate, 353'/674' 3-sulphate	Timmermann et al. (1981)
B. californica	Pat 3-sulphate, 353'/674'	Mues et al. (1979)
B. chlorolepis	353'/674' 3-sulphate	Ulubelen et al. (1980b)
B. glutinosa	574'/363' 4'-sulphate	Goodwin et al. (1984)
B. laciniata	Pat 3-sulphate, 353'/674' 3-sulphate	Timmermann et al. (1979)
B. longifolia	35/674' 3-sulphate, 353'/674' 3-sulphate	El-Sayed et al. (1990)
B. microphylla	Pat 3-sulphate	El-Sayed et al. (1988)
B. veronicaefolia	Pat 3-sulphate; 353'4'/67 3-sulphate, 353'/674' 3-sulphate	Roberts et al. (1980)
Centaurea alexandrina	Q 3,7-bissulphate	Mansour et al. (1988)
Decachaeta haenkeana	353'/674' 3-sulphate	Miski et al. (1985)
Flaveria species	"Flavonol sulphates"	Al-Khubaizi (1977)
F. bidentis	Q and IR $(\text{sulphates})_n$ n = 1–4	Pereyra de Santiago and Juliani (1972), Cabrera and Juliani (1976, 1977, 1979), Cabrera et al. (1985), Varin et al. (1986a,b)
F. chloraefolia	Flavonol $(\text{sulphates})_n$ n = 1 or 2	Barron et al. (1986), Barron and Ibrahim (1986, 1987a,b, 1988)
Haploësthes species	"Flavonol sulphates"	Al-Khubaizi (1977)
Inula britannica	Q 3-sulphate, 35674' 3-sulphate	Öksüz and Topcu (1987)
Iphiona mucronata	Q and IR $(\text{sulphates})_n$ n = 1-4	Ahmed et al. (1988b)
I. scabra	Q and IR $(\text{sulphates})_n$ n = 1-3, 574'/6 7-sulphate	Ahmed and Mabry (1987)
Lasthenia species	K 3-Glc-7-Rhm(sulphate), Q 3-Glc-7-Rhm(sulphate), Pat 3-Glc-7-sulphate, Pat 7-sulphate	Bohm et al. (1974, 1989b)
Neurolaena lobata	5674'/3 7-sulphate, Qt 7-sulphate	Kerr et al. (1981)

Table 4–9. (Continued)

Taxa	Compound[1]	Reference
N. macrophylla	5673'4'/3 7-sulphate	Kerr et al. (1981)
N. oaxacana	5673'4'/3 7-sulphate	Ulubelen et al. (1980a)
Nothobaccharis candolleana	Pat 3-sulphate	El-Sayed et al. (1988)
Pluchea dioscoridis	IR 3-sulphate, K, Q and IR 3,7-bissulphates	Ahmed et al. (1987b)
Pulicaria arabica	Unidentified flavonol sulphate	El-Negoumy et al. (1982)
Sartwellia species	"Flavonol sulphates"	Al-Khubaizi (1977)
Senecio gallicus	IR 3-$(\text{sulphate})_n$	
S. hoggariensis	IR 3-$(\text{sulphate})_n$	
S. vulgaris	IR 3-$(\text{sulphate})_n$ n = 1, 2	Mansour and Saleh (1981)
Silybum marianum	K 3-sulphate	Mericli (1988)

[1] K = kaempferol; Q = quercetin; IR = isorhamnetin; Qt = quercetagetin; Pat = patuletin. Substitutions: Hydroxyl positions/methoxyl positions.

The first type appears to be represented in the family by species of *Lasthenia*. A study of the entire genus (Bohm et al., 1974) revealed the existence of patuletin 7-sulphate and patuletin 3-*O*-glucoside-7-sulphate in some populations of *L. conjugens* and *L. fremontii* but not others. Re-analysis of populations of *L. californica* from its entire range showed that certain individuals have the capacity to make kaempferol and quercetin 3-*O*-glucoside-7-*O*-rhamnosides with a sulphate group attached to a rhamnose hydroxyl (Bohm et al., 1989b; Desrochers and Bohm, 1993) while others apparently do not.

The rest of the reports describe flavonols, and a flavone, that are directly sulphated. Information on known flavonoid sulphates appears in Table 4–9. Species of *Brickellia*, *Decachaeta*, *Inula, Lasthenia, Neurolaena* and *Silybum* yielded monosulfated flavonols. In the case of *Silybum marianum* subsp. *anatolicum*, kaempferol 3-sulphate was reported to be the major flavonoid in aerial parts of the plant (Mericli, 1988). The flavonols identified range from kaempferol to trimethyl ethers of quercetagetin. The only sulphated flavone so far reported from a member of the family is 5,7,4'-trihydroxy-6-methoxyflavone (hispidulin) 7-sulphate which Ahmed and Mabry (1987) isolated from *Iphiona scabra*.

Di-, tri-, or tetrasulphates have been reported from members of *Centaurea*, *Flaveria, Iphiona, Pluchea*, and *Senecio*. Quite complex mixtures of flavonoid sulphates have been identified in some species. Thus, *Iphiona scabra* was shown to accumulate quercetin 3,7-di- and 3,7,4'-trisulphates, isorhamnetin 3-mono, 3,4'-di- and 3,7,4'-trisulphates, and hispidulin 7-sulphate (Ahmed and Mabry, 1987). *Flaveria chloraefolia* achieves a lower level of sulphation but makes these derivatives out of a larger number of flavonols: 6-methoxykaempferol, 6-methoxykaempferol 7-methyl ether (eupalitin), quercetin, quercetin-7,4'-dimethyl ether (ombuin), quercetagetin 6,3'-dimethyl ether (spinacetin), 6,7-dimethyl ether (eupatolitin) and 6,7,4'-trimethyl ether (eupatin) (Barron and Ibrahim, 1986, 1987b). Quercetin 3,7,3',4'-tetrasulphate has been identified in extracts of *Flaveria bidentis* (Pereyra de Santiago and Juliani, 1972) and *Iphiona mucronata* (Ahmed et al., 1988b). Another variant is seen in *Flaveria bidentis* where quercetin-7,3',4'-trisulphate-3-acetate occurs (Varin et al., 1986a,b).

Flavans and C-ring hydroxyflavans

This section encompasses flavans *per se* along with their 3-hydroxy-, 4-hydroxy- and 3,4-dihydroxy derivatives. Some of these compounds have other unusual features, e.g., lack of a 5-hydroxyl group, which have been pointed out in sections above.

Flavans

Other than having the pyran oxygen, flavans have no oxygen substitution on the *C*-ring. Three such compounds have been identified from members of the family. It should perhaps come as no surprise to learn that all three come from *Helichrysum* species. *Helichrysum microcephalus* was shown by Gao et al. (1987) to accumulate 5-hydroxy-7,4'-dimethoxyflavan **[4–91]**, and by Gutierrez and Herz (1988) to have 4'-hydroxy-5,7-dimethoxyflavan **[4–92]**. The only other species involved in making these compounds, so far as we know, is *H. hypocephalum*, which was shown to have 5,7-dihydroxy-8-*C*-prenylflavan (Bohlmann and Abraham, 1979d).

Flavan-3-ols

Flavan-3-ols have been identified from several genera within Asteraceae. One of the earlier reports involved identification of 7-hydroxy-5,4'-dimethoxyflavan-3-ol **[4–93]** as a constituent of *Lagascea rigida* (Bohlmann and Jakupovic, 1978). *Viguiera quinqueradiata* afforded 4'-hydroxy-5,7,3'-

trimethoxyflavan-3-ol (Delgado et al., 1984). Two *C*-prenylated compounds were obtained from *Marshallia tenuifolia* by Herz and Bruno (1987b) and shown to be 5,4'-dihydroxy-5-methoxy-8-*C*-prenylflavan-3-ol and the cyclized derivative involving the prenyl group and the 7-OH function. A recent addition to this short list comes from the work of Stuppner and Müller (1994) who identified 5,7,3'-trihydroxyflavan-3-ol as the 3-isobutyrate **[4–94]** from *Flourensia retinophylla.*

Flavan-4-ols

Flavan-4-ols are among the rarest flavonoids known; only two have been reported from the family. Bohlmann et al. (1980c) reported several unusual flavonoids from species of *Marshallia* one of which, from *M. obovata*, was identified as 5,4'-dihydroxy-7-methoxy-8-*C*-prenylflavan-4-ol **[4–95]**. Wu et al. (1991) identified 7,4'-dihydroxy-4-methoxyflavan 7-*O*-glucoside and the 6"-*p*-hydroxybenzoyl ester as components of *Vernonia* (= *Centratherum*) *anthelmintica* seeds. The stereochemistry of these two flavonoids is different; the *Marshallia* compound has a 4-ß-hydroxyl group whereas in the *Vernonia* compound the 4-methoxy group lies in the α- conformation.

Flavan-3,4-diols

As far as we have determined, the only flavan-3,4-diols in the family are those identified by Bohlmann et al. (1980c) and Jakupovic et al. (1988e) as components of *Marshallia* species. These compounds are accompanied in some of the species by flavanones and/or dihydroflavonols and by some of the other flavan derivatives just mentioned.

Flavonolignans

One of the most thoroughly studied groups of flavonoid compounds in the family are the flavonolignans (Fig. 4–13). These compounds attracted attention because of the antihepatotoxic activity of seeds of *Silybum marianum*, the milk thistle. A flurry of activity in the late 1960s and early 1970s led to the suggestion that the compounds responsible for the pharmacological activity could be derived from condensation of a 3,4-dihydroxyflavonoid (flavones, flavonols, and dihydroflavonols) with coniferyl alcohol. Experimental demonstration of the free radical phenolic

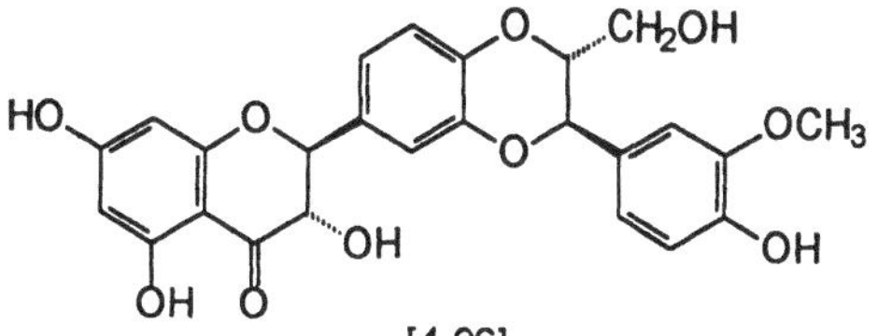

[4-96]

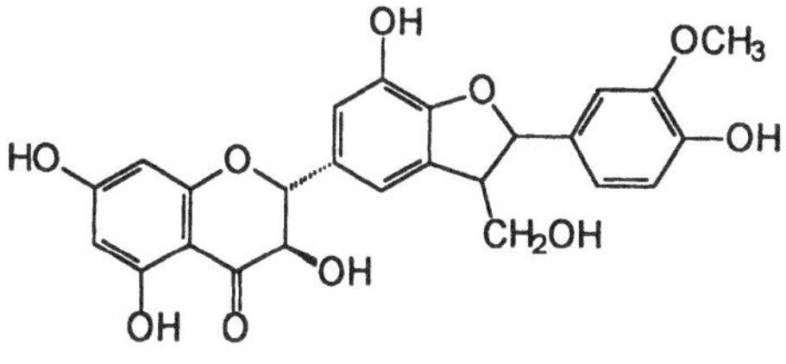

[4-97]

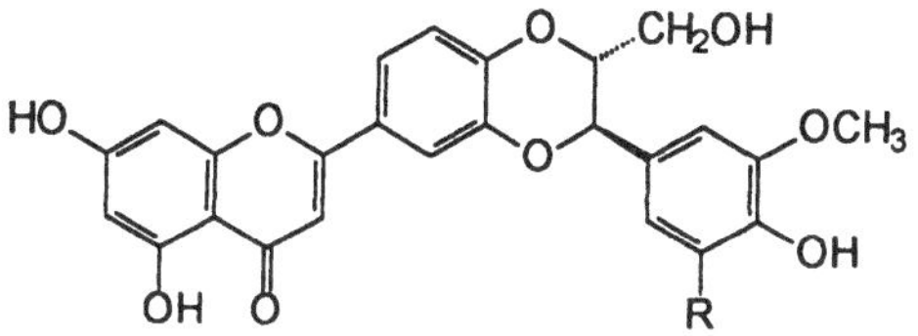

[4-98] R = H

[4-99] R = CH_3

Fig. 4–13. Flavanolignans **[4–96 – 4–99]**

coupling of taxifolin and coniferyl alcohol to produce silybin was provided by Schrall and Becker (1977).

Structural studies (Hänsel and Schöpflin, 1967; Pelter and Hänsel, 1968; Wagner et al., 1971e; Bandopadhyay et al., 1972; Takemoto et al., 1975; Kaloga, 1981a) led to an appreciation that different products were possible based upon differing modes of condensation, and the nature of the flavonoid precursor. Silybin **[4–96]** is characterized by linkage through the two B-ring oxygens of dihydroquercetin (taxifolin) and coniferyl alcohol; isosilybin has the veratryl and hydroxymethyl groups reversed. Dehydrosilybin is the flavonol derivative corresponding to silybin; silandrin is the corresponding derivative based on the flavanone eriodictyol. A second mode of condensation, seen in silychristin **[4–97]**, involves formation of a furan ring linking the flavonoid and coniferyl alcohol. In isosilychristin the two substituents are reversed, as in the case of isosilybin above. Dehydrosilychristin is the flavonol equivalent; silyhermin is the flavanone equivalent. All of the

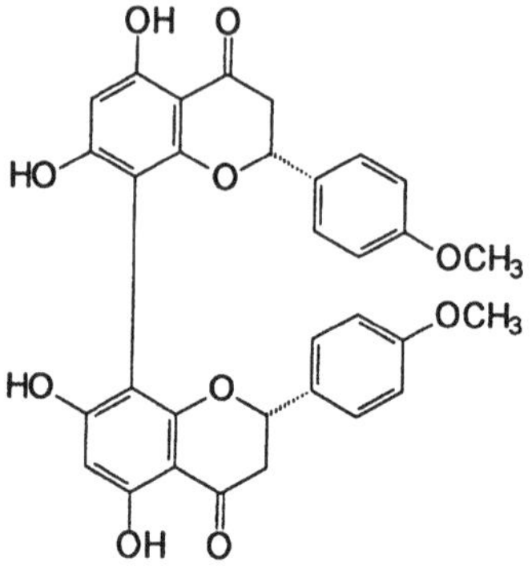

[4-100]

Fig. 4–14. The biflavonoid 4',4'''-dimethyl-cupressuflavanone **[4–100]**

above compounds, plus others, have been isolated from *S. marianum* (for a recent review, see Kurkin and Zapesochnaya, 1987). A recent paper by Cardona et al. (1990) described two flavonolignans from *Onopordum corymbosum*, also a member of Cardueae. The compounds identified were the known hydnocarpin **[4–98]** and the newly described 5"-methoxyhydnocarpin **[4–99]**.

Biflavonoids

Since there appears to have been only one report, biflavonoids must be accorded the status of rarest flavonoid type in the family. Ferraro et al. (1986, 1988b) described the isolation and identification of a biflavonoid from *Eupatorium subhastatum*. (According to King and Robinson, 1987, the proper name for this taxon is *Chromolaena hirsuta*.) The compound can be called 4',4'''-dimethyl-cuppressuflavanone, but formally, it is known as I-5,II-5,I-7,II-7-tetrahydroxy-I-4',II-4'-dimethoxy-I-8,II-8-biflavanone. Perhaps the reader would find it easier to think of two molecules of naringenin-4'-methyl ether (isosakuranetin) joined at the *C*-8 positions **[4–100]**.

Chapter 5

Biosynthesis of Flavonoids

In this chapter we will present a brief overview of flavonoid biosynthesis, along with comments on genetic control of the various steps. Our goal is to demonstrate that the very considerable amount of information that has accumulated on the formation and control of flavonoids in a variety of plant species applies equally well to the pathway in Asteraceae. To this end, we will place particular emphasis on studies done with members of the family. Excellent summary accounts of flavonoid biosynthesis can be found in members of *The Flavonoids* series: Hahlbrock and Grisebach (1975), Ebel and Hahlbrock (1982), and Heller and Forkmann (1988, 1994). Other recent sources include Harborne (1988), Stafford (1990), and Bohm (1998b). Flavonoid genetics has been well reviewed by Gerats et al. (1989), Martin and Gerats (1993), and Forkmann (1994), among others.

The biosynthetic relationships among the main classes of flavonoids are presented in Fig. 5–1 along with enzyme acronyms, which are identified in Table 5–1 (Heller and Forkmann, 1994, p. 501). We have included *O*-methylation and *O*-glycosylation. For the sake of clarity, we have omitted formation of *C*-glycosylated flavones, acylated glycosides, and sulphates.

Chalcone synthase

The initial step in flavonoid biosynthesis involves chalcone synthase (CHS), which serves to join *p*-coumaroyl coenzyme A with three units of malonyl coenzyme A to yield a chalcone. This enzyme has been studied, using a variety of techniques, in preparations from several members of Asteraceae: *Callistephus*, *Centaurea*, *Chrysanthemum*, *Dahlia*, *Gerbera*, *Helichrysum*, and *Zinnia* (as listed in Heller and Forkmann, 1994). The work with *Centauria* involved cultured cells (Kakegawa et al., 1991), whereas the other studies used extracts of floral tissue. Chalcone synthase activity has also been reported from dandelion (*Taraxacum officinale*) callus grown on a cytokinin-rich medium (Akashi et al., 1997). Cells grown in the dark

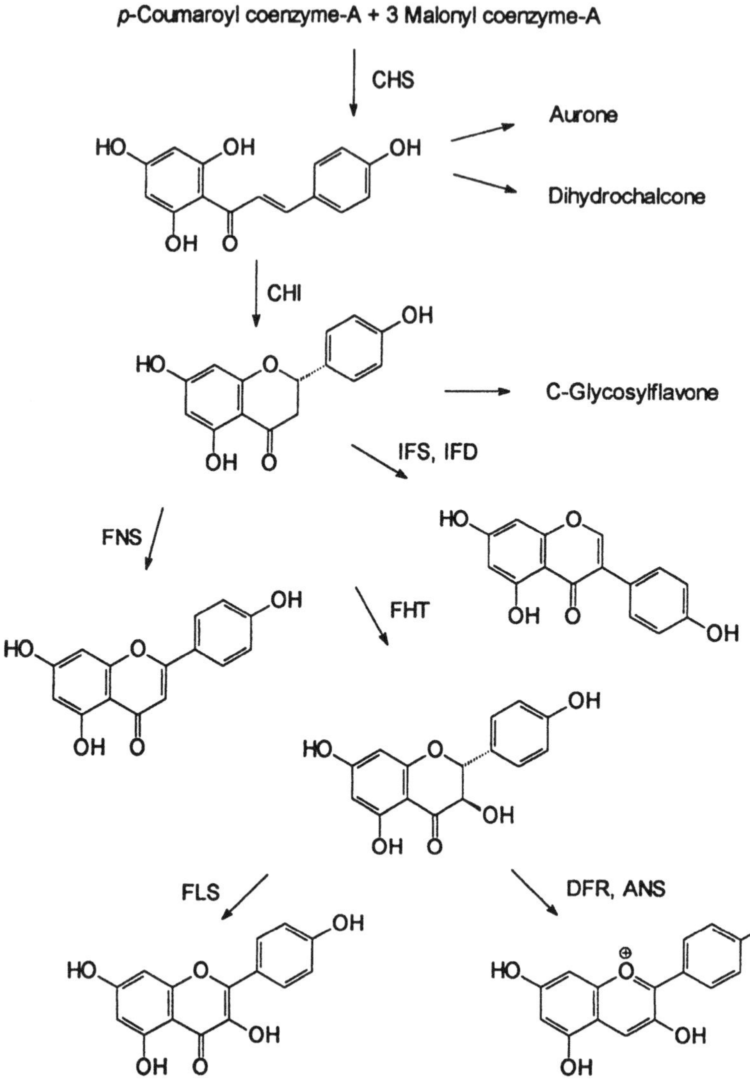

Fig. 5–1. The major steps in flavonoid biosynthesis. See Table 5–1 for full enzyme names and functions

lacked anthocyanins, but a pigment identified as cyanidin 3-*O*-(6"-malonylglucoside) was synthesized in the presence of light.

Several acyanic floral mutants have been studied with regard to the point(s) in the pathway at which the genetic lesion(s) exists. Among these were *Dahlia, Gerbera, Helichrysum*, and *Zinnia*, all of which were found either to have reduced CHS activity or lack the enzyme altogether (Forkmann, 1989).

Table 5–1. Identification of enzyme acronyms used in Fig. 5.1

Acronym	Enzyme name	Function
CHS	Chalcone synthase	Forms chalcone from precursors
CHI	Chalcone isomerase	Chalcone to flavanone
FNS	Flavonesynthases I and II	Flavanone to flavone
FHT	Flavanone 3-hydroxylase	Flavanone to dihydroflavonol
FLS	Flavonol synthase	Dihydroflavonol to flavonol
DFR	Dihydroflavonol 4-reductase	Dihydroflavonol to flavan-3,4-diol
ANS	Anthocyanidin synthase	Flavan-3,4-diol to anthocyanidin
IFS	2-Hydroxyflavanone synthase	Flavanone to 2-hydroxyisoflavanone
IFD	2-Hydroxyisoflavanone dehydratase	2-Hydroxyisoflavanone to isoflavone

Aurones, along with their related chalcones, are major components of the flavonoid profiles of several species in Asteraceae. Information on the biosynthesis of aurones, and their genetic control, however, is very limited, at least in Asteraceae. Some comments on genetic influences on aurone biosynthesis were made by Harborne (1967b) who pointed out that aurone production in snapdragon, *Antirrhinum majus*, is suppressed by the dominant gene **Y**. Yellow pigments accumulate in **yy**. The suppression is not complete, however, as even **YY** plants have a tinge of aurone pigment on the lip of the flower. Only in bottom recessives are aurones, along with all other flavonoids, absent. Comments on the action of the regulatory gene *sulphurea* in *Antirrhinum*, which apparently controls aurone distribution in the flower, were made by Stubbe (1966). Observations of aurone formation in a member of Asteraceae have been made in relation to the next enzyme in the sequence, chalcone isomerase, which we will encounter presently.

Dihydrochalcones, which enjoy a scattered occurrence in the family, are mystery compounds with regard to their biosynthesis and genetic control. Whether these compounds are formed by reduction of chalcones, or possibly even by condensation of a dihydrocinnamoyl coenzyme A derivative with malonyl coenzyme A, is not known.

Chalcone isomerase

The major fate of 2',4',6',4-tetrahydroxychalcone, formed by the action of CHS above, is to serve as substrate for chalcone isomerase (CHI). The product of this reaction is (2*S*)-5,7,4'-trihydroxyflavanone (naringenin). Naringenin is a major branch point in flavonoid biosynthesis serving as substrate for the formation of flavones, isoflavones, dihydroflavonols (3-hydroxyflavanones), flavonols, and flavan-4-ols. Forkmann (1989) listed several cases where CHI mutants were known. Included in his list was the case of the bronze flower mutant of *Helichrysum bracteatum* where chalcone, aurone, and some anthocyanin accumulate. The formation of anthocyanin can be rationalized in terms of the spontaneous isomerization of chalcone to flavanone with subsequent enzymatic conversion of the latter to anthocyanin, which accumulate, on a yellow background. Only in the case of mutants blocked further along in the anthocyanin pathway does one see pure yellow flowers. It is not clear whether the aurone in this system is formed through enzymatically controlled cyclization or whether it is formed spontaneously.

Two types of chalcone isomerase are known, one that will accept both 2',4',6'-trihydroxychalcones (phloroglucinol-type) and 2',4'-dihydroxychalcones (resorcinol-type) as substrates, such as seen in *Cosmos sulphureus* (Chmiel et al., 1983), and one that accepts only the phloroglucinol-type. Plants that contain only flavonoids with phloroglucinol-type A-rings have only the latter enzyme, e.g. *Callistephus* (Kuhn et al., 1978).

Chalcone isomerase has been studied in UV-irradiated callus cells of *Centaurea* (Kakegawa et al., 1991), while its genetic control has been studied in *Callistephus* by Kuhn and coworkers (1978). Plants with the **chch** genotype accumulated isosalipurposide (2',4',6',4-tetrahydroxychalcone 2'-*O*-ß-glucoside), while plants with the **Ch__** genotype exhibited more highly oxidized flavonoids and had normal levels of anthocyanin pigments. Thus, the presence of chalcone in flowers is correlated with the deficiency of the enzyme necessary for its cyclization. Dangelmayr (1982, cited by Forkmann, 1994) and Forkmann (1989) showed that strains of *Helichrysum* with yellow or bronze-colored bracts owed their pigmentation to the accumulation of chalcone and aurones because of lack of CHI activity. Additional mutants lacking CHI were characterized in *Dahlia*, *Gerbera*, and *Zinnia* (Forkmann and Stotz, 1984; Forkmann, 1989).

Flavanone 3-hydroxylase

Flavanone 3-hydroxylase (FHT) catalyzes the stereospecific hydroxylation of naringenin at C-3 to yield (2*R*,3*R*)-3,5,7,4'-tetrahydroxyflavanone (dihydrokaempferol). Forkmann and Stotz (1984) studied FHT in several species among which were cyanic and acyanic forms of *Dahlia variabilis* and *Zinnia elegans*. Administration of naringenin to the acyanic forms failed to yield anthocyanins in these species, whereas use of dihydroquercetin (3,5,7,3',4'-pentahydroxyflavanone) afforded anthocyanins in *Zinnia* but not in *Dahlia*, which indicated that a further block in the pathway existed in the latter. The acyanic forms of both of these species accumulated flavones in their flowers indicating that the flavone synthase system was not influenced by the presence of a mutant FHT gene. Forkmann and Stotz (1984) concluded by saying, "It can be assumed, therefore, that in all species so far investigated the hydroxylation of flavanones in the 3-position to dihydroflavonols is catalyzed by the same type of enzyme." A more recent study of FHT described a partially purified enzyme preparation from flower extracts of *Chrysanthemum* (Stich, cited by Heller and Forkmann, 1994).

Dihydroflavonol 4-reductase

Dihydroflavonol 4-reductase (DFR) catalyzes the stereospecific reduction of (+)-(2*R*,3*R*)-dihydroflavonols to yield (2*R*,3*S*,4*S*)-flavan-2,3-*trans*-3,4-*cis*-diols. DFR has been reported from *Callistephus* (Ruhnau and Forkmann, 1988), *Dahlia* (Fischer et al., 1988), and *Chrysanthemum* (Stich, cited by Heller and Forkmann, 1994). DFR from Asteraceae exhibited properties that were very similar to those of enzymes from members of other plant families.

Formation of anthocyanidins and anthocyanins

Anthocyanin biosynthesis involves loss of the 4-oxygen of flavan-3,4-diols, oxidation of the C-ring to the familiar flavylium ion, and acquisition of one or more sugar moieties by the resulting anthocyanidin. Studies with flower color mutants of *Callistephus chinensis* by Ruhnau and Forkmann (1988) implicated four genes in anthocyanidin biosynthesis: **Ch**, **A**, **F**, and **G**. The gene **Ch**, which controls chalcone isomerase, was mentioned above. The stage of action of the other three genes was not established. Acyanic petals of *C. chinensis*, genotype **AAffGG**, could be induced to form the corresponding anthocyanins upon administration of either leucopelargonidin (3,4,5,7,4'-pentahydroxyflavan) or leucocyanidin (3,4,5,7,3',4'-hexahydroxyflavan). Neither naringenin nor dihydroquercetin led to floral pigmentation with this mutant. In another acyanic line, which carried the **aaFFGG** genotype, none of these precursors lead to floral pigmentation. Cyanic flowers contained dihydroflavonol reductase activity as demonstrated by ready conversion of (+)-dihydrokaempferol to (+)-3,4-*cis*-leucopelargonidin. (+)-Dihydroquercetin and (+)-dihydromyricetin were better substrates for the reductase, which is in line with the normal pattern of pigmentation in *C. chinensis*. It was also pointed out by these workers (Ruhnau and Forkmann, 1988) that the properties of the dihydroflavonol reductase from *C. chinensis* were similar to the corresponding reductases from *Matthiola*, *Dianthus*, *Lycopersicon*, and *Petunia*. Other studies of flower color in Asteraceae include studies of *Dendranthema* (syn. *Chrysanthemum*) (Jordan and Reimann-Philipp, 1983; Teynor et al., 1989a,b), *Parthenium* (Estilai, 1984), and "*Chrysanthemum*" (Hattori, 1992).

Flavanone 4-reductase

Although flavan-4-ols do not figure as major flavonoids in Asteraceae—only a few have been reported from members of the family—the enzyme responsible for their formation from flavanones, flavanone 4-reductase (FNR), has been obtained from *Dahlia* (Fischer et al., 1988).

Flavone synthase

The conversion of flavanones to flavones is catalyzed by two different flavone synthases, FNS I and FNS II. Flavone synthase II, which appears in microsomal preparations, requires NADPH, and is cytochrome-450 dependent, has been reported from several flowering plant species including *Taraxacum officinale* (Stotz and Forkmann, 1981), *Chrysanthemum* (Stich et al., 1988), and *Gerbera* hybrids (Martens and Forkmann, 1998). The *Gerbera* enzyme was detectable only in extracts of genotypes that carried the dominant wild-type allele *Fns*; it was not seen in lines with the recessive condition, *fns fns*. This represents the first correlation between FNS II activity and a specific locus.

Flavonol synthase

Flavonol synthase (FLS), the enzyme responsible for converting dihydroflavonols to flavonols, has apparently not been reported from any member of the family, at least according to the review of Heller and Forkmann (1993). We assume that its existence and operation in members of Asteraceae is typical.

Isoflavone synthesis

The presence of isoflavones in only a few members of Asteraceae is one of the more unusual chemical features of the family. Biosynthesis of isoflavones in these species has apparently not been studied, so it would be interesting to see if they are formed by the same two-step process seen in *Glycine* (Heller and Forkmann, 1988). The first step involves a rearrangement of the aryl group from C-2 to C-3 accompanied by oxidation at C-2 catalyzed by 2-hydroxyisoflavanone synthase (IFS). The product of that reaction is then acted upon by 2-hydroxyisoflavanone dehydratase (IFD) to yield the isoflavone.

Modifications

With the establishment of the fundamental flavonoid classes, which concern the existence of the C-ring, its size, its oxidation level, and the point of attachment of the B-ring, attention can now turn to what one might call the modification reactions. These are the reactions that give us more complex oxygenation patterns, glycosidic derivatives, methyl ethers, and a host of alkylated products. There are excellent enzyme and genetic data available for some of these processes, e.g., *O*-methylation, *O*-glycosylation, and *O*-acylation, but distressingly little for others, particularly on unusual oxygenation patterns. In a few cases, members of Asteraceae have been used in these studies, but most of our understanding of these processes comes from work with corn, petunia, and snapdragon.

Oxygenation pattern

Oxygen atoms on C-5 and C-7 of flavonoids with a six-membered heterocyclic ring originate from acetate, while oxygen at C-4' entered from *p*-coumaroyl CoA. Although caffeoyl CoA can be utilized in the CHS reaction, leading to a flavonoid with 3',4'-oxygenation, the reaction seems to be a minor one. Oxygen at C-3 enters the scene through the action of flavanone 3-hydroxylase, as discussed above. Additional B-ring oxygens can result from the action of two well-defined enzymes. Flavonoid 3'-hydroxylase (F3'H) acts on 4'-hydroxyflavonoids, and can convert naringenin into eriodictyol, dihydrokaempferol into dihydroquercetin, apigenin into luteolin, and kaempferol into quercetin. This enzyme has been reported from flower extracts of *Dahlia* and *Gerbera*, along with species from other families (Stich et al., 1988). A second enzyme, flavonoid 3',5'-hydroxylase (F3',5'H), controls hydroxylation of 4'-hydroxy flavonoids to the 3',4',5'-trihydroxy (myricetin) level. Information on this enzyme in members of Asteraceae does not appear to be available. A recent report by Wimmer et al. (1998) describes the conversion of 2',4',4-trihydroxychalcone (isoliquiritigenin) to

2',4',3,4-tetrahydroxychalcone (butein) by protein preparations from *Dahlia variabilis* floral tissue.

Oxygenation at other points on the B-ring, the 2'- and 6'-positions, has not been demonstrated at the enzyme level, at least in representatives of Asteraceae. Neither has/have the enzyme/s responsible for extra oxygenation at positions 6 and/or 8 on the A-ring been identified. It would be of interest to see whether a situation akin to the F3'H-F3',5'H pair exists such that one enzyme would act at C-6 and the other on both C-6 and C-8. Since flavonoids with the 5,7,8-trioxygenation pattern are known in the family, perhaps there is an enzyme specific for placing oxygen at C-8 as well.

The majority of flavonoids reported from Asteraceae exhibit oxygenation at least at C-5 and C-7. However, a significant number of compounds have been reported that lack oxygen at C-5 (C-4 in aurones; C-6' in chalcones). Although nothing has apparently been reported concerning formation of this group of flavonoids in members of Asteraceae, studies in *Glycine* (Welle et al., 1991; Welle and Schroder, 1992) indicate that a polyketide reductase (PKR) exists that coacts with CHS to produce 2',4',4-trihydroxychalcone (isoliquiritigenin). Early work by Grisebach and Brandner (1961) showed that loss of the oxygen from the 6'-position occurs prior to chalcone formation. Cyclization of the resulting 6'-deoxychalcone would lead either to a series of 4-deoxyaurones (e.g., sulfurein) or 5-deoxyflavanones (e.g., liquiritigenin).

Of some prominence in the family are flavonoids that lack B-ring oxygenation. We are aware of neither enzymatic nor genetic information bearing on the formation of B-ring deoxyflavonoids. The origin of these compounds was the subject of speculation in a recent paper documenting their widespread, but scattered, occurrence in the dicots (Bohm and Chan, 1992). Two possibilities exist to explain the formation of these compounds: (1) activation of cinnamate to form cinnamoyl coenzyme A by a lygase specific for cinnamate; (2) existence of a 4-hydroxycinnamate lygase (4CL) with the capacity to accept either 4-hydroxycinnamate or cinnamate as substrate.

In addition to lacking information on enzymology of extra oxygenation, we have only limited information on the order of substitution of the flavonoid nucleus. What data are available, however, suggest a highly ordered sequence of events. The most detailed study of substitution order comes from R. K. Ibrahim's group in Montreal who worked with *Chrysosplenium americanum* (Saxifragaceae) and examined *O*-methylation, hydroxylation, and *O*-glucosylation (De Luca and Ibrahim, 1985a,b; Ibrahim et al., 1987). These studies dealt with the steps leading to two highly *O*-methylated flavonols: 5,5'-dihydroxy-3,6,7,2',4'-pentamethoxyflavone 5'-*O*-glucoside and

5,2'-dihydroxy-3,7,4',5'-tetramethoxyflavone 2'-*O*-glucoside. The sequence of events for the formation of these flavonoids is: quercetin → 3-methylquercetin → 3,7-dimethylquercetin → 3,7,4'-trimethylquercetin. Establishment of the 6-substitution occurs from either 3,7-dimethyl or 3,7,3'-trimethylquercetagetin.

An unexpected finding emerged from this study. It had generally been assumed that glycosylation occurs as a terminal step in flavonoid biosynthesis. In many simple compounds this is undoubtedly true, but in the case of *Chrysosplenium* a different picture emerged (Khouri and Ibrahim, 1987). An *O*-methyltransferase was obtained that required a glucosylated flavonoid as substrate. The enzyme transferred a methyl group to either the 2'-OH (sugar at 5') or 5'-OH (sugar at 2') of the substrate, depending upon which position was available). We feel that it is reasonable to assume that an orderly process also gives rise to the complex flavonoids present in Asteraceae.

O-Methylation

The sequence of reactions involved in the formation of highly substituted, and highly *O*-methylated, flavonols in *Chrysosplenium* was mentioned above. The high degree of selectivity of methylation site in that system appears to be a useful model for the general process in plant species. There appears to be no information available on flavonoid *O*-methyltransferases in Asteraceae, but it seems reasonable to expect a level of selectivity at least equal to that exhibited by enzymes from other plant species for which data are available.

O-Glycosylation

O-Glycosylation is one of the most thoroughly studied reactions in the flavonoid pathway, in which it is generally considered to be the terminal step. Exceptions to this generalization involve the obvious case where acylation of a flavonoid glycoside occurs, and the not so obvious case that was demonstrated in *Chrysosplenium* where *O*-glycosylation preceded a terminal *O*-methylation.

Enzymatic studies of *O*-glycosylation have involved an assortment of tissues and a wide variety of plant species. Asteraceae is represented in this list by *Flaveria*, in which Ibrahim's group (Varin et al., 1986b; Hannoufa et al., 1991) studied both glycosylation and sulphation reactions. Flavonoid *O*-glycosyltransferases generally exhibit position specificity, but are less sensitive to the class of flavonoid upon which they work. Thus, the same enzyme might be expected to recognize the 7-OH group of a flavone as well as a flavonol as we shall see below.

Genetic studies of *O*-glycosylation with representatives of the family are also scarce, but what is known suggests that the processes are under the same sort of control as has been reported using other species. Thus, Teusch and Forkmann (cited by Forkmann, 1994) reported that the *M* gene in *Callistephus chinensis* controlled anthocyanin 5-*O*-glucosyltransferase whose function it is to transfer a glucose moiety to the 5-OH of anthocyanin 3-*O*-glucosides. Stich et al. (1996) have described an enzyme preparation from floral tissue of *Chrysanthemum segetum* that catalyzes glucosylation of the 7-OH group of quercetin and gossypetin, which are known to be floral pigments in this species. The enzyme would also accept flavanones and flavones as substrates. Evidence was presented by those workers showing that hydroxylation of quercetin at C-8 to produce gossypetin occurred before 7-*O*-glucosylation.

Circumstantial evidence that formation of luteolin 7-*O*-digalactoside is controlled by "dominant allele or alleles" in *Carduus* was discussed by Bain and Desrochers (1988). Artificial crosses between *C. nutans*, which lacks the digalactoside, and *C. acanthoides*, which has it, yielded F_1 progeny all of which possessed the compound. No enzymatic work was done in this system.

Whereas *O*-glycosylation occurs as a late step in flavonoid biosynthesis, the formation of *C*-linked glycosides involves an early event. Administration of labelled flavones to several species yielded *O*-methylated and *O*-glycosylated derivatives but no *C*-glycosyl derivatives (Wallace et al., 1969). However, when labelled naringenin was administered, tracer was found in *O*-methylated and *O*-glycosylated flavones, as in the earlier experiment, but also in the *C*-glycosylated compounds vitexin and orientin, 8-*C*-glucosylapigenin and luteolin, respectively. None of the *O*-glycosyltransferases studied to date has yielded *C*-glycosylated products.

Acylation

Acylated derivatives of anthocyanidin and other flavonoid glycosides occur widely in Asteraceae. Commonly encountered aliphatic acyl groups are derived from acetic, malonic, and succinic acids. Studies of the enzymes involved in acylation have shown a high degree of specificity for both acylating acid and for glycoside substrate. Thus, Ino and Yamaguchi (1993) reported an acetyl-coenzyme A:anthocyanidin 3-*O*-glucoside acetyl transferase from flowers of *Zinnia elegans* that accepted anthocyanidin 3-*O*-

glucosides as substrates but showed no activity toward 3,5-di-*O*-glucosides. A similar selectivity was observed in a malonyl-coenzyme A:anthocyanidin 3-*O*-glucoside malonyltransferase from flowers of *Callistephus chinensis*, which transfers the malonyl group to the 3-*O*-glucosides of pelargonidin, cyanidin, delphinidin, and, at a considerably lower rate, to cyanidin 3,5-di-*O*-glucoside (Teusch and Forkmann, 1987). Malonylation of anthocyanidin 3-*O*-glucosides was also observed using extracts of flower buds of *Dendranthema morifolium* (Ino et al., 1993). The enzyme, present in 14 cyanic and one acyanic cultivar tested, accepted the 3-*O*-glucosides of cyanidin, pelargonidin, delphinidin, and malvidin but did not accept either cyanidin 3,5-di-*O*-glucoside or 3-*O*-rutinoside. The enzyme was specific for malonyl-coenzyme A; acetyl-coenzyme A did not work. A recent study described in some detail a malonyltransferase from *Lactuca sativa* leaves that catalyzes malonylation of anthocyanidin 3-*O*-glucosides (Yamaguchi et al., 1996). These workers reported that the transferase has a wide tolerance for anthocyanidin but not for the glycoside; the enzyme would not accept anthocyanidin 3,5-di-*O*-glucosides as substrates.

Yamaguchi et al. (1995) also described an enzyme from *Centaurea cyanus* that catalyzes the formation of succinylated anthocyanidin 3-*O*-glucosides. The enzyme would accept either cyanidin or pelargonidin 3-*O*-glucoside but would not function with 3,5-di-*O*-glucosides. The enzyme would also accommodate malonate.

Substituted cinnamic acids are also known as acylating acids in a number of cases. Commonly encountered ones are *p*-coumaric, caffeic, and ferulic acids. The corresponding benzoic acids are also met from time to time and 2,3-dihydroxycinnamic acid was reported as the acylating acid involving an apigenin glucoside in *Anthemis nobilis* (Hérisset et al., 1971). Flavonoid glycosides acylated with more than one acid, acetic and *p*-coumaric for example, have been found in several species, e.g., *Bidens pilosa* (Hoffmann and Hölzl, 1988b). An enzyme capable of transferring the *p*-coumaroyl group to anthocyanins in *Silene* was described by Kamsteeg and coworkers (1980). It seems reasonable to assume that a degree of specificity can be expected with enzymes from Asteraceae with regard both to the acid moiety involved as well as the site of acylation. Details of the reactions are wanting, however.

A particularly rich assortment of flavonoid sulphates has been reported from species of *Flaveria* (Varin and Ibrahim, 1989, 1991, 1992). These workers have also examined the enzymatic synthesis of flavonol sulphates of varying degrees of complexity. They found that the enzymes require 3'-phosphoadenosine 5'-phosphosulphate as sulphate donor and will accommodate a range of flavonols. Different proteins are responsible for sulphation at different positions on the flavonol nucleus and a sequence of sulphations appears to exist. Thus, quercetin is first sulphated at the 3-OH, followed by sulphation at the 3'-OH and 4'-OH, respectively. Sulphation occurs next at the 7-OH, although not all species have all of these positions substituted.

C-Alkylation of flavonoids, primarily involving the 3,3-dimethylallyl (prenyl) function, is seen in a number of members of the family. To the best of our knowledge, nothing is known about the enzymology or genetics of this reaction in Asteraceae. Prenyltransferases, which require ß,ß-dimethylallyl pyrophosphate (DMAPP) as prenyl donor, have been identified in members of Fabaceae, however. The reaction appears to be associated with the biosynthesis of phytoalexins (Schröder et al., 1979; Zählringer et al., 1979, 1981). An extensive review of prenylation of isoflavonoids, including structural and biosynthetic considerations, has recently been provided by Tahara and Ibrahim (1995).

Information on the enzyme(s) involved in *C*-methylation appears not to be available for any plant species.

Chapter 6

Biological Functions of Flavonoids

In the earlier days of the 20th Century it was often suggested that plant secondary metabolites, flavonoids included, represented waste materials of plant metabolism. It was also suggested, light-heartedly one assumes, that the major use of these compounds was to challenge the ingenuity of natural product chemists! Even after functions had been identified for some of these compounds, use of the term "secondary metabolite" continued suggesting that they played only a minor role in plant life. Although they are still called secondary metabolites, it has become abundantly clear that many of the processes with which they are involved are critical to the well-being of plants.

The amount of information available on flavonoid functions in plants is large and growing at an impressive rate. A comprehensive review of the subject would far exceed the space available in this book. What we intend in the following is to present a sampling of the activities with which these compounds are involved. We hope that this brief introduction will serve to demonstrate their great versatility. A more detailed discussion can be found in Bohm (1998b).

For an historical view of the subject, the reader is referred to a review by McClure (1975). A more recent treatment by the same author (McClure, 1986) updates the subject. A review by Rao (1990), which deals primarily with root flavonoids, covers more general aspects of the subject as well. Harborne (1988) presents a masterful overview of ecological biochemistry that includes many references to the involvement of flavonoids in Nature. In a recent article, Bennett and Wallsgrove (1994) discuss the role of several classes of secondary compounds, including flavonoids, in plant defense.

Flavonoids function in several ways: (1) through their interaction with other plants; (2) through their influence on micro-organisms; (3) through their capacity to absorb light of various wavelengths; (4) through their capacity to participate in a number of developmental and other processes within the plant itself; and (5) through their capacity to inhibit certain en-

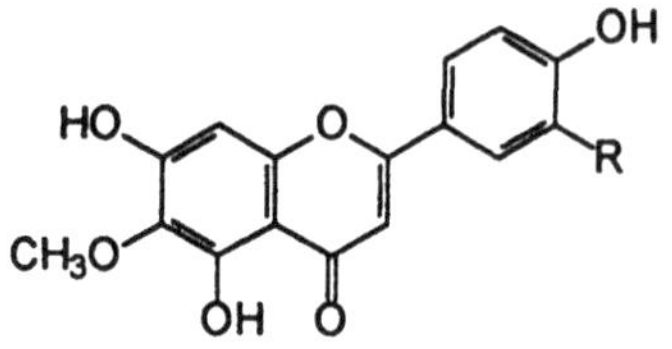

[6-1] R = H

[6-2] R = OH

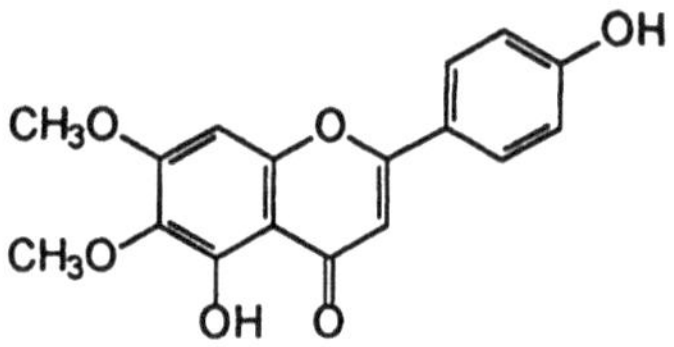

[6-3]

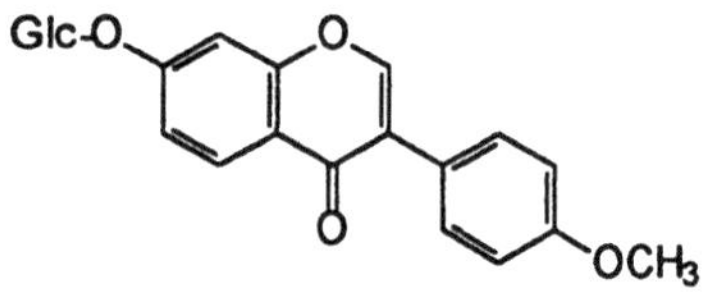

[6-4]

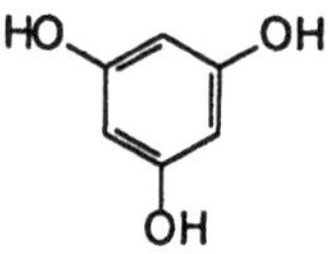

[6-5]

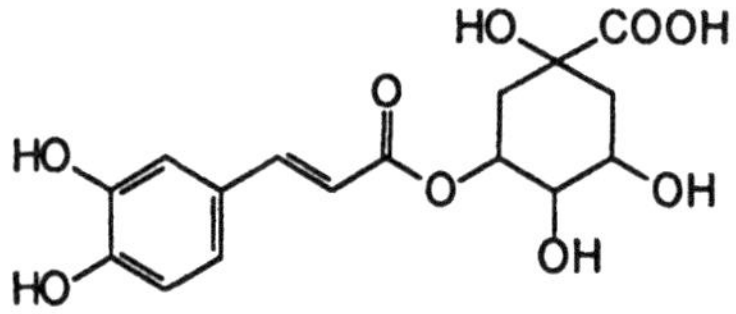

[6-6]

Fig. 6–1. Flavonoids and other phenolic compounds showing possible allelopathic effects. Hispidulin **[6–1]**; eupafolin **[6–2]**; cirsimaritin **[6–3]**; ononin **[6–4]**; phloroglucinol **[6–5]**; a chlorogenic acid **[6–6]**

zyme reactions. Our first examples involve interactions between the producing plant and other plants.

The influence of plant chemicals on the germination of seeds or on the growth of seedlings is referred to as allelopathy. This phenomenon may involve potentially competing species, or seeds and seedlings of the subject species itself. Many examples of allelopathy have been described in the literature involving representatives of a wide variety of plants; in fact, chemical defense against potential competitors appears to be a nearly universal capacity in the plant kingdom. The subject has been thoroughly reviewed by several workers, most notably Rice (1974, 1984, 1995), whose books offer in-depth analyses of the phenomenon. We include in the following a few examples involving members of Asteraceae. An interesting example of allelopathy involving flavonoids comes from the work of Proksch et al. (1982) on the resinous material from the Cape Province endemic *Elytropappus*

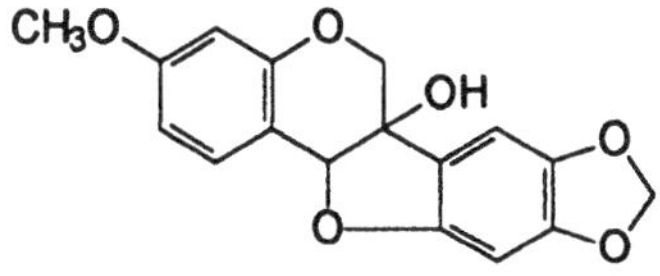

[6-7]

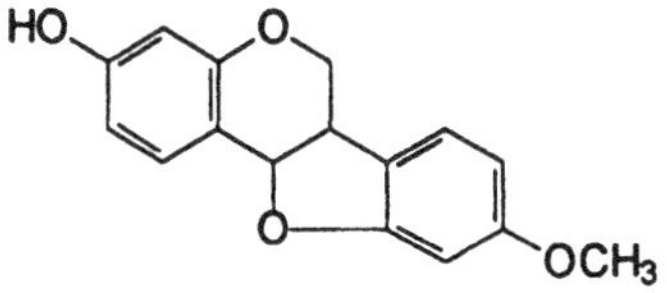

[6-8]

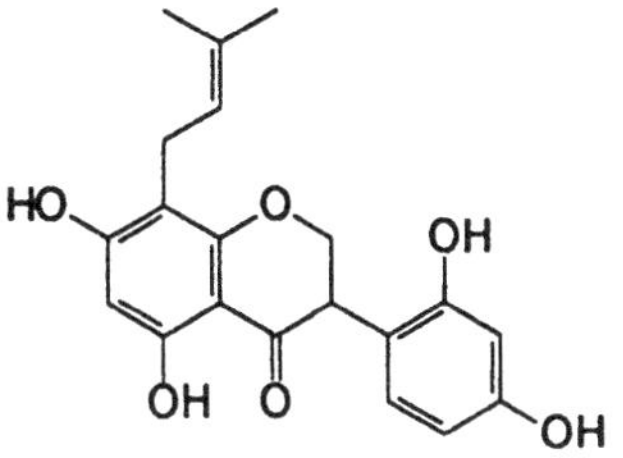

[6-9]

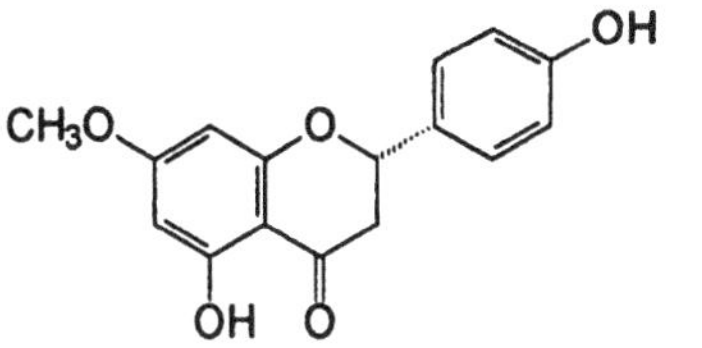

[6-10]

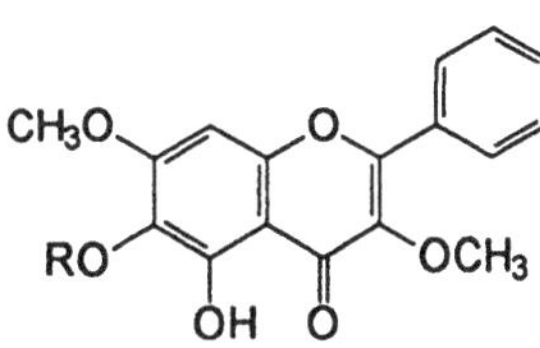

[6-11] R = H
[6-12] R = CH_3

Fig. 6–2. Flavonoids showing activity against microorganisms: pisatin **[6–7]**, medicarpin **[6–8]**, kievitone **[6–9]**, sakuranetin **[6–10]**, and against insects, quercetagetin 3,7-dimethyl ether **[6–11]**, and quercetagetin 3,6,7-trimethyl ether **[6–12]**

rhinocerotis, a member of Gnaphalieae. This species exhibits rapid growth producing nearly pure stands following fires. Allelopathy was suggested in view of the extraordinarily large amounts of phenolic material present in exudates: the lipophilic resin amounted to 20% of the plant's dry weight with 80% of that being contributed by phenolic compounds. Flavonoid components of the mixture (Fig. 6–1) were identified as quercetin, 6-methoxyapigenin (hispidulin) **[6–1]**, 6-methoxyluteolin (eupafolin) **[6–2]**, and 5,4'-dihydroxy-6,7-dimethoxyflavone (cirsimaritin) **[6–3]**. Bioassays of total exudate mixture showed high levels of inhibition of germination of radish and lettuce seed. The authors suggested that the phenolic components might also function to deter herbivores as well as helping to reduce water loss.

Pluchea lanceolata has been known for many years to exert allelopathic influence on other plants. A recent article by Inderjit and Dakshini (1994), which also reviewed their earlier work, described observations on the presence of flavonoids in the soil surrounding roots of this plant. The compounds in question, the isoflavone formononetin 7-*O*-glucoside (ononin) **[6–4]**, phloroglucinol **[6–5]**, a chlorogenic acid, e.g., **[6–6]**, and methylated coumarins, were effective in inhibiting germination of seeds of the test organism *Brassica juncea*. This species of *Pluchea* has become a serious weed in many parts of the world which, despite clearing of above ground parts, continues to exert its influence through its root exudates.

Several flavonoids have been implicated in plant responses to attack by micro-organisms (Fig. 6–2). Attention was focused on the capacity of plants to produce antimicrobial agents by the pioneering work of Cruickshank and Perrin (1960) who observed that isoflavone derivatives, exemplified by pisatin **[6–7]**, were produced by *Pisum sativum* in response to microbial attack. An excellent account of this process is given by Harborne (1988). Some flavonoid derivatives are strongly inhibitory toward nitrifying soil micro-organisms as in the case of medicarpin **[6–8]** and kievitone **[6–9]**, which are produced by several legume species, and are active against *Rhizobium* species (Pankhurst and Biggs, 1980). Other flavonoids exuded by roots play key roles in establishing relationships between roots and nitrogen fixing bacteria, as we shall see below.

A recent study by Kodama et al. (1992) described the isolation and identification of the flavanone sakuranetin **[6–10]** from UV-irradiated rice leaves. Sakuranetin was shown to be active (ED_{50}) at ca. 15 ppm against germination of spores of *Pyricularia oryzae*. It was also noted that sakuranetin concentration was higher in resistant cultivars of rice than in susceptible ones that had been infected by the fungus. Other species have similar responses to fungal infection. For example, Singh and Pandey (1993) found that several flavonoids from *Echinops echinatus* (Asteraceae) were active against spores of *Alternaria tenuissima*. Echinacin, which is the 6''-*p*-coumaroyl derivative of apigenin 7-*O*-glucoside, was the most active of the compounds tested, with apigenin being somewhat less effective. A further example comes from the work of Martin et al. (1994) who studied the isoflavonoids induced in the milkvetch (*Astragalus cicer*) following infection by the pathogenic fungus *Bipolaris zeicola* (syn. *Helminthosporium carbonum*). Considerable variation was found among 41 populations of this vetch with regard to the amount of flavonoid formed, but there appeared to be no geographical pattern.

By no means are all chemical interactions between plants and micro-organisms involving flavonoids negative ones. One of the most interesting and important discoveries in recent years has been the finding that flavonoids exuded by roots of a variety of legume species are intimately involved in helping to establish the association between roots and nitrogen fixing bacteria in the soil. Several flavonoid types have been identified as messengers: chalcones, flavanones, flavones, flavonols, and isoflavones. Representative of these studies is a series of papers from the laboratory of D. A. Phillips of the University of California at Davis on interactions between alfalfa (*Medicago sativa*) root flavonoids and *Rhizobium meliloti* (Hartwig et al., 1989; Dakora et al., 1993; Maxwell et al., 1989). In a concise review of the subject, Phillips and Tsai (1992) pointed out that flavonoids serve in several ways in the establishment of the symbiotic relationship between plant and micro-organism. The first function involves induction of the *nod* genes of the micro-organism. Flavonoids also act to establish a chemotactic field that directs rhizobia toward the root. Additional information has accumulated to suggest that flavonoids have beneficial effects on the growth of soil micro-organisms as well.

Flavonoids are also known to be involved in interactions between plants and a variety of insect pests (Fig. 6–3). For example, *Parthenium argentatum* (Asteraceae) was shown to produce two flavonols, quercetin and quercetagetin 3,7-dimethylether **[6–11]**, that were equally toxic to larvae of *Heliothis zea* and *Spodoptera exigua* (Isman and Rodgriquez, 1983). However, another flavonol derivative from the same plant, 5,4'-dihydroxy-3,6,7-trimethoxyflavone **[6–12]**, was stimulatory toward both larvae. Another example involves kaempferol-7,4'-dimethyl ether (ermanin), which acts as a major insect feeding deterrent in the passion flower, *Passiflora foetida* (Passifloraceae), against *Dione juno* larvae. Larvae of this species consume large quantities of other *Passiflora* species but not *P. foetida* (Echeverri et al., 1991).

Infestation of cassava, *Manihot esculenta*, with mealy-bugs (*Phenacoccus manihoti*) results in changes in the levels of quercetin 3-*O*-rutinoside and kaempferol 3-*O*-glucoside, but not in the concentration of cyanide (actually, cyanogenic glycoside content), in the plant. Flavonoids and a cyanogenic glycoside were detected in the mealy-bug honeydew. Some metabolism by the mealy-bug was suggested by the presence in the honeydew of flavonoid aglycones (Calatayud et al., 1994). At the very least, these organisms possess a glycohydrolase active on the flavonoid glycoside(s). It is well known that release of cyanide from cyanogenic glycosides requires the action of a glycohydrolase. Insect feeding deterrent properties of *Tithonia diversifolia* (Asteraceae) was credited to 6-methoxyapigenin (hispidulin) **[6–1]** and to tagitinins A and C (Dutta et al., 1993). These workers observed a dose-dependent effect when the compounds were evaluated against caterpillars of *Diacrisia obliqua, Phissama transiens, Trabula vishnu,* and grubs of a species of *Epilachna*. Other members of Asteraceae that appear to utilize flavonoids as defense against insect pests include *Helianthus annuus* (Gershenzon et al., 1985; Rieseberg et al.,

1987), and *Artemisia frigida* (Liu and Mabry, 1981a). In both of these cases, *O*-methylated flavonoids located in trichomes or other glandular structures were implicated.

As we have just seen, insect pests can elicit a variety of chemical responses from plants. Interactions between plants and insects, however, can have a very positive impact on the plants as well. Of critical importance in the reproductive cycle of many flowering plants is the attraction of a suitable pollination vector. The disposition of flavonoids in the floral tissues of these plants establishes a pattern recognizable by the pollinating insect. These patterns are visible to the insect because of the flavonoids' capacity to absorb ultraviolet radiation and thus appear dark against a lighter background (often provided by carotenoids). By drawing the insect into the floral tube, where it receives its reward of nectar, deposition or exchange of pollen occurs. In members of Asteraceae the pattern often takes the form of a bulls-eye with pigment deposited in the proximal parts of ray florets. In other species, one finds flavonoids laid down in lines such that the insect is guided into the floral tube. A widespread phenomenon is the change of coloration during floral ontogenesis (Weiss, 1995). In the pre-change situation, the color of the flower, or some specific part thereof, sends a signal to the pollinator that a reward is available, whereas the color after the change does not. Although several different pigment systems are known to be involved, Weiss (1995) reported that the most common one (observed in 68 families) involves anthocyanins. Perhaps the most sophisticated use of flavonoids in attracting insects may be seen in the case of some orchids, which, along with morphological specializations, develop a floral pigment pattern that mimics the coloration of female bees. It thus appears that flavonoids can be very sexy compounds, indeed.

It is of interest to look into the phenomenon of ultraviolet reflectance and absorption by floral parts in a little greater detail. In their description of reflectance patterns in Asteraceae, King and Krantz (1975) listed major contributors to this field starting with Knuth (1891), whose paper on photography of flower color may have been the earliest such report. The adaptive and evolutionary significance of ultraviolet spectral characteristics of plant parts has been discussed from several perspectives (Lutz, 1924; Ornduff and Mosquin, 1970; Kevan, 1972; Scogin et al., 1977). We mentioned above the bulls-eye pattern of floral pigmentation. Studies of the distribution of floral pigments and the existence of patterns have shown a close relationship between pigment-bearing cells and color pattern (Horovitz and Cohen, 1972; Thompson et al., 1972; Scogin, 1976, 1983; Scogin and Zakar, 1976; Scogin et al., 1977; Rieseberg and Schilling, 1985). In a closer inspection of ultraviolet absorption in *Balsamorhiza careyana*, *Lasthenia californica* (cited as *L. chrysostoma*), and *Rudbeckia hirta*, Brehm and Krell (1975) discovered that flavonoids responsible for reflectance patterns in these species were localized in tips of specialized epidermal papillae with resultant unique optical properties. In addition to the species studied by Brehm and Krell (1975), other asteraceous taxa have been examined in some detail: *Bidens* (Scogin, 1976, 1983; Scogin and Zakar, 1976), *Coreopsis* (Scogin, 1976, 1983; Scogin et al., 1977), and *Viguiera* (Rieseberg and Schilling, 1985). In one of his studies, Scogin (1978) examined numerous species from the two tribes Heliantheae and Helenieae.

Plant families other than Asteraceae, of course, have been studied. Kevan's (1972) survey of Arctic plants included a wide variety of families. Specific examples from other families include representatives of Onagraceae (Dement and Raven, 1974), Brassicaceae (Horovitz and Cohen, 1972), Rosaceae (Harborne and Nash, 1983), Scrophulariaceae (Bloom and Vickery, 1973), and Fabaceae (Harborne and Broadley, 1983). A recent paper by Saito and Harborne (1992) discusses correlations between anthocyanin type, pollinator, and flower color in members of Lamiaceae. In all cases, a close relationship between insect and the structure and color of a given flower exists.

The wavelengths of light involved in the above activities are not harmful, whereas other wavelengths of light that fall regularly on plants can be. Ultraviolet-B, in the 280–320 nm range, presents a serious hazard to biological processes. Although wavelengths of light below 295 nm are effectively absorbed by atmospheric ozone (some of which apparently still exists), there is sufficient energy in the remaining transmitted wavelengths to do serious harm to nucleic acids and proteins. It did not go un-noticed for long, in the early days of flavonoid research, that they could serve effectively as filters at these shorter wavelengths, and that one of their major functions might be to protect sensitive processes within the plant cell. The shorter wavelength absorption band of flavonoids, ranging broadly from about 250 to 300 nm, equips them well for such a function. Experimental work with *Oenothera* (Robberecht and Caldwell, 1983) and *Arabidopsis* (Li et al., 1993; Lois, 1994; Lois and Buchanan, 1994), for example, has clearly established that flavonoids do provide a significant measure of protection against shorter wavelength UV radiation. A useful overview of UV-acclimation in higher plants was published by Caldwell et al. (1983).

An example of the level of complexity that can exist in a natural system, and that involves flavonoids, comes from the work of Feeny et al. (1988) on oviposition stimulants for the black swallowtail butterfly (*Papilio polyxenes*). An extract of carrot (*Daucus carota*) leaves that stimulated egg laying behavior in female black swallowtails was shown to contain two major compounds, a chlorogenic acid **[6–6]** and luteolin 7-*O*-(6"-*O*-malonyl)-ß-glucoside. These compounds were inactive when tested alone but in combination exhibited about 70% of the activity of the parent extract. Loss of the malonic acid moiety from the flavone glucoside resulted in loss of biological activity.

The third level at which flavonoids can function in plants involves activities at the molecular level, that is to say, flavonoids interacting with other molecules within the plant, rather than with light or with some other organism. Early work in this area involving flavonoid interactions with indoleacetic acid (IAA) and IAA oxidases was reviewed by McClure (1975). Briefly, it was shown that kaempferol 3-*O*-glucoside and quercetin 3-*O*-glucoside and their *p*-coumaroyl derivatives were involved. The kaempferol derivatives served as cofactors while the quercetin derivatives were inhibitory (Furuya et al., 1962). Further studies involving a variety of flavonoid classes showed that the presence of a flavonoid with a 4'-hydroxyl group invariably led to a stimulation of the enzyme activity whereas 3',4'-dihydroxylation had the reverse effect (Stenlid and Saddik, 1962, 1963). These workers showed, in the same report, that the presence of a 7-hydroxyl group improved the cofactor function of the flavonoids.

In order to exert its effects in the plant, IAA must be transported from the site of its synthesis, apical cells, to target tissues, in a process called polar auxin transport (PAT). Several synthetic compounds, polar auxin transport inhibitors (PATIS), have been found to inhibit PAT, but binding of these compounds does not compete directly with auxin binding. Rather, it appears to involve a separate membrane component, termed the naphthylphthalamic acid (NPA) receptor. Jacobs and Rubery (1988) and Rubery and Jacobs (1988), using a variety of plant tissues, were the first to demonstrate that apigenin and quercetin can effectively bind to the NPA binding sites and significantly inhibit auxin transport. Kinetic studies showed that 50 mM quercetin completely inhibits NPA binding in zucchini hypocotyls. Galangin (3,5,7-trihydroxyflavone) and naringenin were only about one-tenth as effective while caffeic acid had no activity at all. These workers suggested that flavonoids may serve in this regulatory capacity in Nature.

Recent studies using lupines have shown that 5,7,4'-trihydroxy-3'-methoxy-6-*C*-prenylisoflavone (lupisoflavone) **[6–13]**, is active in inducing interconversions of cell wall isoperoxidases. Other tested isoflavones had no activity (Ros Barcelo and Muñoz, 1989; Ferrer et al., 1992).

In an important series of papers, which could well serve as a model for this kind of work, L. P. Taylor and her associates at Washington State University have identified certain flavonoids as key components in early stages of plant reproductive biology. In the lead paper, Taylor and Jorgensen (1992) demonstrated that transgenic petunias with suppressed chalcone synthase gene (CHS) activity produced abnormal anthers that were devoid of flavonoids. This observation is similar to an earlier report (Coe et al., 1981)

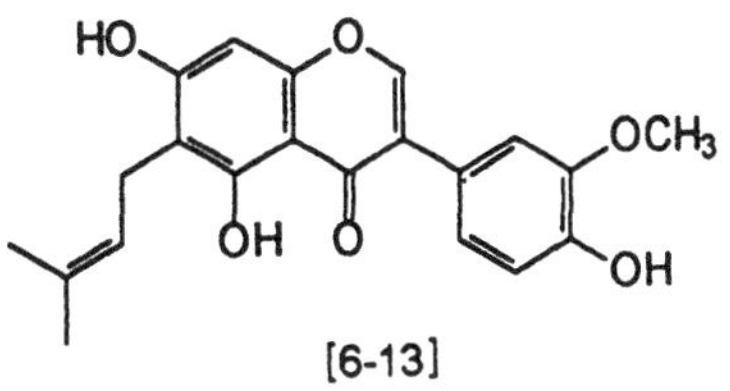

[6-13]

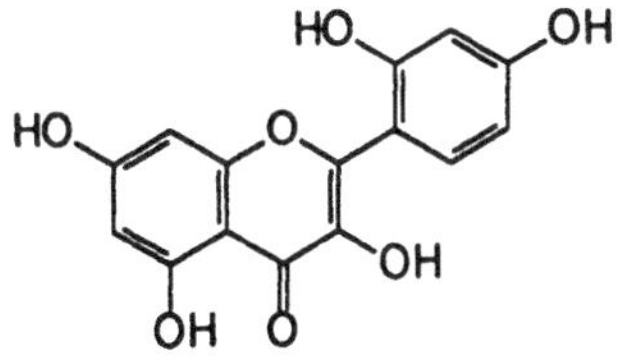

[6-14]

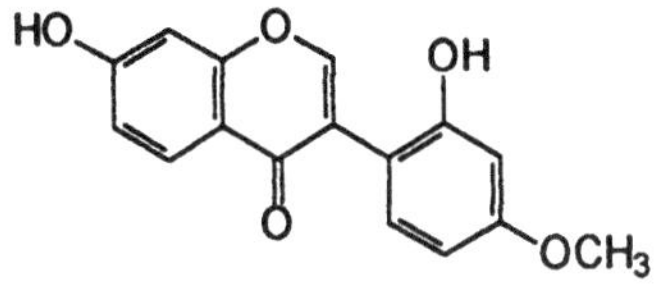

[6-15]

Fig. 6–3. Flavonoids showing a diverse array of biological activities: lupisoflavone **[6–13]**, active in inducing interconversions of cell wall isoperoxidases; morin **[6–14]**, active in growth of pollen tubes; and 7,2'-dihydroxy-4'-methoxyisoflavone **[6–15]**, active in inducing haustorial development

that pollination with CHS-deficient maize pollen was unsuccessful. Further experiments identified kaempferol as a constituent of wild-type stigmas necessary for the germination of the pollen grains and for successful pollen tube growth (Mo et al., 1992). Addition of micromolar amounts of kaempferol to the germination medium or to the surface of the stigma restored full functionality to the system. Experiments described in that report established that only flavonols, as opposed to flavones, dihydroflavonols, and flavanones, could restore biological activity. Kaempferol, 3,5,7-trihydroxyflavone (galangin), and quercetin 3'-methyl ether (isorhamnetin) were active at a concentration of 1 mM. Quercetin and 3,5,7,2',4'-pentahydroxyflavone (morin) **[6–14]** were active at a concentration of 10 mM. Other tested flavonols, including myricetin and 5-deoxyquercetin (fisetin), were less active still. In similar studies of germination and pollen tube growth in tobacco, Ylstra et al. (1992) also found that flavonols were active participants with kaempferol, quercetin, and myricetin all showing significant biological activity.

In an article entitled "Phenolic signals in cohabitation: implications for plant development," Lynn and Chang (1990) summarized information on the (hemi)parasitic genus *Striga* (Scrophulariaceae) involving phenolic compounds, flavonoids included, that exert influence on such processes as germination and haustorial induction. For specific information, the reader should consult the original paper by Lynn and coworkers (1981) in which the identity of one of the first haustorium inducers was established as 7,2'-dihydroxy-4'-methoxyisoflavone **[6–15]**. The physiology and biochemistry of parasitic angiosperms have been reviewed by Stewart and Press (1990).

An interesting function of flavonoids was discussed by Valant-Vetschera and Wollenweber (1994) in their comparative study of exudate flavonoids of *Achillea* sections *Achillea* and *Babounya*. They pointed out that flavonoids in leaf exudates are usually accompanied by terpenes and that the terpenes themselves may play significant roles in plant defense. Flavonoids, known to be excellent antioxidants, may prevent oxidation of the terpene derivatives. Spring (1991) discussed the possibility that the low turnover of sesquiterpene lactones in certain plant specimens may be due to the presence of flavonoids. Antioxidant properties of flavonoids have been discussed by others (Larson, 1988; Puppo, 1992).

Section III

Flavonoid Data

Section Three consists of 11 chapters in which flavonoid data for the family are presented. The information is arranged alphabetically by tribe, and a single tribe constitutes a chapter in most instances. For considerations of space, in two cases chapters contain information on two tribes: Chapter 9 includes information for both Calenduleae and Cardueae; and Chapter 17 includes Vernonieae and Liabeae. The former two tribes clearly are not close relatives; Calenduleae may be allied most closely to Senecioneae (Cronquist, 1955; Jansen et al., 1991a) or somewhere near Gnaphalieae (Inuleae s.l.), Anthemideae, or Asteraceae (Kim and Jansen, 1995b; Jansen and Kim, 1996), but certainly not near Cardueae. Liabeae, however, are closely related to Vernonieae as has been known for some time (Robinson and Brettell, 1973a; Nordenstam, 1977; Turner and Powell, 1977; Jansen and Stuessy, 1980; Bremer, 1996; Jansen and Kim, 1996).

Within each tribal chapter the reader will find an alphabetical presentation of genera. This is done deliberately to facilitate locating data for each genus. This is also done because generic relationships within some tribes are in considerable flux (e.g., Heliantheae) owing to recent, especially cladistic and macromolecular, investigations. In many cases, generic circumscriptions and attendant nomenclature are presented as given by the respective authors. Noteworthy exceptions are Anthemideae and Eupatorieae for which recent comprehensive taxonomic and nomenclatural treatments are available down to the specific level (King and Robinson, 1987; Bremer and Humphries, 1993, respectively); these references have greatly facilitated our placement of these taxa and their flavonoid constituents into a modern context. For all other reports, however, we have checked for taxonomic and nomenclatural changes in revisionary and floristic literature available to us, so that flavonoid data can be presented, and assessed, in the most useful context of modern classifications. This has been quite thorough, but of course, not completely exhaustive. These references are cited where appropriate throughout each chapter. No attempt has been made to confirm original identification of material by examination of voucher specimens.

CHAPTER 7

Flavonoids of Anthemideae

According to Heywood and Humphries (1977), Anthemideae consist of 101 genera with about 1400 species. Bremer and Humphries (1993) see the tribe as comprising 109 genera with 1741 species. Bremer (1994; followed by Arriagada and Miller, 1997) lists 109 genera and 1740 species. We have found flavonoid information for members of 44 genera. By far the largest amount of information available involves *Artemisia*, which is not surprising since it is the largest genus in the tribe with 388 species (Bremer and Humphries, 1993). Species numbers and areas of distribution cited below are those of Bremer and Humphries (1993) unless otherwise indicated.

Achillea: Heywood and Humphries (1977) earlier listed *Achillea* as comprising 85 species, whereas more recently Bremer and Humphries (1993) recognize 115. *Achillea* is a Eurasian genus with some species in North Africa. The widespread *A. millefolium* is a naturalized weed in North America. A high degree of taxonomic and flavonoid interest in the group is revealed by finding flavonoid chemical data for at least 95 named species in the literature (plus many infraspecific taxa). This represents the best sampling of any genus this size in the family. Much credit for this effort goes to K. Valant-Vetschera (née Valant) for her comprehensive studies of the genus (Valant, 1978; Valant et al., 1978, 1980; Valant-Vetschera 1981, 1984, 1985a,b,c, 1987; Valant-Vetschera and Wollenweber, 1986, 1988a,b, 1989, 1994, 1996a,b, 1999).

Members of *Achillea* accumulate a wide variety of flavonoid derivatives including common flavone (apigenin and luteolin) and flavonol (kaempferol, quercetin, and isorhamnetin) glycosides, an array of *C*-glycosylflavones, which will be discussed in more detail below, and a large collection of *O*-methylated flavones, many of which exhibit 6-oxygenation. Many of the *O*-methylated flavonoids reported from the genus were obtained as aglycones from leaf exudates.

Azaleatin (quercetin 5-methyl ether) 3-*O*-galactoside, along with several quercetin 3-O-glycosides, has been reported from *A. urumoffii* by Ivancheva et al. (1987). Several of the more common flavonols have been reported from members of the genus as well; examples include: kaempferol 3-*O*-rhamnoside from *A. micrantha* (Adekenov et al., 1987b), kaempferol and quercetin 3-*O*-glycosides and rutin from *A. cartilaginea* (Neshta and Kaloshina, 1971; Neshta et al., 1973), and rutin from *A. millefolium* (Neshta et al., 1972). Additional examples can be found in Table 7–1.

Table 7–1. Occurrence of simple flavones[1] and flavonols[2] in *Achillea*

Taxon	Api	Lut	Kae	Que	Reference
A. asplenifolia	+	+			Oswiecimska and Miedzobrodzka (1966)
A. biebersteinii		+		+	Shmatova and Komissarenko (1987)
A. cartilaginea	+	+	+	+	Neshta and Kaloshina (1971), Neshta et al. (1973),
	+				Zapesochnaya and Ban'kovskii (1976)
A. chrysocoma		+			Ivancheva and Nedelcheva (1995)
A. clusiana		+			Ivancheva and Stancheva (1996)
A. coarctata				+	Stosic et al. (1988)
A. collina	+	+			Oswiecimska and Miedzobrodzka (1966)
A. fragrantissima	+	+			Seida et al. (1990)
A. holosericea		+		+	Tzakou et al. (1995)
A. kotschyi				+	Ivancheva et al. (1987),
		+			Wollenweber et al. (1987b)
A. krasheninnikovii		+		+	Shmatova and Komissarenko (1987)
A. ligustica	+	+			Tzakou et al. (1995)
A. lingulata		+			Ivancheva and Stancheva (1996)
A. micrantha			+		Adekenov et al. (1987b)
		+			Hatam and Seifert (1994)
A. millefolium		+			Hörhammer et al. (1964)
	+	+			Michaluk (1962)
	+	+			Oswiecimska and Miedzobrodzka (1966)
				+	Neshta et al. (1972)
A. neilreichii				+	Zurabishvili (1974)
A. nobilis	+	+			Solomko et al. (1978)
				+	Ivancheva and Kuzmanov (1986)
A. pannonica	+	+			Oswiecimska and Miedzobrodzka (1966)
A. setacea	+	+			Oswiecimska and Miedzobrodzka (1966)
A. umbellata	+	+			Wollenweber et al. (1987b)
A. urumoffii	+	+			Ivancheva and Kuzmanov (1986)
					Ivancheva et al. (1987)

[1] Api = apigenin; Lut = luteolin. Reported as aglycones and/or 7-*O*-glycosides.

[2] Kae = kaempferol; Que = quercetin. Reported as aglycones and/or 3-*O*- and/or 7-*O*-glycosides.

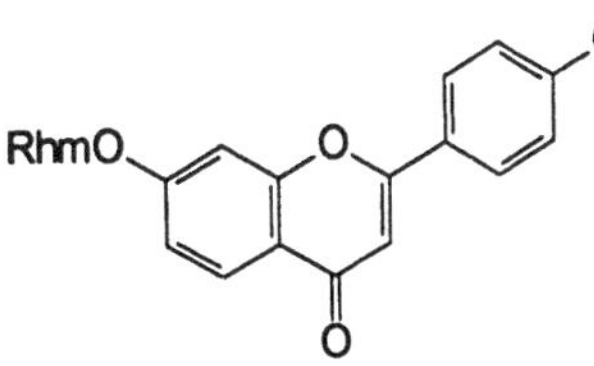

[7-1]

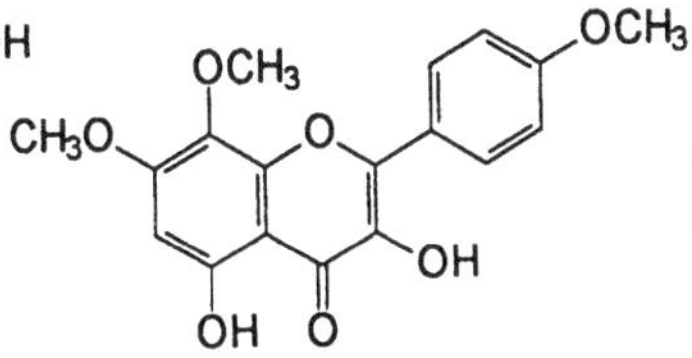

[7-2]

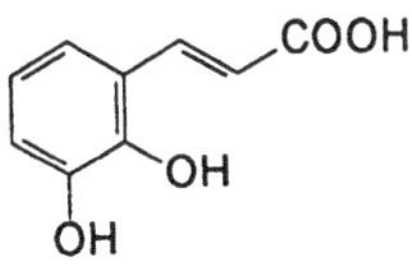

[7-3]

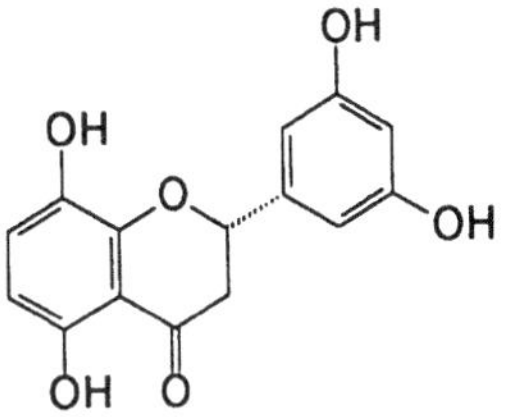

[7-4]

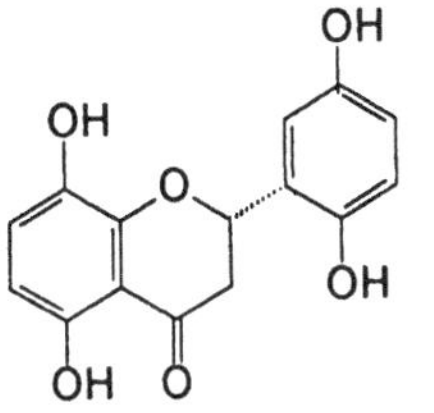

[7-5]

Fig. 7–1. The 5-deoxyflavone **[7–1]** and the 8-substituted flavonol **[7–2]** represent rare reports of these flavonoid types in Anthemideae. 2,3-Dihydroxycinnamic acid **[7–3]** is an unusual acylating acid identified from *Anthemis*. The 3',5'-dioxygenation patterns seen in compounds **[7–4]** and **[7–5]**, respectively, are only rarely found in Asteraceae

5-Deoxyflavonoids are comparatively rare in the family, but two have been identified (Fig. 7–1). 7,4'-Dihydroxyflavone 7-*O*-rhamnoside **[7–1]** was identified as a component of *A. fragrantissima* by Seida et al. (1990) along with 7-*O*-glycosides of apigenin, luteolin, and chrysoeriol. More recently, Ahmad et al. (1995) reported 7-hydroxy-3,6,3',4'-tetramethoxyflavone along with 5-hydroxy-3,6,7,3',4'-pentamethoxyflavone (artemetin) from *A. santolina*.

Glycosylated flavones exhibit sugars at the 7-position, while flavonols have been identified that are glycosylated at either position-3 or -7. The distribution of simple flavone and flavonol glycosides, along with *C*-glycosylflavones, in leaves and flowers of three subspecies of *A. millefolium* was reported by Guédon et al. (1993). The finding of luteolin 7-*O*-glucoside in this species corroborates its earlier report by Hörhammer et al. (1964). Further examples can be found in Table 7–1.

C-Glycosylflavones were reported from 45 species (of 75 examined) (Valant, 1978; Valant et al., 1978). The *C*-glycosylflavone profiles of some species are quite complex as seen, for example, in *A. kotschyi*: vitexin, isovitexin, swertisin, vicenin, orientin, isoörientin, and swertiajaponin (Valant, 1978). *Achillea fragrantissima* yielded vicenin-3, swertisin, swertijaponin, isoörientin, isovitexin 7-*O*-glucoside (saponarin), schaftoside, isoschaftoside, and isovitexin-7,4'-dimethyl ether (embigenin) (Ahmed et al., 1989b).

Table 7–2. *C*-Glycosylflavones from species of *Achillea* published since Valant (1978)

Species	Compound[1]
A. aleppica ssp. *zederbaureri*[2]	ISO, LUC, ORI, SWJ, VIC
A. cretica[3]	I73
A. fragrantissima[2,4]	EMB, ISC, ISO, SAP, SCH, SWE, SWJ, VIC
A. lycaonica[2]	ISO, LUC, ORI, VIC, VIT
A. millefolium[5]	ISC, SCH
A. santolina[4]	ISO, IVX, SCH
A. spinulifolia[2]	ISC, ISO, ORI, SWJ, VIC

1 EMB = embigenin; ISC = isoschaftoside; ISO = isoörientin; IVX = isovitexin; LUC = lucenins; ORI = orientin; SAP = saponarin; SCH = schaftoside; SWA = swertisin 2"-arabinoside; SWE = swertisin; SWJ = swertiajaponin; VIC = vicinens; VIT = vitexin; I73 = isoörientin-7,3'-dimethyl ether.

2 Valant-Vetschera (1985b,c).

3 Valant et al. (1980).

4 Ahmed et al. (1988d).

5 Guédon et al. (1993) (flower heads of ssp. *alpestris*, *ceretanum* & *millefolium*).

Isoörientin-7,3'-dimethyl ether was described from *A. cretica* by Valant et al. (1980). *C*-Glycosylflavone occurrences reported in the genus since the publication of the Valant (1978) survey are summarized in Table 7–2.

O-Methylated flavonoids are also widely distributed in the genus. Referring to the Valant (1978) survey, we see that *O*-methylated compounds were reported from 12 species. These compounds were identified as quercetagetin 6-methyl ether (patuletin) and 6,3'-dimethyl ether (spinacetin). Quercetagetin itself was reported from an additional dozen or so species. Many reports have appeared in the literature since that survey in which these and other *O*-methylated flavonoids are described from the genus. Extra oxygenation at C-6 is common in the genus, but less so at C-8. There appear to be only two reports of the latter type, 3,5-dihydroxy-7,8,4'-trimethoxyflavone (herbacetin-7,8,4'-trimethyl ether) **[7–2]** from *A. depressa* (Tsankova and Ognyanov, 1985) and 5,7-dihydroxy-6,8,4'-trimethoxyflavone from *A. ligustica* (Bruno and Herz, 1988a). Table 7–3 lists other reports of *O*-methylated flavonoids (except *C*-glycosylflavones) identified from members of the genus.

Ajania: *Ajania* is a genus of 34 species that occurs in Japan, China, and south and central Asia. *Ajania fastigiata* has been shown to contain 5,4'-dihydroxy-7,3'-dimethoxyflavone (velutin) (Chumbalov et al., 1972b), 5,4'-dihydroxy-6,7,3'-trimethoxyflavone (fastigenin) (Chumbalov et al., 1974b), and quer-

Table 7–3. *O*-Methylated flavones and flavonols from species of *Achillea*

Species	Compounds[1]	References
A. abrotanoides	573'/364'	Stefanovic et al. (1989), Valant-Vetschera and Wollenweber (1986)
	57/364', 5/3674', 53'/3674', 5/3673'4'	Valant-Vetschera and Wollenweber (1986)
A. ageratifolia s.l.	574'/6, 573'4'/6, 574'/36, 54'/367, 53'4'/367	Valant-Vetschera and Wollenweber (1986), Ivancheva and Stancheva (1996)
	574'/6, 54'/67, 57/64', 5/674', 573'4'/6, 53'4'/67, 53'/674', 574'/36, 54'/367, 57/364', 5/3674', 53'4'/367, 573'/364', 53'/3674', 5/3673'4'	Franzén (1988)
A. ageratum	53'4'/367; 54'/367	Vieira et al. (1997)
A. ambrosiaca	57/64', 573'4'/6, 57/364', 54'/367, 53'4'/367, 573'/364'	Franzén (1988)
A. asiatica	573'/364'	Valant-Vetschera and Wollenweber (1988b)
A. aspleniifolia	573'/364', 53'/3674', 5/3673'4'	Valant-Vetschera and Wollenweber (1988b)
A. atrata ssp. *multifida*	57/364', 573'/364'	Aljancic et al. (1999)
A. barbeyana	573'4'/6, 53'4'/67, 573'/364', 5/3673'4'	Wollenweber cited by Franzén (1988)
A. biebersteinii	3573'4'/6, 574'/363'	Oskay and Yesilada (1984)
A. borealis/lanulosa	57/64', 574'/36, 54'/367, 57/364', 573'4'/36, 53'4'/367, 573'/364', 53'/3674', 5/3673'4'	Valant-Vetschera and Wollenweber (1988b)
A. chrysocoma	573'/4', 574'/3, 54'/37, 573'/364'	Ivancheva and Nedelcheva (1995)

Table 7–3. (Continued)

Species	Compounds[1]	References
A. clusiana	53'4'/367, 574'/363', 54'/3673'	Valant-Vetschera and Wollenweber (1996b)
	574'/6, 57/64', 574'/36, 5/3673'4'	Ivancheva and Stancheva (1996)
A. collina	574'/6, 57/64', 57/34', 54'/367, 57/364', 5/3674', 573'/364', 53'/3674', 5/3673'4'	Valant-Vetschera and Wollenweber (1988b)
A. clavennae	57/64', 573'4'/6, 574'/36, 54'/367, 57/364', 5/3674', 53'4'/367, 573'/364'	Franzén (1988)
A. clypeolata	573'/364', 5/3674', 5/3673'4', 57/364' ?	Ivancheva et al. (1990)
A. coarctica	573'4'/6, 57/34', 54'/37, 5/364', 573'/364'	Ivancheva et al. (1990)
A. cretica	54'/67, 57/64', 5/673'4', 5/3673'4'	Valant-Vetschera and Wollenweber (1986)
A. crithmifolia	574'/363'	Milosavljevic et al. (1991, 1994)
A. cucullata	5/674', 5/673'4'	Öksüz et al. (1991)
A. depressa	5/674', 54'/67, 35/784'	Tsankova and Ognyanov (1985)
A. distans	574'/6, 57/64', 573'4'/6, 57/34', 54'/37, 574'/36, 57/364', 5/3674', 573'/364', 53'/3674'	Valant-Vetschera and Wollenweber (1988b)
A. eriophora	57/4', 5/74', 5/674', 53'/674', 5/673'4', 35/674', 5/3674', 573'4'/36, 53'4'/367	Wollenweber et al. (1997)
A. falcata	54'/67, 5/674', 53'/674', 5/673'4', 53'/3674', 5/3673'4'	Valant-Vetschera and Wollenweber (1986)

Table 7–3. (Continued)

Species	Compounds[1]	References
A. filipendulina	573'/364'	Valant-Vetschera and Wollenweber (1986)
A. fraasii	57/64', 57/364', 53'4'/367, 573'/364'	Franzén (1988)
A. fragrantissima	573'4'/6, 573'4'/3, 573'4'/36	Ahmed et al. (1989b)
	574'/3', 74'/-	Seida et al. (1990)
A. goniocephala	5/674'	Öksüz et al. (1991)
A. grandifolia	573'4'/6	Wollenweber et al. (1987b)
	574'/36, 3573'4'/6, 573'4'/36	Valant-Vetschera and Wollenweber (1988a)
A. holosericea	574'/3', 57/64', 573'4'/3, 574'/33'	Tzakou et al. (1995)
A. kotschyi s.l.	574'/6, 573'4'/6, 574'/3, 53'4'/67, 573'4'/3, 574'/36, 54'/367, 57/364', 5/3674', 573'4'/36, 53'4'/367, 573'/364', 53'/3674', 5/3673'4', 574'/363', 54'/37	Wollenweber et al. (1987b), Valant-Vetschera and Wollenweber (1988a)
A. latiloba	57/34', 574'/33', 57/364', 573'/364'	Valant-Vetschera and Wollenweber (1999)
A. ligustica	574'/36, 57/364'	Valant-Vetschera and Wollenweber (1988a), Tzakou et al. (1995)
	573'4'/36, 573'/364'	Valant-Vetschera and Wollenweber (1988a)
	57/684', 36/574', 53'4'/367	Bruno and Herz (1988a)
A. lingulata	573'4'/6, 57/34'	Ivancheva and Stancheva (1996)
A. lycaonica	54'/67, 57/64', 5/674', 5/673'4', 5/3674', 5/3673'4'	Valant-Vetschera and Wollenweber (1986)
A. micrantha	574'/36, 574'/363'	Hatam and Seifert (1994)
A. millefolium	5/3674', 53'/3674', 5/3673'4'	Falk et al. (1975)

Table 7–3. (Continued)

Species	Compounds[1]	References
A. m. ssp. *lanulosa*	573'4', 574'/36, 57/364', 5/3674', 573'4'/36, 53'4'/367	Wollenweber et al. (1997)
A. m. ssp. *millefolium*	574'/6, 57/64', 574'/36, 57/364', 573'/364'	Valant-Vetschera and Wollenweber (1988b)
A. m. ssp. *sudetica*	574'/6, 57/64', 57/34', 57/364', 573'/364'	Valant-Vetschera and Wollenweber (1988b)
A. moschata	574'/36, 57/364', 573'/364'	Valant-Vetschera and Wollenweber (1986)
A. multifida	57/364', 573'/364'	Valant-Vetschera and Wollenweber (1996a)
A. nana	54'/67, 53'4'/367, 54'/3673', 5/3673'4'	Valant-Vetschera and Wollenweber (1996a)
A. nobilis s.l.	574'/36, 57/364', 5/3674', 53'4'/367, 573'/364', 53'/3674', 5/3673'4'	Valant-Vetschera and Wollenweber (1986), Valant-Vetschera and Wollenweber 1988a)
	35/678	Adekenov et al. (1984)
A. odorata	574'/6, 54'/67, 573'4'/6, 53'/674', 5/3674', 573'4'/36, 53'4'/367, 53'/3674', 5/3674' 5/3673'4'	Valant-Vetschera and Wollenweber (1988a), Barrero et al. (1990a,b)
A. pannonica	574'/6, 57/64', 57/34', 57/364', 573'/364'	Valant-Vetschera and Wollenweber (1988b)
	3574'/3'	Kaloshina and Mazulin (1983b)
A. pindicola	57/64', 573'4'/6, 574'/36, 57/364', 573'/364'	Franzén (1988)
A. roseo-alba	57/64', 5/3674', 573'/364', 53'/3674', 5/3673'4'	Valant-Vetschera and Wollenweber (1988b)
A. santolina	57/64'	Ahmed et al. (1989b)

Table 7–3. (Continued)

Species	Compounds[1]	References
	5/674', 53'/674'	Elgamel et al. (1991a), Balboul et al. (1997)
	57/33'4'	Khafagy et al. (1976)
	5/673'4'	Khafagy et al. (1976), Elgamel et al. (1991a), Balboul et al. (1997)
	5/3673'4'	Ahmed et al. (1989b), Elgamel et al. (1991a), Ahmad et al. (1995)
	7/363'4'	Ahmad et al. (1995)
A. schischkinii	5/674', 57/63'4', 5/673'4', 5/3673'4'	Ulubelen et al. (1987b)
A. serbica	574'/6, 57/64', 574'/36, 5/3673'4'	Ivancheva and Stancheva (1996)
A. setacea	574'/6, 57/64', 54'/67, 5/674', 573'4'/6, 53'4'/67, 57/63'4', 53'/674', 5/673'4', 57/34', 57/364', 5/3674', 53'/3674', 5/3673'4'	Valant-Vetschera and Wollenweber (1988b)
A. sibirica	5/674', 5/3674', 53'/3674', 5,3673'4'	Valant-Vetschera and Wollenweber (1999)
A. sintenisii	5/674', 5/673'4'	Gören et al. (1988)
A. spinulifolia	54'/67, 5/674', 573'4'/6, 53'/674', 5/673'4', 5/3674', 53'4'/367, 5/3673'4'	Valant-Vetschera and Wollenweber (1986)
A. spirokensis	573'/4', 5/674', 5/3673'4'	Ulubelen et al. (1989)
A. umbellata	573'4'/6, 574'/36, 57/364', 573'/364'	Franzén (1988)
A. umbelliformis	57/63'4'	Cappelletti et al. (1986)
A. virescens	54'/67, 5/674', 574'/36, 54'/367, 5/3674', 573'4'/36, 53'4'/367, 53'/3674', 5/3673'4'	Valant-Vetschera and Wollenweber (1988a)

[1] Substitution: Hydroxyl positions/Methoxyl positions.

cetin, luteolin, apigenin, and an apigenin-6,8-di-*C*-glucoside (Chumbalov and Zhubaeva, 1976). Belenovskaya et al. (1977) identified quercetagetin 3,6-

dimethyl ether (axillarin), 3,6,7-trimethyl ether (chrysosplenol-D), and 3,6,7,3',4'-pentamethyl ether (artemetin) from *A. fruticulosa*. *Ajania gracilis* (reported as *Tanacetum gracile*; Shishkin and Bobrov, 1961) was also shown to accumulate artemetin (Shawl, 1993). Two xanthine oxidase inhibitory flavonoids, 5,7-dihydroxy-3,6,4'-trimethoxyflavone and 5,7,3'-trihydroxy-3,6,4'-trimethoxyflavone, have been identified recently from extracts of *A. fruticulosa* by Li et al. (1999).

Anacyclus: *Anacyclus* consists of 12 (Heywood and Humphries, 1977; Humphries, 1979; Bremer and Humphries, 1993) or 13 (Greger, 1978) species and ranges from the western Mediterranean region to southwestern Asia. Glennie and Harborne (1971) reported quercetin 5-*O*-glucoside from *A. pyrethrum*. An extensive study of 13 taxa by Greger (1978) showed the presence of 3-, 5- and, 7-*O*-glycosides of quercetin and isorhamnetin and 7-*O*-glycosides of luteolin and diosmetin although not all compounds were seen in all taxa. From the flowers of *A. radiatus* Greger (1978) also isolated quercetagetin and patuletin 7-*O*-glucosides, quercetagetin-3-methyl ether, and quercetagetin-6,3'-dimethyl ether (spinacetin).

Anthemis: *Anthemis* is a moderately large genus, 211 species according to Bremer and Humphries (1993) and Oberprieler (ca. 210; 1998). Representatives occur in Europe, in the Mediterranean region, and into southwestern Asia (Ghafoor and Al-Turki, 1997). Greger (1969) included a number of species of *Anthemis* in his first survey of the tribe; *A. biebersteinii*, *A. carpatica*, and *A. cotula* had quercetin; *A. cotula* also had kaempferol. (Note that the epithet *biebersteinii* does not occur in *Anthemis* but does occur in *Achillea*.) Studies of the flavonoid glycosides of several species show an array of compounds ranging from simple flavone glycosides to flavonols with extra hydroxylation and some *O*-methylation. Thus, luteolin and/or apigenin 7-*O*-glycosides have been reported (Wagner and Kirmayer, 1957; Chaumont, 1969; Abou-Zied and Rizk, 1973) from *A. nobilis* (= *Chamaemelum nobile*; Tutin et al., 1976; Bremer and Humphries, 1993); *A. nigrescens* (Harborne et al., 1970; probably *A. austriaca*), and *A. pseudocotula* (Saleh and Rizk, 1974). Hérisset et al. (1971) reported an unusual compound, which they called anthenobilic acid, from *A. nobilis* and which they identified as apigenin 7-*O*-glucoside acylated with 2,3-dihydroxycinnamic acid **[7–3]**. Several flavonol derivatives have been identified from members of the genus: quercetin 5-*O*-glucoside from *A. cotula* (Glennie and Harborne, 1971), patuletin 7-*O*-glucoside from *A. carpatica, A. nigrescens* and *A. tinctoria* (Harborne et al., 1970; Kaloshina et al., 1975a), quercetin, patuletin, quercetin 7-methyl ether (rhamnetin), and quercetagetin-3,6-dimethyl ether (axillarin) from *A. melampodina* (Zaghloul et al., 1989) and quercetin 4'-methyl ether (tamarixetin), quercetagetin-6-methyl (patuletin), and 6,3'-dimethyl ether (spinacetin) from *A. tinctoria* (Masterova et al., 1993). Wollenweber and Mayer (1991) reported 6-methoxyapigenin and 6-methoxyluteolin derivatives from the leaf exudate of *A. nobilis* along with a mixture of *O*-methylated derivatives of apigenin, 6-methoxyapigenin, kaempferol, quercetin, and quercetagetin.

Arctanthemum: *Arctanthemum* is a circumarctic genus that consists of four species. Harborne et al. (1970) reported apigenin *O*-glucuronide, an unidentified acyl derivative thereof, and quercetin and patuletin 7-*O*-glucosides from material they identified as *Dendranthema arcticum*. Lee (1967) reported acacetin 7-*O*-rutinoside from *A. sibiricum* (cited as *Chrysanthemum sibiricum*).

Argyranthemum: According to Greger (cited by Wilcox, 1984), *Argyranthemum* accumulates flavonol 5-*O*-glucosides. Bremer and Humphries (1993) list this as a genus of 24 species endemic to Macaronesia.

Artemisia: *Artemisia* is by far the largest genus in Anthemideae (ca. 400 species) and one of the largest in the family. It is a significant element in the floras of Europe, Asia, and North America, with a few species native to southern Africa. Three species are endemic to the Hawaiian Islands (Wagner et al., 1990). Some species of *Artemisia* enjoy a modest degree of fame owing to their use by, or effect on, man (Mabberley, 1987). The oil of *Artemisia absinthium* provided the punch for absinthe, while the flavor of vermouth comes from *A. pontica*. On a more sober note, *A. dracunculus* serves fine diners as the culinary herb tarragon. The various "wormwoods" are *Artemisia* species, the most famous one perhaps being *A. herba-alba* (currently seen as *Serphidium herba-album*; Bremer and Humphries, 1993), the wormwood of Biblical note. Zohary (1982) includes notes on this species in *Plants of the Bible*, and includes the following citation from Jeremiah 23:15: "Therefore thus says the Lord of hosts concerning the prophets: Behold, I will feed them with wormwood, and give them poisoned water to drink." *Artemisia annua* and *A. tilesii* have found application as medicines, while *A. vulgaris* (mugwort) was supposed to possess magical properties. "Sagebrush" is a very common shrub in arid parts of western North America and may be toxic to range animals (Hanks et al., 1973). These plants were traditionally considered to constitute *Artemisia tridentata*, but current taxonomic opinion places them in *Seriphidium* as *S. tridentatum* (Bremer and Humphries, 1993).

Owing to its size, conspicuousness, and the many uses to which members have been put, *Artemisia* has attracted a proportionately large amount

Table 7–4. Common flavanones from species of *Artemisia*

Species	Compounds[1]	References
A. argyi	574'/-	Tan and Jia (1992)
A. campestris ssp. *campestris*	57/4', 54'/73'	Cavaleiro (1986), Rauter et al. (1989)
A. campestris ssp. *glutinosa*	57/-, 5/7, 574'/-, 54'/7, 57/4', 53'4'/7, 574'/	Hurabielle et al. (1982), de Pascual Teresa et al. (1984, 1986), Menichini and Appendino (1987), Appendino et al. (1988), Gonzalez et al. (1983)
A. campestris ssp. *maritima*	54'/73'	Vasconcelos et al. (1998)
A. crithmifolia	574'/-, 57/4', 54'/7, 5/74'	Sanz et al. (1991)
A. dracunculus	574'/-, 53'/74'	Balza and Towers (1984), Balza et al. (1985)
A. halodendron	53'4'/7	Chemesova and Markova (1986)
A. xanthochloa	53'4'/7	Chemesova et al. (1984b, 1987)

[1] Substitution: Hydroxyl positions/Methoxyl positions.

of attention of people interested in its chemical constituents (Marco and Barbera, 1990). One result of these studies is an impressive flavonoid literature from which we have obtained information for at least 70 species. Twenty-eight of these, however, are currently considered to belong to other genera, principally *Seriphidium*, but *Sphaeromeria* and *Stilpnolepis* are also involved (Bremer and Humphries, 1993). Even with the reduction in number of *Artemisia* species for which we have flavonoid information, the overall generic profile is a very rich one. Simple flavones and flavonols, *C*-glycosylflavones, flavanones and dihydroflavonols, high levels of *O*-methylation, and a variety of unusual substitution patterns contribute to a rich array of compounds in the genus.

Flavanones identified from *Artemisia* species are, for the most part, based on pinocembrin, naringenin, or eriodictyol. Pinocembrin and naringenin are known as aglycones, eriodictyol itself has not been reported. Methyl ethers of all three are known from the genus. Occurrences of pinocembrin, naringenin, and eriodictyol derivatives in the genus can be found in Table 7–4. The entries for *A. campestris* ssp. *glutinosus* represent the summation of several reports. The significance of the variation in subspecies *glutinosa* compared to the apparent paucity of these compounds in subspecies *campestris* is unknown.

Table 7–5. *O*-Methylated dihydroflavonols from species of *Artemisia*

Species	Compound[1]	Reference
A. campestris ssp. *glutinosa*	354'/7, 353'4'/7, 354'/73'	Hurabielle et al. (1982), Gonzalez et al. (1983), De Pascual Teresa et al. (1984, 1986)
A. campestris ssp. *maritima*	354'/73'	Rauter et al. (1989)
A. crithmifolia	354'/7, 35/74'	Sanz et al. (1991)
A. dracunculus	354'/7, 354'/73'	Balza and Towers (1984)
A. scoparia	354'/7	Chandrasekharan et al. (1981)
A. xanthochroa	353'4'/7	Chemesova (1987)

[1] Substitution: Hydroxyl positions/Methoxyl positions.

Flavanones based on other substitution patterns have also been reported. Two exceptional compounds, which come from *A. campestris* subsp. *maritima*, lack substitution at C-7, which is an unusual structural feature in flavonoids in general. The compounds were identified as 5,8,4'-trihydroxyflavanone **[4–20]** and 5,6-dihydroxy-4'-methoxyflavanone **[4–21]** (Rauter et al., 1989). Three unusual flavanones were described from *A. subdigitata* (this epithet is not listed by Bremer and Humphries, 1993) by Shi et al. (1992): 5,7,4'-trihydroxy-3',5'-dimethoxyflavanone (the flavanone corresponding to tricin, which also occurs in this species), 5,8,3',5'-tetrahydroxyflavanone **[7–4]**, and 5,8,2'-trihydroxy-5'-methoxyflavanone **[7–5]**.

Dihydroflavonols based on both dihydrokaempferol and dihydroquercetin are known from several *Artemisia* species. The base compounds themselves have not been reported but methyl ethers are known. See Table 7–5 for details. The information for *A. campestris* ssp. *glutinosa* represents the summation from several reports.

Apigenin and luteolin 7-*O*-glycosides and kaempferol and quercetin 3-*O*-glycosides have been reported from a number of species. Although no systematic comparative study of their occurrence appears to have been done, it is likely that they would be found in most taxa. A dedicated search for flavonoid glycosides in *A. judaica* by Saleh et al. (1987) resulted in their finding a complex array of apigenin, luteolin, and chrysoeriol glycosides including apigenin 7-*O*-glucoside, 7-*O*-glucuronide, 4'-*O*-glucoside, 7-*O*-gentiobioside, 7-*O*-diglucuronide and 7-*O*-rutinoside, luteolin 3'-*O*-glucoside, 4'-*O*-glucoside, 7-*O*-gentiobioside, and 7,3'-di-*O*-glucoside. A similar array was seen for chrysoeriol. A quercetin 3,7-di-*O*-glycoside has also been reported from *A. scoparia* (Ryakhovskaya and Alyukina, 1973). So far as we can tell, only a single report of an acylated flavonoid glycoside in *Artemisia* has appeared. Chumbalov et al. (1969) identified isorhamnetin 7-*O*-(*p*-hydroxybenzoyl)galactoside from *A. dracunculus*. Information on flavonoid *O*-glycoside occurrences in the genus are presented in Table 7–6.

Table 7–6. Flavonoid glycosides and simple aglycones from species of *Artemisia*

Species	Compounds[1]	References
A. abrotanum	Q 3-*O*-Rut	Kranen-Fiedler (1956), Oswiecimska et al. (1965)
A. absinthium	Q 3-*O*-Rut	Oswiecimska et al. (1965)
	Q, IR, Pat, and Spin 3-*O*-Glys	Hoffmann and Herrmann (1982)
A. alba	K, Q and IR 3-*O*-Glys	Marco et al. (1987)
A. annua	K, Q, 6OHK and Pat 3-*O*-Glys, Lute 7-*O*-Gly	Marco et al. (1990)
A. argyi	Q	Tan and Jia (1992)
A. campestris	Q 3-*O*-Rut	Oswiecimska et al. (1965)
A. capillaris	Q and IR 3-*O*-Gly	Kiso et al. (1984)
A. dracunculus	Q 3-*O*-Rut	Lukovnikova (1965), Oswiecimska et al. (1965), Chumbalov and Mukhamed'yarova (1970)
	IR 7-*O*-Gal-Acyl	Chumbalov et al. (1969b)
	Lute, Q 3-*O*-Gal-Rhm	Chumbalov and Mukhamed'yarova (1970)
	K, Q, and Pat 3-*O*-Glys	Hoffmann and Herrmann (1982)
	Q and Pat 3-*O*-Glys	Vienne et al. (1989)
A. echegarayi	6-*O*H-Lute 7-*O*-Glc, Qt 7-*O*-Glc	Aguinagalde (1982)
A. frigida	Apig, Lute 7-*O*-Glc	Liu and Mabry (1981b)
A. halodendron	IR 3-*O*-Glc	Chemesova and Markova (1986)
A. hispanica	Apig 7-*O*-Glc, Lute 7-*O*-Glc, Isoëtin 5'-*O*-Glc	Marco et al. (1988)
A. japonica	Apig and Lute 7-*O*-Glc	Kaneta et al. (1978)
A. judaica	Apig, Lute and Chrys-*O*-Glys	Saleh et al. (1987)
A. laciniata	Q	Chemesova et al. (1990)

Table 7–6. (Continued)

Species	Compounds[1]	References
A. ludoviciana	Apig, Lute	Liu and Mabry (1982)
A. palustris	Lute 7-*O*-Glc, Q 3-*O*-Gal	Chemesova et al. (1978)
A. pedemontana	Apig and Lute 7-*O*-Glc	Martinez et al. (1987b)
A. santolinifolia	Lute, K and Q 3-*O*-Rut	Chumbalov and Fadeeva (1974)
A. scoparia	Q 3-*O*-Rut, K and Q 3-Di-*O*-Gly, Q 37 Di-*O*-Gly	Ryakhovskaya and Alyukina (1973)
	Q 3-*O*-Gal	Belenovskaya and Markova (1992)
A. sieversiana	Q 3-*O*-Glys	Berezovskaya and Velikhanova (1968)
A. vulgaris	Q and IR 3-*O*-Glys	Hoffmann and Herrmann (1982)
A. xerophytica	Lute7-*O*-Glc	Belenovskaya et al. (1982)

[1] Apig = apigenin; Lute = luteolin; Chry = chrysoeriol; K = kaempferol; Q = quercetin; IR = isorhamnetin; Pat = patuletin; Spin = spinacetin; Qt = quercetagetin; Glc = glucose; Gal = galactose; Rut = rutinose; Glys = glycosides; Acyl = acylation.

Hoffmann and Herrmann (1982) compared the flavonoid profiles of *A. vulgaris, A. dracunculus*, and *A. absinthium*. The profile of *A. vulgaris* consisted of 3-*O*-glycosides of quercetin and isorhamnetin, that of *A. dracunculus* consisted of 3-*O*-glycosides of kaempferol, quercetin, and patuletin, while that of *A. absinthium* was the most complex consisting of 3-*O*-glycosides of quercetin, isorhamnetin, patuletin, and spinacetin (quercetagetin-6,3'-dimethyl ether). Oswiecimska et al. (1965) had reported rutin from *A. absinthium* and *A. dracunculus* (see also Lukovnikova, 1965) as well as from *A. abrotanum*, and *A. campestris*. An interesting comparative study of simple glycosides in *A. dracunculus* was described recently by Vienne et al. (1989). So-called French tarragon (also known as German tarragon or true tarragon) was compared with Russian tarragon with respect to flavonol glycosides. French tarragon was characterized by the simpler profile consisting of quercetin 3-*O*-glucoside, 3-*O*-rhamnoglucoside, and 3-*O*-rhamnogalactoside, while Russian tarragon had, in addition to these three compounds, free quercetin, patuletin 3-*O*-glucoside, 3-*O*-rhamnoglucoside, and 3-*O*-rhamnogalactoside. These differences parallel differences in the essential oil profile of the two varieties.

C-Glycosylflavones have also been reported from a few species of *Artemisia*. Thus, *A. frigida* accumulates vicenin-2, schaftoside, and

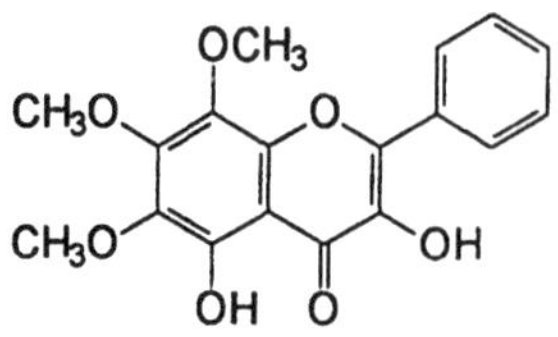

[7-6]

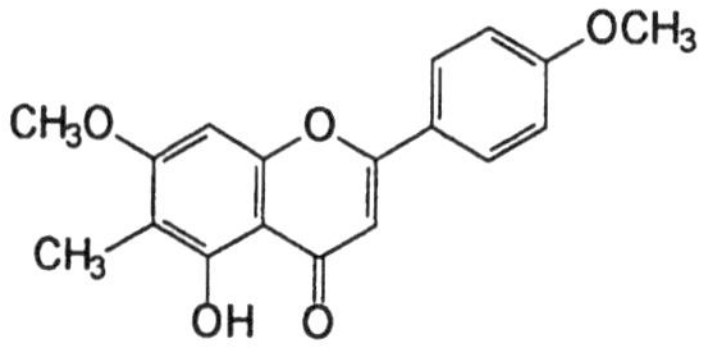

[7-7]

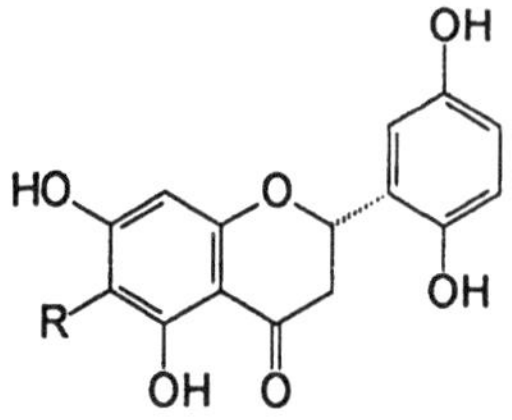

[7-8] R = H

[7-9] R = CH_3

Fig. 7–2. Several flavonoids that lack B-ring substitution, such as **[7–6]**, are known from members of *Artemisia*. Very rare in *Artemisia* are compounds such as **[7–7]**, noteworthy for the presence of the *C*-methyl group. Compounds **[7–8]** and **[7–9]** represent a pair of the more unusual flavanones reported from Asteraceae

isoschaftoside (Liu et al., 1982), *A. judaica* accumulates vicenin-2, and the four schaftosides (schaftoside itself and the iso-, neo-, and neoiso-isomers) (Saleh et al., 1987), and *A. monosperma* has vicenin-2 and lucenin-2 (Saleh et al., 1985). It is interesting to note that all of these are 6,8-di-*C*-glycosylflavones. A similar array will be seen in *Seriphidium*. Again, it is difficult to judge the significance of these findings since there appears not to have been any dedicated search of the genus for these compounds.

O-Methylated flavonoids are a very conspicuous element in the pigment profile of *Artemisia*. Virtually every species tested has afforded at least one such compound. In combination with variable degrees of *O*-methylation, the existence of a variety of oxygenation patterns leads to a particularly varied array of compounds in this category. Extra oxygenation at C-6 is common whereas extra oxygenation at C-8 is seen in only seven compounds, a few of which have substitution at both C-6 and C-8. One of the compounds in the genus that lacks B-ring substitution (Fig. 7–2) is 3,5-dihydroxy-6,7,8-trimethoxyflavone **[7–6]** from *A. klotzschiana* (Dominguez and Cardenas, 1975). 5,7,8,3'-Tetrahydroxy-3,4'-dimethoxyflavone (gossypetin-3,4'-dimethyl ether) was reported from *A. annua* (Shilin et al., 1989). Gossypetin-7,8,3',4'-tetramethyl ether was described from *A. pedemontana* (cited as *A. lanata*) by Esteban et al. (1986). 5,6,7,8,4'-Pentamethoxyflavone has been identified as a component of *A. mesatlantica* (Holeman et al., 1991). Two other 6,8-disubstituted compounds are 5,7-dihydroxy-3,6,8,4'-

tetramethoxyflavone, isolated from *A. ludoviciana* var. *ludoviciana* by Liu and Mabry (1982), and 5,3'-dihydroxy-6,7,8,4'-tetramethoxyflavone (gardenin-D), which was obtained from *A. rupestris* (Liu and Yu, 1985). Naringenin has been identified as a component of *A. ludoviciana* (no varietal status noted) by Jakupovic et al. (1991c). Recently identified for the first time as a natural product is 3,5,6,7,3'-pentahydroxy-4'-methoxyflavone (quercetagetin 4'-methyl ether) from *A. annua* (Yang et al., 1995). *Artemisia annua* continues to attract the attention of various workers likely owing to its production of the anti-malarial sesquiterpene artemisinin. One of these was a study by Sy and Brown (1998) that reported additional sesquiterpenes and six known flavonoids: 5-hydroxy-3,6,7,3',4'-pentamethoxyflavone (artemetin), 5,4'-dihydroxy-3,6,7,3'-tetramethoxyflavone (chrysosplenetin), 5,4'-dihydroxy-3,7,3'-trimethoxyflavone (pachypodol), 5,4'-dihydroxy-3,6,7-trimethoxyflavone (penduletin), and 5-hydroxy-3,7,3',4'-tetramethoxyflavone (retusin).

Whereas *O*-methylation is common in the genus, *C*-methylation is much less so. Adekenov et al. (1987a) reported 5-hydroxy-7,4'-dimethoxy-6-*C*-methylflavone **[7–7]** from *A. austriaca* which appears to be the only member of its class so far reported from the genus.

Diverse B-ring patterns provide for further structural variation in the genus. One of the variants involves the lack of any substitution, a few of which have been reported. Pinocembrin and one of its *O*-methyl ethers were reported from *A. campestris* subsp. *glutinosa*, while 3,5-dihydroxy-6,7,8-trimethoxyflavone **[7–6]** was obtained from *A. klotzschiana* (Dominguez and Cardenas, 1975). The majority of compounds obtained from the genus are based upon 4'-mono or 3',4'-dioxygenation, as is common in the family. The large array of compounds based on these kaempferol- and quercetin-type B-ring patterns is summarized in Table 7–7. Several flavones and a flavonol with 3',4',5'-trioxygenation have been identified from a smattering of species. 5,7,4'-Trihydroxy-3',5'-dimethoxyflavone (tricin) and 6-methoxytricin have been reported from several species. Four flavones oxygenated at the 5,6,7,3',4' and 5'-positions, which exhibit a variety of *O*-methylation patterns different from 6-methoxytricin, have also been reported. The sole flavonol in this group, obtained from *A. monosperma* by Khafagy et al. (1979a), was shown to be 5,7,3',4'-tetrahydroxy-3,5'-dimethoxyflavone (myricetin-3,5'-dimethyl ether). Details of the occurrence of these compounds can be found in Table 7–8.

Five *Artemisia* species accumulate flavonoids based on a B-ring with 2',4',5'-oxygenation. The base compound, 5,7,2',4',5'-pentahydroxyflavone, isoëtin, has been reported from several members of Lactuceae (Chapter 13). Marco et al. (1988) isolated isoëtin 5'-*O*-glucoside from *A. hispanica* along with an array of other compounds including some with 3',4',5'-oxy-

Table 7–7. *O*-Methyl flavones and flavonols with 4'- and 3',4'-B-ring oxygenation in species of *Artemisia*

Species	Compounds[1]	References[2]
A. abrotanum	563'4'/37, 5674'/33', 564'/373', 573'/364', 563'/374', 53'/3674'	Wollenweber et al. (1989b)
	5/3673'4'	Oswiecimska et al. (1965)
A. absinthium	574'/3	Wollenweber et al. (1989b)
	574'/363', 53'4'/367, 54'/3673'	Wollenweber et al. (1989b)
	5/3673'4'	Cekan and Herout (1956), Tunmann and Isaac (1957), Oswiecimska et al. (1965), Akhmedov et al. (1969), Chemesova et al. (1987), Wollenweber et al. (1989b)
A. adamsii	573'4'/6, 574'/63'	Chemesova et al. (1983b)
A. afra	54'/7	Wollenweber et al. (1989b), Wollenweber and Mann (1989)
	5/74'	Jakupovic et al. (1988b), Nkunya et al. (1992)
	53'4'/7, 573'4'/6, 574'/63', 3574'/63', 573'4'/36	Wollenweber et al. (1989b), Wollenweber and Mann (1989)
A. alba	57/4', 574'/3', 57/64', 573'4'/6, 573'4'/3, 574'/33', 573'4'/36, 573'/364'	Wollenweber et al. (1989b)
A. annua	574'/3', 54'/67, 53'/674', 53'4'/67, 54'/673', 354'/7, 54'/367, 3573'/4', 353'4'/7, 573'4'/3, 5673'4'/3, 5783'/34'	Shilin et al. (1989)
	5/3674'	Tu et al. (1985a,b)
	573'4'/36	Shilin et al. (1989), Marco et al. (1990)
	353'/674'	Liu et al. (1981), Misra (1986)
	53'4'/367	Tu et al. (1982), Shilin et al. (1989), Marco et al. (1990)
	35/673'4'	Djermanovic et al. (1975), Bhardwaj et al. (1977), Stefanovic et al. (1979),

Table 7–7. (Continued)

Species	Compounds[1]	References
		Liu et al. (1981), Tu et al. (1982, 1985b), Zheng (1994)
	53'/3674'	Jeremic et al. (1979), Baeva et al. (1988), Shilin et al. (1989), Marco et al. (1990)
	54'/3673'	Baeva et al. (1988), Shilin et al. (1989), Marco et al. (1990)
	5/3673'4'	Liu et al. (1980), Tu et al. (1985a,b), Misra (1986), Baeva et al. (1988), Zheng (1994)
	35673'/4'	Yang et al. (1995)
A. anomala	5/674'	Jakupovic et al. (1987a)
A. arborescens	5/3673'4'	Mazur and Meisels (1955), Appendino and Gariboldi (1982), Betto et al. (1988)
A. arctica ssp. *saxicola*	574'/63'	Herz et al. (1970c)
A. argyi	57/63'4', 5/673'4'	Wu and Tu (1985)
A. assoana	574'/3', 5/674', 574'/63', 54'/673', 5/673'4', 5/3673'4'	Martinez et al. (1987b)
A. aucheri	5/74', 574'/3', 57/64', 54'/67, 5/674', 53'4'/67, 574'/63', 57/63'4', 54'/673', 53'/674', 5/673'4', 573'/364'	Wollenweber et al. (1992)
A. austriaca	574'/3', 573'4'/6, 574'/63', 57/63'4', 5/673'4'	Çubukçu and Melikoglu (1995)
A. campestris	57/64', 574'/63'	Wollenweber et al. (1989b)
	5/374', 53'4'/67, 584'/7	El-Ghazouly and Omar (1984)
A. campestris ssp. *glutinosa*	574'/6	Hurabielle et al. (1982), De Pascual Teresa et al. (1986)
A. campestris ssp. *maritima*	54'/67	Rauter et al. (1989)
A. capillaris	54'/67	Takeda et al. (1976), Komiya et al. (1976), Kiso et al. (1984), Okuno et al. (1988)

Table 7–7. (Continued)

Species	Compounds[1]	References
	54'/7, 54'/673'	Komiya et al. (1976), Kiso et al. (1984)
	354'/7	Komiya et al. (1976), Matsueda et al. (1980), Kiso et al. (1984)
	353'4'/67	Hino et al. (1982)
	52'4'/675'	Namba et al. (1983)
A. copa	574'/353', 54'/3673'	de La Fuente et al. (1986)
A. depauperata	54'/67, 54'/673'	Belenovskaya and Markova (1992)
A. dracunculus	5/3673'4'	Oswiecimska et al. (1965)
A. filifolia	57/4'	Torrance and Steelink (1974)
A. frigida	574'/3', 57/3'4', 574'/6' 573'4'/6, 574'/63', 57/63'4', 57/363'4'	Liu and Mabry (1981b)
A. glacialis	574'/6, 57/64', 574'/63', 57/364'	Wollenweber et al. (1989b)
A. glauca	574'/3, 573'4'/3, 3574'/3, 53'4'/37, 574'/33', 54'/373'	Wollenweber et al. (1989b)
A. globosa	353'5'/7, 53'4'/67	Belenovskaya and Markova (1992)
A. glutinosa	353'4'/7	Gonzalez et al. (1983)
A. gmelinii	57/4', 5/74', 54'/73', 574'/63'	Chemesova et al. (1983a)
A. halodendron	54'/673', 353'4'/7	Chemesova and Markova (1986)
A. hispanica	54'/67	Sanz et al. (1989)
	74'/63', 54'/673', 54'/367, 573'4'/36, 53'4'/367, 54'/3673'	Marco et al. (1988)
A. incanescens	3574'/6, 3574'/3'	Marco et al. (1987)
	573'4'/3, 573'/34', 574'/36, 57/364', 573'4'/36, 573'/364', 53'/3674'	Barbera et al. (1986)
A. japonica	35/673'4'	Gu and Tu (1993)
A. judaica	57/4', 574'/6, 57/64', 54'/67, 574'/63', 57/63'4', 54'/673', 53'/3674'	Saleh et al. (1987)
A. laciniata	3574'/3'	Chemesova et al. (1990)
A. lindleyana	574'/6, 573'4'/6, 354'/67, 574'/3, 574'/36, 3574'/63', 574'/363', 53'4'/367, 57/363'4', 5/3673'4'	McCormick and Bohm (1986)

Table 7–7. (Continued)

Species	Compounds[1]	References
A. ludoviciana		
ssp. *mexicana*	574'/63', 57/63'4'	Ruiz-Cancino et al. (1993)
A. mesatlantica	574'/3', 54'/67, 54'/673', 5/673'4'	Bouzid et al. (1982a,b)
	-/56784'	Holeman et al. (1991)
A. monosperma	57/4', 574'/6, 54'/67, 574'/63'	Saleh et al. (1987)
	573'/4', 57/33'4'	Ismail et al. (1989)
	54'/673'	Khafagy et al. (1979a), Saleh et al. (1987), Elgamel et al. (1991b)
A. pedemontana	35/783'4'	Esteban et al. (1986)
	574'/3', 5/674', 574'/63', 54'/673', 5/673'4', 5/3673'4'	Martinez et al. (1987b)
A. pontica	54'/7, 57/4', 5/74', 53'4'/7, 574'/3', 54'/73', 53'/74', 573'4'/6, 574'/63', 54'/673'	Wollenweber et al. (1989b)
A. roxburghiana	574'/63', 54'/674', 57/63'4', 57/33'4'	Li et al. (1994a,b)
A. rubripes	57/63'4', 54'/673', 574'/63'	Lao et al. (1983)
A. rupestris	53'/6784'	Liu and Yu (1985)
A. rutifolia	574'/3', 54'/673', 54'/3673'	Chemesova et al. (1984a)
A. sacrorum	54'/7	Chandrashekar et al. (1965)
A. salsoloides	54'/7, 5/74', 573'5'/64'	Kaul et al. (1989)
A. schimperi	57/33'4'	Abegaz and Herz (1991)
A. scoparia	54'/67, 354'/7, 354'/67, 353'4'/67	Chandrasekharan et al. (1981)
	574'/33'	Belenovskaya and Markova (1992)
A. sieversiana	54'/3673', 5/3673'4'	Chemesova et al. (1982, 1987)
A. umbelliformis	57/63'4'	Appendino et al. (1983), Cappelletti et al. (1986)
A. vulgaris	53'/374'	Stefanovic et al. (1982)
A. xanthochloa	54'/67, 53'4'/67	Chemesova et al. (1984b)
A. xerophytica	57/64', 54'/67, 574'/63'	Belenovskaya et al. (1982)

1 Substitution: Hydroxyl positions/Methoxyl positions.

2 For a few additional species examined, see Valant-Vetschera and Wollenweber (1995).

genated B-rings. Two of the remaining members of this select group are flavones with 6-methoxylation, the other two are flavonols that lack further A-ring modification. See Table 7–9 for details.

Table 7–8. *O*-Methylated flavones with 3',4',5'-B-ring oxygenation in species of *Artemisia*

Species	Compounds[1]	References
A. anomala	57/63'4'5'	Xiao and Tu (1984)
	574'/3'5'	Xiao and Tu (1986)
A. fragrans	573'4'/65', 573'/64'5'	Belenovskaya et al. (1992)
A. frigida	573'4'/65', 573'/64'5', 574'/63'5'	Liu and Mabry (1981a,1982)
A. giraldii	674'/3'5', 5,5'/83'4'	Zheng et al. (1996)
A. hispanica	573'4'/65', 53'4'/675', 573'/64'5', 57/63'4'5'	Marco et al. (1988)
A. ludoviciana	573'4'/5'	Liu and Mabry (1982)
A. mesatlantica	54'/673'5'	Bouzid et al. (1982a)
	574'/3'5', 574'/63'5'	Bouzid et al. (1982a,b)
A. monosperma	573'4'/35'	Khafagy et al. (1979a)
A. pedemontana	574'/63'5'	Martinez et al. (1987b)
A. rutifolia	574'/3'5'	Chemesova et al. (1984a)
A. salsoloides	573'5'/64'	Kaul et al. (1989)
A. subdigitata	574'/3'5'	Shi et al. (1992)
A. xerophytica	574'/3'5'	Belenovskaya et al. (1982)

[1] Substitution: Hydroxyl positions/Methoxyl positions.

Table 7–9. *O*-Methylated flavonoids with 2',4',5' B-ring oxygenation in species of *Artemisia*

Species	Compounds[1]	References
A. capillaris	52'4'/675'	Hino et al. (1982), Kitagawa et al. (1983), Kiso et al. (1984)
	564'/72'5'	Kitagawa et al. (1983)
A. hispanica	572'4'5'/-	Marco et al. (1988)
A. ludoviciana	572'4'/65'	Liu and Mabry (1982)
A. monosperma	32'5'/574', 2'5'/3574'	Ismail et al. (1989)
A. xanthochloa	572'4'/65'	Chemesova (1987)

[1] Substitution: Hydroxyl positions/Methoxyl positions.

Chamaemelum: Bremer (1994) includes *Chamaemelum* in subtribe Achilleinae and describes it as a genus of six species with an essentially Mediterranean distribution. *Chamaemelum nobile* is one of the included species. This taxon has also been considered as *Anthemis nobilis*. The reader will find flavonoid chemistry published under that name in the section on *Anthemis* above. Reports by Tschan and Sticher (1993) and Tschan et al. (1996) listed apigenin, apigenin 7-*O*-glucoside, and apigenin 7-*O*-(6"-(3"-methyl-3"-hydroxyglutaroyl)glucoside as components of *C. nobile*.

Chrysanthemum: Heywood and Humphries (1977) discuss taxonomic complexity of the group of genera that comprise the *Chrysanthemum* complex (*Argyranthemum, Chrysanthemum, Dendranthema, Leucanthemum*, and *Tanacetum*). More recently, Bremer and Humphries (1993) list only two species in *Chrysanthemum, C. coronarium* and *C. segetum*.

These generic taxonomic difficulties are reflected in the flavonoid literature from which come data for a dozen or so cited *Chrysanthemum* taxa. Among the set are the two species seen as true members of the genus. In the first report, Geissman and Steelink (1957) described isolation of quercetin and gossypetin 7-*O*-glucosides from *C. segetum*. This was soon followed by a report of luteolin, quercetin, and quercetagetin 7-*O*-glucosides in *C. coronarium* (Anyos and Steelink, 1960). Harborne et al. (1970) reported luteolin, chrysoeriol, quercetin, patuletin, and quercetagetin 3'-methyl ether from *C. coronarium*; and luteolin, quercetin, quercetagetin, and gossypetin from *C. segetum*. 7-*O*-Glucosides were identified in all cases, with luteolin 7-*O*-glucuronide and chrysoeriol 7-*O*-rutinoside being identified from *C. coronarium* as well.

Flavonoid information on the other taxa that were published under the generic name *Chrysanthemum* can be found in this chapter under *Arctanthemum, Dendranthema, Heteranthemis, Leucanthemum, Nivellea*, and *Tanacetum*.

Cladanthus: *Cladanthus arabicus*, the sole member of this genus, is native to North Africa, Spain, and the Mediterranean. It was shown by Greger (1969) to have luteolin and quercetagetin (presumably as glycosides). 5-Hydroxy-3,6,7,3',4'-pentamethoxyflavone (artemetin) was reported by Wichlacz et al., (1994).

Coleostephus: This genus of three species (Bremer, 1994) occurs in western Europe and North Africa. Harborne et al. (1976) reported quercetin, quercetagetin, and patuletin 7-*O*-glucosides from *C. myconis*. Wilcox (1984) reported kaempferol, quercetin and, quercetagetin from *C. myconis* and *C. multicaulis*. Greger (1969) reported finding luteolin in *C. myconis* (reported as *Leucanthemum myconis*).

Cotula: *Cotula* contains 55 species in South Africa, Australia, South America, southern oceanic islands, and Mexico. Glennie and Harborne (1971) reported quercetin and isorhamnetin 5-*O*-glucosides from *C. australis* and *C. barbata*, kaempferol 5-*O*-glucoside from *C. goughensis*, and quercetin and luteolin 5-*O*-glucosides from *C. turbinata. Cotula cinerea* accumulates apigenin 7-*O*-rhamnoside, luteolin 7-*O*-mono- and diglucosides, 6-hydroxyluteolin 7-*O*-glucoside, several quercetin glycosides, and quercetagetin-3,6,7-trimethyl ether (chrysosplenol-D) (Ahmed et al., 1987a). Mahran et al. (1975) had reported kaempferol and quercetin glycosides from *C. cinerea.*

Greger (1969) reported isorhamnetin from *C. squalida*, a species that properly belongs in *Leptinella*.

Dendranthema: *Dendranthema* is a genus of 37 species distributed in Asia, especially Japan, Korea, and China and into Europe. *Dendranthema indicum* (cited as *Chrysanthemum indicum*) was shown to accumulate acacetin (Chatterjee et al., 1981; Yu and Xie, 1987; Xu and Liu, 1987) and 5,4'-dihydroxy-3,6,7-trimethoxyflavone (penduletin) (Mladenova et al., 1988). *Dendranthema grandifolium* (cited as *Chrysanthemum morifolium*) afforded apigenin, acacetin, and luteolin 7-*O*-glycosides (Hsu and Hong, 1964; Arisawa et al., 1969; Kaneta et al., 1978; Sinha et al., 1986; Kawase et al., 1970; Yahara et al., 1990) and cyanidin 3-*O*-glucoside (Kawase et al., 1970). Nakayama et al. (1997) recently reported cyanidin 3-*O*-(3",6"-dimalonyl-ß-D-glucoside) as a flower pigment in *D. grandiflorum*. A study of antiviral components of this species (Hu et al., 1994) confirmed the presence of apigenin, acacetin, and luteolin derivatives as well as a 5,6,7-trihydroxyflavone (baicalein) 7-*O*-glycoside. *Dendranthema arctium*, which properly belongs in *Arctanthemum*, accumulates apigenin 7-*O*-glucuronide, an unspecified acyl derivative thereof, and quercetin and patuletin 7-*O*-glucosides (Harborne et al., 1970).

Endopappus: *Endopappus macrocarpus*, native to northern Africa, is the sole member of this genus. Wilcox (1984) reported kaempferol, quercetin, and patuletin from this taxon.

Eriocephalus: *Eriocephalus*, which consists of 26 species, occurs in southern and southwestern Africa. Three species have been examined for flavonoids. *Eriocephalus giessii* (this epithet does not occur in Bremer and Humphries, 1993) was shown to have 5,7-dihydroxy-6,4'-dimethoxyflavone (pectolinarigenin), 5-hydroxy-6,7,4'-trimethoxyflavone (salvigenin), 5,6-dihydroxy-7,3',4'-trimethoxyflavone, and 5-hydroxy-6,7,3',4'-tetramethoxyflavone, while *E. kingesii* afforded only 5,6,4'-trihydroxy-7,3'-dimethoxyflavone (Zdero et al., 1987a). Three other flavones were reported from *E. punctulatus* by Wollenweber and Jay (1988): 5,7,4'-trihydroxy-6-

methoxyflavone (hispidulin), 5,7,4'-trihydroxy-6,3'-dimethoxyflavone (jaceosidin), and 5,7-dihydroxy-6,3',4'-trimethoxyflavone (eupatilin). A second report on *E. punctulatus* (Wollenweber and Mann, 1989) added the following compounds to the list: apigenin, luteolin, luteolin-3',4'-dimethyl ether, 5,7-dihydroxy-6,4'-dimethoxyflavone (pectolinarigenin), quercetin, isorhamnetin, naringenin, and eriodictyol.

Filifolium: *Filifolium* consists of the single species *F. sibiricum*, which occurs in China and Korea (Shishkin and Bobrov, 1961; Bremer and Humphries, 1993). Under the name *Tanacetum sibiricum*, Stepanova et al. (1981) reported an interesting array of flavanones, including two with unusual oxygenation patterns: naringenin, naringenin 4'-methyl ether (isosakuranetin), homoeriodictyol, 6-methoxynaringenin, 5,7,2',5'-tetrahydroxyflavanone **[7–8]**, and 5,7,2',5'-tetrahydroxy-6-methoxyflavanone **[7–9]**.

Glossopappus: *Glossopappus macrotus*, the only member of this genus, occurs in northern Africa and southwestern Europe. It was shown to accumulate luteolin, quercetin, and quercetagetin 7-*O*-glucosides (Harborne et al., 1976).

Gymnopentzia: Bremer and Humphries (1993) consider *Gymnopentzia* to be monotypic with the only species native to southern Africa. *Gymnopentzia bifurcata* was shown to accumulate naringenin 7-methyl ether (sakuranetin) and 4'-methyl ether (isosakuranetin) (Bohlmann and Zdero, 1978).

Heliocauta: *Heliocauta atlantica*, the sole member of this genus, occurs in Morocco. The genus was newly described by Humphries (1977) and consists of individuals formerly put in *Anacyclus*. The flavonoid profile of *A. atlanticus* was reported by Greger (1978) to consist of glycosides of luteolin and quercetin with the possibility of 6-oxygenated luteolin derivatives as well.

Heteranthemis: This monotypic genus occurs in southwestern Europe and North Africa. The flavonoid profile of *H. viscidehirta* was shown to consist of the 7-*O*-glucosides of quercetin, quercetagetin, and patuletin (Harborne et al., 1970). The plant was identified as *Chrysanthemum viscidehirtum* in that paper.

Hippolytia: *Hippolytia* consists of 19 species with a range of occurrence encompassing central Asia, Mongolia, China, and Tibet (Bremer and Humphries, 1993; Bremer, 1994). One of the species, *H. dolichophylla*, has at one time or another been put in *Chrysanthemum* and *Tanacetum*.

Flavonoid information comes from a report by Shawl (1993) that *T. dolichophyllum* accumulates 5-hydroxy-3,6,7,3',4'-pentamethoxyflavone (artemetin).

Inulanthera: This genus has recently been described by Källersjö (1986) based on differences in fruit anatomy for a collection of 10 species of southern Africa formerly residing in *Athanasia* (Källersjö, 1991). We have flavonoid information for only one species, *I. calva* (as *Athanasia calva*) from which Zdero et al. (1991f) obtained naringenin and aromadendrin.

Ismelia: This genus from Morocco consists of the single species *I. versicolor*. Flavonoids of this taxon, described under *Chrysanthemum carinatum*, were identified as quercetin and patuletin 7-*O*-glycosides (Harborne et al., 1970).

Lepidophorum: According to Heywood and Humphries (1977), *Lepidophorum* consists of a single species native to Spain and Portugal. Bremer and Humphries (1993) agree. Greger (1969) reported kaempferol and isorhamnetin from *L. repandum* (cited as *L. grisleyi*). Harborne et al. (1976) reported the 7-*O*-glucosides of quercetin, quercetagetin, patuletin, quercetagetin 3'-methyl ether, and quercetagetin-6,3'-dimethyl ether (spinacetin) from flowers and patuletin 3-*O*-rhamnoside from leaves.

Leptinella: *Leptinella* consists of 33 species that occur in New Guinea, Australia including Tasmania, and New Zealand including the subantarctic islands. One of the species also occurs in southern South America. Several species of *Leptinella* were formerly considered as belonging to *Cotula*. One of these, *C. squalida*, a New Zealand taxon, was reported to accumulate isorhamnetin (Greger, 1969).

Leucanthemopsis: *Leucanthemopsis* consists of nine species mainly in Spain but also in southwestern Russia and Morocco (Bremer and Humphries, 1993). 7-*O*-Glucosides of luteolin, quercetin, quercetagetin, and gossypetin and a 7-*O*-diglucoside of luteolin have been reported from *L. flaveola* (Harborne et al., 1976).

Leucanthemum: *Leucanthemum* consists of 33 species, with a range that includes parts of Europe and northern Africa. Greger (1969) reported kaempferol and isorhamnetin from *L. atratum* and kaempferol in *L. discoideum*. He also reported luteolin from *L. myconis*, but this taxon is now considered to belong to *Coleostephus* according to Bremer and Humphries (1993). Apigenin 7-*O*-glucuronide has been reported from three species, *L. atratum, L. silvaticum, L. vulgare*, by Harborne et al. (1970) and Sagareishvili et al. (1983), while Wilcox (1984) reported various combinations of apigenin, luteolin, kaempferol, quercetin, quercetagetin, and patuletin as well as unspecified glycosides thereof, from additional species. Apiin, the 7-*O*-apiosylglucoside of apigenin, was reported from *Leucanthemum vulgare* and *L. maximum* (cited as *Chrysanthemum leucanthemum* and *C. maximum*, respectively) by Wagner and Kirmayer (1957). They also recorded apiin from *Chrysanthemum uliginosum*, but we cannot locate any taxon with that epithet in Bremer and Humphries (1993).

Leucocyclus: *Leucocyclus* is a monotypic genus native to southwestern Asia in Turkey. Valant-Vetschera (1982) examined both subspecies, *L. formosus* subsp. *formosus* and *L. formosus* subsp. *amanicus*, and reported significant differences in flavonoid profile between them. The profile of subspecies *formosus* was dominated by *C*-glycosylflavones derivatives (isovitexin, isoörientin and glycosides, lucenin-2, vicenin-2, shaftoside, and isoshaftoside) with only a small contribution from flavones or flavonols, whereas subspecies *amanicus* had kaempferol and quercetin glycosides, only a single *C*-glycosylflavone (isoshaftoside), and no flavones. In a second study, Valant-Vetschera and Wollenweber (1989) examined the flavonoid aglycones of leaf exudates of both taxa. Although there were differences between the two subspecies, both exhibited *O*-methylated derivatives of 5,6,7,4'-tetrahydroxyflavone (scutellarein), 6-hydroxyluteolin, and quercetagetin. Only subspecies *amanicus* accumulated 5-hydroxy-3,6,7,4'-tetramethoxyflavone, however.

Lonas: Greger (1969) obtained apigenin and luteolin from *Lonas annua*, the only member of this Mediterranean genus. The name cited in that paper was *L. inodora*, which is a synonym for *L. annua* according to Tutin et al. (1976).

Matricaria: Bremer and Humphries (1993) state that *Matricaria* consists of seven species distributed broadly in the Northern Hemisphere. Nomenclatural concerns by Rauschert (1974) that could have led to a change of name for *Matricaria* have been discussed and argued against by Jeffrey (1979) and Bremer and Humphries (1993). Information on flavonoids is available for six species in all, but the majority of citations deal with *M. recutita* (cited as *M. chamomilla*; synonym fide Bremer and Humphries, 1993).

7-*O*-Glycosides of luteolin and/or apigenin have been reported, either as such or as glycosides, from *M. recutita* by several workers (Sorm et al., 1952, 1953; Tyihak et al., 1962; Harborne et al., 1970; Greger, 1975; Kunde and Isaac, 1979; Redaelli et al., 1980; Pietta et al., 1987). Several acetylated derivatives of apigenin 7-*O*-glucoside have also been reported from this taxon: 2"-acetate (Redaelli et al., 1980), 6"-acetate (Kunde and Isaac, 1979; Redaelli et al., 1980; Pietta et al., 1987), and 2",3"- and 3",4"-diacetates (Redaelli et al., 1982). Simple flavonol glycosides have been reported from *M. recutita* (cited as *M. chamomilla*): quercetin 3-*O*-glucoside and 3-*O*-rutinoside (El-Keiy et al., 1963), quercetin 7-*O*-glucoside (Kunde and Isaac, 1979), and patuletin 7-*O*-glucoside (Hörhammer et al., 1963; Poethke and Bulin, 1969; Kunde and Isaac, 1979).

The presence in *M. recutita* of flavonoids with extra A-ring hydroxylation was noted by Hänsel et al. (1966) who identified 5,4'-dihydroxy-3,6,7,3'-tetramethoxy-flavone (chrysosplenetin). Greger (1975) reported 6-hydroxyluteolin 7-*O*-glucoside as a major component in specimens taken from several locations. Several flavonoids having extra oxygenation and/or *O*-methylation have been reported for this species: luteolin 3'-methyl ether (chrysoeriol), 6-methoxykaempferol, 3,5,4'-trihydroxy-6,7-dimethoxy-flavone (eupalitin), quercetin 3'-methyl ether (isorhamnetin), and quercetagetin-3,6-dimethyl ether (axillarin), 6,7-dimethyl ether (eupatolitin), 6,3'-dimethyl ether (spinacetin), 3,6,7-trimethyl ether (chrysosplenol-D), and 3,6,3'-trimethyl ether (jaceidin) (Exner et al., 1980, 1981). Quercetagetin and glycosides of apigenin and quercetin were present in *M. recutita* (Peneva et al., 1989). This taxon has recently been reported to accumulate apigenin and luteolin 7-*O*-glucosides, kaempferol 3-*O*-glucoside, and chrysosplenetin (Mericli and Korkmaz, 1992). Reichling et al. (1979) studied the flavonoid contents of different organs, including tumors, of *M. recutita* (cited as *M. chamomilla*). They reported apigenin, luteolin, chrysoeriol, quercetin, isorhamnetin, patuletin, and quercetagetin 3,6,7,3'-tetramethyl ether in aerial parts of the plant but no flavonoids at all in roots or tumors.

Matricaria matricarioides (cited as *M. discoidea*; fide Bremer and Humphries, 1993) was shown to accumulate 7-*O*-glycosides of luteolin and quercetin (Greger 1975; Prosovskii and Oleshko, 1985). Quercetagetin 7-*O*-glucoside was shown to be a constituent of *M. matricarioides* by Mathe and Tyihak (1962), while *M. aurea* had the 7-*O*-glucoside of 6-hydroxyluteolin as well (Greger, 1975).

Two flavonoid chemotypes of cultivated *M. recutita* (reported as *Chamomilla recutita*) have been identified by Repcak et al. (1999). Plants of diploid cv. 'Novbona' and tetraploid cv. 'Lutea' were grown in a common garden and their exudate flavonoids determined using HPLC. About 80% of the plants from each cultivar exhibited chrysosplenetin (quercetagetin 3,6,7,3'-tetramethyl ether) in extracts of anthodia, whereas the remaining 20% exhibited jaceidin (quercetagetin 3,6,3'-trimethyl ether). The same flavonoid profile was observed in both diploid and tetraploid forms. A preliminary report appeared in a paper by Repcak and Martonfi (1995).

Nivellea: This genus was described as new recently by Wilcox, Bremer, and Humphries in Bremer and Humphries (1993) as monotypic, with *N. nivellei* native to Morocco (see also Vogt and Oberprieler, 1996). Wilcox (1984) included it in his study of flavonoids of *Leucanthemum* and related North African genera reporting the presence of apigenin, luteolin, and quercetin. The basionym, *Chrysanthemum nivellei* Maire, was included in a flavonoid survey by Harborne et al. (1970) who reported the 7-*O*-glucosides of quercetin, quercetagetin, and patuletin.

Oncosiphon: *Oncosiphon* consists of eight species from southern Africa. There appear to be no flavonoid reports in the literature specifically naming *Oncosiphon*, but we do have information for two species that Bremer and Humphries (1993) consider more correctly placed there. *Oncosiphon grandiflorum* (cited as *Pentzia grandiflora*) afforded eriodictyol (Zdero and Bohlmann, 1990b), while *O. sabulosa*, reported as *Matricaria sabulosa*, accumulates 5,4'-dihydroxy-6,7,3'-trimethoxyflavone, 5-hydroxy-6,7,3',4'-tetramethoxyflavone, and 5-hydroxy-3,6,7,3',4'-pentamethoxyflavone (Zdero and Bohlmann, 1990b). A recent survey paper by Wollenweber et al. (1997) included *O. grandiflorum* (cited as such) from which they obtained *O*-methylated derivatives of scutellarein, 6-hydroxyluteolin, 6-hydroxykaempferol, quercetin, and quercetagetin.

Otanthus: *Otanthus* is a monotypic genus of Europe and the eastern Mediterranean region. *Otanthus maritimus* yielded acacetin (El-Din, 1989) and its 7-*O*-neohesperidoside (Khafagy et al., 1979b; El-Din, 1989), 5-hydroxy-6,7,3',4'-tetramethoxyflavone and 5-hydroxy-6,7,3',4',5'-pentamethoxyflavone (Khafagy et al., 1981c), and the 3,6,4'-trimethyl ether (centaureidin), 3,6,7,4'-tetramethyl ether (casticin) and 3,6,7,3',4'-pentamethyl ether of quercetagetin (artemetin) (Wollenweber et al., 1989b). *C*-Glycosylflavones have also been reported (El-Sayed et al., 1992).

Pentzia: Heywood and Humphries (1977) see *Pentzia* as a genus of 38 species. Bremer and Humphries (1993) cite 23. The genus is distributed in southern and southwestern Africa. Zdero and Bohlmann (1990b) reported 5,4'-dihydroxy-6,7-dimethoxyflavone (cirsimaritin) from *P. calva*. Greger (1977) cites luteolin and quercetin glucosides from the genus.

Plagius: Bremer and Humphries (1993) include two species distributed in Europe. Greger (cited by Wilcox, 1984) reported that a species of *Plagius* contained flavonol 5-*O*-glucosides.

Rennera: This Namibian genus consists of three species. Zdero and Bohlmann (1990b) reported 5,4'-dihydroxy-3,6,7,3'-tetramethoxyflavone (chrysosplenetin) from *R. eenii* (cited as *Pentzia eenii*; taxonomy after Bremer and Humphries, 1993).

Santolina: The genus *Santolina* consists of eight species distributed in the Mediterranean region. Only three species appear to have been studied for flavonoids. *Santolina chamaecyparissus* was shown to accumulate 7-*O*-glycosides of apigenin, luteolin, chrysoeriol (Becchi et al., 1976), quercetagetin 6-methyl ether (patuletin), and 3,6-dimethyl ether (axillarin) (Becchi and Combier, 1976). A further paper from this group (Becchi and Carrier, 1980) reported 6-methoxyapigenin (hispidulin), 5,7-dihydroxy-6,4'-dimethoxyflavone (pectolinaringenin), 6-methoxyluteolin (nepetin), and 5,7,4'-trihydroxy-6,3'-dimethoxyflavone (jaceosidin). A more recent report by Wollenweber et al. (1989b) added 5,4'-dihydroxy-6,7-dimethoxyflavone and luteolin 7-methyl ether to the list. These workers (Wollenweber et al., 1989b) also reported the following compounds from *S. rosmarinifolia*: eriodictyol, 6-methoxyapigenin, 5,4'-dihydroxy-6,7-dimethoxyflavone, luteolin, 6-methoxyluteolin (nepetin), 5,7,4'-trihydroxy-6,3'-dimethoxyflavone (jaceosidin), 5,7-dihydroxy-6,3',4'-trimethoxyflavone, kaempferol, isorhamnetin, 5,7-dihydroxy-3,6,4'-trimethoxyflavone, and quercetagetin-3,6,3'-trimethyl ether (jaceidin). *Santolina pinnata* afforded kaempferol, apigenin, luteolin and its 7-*O*-glucoside, chrysoeriol, and 6-methoxyluteolin (Flamini et al., 1994).

Schistostephium: *Schistostephium* is a genus of 12 species distributed in tropical and southern Africa. Bohlmann et al. (1983b) reported 5-hydroxy-6,7,4'-trimethoxyflavone (salvigenin) from *S. heptalobum*.

Seriphidium: *Seriphidium* consists of 134 species, some of which have in earlier treatments comprised the subgenus *Seriphidium* within *Artemisia*. Flavonoid data are available for 26 species of *Seriphidium*. All of these results appeared in the literature under *Artemisia*.

Occurrence data for simple flavanones and dihydroflavonols in *Seriphidium* appear to be restricted to the report of dihydroquercetin 7,3'-dimethyl ether from *S. pygmaeum* by Rodriguez et al. (1972b). An exceptional flavanone, 5,7,3',4'-tetrahydroxy-6,5'-dimethoxyflavanone **[4–22]**, however, was reported by Belenovskaya et al. (1992) from *S. fragrans* and *S. gurganicum*.

In contrast to the paucity of simple flavanones in *Seriphidium*, the occurrence of common flavones, flavonols, and their *O*-glycosides represents a normal pattern in that they are major components of species whose vacu-

Table 7–10. Common flavonoids from species of *Seriphidium*

Species	Compounds[1]	References
S. cinum	573'4'/3 7-*O*-Glc	Klyshev and Baltabaeva (1978)
	Q 3-*O*-Rut	Ryakhovskaya and Sapko (1985)
S. herba-album	3573'4'/6 Gly	Saleh et al. (1987)
S. lessingianum	Apig and Lute	Ryakhovskaya and Alyukina (1980)
S. oliverianum	Apig	Wollenweber and Rustaiyan (1991)
S. santonicum	K and Q 3-*O*-Glys	Peneva et al. (1988)
S. sublessingianum	Apig, Lute,	
	Q and IR 3-*O*-Rut	Ryakhovskaya et al. (1985),
		Manadilova et al. (1987)
S. tauricum	573'4'/36 7-*O*-Glc	Oganesyan et al. (1976)
S. transiliense	Q 3-*O*-Rut	Chumbalov and Fadeeva (1967)
	573'4'/3 7-*O*-Glc	Chumbalov et al. (1969a)
S. tridentatum	Lute 7-*O*-Glc	Brown et al. (1975)

[1] Substitution: Hydroxyl positions/Methoxyl positions; Apig = apigenin; Lute = luteolin; Q = quercetin; IR = isorhamnetin; Glc = glucoside; Rut = rutinoside; Gly = glycoside.

olar components have been examined. Apigenin, luteolin, kaempferol, quercetin, isorhamnetin, quercetin 3-methyl ether, and quercetagetin occurrences are summarized in Table 7–10. A chromatographic study of "*Artemisia*" sect. *tridentatae* was described by Holbo and Mozingo (1965). One should also note a paper by Wilt and Miller (1992) that describes variation in coumarins and flavonoids in "*A.*" *tridentata* subsp. *wyomingensis*.

C-Glycosylflavones have been reported from four species. *Seriphidium cinum* accumulates vicenin-1 (Baltabaev, as cited by Ryakhovskaya and Sapko, 1985), while Saleh et al. (1985) reported isovitexin, vicenin-2, schaftoside, and isoschaftoside from *S. herba-album*, Ryakhovskaya et al. (1985) reported vicenin from *S. sublessingianum*, and Chumbalov and Fadeeva (1970) reported apigenin-6-*C*-mono- and 6,8-di-*C*-glucosides from *S. transiliense*.

The bulk of information available involves the occurrence of *O*-methylated flavones and flavonols in the genus. We have found information for 22 species. Table 7–11 summarizes the occurrence of compounds that are based upon the 4'- and 3',4'-oxygenation patterns. As can be seen, 6-oxygenation is very common, 8-oxygenation significantly less so. No 6,8-dioxygenated compounds seem to have been observed. It should be pointed out that one of the more unusual compounds obtained (Fig. 7–3) is 7,8,3',4'-tetrahydroxy-3-methoxyflavone (5-deoxygossypetin 3-methyl ether) **[7–10]**, which Chumbalov and Fadeeva (1969) isolated from *S. transiliense*.

Table 7–11. *O*-Methylated flavones and flavonols with 4'- and 3',4'-B-ring oxygenation in species of *Seriphidium*

Species	Compounds[1]	References
S. arbusculum	3574'/6, 54'/367, 53'4'/367	Rodriguez et al. (1972b)
S. canum	54'/3673'	Lee et al. (1969)
S. cinum	574'/6	Ryakhovskaya and Sapko (1985)
S. fragrans	573'/64'5'	Belenovskaya et al. (1992)
S. gorjaevii	574'/6	Musaev and Belenovskaya (1989)
S. gurganicum	574'/3'5', 574'/63', 57/63'4', 573'/64'5'	Belenovskaya et al. (1992)
S. heptapotanicum	574'/33', 54'/373'	Nedel'ko and Nikonov (1987)
S. herba-album	57/4', 574'/6, 54'/67, 574'/63', 54'/673'	Segal et al. (1973), Saleh et al. (1987)
S. incultum	5/3673'4'	Khafagy et al. (1983)
S. lanatum	5/673'4', 35/783'4', 5/3673'4'	Esteban et al. (1986)
S. lessingianum	54'/7, 573'4'/3	Ryakhovskaya and Alyukina (1980)
S. leucodes	57/63'4'	Saitbaeva et al. (1983)
S. longilobum	54'/367	Rodriguez et al. (1972b)
S. mogoltavicum	574'/6, 57/63'4', 573'/64'5'	Belenovskaya et al. (1992)
S. namanganicum	574'/6, 57/63'4', 573'/64'5'	Belenovskaya et al. (1992)
S. oliverianum	54'/7, 57/4', 54'/73', 574'/6, 57/64', 54'/67, 574'/63', 57/63'4', 54'/673'	Wollenweber and Rustaiyan (1991)
S. prasinum	54'/673', 573'/64'5'	Belenovskaya et al. (1992)
S. pygmaeum	57/4', 354'/73'	Rodriguez et al. (1972b)
S. rothrockii	573'4'/36, 53'4'/367	Rodriguez et al. (1972b)
S. sublessingianum	574'/3'	Manadilova et al. (1987), Ryakhovskaya et al. (1987)
	574'/6	Manadilova et al. (1987), Ryakhovskaya et al. (1987)
	574'/63'	Manadilova et al. (1987), Ryakhovskaya et al. (1985, 1987)
S. transiliense	57/4', 54'/7, 573'4'/3, 783'4'/3	Chumbalov and Fadeeva (1969)
S. tridentatum	573'4'/6, 573'4'/36, 53'4'/367	Rodriguez et al. (1972b), Brown et al. (1975)

[1] Substitution: Hydroxyl positions/Methoxyl positions.

Demethylation of the natural product gave 3,7,8,3',4'-pentahydroxyflavone whose melting point agreed with the literature value.

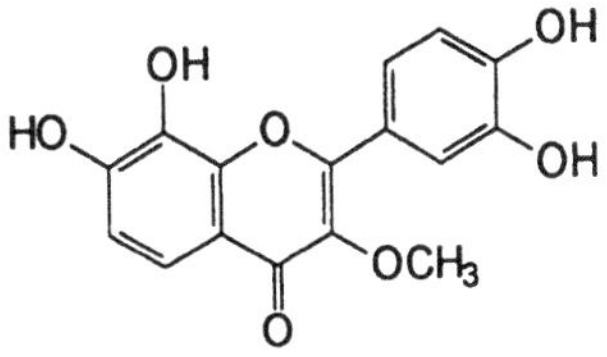

[7-10]

Fig. 7–3. 5-Deoxygossypetin 3-methyl ether **[7–10]** from *Seriphidium transiliense* is unusual with respect to the absence of 5-oxygenation

Several compounds having the 3',4',5'-oxygenation pattern have also been isolated from members of the genus. Saleh et al. (1987) obtained 5,7,3'-trihydroxy-4',5'-dimethoxyflavone from *S. herba-album*. Tricin (5,7,4'-trihydroxy-3',5'-dimethoxyflavone) was obtained from *S. gorjaevii* where it occurred with 5,7,3'-trihydroxy-6,4',5'-trimethoxyflavone (Musaev and Belenovskaya, 1989). This last compound was also obtained from *S. saissanicum* by Musaev and Belenovskya (1989).

Sphaeromeria: This genus of western North America and Mexico consists of nine species. The only information that we have on this genus comes to us under the name *Artemisia argentea* from which El-Emary and Bohlmann (1980) obtained 5-hydroxy-3,6,7,3',4'-pentamethoxyflavone (artemetin).

Stilpnolepis: *Stilpnolepis* consists of two species from Mongolia and China. We have flavonoid data for *S. intricata*, (cited as *Artemisa intricata*) by Belenovskaya and Markova (1992), who reported finding luteolin, 5,4'-dihydroxy-6,7,3'-trimethoxyflavone, 5,7,3',4'-tetrahydroxy-6,5'-dimethoxyflavone, and 5,7,3'-trihydroxy-6,4',5'-trimethoxyflavone.

Tanacetum: *Tanacetum* is a genus of 152 species from Europe and temperate Asia. Some species are important sources of natural insecticide (pyrethrum), and *T. vulgare* is well known as the aromatic herb tansy. Flavonoid reports from this species include luteolin 7-*O*-glucoside and quercetin 5-*O*-glucoside (Michaluk and Sendra, 1964), acacetin (Khvorost et al., 1964; Khvorost, 1966), apigenin, apigenin 7-*O*-glucoside, and luteolin (Khalifa et al., 1991), and the *C*-glycosylflavone orientin (Ivancheva and Behar, 1995). Compounds with extra oxygenation at C-6 and *O*-methylation are common in *Tanacetum*. For details see Table 7–12. One noteworthy point is the presence of quercetin-4'-methyl ether (tamarixetin) 7-acetate in *Tanacetum microphyllum* (Abad et al., 1993). A more recent study of antiinflamatory components in *T. microphyllum* (Martinez et al., 1997) resulted in identification of two active flavonoids, 5,7-dihydroxy-3,6,4'-

Table 7–12. *O*-Methylated flavones and flavonols from species of *Tanacetum*

Species	Compounds[1]	References
T. albipannosum	5/674', 5/673'4'	Gören and Jakupovic (1990)
T. balsamita	Lute, 574'/6, 57/64', 573'4'/6, 574'/63', 573'4'/36	Wollenweber et al. (1997)
T. chiliophyllum	54'/67, 5/674', 574'/63', 57/63'4', 53'4'/367, 573'/364'	Wollenweber et al. (1989b)
T. cilicum	54'/67, 5/674'	Öksüz (1990)
T. cinerariifolium	573'/364'	Glennie and Harborne (1972)
T. coccineum	574'/3' Gly	Harborne et al. (1970)
T. corymbosum	574'/3' Gly	Harborne et al. (1970)
T. densum ssp. *eginense*	574'/6, 574'/36	Gören and Tahtasakal (1994)
T. densum ssp. *sivasicum*	57/64', 56/374', 57/364'	Gören et al. (1992)
T. dolichophyllum	5/3673'4'	Shawl (1993)
T. gracile	5/3673'4'	Shawl (1993)
T. macrophyllum	574'/3' Gly	Harborne et al. (1970)
T. microphyllum	3573'/4'-7-acetate, 573'/364'	Abad et al. (1993),
	57/364'; 57/34'	Martinez et al. (1997)
T. niveum	574'/3' Gly	Harborne et al. (1970)
T. parthenium	564'/37, 56/374', 563'4'/37', 564'/373'	Williams et al. (1995)
T. ptarmicaeflorum	574'/36	Gonzalez et al. (1992b)
T. santolinoides	54'/673', 53'/3674', 5/673'4', 5/3673'4'	El-Din et al. (1985)
T. vulgare	57/4'	Khvorost (1966)
	574'/3'	Khvorost (1966), Wollenweber et al. (1989b)
	573'/4',	Khvorost (1966)
	574'/6, 57/64'	Wollenweber et al. (1989b)
	574'/63'	Ognyanov and Todorova (1983), Wollenweber et al. (1989b)
	57/63'4'	Wagner et al. (1972a)
	3574'/3'	Khvorost (1966)
	573'4'/36, 574'/363'	Ognyanov and Todorova (1983), Wollenweber et al. (1989b)

[1] Substitution: Hydroxyl positions/Methoxyl positions, Gly = glycoside.

trimethoxyflavone (santin) and 5,7-dihydroxy-3,4'-dimethoxyflavone (ermanin). Williams et al. (1995) reported, along with several known flavonoids, the new flavonol "tanetin" from *T. parthenium* (feverfew) and reported its structure as 5,6-dihydroxy-3,7,4'-trimethoxyflavone. A re-investigation of this species (Williams et al., 1999) provided sufficient material of the new compound to confirm its structure, which had, in the original work, been based upon mass spectral and ultraviolet data. Data from NMR now showed that the original structure was incorrect and that tanetin is the known 5-hydroxy-3,6,4'-trimethoxyflavone. Corrected structures for the other four compounds isolated from leaf exudates are 5,7,4'-trihydroxy-3,6-dimethoxyflavone, and the 3,6-di-, 3,6,3'-tri-, and 3,6,4'-trimethyl ethers of quercetagetin. In addition to the major flavonolic components, the leaf exudate afforded only a trace of apigenin. In contrast, the leaf exudate of *T. vulgare* possessed neither of the 6-hydroxykaempferol derivatives and only one quercetagetin derivative, namely the 3,6-dimethyl ether. The major components of tansy leaf exudate were methyl ethers of scutellarein and mono-, di-, and trimethyl ethers of 6-hydroxyluteolin. An interesting observation was the finding of apigenin and luteolin 7-*O*-glucuronides in the glandular trichomes on the lower epidermal surface of the ray florets of *T. parthenium*. The major flavone and flavonol methyl ethers showed significant activity against arachidonate metabolism in leucocytes.

Tanacetum cinerariifolium (cited as *Chrysanthemum cinerariifolium*) was shown by Rao et al. (1973) to accumulate apigenin 4'-*O*-glucoside while Sashida et al. (1983) reported apigenin and luteolin derivatives along with quercetagetin-3,6,3'-trimethyl ether (jaceidin) from the same source.

The list of species in Bremer and Humphries (1993) includes those that have been considered members of *Pyrethrum* in the past. Flavonoids have been reported from two species of *Pyrethrum* by Zemtsova and Molchanova (1978, 1979a): *P. corymbosum* afforded luteolin and its 7-*O*-glucoside while luteolin, diosmetin and its 7-*O*-rutinoside, kaempferol, and quercetin were reported from *P. roseum*, respectively.

Tripleurospermum: *Tripleurospermum* consists of 38 species distributed in Europe, Asia, and North Africa. The genus has been well studied for flavonoids. Several large surveys (Greger 1969, 1975; Harborne et al., 1970) showed the widespread occurrence of apigenin, luteolin, chrysoeriol, quercetin, and isorhamnetin 7-*O*-glucosides. 6-Hydroxyluteolin 7-*O*-glucoside and the 7-*O*-rhamnoglucosides of luteolin and chrysoeriol occur only sporadically. Apigenin and luteolin 4'-*O*-glucosides have a much more restricted, and taxonomically interesting, distribution. Both compounds were observed in one collection (of four) of *T. elongatum* and one collection (of two) of *T. rosellum* and possibly in *T. heterolepis*. Both compounds were major components of *T. disciforme* (all five collections), *T. microcephalum* (four of six collections) and in two unidentified species.

Ursinia: *Ursinia* consists of 38 species distributed in southern and southwestern Africa. In an early survey of Anthemideae, Greger (1969) reported isorhamnetin from *U. anthemoides* and *U. anthemoides* subsp. *versicolor* (cited as *U. versicolor*; Prassler, 1967) and quercetin and apigenin from *U. speciosa*. Bohlmann and Gupta (1982) obtained 5-hydroxy-6,7,4'-trimethoxyflavone (salvigenin) from *U. nana* ssp. *leptophylla*.

CHAPTER 8

Flavonoids of Astereae

Grau (1977) considers that Astereae consist of 135 genera with about 2500 species, Bremer (1994) lists about 170 genera and nearly 3000 species, and Nesom (1994a) gives "at least" 189 genera and about 3020 species. The main centers of distribution are the Americas and southern Africa (Grau, 1977). Recent ITS sequence data suggest that the North American taxa are monophyletic (Noyes and Rieseberg, 1999). We have flavonoid information for 48 genera. Although this sampling is not particularly extensive, the amount of attention paid to several genera is such that we have, nonetheless, a considerable amount of information with which to deal. Information on species numbers and geographic locations are from Bremer (1994), unless otherwise noted.

Amphiachyris: Both species of this Californian genus exhibited "sulphated flavonoids" (Lane, 1979). Bohlmann et al. (1981b) identified quercetin 3,3'-dimethyl ether 4'-*O*-esters along with tetra-, penta-, and hexamethyl ethers of 6,8-dihydroxymyricetin from *A. dracunculoides*.

Aster: This is one of the largest genera in the tribe enfolding some 250, mostly northern hemisphere, species. As is the case with many large genera of the family, *Aster* is currently in the midst of taxonomic controversy. Hence, taxonomic referral of flavonoid reports within the genus is not a simple matter. There are four basic viewpoints on the genus: (1) keeping it together in a broad sense with many internal subdivisions (Jones, 1980; Jones and Young, 1983); (2) keeping most of it together but segregating *Virgulus* (= *Lasallea*; synonymy after Reveal and Keener, 1981); (3) breaking it up into many smaller segregates (Nesom, 1994b); and (4) believing that *Aster* needs to be broken up into smaller natural genera, but that more study is required before formal changes are made (Lane et al., 1996; Xiang and Semple, 1996). Four species, *A. bracteolatus* (= *A. eatonii*), *A. pilosus*, *A. squamatus*, and *A. subulatus*, have been treated in *Aster* by Jones (1980) and Jones and Young (1983), and by Semple and Brouillet (1980), but in different subgeneric placements. As an example to make the point, *A. pilosus* is put in subgen. *Aster* sect. *Dumosi* subsect. *Dumosi* by the former, and in subgen. *Symphyotrichum* sect. *Porteriani* by the latter. Nesom (1994b), however, places these four species in the genus *Symphyotrichum*. Recent cpDNA restriction site studies do provide some support for the recognition of this genus (Xiang and Semple, 1996), but further sampling is needed. For the purposes of this book, however, we retain these taxa in *Aster* pending additional investigations that might recommend transfer.

Despite the size of the genus and its conspicuousness in most floras, we have found flavonoid information for only a few species. Compounds range from simple glycosides of common flavones and flavonols to highly *O*-methylated flavonoids with various degrees of extra oxygenation. Apigenin and luteolin 7-*O*-glycosides have been reported from *A. ageratoides* [cited as *A. leiophyllus* by Kaneta et al. (1978); cited as synonym fide Kitamura (1937) and Ohwi (1965); Makino (1989) however, treats it as distinct], *A. subulatus* (El-Sayed et al., 1987), and *A. bracteolatus* (syn. *A. eatonii*) (Allen, 1988). 3-*O*-Glycosides of kaempferol and/or quercetin have been recorded for *A. pilosus* (Farnsworth et al., 1968), *A. bracteolatus* (Allen, 1988) and *A. subulatus* (El-Sayed et al., 1987). Russian workers (Troshchenko and Limasova, 1966) were the first to report an unusual compound from the genus: 5-hydroxy-6,7,8,3',4'-pentamethoxyflavone from *A. altaicus*. A more comprehensive look at the biosynthetic capacity of an *Aster* species was provided by Ross and co-workers (Ross et al., 1984; Sayed and Ross, 1986). In the earlier paper they described flavonoid constituents of flowers of *A. squamatus*: baicalein (5,6,7-trihydroxy-flavone), luteolin 7-methyl ether, kaempferol, quercetin, quercetin 7-methyl ether, 5,4'-dihydroxy-3,7,8,3'-tetramethoxyflavone, and 5,4'-dihydroxy-3,7,8,3',5'-pentamethoxyflavone (called "squamatin"). In the later paper, the following components of the leaves were described: kaempferol, quercetin and its 3-*O*-diarabinoside, quercetin 3-methyl ether and its 4'-*O*-dirhamnoside, and quercetin 4'-methyl ether. Norbaek et al. (1996) reported the 3-*O*-(6"-malonylglucoside) and 3-*O*-(6"-malonylglucoside)-5-*O*-glucoside derivatives of both cyanidin and delphinidin in flowers of *A. novi-belgii*.

Baccharis: *Baccharis* is another of the large genera in the tribe with about 200 (Hellwig, 1996a), 350 (Nesom, 1994a), or 400 species (Grau, 1977; Bremer, 1994), depending upon taxonomic perspective. The center of distribution of the genus is South America with a few species known from North America. We have flavonoid information for at least 50 species. Members of the genus accumulate flavones, flavonols, a rich array of flavanones (including dihydroflavonols), and a chalcone derivative. Many *O*-methylated compounds, representing all of these structural classes, have been

Table 8–1. Flavones and flavonols from species of *Baccharis*

Species	Compounds[1]	References
B. alaternoides	574'/3	Bohlmann et al. (1982d)
B. articulata	57/4', 54'/7, 5/74', 54'/67, 5/674', 574'/363', 574'/63'	Gianello and Giordano (1984)
B. bigelovii	57/-7-benzoate, 357/6	Arriaga-Giner et al. (1986)
B. boliviensis	53'/6784', 54'/6783'	Morales et al. (1990)
B. boliviensis var. *latifolia*	353'/74', 574'/363', 53'/3674'	Rossomando et al. (1990)
B. chilco	Lute, 57/4', 54'/7,	Bohlmann et al. (1985a)
	5/74', 54'/67, 5/674', 574'/363', 574'/63'	Gianello and Giordano (1984)
B. concava	5/674', 54'/67, 57/64'	Zamorano et al. (1987)
	54'/67	Faini and Castillo (1990)
B. confertifolia	53'/74'	Faini and Castillo (1990)
B. crispa	5/74', 53'/74'	Bandoni et al. (1978)
	Apig, 54'/7	Palacios et al. (1983)
B. decussata	57/64', 57/684'	Morales-Mendez et al. (1984)
B. eleagnoides	54'/67	Mesquita et al. (1986)
B. flabellata	574'/6, 573'4'/6, 574'/63', 53'4'/67	Saad et al. (1988)
B. genistelloides	Apig, 54'/7, 53'4'/67, 54'/67	Kuroyanagi et al. (1985), Suttisri et al. (1994)
	54'/673'	Kuroyanagi et al. (1985)
	53'/674', 5/673'4'	Suttisri et al. (1994)
B. gilliesii	57/4', 54'/7, 5/74', 54'/67, 5/674', 574'/363', 574'/63'	Gianello and Giordano (1984)
B. grandicapitulata	57/4', 357/4'	Bohlmann et al. (1985a)
B. grisebachii	57/64', 57/684', 574'/68	Gianello and Giordano (1987)
B. incarum	54'/36783', 54'/6783'	Faini et al. (1982)
B. intermedia	574'/3, 573'4'/3, 574'/33', 54'/67	Faini et al. (1991)
B. kingii	574'/33'	Bohlmann et al. (1984e)
B. linearis	573'4'/36	Labbe et al. (1986)
	574'/3, 573'4'/3, 574'/33'	Faini et al. (1991)
B. macraei	5/674', 57,64', 54'/67	Faini et al. (1991)
B. magellanica	Apig, 574'/6	Cordano et al. (1982), Reyes et al. (1990)

Table 8–1. (Continued)

Species	Compounds[1]	References
B. microcephala	Lute, 54'/7	Bohlmann et al. (1985a)
B. myrtilloides	57/4', 54'/7, 5/74', 54'/67, 5/674', 574'/363', 574'/63'	Gianello and Giordano (1984)
B. neaei	54'/37, 53'4'/37	Faini et al. (1992)
B. notosergila	Apig, 54'/7	Palacios et al. (1983)
B. ochracea	Q-3-*O*-Glc	Schenkel et al. (1997)
B. paniculata	53'4'/37	Faini and Castillo (1990)
B. patagonica	57/4', 574'/6	Rivera et al. (1988)
B. patens	54'/678, 53'4'/678, 53'/6784'	Silva et al. (1985)
B. pedicellata	Apig, 54'/7, 54'/37	Faini et al. (1987)
B. pedunculata	56/74', 5/674'	Jakupovic et al. (1991b)
B. petiolata	54'/7, 5/74'	Gianello et al. (1990)
	IR, 573'/64'	Faini and Castillo (1990)
B. pingraea[2]	K and Q 3-*O*-Glc	Wagner et al. (1972b)
B. pingraea[3]	54'/7	Gianello and Giordano (1984)
B. pingraea	57/4', 54'/7, 5/74', 54'/67, 5/674', 574'/363', 574'/63'	Gianello and Giordano (1984)
B. quitensis	54'/678, 54'/6783'	Bohlmann et al. (1981f)
B. retamoides	53'4'/37	Rossomando et al. (1990)
B. rhomboidalis	57/4', 5/74'	Silva et al. (1971)
	574'/6, 54'/67, 5/674'	Labbe et al. (1986)
B. salicifolia	54'/73', 574'/3', 54'/673', 54'/6783'	Warning et al. (1986)
B. salicina	573'/364'	Parodi and Fischer (1988),
	573'/364', 57/363'4'	Quijano et al. (1998)
B. sarothroides	573'/34', 573'/364'	Kupchan and Bauerschmidt (1971)
B. solieri	574'/3683', 574'/68, 573'/64', 573'4'/36, 573'/364'	Labbe et al. (1986)
B. spartioides	54'/7, 353'4'/7, 354'/73', 353'/74'	Rossomando et al. (1990)
B. thymifolia	53'4'/678, 54'/6783', 5/6783'4'	Saad et al. (1987)
B. tola	Apig	San Martin et al. (1983)
B. triangularis	53'4'/37	Gianello and Giordano (1989a,b)
B. tricuneata var. *tricuneata*	54'/67, 3'4'/67	Wagner et al. (1978)
B. trimera	Apig, Lute, Q, 574'/6, 573'4'/6	Soicke and Leng-Peschlow (1987)
	53'/674'	Herz et al. (1977)

Table 8–1. (Continued)

Species	Compounds[1]	References
	5/673'4'	Rendon and Vila (1995)
B. trinervis	Apig, 57/4', 54'/7, 5/74', 57/3'4'	Arriaga-Giner et al. (1982)
B. tucumanensis	54'/678, 56/74'	Tonn et al. (1982)
B. ulicina	57/4', 54'/7, 5/74', 54'/67, 5/674', 574'/363', 574'/63'	Gianello and Giordano (1984)
B. umbelliformis	Apig, Apig-7-*O*-Glc, 57/4'	Reyes et al. (1990)
B. vaccinioides	Apig, 57/4', 54'/7, 5/74', 54'/67, 57/64', 574'/6, K, 574'/3, 354'/7, 357/4', 54'/37, 57/34', 35/74', 5/374', 3574'/6, 574'/36, 354'/67, 357/64', 54'/367, 57/364', 5/3674'	Wollenweber et al. (1986)

1 Substitution: Hydroxyl positions/Methoxyl positions; Apig = apigenin; Lute = luteolin; K = kaempferol; Q = quercetin.
2 Reported as *B. angustifolia*; synonymy from Nesom (1988).
3 Reported as *B. medullosa*; synonymy from Nesom (1988).

described, and extra oxygenation at *C*-6 is common. 6,8-Dioxygenation occurs fairly frequently, as well, but no compounds have been reported that have 8-substitution without 6-substitution. B-Ring deoxyflavonoids, including flavones, flavanones, dihydroflavonols, and a dihydrochalcone, have also been reported. Pinocembrin (5,7-dihydroxyflavanone) has been identified from *B. glutinosa* (Miyakado et al., 1976; Dominguez et al., 1986c), *B. viminea*, where it is accompanied by its 7-methyl ether (Bohlmann et al., 1984a), and in *B. bigelovii* where it occurs as the 7-benzoate (Arriaga-Giner et al., 1986). The exudate of *B. bigelovii* yielded several other compounds of interest. Thus, both 3,5,7,-trihydroxy-6-methoxyflavone and flavanone were identified as well as 2,5,7-trihydroxyflavanone (Arriaga-Giner et al., 1986). In another study of exudate chemistry, Wollenweber et al. (1989a) identified ß,2',4'-trihydroxy-6'-methoxychalcone from *B. salicifolia*. Pinobanksin (3,5,7-trihydroxyflavanone) was identified as its 3-acetate in *B. glutinosa* (Dominguez et al., 1986c). Table 8–1 details the occurrences of flavones and flavonols in the genus; Table 8–2 summarizes flavanones and dihydroflavonols.

Table 8–2. Flavanones and dihydroflavonols from species of *Baccharis*

Species	Compounds[1]	References
B. alateroides	Nar, Nar-4'-Me,	
	Nar-5,7-DiMe	Bohlmann et al. (1979b)
	DHK-3-Me	Bohlmann et al. (1982d)
B. bigelovii	357/6, 2857/-	Arriaga-Giner et al. (1986)
B. calvescens	Nar	Metwally (1995)
B. concinna	PC	Bohlmann et al. (1981f)
B. confertifolia	Nar-7-Me,	
	Nar-7,4'-DiMe,	
	Erio-7,3'-DiMe,	
	Erio-3',4'-DiMe,	
	Erio-7,3',4'-TriMe	Faini and Castillo (1990)
B. genistelloides	Erio	Kuroyanagi et al. (1985)
B. glutinosa	PC	Miyakado et al. (1976)
	PC, 357/-3-Ac	Dominguez et al. (1986c)
B. leptocephala	Nar-7-Me, Nar-4'-Me	Bohlmann et al. (1981f)
B. marginalis	Erio-7-Me	Faini et al. (1987)
B. oxydonta	PC, PB	Bohlmann et al. (1981f)
B. paniculata	Nar-7-Me, Erio-7-Me	Faini and Castillo (1990)
B. petiolata	DHQ	Faini and Castillo (1990)
	Nar-7,4'-DiMe,	
	Erio-7-Me	Gianello et al. (1990)
B. polifolia	Nar-4'-Me	Zdero et al. (1986a)
B. pylicoides	Nar-4'-Me,	
	Nar-7,4'-DiMe	Bohlmann et al. (1985a)
B. reticularia	Erio	Bohlmann et al. (1981h)
B. retusa	Nar-7-Me	Herz et al. (1977)
B. salicifolia	Nar-7-Me, 4'/73'	Kavka et al. (1973)
	Nar-7-Me, Nar-4'-Me,	
	Nar-7,4'-DiMe,	
	Erio, Erio-7,3'-DiMe	Zdero et al. (1986a)
B. salzmannii	Nar	Bohlmann et al. (1981h)
B. sternbergiana	Nar-7-Me,	
	Nar-7,4'-DiMe	Bohlmann et al. (1984e)
B. trinervis		
var. *rhexoides*	Nar-7-Me	Bohlmann et al. (1979b)
B. truncata	57/3'	Bohlmann et al. (1981f)
B. varians	Nar, Erio,	
	DHK-3-Ac, DHQ-3-Ac	Bohlmann et al. (1981h)
B. viminea	PC, PC-7-Me	Bohlmann et al. (1984a)
B. vulnearia[2]	Nar-7-Me	Bohlmann et al. (1981h)

1 PC = pinocembrin, PB = pinobanksin, Nar = naringenin, Erio = eriodictyol, DHK = dihydrokaempferol (aromadendrin), DHQ = dihydroquercetin (taxifolin), Ac = acetate, Me = methyl.

2 Reported as *B. serrulata*; synonymy from Nesom (1988).

The bulk of the work described above dealt with nonpolar flavonoids. We are aware of only three papers in which glycosylated forms of flavonoids were reported. Kaempferol and quercetin 3-*O*-glucosides were reported from *B. pingraea* [cited as *B. angustifolia*; see Nesom (1988) for synonymy] (Wagner et al., 1972b), apigenin 7-*O*-glucoside was reported from *B. umbelliformis* (Reyes et al., 1990) and quercetin-3-*O*-mono-, 3-*O*-di- and 3-*O*-triglycosides were reported from *B. thesioides* (Liu et al., 1993).

Recent morphological and taxonomic studies by Hellwig (1990, 1993) have suggested transferral of numerous species out of *Baccharis* and into segregate genera. Of the taxa cited here for flavonoid reports, *B. articulata*, *B. crispa*, *B. marginalis*, *B. oxydonta*, *B. salicifolia*, and *B. rhexioides* have been moved into the reconstructed genus *Pingraea*, and *B. bigelovii*, *B. paniculata*, *B. thesioides*, and *B. ulicina* have been transferred to the new genus *Neomolina*. Nesom (1994a), however, does not accept either of these generic segregates. For the purposes of this book, therefore, we leave these species in *Baccharis* until further studies clarify the generic limits.

Faini et al. (1999) studied the seasonal changes in chemical composition of the epicuticular wax fraction of *Baccharis linearis* leaves. Included in the analyses were wax esters, terpene derivatives, chromenes, and three flavonols, kaempferol 3-methyl ether, quercetin 3-methyl ether, and isorhamnetin 3-methyl ether. Kaempferol 3-methyl ether was observed only in autumn and winter collections, whereas all three compounds were present from those in summer and spring collections.

Bellis: This small genus of eight species (Bremer, 1994), or only seven according to Grau (1977), occurs in Europe and Asia. Flavonoid information seems to be limited to *B. perennis*; Wagner and Kirmayer (1957) reported the apigenin glycoside apiin, while Toki et al. (1991) described malonic acid conjugates of cyanidin 3-*O*-glycosides from a red-flowered form. More recently, Gudej and Nazaruk (1997) reported apigenin 7-*O*-(6"-(*E*)-caffeoyl)-ß-D-glucoside and 7-*O*-ß-D-methylglucuronate.

Callistephus: *Callistephus chinensis*, the sole member of this genus, is native to East Asia. Bogacheva et al. (1971) reported apigenin and apigenin 7-*O*-glucoside from its inflorescences.

Centipeda: *Centipeda*, which consists of four species and enjoys a distribution including parts of east and tropical Asia, Australia, and South America, presents an interesting situation. Heywood and Humphries (1977) included it in their list of genera in Anthemideae but marked it as problematic on the basis of pollen characteristics quoting Skvarla et al. (1977) who placed it in

Astereae. Bremer (1994) placed *Centipeda* in a list of genera not assigned to tribes.

Bohlmann and Mahanta (1979) reported 5,3'-dihydroxy-3,6,7,4',5'-pentamethoxy- and 5-hydroxy-3,6,7,3',4',5'-hexamethoxyflavone from *C. orbicularis*. Wu et al. (1991) reported apigenin and quercetin 3- and 3,3'-methyl ethers from *C. minima*. Bearing in mind the limited sampling, flavonoids do not provide much help in choosing between Astereae or Anthemideae. The combination of 6- and 3',4',5'-oxygenations is seen in Astereae and Anthemideae, *Baccharis*, *Gutierrezia*, and *Gymnosperma* in the former, and *Artemisia* (*Seriphidium*) and *Otanthus* in the latter. The level of *O*-methylation is more reminiscent of the former, however. Sørensen (1977, p. 407) pointed out another chemical feature that relates to the placement of *Centipeda*: the presence of an enepentayne derivative (a polyacetylene) in roots of two *Centipeda* species. The compound, common in the family, is uncharacteristic of acetylene derivatives normally found in Anthemideae.

Ceruana: *Ceruana* is native to Egypt and tropical Africa. The only species in the genus is *C. pratensis* from which El-Karemy et al. (1987) obtained kaempferol 3-*O*-rutinoside.

Chrysopsis: *Chrysopsis* consists of 10 species that occur in southeastern North America. Flavonoid information seems to be limited to an incomplete study of *C. hirtella* (cited as *Bradburia hirtella*) by Semple and Chinnappa (1984) who reported luteolin, quercetin, and "some of their substituted derivatives." Those authors also stated that "...the profiles of *Bradburia* were sufficiently different to make them readily separated from those of *Chrysopsis* species."

Chrysothamnus: *Chrysothamnus* is a genus of 14 (Grau, 1977), 15 (Bremer, 1994), or 12 species (Nesom and Baird, 1993) of shrubby plants native to western North America; many are highly glandular and sticky. The three species that have been studied produce a variety of flavonoids. The earliest work was that of Urbatsch et al. (1975) on *C. viscidiflorus*. These workers identified kaempferol and quercetin 3-methyl ethers, isorhamnetin (quercetin 3'-methyl ether), quercetin 3,3'-dimethyl ether, 6-hydroxykaempferol-3,6-dimethyl ether, quercetagetin-3,6-dimethyl ether, 6,3'-dimethyl ether, 3,6,3'-trimethyl ether, 3,5,6,3'-tetramethyl ether, and the comparatively rare 5,7,4'-trihydroxy-3,6,8,3'-tetramethoxyflavone. *Chrysothamnus pulchellus* accumulates apigenin, luteolin, quercetin, 6-hydroxykaempferol-3,6-dimethyl ether, and quercetagetin-3,6-dimethyl ether (Jakupovic et al., 1990c).

A recent paper by Stevens et al. (1999) treated the surface flavonoids of the three species of *Chrysothamnus* that occur in the dry parts of eastern Oregon, *C. humilis*, *C. nauseosus*, and *C. viscidifolius*. Twenty-six flavonoids were identified from leaf surface extracts of the three species based upon the flavones apigenin and isoscutellarein (8-hydroxyapigenin), the flavonols kaempferol, herbacetin (8-hydroxykaempferol), and quercetin, the flavanones naringenin and eriodictyol, and the dihydroflavonol taxifolin. The flavonoids of *C. nauseosus* were about equally divided between flavone and flavonol derivatives whereas the compounds from *C. viscidiflorus* were, with a single exception (luteolin from a single population), flavonols, flavanones, and the dihydroflavonol. Newly identified from *C. viscidiflorus* were several quercetin and eriodictyol methyl ethers and the set of taxifolin derivatives. The compounds identified from *C. humilis*, here described for the first time, were kaempferol 3-methyl ether, kaempferol 3,5-dimethyl ether, quercetin 3-methyl ether, quercetin 3,3'-dimethyl ether, and naringenin.

Conyza: Bremer (1994) lists *Conyza* as consisting of ca. 60 species, whereas Grau (1977) saw the number closer to 50. The genus occurs mainly in the tropics of both the New and Old Worlds (Nesom, 1990a). Flavonoid data have been gathered for about a dozen taxa, but this limited sampling has revealed a rich and interesting array of structural forms. Variants include simple flavone and flavonol *O*-glycosides, the uncommon quercetin 3'-*O*-glucoside and 3,3'-di-*O*-glucoside, isorhamnetin 3-*O*-rutinoside-4'-*O*-glucoside (El-Sayed et al., 1991) and a series of flavones that exhibit both extra oxygenation and a high degree of *O*-methylation. One of the latter, which has the rare combination of myricetin-type B-ring and extra oxygenation at C-8, is seen in 5,7-dihydroxy-3,8,3',4',5'-pentamethoxyflavone, which Tandon and Rastogi (1977) obtained from *C. stricta*. Herbacetin-3,8,4'-trimethyl ether (herbacetin is 8-hydroxykaempferol) was also reported, from *C. stricta* in the same paper, and more recently from *C. japonica* (Pandey et al., 1984). *Conyza bonariensis* (cited as *Erigeron bonariensis*; synonymy from Cronquist, 1943, and Nesom, 1989b) was shown by Imai and Mayama (1954) to accumulate apigenin, luteolin, and quercetin as aglycones and as common glycosides. A similar array of compounds was reported from *C. canadensis* (cited as *Erigeron canadensis*; synonymy from Cronquist, 1943, and Nesom, 1989b, 1990a) by Glinkowska and Strzelecka (1987). The compounds identified from species of *Conyza* are listed in Table 8–3.

Conyzanthus: Three or four species of *Conyzanthus* are said to occur in South America (Mabberley, 1987). This is an elusive genus. Oddly, it was described in the Flora of the former U.S.S.R. by Tamainschyan [1959; cited

Table 8–3. Flavonoids from species of *Conyza*

Species	Compounds[1]	References
C. aegyptica	Q-3-*O*-Rut	El-Karemy et al. (1987)
	Q-7-*O*-Ara	Ismail et al. (1979)
C. bonariensis	K-3-*O*-Glc, Q-3-*O*-Rut, Q-3,3'-Di-*O*-Glc, Lute-7-*O*-Glc, Scut-7-*O*-Gln, Hisp-7-Di-*O*-Gly	El-Karemy et al. (1987)
	Apig, 57/4', Lute, 574'/3', Q-3-*O*-Glc	Ferraro et al. (1988a)
C. dioscoridis	Q-7-*O*-Rhm, Q-7-*O*-Ara	Ismail et al. (1979)
	Q-4'-*O*-Glc, IR-3-*O*-Rhm, IR-3-*O*-Rut	Sayed et al. (1991)
C. filaginoides	Q-3-*O*-Glc, Q-3-*O*-Rut, Pinostrobin	Mata et al. (1997)
C. incana	54'/367, 54'/3673'	Mossa et al. (1988)
C. ivaefolia	Q	Bohlmann and Grenz (1972)
C. japonica	57/384'	Pandey et al. (1984)
C. linifolia	IR 3-*O*-Rut-4'-*O*-Glc, 54'/67 4'-*O*-Gln, Vitexin, Vicenin II, K-7-*O*-Glc	El-Sayed et al. (1991)
C. pallidiflora	Q, Erio	Zdero et al. (1990a)
C. stricta	57/384', 57/383'4'5'	Tandon and Rastogi (1977)
	56/384', 57/383'4'5'	Sen et al. (1976)
C. stricta var. *pinnatifida*	K-3-*O*-Rut,	
C. sp.	574'/83'	Urzua et al. (1991)

[1] Substitution: Hydroxyl positions/Methoxyl positions; Apig = apigenin; Acac = acacetin; Lute = luteolin; Chry = chrysoeriol; Scut = scutellarein; Hisp = hispidulin; Erio = eriodictyol; Ara = arabinose; Glc = glucose; Gln = glucuronic acid; Rhm = rhamnose; Rut = rutinose; Gly = glycoside.

by Zhang and Bremer (1993)] as being from South America (presumably as an adventive). Nesom (1994a) does not list the genus. Bremer (1994) does list it in his "*Erigeron-Conyza* group" (p. 393), but he does not treat it in the generic synopses. The only flavonoid information available for this genus appears to be the work of El-Karemy et al. (1987) who reported kaempferol 3-*O*-glucoside and 3-*O*-glucuronide from *C. squamatus* collected near Assiut, Egypt.

Dichrocephala: *Dichrocephala* consists of 13 species according to Grau (1977), four according to Bremer (1994), and 10 according to Nesom (1994a), with a range of distribution that includes southern and tropical Africa, Madagascar, south-eastern Asia, and Australia. A recent paper by Khan et al. (1993) described the isolation of kaempferol and quercetin 3-*O*-rutinosides from *D. bicolor*.

Diplostephium: *Diplostephium* consists of about 90 species distributed from Colombia and Venezuela to Bolivia and northern Chile with one species in Costa Rica. A recent report described the identification of naringenin, naringenin 4'-methyl ether (isosakuranetin), eriodictyol, and quercetin 3,3'-dimethyl ether from *D. cinereum* (Urzua et al., 1997).

Egletes: *Egletes* is a genus of 10 (Bremer, 1994; Nesom, 1994a) species that occurs from Texas south through Mexico and Central America to Ecuador, Colombia, and Brazil. Ternatin, 5,4'-dihydroxy-3,7,8,3'-tetramethoxyflavone, was reported from *E. viscosa* by two groups (Melo et al., 1993; Lima et al., 1996).

Ericameria: *Ericameria*, which consists of 27 species according to Bremer (1994) or 31 according to Nesom and Baird (1993), is a genus closely related to *Haplopappus*, or part of it according to Nesom and Baird (1993). Flavonoid information is available for three species. The pigment profiles of *E. diffusa* (Urbatsch et al., 1976) and *E. laricifolia* (Clark and Wollenweber, 1984) are comprised of methyl ethers of apigenin, kaempferol, and quercetin. The only noteworthy compounds reported were 6-methoxyluteolin in *E. laricifolia* (Clark and Wollenweber, 1984) and 5,7-dihydroxy-3-methoxyflavone (galangin 3-methyl ether) from *E. diffusa* (Urbatsch et al., 1976). *Ericameria nauseosa* (cited as *Chrysothamnous nauseosus*; transferred to *Ericameria* by Nesom and Baird, 1993) was shown by Bohlmann et al. (1979a) to have 5,7-dihydroxyflavanone (pinocembrin), 5-hydroxy-7-methoxyflavanone, and 5-hydroxy-6,7-dimethoxyflavanone.

Erigeron: *Erigeron* is a genus of about 200 species distributed mainly in North and South America and Eurasia. Despite the size of the genus, flavonoid data are available for only eight species among which are six from the Juan Fernandez Islands, Chile. Apigenin, luteolin, and quercetin occur as aglycones and as common *O*-glycosides in *E. annuus* (Imai and Mayama, 1953; Kaneta et al., 1978). Scutellarein (and its 7-*O*-glucuronide) were reported from *E. breviscapus* by Zhang et al. (1988). The Juan Fernandez Islands species have been examined for flavonoids by Valdebenito et al. (1992a): *E. fernandezianus*, *E. ingae*, *E. luteoviridis*, *E. rupicola*, *E. stuessyi* and *E. turricola*. Extra A-ring oxygenation, in the form of 6-methoxyluteolin and quercetagetin, not otherwise reported from the genus, was seen in these species. Apigenin, luteolin, and quercetin also were identified, as was a 6-*C*-glucoside of acacetin. All of these compounds occurred as 7-*O*-glucosides in these plants.

Euthamia: *Euthamia* is a genus of six (Nesom, 1994a) or eight (Bremer, 1994) species of North America. The only flavonoid report that applies to *Euthamia* was the finding of schaftoside and isoschaftoside, both *C*-glycosylflavones, from *E. graminifolia* (cited as *Solidago graminifolia* by Budzianowski, 1990). It is interesting to note that when listed with other *bona fide* species of *Solidago*, "*S. graminifolia*" was the only taxon that exhibited *C*-glycosylflavones; all other species had flavonols (and one anthocyanin).

Felicia: This genus consists of about 85 species distributed from southern Africa north to Nigeria and Ethiopia and in Saudi Arabia. The only information on flavonoids of *Felicia* appears to be the recent paper by Bloor (1999) in which he reports the presence of the anthocyanin delphinidin 3-*O*-neohesperidoside-7-*O*-(6-*O*-malonyl glucoside) and the *C*-glycosylflavone swertisin 2''-*O*-rhamnoside-4'-*O*-glucoside as floral components of *F. amelloides*, the blue marguerite daisy. Recombination of the purified pigments at pH 6 recreates the blue flower color.

Grangea: *Grangea* is a small genus of six species native to tropical Asia and Africa. Our knowledge of its flavonoids comes from the work of El-Karemy et al. (1987) who reported quercetin 3-*O*-gentiobioside and quercetin 3-*O*-gentiobioside-7-*O*-glucoside from *G. maderaspatana*.

Grindelia: *Grindelia* is a moderately large genus; some 55 (Bremer, 1994) to 60 (Grau, 1977) species occur in North America and in southern South America. Flavonoid data are available for seven species. The majority of compounds identified are flavone and flavonol methyl ethers; luteolin and quercetin were also reported, and dihydroquercetin 7,3'-dimethyl ether was described from *G. columbiana* [cited as *G. discoidea*; synonymy from Cronquist (1994)] (Timmermann et al., 1986). Several kaempferol and quercetin methyl ethers, in various combinations, were reported from *G. aegialitis*, *G. chiloensis*, and *G. pulchella* by Ruiz et al. (1981). A rich array of *O*-methylated flavones and flavonols based on apigenin, kaempferol, and quercetin, and their 6-oxygenated derivatives, was reported from *G. tarapacana* (Wollenweber et al., 1993a) and from *G. glutinosa* and *G. robusta* (Timmermann et al., 1994). Details can be found in Table 8–4.

Table 8–4. Flavonoids from species of *Grindelia*

Species	Compounds[1]	References
G. aegialitis	574'/3, 57/34'	Ruiz et al. (1981)
G. camporum	57/4', 54'/37, Q, 574'/33'	Timmermann et al. (1985)
G. chiloensis	Q	Ruiz et al. (1981)
G. columbiana[2]	574'/33', 54'/373', 354'/73'-Flavanone	Timmermann et al. (1986)
G. glutinosa	57/4', 57/64', 57/34', 5/374', 57/364', 5/3674'	Timmermann et al. (1994)
G. nuda var. *aphanactis*[3]	Lute, Q, 573'4'/3, 574'/33'	Torck et al. (1976) Pinkas et al. (1977)
G. pulchella	Q, 573'4'/3, 574'/33', 54'/373', 574'/3	Ruiz et al. (1981)
G. robusta	574'/3, 57/34', 574'/33', 54'/373', 57/33'4', 574'/36, 54'/367, 573'/364', 53'/3674'	Timmermann et al. (1994)
G. squarrosa	574'/3'-7-Gln	Wagner et al. (1972j)
	Lute, 574'/3, 54'/37, Q, 573'4'/3, 574'/33'	Pinkas et al. (1977, 1978)
G. tarapacana	574'/3, 54'/37, 54'/367, 57/364', 53'4'/37, 574'/33', 573'/34', 54'/373', 573'/364', 54'/3673'	Wollenweber et al. (1993a)

1 Substitution: Hydroxyl positions/Methoxyl positions; Lute = luteolin; K = kaempferol; Q = quercetin; Gln = glucuronic acid.
2 Cited as *G. discoidea*; synonymy after Cronquist (1994).
3 Cited as *G. aphanactis*; submerged as variety into *G. nuda* by Nesom (1990b).

Gutierrezia: *Gutierrezia* is a genus of 19 (Mabberley, 1987), 20 (Grau, 1977), or 27 species (Bremer, 1994) native to North America and South America south of the tropics. Flavonoid data are available for nine species. The group is characterized by a combination of extra oxygenated A-rings, unusual B-ring oxygenation patterns, and high levels of *O*-methylation. The biosynthetic capacity of some members of the genus is quite impressive: no fewer than 51 compounds were reported from *G. microcephala*, which involved

Table 8–5. *O*-Methylated flavonols from species of *Gutierrezia*

Parent Flavonol and Derivatives	Species	References
6-Hydroxykaempferol (35674')		
3,6-Dimethyl	*G. grandis*	Fang et al. (1985a)
	G. wrightii	Fang et al. (1986d)
	G. microcephala	Fang et al. (1986b)
8-Hydroxykaempferol (35784')		
3-Methyl	*G. microcephala*	Roitman and James (1985)
3,8-Dimethyl	*G. microcephala*	Roitman and James (1985), Fang et al. (1986b)
6,8-Dihydroxykaempferol (356784')		
3,6-Dimethyl	*G. microcephala*	Fang et al. (1986b)
3,6,8-Trimethyl	*G. microcephala*	Roitman and James (1985) Fang et al. (1986b)
	G. grandis	Fang et al. (1985a)
	G. wrightii	Fang et al. (1986d)
3,6,7,8-Tetramethyl	*G. microcephala*	Fang et al. (1986b)
3,6,8,4'-Tetramethyl	*G. wrightii*	Fang et al. (1986d)
Quercetin (3573'4')		
3-Methyl	*G. microcephala*	Roitman and James (1985), Fang et al. (1986b)
	G. sarothrae	Hradetzky et al. (1987)
3-Methyl and 7-*O*-Glucoside	*G. texana*	Li et al. (1987)
3'-Methyl 3-*O*-Galactoside	*G. microcephala*	Fang et al. (1986b)
3'-Methyl 3-*O*-Glucoside	*G. wrightii*	Fang et al. (1986b)
3,3'-Dimethyl	*G. microcephala*	Dong et al. (1987),
	G. sarothrae	Hradetzky et al. (1987)
3,4'-Dimethyl	*G. texana*	Li et al. (1987)
3,7-Dimethyl	*G. microcephala*	Dong et al. (1987)
	G. alamanii var. *megalocephala*	Lenherr et al. (1986)
3,7,4'-Trimethyl	*G. alamanii* var. *megalocephala*	Lenherr et al. (1986)
Myricetin (3573'4'5')		
3,7,4'-Trimethyl	*G. alamanii* var. *megalocephala*	Lenherr et al. (1986)
3,3',4',5'-Tetramethyl	*G. alamanii* var. *megalocephala*	Lenherr et al. (1986)
Gossypetin (35783'4')		
3-Methyl	*G. microcephala*	Roitman and James (1985)
3,3'-Dimethyl	*G. microcephala*	Roitman and James (1985)

Table 8–5. (Continued)

Parent Flavonol and Derivatives	Species	References
3,8-Dimethyl	*G. microcephala*	Roitman and James (1985)
	G. grandis	Fang et al. (1985a)
3,8,3'-Trimethyl	*G. grandis*	Fang et al. (1985a)
	G. microcephala	Roitman and James (1985)
Quercetagetin (35673'4')		
3-Methyl	*G. texana*	Li et al. (1987)
6-Methyl	*G. wrightii*	Fang et al. (1986d)
	G. sphaerocephala	Li et al. (1988)
3,6-Dimethyl	*G. grandis*	Fang et al. (1985a)
	G. microcephala	Fang et al. (1986a)
6,3'-Dimethyl	*G. wrightii*	Fang et al. (1986a)
	G. sarothrae	Hradetzky et al. (1987)
	G. sphaerocephala	Li et al. (1988)
3,6,3'-Trimethyl	*G. microcephala*	Roitman and James (1985), Fang et al. (1986b)
	G. sarothrae	Hradetzky et al. (1987)
3,6,4'-Trimethyl	*G. grandis*	Fang et al. (1985a)
3,6,7,4'-Tetramethyl	*G. microcephala*	Fang et al. (1986b)
6-Hydroxymyricetin (35673'4'5')		
3,6-Dimethyl	*G. microcephala*	Fang et al. (1986b)
3,6,4'-Trimethyl	*G. grandis*	Fang et al. (1985a)
3,6,5'-Trimethyl	*G. microcephala*	Fang et al. (1986b)
3,6,4',5'-Tetramethyl	*G. microcephala*	Fang et al. (1986b)
8-Hydroxymyricetin (35783'4'5')		
3,8,4'-Trimethyl	*G. microcephala*	Fang et al. (1986b)
3,8,3',4',5'-Pentamethyl	*G. microcephala*	Roitman and James (1985)
6,8-Dihydroxyquercetin (356783'4')		
3,6-Dimethyl	*G. microcephala*	Roitman and James (1985)
6,8-Dimethyl	*G. microcephala*	Roitman and James (1985), Fang et al. (1986b)
3,6,8-Trimethyl	*G. grandis*	Fang et al. (1985a)
	G. microcephala	Roitman and James (1985)
6,7,8-Trimethyl	*G. microcephala*	Fang et al. (1986b)
6,8,3'-Trimethyl	*G. microcephala*	Roitman and James (1985) Fang et al. (1986b)
3,6,7,8-Tetramethyl	*G. resinosa*	Bittner et al. (1983)
	G. microcephala	Fang et al. (1986b)
3,6,8,3'-Tetramethyl	*G. wrightii*	Fang et al. (1986a)
	G. sarothrae	Hradetzky et al. (1987)
	G. microcephala	Roitman and James (1985),

Table 8–5. (Continued)

Parent Flavonol and Derivatives	Species	References
		Fang et al. (1986b)
3,6,8,4'-Tetramethyl	*G. grandis*	Fang et al. (1985a)
	G. microcephala	Fang et al. (1986b)
	G. sarothrae	Hradetzky et al. (1987)
3,6,7,8,3'-Pentamethyl	*G. resinosa*	Bittner et al. (1983)
	G. microcephala	Roitman and James (1985)
	G. sarothrae	Hradetzky et al. (1987)
3,6,7,8,4'-Pentamethyl	*G. microcephala*	Fang et al. (1986b)
	G. sarothrae	Hradetzky et al. (1987)
6,8-Dihydroxymyricetin (356783'4'5')		
3,6,8-Trimethyl	*G. grandis*	Fang et al. (1985a)
	G. microcephala	Fang et al. (1986b)
3,8,4'-Trimethyl	*G. grandis*	Fang et al. (1985a)
3,6,8,4'-Tetramethyl	*G. microcephala*	Fang et al. (1986b)
	G. grandis	Fang et al. (1985a)
3,6,8,5'-Tetramethyl	*G. microcephala*	Fang et al. (1986b)
3,7,8,4'-Tetramethyl	*G. microcephala*	Fang et al. (1986b)
3,6,7,8,4'-Pentamethyl	*G. grandis*	Fang et al. (1985a)
	G. microcephala	Fang et al. (1986b)
3,6,8,3',5'-Pentamethyl	*G. microcephala*	Roitman and James (1985)
	G. grandis	Fang et al. (1985a)
3,6,8,4',5'-Pentamethyl	*G. microcephala*	Roitman and James (1985), Fang et al. (1986b)
3,6,7,8,4',5'-Hexamethyl	*G. microcephala*	Fang et al. (1986b)
3,6,8,3',4',5'-Hexamethyl	*G. microcephala*	Roitman and James (1985), Fang et al. (1986b)
	G. grandis	Fang et al. (1985a)
3,6,7,8,3',4',5'-Heptamethyl	*G. microcephala*	Fang et al. (1986b)
3,5,7,2',4',5'-Hexahydroxyflavone (3572'4'5')		
3,4'-Dimethyl and 5'-Acetate	*G. texana*	Li et al. (1987)
3,5,6,7,2',4',5'-Heptahydroxyflavone (35672'4'5')		
3,6,4'-Trimethyl	*G. grandis*	Fang et al. (1985a)
	G. texana	Li et al. (1987)
3,6,5'-Trimethyl	*G. grandis*	Fang et al. (1985a)
3,6,4',5'-Tetramethyl	*G. microcephala*	Fang et al. (1986b)
3,5,7,8,2',4',5'-Heptahydroxyflavone (35782'4'5')		
3,8,4'-Trimethyl	*G. microcephala*	Fang et al. (1986b)
3,8,5'-Trimethyl	*G. microcephala*	Fang et al. (1985b)
3,5,6,7,8,2',4',5'-Octahydroxyflavone (356782'4'5')		
3,6,8,4'-Tetramethyl	*G. microcephala*	Fang et al. (1985b, 1986b)

Table 8–5. (Continued)

Parent Flavonol and Derivatives	Species	References
	G. grandis	Fang et al. (1985a)
3,6,8,5'-Tetramethyl	*G. microcephala*	Roitman and James (1985), Fang et al. (1985b, 1986b)
	G. grandis	Fang et al. (1985a)
3,6,8,4',5'-Pentamethyl	*G. microcephala*	Fang et al. (1985b, 1986b)
3,6,7,8,4',5'-Hexamethyl	*G. microcephala*	Fang et al. (1985b, 1986b)

Table 8–6. *O*-Methylated flavones and flavanones from species of *Gutierrezia*

Species	Compounds[1]	References
G. grandis	573'4'/68	Fang et al. (1985a)
G. microcephala	573'4'/68, 572'/5'/84', 572'5'/684', 52'5'/6784'	Fang et al. (1986b)
G. sarothrae	574'/3', 573'/4', 573'4'/6, 574'/683', 53'4'/7[2]	Hradetzky et al. (1987)
G. sphaerocephala	573'4'/6, 574'/63', 57/64', 57/63'4', 574'/63'5', 53'4'5'/67, 573'4'/6[2], 574'/63[2]	Li et al. (1988)
G. wrightii	574'/3'	Fang et al. (1986d)

[1] Substitution: Hydroxyl positions/Methoxyl positions.
[2] Flavanone.

no fewer than 16 different substitution patterns among the flavonols, eight different patterns within the flavones and two within the flavanones. Among the substitution patterns are the unusual 2',4',5'-system. The overall complexity of the flavonoid profile of the genus arises from the combination of oxygenation pattern and *O*-methylation, the nature of which can be appreciated by referring to Table 8–5, which gives details on the occurrence of flavonols in the genus and Table 8–6, which summarizes the occurrence of flavones and flavanones. Responsible for documenting this impressive array of compounds is Fang and his coworkers (Fang et al., 1985a,b, 1986b,c,d).

Gymnosperma: This genus consists of the single species *G. glutinosum*, which occurs from Texas to central South America. *Gymnosperma* presents an extremely complex array of flavonols, plus one flavone, all of which have extensive extra oxygenation and high levels of *O*-methylation. The first paper describing flavonoids of this species, by Dominguez and Torre (1974), presented evidence for two highly unusual compounds, 5,7-dihydroxy-6,8,3',4',5'-pentamethoxyflavone (called luiselizondin), and 5-hydroxy-3,6,7,8,3'-pentamethoxyflavone **[4–16]**. The first was unusual because of the combination of A-ring oxygenation at both C-6 and C-8 coupled with a myricetin type B-ring. The second compound is novel by virtue of the absence of any substituent at C-4'. This rare B-ring substitution pattern was confirmed by isolation of 3-methoxybenzoic acid **[4–17]** from a micro alkaline fusion of the pentamethoxy flavonoid.

The second paper (Yu et al., 1988) described 21 compounds including luteolin, its 7-*O*-glucoside, and an apigenin-8-*C*-glycoside. The other 18 compounds exhibited extra oxygenation and high levels of *O*-methylation. Two unusual B-ring oxygenation patterns characterize many of the compounds: 3',4',5'-trioxygenation (the myricetin pattern); and 2',4',5'-trioxygenation (the isoëtin pattern). In many instances these patterns are combined with extra oxygenation at C-6 and/or C-8 to provide some of the most highly oxygenated and *O*-methylated flavonoids known from the family. The oxygenation and *O*-methylation patterns reported from *Gymnosperma* are shown in Table 8–7.

Haplopappus: According to Bremer (1994) and Nesom (1994a), *Haplopappus* consists of about 70 species, whereas Grau (1977) saw a larger group, some 160 species, that encompasses *Ericameria* and *Hazardia*. We

Table 8–7. Levels of oxygenation and *O*-methylation of flavonoids from *Gymnosperma glutinosum* (Yu et al., 1988)

Oxygenation Pattern	*O*-Methylation Patterns
356782'4'5'	3682'; 3685'; 36782'; 3682'5'; 3684'5'; 3682'4'5'
356783'4'5'	368; 3684'; 3684'5'; 36784'5'
356783'4'	68; 368; 3683'; 6783'
35672'4'5'	362'
3572'4'5'	32'
35673'4'5'	364'
356784'	368
356783'	36783'
56783'4'5'	683'4'5'

include the latter two in this chapter as genera in their own right (Urbatsch, 1978; Clark, 1979; Bremer, 1994). Nonetheless, *Haplopappus* is a moderately large genus well known in the floras of North and South America. [In the narrow sense, Nesom and Morgan (1990) see *Haplopappus* as strictly South American.] In the broader context, the genus is highly diverse in life forms and also has an impressive array of chromosome numbers, with n = 2–10, 12, 18, and 45 (Grau, 1977).

Flavonoid data are available for 16 species. Although flavones have been reported from a few, the majority of species accumulate flavonol derivatives. Anthocyanins based on cyanidin have been reported from *H. gracilis* (Constabel et al., 1971; Stickland and Sunderland, 1972), *C*-glycosylflavones from *H. scrobiculatus* and *H. sericeus* (Ates et al., 1982), and flavanones from two species. Zdero et al. (1991d) reported three flavanone derivatives from *H. remyanus*, eriodictyol and two that lack B-ring oxygenation, namely, 5-hydroxy-7-methoxyflavanone and 3,5-dihydroxy-7-methoxyflavanone 3-acetate. Another dihydroflavonol, 3,5,4'-trihydroxy-7,3'-dimethoxy 3-acetate, was identified by Marambio and Silva (1989) from *H. taeda*. The 7,4'-dimethyl ethers of naringenin and dihydrokaempferol and the 7,3'-dimethyl ethers of eriodictyol and dihydroquercetin, along with some common *O*-methylated flavonols have been identified from *H. hirtellus* (Maldonado et al., 1993).

Flavones have been described from two species. *Haplopappus canescens* accumulates scutellarein 6-*O*-glucoside, scutellarein 6-methyl ether, and luteolin 3'- and 4'-methyl ethers, chrysoeriol and diosmetin, respectively. Ulubelen and Mabry (1981a) identified apigenin, luteolin, and scutellarein 6-*O*-glucoside as components of *H. rengifoanus*, along with several flavonols.

Flavonols from *Haplopappus* species include a variety of compounds based upon kaempferol, quercetin, myricetin, and quercetagetin. Evidently, 8-oxygenation does not occur in this genus. A modest level of *O*-methylation can be seen in several species with the extremes being quercetagetin 3,6,4'-trimethyl ether from *H. canescens* (Öksüz et al., 1981) and myricetin-3,3',4'-trimethyl ether from *H. integerrimus* var. *punctatus*. Full details of flavonols and their occurrence in the genus are given in Table 8–8.

Hazardia: *Hazardia* consists of 13 species that range in occurrence from southern Oregon to northern Baja California (Clark, 1979). Until Clark's study, *Hazardia* was considered as a section within the large genus *Haplopappus*. Although agreeing that *Hazardia* is distinct, Nesom and Morgan (1990) suggest that it might be congeneric with *Haplopappus* sect. *Polyphylla* of South America. The flavonoid chemistry of all 13 species (15 taxa) was the subject of a study by Clark and Mabry (1978). Variously

Table 8–8. Flavonoids of species of *Haplopappus*

Species	Compounds[1]	References
H. baylahuen	K, Q, Q3Me	Hörhammer et al. (1973)
H. bezanillanus	54'/3673'	Maldonado et al. (1993)
H. canescens	Scut-6-*O*-Glc, Scut6Me, Chry, Dios, M, 573'/364'	Öksüz et al. (1981)
H. foliosus	K and Q 3-*O*-Glys, IR-3-*O*-Glc, K3Me, Q3Me, 353'4'/67	Ulubelen et al. (1982)
H. hirtellus	35/74', 5/374', 54'/373'	Maldonado et al. (1993)
H. integerrimus var. *punctatus*	Q-3-*O*-Glc, Q3Me, IR, 53'4'/37, 574'/33', 354'/73', 3575'/3'4', 575'/33'4'	Ayanoglu et al. (1981)
H. multifolius	Q, Q-3-Me	Nuñez-Alarcón and Quiñones (1995)
H. remyanus	Pinostrobin, Pinob-7-Me-3-Ac, Erio	Zdero et al. (1991d)
H. rengifoanus	Apig, Lute, Scut-6-*O*-Glc, IR-3-*O*-Glys, 5673'4'/3, 563'4'/37	Ulubelen and Mabry (1981a)
H. scrobiculatus	Q and IR 3-*O*-Glc, K3Me, 3573'4'/6, 353'4'/67, Vitexin, Isovitexin, Vicenin-2, Isoschaftoside	Ates et al. (1982)
H. sericeus	Q-3-*O*-Gln, Vitexin, Isovitexin, Vicenin-2	Ates et al. (1982)
H. taeda	Tax-73'-DiMe-3-*O*-Ac	Marambio and Silva (1989)
H. velutinus	K3Me, K37DiMe	Urzua et al. (1991), Urzua and Mendoza (1993)

[1] Substitution: Hydroxyl positions/Methoxyl positions; Apig = apigenin; Lute = luteolin; Chry = chrysoeriol; Dios = diosmetin; Scut = scutelarein; K= kaempferol; Q = quercetin; IR = isorhamnetin; M = myricetin; Tax = taxifolin (dihydroquercetin); Pinob = pinobanksin; Erio = eriodictyol; Cyan = cyanidin; Me = methyl; Ac = acetate; Glc = glucose; Gal = galactose; Rut = rutinose.

Table 8–9. Leaf exudate flavonoids from *Hazardia squarrosa* var. *grindelioides*[1]

- Flavanones:
  - Eriodictyol
    - 7-methyl
    - 7,3'-dimethyl
    - 5,4',5'-Trihydroxy-7,3'-dimethoxyflavanone
- Flavones:
  - Apigenin
  - Scutellarein-6-methyl ether
    - 6,4'-dimethyl ether
  - Luteolin
    - 7-methyl ether
    - 3'-methyl ether
  - 6-Methoxyluteolin
- Flavonols:
  - Kaempferol
    - 7-methyl ether
    - 4'-methyl ether
    - 3,4'-dimethyl ether
    - 7,4'-dimethyl ether
    - 3,7,4'-trimethyl ether
- 6-Hydroxykaempferol 3,6,4'-trimethyl ether
  - Quercetin
    - 3-methyl ether
    - 3'-methyl ether
    - 7-methyl ether
    - 3,7-dimethyl ether
    - 3,3'-dimethyl ether
    - 7,3'-dimethyl ether
    - 3,7,3'-trimethyl ether
    - 3,3',4'-trimethyl ether
    - 7,3',4'-trimethyl ether
    - 3,7,3',4'-tetramethyl ether
- Quercetagetin 6-methyl ether (patuletin)
    - 3,6-dimethylether
    - 6,3'-dimethylether
    - 3,6,4'-trimethylether

[1] From Clark and Wollenweber (1985).

distributed in the genus are *O*-glycosides of apigenin, luteolin, kaempferol, quercetin, and isorhamnetin. Also identified from many species were *C*-glycosylflavones based on both apigenin and luteolin. Details of the distribution can be found in Chapter 22. In addition, Clark and Mabry (1978) observed four flavonol methyl ethers in several of the species but did not provide structural data on them. In a later work, Clark and Wollenweber (1985) described 33 flavonoid aglycones from the external stem and leaf resin of *H. squarrosa* var. *grindelioides*. These have been listed in Table 8–9.

Heteromma: This genus of three species is native to South Africa. A paper by Bohlmann and Fritz (1979) reported three flavonols from *H. simplicifolium*: 3,5-dihydroxy-7,8-3',4'-tetramethoxyflavone, 3,5-dihydroxy-7,8,3',4',5'-pentamethoxyflavone, and 3,5,4'-trihydroxy-7,8,3'-tri-methoxyflavone. A re-examination of the data led Horie et al. (1988) to correct these structures to, respectively, 5,7-dihydroxy-3,8,3',4'-tetramethoxyflavone, 5,7-dihydroxy-3,8,3',4',5'-pentamethoxyflavone, and 5,7,4'-trihydroxy-3,8,3'-trimethoxyflavone. The noteworthy feature of these compounds is extra oxygenation at C-8 (not C-6) and the presence of a moderately high level of *O*-methylation.

Heteropappus: *Heteropappus* is an East Asian genus with species numbers varying from three in Grau's view (1977) to about 20 (Bremer, 1994). Bohlmann et al. (1985e) reported 5-hydroxy-6,7,8,3',4'-pentamethoxyflavone (5-desmethylnobiletin) from *H. altaicus*. Those authors pointed out that this taxon was once recognized as an aster, *Aster altaicus* Willd. (e.g., Kitamura, 1937). A recent note by Bader et al. (1993) described kaempferol, quercetin, and isorhamnetin 3-*O*-rutinosides from *H. altaicus* and *H. biennis*.

Heterothalamus: *Heterothalamus* consists of eight South American species. Our only flavonoid information comes from a recent paper by Kerber et al. (1993) who reported 5-hydroxy-7-methoxyflavanone (pinostrobin) and 3,5,7-trihydroxyflavone (galangin) from *H. psiadioides*.

Heterotheca: This genus consists of about 15 (Grau, 1977), 20 (Mabberley, 1987), 25 (Bremer, 1994; Nesom, 1994a) or 28 (Semple, 1996) species ranging from southern North America to Mexico. Three species have been examined in detail for leaf exudate flavonoids, *H. grandiflora* and *H. psammophila* by Wollenweber et al. (1985b), and *H. inuloides* by Jerga et al. (1990). Table 8–10 summarizes these observations. Note the variation in the patterns of the first two species.

Table 8–10. Flavonoids from three species of *Heterotheca*[1]

Compound[2]	Species GRA	PSA	INU
Kaempferol:	+/–	–	+
3-Me	+/–	+/–	–
7-Me	+/–	–	–
4'-Me	+/–	–	–
3,7-DiMe	+	+/–	+
3,4'-DiMe	+/–	+/–	–
3,7,4'-TriMe	–	+/–	+
6-OH-Kaempferol:			
6-Me	+/–	–	–
3,6-DiMe	+/–	+/–	–
6,7-DiMe	+	+	–
3,6,7-TriMe	+	+	+
3,6,4'-TriMe	+	–	–
6,7,4'-TriMe	–	+/–	–
3,6,7,4'-TetraMe	+/–	+/–	–
Quercetin:	+/–	–	–
3-Me	+/–	–	–
3,3'-DiMe	–	–	+
3,4'-DiMe	–	–	+
3,7-DiMe	–	–	+
7,3'-DiMe	+/–	–	–
3,7,3'-TriMe	+	–	+
3,7,4'-TriMe	–	–	+
3,7,3',4'-TetraMe	–	–	+
Quercetagetin:			
3,6-DiMe	+/–	–	–
6,3'-DiMe	+/–	–	–
3,6,7-TriMe	+	+/–	–
6,7,4'-TriMe	+/–	–	–
3,6,7,4'-TetraMe	–	+/–	+
Scutellarein:			
6,7-DiMe	+	–	–
6,4'-DiMe	–	+/–	–
6,7,4'-TriMe	–	+/–	–
Luteolin:			
7,3'-DiMe	+/–	–	–

Table 8–10. (Continued)

Compound[2]	Species GRA	PSA	INU
6-OH-Luteolin:			
6,7,3'-TriMe	+/–	–	–
Eriodictyol:			
7-Me	+/–	–	–
3'-Me	–	–	+
7,3'-DiMe	–	–	+

[1] *H. grandiflora* (GRA) and *H. psammophila* (PSA) data from Wollenweber et al. (1985b); *H. inuloides* (INU) data from Jerga et al. (1990).

[2] Me = methyl; +/– = present in one population; + = present in both; only one population of *H. inuloides* studied.

Isocoma: Bremer (1994) and Nesom (1991, 1994a) consider that *Isocoma* consists of 16 species. Formerly considered part of *Haplopappus*, members of *Isocoma* occur throughout southwestern United States and adjacent Mexico (Clark et al., 1987). While Dominguez et al. (1986b) reported naringenin 7-methyl ether (sakuranetin) from *I. coronopifolia*, the bulk of our information comes from a survey of five taxa done by Clark et al. (1987). They found a variety of *O*-methylated compounds based on apigenin, 6-hydroxyapigenin (scutellarein), luteolin, kaempferol, 6-hydroxykaempferol, and quercetin. Seasonal variation in exudate flavonoids of *I. acradenia* has been studied by Clark and Clark (1990). The data are summarized in Table 8–11.

Kalimeris: Bremer (1994) lists *Kalimeris* as a genus of 10 species native to Central, East, and Southeast Asia, but Nesom (1994a) and Gu and Hoch (1997) list only eight. *Kalimeris* is closely allied to Asian and may be at least biphyletic (Ito et al., 1998). *Kalimeris pinnatifida* accumulates apigenin 7-*O*-glucoside and isorhamnetin 3-*O*-galactoside (Kaneta et al., 1978).

Lepidophyllum: *Lepidophyllum* consists of a single species native to Patagonia. Marini-Bettolo et al. (1957) identified 5,4'-dihydroxy-3,6,7,3'-tetramethoxyflavone from *L. quadrangulare*.

Machaeranthera: *Machaeranthera* is a North American genus with 36 species. This genus was "reconstituted" out of *Aster* by Cronquist and Keck (1957), a taxonomic perspective that has been followed by nearly all sub-

Table 8–11. Simple and *O*-methylated flavonoids from species of *Isocoma*[1]

Species	Compounds[2]
I. acradenia	Apig, Lute, 53'4'/7, 574'/3, 54'/37, 573'4'/3, 574'/33', 53'/374'
I. acradenia var. *eremophila*[3]	Apig, Genk, Chry, K, 574'/3, 54'/37, 5/374', 573'4'/3, 574'/33'
I. drummondii	Apig, Acac, 5/74', Lute, Chry, Dios, 54'/73', K, 354'/7, 357/4', 54'/37, 35/74', 5/374', Q, IR, 353'4'/7, 53'4'/37, 574'/33', 354'/73', 54'/373', 53'/374'
I. tenuisecta	Apig, Lute, Chry, 574'/3, 54'/37, 53'/374', 54'/367, 5/3674', 573'4'/3, 574'/33'
I. veneta	Apig, Genk, Acac, 5/74', 574'/6, Chry, 574'/3, 54'/37, 5/374', 573'4'/3, 53'4'/37, 574'/33', 53'/374', 5/373'4'

1 From Clark et al. (1987).
2 Substitution: Hydroxyl positions/Methoxyl positions; Apig = apigenin; Acac = acacetin; Genk = genkwanin; Lute = luteolin; Chry = chrysoeriol; Dios = diosmetin; K = kaempferol; Q = quercetin; IR = isorhamnetin.
3 Reported as *I. eremophila*; varietal status after Nesom (1991).

sequent workers (e.g., Jones and Young, 1983; Hartman, 1990; Morgan and Simpson, 1992; Cronquist, 1994; Nesom, 1994a; Morgan, 1997). Nesom (1994a) has gone further and created a new subtribe, Machaerantherinae, to include the genus plus additional presumptive relatives, such as *Haplopappus*. Cyanidin derivatives were reported from *M. gracilis* (cited as *Haplopappus gracilis*) by Constabel et al. (1971) and Stickland and Sunderland (1972). 5-Hydroxy-6,7,3',4'-tetramethoxyflavone has been reported from *M. tanacetifolia* (cited as *Aster tanacetifolius*) by Dominguez et al. (1986a).

Microglossa: *Microglossa*, with ca. 10 (Bremer, 1994) to 18 species (Nesom, 1994a), occurs from southern to tropical Africa and in eastern Asia. Flavonoid information is available for two species, *M. zeylandica*, from which 5,4'-dihydroxy-6,7,8,3'-tetramethoxyflavone was obtained (Gunatilaka et al., 1987), and *M. pyrifolia*, from which mono- and diacetylated derivatives of the aurone maritimein were isolated (Rücker et al., 1994).

Nidorella: *Nidorella* ranges from southern Africa to the Middle East and consists of 11 (Grau, 1977) to about 15 (Bremer, 1994; Nesom, 1994a) species. We have information on flavonoids from only a single one, however. Bohlmann et al. (1982g) reported 5,4'-dihydroxy-3,7,3'-trimethoxyflavone from *N. hottentotica*.

Olearia: This large genus of about 100 (Nesom, 1994a) to 130 species (Bremer, 1994) occurs mainly in Australia with about 30 known from New Zealand and a few from New Guinea and Lord Howe Island. Flavonoid data are available for five species, sufficient to show us that *O*-methylated flavones and flavonols, plus one flavanone, accumulate. Extra oxygenation can be seen at either C-6 or C-8; in one compound, 5,4'-dihydroxy-6,7,8-trimethoxyflavone (xanthomicrol) from *O. heterocarpa*, both positions have been oxygenated (Pinhey et al., 1972). Naringenin 7-methyl ether (sakuranetin) was isolated from *O. brachyphylla* (Warning et al., 1988). The rest of the compounds identified are flavones (or flavonols): 5,7,4'-trihydroxy-8,3'-dimethoxyflavone from *O. axillaria* (Warning et al., 1988); quercetagetin-3,6-dimethyl and 3,6,4'-trimethyl ethers from *O. muelleri* (Jefferies et al., 1974); and 5,7-dihydroxy-6,4'-dimethoxyflavone from *O. paniculata* (Chivers et al., 1966).

Parastrephia: *Parastrephia* is a genus of three (Nesom, 1993, 1994a) or five (Marticorena and Quezada, 1985; Bremer, 1994) species that occur in southern South America. Marini-Bettolo et al. (1957) reported 5,4'-dihydroxy-3,6,7,3'-tetramethoxyflavone from *P. quadrangularis* (cited as *Lepidophyllum quadrangulare*; synonymy according to Cabrera, 1954, and Nesom, 1993). Loyola et al. (1985) reported 5,7-dihydroxy-3,8,3',4'-tetramethoxyflavone from *P. quadrangularis*, and Nieto et al. (1994) reported apigenin 7-methyl ether (genkwanin), apigenin 7,4'-dimethyl ether, hispidulin (6-methoxyapigenin), and 5-hydroxy-6,7,8,4'-tetramethoxyflavone (gardenin B) from *P. lucida*. A recent article by Loyola et al. (1995) reported 5,3',4'-trihydroxy-7-methoxy- and 5,4'-dihydroxy-7,3'-dimethoxyflavanones along with 5,4'-dihydroxy-7-methoxyflavone (genkwanin), 5,4'-dihydroxy-7,8-dimethoxyflavone, and 5,7,4'-trihydroxy-3,8,3'-trimethoxyflavone from *P. lucida*, *P. phylicaeformis*, and *P. teretiuscula*.

Prionopsis: *Prionopsis* consists of the single species, *P. ciliata*, which occurs in central and southwestern United States. Originally described as a distinct genus by Nuttall (1841), it was treated by most workers in *Haplopappus* until just recently resurrected (Correll and Johnston, 1970; Howe, 1975; Bremer, 1994; Lane and Hartman, 1996). A paper by Bittner and Watson (1982) reported quercetin, quercetin 3-methyl ether, quercetin 3'-methyl ether (isorhamnetin), and 5,7,4'-trihydroxy-3,6-dimethoxyflavone from *P. ciliata* (cited as *Haplopappus ciliatus*).

Table 8–12. *O*-Methylated flavonoids from species of *Psiadia*

Species	Compounds[1]	References
P. arabica	573'/2'4'5'	Mossa et al. (1992)
	53'/672'4'5';	Al-Yahya et al. (1987)
	53'/672'4'5'; 53'/72'4'5';	
	5/72'3'4'5'	El-Feraly et al. (1990)
P. punctata	5/6784'; 5/3674'; 54'/673';	
	53'/674'5'; 53'/6784'5';	
	5/673'4'5'	Abou-Zaid et al. (1991)
P. trinerva	54'/37; 54'/367; 53'4'/37;	
	574'/33'; 53'/374'; 567/33'4';	
	53'/3674'; 5/3673'4';	
	54'/3783'; 57/383'4'	Wang et al. (1989b)
	573'4'/3; 574'/38; 574'/3	Wang and Hostettmann (1990)

[1] Substitution: Hydroxyl positions/Methoxyl positions.

Psiadia: *Psiadia* is a genus of about 60 species found in southern to tropical Africa, Madagascar and in tropical Asia. Four species have been studied for their flavonoid chemistry. In the earliest study apigenin, aromadendrin (dihydrokaempferol), and aromadendrin 7-methyl ether were described from *P. altissima* (Canonica et al., 1969). Apigenin, acacetin, luteolin, and chrysoeriol were isolated from *P. punctulata*, along with the *C*-glycosylflavones orientin and isoörientin and several highly *O*-methylated flavones (Abou-Zaid et al., 1991). *Psiadia arabica* yielded two flavones that had 2',3',4',5'-tetraoxygenation (Al-Yahya et al., 1987; Mossa et al., 1992). A rich array of *O*-methylated compounds based on kaempferol, 6-hydroxykaempferol, quercetin, quercetagetin, and gossypetin has been reported by Wang (1989a) from *P. trinerva* (Wang et al., 1989a; Wang and Hostettmann, 1990). The *O*-methylated flavonoids from *Psiadia* are listed in Table 8–12. Ibrahim et al. (1997) observed the glucosylation of 5,3'-dihydroxy-7,2',4',5'-tetramethoxyflavone and 5,3'-dihydroxy-6,7,2',4',5'-pentamethoxyflavone at postion-3' by cultures of *Cunninghamella elegans*. Wang et al. (1989b) reported positive antifungal activity for 5,3'-dihydroxy-3,7,4'-trimethoxyflavone (ayanin), 5,3'-dihydroxy-3,6,7,4'-tetramethoxyflavone (casticin), 5,3',4'-trihydroxy-3,6,7-trimethoxyflavone (chrysosplenol-D), and 5,7,4'-trihydroxy-3,8-dimethoxyflavone, all isolated from *P. trinervia*. In the same paper, Wang et al. (1989b) reported antibacterial activity for chrysosplenol-D, kaempferol 3-methyl ether (isokaempferide), 5,7,4'-trihydroxy-3,3'-dimethoxyflavone, and 5,7,4'-trihydroxy-3,8-dimethoxyflavone, also from *P. trinervia*. An *in-vitro* screening of several *Psiadia* species for antipoliovirus activity resulted in identification of two active flavonoids, 5,7-dihydroxy-3,4'-

dimethoxyflavone (ermanin) and kaempferol 3-methyl ether (kaempferide), in extracts of *P. dentata* (Robin et al., 1998).

Pteronia: This genus of about 80 species is native to South Africa (Bremer, 1994; Nesom, 1994a). Wollenweber and Jay (1988) and Wollenweber and Mann (1989) reported a series of 6-methoxy and 6,8-dimethoxy flavones along with kaempferol-3,7-dimethyl ether from *P. incana*, while Mossa et al. (1988) reported 5,3',4'-trihydroxy-3,6,7,-trimethoxyflavone (chrysosplenol-D) from the same species. Taxifolin (dihydroquercetin) has also been reported from *P. incana* (Zdero et al., 1990f).

Rigiopappus: This monotypic genus is native to southwestern North America. *Rigiopappus leptocladus* accumulates the 3-*O*-glucosides and 3-*O*-rutinosides of kaempferol, quercetin, and isorhamnetin (Ornduff and Bohm, 1975).

Solidago: The center of distribution of this large genus of perhaps 100 (Grau, 1977; Nesom, 1994a) to 150 (Bremer, 1994) species is North America. A few species occur in South America, Macronesia, and Eurasia (Mabberley, 1987). Cyanidin 3-*O*-gentiobioside was identified as a constituent of *S. virgaurea* (Bjorkman and Holmgren, 1958). However, flavonols predominate in the profiles of the few species for which information is available. 3-*O*-Glycosides of kaempferol, quercetin, quercetin 3'-methyl ether (isorhamnetin), and quercetin 7-methyl ether (rhamnetin) have been reported from *S. canadensis* (Fuchs and Iliev, 1949; Batyuk and Kol'tsova, 1968, 1969; Batyuk and Kovaleva, 1985). A paper by Hirose et al. (1962) reporting kaempferol, quercetrin, and quercetin 3-*O*-glycosides can be added to the *S. canadensis* list, although those workers identified their plant material as *S. altissima* (synonymy by Gleason and Cronquist, 1991, and Cronquist, 1994). Quercetin 3-*O*-glycosides have also been reported from *S. chilensis* by Torres et al. (1987) who described their material as *S. microglossa* and from *S. gigantea* (Kustrak and Males, 1993). Fuchs and Iliev (1949) reported quercetin 3-*O*-rhamnoside from this latter taxon but called their plant material *S. serotina* (synonymy by Gleason and Cronquist, 1991, and Cronquist, 1994). Glycosides of all four of the above flavonols have been identified from *S. virgaurea* (Fuchs and Iliev, 1949; Skrzypczakowa, 1961; Hiller et al., 1979; Batyuk and Kovaleva, 1985).

Stephanodoria: *Stephanodoria* is a monotypic genus native to Mexico (Nesom, 1997). Fang et al. (1986a) reported kaempferol and quercetin 3-*O*-monoglycosides, quercetin 3-methyl ether and two *C*-glycosylflavones, vitexin and vicenin-2, from *S. tomentella*.

Tracyina: *Tracyina rostrata*, the only species in this endemic Californian genus, contains kaempferol and quercetin 3-*O*-glucosides and 3-*O*-rutinosides (Ornduff and Bohm, 1975).

Trimorpha: *Trimorpha* is a genus of 40–45 species of North America and Eurasia (Nesom 1989a). Often treated as a section of *Erigeron* (e.g., Cronquist, 1947), a close relationship with *Conyza* has also been mentioned (Cronquist, 1943). Nesom (1989a,b) has argued for generic distinction for *Trimorpha*, and this perspective has been followed by Bremer (1994). More recently, however, Nesom (1994a) has re-submerged *Trimorpha* into *Erigeron* as a subgenus based on new biosystematic evidence from W. Huber and colleagues (e.g., Huber and Leuchtmann, 1992; other references in Nesom, 1994a) that show close similarity between European and Eurasian species of the two taxa. We follow Nesom (1989a,b) and Bremer (1994) and treat *Trimorpha* as a distinct genus pending further comprehensive investigations on the two genera (as well as on *Conyza*). Kaneta et al. (1978) reported apigenin, luteolin, quercetin, and common glycosides thereof from *T. acris* (cited as *Erigeron acris*).

Vanclevea: *Vanclevea stylosa*, the sole member of this southwestern North American genus, afforded kaempferol 3,7,4'-trimethyl ether, quercetin 3,3'-dimethyl ether, and quercetin 3,7,3',4'-tetramethyl ether (Jolad et al., 1988).

Xanthisma: This genus, native to the southern United States, consists of the single species *X. texanum*, although two subspecies and two varieties have been described. Semple and Averett (1975) reported variable distributions of luteolin, quercetin, and some unidentified compounds in all four taxa.

Xanthocephalum: This genus, which consists of five species, occurs in Mexico and parts of southwestern United States. Yu et al. (1987a) reported 5,7,4'-trihydroxy-3',5'-dimethoxyflavone (tricin), kaempferol 3-methyl ether, quercetin, quercetin-3,7-dimethyl ether, quercetin-3,3'-dimethyl ether, myricetin-3,7,4'-trimethyl ether, and myricetin-3,3',5'-trimethyl ether from *X. gymnospermoides* var. *gymnospermoides*.

Xylothamnia: *Xylothamnia* is a genus of eight species concentrated in the Chihuahuan and Sonoran deserts of Mexico (Nesom et al., 1990). Previously placed in *Haplopappus* and *Ericameria*, these species have recently been judged generically distinct based on morphological and macromolecular evidence (Nesom et al., 1990; Bremer, 1994). The one flavonoid report from *X. diffusa* (cited as *Haplopappus sonoriensis*) shows the presence of kaempferol 3,7,4'-trimethyl ether (Gajhede et al., 1989).

Chapter 9

Flavonoids of Calenduleae and Cardueae

Calenduleae

Norlindh (1977) reckoned that Calenduleae are composed of only eight genera. With some taxonomic adjustment the tribe is still thought to comprise eight genera (Nordenstam, 1994, 1996). Within Bremer's (1994) treatment of the family, Calenduleae are one of the smallest tribes with about 110 species. Flavonoid data are available for a single species of each of three genera.

Calendula: *Calendula* consists of ca. 15 species but Nordenstam (1994) points out that this may vary from 10 to 25 depending upon taxonomic viewpoint. Species occur in most regions surrounding the Mediterranean, in Macaronesia, and into Iraq, Iran, and Yemen. The flavonoid profile of *Calendula officinalis* appears to consist solely of glycosides of quercetin and isorhamnetin. In addition to the usual monoglycosides and rutinosides (Friedrich 1962a, 1962b; Biryuk and Chernobai, 1972; Gora et al., 1979; Komissarenko et al., 1988), several unusual triglycosides of both quercetin and isorhamnetin have been reported. Vidal-Ollivier et al. (1989, 1991) reported, among other common glycosides, quercetin and isorhamnetin 3-*O*-2^{G}-rhamnosylrutinosides, while Masterova et al. (1991) described isorhamnetin 3-*O*-(2^{R}-rhamnosylrutinoside). Pinkas et al. (1975) isolated quercetin 3-*O*-glucoside, 3-*O*-rutinoside, and isorhamnetin 3-*O*-rutinoside from *C. arvensis*.

Dimorphotheca: According to Nordenstam (1996), this genus consists of 19 species native to southern Africa north to Angola and Zimbabwe. The sole report of flavonoids from *Dimorphotheca* involved isolation of cyanidin and delphinidin 3-*O*-glucosides from *D. sinnata* (Ball et al., 1972).

Osteospermum: *Osteospermum* consists of 45 species with a range that includes southern and tropical Africa, St. Helena, Somalia, and southwestern Arabia. Valedon and Mummery, cited by Valadon (1977), failed to obtain flavonols from *Osteospermum jucundum* but did report having found apigenin *O*-glycosides. It should be noted that Nordenstam (1994, 1996), based on a cladistic analysis of genera of Calenduleae, has suggested a merger of sect. *Blaxium* of *Osteospermum* (in which *O. jucundum* belongs; Norlindh, 1943) with *Dimorphotheca*. We have not yet seen the formal nomenclatural transfers, however.

Cardueae

This is the tribe wherein dwell the thistles, probably one of the most conspicuous members of the family. According to Dittrich (1977), the tribe consists of 79 genera with about 2500 species. Bremer (1994) agrees with that number of species but lists 83 genera. According to the treatment of Dittrich (1977) four major groups exist (partitioned into three separate tribes): Echinopeae, Carlineae, and Cardueae, the latter of which were further divided into Carduinae and Centaureinae. Bremer (1994) rejects multiple tribal recognition for portions of Cardueae (as has nearly everyone else), and he views the same four units as coordinate subtribes: Echinopsidinae, Carlininae, Carduinae, and Centaureinae. Petit et al. (1996) and Petit (1999), however, based on cladistic analysis of morphological and palynological features, re-elevated *Echinops* and relatives to tribe Echinopeae and recognized only two subtribes within Cardueae: Carlininae and Carduinae (including also former subtribe Centaureinae). Within Carduinae they list two "series," the first containing five genera centering around *Xeranthemum*, and the second all the remaining genera. We see no compelling reason to elevate the *Echinops* complex to tribal status; however, we do believe the Centaureinae and Carduinae subtribal distinctions useful. Hence, for purposes of this book, the more traditional treatment of four subtribes is accepted. For recent phylogenetic treatments, based upon morphological and molecular data, see Susanna et al. (1995) and Hellwig (1996b).

We have flavonoid data for members of 26 genera. The largest samplings have been from two genera, *Centaurea* and *Cirsium*, which is not surprising considering that these are among the largest genera in the tribe and are conspicuous floristic elements throughout their respective ranges. Owing to the amount of information available for these two genera, the following presentation will naturally have disproportionately larger sections on them than on the other genera. Information on species number and ranges of occurrence are from Dittrich (1977) and Bremer (1994).

Acroptilon: *Acriptilon* is a monotypic genus native to eastern Europe and adjacent areas in Asia. Zheng et al. (1990a) reported apigenin 5-*O*-glucoside from *A. repens* (cited as *Centaurea repens*; generic referral from Dostál, 1976).

Arctium: *Arctium* consists of nine species at home in temperate Eurasia (Duistermaat, 1996). Flavonoid information is available for three species. Rutin and quercetin 3-*O*-galactoside were reported for *A. leiospermum* and *A. tomentosum* (Kharlamov et al., 1968). *Arctium minus*, "common burdock," is a weedy European species widely dispersed in North America. It was shown to accumulate a series of common 3-*O*-mono- and diglycosides of kaempferol and quercetin (Saleh and Bohm, 1971).

Atractylis: This genus, which consists of ca. 30 species, enjoys an interesting distributional pattern: Mediterranean, Macaronesia, and S. W. Asia. *Atractylis gummifera*, the only species for which we have flavonoid data, has been shown to accumulate *C*-glycosylflavones (Fassina and Veronese, 1961; Raynaud and Dombris, 1978; Chaboud et al., 1988, 1990; Proliac et al., 1991).

Carduus: *Carduus*, with about 95 species, is native to Eurasia and parts of Africa, but is represented by weedy species in many parts of the world. Flavonoid data are available for only a few species, however, *C. acanthoides, C. assoi, C. nutans*, and *C. pycnocephalus*. Reports have shown the existence of a variety of flavone and flavonol *O*-glycosides based on 4'- and 3',4'-oxygenaton patterns, except that *C. assoi*, endemic to the Iberian Peninsula, also accumulates 5,7,4'-trihydroxy-3',5'-dimethoxyflavone (tricin) (Fernandez et al., 1991). The only report of extra oxygenation of the A-ring was also reported by Fernandez et al. (1991) who found 6-hydroxyapigenin (hispidulin) 7-*O*-glucoside in *C. assoi*. *O*-Methylation appears to be limited to the formation of isorhamnetin, found in *C. nutans* (Kaloshina and Mazulin, 1988), and tricin, found in *C. assoi* as noted above. Kaempferol-3,4'-dimethyl ether 7-*O*-glucoside was reported from *C. pycnocephalus* (Amer et al., 1985). A more recent paper by El-Lakany et al. (1995) described isolation of kaempferol 3-*O*-rhamnoside and diosmetin from the same taxon. Table 9–1 summarizes the flavonoid occurrence data for this genus.

Carlina: *Carlina* is a genus of about 28 species found in the Mediterranean area, Central Europe, and southwestern Asia. We have flavonoid data for only five. The principal compounds reported are *C*-glycosylflavones in *C. acanthifolia* (Raynaud and Rasolojaona, 1978; Proliac and Raynaud, 1983), *C. acaulis* (Raynaud and Rasolojaona, 1976, 1979), *C. corymbosa* (Besson et al., 1979; Dombris and Raynaud, 1977, 1981b; Raynaud et al., 1981), *C. macrocephala* (Dombris and Raynaud, 1980, 1981a), and *C. vulgaris*

Table 9–1. Flavonoids of *Carduus*

Species	Compounds[1]	References
C. acanthoides	Lute-7-*O*-Glys	Bain and Desrochers (1988)
C. assoi	Apigenin, Luteolin, Tricin, Lute-3'-*O*-Glc, Hisp-7-*O*-Glc, K-3-*O*-Glys	Fernandez et al. (1991)
C. nutans	K-3-*O*-Rhm	Kaloshina et al. (1975b), Abdallah et al. (1989)
	Acac-7-*O*-Glc, Apig-7-*O*-Glys, Lute-7-*O*-Glys	Kaloshina et al. (1975b); Bain and Desrochers (1988)
	K and Q-3-*O*-Glys, IR, Apigenin	Kaloshina and Mazulin (1988)
	K-3-*O*-Glc-7-*O*-Rhm, Apig-7-*O*-Neohesp	Abdallah et al. (1989)
C. pycnocephalus	Apig-7-*O*-Glc, Kaempferol, K-34'-DiMe-7-*O*-Glc	Amer et al. (1985)
	K-3-*O*-Rhm, Diosmetin	El-Lakany et al. (1995)

[1] Apig = apigenin; Acac = acacetin; Lute = luteolin; Hisp = hispidulin; K = kaempferol; Q = quercetin; IR = isorhamnetin; Gly = glycoside; Glc = glucoside; Rhm = rhamnoside; Neohesp = neohesperidoside; Me = methyl.

(Rasolojaona and Raynaud, 1979). There is a report that apigenin 7-*O*-glucoside occurs in *C. acaulis* (Raynaud and Rasolojaona, 1979).

Carthamus: *Carthamus*, a genus of 17 species (Bremer, 1994), occurs in the Mediterranean region, parts of Africa, and in parts of Asia. *Carthamus tinctorius* is cultivated in Asia where the flowers serve as an important commercial source of pigments. Although the main interest in the flavonoid chemistry of *Carthamus* continues to involve structural studies of the floral pigments, information is available for foliar flavonoids from two species in addition to *C. tinctorius*.

One of the earliest reports of a flavonoid form this genus was the description of scutellarein 7-*O*-glucuronide from *C. scabiosa* by Charaux and Rabaté (1940). *Carthamus oxycantha* yielded luteolin 7-*O*-glucoside (Kapoor et al., 1970), but the bulk of work on foliar flavonoids involves *C. glaucus* subsp. *alexandrinus*, which has yielded a number of interesting and unusual compounds. Three B-ring deoxyflavonoids have been described, 5,7-dihydroxy-6-methoxyflavone, its 7-*O*-glucoside and 5-hydroxy-

6,7-dimethoxyflavone (Khafagy et al., 1979c, 1980; El-Din, 1987). The results of extra oxygenation at C-8 can be seen in the accumulation of 8-methoxyapigenin (5,7,4'-trihydroxy-8-methoxyflavone) (Khafagy et al., 1979, 1980). *O*-Methylation results in the accumulation of 5,4'-dihydroxy-7,3'-dimethoxyflavone (velutin) (El-Din, 1987), quercetin 3-methyl ether (as the 7-*O*-glucoside) and quercetin-3,3'-dimethyl ether (Khafagy et al., 1979c, 1980). The 7-*O*-glucosides of apigenin and quercetin were reported by El-Din (1987). The foliar flavonoid profile of *C. tinctorius* appears to be much simpler; only luteolin 7-*O*-glucoside and a kaempferol 3-*O*-rhamnosylglucoside have been reported (Nakaoki and Morita, 1960b; Murti et al., 1962; Homma et al., 1990). However, a complex array of flavonoids has been reported from "petals" of *C. tinctorius* by Hattori et al. (1992). The array was dominated by 6-hydroxykaempferol and a series of glycosides including 3-*O*-mono-, 3,6-*O*-di-, and 3,6,7-*O*-triglucosides and 3-*O*-rutinoside-6-*O*-glucoside. Simple *O*-glycosides of kaempferol and quercetin were also present as were eriodictyol and 5,6,7,4'-tetrahydroxyflavone (scutellarein).

The yellow to red floral flavonoids of *C. tinctorius* have attracted a good deal of attention owing to their commercial importance. These compounds have been shown to be complex *C*-glucosyl chalcones whose structures were discussed in Chapter 4.

Centaurea: *Centaurea* is a genus with as many as 600 species distributed mainly in the Mediterranean region and southwestern Asia with a few species native to Africa and the New World. *Centaurea* has been the subject of numerous chemical studies many of which have been tabulated by Al-Easa and Rizk (1992) in a detailed review. Some of the information used in the present work was taken from their survey.

A major effort has been made to determine the structure of the remarkable blue pigment from *C. cyanus*, the cornflower. Willstätter and Everest (1913) described the pigment from cornflower as cyanidin 3,5-di-*O*-glucoside. Re-examination of the pigment by Japanese workers (Takeda and Tominaga, 1983; Tamura et al., 1983) led to a revision of the structure to cyanidin 3-*O*-(6"-*O*-succinoylglucoside)-5-*O*-glucoside. A recent report by Kondo et al. (1994) describes the structure of "protocyanin" as consisting of six molecules each of succinoylcyanin and malonyl flavone, and one equivalent each of ferrous and magnesium ions. This "supermolecule," which has a molecular weight of 8511, has the capacity for self-organization. A survey of several additional species of *Centaurea* suggests that this pigment may occur widely in the genus (Sulyok and Laszlo-Bencsik, 1985). The anthocyanidin derivatives obtained from members of the genus, including results of that survey, are summarized in Table 9–2.

Table 9–2. Anthocyanin derivatives in *Centaurea*

Source	Derivative[1]	References
C. apiculata		
ssp. *spinulosa*[2]	Cyan-3-(6"-Succ)-Glc-5-Glc	Sulyok and Laszlo-Bencsik (1985)
C. biebersteinii		
ssp. *biebersteinii*[3]	Cyan-3-(6"-Succ)-Glc-5-Glc	Sulyok and Laszlo-Bencsik (1985)
C. calcitropa	Cyan-3-Glc,	
	Cyan-3-DiGlc	Frontera et al. (1991)
C. cyanus	Cyan-3,5-DiGlc,	
	Pel-3,5-DiGlc,	
	Pel-3,5-DiGlc-Caff	Asen (1967)
	Cyan-3-(6"-Succ)-Glc-5-Glc	Tamura et al. (1983),
		Takeda and Tominaga (1983),
		Sulyok and Laszlo-Bencsik (1985)
	Pel-3-(6"-Succ)-Glc-5-Glc	Tamura et al. (1983)
	Cyan-3-(6"-Mal-Gal)	Kakegawa et al. (1987)
C. grinensis		
ssp. *fritschii*	Cyan-3-(6"-Succ)-Glc-5-Glc	Sulyok and Laszlo-Bencsik (1985)
C. jacea	Cyan-3-(6"-Succ)-Glc-5-Glc	Sulyok and Laszlo-Bencsik (1985)
C. macroptilon		
ssp. *oxylepis*	Cyan-3-(6"-Succ)-Glc-5-Glc	Sulyok and Laszlo-Bencsik (1985)
C. pannonica	Cyan-3-(6"-Succ)-Glc-5-Glc	Sulyok and Laszlo-Bencsik (1985)
C. sadlerana	Cyan-3-(6"-Succ)-Glc-5-Glc	Sulyok and Laszlo-Bencsik (1985)
C. triumfettii		
ssp. *lugdunensis*[4]	Cyan-3-Glc	Kamanzi and Raynaud (1977)

[1] Cyan = cyanidin; Pel = pelargonidin ; Glc = glucoside; Gal = galactose; Mal = malonyl; Succ = succinoyl.
[2] Reported as *C. spinulosa*; synonymy following Dostál (1976).
[3] Reported as *C. micranthus*; synonymy following Dostál (1976) and Adler et al. (1994).
[4] Reported as *C. lugdunensis*; synonymy following Dostál (1976).

Whereas the majority of glycosylated, non-anthocyanic, flavonoids found in *Centaurea* species are 7-*O*-glycosides, occasional variants have been reported such as the 4'-*O*-glucosides of apigenin and luteolin that have been reported from *C. montana* (Gonnet, 1992). Quercetin and patuletin 7-*O*-glucosides, along with quercetin-3-*O*-glucoside, were identified as components of *C. collina* by Kamanzi et al. (1982b). Quercetin 3-*O*-glucoside-7-*O*-rhamnoside was reported from four species of *Centaurea* by Plouvier (1970): *C. kerria, C. rhus,* and *C. scabiosa*. Other early surveys (Ahmed et al., 1970a,b) recorded occurrences of apigenin 7-*O*-glucuronide, kaempferol 3-*O*-glucoside, and rutin in *C. alexandrina, C. calcitropa, C. glomerata* and *C. pallescens*. Quercetin 7-*O*-glucoside was reported from several species (Bandyukova, 1967; Bandyukova et al., 1969). Naringenin was also reported from *C. calcitropa* and *C. glomerata* in this paper. Other flavanone derivatives reported from *Centaurea* include eriodictyol and taxifolin (dihydroquercetin) from *C. triumfettii* subsp. *adscendens* (cited as *C. floccosa*; synonymy from Dostál, 1976) by Negrete et al. (1987).

C-Glycosylflavone occurrences in the genus are detailed in Table 9–3. Several of the compounds reported by Gonnet (1992) from *C. montana* were acylated or secondarily glycosylated derivatives of vitexin, isovitexin, and isoörientin.

Whereas B-ring oxygenation in *Centaurea* flavonoids is predominantly of the 4'- and 3',4'-types, several exceptions are known. These are the B-ring deoxyflavonoids pinocembrin (5,7-dihydroxyflavanone) 7-*O*-diglucoside, reported from *C. ragusina* by Mahmoud et al. (1986), baicalein (5,6,7-trihydroxyflavone) 7-*O*-glucuronide from *C. scabiosa* (Marsh, 1957), and 5,6-dihydroxy-7-methoxyflavone, its 6-*O*-glucoside, and pinocembrin and its 5,7-di-*O*-glucoside from *C. eryngioides* (Sarg et al., 1993) and 5,7-dihydroxyflavone (chrysin) from the roots of *C. scabiosa* (Tesevic et al., 1994). As we shall see below in the discussion of flavone and flavonols, the primary modification of the A-ring is 6-oxygenation which occurs widely in the genus. A single 8-oxygenated compound has been reported; 5,7,4'-trihydroxy-8-methoxyflavone (8-methoxyapigenin) was reported by El-Emary et al. (1983) from *C. cineraria*. Chalcone itself, lacking oxygenation in both rings, was identified from *C. calcitrapa* by Dawidar et al. (1989). Both isomers were evident from the spectral data but they could not be separated. Mansour et al. (1988) reported the presence of quercetin 3,7-bissulphate from *C. alexandrina*, which appears to be the only report of a sulphate from a member of this tribe.

The most commonly reported flavonoids from *Centaurea* species are flavones and flavonols that have both oxygenation at C-6 and *O*-methylation. Di- and tri-*O*-methylation are common and 5-hydroxy-6,7,3',4'-tetramethoxyflavone has been reported from several species. A rich array of flavone methyl ethers, some with 6-substitution, was described from the leaf exudate of *C. macrocephala* (Wollenweber, 1991). The distribution of flavones, flavonols, and their *O*-methyl derivatives is summarized in Table 9–4. A recent paper by Sepulveda et al. (1994b) reported taxifolin (dihydroquercetin) along with the 8-methoxy derivatives of both apigenin and luteolin from *C. chilensis*.

Cheirolophus: *Cheirolophus* is a genus of 23 species (Susanna et al., 1999) native to the Canary Islands, southwestern Europe, and adjacent northwestern Africa. Three species have been examined for flavonoids under the

Table 9–3. *C*-Glycosylflavones from species of *Centaurea*

Species	Compounds	References
C. cyanus	Swertisin	Asen and Jurd (1967)
	Isoswertisin	Asen (1967)
		Asen and Jurd (1967)
C. incana	6-*C*-Glc-Apig 2"-*O*-Glc,	
	Apig-6,8-di-*C*-Glc	Akkal et al. (1997)
C. macrophylla	Isovitexin,	
	Isoscoparin	Reynaud et al. (1992)
C. melitensis	Orientin,	
	Homoörientin,	
	Schaftoside,	
	Viccnin-2	Kamanzi et al. (1983b)
C. montana	Isovitexin,	
	Isovitexin-2"-O-Ara,	
	Isovitexin-2"-O-Glc[1],	
	Isoörientin-2"-O-Glc[1],	
	Isoörientin-2"-O-Ara,	
	Scoparin,	
	Isoscoparin,	
	Isoscoparin-2"-O-Glc,	
	Vicenin-2,	
	Swertisin	Gonnet (1992)
C. ragusina		
ssp. *ragusina*	Vitexin	Mahmoud et al. (1986)
C. regia	Vitexin	Ulubelen et al. (1988)
C. thracica	Vitexin,	
	Isovitexin,	
	Orientin	
	Isoörientin	Putun et al. (1989)
C. virgata[2]	Isovitexin,	
	Isoschaftoside	Öksüz et al. (1984)

1 Also occurs as a caffeoyl derivative.

2 We follow Wagenitz (1972, 1986) in use of this name; Dostál (1976) places it in synonymy with *C. ovina* Pallas ex Willd. ssp. *besserana* (DC.) Dostál.

generic title of *Centaurea*. *Cheirolophus arbutifolius* (cited as *Centaurea arbutifolia*; generic placement of this and the following two taxa from Hansen and Sunding, 1979) was shown by Breton Fumes et al. (1969) to accumulate chrysoeriol and kaempferol and quercetin 3-methyl ethers. *Cheirolophus arguta* (cited as *Centaurea arguta*) was also shown to accumulate quercetin 3-methyl ether (as the 7-*O*-rutinoside) but had 6-methoxyapigenin, 6-methoxyluteolin, and 5,7,4'-trihydroxy-6,3'-dimethoxyflavone as well

Table 9–4. Substituted flavonoids from species of *Centaurea*

Species	Compounds[1]	References
C. aggregata	53'/674', 5/673'4'	Zapesochnaya et al. (1977)
C. alexandrina	574'/63', 57/63'4'	Amer et al. (1984)
	574'/6-7-*O*-Glc,	
	57/64'-7-*O*-Glc	Harraz et al. (1989)
	574'/363'-7-Glc	Gavina et al. (1974)
	573'4'/-, 574'/3',	
	574'/63', 574'/3,	
	357/64'	Ismail et al. (1986)
C. araneosa	574'/6, 54'/73', 54'/67	Farrag et al. (1993)
C. aspera	574'/6, 573'4'/6,	
	574'/63'	Ferreres et al. (1980)
		Ferreres and Tomas (1980)
var. *subinermis*	57/64', 574'/6,	
	573'4'/6, 574'/63'	Cardona et al. (1991)
C. aspera		
var. *stenophylla*	573'4'/6	Picher et al. (1984)
C. behen	54'/67	Öksüz et al. (1982)
C. bruguierana	5/673'4'	Rustaiyan et al. (1982)
	574'/6-7-*O*-Rut,	
	54'/67, 54'/673'	Harraz et al. (1994)
C. cariensis	57/64', 53'/674',	
	354'/67, 573'/364'	Halfon et al. (1989)
C. chilensis	574'/3' and 7-*O*-Glc,	
	574'/6	Negrete et al. (1988)
	574'/8, 573'4'/8	Sepulveda et al. (1994b)
C. cineraria	574'/8	El-Emary et al. (1983)
C. cineraria		
ssp. *umbrosa*	57/63'4', 574'/63',	
	5/674', 5/673'4'	Bruno and Herz (1988b)
C. clementei	56/7, 56/74',	
	574'/6, 574'/3	Collado et al. (1985)
C. collina	57/64', 3573'4'/6,	
	353'4'/7	Kamanzi et al. (1982a,b,c),
	5674'/-, 57/64',	
	573'4'/6	Fernandez et al. (1989)
C. coronopifolia	574'/6, 573'4'/6,	
	573'4'/36 and-7-*O*-Glc	Öksüz and Ayyildiz (1986)
C. cuneifolia	56/74', 5/674',	
	57/63'4', 574'/63',	
	53'/674', 57/4'	Öksüz et al. (1988)
C. cyanus	573'4'/6,	
	3574'/3' and 7-*O*-Glc	Litvinenko and Bubenchikova (1988)

Table 9–4. (Continued)

Species	Compounds[1]	References
C. dealbata	574'/3	Christensen and Lam (1991c)
C. depressa	5674'/-	Bandyukova et al. (1967)
C. glomerata	573'4'/3, 354'/73', 3574'/63'	Zaghloul et al. (1990)
C. granata	53'/674', 5/673'4'	Medjroubi et al. (1998)
C. hyssopifolia	574'/363' and 7-*O*-Glc	Gonzalez et al. (1971)
C. iberica	5/673'4'	Ahmed and Bibi (1979)
C. incana	574'/6, 573'4'/6, 54'/73'5', 5/73'4'5', 3574'/6, 3574'5'/3', 573'4'/35', 3573'4'/6	Akkal et al. (1997)
C. inermis	574'/6, 573'4'/6, 574'/63', 53'/674', 574'/3	Öksüz et al. (1984)
C. jacea	574'/63'-7-*O*-Glc	Farkas et al. (1964a,b), Wagner et al. (1969)
	573'/364'-7-*O*-Glc	Farkas et al. (1964a,c), Wagner et al. (1973a)
	574'/36 and 7-*O*-Glc	Rosler et al. (1971)
C. kilaea	574'/63', 5/673'4'	Öksüz et al. (1984)
C. kotschyi var. *kotschyi*	3574'/3', 3574'/63', 35674'/3'-7-*O*-Glc	Öksüz and Putun (1987)
C. macrocephala	5674'/-	Monea (1986)[2]
	574'/-, 57/4', 54'/67, 5/674', 573'4'/-, 574'/3', 573'4'/6, 53'4'/67, 574'/63', 54'/673', 5/673'4'	Wollenweber (1991)
C. macrophylla	3574'/3'-3-*O*-Glc	Reynaud et al. (1992)
C. melitensis	573'4'/6-7-*O*-Glc, 3573'4'/6-7-*O*-Glc	Kamanzi et al. (1982d)
	574'/6	Negrete et al. (1989)
C. montana	574'/3'-7-*O*-Glc, 3573'/4'-7-*O*-Glc	Gonnet (1992)
C. nigrescens	573'/64'	Bohlmann and Zdero (1967)
C. ornata	57/64'	Bastos et al. (1994)
C. pallescens	57/4', 574'/63', 574'/363'	Ali et al. (1987)
C. phyllocephala	574'/6, 573'4'/6, 53'4'/67, 574'/63'	Twaij et al. (1983), Kery et al. (1985)

Table 9–4. (Continued)

Species	Compounds[1]	References
	5/673'4'	Ahmed and Bibi (1981)
C. phrygia		
ssp. *pseudophrygia*[3]	573/64'	Bohlmann and Zdero (1967)
C. pichleri		
ssp. *pichleri*	574'/6-7-*O*-Glc, 57/64'-7-*O*-Glc, 574'/6, 54'/67, 573'4'/6	Putun and Putun (1990)
C. pumila	574'/6, 54'/67, 574'/683', 53'4'/67, 54'/673', 5/673'4'	El-Masry et al. (1980, 1981)[4]
C. regia	574'/3'	Ulubelen et al. (1988)
C. rhenana		
ssp. *pseudomaculosa*[5]	5/673'4'	Zapesochnaya et al. (1977), Adekenov et al. (1986)
C. rothmalerana	5/673'4', 57/63'4', 54'/673'	Santos et al. (1992)
C. salicifolia	574'/363'	Zapesochnaya et al. (1977)
C. scoparia	574'/6, 54'/67, 5/674'	Youssef and Frahm (1995)
C. sinaica	53'4'/67, 3574'/3', 5/674', 574'/33'	Al-Easa (1990), Al-Easa et al. (1992)
C. solstitialis	574'/36, 3573'4'/6, 3574'/63', 573'4'/36	Kamanzi et al. (1983c,d)
C. sulphurea	57/64'	Gonzalez et al. (1984a)
C. trumfettii		
ssp. *adscendens*[6]	574'/3', 574'/6	Negrete et al. (1987)
C. urvillei	574'/6, 53'4'/67, 573'4'/6, 54'/67, 5/674', 54'/7	Ulubelen and Öksüz (1982)
C. virgata[7]	574'/6, 573'4'/6, 53'/674', 574'/63', 574'/3	Öksüz et al. (1984)

1 Substitution: Hydroxyl positions/Methoxyl positions, Glc = glucose, Rut = rutinose.
2 Reported as *Grossheimia macrophyla.*
3 Treated as *C. pseudophrygia* by Hellwig (1996b).
4 Reported as *Aegialophia pumila.*
5 Reported as *C. pseudomaculosa*; nomenclature after Dostál (1976).
6 Reported as *C. floccosa*; synonymy after Dostál (1976).
7 We follow Wagenitz (1972, 1986; Wagenitz and Gamal-Eldin, 1985) in use of this name; Dostál (1976) places it in synonymy with *C. ovina* ssp. *besserana*.

(Breton Fumes et al., 1968; Gadeschi et al., 1989). Plouvier (1970) reported quercetin 3-*O*-glucoside-7-*O*-rhamnoside from *C. sempervirens* (cited as *Centaurea sempervirens*).

Cirsium: *Cirsium*, the thistle genus, consists of 250-300 species native to Eurasia and North America. Although there have appeared a few reports of flavonols in members of the genus, flavones are the major compounds present in almost all species examined. Mono- and di-*O*-methyl derivatives are known; higher levels of *O*-methylation apparently do not exist. Extra oxygenation at C-6 is common. A report of an 8-oxygenated flavone (5,7-dihydroxy-8,4'-dimethoxyflavone) and its 7-*O*-glucoside in *C. japonicum* var. *takaoense* (Lin et al., 1978) was challenged by Markham (1983) who argued for 5,4'-dihydroxy-6,7-dimethoxyflavone 4'-*O*-glucoside as the correct structure for the glycoside. The corrected structure was corroborated by Nakayama et al. (1983) who synthesized the 8-methoxy flavone. Iwashina and coworkers (Iwashina et al., 1988, 1989, 1995) have investigated the flavonoid chemistry of a large number of Japanese species of *Cirsium*. These and other occurrences of flavone and flavonol derivatives in the genus are summarized in Table 9–5.

Hispidulin and nepetin (6-methoxyapigenin and 6-methoxyluteolin, respectively) 4'-*O*-glucosides were reported as new compounds from *C. oligophyllum* by Iwashina et al. (1999) along with the known compounds hispidulin 7,4'-*O*-diglucoside, vicenin-2, hispidulin, and nepetin.

Cnicus: According to Dittrich (1977) and Wagenitz and Hellwig (1996), *Cnicus* consists of a single species, *C. benedictus*, which occurs naturally from the lands of the Mediterranean to southwestern Asia. Bremer (1994) includes *Cnicus* in *Centaurea*, and even Wagenitz and Hellwig (1996, p. 496) state, "There is clearly more study necessary to determine the exact position of *Cnicus*, but there is no doubt that it is part of the *Jacea* group of *Centaurea*."

A study of *C. benedictus* by Sendra and Walczak (1967) led to the discovery of apigenin 7-*O*-glucoside, a luteolin *O*-diglucoside and kaempferol 3-*O*-glucoside.

Cousinia: This very large genus, perhaps 600 species, occurs in southwestern and central Asia. Despite its size we have information on flavonoids for only one species, *C. eriocephala*, from which Ulubelen and Tuzlaci (1988) isolated apigenin and its 4-methyl ether (acacetin), luteolin 3'-methyl ether (chrysoeriol), and 5,4'-dihydroxy-6,7-dimethoxyflavone (cirsimaritin).

Cynara: This genus consists of eight species distributed in southern Europe, the Middle East to Iran, and in North Africa. Flavonoids have been

Table 9–5. Flavonoids from species of *Cirsium*

Species	Compounds[1]	References
C. arvense	Apig-7-*O*-Glys,	
	Lute	Shelyuto et al. (1970)
	Tricin 5-*O*-Glc	Wallace (1974)
	Apig-7-*O*-Gln,	
	Acac-7-*O*-Gln	Shelyuto et al. (1972b)
	K3Me	Shelyuto et al. (1972a)
	Acac	Rendyuk et al. (1977)
C. babanum	K & Q-3-O-Glys	Iwashina et al. (1995)
C. bitchuense	Pec	Iwashina et al. (1995)
C. brevistylum	Pec	Wallace and Bohm (1971)
C. carolinianum	Apig, 5674'/-	Bohlmann and Abraham (1981)
C. chikushiense	Apig & Lute-7-*O*-Gly,	
	Chry-7-*O*-Gly	Iwashina et al. (1995)
C. gratiosum	Pec, Aca, Vicenin-2	Iwashina et al. (1995)
C. hachijoense	Apig-7-*O*-Glys,	
	Lute-7-*O*-Glys,	
	Apig-6,8-di-*C*-Gly	Iwashina and Ootani (1998)
C. helenioides[2]	Lute-7-*O*-GLc,	
	Lute-4'-*O*-Glc	Ogurtsova and Syuzeva (1974)
C. inundatum	Pec	Nakaoki and Morita (1960a)
	Lute-7-*O*-Glc,	
	Lute-4'-*O*-Rut	Iwashina et al. (1988)
C. japonicum	Pec	Iwashina et al. (1989)
	574'/6-7-*O*-Rut	Miyaichi et al. (1995)
	Pec, Aca	Iwashina et al. (1988, 1995)
C. japonicum		
var. *takaoense*	Cir and 4'-*O*-Glc	Markham (1983)
var. *ussuriense*	Acac,	
	Hisp-7-*O*-Neohesp,	
	54'/67-4'-*O*-Glc	Park et al. (1995)
C. kagamontanum	57/64'-7-*O*-Rut	Nakaoki and Morita (1960a)
	Apig & Lute-5-*O*-Glc	Iwashina et al. (1989)
	Pec, Aca	Iwashina et al. (1995)
C. kamtschaticum	Pec	Morita et al., (1973), Iwashina et al. (1988)
C. lineare	53'4'/67-4'-*O*-Glc,	
	54'/673'-4'-*O*-Glc	Morita et al. (1973)
C. longepedunculatum	Apig & Lute-5-*O*-Glc	Iwashina et al. (1989)
C. lucens var. *opacum*	Pec, K-3-*O*-Glc	Iwashina et al. (1988)
C. magofukui	Apig & Lute-5-*O*-Glc	Iwashina et al. (1989, 1995)
C. matsumurae		
var. *pubescens*	Lute-7-*O*-Glc	Nakaoki and Morita (1960a)

Table 9–5. (Continued)

Species	Compounds[1]	References
C. microspicatum	Pec	Nakaoki and Morita (1959)
	Pec, Aca	Iwashina et al. (1988,1995)
C. m. var. *kiotense*	Pec, Aca	Iwashina et al. (1995)
C. nipponicum	Apig-7-*O*-Glc,	
	Lute-7-*O*-Glc	Kaneta et al. (1978)
	Pec	Do et al. (1994)
C. n. var. *incomptum*	Pec, Aca	Iwashina et al. (1988, 1995)
C. oleraceum	Pec, Aca	Wagner et al. (1960)
	Apig, Lute,	
	K-3-Me	Shelyuto et al. (1971a)
	Apig-7-*O*-Glc,	
	Lute-7-*O*-Glc,	
	Pec, Aca	Shelyuto et al. (1971b)
C. otayae	Pec	Iwashina et al. (1995)
C. palustre	Lute-7-*O*-Glc	Shelyuto et al. (1972c)
C. pectinellum	Linarin	Morita et al. (1973)
C. purpuratum	Pec	Nakaoki and Morita (1959)
	Apig-7-*O*-Glc,	
	Acac-7-*O*-Glc,	
	Q	Kaneta et al. (1978)
C. rhinoceros	Apig, Lute,	
	Acac-7-*O*-Glc,	
	Hisp-7-*O*-Rut,	
	Pec & 7-*O*-Glys	Lee et al. (1994)
C. scariorum[3]	Pec, Aca	Gardner (1973)
C. senjoense	Pec, Aca	Iwashina et al. (1995)
C. setosum	Apig, K-3-*O*-Glc	Syrchina et al. (1997)
C. sieboldii	Apig & Lute-5-*O*-Glc	Iwashina et al. (1989, 1995)
C. spicatum	Pec, Aca	Iwashina et al. (1995)
C. suffultum	Lute-7-*O*-Glc	Morita et al. (1973),
		Iwashina et al. (1995)
C. syriacum	Apig-7-*O*-Glc,	
	Apig-7-*O*-DiGlc,	
	Lute, K	Khafagy et al. (1981b)
C. yegoense[4]	Pec	Nakaoki and Morita (1959)
C. yezoense	Pec, Aca	Iwashina et al. (1988, 1995)
C. vulgare[5]	Apig-7-*O*-DiGlc,	
	Genk-4'-*O*-Glc,	
	K and Q-3-*O*-Glc,	
	Q-3-*O*-Gal	McGowan and Wallace
		(1972)

Table 9–5. (Continued)

Species	Compounds[1]	References
C. wallichii[6]	Lute 7-*O*-Glc, IR 7-*O*-Glc	Singh and Pandey (1986), Singh et al. (1988)
C. yakushimense	Lute-7-*O*-Glc	Iwashina et al. (1995)

[1] Apig = apigenin; Aca = acacetin-7-O-rutinoside; Cir = cirsimaritin; Genk = genkwanin, Hisp = hispidulin; Lute = luteolin; Pec = pectolinarigenin-7-O-Rutinoside; K = kaempferol; Q = quercetin; IR = isorhamnetin; Me = methyl; Glc = glucose; Gal = galactose; Rut = rutinose.
[2] Reported as *C. otoyae*; taxonomic perspective from Ohwi (1965).
[3] Reported as *C. heterophyllum*; synonymy from Werner (1976).
[4] Reported as *C. coloradense*; synonymy from Cronquist (1994).
[5] Reported as *C. yoshizawae*; synonymy from Ohwi (1965).
[6] Reported as *C. lanceolatum*; synonymy from Werner (1976).
[7] Reported as *Cnicus walichii*; generic referral after Petrak (1979).

reported from two species, *C. sibthorpiana* and *C. scolymus* (*C. cardunculus*, according to Bremer, 1994). Several flavone glycosides, among which are the 7-*O*-glucoside, 7-*O*-rutinoside, 7,4'-di-*O*-glucoside, 3'-*O*-glucoside, and 4'-*O*-glucoside, have been described from both species. El-Ansari et al. (1988) reported apigenin and luteolin *O*-glycosides, along with naringenin 7-O-glycosides, from the former. Reports of glycosides of apigenin and/or luteolin came from the work of Masquelier and Michaud (1956), Dranik et al. (1964), Dranik (1966), Bombardelli et al. (1977), and El-Negoumy et al. (1987). Dranik and Chernobai (1966) also reported a luteolin 7-*O*-triglycoside from this species. An interesting finding in *C. scolymus* was the aurone maritimein (Hinou et al., 1989). Cynarin has been reported from *C. cardunculus* (*C. scolymus*) by Grancai et al. (1994).

Echinops: *Echinops*, a genus of 100–120 species, is represented in the flavonoid literature by only four species, *E. echinatus, E. niveus, E. ritro*, and *E. spinossimus*. With one exception, *E. ritro* was shown to accumulate quercetin derivatives (Chevrier and Raynaud, 1973), the flavonoid profile of the genus is dominated by simple flavones. Apigenin and its 7-*O*-glucoside have been reported in *E. echinatus* by several groups (Chaudhuri and Thakur, 1986; Hogale and Uthale, 1988; Singh and Pandey, 1989, 1993; Ram et al., 1996) and in *E. spinossimus* by Omar (1980). In addition to the simple glucoside, Chaudhuri and Thakur (1986) found apigenin 7-*O*-(4"-*O*-*cis*-*p*-coumaroyl)glucoside. Two *p*-coumaroyl esters were reported by Singh et al. (1986); "echinacin" was shown to be the 6"-*O*-*p*-coumaroyl derivative and "echinaticin" the 4" analogue. *Echinops niveus* has apigenin and luteolin (Singh et al.,1990) and quercetin 4'-methyl ether (tamarixietin) (Singh and Pandey, 1994), while *E. ritro* was shown by Chevrier and Raynaud (1973) to have apigenin, luteolin, and luteolin 3'-methyl ether (chrysoeriol), as well as quercetin, as noted above.

Leuzea: *Leuzea* is a small genus, three species, that occurs in southwestern Europe and in North Africa. Varga et al. (1990) reported quercetin 5-*O*- and 7-*O*-glucosides, isorhamnetin 5-*O*-glucoside, and quercetin 3,3'-dimethylether from *L. carthamoides*.

Notobasis: *Notobasis syriaca*, the only member of this genus, occurs over a wide area that includes southern Europe, Turkey, the Middle East to Iran, central Asia, and North Africa. A study by Mericli and Dellamonica (1983) resulted in the description of different profiles of flavonoids in aerial parts of the plant vs. fruits. In the former, these workers found apigenin and luteolin, both as aglycones and as their 7-*O*-glucosides, chrysoeriol, kaempferol, and isorhamnetin. They failed to find the flavonols in the fruits, but the flavones were present, as were two *C*-glycosylflavones, schaftoside and isoschaftoside. Khafagy et al. (1981b) also reported apigenin and luteolin 7-*O*-glycosides from this taxon (cited as *Cirsium syriacum*).

Onopordum: *Onopordum* is a moderately large genus with some 60 species in southwestern and central Asia. Although there are flavonoid data for only about a half-dozen species, the profiles show a fair degree of biosynthetic diversity. The earliest paper was that of Bogs and Bogs (1967) who reported apigenin, luteolin, isorhamnetin, and a quercetin glucoside from *O. acanthium*. *Onopordum macranthum* accumulates apigenin, acacetin and its 7-*O*-methylglucuronide, luteolin, chrysoeriol, and 6-methoxyluteolin (eupafolin) (Cardona et al.,1987). In addition to the simple flavones just listed, two more examples of A-ring extra oxygenation were reported from *O. corymbosum* (Cardona et al., 1990). These workers reported 6-methoxyapigenin (hispidulin) and 5,7-dihydroxy-6,4'-dimethoxyflavone. The most remarkable find in this species, however, was the presence of hydnocarpin and 5"-methoxyhydnocarpin, compounds of the flavonololignan group normally associated, in Asteraceae, with *Silybum marianum* (see below). Apigenin, luteolin, and their 7-*O*-glucosides, acacetin, 6-methoxyapigenin (hispidulin), and 6-methoxyluteolin (nepetin) were reported from *O. tauricum* (Mericli et al., 1991), while apigenin, luteolin, chrysoeriol, and hispidulin were reported from *O. acaulon* (Cardona et al., 1992). Studies of *O. alexandrinum* have shown the presence of apigenin, luteolin, and quercetin 7-*O*-glucosides (Wassel, 1975) and scutellarein 4'-methyl ether (6-

hydroxyapigenin 4'-methyl ether) (Ayyad et al., 1994). Recent reports have described moderately complex arrays of compounds from several species. Garcia et al. (1996), for example, identified apigenin, luteolin, chrysoeriol, hispidulin, sorbifolin (5,6,4'-trihydroxy-7-methoxyflavone), onopordin (5,7,3',4'-tetrahydroxy-8-methoxyflavone), eupafolin (5,7,3',4'-tetrahydroxy-6-methoxyflavone, the flavanone eriodictyol, and the *C*-glycosylflavone vitexin from *O. myriacanthum*. Lazari et al. (1998) described isolation of luteolin, chrysoeriol, hispidulin, sorbifolin, and eriodictyol from *O. sibthorpianum*, and apigenin, luteolin, and chrysoeriol and their 7-*O*-glucosides, acacetin, hispidulin, pilloin, onopordine, and vitexin from *O. laconicum*.

Picnomon: *Picnomon acarna*, the sole species in this genus, occurs in southern Europe, southwestern Asia, and in North Africa. Souleles and Laskaris (1988) reported that it accumulates luteolin-3'-methyl ether (chrysoeriol), 5,4'-dihydroxy-7,3'-dimethoxyflavone, and 3,5,7-trihydroxy-8,3',4'-trimethoxyflavone (gossypetin 8,3',4'-trimethyl ether). This last named compound was a new natural product at the time of its isolation. A more recent paper (Laskaris et al., 1995) recorded luteolin, quercetin, apigenin 4'-methyl ether (acacetin) 7-*O*-rutinoside, 6-methoxyapigenin (hispidulin) and its 7-*O*-rhamnoside and 7-*O*-rutinoside.

Saussurea: *Saussurea* is one of the larger genera in Cardueae having at least 300 species widely distributed in Eurasia (Lipschitz, 1979). One species extends into northern Australia. We have flavonoid information for eight species. Apigenin, luteolin, and/or chrysoeriol and various *O*-glycosides thereof occur in *S. gossypiphora* (Qiu et al., 1989; Zheng et al., 1990b, 1991), *S. hieracioides* (Liu et al., 1989), *S. medusa* (Jia et al., 1986, 1990), and *S. pulchella* (Basargin and Tsiklauri, 1990). Hispidulin and jaceoside have been reported in tissue-cultured *S. medusa* by Wang (1997). Quercetin and/or its 3-*O*-glycosides have been reported from *S. gossypiphora* (Zheng et al., 1990b), *S. involucrata* (Jia et al., 1988), *S. medusa* (Jia et al., 1986, 1990) and *S. controversa* (Syrchina et al., 1993). Shi et al. (1991) identified kaempferol and quercetin 3-*O*-glucosides and their 6''-*O*-crotonoyl derivatives as components of *S. japonica*. Cyanidin 3-*O*-glucoside was reported from *S. pulchella* (Basargin and Tsiklauri, 1990). The only extra-oxygenated flavonoids reported from a member of the genus are 6-methoxyapigenin hispidulin and 5,7-dihydroxy-6,4'-dimethoxyflavone (pectolinarigenin) described from *S. elegans* by Sham'yanov et al. (1983).

Schmalhausenia: This genus consists solely of *S. nidulans*, which is endemic to the Tien Sian Mountains of Kazakhstan. In a recent paper, Iwashina and Kadota (1999) reported apigenin, its 7-*O*-glucoside and 7-*O*-rutinoside, luteolin, its 7-*O*-glucoside, 7-*O*-rutinoside, and 7-*O*-glucuronide, and quercetin 3-*O*-glucoside and 3-*O*-rutinoside from this taxon. The authors argued that the difference in flavonoid profiles of this taxon and *Cirsium* add to the features by which these two genera may be distinguished.

Serratula: This genus contains some 70 species in Eurasia and North Africa. We have flavonoid information on only two species, *S. radiata* (cited as *S. bracteifolia*; synonymy from Cannon and Marshall, 1976 for this and the following species) and *S. tinctoria* (cited as *S. inermis*). Kaempferol and quercetin were described from the former (Yatsyuk et al., 1970), quercetin 3-methyl ether from the latter (Glyzin et al., 1982). Yatsyuk and Lyashenko (1969) also reported apigenin and luteolin from *S. tinctoria* (cited as *S. inermis*).

Silybum: *Silybum* has attracted a considerable amount of interest despite the existence of only two species owing in part at least to the weedy nature and cosmopolitan distribution of *S. marianum* (e.g., in Chile; Matthei, 1995). The seeds of *S. marianum* have long been utilized for their antihepatotoxic activity (Hahn et al., 1968). Intensive searches for the active component(s) led to the discovery of an entire class of flavonoids, the flavonololignans, e.g., compound **[4–93]**, wherein a normal fifteen-carbon flavonoid unit has condensed with an equivalent of coniferyl alcohol. As well as structural work on the compounds themselves (Hänsel and Rimpler, 1968; Wagner et al., 1965, 1968; Abraham et al., 1970; Mishima et al., 1971; Hänsel et al., 1972), some workers have looked into variation in flavonolignan content as in the case of Dranik et al. (1992, 1993) who compared white-flowered and purple-flowered forms of *S. marianum*. In addition to flavonololignans, ordinary flavonoids have also been reported: 5,7,4'-trihydroxy-3',5'-dimethoxyflavanone (Boehrer, 1968), simple flavone and flavonol glycosides (Halbach and Winkler, 1971; Khafagy et al., 1981a; Fiebig and Wagner, 1984; Mericli, 1988; Ahmed et al., 1989c), flavanones and dihydroflavonols (Kaloga, 1981b; Fiebig and Wagner, 1984; Hikino et al., 1984; Mericli, 1988). Unusual compounds of note are the flavanone with 3',4',5'-trioxygenation (Boehrer, 1968), and kaempferol 3-sulphate, which Mericli (1988) reported as the major component of aerial parts of *S. marianum* ssp. *anatolicum*. An extensive review of the chemistry and pharmacology of *S. marianum*, with over 200 citations, was recently published by Morazzoni and Bombardelli (1995).

Staehelina: Flavonoid information is available for only *S. dubia* from this group of eight Mediterranean species. Raynaud et al. (1971) reported 6-methoxyluteolin and a 6-*C*-glycosylluteolin.

Stemmacantha: According to Dittrich (1977), *Stemmacantha* (as *Rhaponticum*; see Dittrich, 1984, for nomenclatural changes) consists of 24 species distributed in southern Europe, Asia, North Africa, and in Australia (1 species); Bremer (1994) lists 20 species. Flavonoid information is available for *S. carthamoides* (cited as *Rhaponticum carthamoides*). A hesperidin derivative, which gave glucose and rhamnose upon hydrolysis, was reported by Bukharov and Shcherbak (1967). Varga et al. (1982) reported quercetin and isorhamnetin 5-*O*-glucosides, quercetin 7-*O*-glucoside, and quercetin 3,3'-dimethyl ether from roots of the same species.

Tricholepis: Quercetin 3-*O*-rutinoside was reported from *T. glaberrima* by Manerikar and Kulkarni (1978). This appears to be the only information available for this genus of 18 species which occur from Afghanistan through India to Burma.

Volutaria: *Volutaria* consists of 16 species, which occur in the Mediterranean region, southwestern Asia, and in tropical Africa. Our only flavonoid information is the report of 5,7,4'-trihydroxy-6,3'-dimethoxyflavone (jaceosidin) from *V. divaricata* (Forgacs et al., 1981).

Xeranthemum: *Xeranthemum* is a genus of 5–6 species native to the Mediterranean region and southwestern Asia. The only flavonoid work on the genus is apparently the report by Zemtsova and Molchanova (1979b) that *X. annuum* contains luteolin, its 7-*O*-glucoside, and quercetin.

CHAPTER 10

Flavonoids of Eupatorieae

According to Robinson and King (1977) and King and Robinson (1987), Eupatorieae comprise approximately 2000 species arrayed in 180 genera. Bremer et al. (1994) recognize only 170 genera but put the species number nearer 2400. Before we proceed with description of flavonoids of the tribe, it is necessary to attend to some nomenclatural problems. One of the results of the recent monographic treatment of the tribe (King and Robinson, 1987) is a generic realignment of taxa. Many of the flavonoid reports in the literature involve plants whose taxonomic positions or nomenclature have been changed to correspond to the newer view of the tribe. As an aid to the reader, therefore, we list in Table 10–1 name changes that affect our presentation of flavonoid data. Our use of the generic concepts presented in King and Robinson (1987) is motivated more by wanting to have a recent taxonomic and nomenclatural reference point for all flavonoid reports than by believing in phyletic reality of all generic concepts presented in that comprehensive study and earlier publications, some of which are still controversial (e.g., Gage, 1985; Scott, 1985; Sundberg, 1985; Turner, 1985a,b, 1997; Nesom 1989c; Bremer et al., 1994). Flavonoid information exists for 32 genera. Notes on size of taxa and their distribution come from Bremer (1994) unless stated otherwise.

Acanthostyles: Two species are recognized by King and Robinson (1987) in this South American genus. Flavonoids of *A. buniifolius* (cited as *Eupatorium buniifolium* Hook. and Arn.) are as follows: 5,7-dihydroxy-4'-methoxyflavone (acacetin), 5-hydroxy-6,7,4'-trimethoxyflavone (salvigenin), 5,7,3'-trihydroxy-3,6,4'-trimethoxyflavone (quercetagetin-3,6,4'trimethyl ether, centaureidin), and 5,7,4'-trihydroxy-3,6,3'-trimethoxyflavone (quercetagetin-3,6,3'-trimethyl ether, jaceidin) (Caula et al., 1991). More recent work by Muschietti et al. (1993, 1994) resulted in the identification of 5,2'-dihydroxy-3,6,7,4',5'-pentamethoxyflavone (brickellin) (Fig. 10–1) **[10–1]** and 5,7,5'-trihydroxy-3,6,2',4'-tetramethoxyflavone **[10–2]** from *A. buniifolius* (reported as *Eupatorium*). The authors note that this is the first report of flavonoids with the 2',4',5'-oxygenation pattern in *Eupatorium*.

Table 10–1. Nomenclatural and taxonomic changes in Eupatorieae following King and Robinson (1987)

Name in flavonoid paper	Reference	Name according to King and Robinson
Ageratina stevioides	Morales-Mendez and Rosquete (1988)	*Ageratina jahnii* (B. Robinson) King and Robinson
Ageratum corymbosum f. *corymbosum*	Quijano et al. (1992)	*Ageratum corymbosum* Zucc.
Ageratum mexicanum	Mionskowski and Gill (1973)	*Ageratum houstonianum* Mill.
Ageratum strictum	Quijano et al. (1982a)	*Ageratum corymbosum* Zucc.
Brickellia guatemaliensis	Bohlmann and Zdero (1969)	*Brickellia argyrolepis* B. Robinson
Brickellia microphylla	El-Sayed et al. (1988)	*Nothobaccharis candolleana* (Steudel) King and Robinson
Brickellia squarrosa	Flores and Herran (1960)	*Brickellia cavanillesii* (Cass.) A. Gray
Eupatorium areolare	Yu et al. (1986c)	*Piptothrix areolaris* (DC.) King and Robinson
Eupatorium buniifolium	Caula et al. (1991)	*Acanthostyles buniifolius* (Hook. and Arn.) King and Robinson
Eupatorium coelestinum	Le-Van and Pham (1979), Wagner et al. (1972f)	*Conoclinium coelestinum* (L.) DC.
Eupatorium glandulosum	Nair and Sivakumar (1990)	*Ageratina adenophora* (Sprengel) King and Robinson
Eupatorium inulaefolium	Ferraro et al. (1977)	*Austroeupatorium inulaefolium* (H.B.K.) King and Robinson
Eupatorium ligustrinum	Quijano et al. (1970)	*Ageratina ligustrina* (DC.) King and Robinson
Eupatorium littorale	Sato and Moreira (1985)	*Disynaphia littoralis* (Cabrera) King and Robinson
Eupatorium macrocephalum	Gonzalez et al. (1973)	*Campuloclinium macrocephalum* (Less.) DC.
Eupatorium micranthum	Sagareishvili (1985)	*Ageratina ligustrina* (DC.) King and Robinson
Eupatorium odoratum	Bose et al. (1973), Talapatra et al. (1974), Arene et al. (1978), Metwally and Ekejiuba (1981) Nguyen et al. (1993)	*Chromolaena odorata* (L.) King and Robinson

Table 10–1. (Continued)

Name in flavonoid paper	Reference	Name according to King and Robinson
Eupatorium recurvum × *rotundifolium*	Wagner et al. (1972f)	*E. mohrii* E. Greene X *E. rotundifolium* L.
Eupatorium rugosum	Wagner et al. (1972f)	*Ageratina altissima* (L.) King and Robinson
Eupatorium salvia	Gonzalez et al. (1990)	*Aristeguietia salvia* (Colla) King and Robinson
Eupatorium sternbergianum	D'Agostino et al. (1994)	*Ageratina sternbergiana* (DC.) King and Robinson
Eupatorium stoechadosmum	Nakaoki and Morita (1958)	*Eupatorium fortunei* Turcz.
Eupatorium subhastatum	Ferraro and Coussio (1973), Ferraro et al. (1986, 1987, 1988b)	*Chromolaena hirsuta* (Hook. and Arn.) King and Robinson
Eupatorium tinifolium	Moreno et al. (1979), D'Agostino et al. (1990b, 1991a),	*Ageratina tinifolia* (H.B.K.) King and Robinson
Stevia nepetaefolia	Rajbhandari and Roberts (1984)	*Stevia suaveolens* Lagasca
Stevia purpurea	Bohlmann et al. (1976)	*Stevia eupatoria* (Sprengel) Willd.
Symphyopappus polystachyus	Mesquita et al. (1986)	*Symphyopappus compressus* (Gardner) B. Robinson

Acritopappus: This Brazilian genus consists of 13 species. The only report of flavonoids in this group is that of kaempferol 3-methyl ether from *A. hagei* (Bohlmann et al., 1980g).

Ageratina: *Ageratina* is a large genus, with 248 (King and Robinson, 1987) to ca. 290 (Bremer et al., 1994) species, distributed in the Americas and in the West Indies. Flavonoids with 6-oxygenation, moderate levels of *O*-methylation, and, in some cases, combinations of *O*-methylation and *O*-glycosylation have been identified. Details of the occurrences of these compounds can be found in Table 10–2. Compounds of note include dihydrokaempferol 7-methyl ether and naringenin 7-methyl ether from *A. tristis* (Bohlmann et al., 1985b), quercetagetin-6,7-dimethyl ether 3-*O*-apioside from *A. calophylla*, and, from the same source, 6-*C*-glucosylquercetin **[10–3]** (Fang et al., 1986e). This was the first report of *C*-glucosylquercetin from Nature. More recently, a flavonoid with a rare A-ring substitution

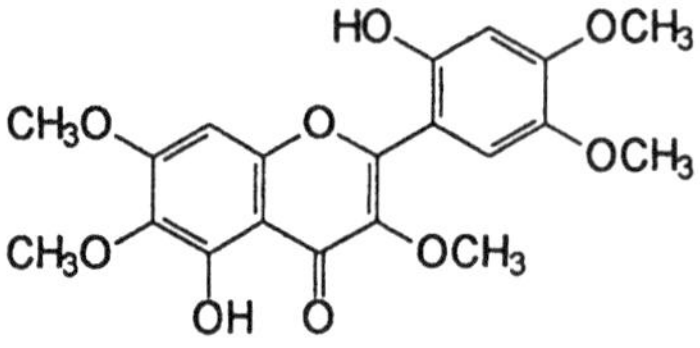

[10-1]

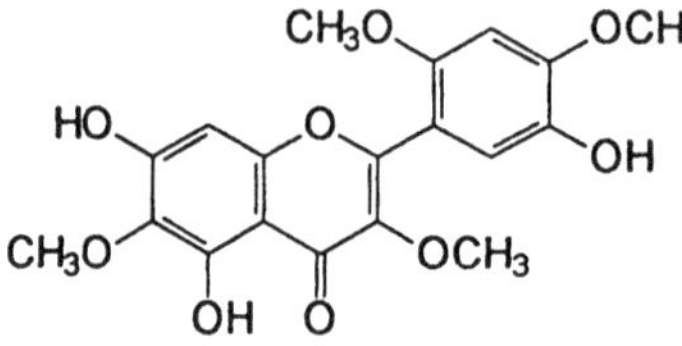

[10-2]

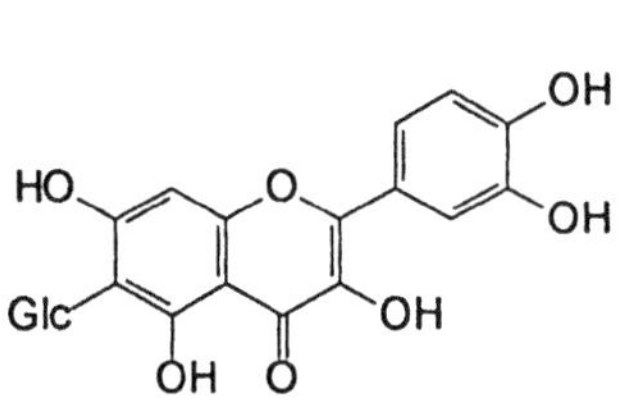

[10-3]

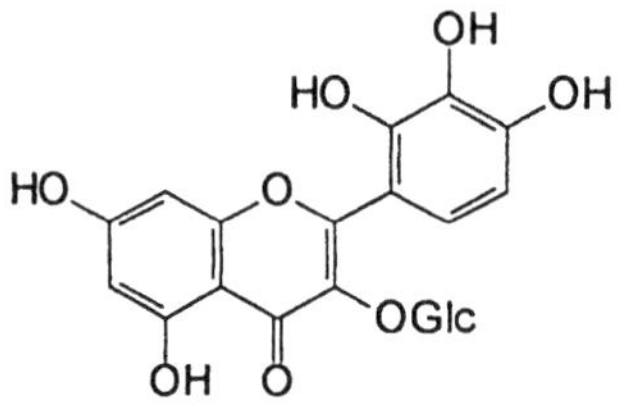

[10-4]

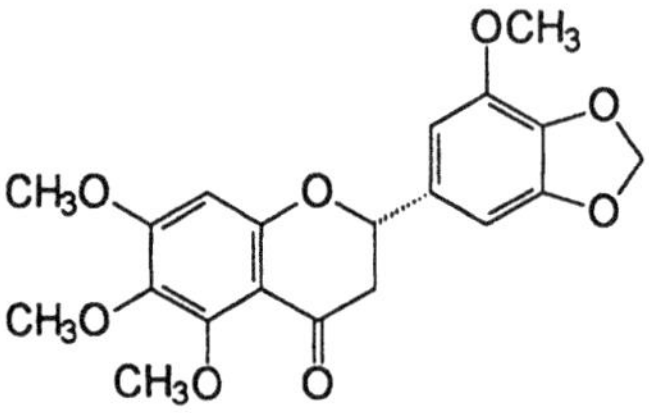

[10-5]

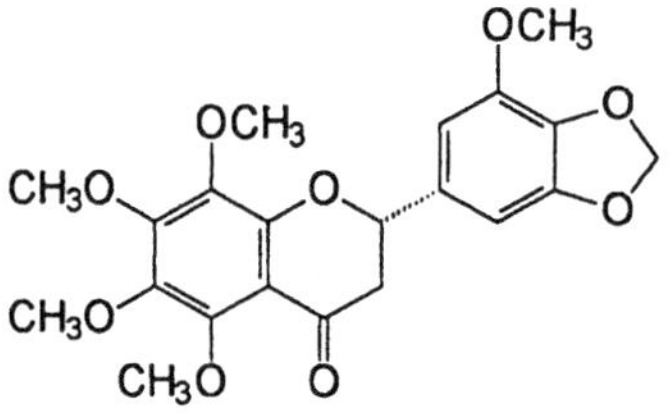

[10-6]

Fig. 10–1. Compounds **[10–1]** and **[10–2]** from *Acanthostyles buniifolium* exhibit the unusual 2',4',5'-trioxygenation pattern. Other unusal compounds from Eupatorieae include 6-*C*-glucosylquercetin **[10–3]** and the 2',3',4'-trihydroxyflavonol **[10–4]**. High levels of *O*-alkylation are exemplified by compounds **[10–5]** and **[10–6]**, both of which come from *Ageratum corymbosum*

pattern was isolated from *A. sternbergiana* (reported as *Eupatorium sternbergianum*). D'Agostino et al. (1994) determined the structure of the new compound to be 3,5,7,2',3',4'-hexahydroxyflavone 3-*O*-glucoside **[10–4]**.

Ageratum: This well known genus from tropical America consists of about 40 species only a few of which have been studied for their flavonoids. Nonetheless, some of the compounds identified are among the more unusual of the family, both with regard to oxygenation pattern and level of *O*-alkylation. The simplest compounds reported are kaempferol and quercetin *O*-glycosides from *A. mexicanum* (Mionskowski and Gill, 1973). One of the early hints of the complexity of *Ageratum* flavonoids came from the

Table 10–2. Flavones and flavonols from species of *Ageratina*

Species	Compounds[1]	References
A. adenophora[2]	357/64'-3-*O*-Gal	Nair and Sivakumar (1990)
A. altissima[3]	K & Q-3-*O*-Rut	Wagner et al. (1972f)
A. angustifolia	574'/6, 57/64'	de Villarraga and de Perez (1994)
A. arbutifolia	Q & Pat-3-*O*-Gal	Espitia de Perez and Villegas Molina (1995)
A. calophylla	574'/6, 57/64', 57/63'4', Q, Pat, Q-6-*C*-Glc, 3574'/6-3-*O*-Glc, 354'/67-3-*O*-Glys, 353'4'/67-3-*O*-Glys	Fang et al. (1986e)
A. deltoidea	53'/374', 53'/3674', 574'/36, 3574'/6, 53'4'/67, 563'4'/37, 53'4'/367, 5674'/3, 53'4'/37, 354'/67, 57/33'4', 574'/33', 564'/37	Yang et al. (1990)
A. fastigiata	5/74', 5/374'	Torrenegra et al. (1995)
A. gilbertii	5/674'	Herz and Gibaja (1972)
A. gracilis[4]	Q, Pat, 35784'/-, 3574'/8, 354'/78, 35673'4'/-	Torrenegra et al. (1984)
A. glyptophlebia	574'/6, 573'4'/6, 574'/63'	Espitia de Perez and Salgado de Pinilla (1990)
A. jahnii[5]	Acac, 5/74', 5/674', 5/374', 5/3673'	Morales-Mendez and Rosquete (1988)
A. ligustrina[6]	354'/67-3-*O*-Rhm, 353'4'/67-3-*O*-Rhm	Quijano et al. (1970)
	354'/7	Sagareishvili (1985)
	54'/67	Tamayo-Castillo et al. (1988)
A. saltillensis	Q & Pat-3-*O*-Glc, Q3Me, 3574'/6 3-*O*-Glys 354'/67 3-*O*-Rhm, 353'4'/67 3-*O*-Rhm, 573'4'/36; 573'/634'	Yu et al. (1986a)
A. tinifolia[7]	Acac; 5/74'; 574'/3; 57/34'; 5/374'	Moreno et al. (1979)
	IR-3-*O*-Glc, Pat-3'-*O*-Glc, Pat-3-*O*-Gal	D'Agostino et al. (1990b)
	34'/567-3-*O*-Glc	D'Agostino et al. (1991a)

isolation of 5,6,7,5'-tetramethoxy-3',4'-methylenedioxy **[10–5]** and 5,6,7,8,5'-pentamethoxy-3',4'-methylenedioxy flavanones **[10–6]** from *A. corymbosum* (Quijano et al., 1980). That work was followed by papers in which these

Table 10–2. (Continued)

Species	Compounds[1]	References
A. tomentella	574'/6, 573'4'/6, 574'/63', 53'/674', 54'/673', 5/673'4'	Fang and Mabry (1986)
A. viscosa	5/673'4', 5/783'4'	Torrenegra et al. (1990)

[1] Substitution: Hydroxyl positions/Methoxyl positions; Acac = acacetin; Q = quercetin; IR = isorhamnetin; Pat = patuletin; Glc = glucose; Gal = galactose; Rhm = rhamnose; Gly = glycoside; Me = methyl.
[2] Reported as *Eupatorium glandulosum.*
[3] Reported as *Eupatorium rugosum.*
[4] Reported as *Eupatorium gracile.*
[5] Reported as *Ageratina stevioides.*
[6] Reported as *Eupatorium ligustrinum.*
[7] Reported as *Eupatorium micranthum.*
[8] Reported as *Eupatorium tinifolium.*

workers (Quijano et al., 1982a, 1992) reported additional chalcones and flavanones from *A. corymbosum* (cited as *A. strictum*) including 3',6'-dihydroxy-2',4'-dimethoxy-3,4-methylenedioxychalcone, the corresponding flavanone, 6,4'-dihydroxy-5,7,3'-trimethoxyflavanone, 6-hydroxy-5,7,3',4'-tetramethoxyflavanone, and related compounds. Peralkylated flavones were first reported by Adesogan and Okunade (1979), who identified 5,6,7,8,3'-pentamethoxy-4',5'-methylenedioxyflavone, which is the flavone corresponding to **[10–6]**, and 5,6,7,8,3',4',5'-heptamethoxyflavone from *A. conyzoides*. The structure of the heptamethoxylated flavone was recently confirmed by synthesis (Shaw et al., 1990). Other studies by Quijano et al. (1980, 1982b, 1985, 1987) have resulted in identification of several flavones with 2',4',5'- and 2',3',4',5'-B-ring oxygenations. It is of interest to note that flavonols (as opposed to flavones) with extra oxygenation and high levels of *O*-alkylation do not appear to exist in this genus. Gonzalez et al. (1991) reported, among a number of other compounds, one that they identified as 5,6,8,3',4',5'-hexamethoxyflavone, which is noteworthy for its lack of a substituent at C-7. A recent study of this compound, however, led to a reassignment of its structure as 5,6,7,3',4',5'-hexamethoxyflavone (Horie et al., 1993). Details of the occurrence of flavones and flavonols in the genus can be found in Table 10–3.

Aristaguietia: *Aristaguietia* is an Andean genus of 21 species. Information on flavonoids of the genus has been reported for two species. 5,4'-Dihydroxy-6,7-dimethoxyflavone (cirsimaritin) was reported from *A. buddleaefolia* by Bohlmann et al. (1979b, 1980f), while 5,7-dihydroxy-

Table 10–3. Flavones and flavonols from species of *Ageratum*

Species	Compounds[1]	References
A. conyzoides	4'/5673'; 4'/5673'5';	
	4'/56783'; 4'/56783'5'	Vyas and Mulchandani (1986)
	8/5673'4'5'	Gonzalez et al. (1991)
	-/5673'4'	Vyas and Mulchandani (1986), Gonzalez et al. (1991)
	-/56783'4'	Vyas and Mulchandani (1986), Gonzalez et al. (1991)
	-/5673'4'5'	Vyas and Mulchandani (1986), Gonzalez et al. (1991)
	-/56783'4'5'	Adesogan and Okunade (1979), Vyas and Mulchandani (1986), Gonzalez et al. (1991)
	-/567/3'4'-MDO	Vyas and Mulchandani (1986),
	-/5678/3'4'-MDO	Vyas and Mulchandani (1986), Gonzalez et al. (1991)
	-/5675'/3'4'-MDO	Gonzalez et al. (1991)
	-/5673'/4'5'-MDO	Vyas and Mulchandani (1986)
	-/56783'/4'5'-MDO	Adesogan and Okunade (1979), Vyas and Mulchandani (1984)
	-/56783'/4'5'-MDO	Vyas and Mulchandani (1986)
	-/568785'/3'4'-MDO	Gonzalez et al. (1991)
A. corymbosum	52'4'/6785'	Quijano et al. (1980)
	-/56782'4'5'	Quijano et al. (1980, 1992)
	53'/672'4'5'; 5/672'4'5'; 5/672'4'5'; 5/672'3'4'5'; 3'/5672'4'5'; -/5672'3'4'5'; -/567/3'4'-MDO; -/5678/3'4'-MDO	Quijano et al. (1992)
A. houstonianum	-/5672'3'4'5'; -/56782'3'4'5';	Quijano et al. (1982b, 1987)
	53'/6782'4'5'; 3'/56782'4'5'	Quijano et al. (1985, 1987)
	-/5673'4'5'; 5/672'4'5'; 5/6782'4'5'; 52'/6784'5'	Quijano et al. (1987)
	K and Q Glys	Mionskowski and Gill (1973)[2]

1 Substitution: Hydroxyl positions/Methoxyl positions; MDO = methylenedioxy; K = kaempferol; Q = quercetin Gly = glycoside.

2 Reported as *Ageratum mexicanum*.

6,4'-dimethoxyflavone and 5-hydroxy-6,7,4'-trimethoxyflavone were reported (Gonzalez et al., 1990) from *A. salvia* (reported as *Eupatorium salvia*).

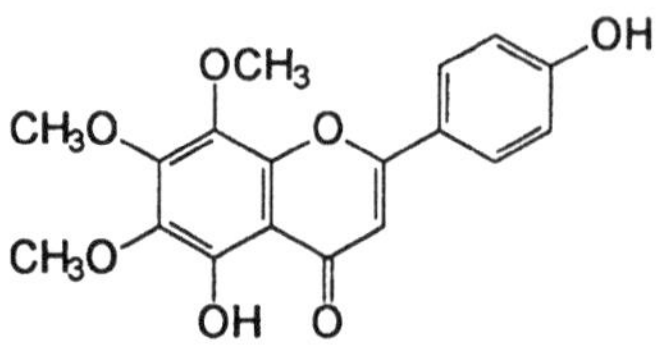

[10-7]

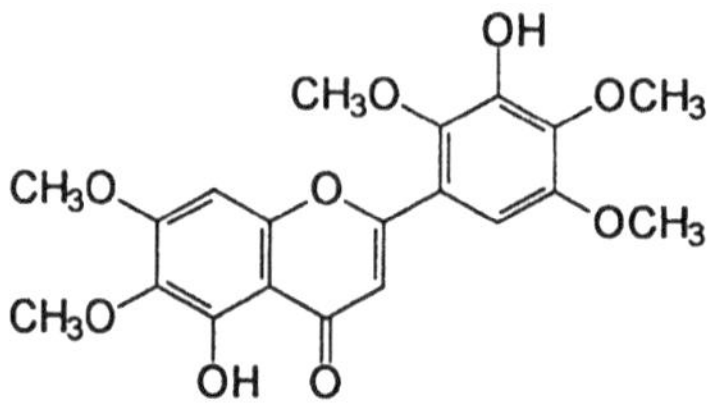

[10-8]

Fig. 10–2. Compound **[10–7]** appears to be the only 8-substituted flavonoid known from *Brickellia*. Compound **[10–8]**, also known from a species of *Brickellia*, exhibits the rare 2',3',4',5'-tetraoxygenation pattern

Austroeupatorium: *Austroeupatorium*, a genus of 13 species, occurs in South America but extends northward into Panama and Trinidad. Flavonoids of *A. inulaefolium* (cited as *Eupatorium inulaefolium* var. *suaveolens*) were described by Ferraro et al. (1977) as 5,6,3'-trihydroxy-7,4'-dimethoxyflavone and 5,6,3',4'-tetrahydroxy-7-methoxyflavone.

Bejaranoa: According to Bremer (1994), this South American genus consists of two species distributed in Bolivia, Brazil, and Paraguay. A recent study of both species for sesquiterpene lactones by Morgenstern et al. (1996) also resulted in the identification of 6-methoxyapigenin (hispidulin) from *B. balansae*.

Brickellia: *Brickellia* consists of 98 (Robinson and King, 1977) to about 110 (Bremer et al., 1994) species distributed throughout western North America, Mexico, and Central America with one, *B. diffusa*, known from throughout the Neotropical Region. We have flavonoid information for about 20 species. Members of the genus accumulate a variety of flavones and flavonols many of which exhibit 6-oxygenation. The only flavanones reported from the genus are naringenin 7-methyl ether, eriodictyol, eriodictyol 4'-methyl ether, and eriodictyol 7,3'-dimethyl ether from *B. vernicosa* (Ahmed et al., 1986). The report of 5,7-dihydroxyflavone (chrysin) from *B. laciniata* by Jakupovic et al. (1989b) represents the only record of a B-ring deoxyflavonoid from a member of this genus. *Brickellia eupatorioides* (cited as *Kuhnia eupatorioides*) was shown to accumulate 5,4'-dihydroxy-6,7,8-trimethoxyflavone (Fig. 10–2) **[10–7]**, which appears to be the only 8-substituted compound from the genus (Bohlmann et al., 1982c), and the more common 5-hydroxy-6,7,3',4'-tetramethoxyflavone (Herz, 1961). An interest in cytotoxic flavonoids led Mabry et al. (1980) to a study of a large number of *Brickellia* species. Well over 40 compounds were observed (although not listed in that paper) with the 3-*O*-rhamnosylglucosides of 6-

methoxykaempferol, quercetagetin 6,3'-dimethyl ether (spinacetin), and quercetagetin 6,7,3'-trimethyl ether, all from *B. chlorolepis*, reported as new at that time.

Flavonoid B-rings in *Brickellia* are commonly 4'- or 3',4'-based. A rare exception is the report of 5,3'-dihydroxy-6,7,2',4',5'-tetramethoxyflavone **[10–8]** from *B. glutinosa* (Goodwin et al., 1984). The report by Roberts et al. (1984) that 5,6'-dihydroxy-6,7,2',3',4'-pentamethoxyflavone (brickellin) existed in *B. chlorolepis* and *B. veronicaefolia* was challenged by Iinuma et al. (1985) who showed the correct structure to be 5,2'-dihydroxy-3,6,7,4',5'-pentamethoxyflavone. Eight species have been shown to accumulate flavonol sulphates. Occurrences of flavones and flavonols in species of *Brickellia* are listed in Table 10–4.

Brickelliastrum: *Brickelliastrum fendleri*, the sole member of the genus, occurs in New Mexico and western Texas. It is closely related to *Brickellia coulteri* with which it shares a suite of simple kaempferol and quercetin 3-*O*-glycosides (Bulinska-Radomska et al., 1985b; El-Sayed et al., 1988). *Brickelliastrum* also has isorhamnetin 3-*O*-glucoside and 5,4'-dihydroxy-3,6,7-trimethoxyflavone (Bulinska-Radomska et al., 1985b).

Campuloclinium: *Campuloclinium* is a genus of 14 species centered in Brazil but with representatives as far afield as Central America and Mexico. Gonzalez et al. (1973) reported rhamnocitrin 3-*O*-glucoside and dihydrokaempferol 7-methyl ether 3-*O*-glucoside from *C. macrocephalum* (reported as *Eupatorium macrocephalum*).

Carphochaete: Seven species according to Turner (1987) and Bremer et al. (1994) comprise this principally Mexican genus with one, *C. bigelovii*, extending into the southwestern part of the United States. Meuer and Mabry (1987) reported 5,7,4'-trihydroxy-6-methoxyflavone (hispidulin), 5,7,4'-trihydroxy-6,3'-dimethoxyflavone, 5,3'-dihydroxy-6,7,4'-trimethoxyflavone, 5,7,4'-trihydroxy-6,3',5'-trimethoxyflavone, and 5,3'-dihydroxy-6,7,4',5'-tetramethoxyflavone from *C. bigelovii*.

Chromolaena: King and Robinson (1987) recognize 165 species in this genus, which is centered in Brazil but occurs in other parts of South America and extends northward into Mexico and the southeastern United States. We have flavonoid data for several of their listed species including *C. odorata* (which appears in the literature as *Eupatorium odoratum*) and *C. hirsuta* (appearing as *E. subhastatum*). A diverse array of chalcones, flavanones, and flavones has been reported as well as a biflavanone. Structural features include B-ring deoxy flavonoids, oxygenation at C-8 and, more prevalently

Table 10–4. Flavones and flavonols from species of *Brickellia*

Species	Compounds[1]	References
B. arguta var. *odontolepis*	IR 3-*O*-Glc, 574'/33', 3574'/6 3-*O*-Rob, 3573'4'/6 3-*O*-Gal, 54'/3673', 3573'4'/6 3-*O*-Rob,	Rosler et al. (1984)
B. argyrolepis[2]	54'/367	Bohlmann and Zdero (1969)
B. baccharidea	574'/6, 5/674', 573'/64', 53'/674', 3573'4'/6 3-*O*-Glys, 3-Sulf, 353'4'/67 3-*O*-Glc, 353'/674' & 3-Sulf, 573'/364', 53'/3674', 5/3673'4'	Timmermann et al. (1981)
B. californica	574'/6, 54'/67, 57/64', 573'4'/6, 53'4'/67, 573'/64', 3573'4'/6 3-*O*-Glys & 3-Sulf, 353'4'/67 3-*O*-Gal, 3574'/63', 353'/674' & 3-Sulf	Mues et al. (1979)
B. cavanillesii[3]	54'/367-4'-*O*-Glc	Flores and Herran (1960)
B. chlorolepis	Q-3-*O*-Gly, IR 3-*O*-Gal, 3574'/6 3-*O*-Gly, 3573'4'/6, 3574'/63' 3-*O*-Gly, 353'4'/67 3-*O*-Gly, 353'/674' 3-Sulf, 354'/673' 3-*O*-Gly 53'/3674', 5/3673'4'	Ulubelen et al. (1980b)
B. coulteri	K & Q 3-*O*-Glcs	El-Sayed et al. (1988)
B. cylindracea	573'/364', 57/363'4', 5/3673'4', Pat 3-*O*-Glc	Timmermann and Mabry (1983)
B. dentata	57/64', 5/674', 573'4'/6, 53'4'/67, 53'/674', 354'/67 3-*O*-Gal, 353'4'/67 & 3-*O*-Glys, 353'/674'	Ulubelen et al. (1980b)
B. diffusa	K & Q 3-*O*-Glys, 353'4'/7-3-*O*-Rhm	Ahmed et al. (1988c)
B. eupatorioides	54'/678	Bohlmann et al. (1982c)
B. glutinosa	Q, 573'4'/3, 53'4'/37, 54'/367, 573'4'/36-7-*O*-Glc, 353'4'/67, 573'/364'-7-*O*-Glc, 574'/363'-4'-Sulf, 53'4'/367, 57/363'4', 53'/672'4'5'	Goodwin et al. (1984)
B. laciniata	57/- 573'/4', 57/64', 5/674', 354'/67-3-*O*-Gal, 3573'4'/6-3-*O*-Glys, 3-Sulf, 353'4'/67 & 3-*O*-Gal, 3574'/63', 3573'/64', 573'/634', 353'/674' & 3-Sulf, 53'/3674'	Jakupovic et al. (1989b) Timmermann et al. (1979)
B. longifolia	354'/67, 35/674'-3-Sulf, 353'4'/67-3-*O*-Gal, 353'/674'-3-Sulf	El-Sayed et al. (1990)
B. monocephala	354'/67-3-*O*-Glys, 573'4'/3,	

Table 10–4. (Continued)

Species	Compounds[1]	References
	353'4'/67-3-*O*-Glys	Norris and Mabry (1985)
B. pendula	54'/367-4'-*O*-Glc	Flores and Herran (1958)
B. schaffneri	5/3674', 5/3673'4'	Mericli et al. (1989)
B. scoparia	3574'/6-3-*O*-Glc, 354'/67-3-*O*-Glc, 57/364', 5/3674', 3573'4'/6-3-Glc, 574'/36, 573'4'/36, 573/364'	Li et al. (1986)
B. vernicosa	54'/7, K & Q 3-*O*-Glc, 54'/37, 3574'/6, & 3-*O*-Rhm, 354'/67-3-*O*-Rhm, 573'4'/3, 353'4'/7, 3573'4'/6-3-*O*-Glys, 353'4'/67-3-*O*-Glys	Ahmed et al. (1986)
B. veronicaefolia	3573'4'/6-3-Sulf, 353'4'/67 3-Glc, 3-Sulf, 354'/673'-3-DiGal, 3-Sulf, 353'/674' & 3-Sulf, 53'4'/367, 53'/3674', 5/3673'4'	Roberts et al. (1980)

[1] Substitution: Hydroxyl positions/ Methoxyl positions; K = kaempferol; Q = quercetin; IR = isorhamnetin; Glc = glucose; Gal = galactose; Rhm = rhamnose; Rob = robinoside; Gly = glycoside; Sulf = sulfate.

[2] Reported as *B. guatemaliensis* in Bohlmann and Zdero (1969).

[3] Reported as *B. squarrosa* in Flores and Herran (1960).

at C-6, and a high level of *O*-methylation. The reader can get a good idea of the flavonoid richness of the species by consulting Wollenweber et al. (1995a) who were interested in the exudate chemistry of this weedy species in relationship to its interaction with the possible biological control agent *Pareuchaetes pseudoinsulata*. They reported a series of common mono- and dimethyl ethers of kaempferol and quercetin along with acacetin and quercetagetin 6,4'-dimethyl ether (laciniatin). Naringenin and taxifolin derivatives from these species are listed in Table 10–5. A recent follow-up on that work (Wollenweber and Roitman, 1996) described identification of the novel quercetagetin derivative 3-hydroxy-5,6,7,3',4'-pentamethoxyflavone from *C. odorata*. Baez et al. (1998) reported kaempferol and kaempferol and quercetin 3-*O*-rutinosides from *C. moritziana* but found no flavonoids in the dichloromethane leaf wash, which eliminated flavonoids as the causative agents for the antibacterial activity of the leaf wash extract.

The biflavanone, described by Ferraro et al. (1986, 1988b), is formally known as I-5,II-5,I-7,II-7-tetrahydroxy-I-4',II-4'-dimethoxy-I-8,II-8-biflavanone. It can be more efficiently pictured, perhaps, as two

Table 10–5. Flavanones and dihydroflavonols from species of *Chromolaena*

Species	Compounds[1]	References
C. chaseae	57/-, 357/-, 357/6	Bohlmann et al. (1982e)
C. connivens	574'/-, 54'/7	Tamayo-Castillo et al. (1989)
C. heteroclinia	64'/573', -/5673'4', 4'/5673', 6/573'4'	Boeker and Bohlmann (1986)
C. hirsuta[2]	573'4'/-	Ferraro and Coussio (1973)
	573'4'/6	Ferraro et al. (1987)
C. meridensis	53'4'/7	Amaro and Morales-Mendez (1983)
C. odorata[3]	54'/7	Bose et al. (1973)[3]
	5/74', 54'/73'	Arene et al. (1978)[3]
	4'/567	Barua et al. (1978)
	54'/7, 57/4'	Metwally and Ekejiuba (1981)[3], Iwu and Chiori (1984),
	354'/7	Wollenweber et al. (1995a)
C. subscandens	54'/7, 54'/67, 57/84'	Amaro and Delgado-Mendez (1993)
C. tunariensis	4'/567, -/5674'	Bohlmann et al. (1979d)

[1] Substitution: Hydroxyl positions/Methoxyl positions.
[2] Reported as *Eupatorium subhastatum*.
[3] Reported as *Eupatorium odoratum*.

naringenin-4'-methyl ether molecules joined through their C-8 positions **[4–97]**.

Bohlmann et al. (1982e) reported 2',4',6'-trihydroxychalcone from *C. chasae*. Several tri- and tetramethyl ethers have been reported from *C. odorata* (reported at *Eupatorium odoratum*) based on 2',4',5',6',4-pentahydroxychalcone (Bose et al., 1973; Arene et al., 1978; Barua et al., 1978; Nguyen et al., 1993). In the paper by Barua et al. (1978) 2'-hydroxy-4',5',6',3,4-pentamethoxychalcone was also reported. The array of flavanones and dihydroflavonols in *Chromolaena* is more complex and can be appreciated best by referring to Table 10–5. Flavones and flavonols from *Chromolaena* are listed in Table 10–6.

Conoclinium: *Conoclinium* is a genus of three species native to southwestern United States and adjacent Mexico. Three groups of workers, Le-Van and Pham (1979), Herz et al. (1980), and Martinez-Vasquez et al. (1993), have identified flavonoids from two of them, *C. coelestinum* (cited as *Eupatorium coelestinum*) and *C. greggii*. The pigment profiles of the two species are quite similar consisting of extensively *O*-alkylated derivatives of four aglycones, 5,6,7,3',4'-pentahydroxyflavone, 5,6,7,3',4',5'-hexahydroxyflavone,

Table 10–6. Flavones and flavonols from species of *Chromolaena*

Species	Compounds[1]	References
C. arnottiana	53'/674', 5/673'4', 5/673'4'5'	de Gutierrez et al. (1995)
C. chaseae	357/-, 357/6	Bohlmann et al. (1982e)
C. heteroclinia	-/5673'4'	Boeker and Bohlmann (1986)
C. hirsuta[2]	Q-3-*O*-Glys, 573'4'/6	Ferraro and Coussio (1973)
	57/4', 574'/6, 53'/674'	Ferraro et al. (1987)
C. meridensis	54'/73', 353'4'/7, 353'4'/67	Amaro and Morales-Mendez (1983)
C. morii	54'/373', 54'/3783'	Bohlmann et al. (1981d)
C. moritziana	K & Q 3-*O*-Glys	Baez et al. (1998)
C. odorata[3]	5/674'	Talapatra et al. (1974)
	-/5674', -/5673'4'	Barua et al. (1978)
	357/4', 3573'/4'	Metwally and Ekejiuba (1981), Iwu and Chiori (1984), Wollenweber et al. (1995a)
	57/4', 354'/7, 35/74', Q-3-Rut, 353'4'/7, 353'/74', 35/73'4', 3573'/64'	Wollenweber et al. (1995a)
	64'/3573'	Iwu and Chiori (1984)
C. subscandens	K-3-O-Rhm	Amaro and Delgado-Mendez (1993)

1 Substitution: Hydroxyl positions/Methoxyl positions; K = kaempferol; Q = quercetin; Gly = glycoside; Rhm = rhamnose.
2 Reported as *Eupatorium subhastatum*.
3 Reported as *Eupatorium odoratum*.

5,6,7,8,3',4'-hexahydroxyflavone, and 5,6,7,8,3',4',5'-heptahydroxyflavone. Among the more highly *O*-alkylated compounds (Fig. 10–3) identified were 3'-hydroxy-5,6,7,8-tetramethoxy-4',5'-methylenedioxyflavone **[10–9]** from *C. coelestinum* and 5,6,7,8,3'-pentamethoxy-4',5'-methylenedioxyflavone **[10–10]** from both taxa.

Critonia: *Critonia* consists of 40 or so species distributed from Mexico to southern Argentina. Only *C. daleoides* has been studied for flavonoids from which Yang et al. (1989) reported kaempferol and quercetin 3-*O*-galactosides and 3-*O*-rhamnosylgalactosides.

Decachaeta: Seven species constitute this Mexican and Central American genus. Flavonoid data are available for three. Features of this group include the presence of both flavones and flavonols, the presence of 6-oxygen-

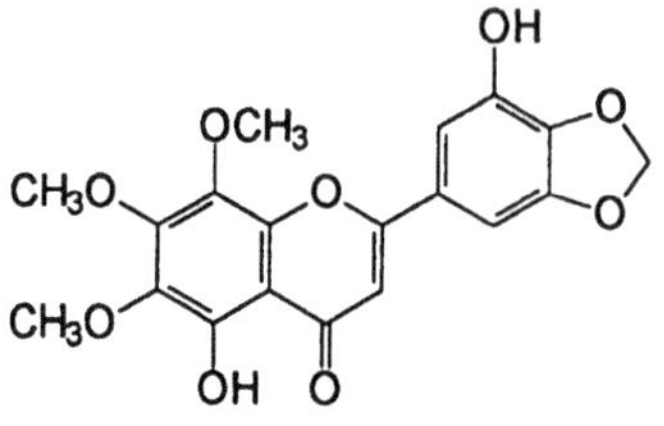

[10-9]

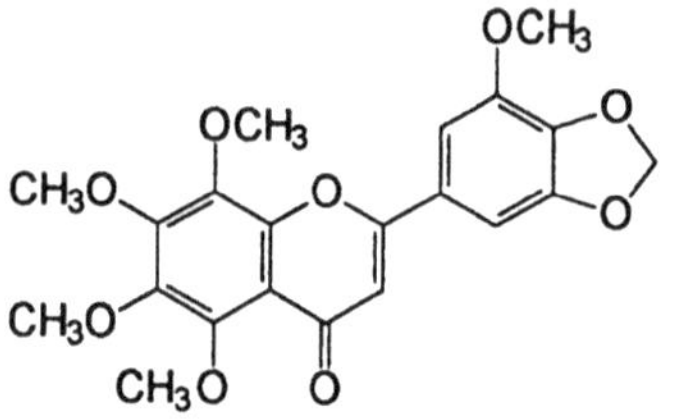

[10-10]

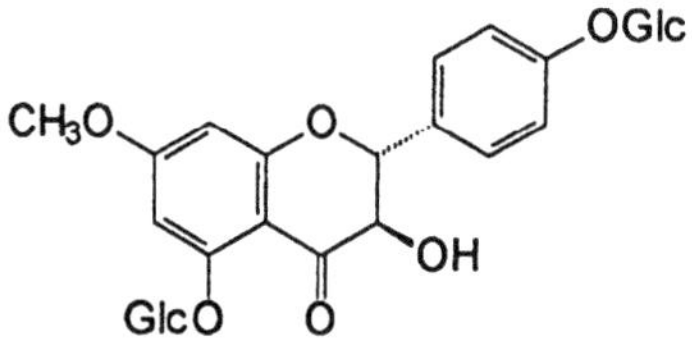

[10-11]

Fig. 10–3. Other compounds exhibiting high levels of *O*-alkylation include **[10–9]** and **[10–10]** known from species of *Conoclinium*. Compound **[10–11]** is unusual in its 5,4'-di-*O*-glycosylation pattern

Table 10–7. Flavones and flavonols from species of *Decachaeta*

Species	Compounds[1]	References
D. haenkeana	3574'/63'; 357/63'4'	
	353'/674' & 3-sulphate	Miski et al. (1985)
D. ovatifolia	574'/6; 57/64'; 54'/67;	
	3574'/6 3-*O*-Glc;	
	3573'4'/6 3-*O*-Glys	de Luengo and Mabry (1986)
D. scabella	574'/6; 54'/67	Miski et al. (1987)

[1] Substitution: Hydroxyl positions/Methoxyl positions; Glc = glucose; Gly = glycoside.

ation, a modest level of *O*-methylation, and the presence of a sulphated flavonol. Details appear in Table 10–7.

Disynaphia: Another South American genus, *Disynaphia*, is considered by King and Robinson (1987) to comprise 16 species. Rutin was reported by Sato and Moreira (1985) as a component of *D. littoralis* (cited as *Eupatorium littorale*).

Table 10–8. Flavones and flavonols from species of *Eupatorium*

Species	Compounds[1]	References
E. adenophorum	54'/67, 35674'-7-*O*-Glc, 354'/67-3-*O*-Glc, Qt-7-*O*-Glc, 3573'/4'-7-*O*-Glc	Li et al. (1997)
E. africanum	57/6 7-*O*-Rhm-Glc, 53'4'/67 4'-*O*-Glc, 353'4'/67 3-*O*-Gal, 354'/673' 3-*O*-Gal-Glc, 54'/6783' 4'-*O*-Gal	Aqil (1995)
E. album	K & Q-3-*O*-Rut	Wagner et al. (1972f)
E. altissimum	53'/674',5/673'4'	Dobberstein et al. (1974, 1977), Herz et al. (1978)
E. buniifolium	Q 3-*O*-Gal, Q-3-*O*-Rut	Muschietti et al. (1990)
E. cannabinum	K & Q-3-*O*-Glys	Grzybowski and Jerzmanowska (1954), Stefanovic et al. (1986), Aquino et al. (1988)
	574'/6, 57/64', 573'4'/6, 574'/63', 57/364', 573'/364'	Stevens et al. (1995)
E. capillifolium	K & Q-3-*O*-Glc	Wagner et al. (1972f)
E. capillifolium × *E. perfoliatum*	54'/7, 573'4'/3	Herz et al. (1972c)
E. coelestinum	K & Q 3-*O*-Glcs & 3-O-Ruts	Wagner et al. (1972f)
E. cuneifolium	K 3-*O*-Glc, Q 3-*O*-Gal	Wagner et al. (1972f),
	574'/6, 573'4'/6	Kupchan et al. (1969a)
E. fortunei	353'/674'	Nakaoki and Morita (1958)
E. glandulosum	35674'/- 7-*O*-Glc(Caff)	Nair et al. (1993)
	35674'/- 7-*O*-Glc, Qt-7-*O*-Glc(Acetyl), 35673'/4'-7-*O*-Glc(Caff), Qt-7-*O*-Glc, Q-7-*O*-Glc	Nair et al. (1995)
E. grayanum	K & Q 3-*O*-Glcs & 3-O-Ruts	D'Agostino et al. (1990a)
E. hyssopifolium	K & Q 3-*O*-Glys	Wagner et al. (1972f)
E. laeve	K & Q 3-*O*-Glys	Schmeda et al. (1983)
E. leptophyllum	IR-3-*O*-Glys	Wagner et al. (1972f, 1972l)
E. leucolepis	574'/63', 4'/56783', 4'/56783'5', 3'/56784'5', -/5678/3'4'-MDO, -/56785'/3'4'-MDO, -/56783'4'	Herz and Kulanthaivel (1982)
E. mohrii × *rotundifolium*	K & Q 3-O-Glys	Wagner et al. (1972f)
E. perfoliatum	K & Q 3-O-Glys	Wagner et al. (1972f)

Table 10–8. (Continued)

Species	Compounds[1]	References
E. semiserratum	53'/674'	Kupchan et al. (1965)
	353'/674', 33'/5674'	Kupchan et al. (1969a)
	57/64', 53'/674'	Kupchan et al. (1969b)
E. semiserratum var. *lancifolium*	574'/6, 57/64', 5/674', 53'/674'	Herz et al. (1981)
E. serotinum	574'/6, 57/64'	Herz et al. (1979)
E. tinifolium	34'/567 4'-*O*-Glc	D'Agostino et al. (1990b)

[1] Substitution: Hydroxyl positions/Methoxyl positions; K = kaempferol; Q = quercetin; IR = isorhamnetin; Qt = quercetagetin; Gly = glycoside; Glc = glucose; Gal = galactose; Rut = rutinose; MDO = methylenedioxy; Caff = caffeoyl.

Eupatorium: *Eupatorium*, a genus of 48 species, is North American, occurring from the Rocky Mountains eastward. *Eupatorium* has been fairly well studied chemically so that we have information on flavonoids from over 30 species. A taxonomic and nomenclatural problem exists, however, insofar as at least half of the names taken from the literature do not appear as species of *Eupatorium* recognized by King and Robinson (1987). Individual situations will be commented upon as they arise, e.g., *Chromolaena odorata* = *Eupatorium odoratum*. The reader should refer to the following generic entries for information on taxa that had at one time been considered members of *Eupatorium*: *Acanthostyles*, *Ageratina*, *Aristeguietia*, *Austroeupatorium*, *Campuloclinium*, *Disynaphia*, and *Piptothrix*.

The flavonoid picture in *Eupatorium* is a complex one. A-Ring extra oxygenation is seen in many compounds and a variety of *O*-methylated compounds exist (e.g., Wagner et al., 1971a). Flavone and flavonol derivatives of *Eupatorium* sensu King and Robinson are listed in Table 10–8. Included in that table is the report of quercetagetin 4'-methyl ether 7-*O*-(6"-*E*-caffeoyl) glucoside from *E. glandulosum* (Nair et al., 1995). 6,8-Di-*C*-glucosylapigenin was reported from *E. serotinum* by Wagner et al. (1972f). *O*-Methylated derivatives of naringenin and eriodictyol have been isolated from three species, *E. buniifolium* (Muschietti et al., 1990), *E. havanense* (Dominguez and de la Fuente, 1973; see also Yu et al., 1987b), and *E. sternbergianum* (Gonzalez et al., 1982, 1984b). Dihydroflavonol derivatives have been reported from *E. capillifolium* and *E. capillifolium* × *E. perfoliatum* (Herz et al., 1972c) and from *E. micranthum* (Sagareishvili et al., 1981). These last workers identified dihydrokaempferol 7-methyl ether 5,4'-di-*O*-glucoside **[10–11]**, which represents a very unusual glycosylation pattern.

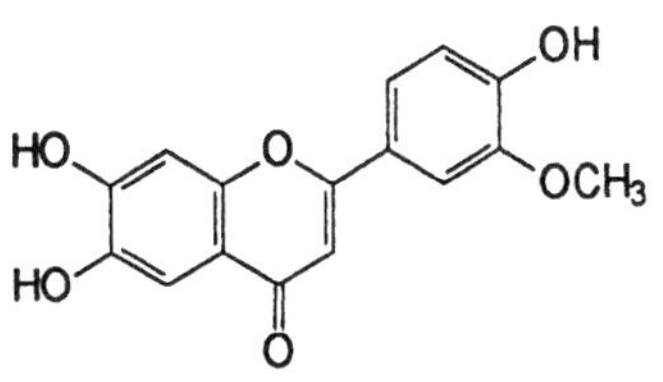

[10-12]

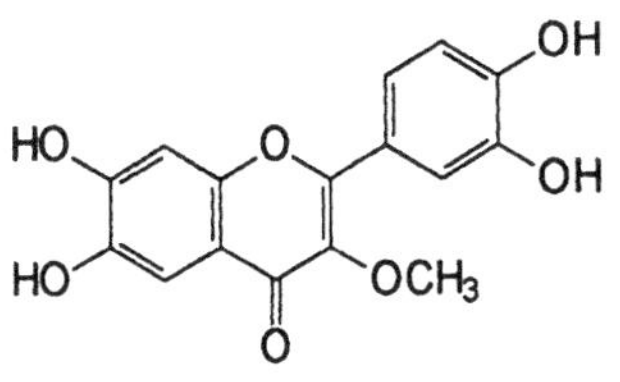

[10-13]

Fig. 10–4. Compounds **[10–12]** and **[10–13]**, both from *Graziela mollissima*, are characterized by the absence of 5-oxygenation

Fleischmannia: *Fleischmannia* consists of about 80 species distributed from southeastern United States south to Argentina with a concentration of species in Mexico, Central America, and the Andes. We have flavonoid information on only one species. *Fleischmannia viscidipes* from Guatemala yielded 5-hydroxy-7,3',4'-trimethoxyflavone (Bohlmann et al., 1984b).

Flyriella: *Flyriella* is a genus of six species distributed in northern Mexico and southern Texas. Mabry et al. (1981) described the flavonoid profiles of five of them. Two quercetin 3-*O*-glycosides plus quercetin 4'- and 7,4'-methyl ethers (present as diglycosides) were identified, although not all specimens had all these flavonoids. *Flyriella parryi* had all four compounds, *F. leonensis* and *F. chrysostyla* had the later three compounds, *F. sphenopoda* had only the two methylated compounds, and *F. stanfordii* had the simplest profile with only quercetin 3-*O*-arabinoside.

Goyazianthus: *Goyazianthus tetrastichus*, the sole member of this genus, occurs in Brazil. Bohlmann et al. (1982b) reported apigenin and naringenin from this species.

Graziela: *Graziela mollissima* is the only one of 10 species of this South American genus, known from Argentina, Brazil, and Uruguay (Bremer, 1994), that has been studied for flavonoids. Nakashima et al. (1994) reported two unusual compounds (Fig. 10–4), 6,7,4'-trihydroxy-3-methoxyflavone **[10–12]** and 6,7,3',4'-tetrahydroxy-3-methoxyflavone **[10–13]**. Note the lack of C-5 oxygenation in these compounds.

Lasiolaena: *Lasiolaena* is a Brazilian genus of five species. Bohlmann et al. (1981e) reported 5-hydroxy-6,7,3',4'-tetramethoxyflavone from *L. santosii*.

Liatris: The names of 42 recognized species and hybrids belonging to *Liatris* are listed by King and Robinson (1987). The genus occurs largely in eastern North America with some representatives in Colorado and New Mexico. Flavonoid information is available for several species. A survey of ten species (Wagner et al., 1973b) disclosed kaempferol and/or quercetin glycosides in most and vicenins (*C*-glycosylflavones) in half. *Liatris aspera* contains 5-hydroxy-6,7,4'-trimethoxyflavone, 5,3'-dihydroxy-6,7,4'-trimethoxyflavone, and 5-hydroxy-6,7,3',4'-tetramethoxyflavone (Herz and Kulanthaivel, 1983). Iyengar et al. (1976a) reported myricetin 4'-methyl ether (3,5,7,3',5'-pentahydroxy-4'-methoxyflavone) from *L. pauciflora*. Quercetin 3-*O*-glucoside, 3-*O*-rutinoside, and 3-*O*-glucoside-7-*O*-rhamnoside were reported from *L. spicata* by Kagan (1968), quercetin and myricetin from *L. pauciflora* (Iyengar, 1975).

Lourteigia: Nine species constitute this South American genus. 6-Methoxyapigenin, 6-methoxyluteolin, and quercetin 3-methyl ether were reported from *L. ballotaefolia* by Triana (1984). Kaempferol, quercetin, rutin and, 5-hydroxy-6,7,3',4'-tetramethoxyflavone were reported from *L. microphylla* [cited as *Eupatorium micropillum* (sic)] (Torrenegra et al., 1985).

Mikania: *Mikania* is a genus of 415 (Robinson and King, 1977) to 430 (Bremer et al., 1994; Holmes, 1995) species, pantropical in distribution, but best known from South America. Despite the size of the genus, we have flavonoid data for only a few species. *Mikania batatifolia* afforded 5,6,7,4'-tetrahydroxy-3'-methoxyflavone (Herz et al., 1969, 1970d), while *M. cordata* provided 3,5-dihydroxy-6,7,4'-trimethoxyflavone (Kiang et al., 1965; Wagner et al., 1967). Bohlmann et al. (1982f) reported simple flavanones from two species, naringenin 7-methyl ether from *M. purpurascens*, and naringenin and naringenin 4'-methyl ether from *M. grazielae* (Bohlmann et al., 1982a). *Mikania cordifolia* accumulates quercetin 3-*O*-glucoside (D'Agostino et al., 1991c).

Nothobaccharis: *Nothobaccharis candolleana* is the sole member of this Peruvian genus. The flavonoids of this taxon, reported as *Brickellia microphylla* by El-Sayed et al. (1988), are patuletin 3-*O*-mono- and di-*O*-glycosides, patuletin 3-sulphate, quercetagetin 6,3,3'-trimethyl ether and quercetagetin-6,3',4'-trimethyl ether and its 7-*O*-glucoside.

Ophryosporus: This genus of 37 species is found only in South America. Flavonoid information is available for four species. Ferracini et al. (1989) reported apigenin 7-methyl ether (genkwanin) and apigenin 7,4'-dimethyl ether from *O. heptanthus*, Zdero et al. (1990e) reported naringenin and eriodictyol 7-methyl ethers from *O. floribundus*, Sigstad et al. (1993) reported naringenin 7-methyl ether from *O. macrodon*, and Favier et al. (1997) reported apigenin 7,4'-dimethyl ether, cirsimaritin, salvigenin, and eriodictyol 7-methyl ether from *O. charrua*.

Piptothrix: *Piptothrix* consists of five species, four of which occur only in Mexico. *Piptothrix areolaris*, the only species for which flavonoid data are available, occurs in Guatemala as well as in Mexico. A moderately complex assemblage of compounds was reported from this taxon (recorded as *Eupatorium areolare* var. *leiocarpum*) by Yu et al. (1986c). The flavonoids reported were: quercetin and its 3-*O*-glucoside and 3-*O*-galactoside, 6-methoxykaempferol and its 3-*O*-glucoside and 3-*O*-galactoside, 6-methoxyquercetin (patuletin) and its 3-*O*-glucoside and 3-*O*-galactoside, quercetagetin 6,7-dimethyl ether and its 3-*O*-glucoside and 3-*O*-galactoside, quercetin-7,4'-dimethyl ether (ombuin), 3,5,4'-trihydroxy-6,7-dimethoxyflavone, and 5,4'-dihydroxy-3,6,7-trimethoxyflavone.

Stevia: *Stevia* is a large genus, 230 species according to King and Robinson (1987), distributed from southwestern North America south through Central America to central South America. We have flavonoid chemistry for 18 species from King and Robinson's (1987) list. The flavonoid chemistry of *Stevia* consists of flavones and flavonols, many of which occur as *O*-glycosides. 6-Oxygenation is common. 8-Oxygenation, in contrast, is seen only in 5,6-dihydroxy-7,8,4'-trimethoxyflavone, which occurs in *S. berlandieri* (Dominguez et al., 1974) and *S. mercedensis* (Mendiondo and Seeligmann, 1985). No other compounds were reported from either of these species. *O*-Methylation is common in members of the genus. The only flavanone reported from the genus to date is naringenin 7-methyl ether, which Bohlmann et al. (1976) isolated from *S. purpurea* (= *S. eupatoria*). *Stevia nepetifolia* (= *S. suaveolens*) has an array of flavones and flavonol *O*-glycosides entirely in line with other species of *Stevia*. Details of flavonoid occurrences in this genus are given in Table 10–9.

Steviopsis: This Mexican genus consists of seven species (Turner, 1988a) only one of which, with two varieties, has been examined for flavonoids (Yu et al., 1986b). *Steviopsis thyrsiflora* var. *solidaginifolia* (cited as *Asanthus solidaginifolia*; see Turner, 1988a, for synonymy] afforded 6-hydroxyapigenin and 6-hydroxyluteolin mono and dimethyl ethers, kaempferol, quercetin, and isorhamnetin 3-*O*-glycosides, and quercetin 3-methyl ether 7-*O*-glucoside. *Steviopsis thrysiflora* var. *thyrsiflora* (cited as *Asanthus thyrsiflora*), on the other hand, had a much simpler pigment profile consisting of 6-methoxyapigenin, 6-methoxyluteolin, and two quercetin 3-*O*-glycosides.

Table 10–9. Flavonoids from species of *Stevia*

Species	Compounds[1]	References
S. berlandieri	56/784'	Dominguez et al. (1974)
S. cuzcoensis	Q 3-*O*-Glys, 574'/363', 573'/364'	Rajbhandari and Roberts (1985b)
S. eupatoria[2]	Nar 7-*O*-Me	Bohlmann et al. (1976)
S. galeopsidifolia	Q 3-*O*-Glys, 573'/364'	Rajbhandari and Roberts (1985b)
S. jujuyensis	574'/63'	Gil et al. (1992)
	5/3674', 5/3673'4'	De Gutierrez et al. (1992)
S. lucida	57/64', 57/63'4'	Amaro and Hung (1985)
S. mercedensis	56/784'	Mendiondo and Seeligmann (1985)
S. microchaeta	Lute *O*-Glys, Q 3-*O*-Glys	Rajbhandari and Roberts (1985a)
S. monardifolia	Q 3-*O*-Glys, 574'/36', 573'/364'	Rajbhandari and Roberts (1985a)
S. origanoides	Q 3-*O*-Glys, 574'/6, 54'/67' 4'-*O*-Glc, 53'/674', 57/364', 573'/364'	Rajbhandari and Roberts (1985a)
S. procumbens	53'/674', 5/3673'4'	Sosa et al. (1985)
S. rebaudiana	Apig *O*-Glys	Rajbhandari and Roberts (1983), Matsuo et al. (1986)
	Lute 7-*O*-Glc, K 3-*O*-Rhm, Q 3-*O*-Glys, 573'/364'	Rajbhandari and Roberts (1983)
	Q-3-*O*-Glys	Komissarenko et al. (1994)
S. sanguinea	574'/6	Gil et al. (1989)
S. satureifolia	54'/67, 53'/674'	Sosa et al. (1984)
S. serrata	Q 3-*O*-Glys, 57/363'4'	Rajbhandari and Roberts (1985b)
S. soratensis	Apig 7-*O*-Glc, Q 3-*O*-Gal	Rajbhandari and Roberts (1985b)
S. suaveolens[3]	Lute 7-*O*-Glc, Q 3-*O*-Glys, 573'/364'	Rajbhandari and Roberts (1984)
S. vaga	57/63'4', 53'/3674'	Hernández et al. (1996)

[1] Substitution: Hydroxyl positions/Methoxyl positions; Apig = apigenin; Lute = luteolin; Nar = naringenin; Q = quercetin; Gly = glycoside; Glc = glucoside; Gal = galactose; Rhm = rhamnoside; Me = methyl.

[2] *Stevia purpurea* in Bohlmann et al. (1976).

[3] *Stevia nepetifolia* in Rajbhandari and Roberts (1984).

The pigment profile of the latter variety was a subset of the profile of the former.

Symphyopappus: Eleven species restricted to Brazil comprise this genus. We have flavonoid data for two. 5-Hydroxy-3,6,7,3',4'-pentamethoxyflavone (artemetin) has been identified from *S. reticulatus* (Bohlmann et al.,

1981i). *Symphopappus compressus* (reported as *S. polystachyus*) afforded apigenin 7-methyl ether (genkwanin) (Mesquita et al., 1986).

Urolepis: *Urolepis hecatantha*, the sole member of this genus, occurs in Argentina, Bolivia, Brazil, Paraguay, and Uruguay. A recent paper by de Gutierrez et al. (1995) recorded 5-hydroxy-3,7,3',4'-tetramethoxyflavone (retusin), and 5,6,3'-trihydroxy-3,7,4'-trimethoxyflavone (oxyayanin) from this species.

CHAPTER 11

Flavonoids of Heliantheae s.l.

The tribe Heliantheae is one of the largest and most morphologically variable of Asteraceae. This size and complexity has led to two different subtribal classifications in recent years (Robinson, 1981; Stuessy, 1977), which in turn have stimulated detailed morphological cladistic analyses and discussions of the entire tribe (Karis, 1993b; Karis and Ryding, 1994a,b) and its parts (e.g., Bolick, 1983; Ryding and Bremer, 1992; Karis, 1995; Strother, 1999) plus recent molecular analyses (Panero et al., 1999).

Central to classification of Heliantheae is the relationship to and disposition of Helenieae. Traditionally treated as a distinct, yet related, tribe (e.g., Bentham, 1873), it was submerged into Heliantheae s.s. by Cronquist (1955). Turner and Powell (1977) continued this perspective and suggested specific referrals of taxa of Helenieae to other tribes, especially Heliantheae and also to Senecioneae, even though some resistance was expressed regarding the latter (e.g., Nordenstam, 1977). Recent macromolecular cladistic analyses of representative genera within Heliantheae s.l. (e.g., Kim and Jansen, 1995b; Jansen and Kim, 1996; see our Fig. 2–7) provide some support for separation of Helenieae as a distinct unit, even perhaps as a tribe, and this is the perspective adopted by Karis and Ryding (1994a) in Bremer's (1994) book. Turner (in Turner and Zippin, 1992) has also recently agreed to tribal recognition. We, however, present flavonoid data for Heliantheae in a broad sense including Helenieae. This is done because we believe that much still needs to be learned about the proper disposition of subtribes within this complex (a perspective also echoed by Karis, 1993). Central to problems are the large number of genera that makes complete sampling and comprehensive analysis difficult, and the wide amount of morphological variation among genera that complicates interpretations of characters, states, and homologies (Panero et al., 1999). Other units also appear to be quite distinct, such as Coreopsidinae and Madiinae, treated earlier by some also as distinct tribes (e.g., Jepson, 1925; Ryding and Bremer, 1992). Achieving a classification of subtribes and tribes within Heliantheae s.l. of high information content and predictive value, therefore, will require additional detailed studies. For the convenience of the reader, however, we also mark genera treated as Helenieae by Karis and Ryding (1994a) with "Hele". Discussion of the significance of flavonoid data as they relate to various views of tribes and subtribes within Heliantheae will be found in Chapters 19 and 20.

Heliantheae s.l. consist of nearly 300 genera and 3300 species (Karis and Ryding, 1994a,b) distributed throughout the New World. They are abundant in Latin America with particularly strong diversity in Mexico. Flavonoid reports exist for 110 genera (including 32 Helenieae). Species number and information on geographic distribution in the comments for each genus are taken from Karis and Ryding (1994a,b) unless noted otherwise.

Acanthospermum: *Acanthospermum* consists of six species from tropical and subtropical America. Flavonoids have been reported for three species, *A. australe, A. glabratum*, and *A. hispidum*, the latter being a widespread weed. Compounds identified from the first two are *O*-methylated flavonols based on 6-hydroxykaempferol and 6-hydroxyquercetin (quercetagetin). The 3,6-di- and 3,6,7-trimethyl ethers of these compounds have been described from both species (Saleh et al., 1976, 1980; Debenedetti et al., 1987a). *Acanthospermum hispidum* was shown by Nair et al. (1985) to accumulate two flavones, 5-hydroxy-6,7,4'-trimethoxyflavone and 5,7-dihydroxy-6,8,4'-trimethoxyflavone.

Acmella: *Acmella* consists of 30 species mainly from tropical America, Africa, the Malay peninsula, Borneo, and northern Australia. Our knowledge of its flavonoids involves only *A. oleracea* (cited as *Spilanthes oleracea*; synonymy from Jansen, 1985), from which Verykokidou-Vitsaropoulos and Becker (1983) obtained quercetin 3-*O*-glucoside and 3-*O*-rutinoside and the 7-*O*-glucoside and 7-*O*-neohesperidoside of apigenin.

Adenothamnus (Hele): *Adenothamnus validus*, the sole member of this genus of tarweeds, is an endemic species of northern Baja California, Mexico. Flavonoids of a bulked sample were identified by Crins and Bohm (1988). Nonpolar flavonoids identified were naringenin, naringenin 4'-methyl ether, naringenin 7,4'-dimethyl ether, eriodictyol 7-methyl ether, eriodictyol 7,4'-dimethyl ether, dihydroquercetin 7-methyl ether, 6-methoxyapigenin, luteolin 7-methyl ether, 6-methoxyluteolin, kaempferol 3-methyl ether, 6-methoxykaempferol 3-methyl ether, and rhamnetin. Polar constituents were identified as quercetin and patuletin 3-*O*-glucosides, eriodictyol 7-*O*-monoglycoside (probably glucoside), 6-methoxyapigenin 7-*O*-glucoside, and a luteolin dimethyl ether 7-*O*-glucoside.

Amauria (Hele): *Amauria* consists of three species native to northwestern Mexico. The flavonoid fraction of *A. rotundifolia* was shown by Crins et al.

(1988a) to include kaempferol 3,4'-dimethyl ether, the 3,6-dimethyl and 3,6,7-trimethyl ethers of 6-hydroxykaempferol, and quercetagetin 3,6,3',4'-tetramethyl ether.

Amblyolepis (Hele): *Amblyolepis setigera*, the sole member of this genus, occurs in southwestern North America and adjacent Mexico. Bierner (1979) reported quercetin and isorhamnetin 3-*O*-galactosides from this species.

Ambrosia: *Ambrosia* is native to southwestern North America and consists of 43 species (including those from *Franseria*, Payne, 1964, 1976). Miao et al. (1995c), in a cpDNA study of *Ambrosia*, *Franseria*, and *Hymenoclea*, concluded that all three genera should be merged into a single larger genus *Ambrosia*. Morphological cladistic investigations by Karis (1995) support this view, but he would also recommend inclusion of *Xanthium*. We would have serious reservations about such a drastic merger without many additional comparative studies among all these genera of Ambrosiinae. We note the numerous autapomorphies separating *Xanthium* from species of *Ambrosia* (see his Fig 2B), a fact that we believe must be taken into consideration in formal classification (Stuessy, 1997). Payne (1964) earlier recommended combining *Franseria* with *Ambrosia*. The cpDNA data do provide support for a recognizable evolutionary unit representing most of the former *Franseria*, but this is treated at the level of subgenus.

Flavonoids have been described from nine species. The compounds range from common flavonols, quercetin 3-*O*-glucoside and isorhamnetin in the pollen of *A. artemisiifolia* (Hehl, 1919) and quercetin 3-*O*-glucoside in *A. trifida* (Bhatia and Kagan, 1971), to flavonols with substitution at both C-6 and C-8 and a moderately high level of *O*-methylation. The work of Herz and Sumi (1964) on *A. hispida* resulted in the first report of 5,7,4'-trihydroxy-6-methoxyflavone, named "hispidulin" by these workers, from a natural source. Jakupovic et al. (1988a) reported eriodictyol, aromadendrin (dihydrokaempferol), and aromadendrin 3-methyl ether from *A. artemisioides*. Details on flavonoids from the genus can be found in Table 11–1.

Argyroxiphium (Hele): This endemic Hawaiian genus, the silverswords, consists of five species. The glandular exudate of *A. sandwicense* was shown to have two flavonoids, 5,4'-dihydroxy-7,8-dimethoxy- and 5,4'-dihydroxy-6,7,8-trimethoxyflavones (Bohm and Fong, 1990).

Arnica (Hele): *Arnica* consists of about 30 species with a northern temperate and Arctic distribution. We have flavonoid information for about a dozen taxa including infraspecific data for a few. In general, members of the genus have been shown to accumulate derivatives of the common flavones

Table 11–1. Flavones and flavonols from species of *Ambrosia*

Species	Compounds[1]	References
A. artemisiifolia	Q-3-*O*-Glc, IR	Hehl (1919)
A. artemisioides	Apig	Jakupovic et al. (1988a)
	K, 3574'/6	Silva et al. (1992b)
A. deltoidea	574'/68	Wollenweber and Jay (1988)
A. dumosa	574'/6	Carman et al. (1972)
	574'/83'	Seaman et al. (1972)
A. eriocentra	54'/37	Herz et al. (1973b)
A. grayi	57/3684'	Herz et al. (1973a)
A. hispida	574'/6	Herz and Sumi (1964), Alvarez et al. (1992)
	563'4'/7	Alvarez et al. (1992)
A. tenuifolia	574'/6	Silva et al. (1992b)
A. trifida	Q-3-*O*-Glc	Bhatia and Kagan (1971)

[1] Substitution: Hydroxyl positions/Methoxyl positions; Apig = apigenin; K = kaempferol; Q = quercetin; IR = isorhamnetin; Glc = glucose.

apigenin and luteolin and flavonols based on kaempferol and quercetin as well as on aglycones that have undergone extra hydroxylation at C-6 and various degrees of *O*-methylation. Particularly rich accumulations of extra oxygenated and *O*-methylated flavonoids have been reported from *A. acaulis* (Herz and Sosa, 1988; Ebert et al., 1988; Merfort, 1994) and from *A. nevadensis* and *A. viscosa* (Ebert et al., 1988). Included in the *O*-methylated compounds from these species are 6-methoxyapigenin (hispidulin) and 6-methoxyluteolin (nepetin), luteolin 7,3'-dimethyl and 7,4'-dimethyl ethers, luteolin 7,3',4'-trimethyl ether and 5,3'-dihydroxy-6,7,4'-trimethoxyflavone. The study by Merfort (1994) also included O-methylated flavonols (6-methoxykaempferol and 6-methoxyquercetin). This richness in flavonoid expression is also seen in *A. longifolia* from which Skibinski et al. (1994) reported no less than 19 aglycones and 18 glycosides. *O*-Glycosylation is common involving both glucosides and glucuronides of common flavonoids as well as of *O*-methylated flavonoids. Acetylated glucosides have also been reported in *A. chamissonis* (Merfort, 1986, 1988a,b; Merfort and Wendisch, 1987) as has an unusual 2-methylbutyryl glucoside of eupafolin (Merfort, 1988b). Table 11–2 gives details of the occurrence of flavonoids in taxa other than *A. chamissonis* and *A. montana*, which will be considered separately below.

Arnica chamissonis and *A. montana* have been the subject of numerous chemical studies owing, in part at least, to their pharmaceutical properties. Luteolin and quercetin glycosides have been described from *A. chamissonis*

Table 11–2. Flavonoids from species of *Arnica* (other than *A. chamissonis* and *A. montana*)

Species	Compounds[1]	References
A. acaulis	53/74', 5/674'	Herz and Sosa (1988)
	574'/6, 573'4'/6, 3574'/6, 54'/73', 5/73'4', 54'/67, 53'/674'	Ebert et al. (1988)
A. angustifolia "group"	K & Q 3-*O*-Glys, Apig & Lute 7-*O*-Glc	Downie (1988), Downie and Denford (1988)
A. cernua	K & Q 3-*O*-Glys	Wolf and Denford (1984b)
	3574'/6 3-*O*-Glc	Wolf and Whitkus (1987)
A. cordifolia	K & Q 3-*O*-Glys, Q3Me, 3574'/6 3-*O*-Glc, Lute, 54'/7, 574'/6, 573'4'/6 7-*O*-Glc	Wolf and Denford (1983, 1984a), Wolf and Whitkus (1987)
A. discoidea	K & Q 3-*O*-Glys, 54'/7, Lute 7-*O*-Glc, 574'/6, 54'/67, 573'4'/6 7-*O*-Glc, 573/64', 3574'/6 3-*O*-Glc	Wolf and Denford (1984b), Wolf and Whitkus (1987)
A. frigida subspecies	Apig 7-*O*-Glc, K & Q 3-*O*-Glys	Downie and Denford (1986b, 1988)
A. fulgens	K & Q 3-*O*-Glys, Apig, Lute, 573'4'/6 7-*O*-Glc	Downie and Denford (1987, 1988)
A. gracilis	K, Q, and Pat 3-*O*-Glys, Apig & Lute 7-*O*-Glc, 574'/6, 573'/4', 573'4'/6 7-*O*-Glc, 3574'/6 3-*O*-Glc	Wolf and Whitkus (1987)
A. latifolia	K, Q, and Pat 3-*O*-Glys, 574'/6	Wolf and Denford (1984a), Wolf and Whitkus (1987)
A. lonchophylla	K & Q 3-*O*-Glys, Lute 7-*O*-Glc	Downie and Denford (1988)
A. louiseana	K & Q 3-*O*-Glys, Apig 7-*O*-Glc	Downie and Denford (1986b, 1988)
A. nevadensis	K & Q 3-*O*-Glys, 574'/6, 573'/4', 573'4'/6	Wolf and Denford (1984b)
	573'/4'	Wolf and Whitkus (1987)
	57/64', 54'/67, 574'/63', 3574'/6, 3574'/63'	Ebert et al. (1988)
A. rydbergii	K & Q 3-*O*-Glys, Q 3,7-*O*-DiGlc, Lute 7-*O*-Glc,	

Table 11–2. (Continued)

Species	Compounds[1]	References
	3574'/6 3-*O*-Glc	Downie and Denford (1988)
A. sachalinensis	IR & Pat 3-*O*-Glc, K, Q	
	3574'/63' 3-*O*-Glc	Passreiter et al. (1990)
A. sororia	K & Q 3-*O*-Glys,	
	Apig, Lute, 573'4'/6 7-*O*-Glc	Downie and Denford (1987, 1988)
A. spathulata	Q 3-*O*-Glys, 574'/6, 54'/7,	
	573'/64', 3574'/6 3-*O*-Glc	Wolf and Denford (1984b), Wolf and Whitkus (1987)
A. venosa	K & Q 3-*O*-Glys, 574'/6, 54'/7,	
	3574'/6 3-*O*-Glc	Wolf and Denford (1984b), Wolf and Whitkus (1987)
A. viscosa	K, Q, and Pat 3-*O*-Glys,	
	574'/6, 54'/7, 54'/67, 5673'/4',	
	5673'/64', 4'/673'	Wolf and Denford (1984b), Wolf and Whitkus (1987), Ebert et al. (1988)
	5/74', 54'/73', 5/73'4',	
	Nar 7-Me, Nar 7,4'-DiMe,	
	Erio 7,4'-DiMe	Ebert et al. (1988)

[1] Substitution: Hydroxyl positions/Methoxyl positions; Apig = apigenin; Lute = luteolin; K = kaempferol; Q = quercetin; Pat = patuletin; Nar = naringenin; Erio = eriodictyol; Glc = glucose; Gly = glycoside; Me = methyl.

(Borkosky et al., 1966; Schulte et al., 1969) while glycosides of kaempferol and quercetin have been recorded from *A. montana* by several workers (Friedrich, 1962c; Saner and Luepin, 1966; Dombrowicz and Greiner, 1968; Gora et al., 1979). The bulk of flavonoid data on these species, however, come from the work of Merfort and Wendisch. A complex array of compounds has been compiled from comparisons of *A. chamissonis* ssp. *chamissonis* (Merfort, 1985) and *A. montana*, and *A. chammissonis* ssp. *foliosa* var. *incana* (Poplawski et al., 1971; Marchishyn and Komissarenko, 1981; Merfort, 1984, 1985, 1986, 1988a, 1988a,b; Merfort and Wendisch, 1987, 1988, 1992). The profiles consist of flavones and flavonols some of which exhibit extra oxygenation at C-6 and/or a moderate level of *O*-methylation. Many of the compounds occur as glucosides, glucuronides, or both. In *A. chammissonis* ssp. *foliosa* var. *incana* some of the glucosides have been acylated with acetic or 2-methylbutyric acids while flavonol 3-*O*-glucuronides appear to be restricted to *A. montana*. Compounds identified from these taxa are listed in Table 11–3.

Table 11–3. Flavonoids[1] of *Arnica chamissonis* and *A. montana* (see text for citations)

Flavones	ACC	Taxon[2] ACF	AMT
Apigenin		+[3]	+
Acacetin	+	+	
Luteolin	+	+[3]	
Chrysoeriol (574'/3')[4]	+		+
Diosmetin (573'/4')		+	
Hispidulin (574'/6)	+	+	+
Pectolinarigenin (57/64')	+	+	+
Eupafolin (syn. nepetin) (573'4'/6)	+	+	+
Jaceosidin (574'/63')	+	+	+
Tricin (574'/3'5')			+
Flavonols			
Kaempferol	+	+	+
6-Methoxykaempferol	+	+	+
Betuletol (357/64')	+	+	+
Quercetin	+	+	+
Isorhamnetin (3574'/3')	+	+	+
Dillenetin (357/3'4')	+		
Patuletin (3573'4'/6)	+	+	+
Laciniatin (3573'/64')			+
Spinacetin (3574'/63')		+	+
357/63'4'			+

[1] Includes aglycones and glycosides.
[2] ACC = *A. chamissonis* ssp. *chamissonis*; ACF = *A. chamissonis* ssp. *foliosa*; AMT = *A. montana*.
[3] Apigenin, luteolin, kaempferol, and quercetin reported by Shelyuto et al. (1976b) in *A. chamissonis* ssp. *foliosa* (cited as *A. foliosa*; synonymy after Cronquist, 1994).
[4] Substitution: Hydroxyl positions/Methoxyl positions.

Bahia (Hele): *Bahia* consists of 11 (Stuessy, 1977) species distributed in western United States, Mexico, and Chile. Karis and Ryding (1994a) listed 13 species, following Ellison (1964). This same author (1966), however, later segregated two species as *Picradeniopsis* (following Rydberg, 1914), and we follow that perspective here. Flavonoids have been reported from six species. The predominant compounds are quercetagetin methyl ethers. *Bahia pringlei*, for example, afforded the 3,6-dimethyl and 3,6,3'-trimethyl ethers, along with kaempferol and quercetin. The only exception was seen with *B. xylopoda* from which 5,7-dihydroxy-3,6,4'-trimethoxyflavone was also obtained (Perez et al., 1990). See Table 11–4 for details.

Table 11–4. Flavonoids from species of *Bahia*

Species	Compounds[1]	References
B. absinthifolia	574'/363'	Perez et al. (1987)
B. ambrosioides	5/3673'4'	Zdero et al. (1990c)
B. glandulosa	54'/73', 53'/374'	Perez-Castanorena et al. (1997b)
B. pringlei	K, Q, 573'4'/36, 574'/363'	Perez et al. (1990)
B. schaffneri	573'/364', 57/363'4'	Perez and Romo de Vivar (1989)
B. xylopoda	57/364', 57/363'4'	Perez et al. (1990)

[1] Substitution: Hydroxyl positions/Methoxyl positions; K = kaempferol; Q = quercetin.

Baileya (Hele): This small genus of southwestern North American and Mexican plants consists of four species only one of which appears to have been studied for flavonoids. From *B. multiradiata* Dominguez et al. (1976) reported isolation of "morelosin-IV," 3,5,7-trihydroxy-3',5'-dimethoxyflavone **[4–18]**! The absence of substitution at C-4' was confirmed by alkaline cleavage of the compound, which yielded 3,5-dimethoxybenzoic acid **[4–19]**.

Balduina (Hele): *Balduina*, a genus of three species, is at home in the southern United States. All three species have been examined for flavonoids. Lee et al. (1972) reported 5,7,4'-trihydroxy-6-methoxyflavone, along with the isoflavonoid derivative rotenone **[4–75]**, from *B. angustifolia*, while Parker (1972) reported luteolin and the *C*-glycosylflavones orientin, isoörientin, and isovitexin from *B. angustifolia*, *B. atropurpurea*, and *B. uniflora*, apigenin from the first two and acacetin 7-*O*-glucoside from *B. atropurpurea*.

Balsamorhiza: *Balsamorhiza* is a genus of 14 species distributed widely over western North America. Five species have been examined for their leaf exudate flavonoids; data for vacuolar flavonoids are available for only one of these. Whereas the vacuolar components appear to be very simple, kaempferol and quercetin 3-*O*-glycosides in *B. deltoidea* (Bohm and Choy, 1987), the glandular flavonoid fraction is more elaborate. Extra oxygenation at C-6 occurs in all five species examined and *O*-methylation is common. A possible instance of 8-oxygenation was recorded for *B. hookeri* but confirmation is required (Bohm et al., 1989a). Occurrences of *O*-methylated flavonoids are given in Table 11–5. The reader should note that two different reports are given for *B. sagittata* and *B. deltoidea* indicating the existence of interpopulational variation in this genus.

Table 11–5. *O*-Methylated flavonoids from species of *Balsamorhiza*

Species	Compounds[1]	References
B. careyana	574'/6, 357/64', 57/364', 574'/363'	Robson and McCormick (1988)
B. deltoidea (pop'n. 1)	357/4', 3564'/7, 574'/36, 573'4'/3, 3573'/4', 573'/34', 357/3'4', 3563'4'/7, 573'4'/36, 53'4'/367, 574'/363'	Bohm and Choy (1987)
B. deltoidea (pop'n. 2)	3574'/6, 357/64', 3573'4'/6	Robson and McCormick (1988)
B. hookeri	573'4'/3, 574'/33', 573'4'/36	Bohm et al. (1989a)
B. macrophylla	573'4'/3, 3573'/4', 573'/34', 573'4'/36, 3574'/63', 573'/364'	McCormick et al. (1985b)
B. sagittata (pop'n. 1)	3654'/7	Bohm et al. (1989a)
B. sagittata (pop'n. 2)	3574'/6, 357/64', 3573'4'/6, 573'4'/36, 574'/363'	Robson and McCormick (1988)

[1] Substitution: Hydroxyl positions/Methoxyl positions.

Berlandiera: *Berlandiera* consists of four species native to the southern United States and Mexico. Flavonoid data are available for two of these. Wagner et al. (1972b) reported kaempferol and quercetin 3-*O*-glucosides from *B. pumila*, while Bulinska-Radomska et al. (1985a) reported those two compounds as well as apigenin, 6-methoxyapigenin (hispidulin), and 6-methoxyluteolin from *B. texana* (cited as *B. texana* var. *texana*).

Bidens: *Bidens*, which enjoys a worldwide distribution, consists of 240 species, making it the largest genus of Heliantheae. Despite its size and conspicuous nature, most of what we know about the genus in terms of flavonoids comes from a comparatively small number of species. Nonetheless, the flavonoid profile of the genus is a complex one that includes aurones, chalcones, flavanones, flavones, and flavonols with a wide variety of *O*-methylation patterns and glycosylations, although not all taxa have all aglycone types nor all structural variants.

Flavones and flavonols identified from members of *Bidens* are for the most part commonly encountered compounds, i.e., glycosides of apigenin, luteolin, kaempferol, and quercetin, while diosmetin (luteolin 4'-methyl

Table 11–6. Flavonoid occurrences in species of *Bidens*

Species	Compounds[1]	References
B. bipinnata	Q 3-*O*-Gal, isoökanin, Common anthochlors	Wang et al. (1992)
B. cernua	Sulfuretin, maritimetin, Butin 7-*O*-Glc, 8-OH-butin	Borisov et al. (1980)
	Common anthochlors, Flavones, quercetin	Isakova et al. (1986)
B. campylotheca	Complex okanin derivatives	See Table 12-7
B. chrysanthemoides	Common anthochlors, quercetin	Isakova et al. (1986)
B. clavata	Apig & Lute Glys	Ballard (1988)
B. cordylocarpa	Nar 7-*O*-Gly, Q 3-*O*-Glc, common anthochlors	Crawford (1971)
B. dahlioides	Common anthochlors, quercetin	Isakova et al. (1986)
B. ferulaefolia complex	K & Q 3-*O*-glys, Apig & Lute 7-*O*-Glys, common anthochlors	Hart (1979), Isakova et al. (1986)
B. frondosa	Luteolin, common anthochlors	Romussi and Pagani (1970)
	Common anthochlors, quercetin	Isakova et al. (1986)
	Okanin derivatives	Karikome et al. (1992)
B. laevis	Anthochlors	Scogin and Zakar (1976)
B. maximowicziana	Common anthochlors, quercetin	Isakova et al. (1986)
B. parviflora	Common aurones, quercetin	Isakova et al. (1986)
	K & Q 3-*O*-glc, common anthochlors	Ma et al. (1991)
	572'5'/-[2], Q 3-*O*-Rut, Nar 7-*O*-Rut	De Tommasi et al. (1992)
B. pilosa	Complex okanin & maritimein derivatives	See Table 11-7
B. radiata	Common anthochlors, quercetin	Isakova et al. (1986)
B. torta	Complex okanin derivatives	See Table 11-7
B. tripartita	Lute 7-*O*-Glc	Baranska (1963)
	Common anthochlors	Serbin et al. (1972a,b, 1973, 1975)
Bidens in Hawaii	See Chapter 21	Ganders et al. (1990)

[1] Apig = apigenin; Lute = luteolin; K = kaempferol; Q = quercetin; Nar = naringenin; Gal = galactose; Glc = glucose; Rut = rutinose;

[2] Substitution: Hydroxyl positions/Methoxyl positions.

ether) has been reported from *B. sandvicensis* ssp. *confusa* (Ganders et al., 1990). Luteolin and luteolin 7-*O*-glucoside were reported from *B. tenuifolia* (cited as *Glossogyne tenuifolia*; synonymy from Veldkamp and Kreffer, 1991) by Lin et al. (1975). Quercetin 3,3'-dimethyl ether 7-*O*-(6-rhamnosylglucoside), a new compound, and the corresponding 7-*O*-glucoside were identified as root components of *B. pilosa* by Brandao et al. (1998). Another interesting compound, recently reported, is 5,7,2',5'-tetrahydroxyflavone from *B. parviflora* (De Tommasi et al., 1992). Eriodictyol 7-*O*-glucoside was identified in several Hawaiian species of *Bidens* (Ganders et al., 1990). 7,3',4'-Trihydroxyflavanone, the flavanone corresponding to butein, was identified as a component of *B. tripartita* by Serbin et al. (1972a,b, 1975). 7,8,3',4'-Tetrahydroxyflavanone was identified by Hart (1979) in the three species comprising the *B. ferulaefolia* complex (*B. ferulaefolia*, *B. carpodonta* and *B. longistyla*; see Hart, 1978). Table 11–6 lists general information on flavonoid occurrences in *Bidens*. Studies of variation patterns in the Hawaiian *Bidens* system, and Hart's (1979) study of the *B. ferulaefolia* complex have associated with them large data arrays which will be dealt with in later chapters.

Aurones and chalcones have been reported from all members of the genus so far examined. The aurones commonly reported are sulfurein **[4–48]** (sulfuretin 6-*O*-glucoside, 6,3',4'-trihydroxyaurone 6-*O*-glucoside) and maritimein **[4–52]** (maritimetin 6-*O*-glucoside, 6,7,3',4'-tetrahydroxyaurone 6-*O*-glucoside). The chalcones commonly reported are coreopsin **[4–46]** (butein-4'-*O*-glucoside, 2',4',3,4-tetrahydroxychalcone 4-*O*-glucoside) and marein **[4–50]** (okanin 4'-*O*-glucoside, 2',3',4',3,4-pentahydroxychalcone 4'-*O*-glucoside). These are frequently reported on the basis of chromatographic behavior and are often reported in the literature as occurring in pairs, e.g., coreopsin/sulfurein. Scogin and Zakar (1976) reported positive anthochlor tests (color change with ammonia) for floral tissue for several *Bidens* species and discussed the results in terms of pollinator attraction (see Chapter 6).

Acetylated aurone and chalcone glycosides and a variety of *O*-methylated compounds have also been reported. Several of the compounds reported by Ganders et al. (1990) display both features. For example, *Bidens torta* accumulates a series of okanin derivatives (Fig. 11–1): okanin 3,4,3',4'-tetramethyl ether **[11–1]**, okanin 3,4,3'-trimethyl ether 4'-*O*-glucoside **[11–2]**, okanin 3,4-dimethyl ether 4'-*O*-glucoside **[11–3]** and okanin 4-methyl ether 4'-*O*-glucoside **[11–4]** and an acetylated derivative thereof (McCormick et al., 1984). Several complex acylated aurone and chalcone glycosides have been reported by Hoffmann and Hölzl (1989a,b), who found 4",6"-diacetyl- and 2",4",6"-triacetylmaritimein in *B. pilosa*. These workers have also reported a series of derivatives of the chalcone okanin from *B. pilosa*.

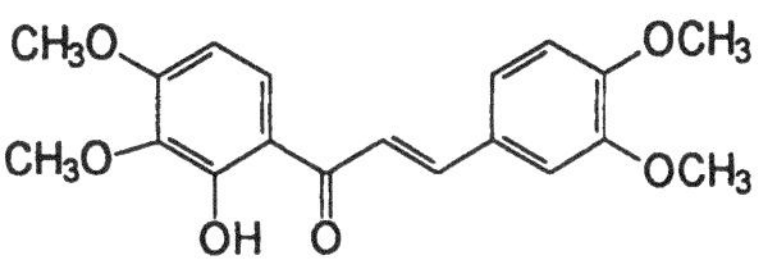

[11-1]

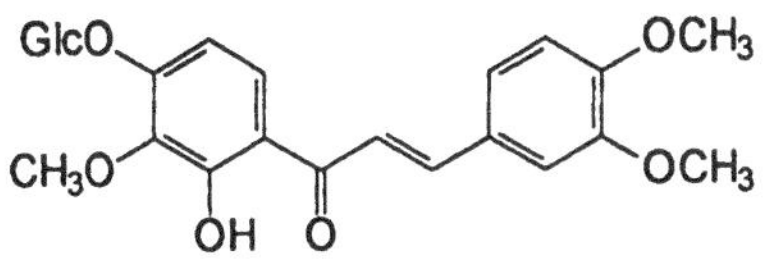

[11-2]

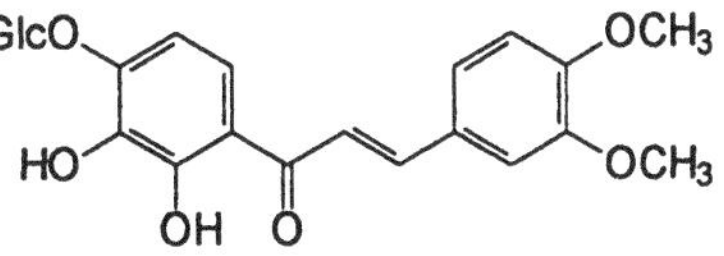

[11-3]

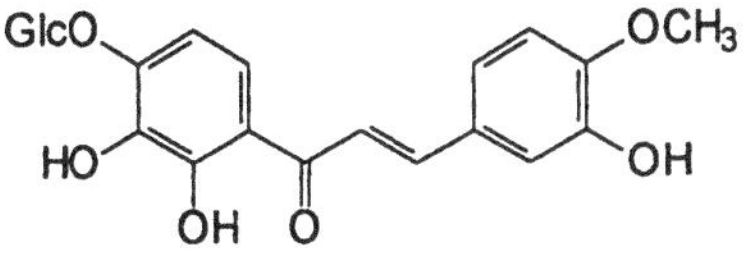

[11-4]

Fig. 11–1. Okanin derivatives **[11–1 – 11–4]** isolated from *Bidens torta*

Redl et al. (1993) have recently described three derivatives of okanin 4-methyl ether from *B. campylotheca* subsp. *campylotheca*, the 4'-*O*-glucoside **[11–4]**, the 4'-*O*-(6"-*O*-acetyl)glucoside, and the 4'-*O*-primoveroside. More recently still, Wang et al. (1997) described several new anthochlor glycosidic derivatives from *B. pilosa* var. *radiata*, among which were the novel (*Z*)-6,7,3',4'-tetrahydroxyaurone-6-*O*-(3",4",6"-triacetyl-ß-D-glucoside) and corresponding 2",4",6"-triacetate. Compounds of this sort, and their species of origin, are listed in Table 11–7.

Shimokoriyama (1962) recorded a positive anthochlor test for an (unnamed) species of *Bidens* (cited as *Microlecane*).

Calea: *Calea* consists of approximately 110 Latin American species. Despite its size, we have flavonoid information for only a few. Apigenin 4'-methyl ether (acacetin) has been reported from: *C. jamaicensis* (Ober et al., 1985b, 1986; cited as *C. berteriana* and *C. prunifolia*; synonymy after Wussow et al., 1985), *C. divaricata* (Ober et al., 1985a), *C. solidaginea* (Ober et al., 1985b) and *C. ternifolia* (cited as *C. zacatechichi*, Herz and Kumar, 1980; Martinez et al., 1987, 1988a; synonymy from Wussow et al., 1985). *Calea crocinervosa* afforded apigenin 7-methyl ether (genkwanin) and apigenin 7,4'-dimethyl ether (Ortega et al., 1989). Apigenin-7,4'-dimethyl ether was also reported from *C. jamaicensis* (Ober et al., 1986), *C. divaricata* (Ober et al., 1985a) and by all three groups who studied *C. ternifolia* (cited as *C. zacatechichi*).

Calycadenia (Hele): *Calycadenia* consists of 11 species distributed in western United States with a concentration of species in California. Studies of the exudate flavonoids of six species have been done (Emerson et al., 1986;

Table 11–7. *O*-Methylated, glycosylated, and acylated anthochlor derivatives from species of *Bidens*

Species	Compounds[1]	References
B. campylotheca	4-*O*-Methyl okanin, 4-*O*-Methyl okanin 4'-*O*-(6"-acetyl)glucoside, 4-*O*-Methyl okanin 4'-*O*-primoveroside	Redl et al. (1993)
B. frondosa	Okanin 4-*O*-(2"-caffeoyl-6"-acetyl)glucoside, Okanin 4-*O*-(2"-caffeoyl-6"-*p*-coumaroyl)glucoside, 4-*O*-Methyl okanin 4'-*O*-(acetyl)glucoside, 4-*O*-Methyl okanin 4'-*O* -(6"-*p*-coumaroyl)glucoside 4-*O*-Methyl okanin 4'-*O*-(2"-caffeoyl-6"-acetyl)glucoside	Karikome et al. (1992)
B. pilosa	2",4",6"-Triacetyl maritimein, 4",6"-Diacetylmaritimein	Hoffmann and Hölzl (1989b)
	Okanin 3'-*O*-glucoside, Okanin 4'-*O*-(2",4",6"-triacetyl)glucoside, Okanin 4'-*O*-(6"-*p*-coumaroyl)glucoside, Okanin 4'-*O*-(4"-acetyl-6"-*p*-coumaroyl)glucoside, Okanin 4'-*O*-(2",4"-diacetyl-6"-*p*-coumaroyl)glucoside, Okanin 4'-*O*-(3",4"-diacetyl-6"-*p*-coumaroyl)glucoside	Hoffmann and Hölzl (1988a,b)
	4-*O*-Methyl okanin 3'-*O*-glucoside	Hoffmann and Hölzl (1988c)
	Okanin 3'-*O*-glucoside, Okanin 4'-*O*-glucoside, Okanin 3',4'-di-*O*-glucoside, Okanin 4'-*O*-gentiobioside	Hoffmann and Hölzl (1989a)
B. pilosa var. *radiata*	3",4",6"-Triacetylmaritimein, 2",4",6"-Triacetylmaritimein, Okanin 4'-*O*-(4",6"-diacetyl)glucoside,	

Table 11–7. (Continued)

Species	Compounds[1]	References
	Okanin 4'-*O*-(3",4",6"-triacetyl)glucoside	Wang et al. (1997)
B. torta	4-*O*-Methyl okanin 4'-*O*-glucoside, 4-*O*-Methyl okanin 4'-*O*-(acetyl)glucoside, 3,4-Di-*O*-methyl okanin 4'-*O*-glucoside, 3',3,4-Tri-*O*-methyl okanin 4'-*O*-glucoside, 3',4',3,4-Tetra-*O*-methyl okanin	McCormick et al. (1984)

[1] Okanin = 2',3',4',3,4-Pentahydroxychalcone; Maritimein = 6,7,3',4'-Tetrahydroxyaurone 6-*O*-glucoside.

Table 11–8. Leaf exudate flavonoids from species of *Calycadenia*[1]

Species	Compounds[2]
C. ciliosa	54'/3678, 53'4'/3678, 53'4'/378, 563'4'/378, 54'/36783'
C. mollis	573'/684',573'4'/36, 573'4'/368, 563'/3784', 53'/36784'
C. multiglandulosa	54'/6783',574'/36, 53'4'/3678, 53'4'/378, 54'/36783'
C. pauciflora	573'4'/36, 573'4'/368, 563'/3784'
C. truncata	567/84', 573'/684', 53'/6784', 573'4'/36, 573'4'/368, 573'/34', 563'/3784', 53'/36784'
C. villosa	573'/684', 53'/6784',574'/36, 563'/3784'

[1] Data from Emerson et al. (1986) and Bohm et al. (1992).
[2] Substitution: Hydroxyl positions/Methoxyl positions.

Bohm et al., 1992). The profiles are dominated by a high frequency of compounds having both 6- and 8-oxygenation plus consistently high levels of *O*-methylation; both flavones and flavonols have been identified. The distribution of 15 compounds in members of the genus so far studied can be found in Table 11–8.

Chrysanthellum: This Mexican genus consists of 13 species, two of which appear to have been examined for flavonoids. Brasseur et al. (1987) reported luteolin 7-*O*-glucoside, rutin, eriodictyol 7-*O*-glucoside, isoökanin 7-*O*-glucoside (isoökanin is 6,7,3',4'-tetrahydroxyflavanone), and the chalcone/aurone pair marein and maritimein in *C. indicum* as part of a study of antiradical effects of extracts of the plant. Shimokoriyama (1962) recorded positive anthochlor tests for six species (not named). Crawford and Stuessy (1981) reported butein/sulfuretin and okanin/maritimetin glucoside pairs in *C. pinnatisectum* (cited as *Eryngiophyllum pinnatisectum*; synonymy after Veldkamp and Kreffen, 1991).

Clibadium: *Clibadium* consists of 29 species distributed in Mexico, Central, and South America (Arriagada, 1994). Flavonoid information is available for 20 species. Kaempferol and quercetin 3-*O*-mono- and 3-*O*-diglycosides are widespread in the genus, glycosides of *O*-methylated flavonols much less so. Making up this latter group are quercetagetin, patuletin, quercetagetin 3'-methyl ether, and quercetagetin-3,3'-dimethyl ether; all occur as 7-*O*-glucosides (Bohm and Stuessy, 1981a, 1985). Two unusual findings are of interest. Of fifty populations of *C. surinamense* studied, only two had kaempferol and quercetin 3,7-di-*O*-rhamnosides, which was also identified in *C. pilonicum* (Bohm et al., 1983; recently synonymized into *C. anceps* by Arriagagda, 1994). A further distinctive populational difference was observed in *C. trianae* where six populations had only kaempferol and quercetin 3-*O*-glycosides, one population added quercetagetin and patuletin 7-*O*-glucosides, and one population had the simple flavonol glycoside profile plus 4,6,3',4'-tetrahydroxyaurone 6-*O*-glucoside and three chalcones, 2',3',3,4-tetrahydroxy-chalcone 4'-*O*-glucoside (coreopsin), 2',4',4-trihydroxychalcone 4'-*O*-glucoside and 2',4',4-trihydroxy-3-methoxychalcone 4'-*O*-glucoside (Bohm et al., 1983). A recent note discusses the ethnobotanical uses of *Clibadium* in Latin America (Arriagada, 1995).

Coreocarpus: *Coreocarpus* consists of 10 species that range from California to northwestern Mexico. Crawford and Stuessy (1981) reported a butein/sulfuretin glycoside pair from three species: *C. arizonicus*, *C. dissectus*, and *C. parthenioides* (cited as *C. shrevei*; synonymy from Smith, 1989).

Coreopsis: *Coreopsis* is a large genus with over 100 species distributed mainly in North and South America and Africa. Despite the impressive taxonomic contributions of Sherff on *Coreopsis* (1936, 1955b) and *Bidens* (1937), considerable difficulties still remain on generic separation between the two. Based on comparative morphological studies, Mesfin (1986) recently recommended inclusion of all African species of *Coreopsis* into *Bidens*, but further studies (Mesfin et al., 1995) have suggested that this may have been premature; molecular ITS sequence data suggest that *Bidens* was derived twice from out of *Coreopsis* (Kim et al., 1999b). Many additional comparisons, especially involving South American taxa, are needed before the generic limits can be resolved satisfactorily.

Members of the genus have figured significantly in the history of flavonoid chemistry, particularly through the pioneering efforts of Ted Geissman at the University of California at Los Angeles. He isolated butein (2',4',3',4-tetrahydroxychalcone, naringenin chalcone) from *C. douglasii* (Geissman, 1941a) and *C. gigantea* (Geissman, 1941b). Soon thereafter, came 6,3',4'-trihydroxy-7-methoxyaurone 6-*O*-glucoside (leptosin) from *C. grandiflora* (Geissman and Heaton, 1944a,b), 2',4',5',3,4-pentahydroxychalcone (stillopsidin) from *C. stillmanii* (Seikel and Geissman, 1950), a total synthesis of leptosidin and leptosin (Geissman and Moje, 1951), 2',3',4',3,4-pentahydroxychalcone 4'-*O*-glucoside (marein) and 6,7,3',4'-tetrahydroxyaurone 6-*O*-glucoside (maritimein) from *C. maritima* (Harborne and Geissman, 1956), and a suite of compounds from *C. gigantea* that included butein, coreopsin, marein, maritimein, sulfurein, and luteolin 7-*O*-glucoside.

Other work that appeared during this time included identification of leptosin **[4–56]** and lanceolin **[4–54]** in *C. lanceolata* and *C. saxicola* (Shimokoriyama and Hattori, 1953), the report of marein and maritimein from *C. tinctoria* by Shimokoriyama (1957), and a suite of compounds including butein, butin, isocoreopsin (the latter two are flavanones), and luteolin from *C. basalis* (cited as *C. drummondii*; synonymy from Correll and Johnston, 1970) and *C. tinctoria* by Puri and Seshadri (1954). Reports of many of these compounds from new sources continued to appear. Thus, Julian and Crawford (1972) found sulfuretin 6-*O*-diglucoside, a caffeoyl derivative thereof, and a caffeoylated triglucoside in *C. mutica*. *Coreopsis mutica* var. *simplicifolia* gave mono-, di- and triglucosides of simple chalcones and aurones plus apigenin and luteolin 7-*O*-glucosides and kaempferol, quercetin, and isorhamnetin 3-*O*-monoglycosides (Crawford, 1981). Crawford (1978) found okanin 4'-*O*-diglucoside in the ray florets of *C. petrophiloides*. Marein and lanceolin were reported from *C. nuecensis* (Harborne, 1977a), while several common chalcones and aurones were reported, along with luteolin 7-*O*-glucoside and 6-hydroxyluteolin 7-*O*-glucoside, from *C. bigelovii* (Harborne and Smith, 1978; Nicholls and Bohm, 1979). The *C. nuecensis* and *C. nuecensoides* species pair was studied by Smith and Crawford (1981) who reported apigenin and luteolin 7-*O*-glucosides, quercetin 3-*O*-glycosides, 6-hydroxy and 6-methoxyluteolin 7-*O*-glucosides, quercetagetin and patuletin 7-*O*-glucosides and the marein/

maritimein and lanceolin/leptosin chalcone/aurone pairs. Much the same profile of flavonoids was reported from *Coreopsis* section *Palmatae* (Crawford et al., 1980), and from the four varieties of *C. grandiflora* (Crawford and Smith, 1980). Similar results were seen in a study of *C. mutica* (Crawford 1970) except that *C. mutica* var. *leptomera* also had the *C*-glycosylflavone vicenin. *C*-Glycosylflavones were also identified from *C. cyclocarpa*, *C. calliopsidea*, *C. douglasii* and *C. californica* (Crawford and Smith, 1983b). Anthochlors, in the form of the butein/sulfuretin pair, were reported for *C. buchii* (cited as *Selleophyton buchii*; synonymy from Blake, 1924) by Crawford and Stuessy (1981).

Cosmos: *Cosmos* is a New World genus consisting of 26 species with a concentration of representatives in Mexico (Mabberley, 1987). *Cosmos bipinnatus* is a widely cultivated garden ornamental, which figures significantly in the history of flavonoid research. Hayashi (1941) reported cyanidin 3-*O*-rutinoside from the species, while Geissman (1942), Shimokoriyama and Hattori (1953), and Puri and Seshadri (1954) were concerned with non-anthocyanic floral pigments which involved several aurones and chalcones, including butein and its glucoside coreopsin, sulfuretin and its glucoside sulfurein, the flavanone butin, and glycosides of luteolin and quercetin. Common flavone and flavonol glycosides have been reported from *C. bipinnatus* (Nakaoki, 1936; Nakaoki et al., 1961; Saito, 1976; Kaneta et al., 1978).

Dahlia: *Dahlia* is a genus of 30 species (Sorensen, 1969, 1980, 1987) that occurs naturally from Mexico to Colombia. According to Giannasi (1975a,b; see also Sorensen, 1969), whose chemical monograph treats the genus in detail, the dahlias of commerce are lumped under *D. variabilis*, which is a tetraploid that arose through hybridization between *D. coccinea* and *D. pinnata*, both of which are Mexican species. Much of the early flavonoid chemistry was performed on material identified as *D. variabilis* (now placed into synonymy of *D. pinnata* Cav. by Sorensen, 1969). Giannasi's work involved 24 species. An additional species, *D. imperialis* (cited as *D. lehmanni*; synonymy from Sorensen, 1969) was examined by Kaufmann and Lam (1967).

Several of the earlier papers on naturally occurring flavonoids involved analyses of *D. variabilis* (see comment in preceding paragraph). Among the earliest were the papers by Hayashi (1933) who identified anthocyanin 3,5-di-*O*-glucosides, Nakaoki (1938), who reported luteolin and diosmetin, and Price (1939), who reported the presence of butein (2',4',3,4-tetrahydroxychalcone). The next generation of workers extended these studies showing the presence of a variety of flavone and anthocyanin *O*-glycosides (Nordström and Swain, 1953), aurone and chalcone glycosides,

Table 11–9. Flavonoids from species of *Dahlia* reported by Giannasi (1975a,b)

Flavonoid type	Derivative
Anthocyanins:	
Cyanidin	Glycosides (Unident.)
Aurones:	
Sulfuretin (63'4'/-)[1]	Aglycone
	6-*O*-Glucoside
	6-*O*-Galactoside
	6-*O*-Arabinoside
	6-*O*-Diglucoside
	6-*O*-Arabinosylgalactoside
Chalcones:	
Isoliquiritigenin (2'4'4/-)	Aglycone
	4'-*O*-Glucoside
	4'-*O*-Diglucoside
	4'-*O*-Diglucoside-4-*O*-glucoside
	4'-*O*-Diglucoside-4-*O*-diglucoside
	4-*O*-Glucoside
Butein (2'4'34/-)	Aglycone
	4'-*O*-Glucoside
	4'-*O*-Galactoside
	4'-*O*-Diglucoside
	4'-*O*-Glucoside-4-*O*-glucoside
	4'-*O*-Diglucoside-4-*O*-glucoside
	4'-*O*-Diglucoside-4-*O*-diglucoside
	4-*O*-Glucoside
Butein 3-methyl ether	Aglycone
	4'-*O*-Glucoside
	4'-*O*-Diglucoside
	4'-*O*-Glycosides (Unident.)
Methoxychalcone (Unident.)	
Dimethoxychalcone (Unident.)	
Flavones:	
Apigenin	Aglycone
	7-*O*-Glucoside
	7-*O*-Diglucoside
	7-*O*-Glycosides (Unident.)
	4'-*O*-Glucoside
Acacetin (57/4')	Aglycone
	7-*O*-Glucoside
	7-*O*-Rutinoside
Luteolin	Aglycone
	7-*O*-Glucoside
	7-*O*-Diglucoside
	7-*O*-Glycoside (Unident.)

Table 11–9. (Continued)

Flavonoid type	Derivative
	4'-*O*-Glucoside
Diosmetin (573'/4')	Aglycone
	7-*O*-Glucoside
	7-*O*-Diglucoside
	7-*O*-Glycoside (Unident.)
Hispidulin (574'/6)	Aglycone
	7-*O*-Glycoside (Unident.)
Pectolinarigenin (57/64')	Aglycone
	7-*O*-Glucoside
	7-*O*-Arabinosylgalactoside
6-Methoxyflavone (Unident.)	
6-Methoxyluteolin (573'4'/6)	7-*O*-Glycoside (Unident.)
573'/64'	Aglycone
	Glycoside (Unident.)
Flavonols:	
Kaempferol	Aglycone
	3-*O*-Glucoside
	3-*O*-Rutinoside
Quercetin	Aglycone
	3-*O*-Diglucoside
	3-*O*-Rutinoside
	3,7-*O*-Diglycosides (Unident.)
C-Glycosylflavone:	
Apigenin	6,8-Di-*C*-glycosides (Unident.)

[1] Substitution: Hydroxyl positions/Methoxyl positions.

and assorted other flavonoids (Bate-Smith and Swain, 1953; Puri and Seshadri, 1954; Bate-Smith et al., 1955; Nordström and Swain, 1956; Harborne and Sherratt, 1957; Saito et al., 1970).

The bulk of our knowledge concerning *Dahlia* flavonoids, however, comes from the survey of the genus by Giannasi (1975a,b). He listed 64 compounds which included three xanthones and several unknown, presumably phenolic, compounds. The compound types identified included anthocyanins, aurones, chalcones, flavanones, flavones, flavonols, and *C*-glycosylflavones. Table 11–9 lists the specific members of these groups reported from the genus by Giannasi. Three additional studies will round out our look at the genus. Kaufmann and Lam (1967) reported 5,7,4'-trimethoxyflavanone (naringenin trimethyl ether) from *D. imperialis* (cited as *D. lehmannii*; synonymy after Sorensen, 1969). *Dahlia tenuicaulis* afforded 5,7,4'-trimethoxyflavanone, 5-hydroxy-7,4'-dimethoxyflavanone, 2'-hydroxy-4',6',4-trimethoxychalcone and a flavan derivative, 4-hydroxy-5,7,4'-trimethoxyflavan, from the leaves, and 2',4',4-trihydroxychalcone, 2',4',3-trihydroxy-4-methoxychalcone and 5-hydroxy-7,4'-dimethoxyflavanone from floral tissue. Kaneta et al. (1978) identified apigenin 7-*O*-glucoside from *D. pinnata* and Harborne et al. (1990) identified butein 4'-*O*-malonylglucoside and 4'-*O*-malonylsophoroside from *D. pinnata* (cited as *D. variabilis*; synonymy from Sorensen, 1969) and *D. coccinea.*

Desmanthodium: *Desmanthodium* is a Mexican genus of seven species (Turner, 1996), three of which have been examined for flavonoids (Bohm and Stuessy, 1981b). Field-collected material of *D. fruticosum* and *D. hondurense* (recently referred to synonymy under *D. guatemalense* Hemsl. by Turner, 1996) and both field-collected and greenhouse-grown material of *D. perfoliatum* were studied. Kaempferol and quercetin 3-*O*-mono- and diglycosides were evenly distributed between the three species except that quercetin 3-*O*-rhamnoside was only found in *D. perfoliatum*. Four *O*-methylated flavonol derivatives were obtained from all three species: patuletin 3-*O*-glucoside, quercetagetin 3-methyl ether 7-*O*-glucoside, quercetin-3,3'-dimethyl ether 7-*O*-glucoside, and quercetagetin 3,6,3'-trimethyl ether 7-*O*-glucoside. In greenhouse-grown specimens of *D. perfoliatum* the quercetagetin trimethyl ether was absent but quercetagetin 7-*O*-glucoside and quercetagetin 3'-methyl ether 7-*O*-glucoside were seen instead.

Dicranocarpus: *Dicranocarpus parviflorus*, the only member of this Mexican genus, gave a positive anthochlor test (Crawford and Stuessy, 1981).

Dubautia (Hele): According to the monograph of Carr (1985), *Dubautia* consists of 26 species endemic to the Hawaiian Islands. Flavonoid data for this group come from a study of hybridization between *D. scabra* subsp. *scabra* and *D. ciliolata* ssp. *glutinosa* (Crins et al., 1988c). The vacuolar flavonoid profile was shown to consist of 3-*O*-glycosides based upon kaempferol, quercetin, 6-hydroxykaempferol, and quercetagetin plus eriodictyol 7-*O*-glucoside. Leaf surface flavonoids consisted of flavones and flavonols based upon apigenin, luteolin, kaempferol, and quercetin. The only compound from leaf surface extracts with extra oxygenation was 6-hydroxyapigenin. Naringenin and naringenin 7-methyl ether were also present in the exudate fraction. *Dubautia arborea* (Madiinae), an alpine and desert shrub endemic to the Island of Hawaii, has been shown to accumulate apigenin, luteolin 7-methyl ether, luteolin 4'-methyl ether (diosmetin), 6-methoxyluteolin (hispidulin), naringenin 7-methyl ether (sakuranetin), eriodictyol 7-methyl ether, and 5,4'-dihydroxy-6,7-dimethoxyflavone (cirsimaritin) in its leaf exudate (Bohm, 1999).

Dugaldia (Hele): Recognized by some (Bierner, 1974; Stuessy, 1977; Karis and Ryding, 1994a) as distinct from *Helenium*, but treated as belonging within *Hymenoxys* by others (Bierner, 1994b; Bierner and Jansen, 1998), *Dugaldia* Cass. consists of three species native to western North America and Mexico. *Dugaldia integrifolia* yielded quercetin and 6-methoxyluteolin (nepetin) (Bohlmann et al., 1985d), while a recent report by Bierner (1994a) described isolation of 5,7-dihydroxy-6,4'-dimethoxyflavone from *D. pinetorum*. Kaempferol and quercetin 3-O-glucosides and quercetin 3-*O*-(acetylglucoside) were identified as components of *D. hoopesii* (cited as *Helenium hoopesii*; taxonomy from Bierner, 1974) by Wagner et al. (1971c).

Echinacea: *Echinacea* consists of nine species native to the eastern United States. The only flavonoid information available comes from two reports, cyanidin 3-*O*-glucoside and cyanidin 3-*O*-(6-*O*-malonyl)glucoside from *E. pallida* and *E. purpurea* (Cheminat et al., 1989) and quercetin, apigenin, luteolin, the flavone 7-*O*-glucosides, and apigenin 7-*O*-(4"-*O*-p-coumaroyl) glucoside from "*E. echinacea*" (Chaudhuri, 1988).

Eclipta: *Eclipta*, a genus of only four species, occurs in warm tropical regions. The earliest report of a flavonoid derivative from *Eclipta* seems to be that of Govindachari et al. (1956b) who identified the coumestan derivative wedelolactone **[4–71]** from *E. alba*. Some years later Bhargava et al. (1970) reported demethylwedelolactone **[4–72]** and its glucoside **[4–73]** from the same source. Wagner et al. (1986) reported apigenin from *E. alba*.

Encelia: Fifteen species (17 according to Proksch et al., 1988, but two are marked as undescribed) constitute the New World genus *Encelia*, which is also represented on the Galapagos Islands. Seventeen taxa were studied with regard to their epicuticular flavonoids (Proksch et al., 1988) twelve of which yielded identifiable compounds. Occurrence of all compounds was highly variable within the genus. Kaempferol 3-methyl, 3,7-dimethyl, and 3,4'-dimethyl ethers were identified as were quercetin and its 3-methyl-, 7-methyl-, 3,7-dimethyl, 3,3'-dimethyl-, 3,3',4'-trimethyl-, and 3,7,4'-trimethyl ethers. Two herbacetin (8-hydroxykaempferol) derivatives were identified, the 3,8-dimethyl and 3,8,4'-trimethylethers. The following gossypetin (8-hydroxyquercetin) derivatives were also identified: 3,8-dimethyl ether, 3,8,3'-trimethyl ether, and 3,8,3',4'-tetramethyl ether. Two 6,8-dioxygenated compounds was identified, 5,7,4'-trihydroxy-3,6,8-trimethoxyflavone and 5,7,3'-trihydroxy-3,6,8,4'-tetramethoxyflavone. The 6,8-dioxygenated flavonols, along with gossypetin-3,8,3'-trimethyl ether, were seen only in material of "*E. densifolius*," an undescribed species. Simple flavones identified were luteolin, luteolin 7-methyl ether, and luteolin 3'-methyl ether (chrysoeriol).

Luteolin occurred in several species whereas the two methyl ethers were seen only in *E. stenophylla*. *Encelia stenophylla* was also the only source of two *C*-prenylated flavanones, 8-*C*-prenylpinocembrin and 8-*C*-prenyleriodictyol. Eriodictyol, its 7-methyl ether, and naringenin 7-methyl ether were observed in several species, whereas the 8-methoxy derivatives of both naringenin and eriodictyol plus the 3'-methyl ether of the latter were found to occur in *E. frutescens* var. *resinosa*. A simple but unusual arrangement of hydroxyls (Fig. 11–2) was observed in 5,7,3'-trihydroxyflavanone **[11–5]**, which was obtained from *E. stenophylla*. Its structural assignment was based on proton magnetic resonance data and is certainly correct.

Enceliopsis: *Enceliopsis* consists of four species native to southwestern North America. Proksch et al. (1987) studied leaf and root flavonoids of *E. nudicaulis*. Leaves of *E. nudicaulis* var. *nudicaulis* afforded herbacetin 8-methyl and 3,8-dimethyl ethers, gossypetin-3,8,3'-trimethyl ether, dihydroherbacetin 8-methyl ether and 8,4'-dimethyl ether, and dihydrogossypetin-8,4'-dimethyl ether. Roots of this variety afforded only benzofurans. Leaves of *E. nudicaulis* var. *corrugata* had the same array of flavonoids as var. *nudicaulis* plus gossypetin 3,8-dimethyl ether. Roots of this variety had the same array of benzofurans plus dihydroherbacetin-8,4'-dimethyl ether **[11–6]** and dihydrogossypetin-8,4'-dimethyl ether **[11–7]**.

Ericentrodea: Six species comprise this South American genus (Robinson, 1993b). A butein/sulfuretin glycoside pair was observed by Crawford and Stuessy (1981) in two species, *E. corazonensis* and "*E. campii*."

Eriophyllum (Hele): Eleven species, native to western North America, comprise the genus *Eriophyllum*. Seven species have been studied for flavonoids by Harborne and Smith (1978): *E. ambiguum, E. confertiflorum, E. lanatum* var. *obovatum, E. multicaule, E. pringlei, E. staechadifolium*, and *E. wallacei*. Quercetin 5-*O*-glucoside, quercetagetin 7-*O*-glucoside, and patuletin 7-*O*-glucoside were identified, although not all taxa exhibited all three compounds.

Espeletia: *Espeletia* comprises about 15 species with a range that includes Venezuela, Colombia, Peru, and Ecuador. Recent reports by Torrenegra et al. (1994) and Torrenegra and Tellez (1995, 1996) reported luteolin, quercetin, and quercetin 3-methyl ether from two species, *E. killipii* and *E. tunjana*.

Flaveria (Hele): *Flaveria* is represented in North America by 22 species (Powell, 1979; Turner, 1995) with an additional one known from Australia. Five have

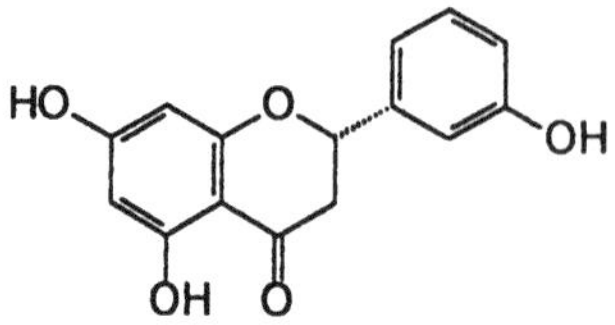

[11-5]

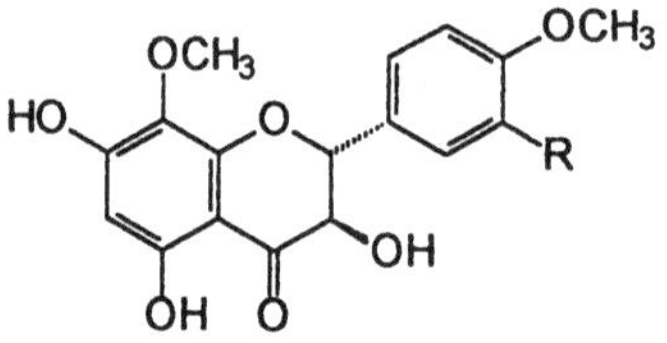

[11-6] R = H

[11-7] R = OH

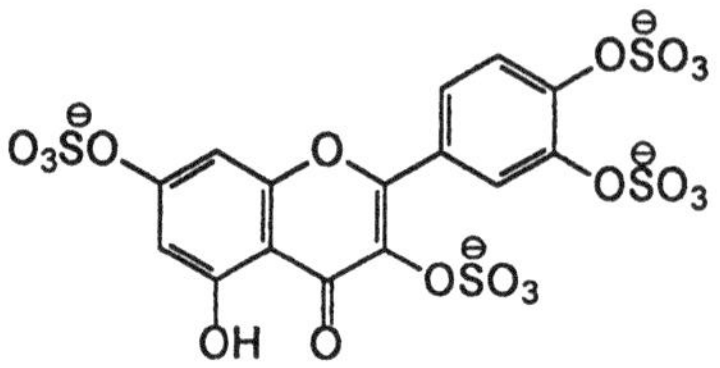

[11-8]

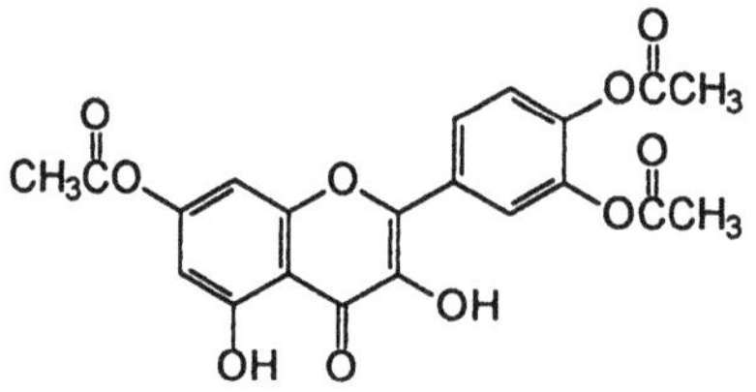

[11-9]

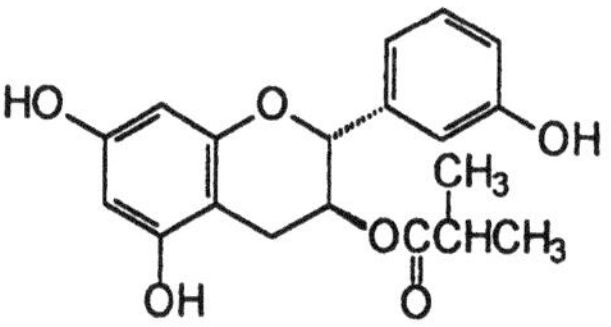

[11-10]

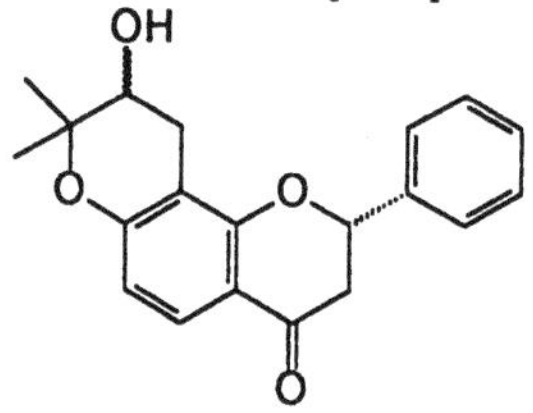

[11-11]

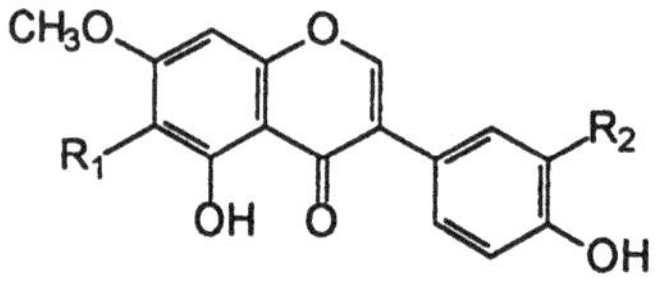

[11-12] R_1 = H, R_2 = OCH_3

[11-13] R_1 = OH, R_2 = H

[11-14] R_1 = OH, R_2 = OCH_3

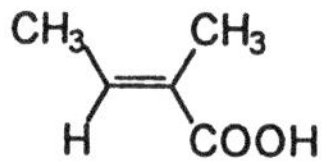

[11-15]

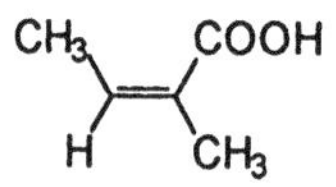

[11-16]

Fig. 11–2. An assortment of flavonoids from members of Heliantheae s.l. including **[11–5]** from *Encelia stenophylla*, the unusual dihydroflavonols **[11–6, 11–7]** from *Enceliopsis nudicaulis* var. *nudicaulis*, the sulfated **[11–8]** and polyacetylated **[11–9]** flavonols from *Flaveria* species, and the flavan-3-ol isobutyrate **[11–10]** from species of *Flourensia*. Compound **[11–11]** illustrates a further modification, the hydroxyl group on the alkyl side chain, that can occur with *C*-prenyl flavonoids. The isoflavones **[11–12 – 11–14]** make the flavonoid profile of *Gaillardia* highly uncommon in the family. Compounds **[11–15]**, tiglic acid, and **[11–16]**, angelic acid, have been encountered as acylating acids in *Geraea canescens* been subjected to flavonoid analysis. Wagner et al. (1971b) reported patuletin 3-*O*-glucoside from *F. linearis* and *F. trinervia*; Suarez et al. (1979) reported kaempferol, quercetin, isorhamnetin, and *O*-glycosides thereof from *F. bidentis*; Al-Khubaizi et al. (1978) added 6-methoxykaempferol 3-*O*-glucoside, which they found in *F. brownii*. The first report of sulphated flavonoids in the genus was apparently that of Pereyra de Santiago and Juliani (1972) who identified quercetin-3,7,3',4'-tetrasulphate **[11–8]** from *F. bidentis*. This paper was followed by a series of others in which additional sulphated flavonols were described from this species: quercetin-7,3',4'-trisulphate-3-acetate **[11–9]** (Cabrera and Juliani, 1976), isorhamnetin-3,7-disulphate (Cabrera and Juliani, 1977), quercetin-3,4'-di- and 3,7,4'-trisulphates (Cabrera and Juliani, 1979), and quercetin-3,7,3'-trisulphate (Cabrera et al., 1985). Confirmation of these observations as well as biosynthetic studies using *F. bidentis* were described by Varin et al. (1986a,b). Detailed studies by Barron and Ibrahim and coworkers (1986, 1987a,b, 1988; Barron et al., 1986) have demonstrated the existence of a complex assemblage of sulphated flavonols based on quercetin, quercetin-7,4'-dimethyl ether (ombuin), and a series of quercetagetin methyl ethers in *F. chloraefolia*. Recent biosynthetic work dealing with these compounds was discussed in Chapter 5.

Flourensia: *Flourensia* consists of at least 30 species native to Latin America. Flavonoid data are available for eight species. A rich array of flavonoid types characterizes the group including chalcones, flavanones, flavones, flavonols, and *C*-glycosylflavones. Flavonoids with extra hydroxylation at C-6 were reported from several species, *C*-prenylated compounds have been reported from two and B-ring deoxyflavonoids are known from six species. A series of 6,8-di-*C*-glycosyl flavones was reported by Dillon et al. (1976) from *F. cernua*. 2',4'-Dihydroxychalcone and the isomeric 7-hydroxyflavanone were identified in *F. heterolepis* (Bohlmann and Jakupovic, 1979) and *F. oolepis* (Guerreiro et al., 1979). *Flourensia heterolepis* also afforded 7-methoxyflavanone (Bohlmann and Jakupovic, 1979). 5,7-Dihydroxyflavanone (pinocembrin) has been reported from two species, *F. ilicifolia* and *F. retinophylla* (Dillon and Mabry, 1977), and 3,5,7-trihydroxyflavanone (pinobanksin) was reported from the resin of *F. resinosa* by Wollenweber and Yatskievych (1985). Reinvestigation of *F. retinophylla* (Stuppner and Müller, 1994) resulted in identification of three flavonoids with the unusual 3'-hydroxy B-ring substitution pattern: 5,7,3'-trihydroxyflavanone **[11–5]**, its 3'-methyl ether, and 5,7,3'-trihydroxyflavan-3-ol 3-isobutyrate **[11–10]**.

B-Ring deoxyflavones (and flavonols) have been reported from other members of the genus: 5,7-dihydroxyflavone (chrysin) from *F. laurifolia* (Guerrero and de Lourdes Flores, 1990), chrysin, 5-hydroxy-7-methoxyflavone (tectochrysin), 3,5,7-trihydroxyflavone (galangin), galangin 3-methyl ether and galangin 7-methyl ether from the leaf resin of *F. resinosa* (Wollenweber and Yatskievych, 1985). Two further elaborations on the B-ring deoxy theme are found in *F. laurifolia* in the form of 5,7-dihydroxy-6-methoxyflavone and an 8-*C*-prenylated flavanone that has undergone cyclization with the 7-OH group, as well as further modification to yield **[11–11]** (Guerrero and de Lourdes Flores, 1990). A *C*-prenylated flavanone was reported from *F. heterolepis* by Bohlmann and Jakupovic (1979) and shown to be 5,3',4'-trihydroxy-7-methoxy-8-*C*-prenylflavanone. Flavones and flavonols having B-ring oxygenation are listed in Table 11–10.

Gaillardia (Hele): Twenty-eight species comprise this North American genus. Flavonoid data are available for only 10 of these, but the work has resulted in identification of a surprising variety of flavonoid types. Stoutamire (1960) reported the presence of cyanidin *O*-glycosides in six species (not named in abstract). The Carman et al. (1972) survey of the Asteraceae for 6-methoxyapigenin (hispidulin) reported sightings in seven species (*G. aestivalis, G. mexicana, G. multiceps, G. pinnatifida, G. powellii, G. pulchella* and *G. suavis*). Hispidulin has also been reported from *G. fastigiata* (Herz et al., 1966b; Averett and Beaman, 1975) and *G. aestivalis* (Averett and Beaman, 1975; cited as *G. serotina*; synonymy from Correll and Johnston, 1970). The related 6-methoxyluteolin (nepetin) has been reported from the genus as well: *G. aestivalis, G. fastigiata* (Averett and Beaman, 1975; Herz et al., 1991), *G. powellii* (Herz and Bruno, 1987a), and *G. cabrerae, G. megapotamica* and *G. tontalensis* (Petenatti et al., 1996). This

Table 11–10. B-Ring oxygenated flavones and flavonols from species of *Flourensia*

Species	Compounds[1]	References
F. campestris	574'/36	Dillon and Mabry (1977)
F. cernua	574'/6, 54'/67	Rao et al. (1970)
	54'/67, 54'/37, 573'4'/3	Dillon et al. (1976)
F. ilicifolia	574'/3, 3574'/6, 574'/36, 573'4'/36, 574'/363'	Dillon and Mabry (1977)
F. resinosa	574'/-, 54'/7, 5/74', 574'/3', 54'/37, 57/34', 574'/33'	Wollenweber and Yatskievych (1985)
F. retinophylla	54'/37	Dillon and Mabry (1977)
F. thurifera	353'4'/7	Faini et al. (1997)

[1] Substitution: Hydroxyl positions/Methoxyl positions.

latter cited paper also reported the presence of luteolin in *G. megapotamica* and eriodictyol and 5,7,4'-trihydroxy-6,3'-dmethoxyflavone (jaceosidin) in *G. tontalensis*.

C-Glycosylflavones based upon both apigenin and luteolin have been reported from several species: *G. aristata* (Zielinska and Gill, 1974; Zielinska-Stasiek and Gill, 1977); *G. pulchella* (Wagner et al., 1972d); *G. aestivalis* (cited as *G. serotina*; see above) and *G. fastigiata* (Averett and Beaman, 1975). The most unusual flavonol sighting in this genus, however, must be credited to Herz et al. (1991) who identified three isoflavones from *G. suavis*, 5,4'-dihydroxy-7,3'-dimethoxyisoflavone (orobol 7,3'-dimethyl ether) **[11–12]**, 5,6,4'-trihydroxy-7-methoxyisoflavone **[11–13]**, and 5,6,4'-trihydroxy-7,3'-dimethoxyisoflavone **[11–14]**. Isoflavones are among the rarest of flavonoids in the family.

Galeana (Hele): This small genus of three species occurs in Mexico and Central America. Two unusual flavonoid esters were identified from *G. pratensis* by Maldonado et al. (1992b): 5,8,4'-trihydroxy-3,6,7-trimethoxyflavone 8-(2-methyl)butyrate and 5,7,8-trihydroxy-3,6,4'-trimethoxyflavone 8-tiglate. (Tiglic acid is 2-methylbut-2(*E*)-enoic acid **[11–15]**.)

Galinsoga: *Galinsoga* consists of 13 species native to temperate and subtropical Central and South America. Only one species has apparently been examined for flavonoids; Plekhanova et al. (1977) reported apigenin and luteolin 7-*O*-glucosides from *G. parviflora*.

Geraea: Both species of this genus, native to southwestern North America and Mexico, have been studied for their flavonoids. Two papers by Harborne et al. (Harborne and Smith, 1978; Harborne et al., 1978) focused on floral pigments. The main flavonoid compound identified in that work was gossypetin 8-methyl ether 3-*O*-monoglucoside. Detailed studies of resin components yielded an array of *O*-methylated flavonols. Resin of *G. viscida* had the simpler profile with the 3-methyl ethers of kaempferol, quercetin, and isorhamnetin (Proksch et al., 1986). *Geraea canescens* leaf resin afforded 8-methoxyapigenin, herbacetin 3,8-dimethyl ether, and gossypetin 3,8-dimethyl and 3,8,3'-trimethyl ethers (Proksch et al., 1986). A later paper by Proksch et al. (1987) reported the presence of 3,5,7,4'-tetrahydroxy-8-methoxyflavanone 3-acetate and 3-angelate in the leaf resin of the same species. (Angelic acid is 2-methylbut-2(*Z*)-enoic acid **[11–16]**.)

Glossocardia: This genus consists of 12 species native to India, Vietnam, Australia, and the Pacific. Two taxa gave positive anthochlor tests (Crawford and Stuessy, 1981): *G. bosvallia* and *G. tridenta* (cited as *Guerreroia macrocephala* (sic = *G. monocephala*); synonymy after Veldkamp and Kreffer, 1991).

Goldmanella: *Goldmanella sarmentosa*, native to Mexico, is the sole member of this genus. Okanin/maritimetin and butein/sulfuretin glucoside pairs were reported from this taxon (Crawford and Stuessy, 1981).

Guardiola: This genus of 10 species occurs in Mexico and southwestern United States. It was unusual in Coreopsidinae in that all five species tested by Crawford and Stuessy (1981), *G. angustifolia, G. arguta, G. mexicana, G. platyphylla* and *G. tulocarpus*, failed to give positive anthochlor tests.

Guizotia: *Guizotia* consists of approximately six species native to tropical Africa. Shimokoriyama (1962) reported positive anthochlor tests for six species (names not given).

Haploësthes (Hele): *Haploësthes* consists of three species native to southwestern North America. Al-Khubaizi (1977), describing a preliminary study of the genus, reported the presence of sulphated flavonoids based upon kaempferol, quercetin, isorhamnetin, patuletin, or spinacetin. Details were not available.

Helenium (Hele): The genus *Helenium*, native to the United States and Mexico, consists of 40 species. Cyanidin 3,5-di-*O*-glucoside was reported from *H. autumnale* in 1916 (Willstätter and Bolton, 1916). More recent studies include finding 6-methoxyapigenin (hispidulin) in 20 species (Carman et al., 1972) and *C*-glycosylflavones based on both apigenin and luteolin in *H. alternifolium*, *H. autumnale* and *H. brevifolium* (Wagner et al., 1972c, 1972g). More recent still are the reports of 5,7-dihydroxy-6,4'-dimethoxyflavone (pectolinarigenin) in *H. integrifolium* (Ozawa et al., 1983), 5,7,3',4'-tetrahydroxy-6-methoxyflavone in *H. scorzoneraefolium* (Maldonado et al., 1986), and 5,7,4'-trihydroxy-6,3'-dimethoxyflavone (jaceosidin) in *H. donianum* (Silva et al., 1992a).

Helenium chihuahuensis and three putatively closely related species, *H. arizonicum, H. mexicanum*, and *H. laciniatum*, were studied by Bierner et al. (1977) in an effort to assess their inter-relatedness. Analysis of the flavonoid profiles of the four taxa showed a rich array of *C*-glycosylflavones, involving both apigenin and luteolin types, and 6-methoxy derivatives of both apigenin and luteolin.

Helianthopsis: *Helianthopsis* was described as a Peruvian-Ecuadorian genus of 20 species (Robinson, 1979). Karis and Ryding (1994b) include these taxa in *Pappobolus*. The only reference to flavonoids in *Helianthopsis* is a comment made by Schilling and Spooner (1988) that the genus tested positive for anthochlors. No mention was made of what taxa were examined.

Helianthus: *Helianthus* is a North American genus of about 50 species the "flagship" member of which is the common sunflower *H. annuus*. Early work by Sando (1926a,b) described quercetin 7-*O*-glucoside from *H. annuus*, which species, along with *H. tuberosus*, was shown to possess a "rutin-like" compound by Beal (1949). Waddell (1973) identified 5,7-dihydroxy-6,8,3',4'-tetramethoxyflavone (hymenoxin) from *H. angustifolius*. Numerous reports of *O*-methylated compounds, some of which featured extra oxygenation, soon appeared in the literature. These reports are detailed in Table 11–11.

Table 11–11. Less common flavones and flavonols from species of *Helianthus*

Species	Compounds[1]	References
H. series *Corono-solis*		
14 of 15 species	574'/6,	
13 of 15 species	573'4'/6	
13 of 15 species	57/-	Schilling and Mabry (1981)
H. angustifolius	57/683'4'	Waddell (1973); Ohno et al. (1981)
H. annuus	574'/6, 573'4'/6,	
	574'/63', 57/684'	Rieseberg et al. (1987)
	57/684'	Alfatafta and Mullin (1992)
H. argophyllus	57/684'	Watanabe et al. (1982)
H. gracilentus	57/64'	Melek et al. (1985)
H. grosseserratus	574'/6, 57/64'	Herz and Kumar (1981a)
H. hirsutus	57/683'4'	Melek et al. (1985)
H. microcephalus	564'/7, 353'/674',	
	53'/3674', 35/674'	Gao et al. (1987)
	574'/6, 573'/684',	
	573"4'/6, 574'/63'	
	57/683'4', 57/684'	Schilling et al. (1987)
	57/683'4', 57/684'	Gutierrez and Herz (1988)
H. pumilus	57/684'	Herz and DeGroote (1977); Herz and Kumar (1981b)
H. resinosus	57/684'	Schilling and Mabry (1981)
H. schweinitzii	57/683'4'	Schilling and Mabry (1981)
H. simulans	57/683'4'	Herz et al. (1983)
H. strumosus	57/684', 57/683'4',	
	574'/683', 573'/684'	Herz and Kulanthaivel (1984)

[1] Substitution: Hydroxyl positions/Methoxyl positions.

A flavan, a comparatively rare flavonoid type characterized by a completely reduced C-ring, has been isolated from a member of *Helianthus*; 5-hydroxy-7,4'-dimethoxyflavan **[4–88]** has been reported from *H. microcephalus* (Gao et al., 1987; Gutierrez and Herz, 1988). Similarly rare, particularly in Asteraceae, are 5-deoxyflavonols. It is of interest, then, that Harborne and Smith (1978) reported 5-deoxyquercetin (fisetin) as a constituent of *H. gracilentus*.

Most species of *Helianthus*, when specifically examined for flavonols, have been shown to accumulate kaempferol and quercetin mono- and diglycosides. Both 3-*O*-glycosides and 7-*O*-glycosides have been identified. Occurrences of these compounds can be found in Schilling and Mabry (1981); Schilling (1983); and in Schilling et al. (1987). Rieseberg et al. (1987) reported luteolin and other compounds (see Table 11–11) as components of leaf glands of *H. annuus*.

Two groups of workers have described studies of the pigments responsible for the purple color of seeds of *H. annuus*. Vaccari et al. (1982) identified acylated derivatives of peonidin, cyanidin, and malvidin 3-*O*-glycosides. More recently, Mazza and Gao (1994) reported cyanidin 3-*O*-xyloside and 3-*O*-glucoside and their respective malonyl derivatives from purple sunflower seeds.

The first report of anthochlors in *Helianthus* came from the study of nectar guides in the family (Harborne and Smith, 1978): coreopsin and sulfurein were identified from *H. atrorubens*, *H. glaucophyllus*, and *H. gracilentus* while anthochlors were detected in the ray florets of 28 of 37 additional species sampled from herbarium specimens. All surveys of the genus since have reported anthochlors in at least some of the species, usually as components of floral tissue: *Helianthus* series *Corono-solis* (Schilling and Mabry, 1981), *Helianthus* series *Angustifolii* (Schilling, 1983), *Helianthus* series *Microcephali* (Schilling et al., 1987). An unusual aurone (Fig. 11–3) was isolated from an epicuticular extract of flowers of *H. annuus* by Alfatafta and Mullin (1992) and shown to be 5-hydroxy-4,6,4'-trimethoxyaurone **[11–17]**. These workers reported the presence of 5,7-dihydroxy-6,8,4'-trimethoxyflavone (nevadensin) in the same tissue as well. An array of chalcones, the related flavanones, and a flavonol have been identified as allelopathic constituents of *H. annuus* cult. VYP (Macias et al., 1997). The chalcones were shown to be 2',4',4-trihydroxy-3'-methoxychalcone and the hitherto unknown (in Nature) 2',4-dihydroxy-3',4'-dimethoxychalcone **[11–18]**, which was named "heliannone A" by those authors. The flavanone corresponding to the new chalcone **[11–19]**, named "heliannone C" also appears to be a new natural product. The flavonol was identified as tambulin, 3,5-dihydroxy-7,8,4'-trimethoxyflavone. All of the compounds showed some level of activity in germination tests.

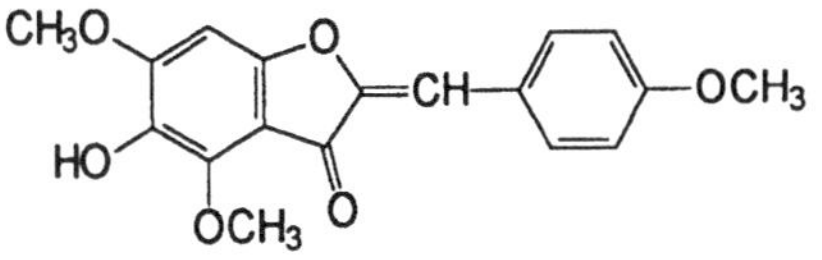

[11-17]

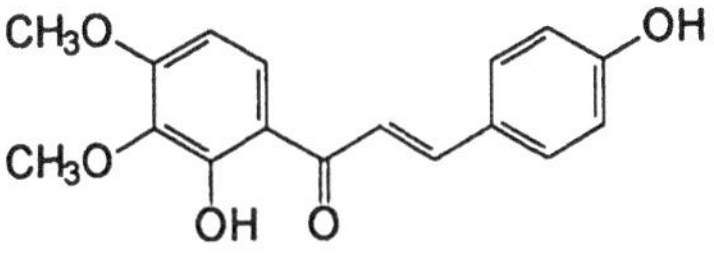

[11-18]

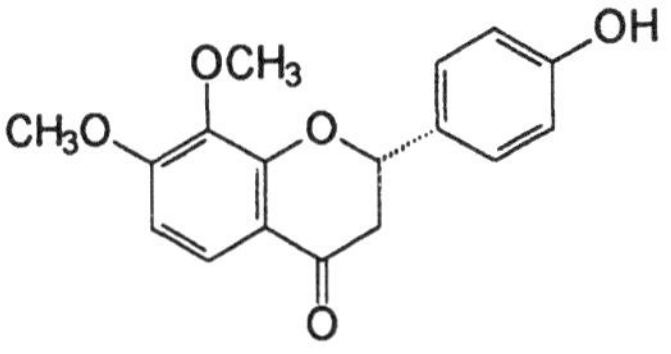

[11-19]

Fig. 11–3. An aurone **[11–17]**, a chalcone **[11–18]**, and a flavanone **[11–19]** isolated from members of Heliantheae s.l.

Three recent papers should be consulted for information on the application of macromolecular methods to problems of relationships within the genus, Kahler and Lay (1985) on the genetics of electrophoretic variants in the cultivated sunflower, Rieseberg et al. (1991) on chloroplast DNA applied to hybridization within *Helianthus* section *Helianthus*, and Gentzbittel et al. (1992) on a study of a molecular phylogeny of the genus using restriction fragment length polymorphism (RFLP).

Heliomeris: *Heliomeris* consists of five species native to western United States and Mexico. Floral flavonoids of *H. multiflora* (reported as *Viguiera multiflora*) were shown by Shimokoriyama and Geissman (1960) to be butein, coreopsin, sulfuretin, sulfurein, and quercetin 7-*O*-glucoside.

Heliopsis: *Heliopsis* consists of 15 species native to North America. Jakupovic et al. (1988d) reported 5,3'-dihydroxy-3,6,7,8,4'-pentamethoxyflavone from *H. helianthoides* in the only report we have for flavonoids from the genus.

Hemizonia (Hele): *Hemizonia* consists of at least 30 species restricted to California and Baja California (Proksch et al., 1984; Tanowitz et al., 1987). Two studies of foliar exudate flavonoids in this genus have been published, the first a report of such compounds from four species (Proksch et al., 1984), the second a study of 18 taxa with accompanying taxonomic discussion (Tanowitz et al., 1987). Three flavanones were identified, naringenin, eriodictyol, and eriodictyol 7-methyl ether. Flavones identified were luteolin and 5,7,8,3',4'-pentahydroxy-6-methoxyflavone. Simple flavonols identified were kaempferol 4'-methyl ether, quercetin, and quercetin 3-methyl and 3,7-dimethyl ethers. Extra hydroxylation at C-6 occurs in all taxa studied as evidenced by the presence of the following flavonols: patuletin, quercetagetin-3,6-dimethyl ether, and quercetagetin-3,6,7-trimethyl ether. Three compounds were identified that were characterized by substitution at C-8: 5,7,3',4'-tetrahydroxy-3,8-dimethoxyflavonol (gossypetin 3,8-dimethyl ether), 5,3',4'-trihydroxy-3,7,8-trimethoxy-flavonol (gossypetin 3,7,8-trimethyl ether), and 5,7,3',4'-tetrahydroxy-3,6,8-trimethoxyflavone.

Henricksonia: *Henricksonia mexicana*, the sole member of this genus (Turner, 1977c), gave a positive anthochlor test (Crawford and Stuessy, 1981).

Heterosperma: This genus ranges from Mexico to Central America and consists of five species. Shimokoriyama (1962) reported positive anthochlor tests for all five.

Hidalgoa: *Hidalgoa* consists of five species ranging from Mexico to Central America. Shimokoriyama (1962) reported a positive anthochlor test for one (un-named) species.

Holocarpha (Hele): *Holocarpha* is a genus of four endemic Californian species: *H. heermanii, H. macradenia, H. obconica*, and *H. virgata*. All four have been examined for leaf surface flavonoids (Bohm et al., 1994). Present in all four taxa are naringenin 7,4'-dimethyl ether, eriodictyol 7-methyl ether, taxifolin 7-methyl ether, kaempferol 3-methyl ether, and quercetin 7-methyl ether (rhamnetin). Variously distributed are apigenin, luteolin 7-methyl ether, kaempferol 7-methyl ether, quercetin 3-methyl ether, quercetin 3,7-dimethyl and 3,3'-dimethyl ethers, and quercetagetin 3,6-dimethyl and 3,7-dimethyl ethers.

Hulsea (Hele): *Hulsea* consists of seven species all of which occur in western North America. Flavonoids of the genus have been studied by Wilken (1975). The flavone diosmetin (luteolin 4'-methyl ether) and several quercetin 3-*O*-glycosides were present in all seven species. Apigenin 7-*O*-glucoside was more restricted and cyanidin 3-*O*-glycosides were identified in about half of the species studied.

Hymenoclea: *Hymenoclea* is a small genus, only three species, native to southwestern United States (Peterson and Payne, 1973; Welsh, 1993; Holmgren, 1994). Miao et al. (1995c), based on cpDNA work, have concluded that the species are best treated within *Ambrosia*. As none of the taxa at this time appear to have a name available in the latter genus, however, we leave *Hymenoclea* separate here following Karis and Ryding (1994b). Zdero et al. (1989a) reported naringenin and 3,5,7-trihydroxyflavanone (pinobanksin) from *H. monogyra*.

Hymenoxys (Hele): *Hymenoxys* consists of 28 species distributed in western North America. Flavonoid data are available for 15 species. Eight species (*H. anthemoides, H. argentea, H. bigelovii, H. odorata, H. quinquesquamata, H. richardsonii, H. rusbyi, H. subintegra*) were surveyed for common flavonol glycosides by Wagner et al. (1972h) who found quercetin 3-*O*-glucoside in most but kaempferol glycosides, quercetin 3-*O*-rutinoside and isorhamnetin 3-*O*-galactoside more restricted in distribution. Quercetagetin and patuletin 7-*O*-glucosides and patuletin 3-*O*-glucoside and 3-*O*-rutinoside have been reported from *H. scaposa* (Thomas and Mabry, 1968a). 6-Methoxyapigenin (hispidulin) was reported from *H. odorata* and *H. richardsonii* by Herz et al. (1970a,b), from *H. anthemoides, H. cabrerae, H. robusta, H. rusbyi* and *H. tweediei* by Carman et al. (1972) and *H. turneri* by Gao et al. (1990). Several highly *O*-methylated flavonoids have been identified from *Hymenoxys* including, of course, hymenoxin, which Thomas and Mabry (1967) obtained from *H. scaposa* and showed to be 5,7-dihydroxy-6,8,3',4'-tetramethoxyflavone. These workers (Thomas and Mabry, 1968b) isolated 5,7,5'-trihydroxy-6,8,3',4'-tetramethoxyflavone from the same source. Other reports of *O*-methylated flavones from the genus are hymenoxin from *H. linearifolia* (Herz et al., 1970a,b; Dominguez et al., 1980), 5,7,4'-trihydroxy-3,3'-dimethoxylflavone from *H. acaulis*, 5,7-dihydroxy-6,4'-dimethoxyflavone (pectolinarigenin) from *H. subintegra*, and 5,7-dihydroxy-6,7,8,4'-tetramethoxyflavone from *H. turneri* (Gao et al., 1990). These latter workers also isolated the known dihydrokaempferol 3-acetate and a series of 6-methoxydihydroflavonols, some of which occurred as the 3-acetates. A recent paper by Ferracini et al. (1994) reported the identification of 5,4'-dihydroxy-6,7,8-trimethoxyflavone (xanthomicrol) from *H. scaposa* var. *linearis*.

Diaz et al. (1992) described several compounds that they obtained from an investigation of *H. linearifolia* var. *arenicola* (cited as *Tetraneuris linearifolia* var. *arenicola*; generic synonymy after Cronquist, 1994; Karis and Ryding, 1994b). The workers identified a series of 6- and 6,8-oxygenated flavones in addition to kaempferol. Known compounds identified were 5,7,4'-trihydroxy-6,3'-dimethoxyflavone 7-*O*-glucoside, 5,7,4'-trihydroxy-6,8-dimethoxyflavone (demethoxysudachitin), 5,7-dihydroxy-6,8,4'-trimethoxyflavone (nevadensin), 5,7,3',4'-tetrahydroxy-6,8-dimethoxyflavone, 5,7,4'-trihydroxy-6,8,3'-trimethoxyflavone (sudachitin), and 5,7-dihydroxy-6,8,3',4'-tetramethoxyflavone (hymenoxin). Newly reported in that paper was 5,7,3',4'-tetrahydroxy-6,8,5'-trimethoxyflavone.

Recent DNA restriction site data (Bierner and Jansen, 1998) suggest that *Tetraneuris* Greene should be segregated from *Hymenoxys*, which would refer to *H. acaulis*, *H. linearifolia*, *H. scaposa*, and *H. turneri* reported here for flavonoids. The flavoniods of this group of species do not provide any additional insights into this taxonomic discussion.

Ichthyothere: Approximately 20 species constitute this South American genus (Karis and Ryding, 1994b; Robinson, 1995). The only flavonoid report in the literature concerns *I. terminalis* (Bohm and Stuessy, 1982). The glycosidic fraction consisted of quercetin 3-*O*-mono-, 3-*O*-di- and 3-*O*-triglycosides. The aglycone fraction consisted of small quantities of kaempferol, quercetin, and dihydroquercetin in addition to a large amount of quercetin 3-methyl ether.

Iostephane: This is a small genus, with perhaps four species, endemic to Mexico (Strother, 1983; Schilling and Panero, 1991). The flavonoid profile of the species, *I. heterophylla, I. madrensis, I. papposa* and *I. trilobata*, is a simple one consisting of quercetin 3-*O*-glucoside and 3-*O*-glucuronide and the anthochlor pair coreopsin/sulfurein.

Isostigma: *Isostigma* is a Mexican genus consisting of 11 species. Shimokoriyama (1962) reported positive anthochlor tests for four species which he did not name.

Iva: *Iva* is a North American genus of 15 species, five of which have been examined for flavonoids. 6-Oxygenation is common, 8-oxygenation occurs in two compounds (three species) and moderate levels of *O*-methylation are seen in *Iva* flavonoids. Early studies of *I. acerosa* and *I. nevadensis* (Farkas et al., 1966; Nogradi and Farkas, 1968) resulted in isolation and identification of the new flavonoid 5,7-dihydroxy-6,8,4'-trimethoxyflavone (nevadensin) from the former, and nevadensin, 6-methoxyapigenin (hispidulin) and 5,7-dihydroxy-6,4'-dimethoxyflavone (pectolinaringin) from the latter. Another new flavonoid obtained from *I. acerosa* by these workers (Farkas et al., 1967) was identified as 5,7,3'-trihydroxy-6,8,4'-trimethoxyflavone (acerosin). *Iva axillaris* afforded the new compound 5,7,3',4'-tetrahydroxy-3,6-dimethoxyflavone (axillarin) (Herz et al., 1966a,c; Hörhammer et al., 1966; Nogradi and Farkas, 1968). The last mentioned authors also described isolation of 5,7,3'-trihydroxy-6,8,4'-trimethoxyflavone (acerosin) from the same source. Herz et al. (1972a) isolated 5,7,3'-trihydroxy-3,6,4'-trimethoxyflavone from *I. frutescens* ssp. *frutescens* and hispidulin and pectolinaringen from *I. frutescens* ssp. *oparia*. Herz and Sudarasanam (1970) reported hispidulin and axillarin as components of *I. hayesiana*.

A comment on generic delimitation with respect to *Iva* is in order. It is controversial whether to recognize one genus (following Jackson, 1960) or several genera (following Rydberg, 1922; Bolick, 1983, 1985; Karis, 1995). Taking the genus in a broad sense, there are three main units within *Iva* (after Jackson, 1960): sect. *Iva*, sect. *Linearbractea*, and sect. *Cyclachaena*. All workers agree that the first two units belong in *Iva* s.s. Miao et al. (1995b), based on cpDNA restriction site data, recently elevated *I. axillaris* from out of sect. *Iva* to its own new section *Rhizoma*. Bolick (1983, 1985) and Karis (1995) recommend splitting up Jackson's sect. *Cyclachaena* into four genera: *Chorisiva*, *Leuciva*, *Oxytenia*, and *Cyclachaena* (all monotypic except the last which would be bitypic).

The issue that *Iva* highlights, at least in part, is whether one should employ strict cladistic rules for formal taxonomic recognition or an explicit phyletic approach that also takes into account apomorphic changes within lineages (e.g., Stuessy, 1990, 1997). If one examines the more recently generated morphological cladogram of Karis (1995, his Fig. 2A), we see holophyletic *Iva* s.l. in which are nested *Dicoria* (four species, but two only used in the analysis) and *Euphrosyne* (monotypic). Seven apomorphic character state changes (one non-homoplasious) define *Dicoria*, and suggest strongly that it should be recognized as a distinct genus. No one, in fact, has suggested that this should be otherwise. The data also suggest that *Dicoria* may have evolved from out of *Iva*. Such a possibility cannot be demonstrated by cladistic analysis and hence is dismissed in any conclusion derived from strict cladistic principles. *Euphrosyne* is less well defined with only four autapomorphies and none of them non-homoplasious. If in a strict cladistic approach *Dicoria* is treated as a distinct genus, then *Iva nevadensis* must also be recognized as a distinct genus (as *Chorisiva*), *I. acerosa* as *Oxytenia*, and *I. dealbata* as *Leuciva*. *Cyclachaena* (with two species) does appear to be biphyletic and problematic, meriting further study.

For continuity in this book, as in most all other instances, we follow Karis and Ryding (1994b) in treating *Iva* in the broad sense. Although these authors opt for a phyletic perspective with *Iva* (they only explicitly include *Oxytenia* in synonymy, however), they make it clear (p. 578) that this is only because they believe more study is needed and that "...some of the segregate genera recognized by Rydberg (1922) may be resurrected." Additional macromolecular data will also be helpful here; Miao et al. (1995b) did apply such data to the complex, but they focused on interrelationships within *Iva* s.s. and did not examine taxa of sect. *Cyclachaena*, the controversial ones, in their study.

Jaumea (Hele): *Jaumea* is a genus of three species from the west coast of North and South America, plus Argentina and Uruguay (McVaugh, 1984). Crawford and Stuessy (1981) reported the butein/sulfuretin glucoside pair from *J. carnosa* and *J. peduncularis*.

Kingianthus: *Kingianthus* is a genus of two species native to Ecuador. Kaempferol and quercetin 3-methyl ethers have been reported from *K. sodiroi* (cited as *Zaluzania sodiroi*; synonymy from Robinson, 1978a) by Olsen and Mabry (1980). Although we include *Kingianthus* here following Karis and Ryding (1994b), the monographer of *Zaluzania* rejects this generic segregate (Olsen, 1979).

Lagascea: *Lagascea* is a Mexican genus consisting of eight species. Flavonoid information is available for all of them. Bohlmann and Jakupovic (1978) reported 3,7-dihydroxy-5,4'-dimethoxyflavan (7-hydroxy-5,4'-dimethoxyflavan-3-ol) from *L. rigida*. El-Naggar and Doskotch (1979) identified patuletin 7-*O*-glucoside and an acetylated derivative from *L. mollis*. The position of acetylation, presumably on the sugar, was not determined. A survey of flavonoids of the entire genus has been described by La Duke et al. (unpubl. ms.) who reported kaempferol, quercetin, isorhamnetin, and patuletin *O*-glycosides.

Lagophylla (Hele): *Lagophylla* consists of five species native to western United States, only one of which has been examined for flavonoids. Leaf exudate of *L. glandulosa* contained *O*-methylated flavones and flavonols, all but one of which exhibited 6-oxygenation (Bohm et al., 1992). The compounds identified were apigenin-7-methyl ether (genkwanin), 5,4'-dihydroxy-6,7-dimethoxyflavone, the 6,3'-dimethyl-, 6,7,3'-trimethyl-, and 6,7,3',4'-tetramethyl ethers of 6-hydroxyluteolin, and quercetagetin 3,6,3'-tri- and 3,6,7,4'-tetramethyl ethers.

Lasthenia (Hele): *Lasthenia* is a genus of 16 or 17 species. One occurs in Chile, the rest in western North America with a concentration in California. One species, *L. maritima*, occurs on bird-islands and can be found as far north as the northwestern tip of Vancouver Island, British Columbia. The entire genus has been subjected to detailed flavonoid analysis (Saleh et al., 1971; Bohm et al., 1974; Ornduff et al., 1973a, 1974; Bohm and Banek, 1986; Bohm et al., 1989b; Desrochers and Bohm, 1993; Rajakaruna and Bohm, unpubl.). Luteolin 7-*O*-glucoside, kaempferol, quercetin, and patuletin 3-*O*-glycosides, sulphated flavonol glycosides, eriodictyol 7-*O*-glycosides, the aurones sulfurein and maritimein, and the chalcones coreopsin and marein have been identified in various combinations in the genus. The aurones and chalcones were observed only in *L.* (*chrysostoma*) *californica*, *L. macrantha* and *L. debilis*.

Layia (Hele): *Layia* consists of 14 species native to California (Baldwin and Bainbridge, 1993). Crins et al. (1988b) studied the phenolic chemistry of

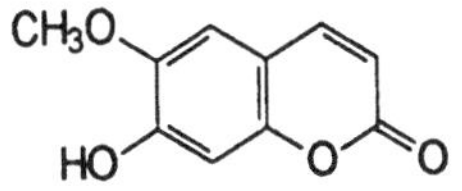

[11-20]

Fig. 11–4. The coumarin scopoletin **[11–20]** was observed in all species of *Layia* studied

Table 11–12. Leaf surface flavonoids of species of *Layia*[1]

Species	Compounds[2]
L. glandulosa	574'/36
L. hieracioides[3]	Apigenin, 574'/6, 54'/67, 573'4'/6, 574'/36, 57/63'4'
L. platyglossa	574'/36
L. septentrionalis	574'/6, 54'/67

1 Data from Crins et al. (1988c).
2 Substitution: Hydroxyl positions/Methoxyl positions.
3 Includes flavonoids reported from *L. paniculata*; synonymy after Baldwin and Bainbridge (1993).

leaf surface extracts of seven of them. *O*-Methylated flavonoids based on apigenin, luteolin, and kaempferol were identified from five of the species; the coumarin scopoletin (6-methoxy-7-hydroxycoumarin) (Fig. 11–4) **[11–20]** was seen in all. *Layia carnosa* and *L. chrysanthemoides* afforded only the coumarin. The occurrence of flavonoids is detailed in Table 11–12.

Lecocarpus: *Lecocarpus* consists of three species endemic to the Galapagos Islands. Macias and Fischer (1992) described 5,4'-dihydroxy-3,6,7-trimethoxyflavone (penduletin) from *L. pinnatifidus* and more recently Macias et al., 1993) reported 5-hydroxy-3,6,7,4'-tetramethoxyflavone from *L. lecocarpoides*.

Lipochaeta: The 23 species (Gardner, 1979) that comprise *Lipochaeta* are endemic to the Hawaiian Islands. A study by Gardner (1976) showed the presence of apigenin, luteolin, and quercetin *O*-glycosides. However, the distribution was not uniform; diploid species contained only quercetin derivatives while tetraploid species contained both flavone and flavonol derivatives. Additional morphological and cytogenetic studies by Rabakonandrianina (1980) and Rabakonandrianina and Carr (1981) have shown convincingly that the diploid and tetraploid sections of *Lipochaeta* have had separate ancestry from *Wedelia* [and/or *Thelechitonia* (= *Complaya*; Robinson and Cuatrecasas, 1992) and *Wollastonia*; see Strother, 1991, for these generic concepts], hence underlining the documented flavonoid differences. Further phyletic analysis is obviously warranted that will doubtless lead to future taxonomic changes.

Madia (Hele): *Madia* is a genus of 18 species, primarily western North American in distribution, but also represented in Chile. *Madia sativa*, the Chilean species now naturalized in California, was examined for flavonoids by Bohm et al. (1992) who reported 5,4'-dihydroxy-7-methoxydihydroflavonol (dihydrorhamnocitrin), naringenin, three eriodictyol methyl ethers, and two flavonols, kaempferol 7-methyl ether (rhamnocitrin) and 5,3',4'-trihydroxy-3,6,7-trimethoxyflavone (chrysosplenol-D).

Marshallia (Hele): *Marshallia,* native to the southern United States, consists of seven species. Long problematic as to tribal placement, recent cpDNA restriction site data (Watson et al., 1991) recommend positioning in Heliantheae s.l. Karis and Ryding (1994a) place it in Helenieae, following Watson et al. (1991) who suggest ties to *Bahia* and *Palafoxia*. The generic sample for cpDNA comparison is small, however; further, a tie has also been revealed with *Galinsoga*. Broader generic sampling is obviously needed to determine the proper affiliation of *Marshallia* within Heliantheae s.l.

The earliest report of flavonoids in the group is that of Bohlmann et al. (1979e) who isolated, from *M. grandiflora*, 8-*C*-prenylnaringenin, 8-*C*-prenylapigenin, and two 3,4-flavandiols identified as 3,4,7,4'-tetrahydroxy-5-methoxy-8-*C*-prenylflavan and 3,7,4'-trihydroxy-4,5-dimethoxy-8-*C*-prenylflavan. The following year a flavan-4-ol, several flavan-3,4-diols, plus two dihydroflavonols, all bearing a prenyl group at C-8 and various levels of *O*-methylation, were described from *M. obovata* (Bohlmann et al., 1980c). In two of these compounds, a flavan-3,4-diol and a dihydroflavonol, the *C*-prenyl group at C-8 has undergone cyclization with the phenolic group at C-7 to yield 6",6"-dimethylchromeno derivatives. Jakupovic et al. (1988e) expanded the study to include five additional taxa, *M. graminifolia*, *M. graminifolia* var. *tenuifolia* (cited as *M. tenuifolia*; synonymy after Watson and Estes, 1990), *M. mohrii*, *M. ramosa*, and *M. trinervia*, all of which contained flav-3,4-diols. Newly described in that paper were 7-*O*-glucosides of two of these plus a 4-*O*-xylosylglucoside of one. Distribution of these compounds is detailed in Table 11–13. Herz and Bruno (1987b) added further to the group describing two 8-*C*-prenylated flavan-3-ols from *M. graminifolia* var. *tenuifolia* (cited as *M. tenuifolia*; see comment above), along with 3,5,4'-trihydroxy- 6",6"-dimethylchromeno-[2",3":7,6]-flavone.

Table 11–13. 8-*C*-Prenylflavanones, -dihydroflavanols, and -flavans from species of *Marshallia*[1]

Base compounds	Species[2] GRA	GRM	MOH	OBO	RAM	TEN	TRI
Naringenin	–	–	+[3]	–	–	+	–
Aromadendrin	–	–	–	+[3]	–	–	–
Aromadendrin 5-methyl ether	+	–	–	–	–	–	–
4,5,4'-Trihydroxy-7-methoxyflavan	–	–	–	+	–	–	–
3,4,7,4'-Tetrahydroxy-5-methoxyflavan	+	+	–	–	+	–	–
" 7-*O*-glucoside	–	–	+	–	–	–	–
4,7,4'-Trihydroxy-3,5-dimethoxyflavan	–	–	–	+	–	–	–
3,7,4'-Trihydroxy-4,5-dimethoxyflavan	+	+	–	+	+	+	+
" 7-*O*-glucoside	–	–	+	–	–	–	–
" 4-*O*-xylosylglucoside	–	–	+	–	–	–	–
3,4,4'-Trihydroxy-5-methoxyflavan	–	–	–	+[3]	–	–	–

[1] *M. grandiflora* data from Bohlmann et al. (1979e) and Jakupovic et al. (1988e), *M. obovata* data from Bohlmann et al. (1980c), *M. tenuifolia* data from Herz and Bruno (1987b), the remainder from Jakupovic et al. (1988e).

[2] GRA = *M. grandiflora*; GRM = *M. graminifolia*; MOH = *M. mohrii*; OBO = *M. obovata*; RAM = *M. ramosa*; TEN = *M. tenuifolia*; TRI = *M. trinervia*.

[3] Prenyl group cyclized with 7-OH.

Megalodonta: *Megalodonta beckii* is the sole member of this genus. Its flavonoid chemistry and relationship with *Bidens* have been discussed by Roberts (1980). The flavonoid profile of *M. beckii* consists of common cyanidin and quercetin 3-*O*-glycosides, the anthochlor pairs marein/maritimein and coreopsin/sulfurein (and their respective aglycones), and the newly reported 2',4',3-trihydroxy-4-methoxychalcone 4'-*O*-glucoside (and aglycone).

Melampodium: *Melampodium* is a Mexican and Central American genus consisting of 39 species (Stuessy, 1972; Turner, 1988b, 1993a). We have flavonoid data for five of them. Bohm and Stuessy (1991) reported simple arrays of kaempferol and quercetin 3-*O*-mono- and diglycosides as the sole flavonoid components of *M. aureum, M. divaricatum*, and *M. longipilum*. From the latter work we also learned that *M. bibracteatum* has simple kaempferol and quercetin 3-*O*-monoglycosides plus quercetagetin 3-methyl ether, and that *M. americanum* has, again in addition to simple flavonol 3-*O*-glycosides, a series of quercetagetin derivatives, viz. 3-methyl ether, 3,7-dimethylether (tomentin), 3,6,7-trimethyl ether (chrysosplenol-D), 3,6,7,3'-tetramethyl ether (chrysosplenetin), and 3,6,7,3',4'-pentamethyl ether (artemetin).

Montanoa: *Montanoa* consists of 25 species in Mexico and Central America. Kaempferol and quercetin glycosides were reported from *M. tomentosa* by Oshima et al. (1986). A recent paper by Quijano et al. (1994) reported four *O*-methylated flavonoids from *M. leucantha* ssp. *leucantha*: apigenin 7,4'-dimethyl ether, kaempferol 3,7,4'-trimethyl ether, 5-hydroxy-6,7,4'-trimethoxyflavone and 5-hydroxy-6,7,8,4'-tetramethoxyflavone.

Moonia: *Moonia* consists of a single species, *M. heterophylla*, native to southern India and Sri Lanka (Stuessy, 1975). Crawford and Stuessy (1981) recorded a positive anthochlor test for a specimen of this taxon.

Narvalina: South America and the West Indies are home to the four species that constitute this small genus. Crawford and Stuessy (1981) reported the presence of okanin/maritimetin glycoside and coreopsin/sulfuretin glycoside pairs in *N. domingensis* from Haiti.

Neurolaena: *Neurolaena* consists of 13 species native to the West Indies and tropical America. Detailed flavonoid information is available for five (Ulubelen et al., 1980a, 1981; Ulubelen and Mabry, 1981b; Kerr et al., 1981). Profiles are based primarily on flavonols but a few flavones have been reported. 6-Hydroxylated compounds were present in all five pro-

Table 11–14. Flavonoid aglycones of species of *Neurolaena*

Species	Compounds[1]	Reference
N. lobata	Qt, 5673'4'/-, 5674'/3', 35674'/-, 564'/37, 573'4'/3, 573'4'/36, 563'4'/37	Kerr et al. (1981)
N. macrocephala	Qt, 5673'4'/3, 3573'4'/6, 573'4'/36, 563'4'/37	Kerr et al. (1981)
N. macrophylla	Lute, K, Qt, 5673'4'/-, 35674'/-, 3574'/6, 57/364', 5673'4'/3, 3573'4'/6, 573'4'/36, 563'4'/37	Ulubelen and Mabry (1981b)
N. oaxacaca	K, Qt, 573'4'/3, 5673'4'/-, 5674'/3', 5674'/3, 564'/37, 5673'4'/3, 3573'4'/6, 573'4'/36, 563'4'/37	Ulubelen et al. (1980a)
N. venturana	Qt, 53'4'/67, 573'4'/36, 563'4'/37	Ulubelen et al. (1981)

1 Lute = luteolin; K = kaempferol; Qt = quercetagetin; Substitution: Hydroxyl positions/Methoxyl positions.

files, mono- and di-*O*-methylation are common, a trimethyl ether is known, and sulphated compounds have been identified in three of the species. Table 11–14 details the occurrences of compounds in the genus.

Oparanthus: *Oparanthus* consists of four (Stuessy, 1977) species endemic to French Polynesia (Shannon and Wagner, 1997). Crawford and Stuessy (1981) reported a positive anthochlor test for *O. rapensis*. Stuessy (1988), based on comparative morphological studies, referred *Oparanthus* to *Bidens*, but Shannon and Wagner (1997) have recently resuscitated it.

Parthenium: *Parthenium*, which consists of 16 species, is distributed in the western United States and Mexico. The genus has been the subject of intensive chemical analysis owing in part to the importance of *P. argentatum* (guayule), which has proven to be a useful source of latex for rubber production (e.g., Lloyd, 1911) and, on the negative side, because of severe dermatitis caused by *P. hysterophorus* (Mears, 1980a). The entire genus, including several infraspecific taxa, has been examined for flavonoids. Profiles are based on kaempferol, quercetin, 6-hydroxykaempferol, and quercetagetin with abundant *O*-methylation with *O*-glycosides of many of the compounds known as well. The occurrence of aglycones in *Parthenium* is given in Table 11–15. This table represents a condensation of data arising from detailed surveys (Mears, 1973, 1979, 1980a). Mears distinguished between major and minor compounds; we present simply presences or absences. We have also presented total species chemistries in those cases where infraspecific taxa were listed in his table. These data compressions will be expanded in a later chapter where the systematic significance of the patterns of occurrence will be discussed. Earlier papers on *P. tomentosum* (Rodriguez et al., 1972a) and on *P. bipinnatifidum, P. glomeratum, P. hysterophorus* and *P. rollinsianum* (Shen et al., 1976) presented data that are in general agreement with Mears's findings.

Pericome (Hele): *Pericome* consists of four species of southwestern North America, only one of which has been examined for flavonoids. Leaf surface flavonoids of *P. caudata* have been shown to be 5,7-dihydroxy-3,4'-dimethoxyflavone, 5,7,4'-trihydroxy-3,6-dimethoxyflavone, 5,7-dihydroxy-3,6,4'-trimethoxyflavone, and 5,7,3'-trihydroxy-3,6,4'-trimethoxyflavone (Wollenweber and Mann, 1987; Crins et al., 1988a). Wollenweber and Mann (1987) also analyzed the vacuolar flavonoids of this species and showed that the profile consists of, after acid hydrolysis, quercetin, isorhamnetin, rhamnetin, the 7,3'-dimethyl and 7,3',4'-trimethyl ethers of quercetin, and 6,7,4'-trimethyl and 6,7,3',4'-tetramethyl ethers of quercetagetin.

Table 11–15. Flavonoids of species of *Parthenium*[1]

Taxon	Compounds[2]
Sect. *Argyrochaeta*	
P. bipinnatifidum	K, Q & IR 3-O-Gly, K3Me[3], Q33'Me, 574'/36, 564'/37, Qt36Me, Qt37Me, Qt373Me', Qt3674'
P. confertum	K & Q 3-*O*-Glys, K3Me, Q33'Me, 564'/37, Qt36Me, Qt37Me, Qt373'Me
P. densipilum	K & Q 3-*O*-Glys, K3Me, 564'/37, Qt36Me, Qt37Me
P. hysterophorus	K, Q & IR 3-*O*-Glys, K3Me, Q33'Me, 574'/36, 564'/37, Qt36Me, Qt37Me, Qt3674'
Sect. *Bolophytum*	
P. alpinum	Qt373'Me, Qt3674'Me, Qt3673'Me
P. ligulatum	Qt363'4'Me, Qt3674'Me
P. tetraneuris	574'/36, Qt36Me, Qt373'Me, Qt3674'Me, Qt3673'Me
Sect. *Partheniastrum*	
P. integrifolium	K, Q & IR 3-*O*-Glys, K3Me, K7Me, Q33'Me, Q373'4', 574'/36, 564'/37, Qt373'Me
P. radfordii	K, Q & IR 3-*O*-Glys, K7Me, Q33'Me, 564'/37
Sect. *Parthenichaeta*	
P. argentatum	K & Q 3-*O*-Glys, K3Me, K7Me, Q33'Me, 564'/37, 54'/367, Qt36Me, Qt37Me
P. incanum	K & Q 3-*O*-Glys, K3Me, Q33'Me, Qt36Me, Qt36Me
P. fruticosum	K & Q 3-*O*-Glys, K3Me, K7Me, Q33'Me, 564'/37, 54'/367, Qt37Me
P. lozanianum	K 3-*O*-Glys, Q33'Me, 564'/37, Qt36Me, Qt37Me
P. rollinsianum	Q33'Me, 54'/367, Qt37Me, Qt367Me, Qt3673'Me, Qt3673'4'Me
P. schottii	Qt37Me
P. tomentosum	K & Q 3-*O*-Glys, K7Me, Qt36Me, Qt37Me

[1] Data (condensed) from Mears (1980a).
[2] K = kaempferol; Q = quercetin; IR = isorhamnetin; Qt = quercetagetin; Gly = glycoside; Me = methyl; Substitution: Hydroxyl positions/Methoxyl positions;
[3] Number(s) preceding Me represents positions of *O*-methyl group(s).

Perityle (Hele): *Perityle* (including *Laphamia*; Powell, 1968) is a large genus involving about 63 species from western North America. Flavonoid information for the genus comes from a report of isorhamnetin (Pfeil et al., 1987) and 5-hydroxy-3,6,7,4'-tetramethoxyflavone (Southwick et al., 1972) in *P. vaseyi* and a survey of nine species by Crins et al. (1988a). 5,7,4'-Trihydroxy-3,6-dimethoxyflavone and 5,7-dihydroxy-3,6,4'-trimethoxyflavone

Table 11–16. *O*-Methylated flavonols (and two flavanones) from species of *Perityle*

Taxon	Compound[1]	Reference
Sect. *Laphamia*		
P. bisetosa var. *scalaris*	574'/36, 57/364'	Crins et al. (1988a)
P. lemmonii	574'/3, 574'/36, 57/364'	Crins et al. (1988a)
P. villosa	574'/36, 57/364'	Crins et al. (1988a)
Sect. *Pappothrix*		
P. cinerea	574'/36, 57/364', 574'/363'	Crins et al. (1988a)
P. fosteri	574'/3, 574'/36, 57/364'	Crins et al. (1988a)
P. rupestris var. *albiflora*	574'/36, 57/364', 53'4'/367	Crins et al. (1988a)
Sect. *Perityle*		
P. californica	57/34', 574'/36, 54'/367, 57/364'	Crins et al. (1988a)
P. emoryi	574'/3, 57/34', 574'/36, 54'/367, 57/364', 573'4'/36, 53'4'/367, 574'/363', 57/363'4', 573'/364', 53'4'/7[2]	Crins et al. (1988a), Zdero et al. (1990d)
P. vaseyi	57/364', 573'4'/36, 53'4'/367, 57/364', 5/3674', 3574'/3', 53'4'/367, 574'/363', 57/363'4', 5/74'[2]	Southwick et al. (1972), Pfeil et al. (1987), Crins et al. (1988a)

[1] Substitution: Hydroxyl positions/Methoxyl positions.
[2] Flavanone.

occurred in all species examined (the latter compound was missing from one population of *P. emoryi*). The other compounds identified, also highly *O*-methylated flavones (flavonols, actually), and two flavanones, were more scattered. Table 11–16 details these occurrences. The data for *P. emoryi* are the sum of observations for material from four widely distant sites; the actual situation is more complex.

Perymenium: *Perymenium* consists of 41 species native to Mexico, Central America, Colombia, and Peru. It appears that only one has yielded any flavonoid information, however. Bohlmann et al. (1985c) reported naringenin 7,4'-dimethyl ether from *P. jelskii* (cited as *P. serratum*; synonymy from Turner, 1993b).

Petrobium: *Petrobium* is a monotypic genus endemic to St. Helena. A positive anthochlor test was recorded for *P. arboreum* by Crawford and Stuessy (1981). Although relegated to synonymy in *Bidens* by Stuessy (1988), Karis and Ryding (1994b) keep the genus distinct. The appropriate disposition of these segregate genera relative to the large *Bidens* awaits a macromolecular phylogenetic study with strategic sampling of taxa.

Phoebanthus: *Phoebanthus* is a North American genus of two species both of which have been examined for flavonoids (Schilling and Rieseberg, 1985). The leaf surface extract of both *P. tenuifolius*, the diploid member of the genus, and *P. grandiflorus*, the tetraploid, contain 5,7,3',4'-tetrahydroxy-6-methoxyflavone (6-methoxyluteolin), 5,7,4'-trihydroxy-6,3'-dimethoxyflavone (jaceosidin), and 5,7-dihydroxy-6,8,3',4'-tetramethoxyflavone (hymenoxin).

Picradeniopsis (Hele): *Picradeniopsis* consists of two species restricted to the southwestern United States (Ellison, 1966; Stuessy et al., 1973). Placed in *Bahia* by some workers (e.g., Ellison, 1964), we recognize *Picradeniopsis* following Ellison (1966). The genus is of interest evolutionarily because *P. woodhousei* is a diploid (n = 12) and *P. oppositifolia* a tetraploid (n = 24), with the former being one parent involved with the alloploid origin of the latter (see Stuessy et al., 1973, for details). Spot patterns of phenolics have been observed from 191 individuals from 40 populations of these taxa, and although the profiles are quite similar, four species-specific spots occur (Stuessy et al., 1973). Herz and Bhat (1972) reported 5,7,4'-trihydroxy-3,6,3'-trimethoxyflavone from *P. woodhousei* (cited as *Bahia woodhousei*); Herz et al. (1972b) reported 5,7-dihydroxy-3,6,3',4'-tetramethoxyflavone from *P. oppositifolia* (cited as *Bahia oppositifolia*).

Plummera (Hele): *Plummera*, which consists of only two species, is a segregate from *Hymenoxys*, and, in fact, is most recently treated as belonging to that genus (Bierner, 1994b; Bierner and Jansen, 1998). Wagner et al. (1971c, 1972i) described kaempferol, quercetin, and patuletin 3-*O*-glycosides from both *P. ambigens* and *P. floribunda*. In a survey of Asteraceae for the occurrence of hispidulin, Carman et al. (1972) reported positive sightings in both species. Bierner (1978) examined both species in detail and reported the presence, in both, of 6-methoxyapigenin (hispidulin), 5,7-dihydroxy-6,4'-dimethoxyflavone (pectolinarigenin), 5,7,4'-trihydroxy-6,3'-dimethoxyflavone, quercetin, quercetin 3-*O*-glucoside, and an acetate of quercetin 3-*O*-glucoside. He did not detect kaempferol 3-*O*-glucoside in *P. floribunda* as earlier reported (Wagner et al., 1971c), nor the patuletin derivatives in *P. ambigens* (Wagner et al., 1972i).

Polymnia: *Polymnia* consists of only two species from eastern North America. Herz and Bhat (1970) identified 5-hydroxy-3,6,7,3',4'-pentamethoxyflavone in a study of *P. canadensis*.

Psathyrotes (Hele): *Psathyrotes* consists of three species native to southwestern North America (Turner, 1993c). The only flavonoid information available comes from the study of Finnegan et al. (1972) who isolated 5-hydroxy-6,7,4'-trimethoxyflavone from *P. annua*.

Ratibida: There are seven species in the North American genus *Ratibida*, but flavonoid data are available for only one. Ellmauerer et al. (1987) reported 5,7-dihydroxyflavanone (pinocembrin) from *R. columnifera*.

Rudbeckia: *Rudbeckia* is a North American genus consisting of 16 species, six of which appear in the flavonoid literature. Herz and Kulanthaivel (1985) identified kaempferol 3-*O*- and 7-*O*-monoglucosides, kaempferol 7-*O*-rhamnoside, quercetin 3-*O*-glucoside, and 3-*O*-rhamnoside from *R. fulgida* var. *fulgida*. Isoörientin and vitexin 2"-xyloside were reported from most members of *Rudbeckia* subgen. *Macrocline* and herbacetin 3-*O*-rhamnoside and "quercetagetin derivatives" from *R. hirta* of subgen. *Rudbeckia* in a preliminary survey of the genus for flavonoids (Cox and Urbatsch, 1987). All other reports involved molecules that have extra hydroxylation at C-6 and a modest level of *O*-methylation. A 1979 report of the existence of 3,5,4'-trihydroxy-7,8-dimethoxyflavone 3-*O*-rhamnoside (Jauhari et al., 1979) was corrected by Horie et al. (1988) to 3,5,4'-trihydroxy-6,7-dimethoxyflavone 3-*O*-rhamnoside. This compound had already been reported from *R. hirta* var. *pulcherrina* (cited as *R. serotina*; nomenclature following Cronquist, 1980, and Gleason and Cronquist, 1991) where it occurred with 3,5,3',4'-tetrahydroxy-6,7-dimethoxyflavone (eupatolitin 3-*O*-rhamnoside) and 3,5,4'-trihydroxy-6,7,3'-trimethoxyflavone (veronicafolin) (Waddell et al., 1985). Eupatolitin rhamnoside has also been identified as a constituent of *R. mollis* (Iyengar et al., 1976b) and *R. subtomentosa* (Vasquez et al., 1990). Eupatolitin 3-*O*-glucoside and quercetagetin and patuletin 7-*O*-glucosides were described as UV-absorbing nectar guide components of *R. hirta* (Thompson et al., 1972). The presence of these compounds in *R. hirta* was recently confirmed by Cisowski et al. (1993) who added 3,5,4'-trihydroxy-6,7-dimethoxyflavone 3-*O*-rhamnoside (eupalitin rhamnoside) to the list for this species.

Sartwellia (Hele): *Sartwellia* consists of four species native to southern North America. Al-Khubaizi (1977), describing a preliminary study of the genus, reported the presence of sulphated flavonoids based upon kaempferol, quer-

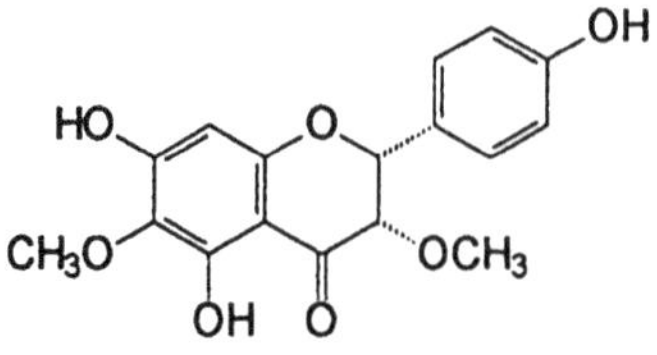

[11-21]

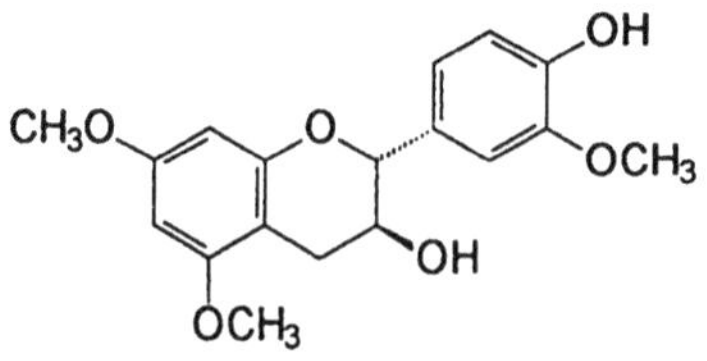

[11-22]

Fig. 11–5. The unusual dihydroflavonol **[11–21]** was isolated from *Scalesia*, an endemic genus of the Galapagos Islands. The catechin derivative **[11–22]** occurs in *Viguiera quinqueradiata*

cetin, isorhamnetin, patuletin, or spinacetin. No further information was provided.

Scalesia: This genus consists of 11 species endemic to the Galapagos Islands. Spring et al. (1997) described naringenin and 6-methoxynaringenin from *S. baurii* ssp. *hopkinsii*, and observed an unidentified flavonoid in extracts of *S. stewartii*. Extraction of combined plant material from several species afforded the two flavanones plus a small amount of a flavonoid that was identified as the unusual dihydroflavonol **[11–21]** (Fig. 11–5). An expanded survey (Spring et al., 1999) revealed these compounds in *S. aspera*, *S. baurii* ssp. *baurii*, *S. crockeri*, *S. helleri*, and *S. retroflexa*.

Schkuhria (Hele): *Schkuhria* is a Latin American genus of 10 species only, three of which appear to have been examined for flavonoids. Ganzer and Jakupovic (1990) reported 5,7-dihydroxy-6,4'-dimethoxyflavone (pectolinarigenin) from *S. pinnata*. This report was corroborated by Pacciaroni et al. (1995) who identified their plant material as var. *abrotanoides*. Mendiondo et al. (1990a) reported quercetin 3-*O*-glucoside, isorhamnetin 3-*O*-glucoside, and the 4'-methyl ether 3-*O*-glucosides of quercetin and patuletin from floral tissue of *S. pinnata*. In a subsequent paper from that laboratory, De Israilev et al. [1992 (1993)] reported quercetin and isorhamnetin 3-*O*-glucosides and quercetin 4'-methyl ether 3-*O*-glucoside from aerial parts of *S. multiflora*.

Simsia: *Simsia* is a New World genus of 18 species whose range extends from southwestern United States to Argentina with one species endemic to Jamaica (Schilling and Spooner, 1988; Spooner, 1990). Nearly all have been examined for flavonoids (Schilling and Spooner, 1988). The compounds identified were *O*-glycosides of apigenin, luteolin and quercetin and the anthochlor pairs coreopsin/sulfurein and marein/maritimein. The quercetin glycosides were most uniformly distributed while the flavones and anthochlors were predominantly present in members of the "*S. amplexicaulis* group" of species. The anthochlors were seen in *S. ghiesbreghtii*, which is one of the "*S. fruticulosa* group" of species. 5,7-Dihydroxy-6,8,4'-trimethoxyflavone (nevadensin) was described by Maldonado et al. (1992a) as a component of *S. foetida* var. *megacephala* (cited as *S. cronquistii*; synonymy after Spooner, 1990).

Smallanthus: This genus consists of about 19 species distributed throughout Latin America. All taxa were previously treated within *Polymnia* (Wells, 1965), but Robinson (1978b) more recently has generically separated the type species *P. canadensis* and its close relative *P. laevigata*, leaving the bulk of the species in a separate genus with the already available name *Smallanthus* MacKenzie. Naringenin and naringenin 7-methyl ether (sakuranetin) were isolated from *Smallanthus fruticosus* (cited as *Polymnia fruticosa*) by Bohlmann and Zdero (1977). More recently, Beutler et al. (1993) isolated a cytotoxic flavonoid from *S. fruticosus* and showed it to be 5,7,3'-trihydroxy-3,6,4'-trimethoxyflavone (centaureidin).

Staurochlamys: *Staurochlamys burchellii* is the only member of this Brazilian genus and is one of the species from which Crawford and Stuessy (1981) obtained a positive anthochlor test.

Steiractinia: A dozen species comprise this South American genus. Flavonoid information consists of the report of naringenin from *S. mollis* (Bohlmann et al., 1980d).

Tetragonotheca: *Tetragonotheca* is a genus of four species native to southern United States and Mexico all of which have been examined for flavonoids (Bacon et al., 1978). Kaempferol, quercetin, and isorhamnetin 3-methyl ethers were identified along with a series of 6-oxygenated compounds. The latter group consisted of 6-hydroxykaempferol and its 6-methyl and 3,6-dimethyl ethers, quercetagetin and its 6-methyl, 3,6-dimethyl, 6,3'-dimethyl, 3,6,3'-trimethyl, and 3,6,4'-trimethyl ethers. 7-*O*-Glucosides of several of these flavonols were also identified. Not all species had the same combination of pigments.

Thelesperma: Fifteen species native to western North America comprise this genus. It appears that flavonoids have been reported for only one, *T. megapotamicum*, from which Ateya et al. (1982) obtained luteolin, luteolin 7-*O*-glucoside, and 2',3',4',3,4-pentahydroxychalcone 4'-*O*-glucoside (marein).

Tithonia: *Tithonia* consists of 11 species native to Mexico and Central America. We have information on four of them. Correa and Cervera (1971) isolated an unusual compound, 3'-hydroxy-7,4'-dimethoxyflavone from *T. tubaeformis*, which qualifies as one of the few 5-deoxy flavonoids known from Asteraceae. Perez et al. (1988, 1992) reported several *O*-methylated flavonoids from members of the genus: 5,7-dihydroxy-6,8,4'-trimethoxyflavone (nevadensin) and 5,7,3'-trihydroxy-6,8,4'-trimethoxyflavone (acerosin) from *T. pedunculata* and nevadensin and 5,7-dihydroxy-6,8,3',4'-tetramethoxyflavone (hymenoxin) from *T. longiradiata*. 6-Methoxyapigenin (hispidulin) was identified as an insect feeding deterent in *T. diversifolia* by Dutta et al. (1993). The presence of hispidulin in *T. diversifolia* has been confirmed recently by Pereira et al. (1997).

Unxia: *Unxia* is a genus of three species (Stuessy, 1969; Robinson, 1980a) distributed in Panama and northern South America. Jacobs et al. (1986b) reported apigenin 7-methyl ether (genkwanin) and apigenin-7,4'-dimethyl ether from *U. camphorata* (cited as *Melampodium camphoratum*; synonymy from Stuessy, 1969).

Venegasia (Hele): *Venegasia* is a monotypic genus native to California and northwestern Mexico. It appears related to *Arnicastrum* and *Jamesianthus* (Turner and Zippin, 1992) of Robinson's (1981) Chaenactidinae. The lack of anthochlors in *Venegasia* (Crawford and Stuessy, 1981) would not be inconsistent with this taxonomic perspective.

Verbesina: *Verbesina*, with 200 or 300 species native to the Americas (Olsen, 1985; Karis and Ryding, 1994b), is one of the larger genera of Heliantheae. Despite its size, flavonoids appear to have been reported from only two species, *V. encelioides* and *V. myriocephala*. Glennie and Jain (1980) reported a series of quercetin *O*-glycosides involving substitution at positions-3 and/or -7 from the former species, while Wagner et al. (1974) isolated kaempferol 7-methyl ether (rhamnocitrin) 3-*O*-glucuronide from the latter.

Viguiera: *Viguiera* is a large New World genus consisting of 180 species. Flavonoid sampling has been fairly extensive. One of Geissman's "Anthochlor Pigments" series featured *V. multiflora* from which quercetin 7-*O*-glucoside, butein, coreopsin, sulfuretin, and sulfurein were isolated (Shimokoriyama and Geissman, 1960). In their study of nectar guides in Asteraceae, Harborne and Smith (1978) reported coreopsin and sulfurein from floral tissue of *V. laciniata*. One of the more unusual compounds reported from a member of the genus, or from the family for that matter, is

3,4'-dihydroxy-5,7,3'-trimethoxyflavan (catechin 5,7,3'-trimethyl ether) **[11–22]** which Delgado et al. (1984) obtained from *V. quinqueradiata*. Other reports include identification of eriodictyol 3'-methyl and 7,3'-dimethyl ethers from *V. laciniata* (Gao et al., 1989), and 5,7-dihydroxy-6,8,4'-trimethoxyflavone (nevadensin) from *V. procumbens* (Tamayo-Castillo et al., 1990). The rest of our information on *Viguiera* flavonoids comes from another set of "Schilling surveys." In the earliest of these (Rieseberg and Schilling, 1985) luteolin 7-*O*-glycosides, quercetin 3-*O*- and 7-*O*-glucosides, and a set of quercetin 3-methyl ether 7-*O*-glycosides were reported from nine members of *Viguiera* series *Viguiera* (12 species are recognized). In the same paper, coreopsin and sulfurein were also recorded for all nine species, while marein and maritimein were seen only in *V. dentata*. The external flavonoid aglycone fraction of all members of series *Viguiera* was described by Schilling (1989). The compounds identified from 10 species (*V. dentata* and *V. potosina* produce no aglycones) were derivatives of apigenin, luteolin, 6-hydroxyapigenin and 6-hydroxyluteolin, quercetin, quercetagetin, naringenin, isoliquiritigenin (2',4',4-trihydroxychalcone), butein, and sulfuretin. The arrays of compounds are complex and will be discussed in relation to specific problems in the genus in a later chapter.

Viguiera series *Brevifoliae* consists of three species which were examined for both foliar aglycones and floral flavonoid glycosides (Schilling and Panero, 1988). The floral profiles were very similar among the three species consisting of glycosides of quercetin, quercetagetin, isoliquiritigenin, butein, and sulfuretin. *Viguiera brevifolia* yielded no foliar aglycones and *V. bicolor* only 5,7-dihydroxy-6,8,4'-trimethoxyflavone (nevadensin), while *V. greggii* afforded eight flavones: 6-methoxyapigenin (hispidulin), 5,6,4'-trihydroxy-7-methoxyflavone (sorbifolin), 6,8-dimethoxyapigenin, nevadensin, luteolin, 6-methoxyluteolin (nepetin), 5,3',4'-trihydroxy-6,7-dimethoxyflavone (cirsiliol), and 5,7-dihydroxy-6,8,3',4'-tetramethoxyflavone (hymenoxin). De La Fuente et al. (1994) reported 5,7,3'-trihydroxy-6,8,4'-trimethoxyflavone (acerosin) and nevadensin from *V. mollis*.

A study of *Viguiera* section *Maculatae* (11 taxa representing 10 species were examined) showed leaf profiles to consist of luteolin, hispidulin, 6-hydroxyluteolin, nepetin, and derivatives of isoliquiritigenin (Schilling et al., 1988). The isoliquiritigenin derivatives were seen only in *V. splendens*, however. Neither *V. oaxacana* nor *V. neocronquistii* afforded any foliar flavonoids. Floral pigment profiles in this group of species were based on common anthochlors (derivatives of butein and sulfuretin) plus glycosides of quercetin. Again, *V. splendens* stood out from the group in having only butein and sulfuretin and their glucosides as floral flavonoids. This simple floral profile was also recorded for *V. oaxacana* and *V. neocronquistii*.

Wedelia: *Wedelia* is a comparatively large genus, some 100 species, distributed in tropical and subtropical regions of the world. The coumestan derivative wedelolactone was reported from *W. calendulacea* (assigned with hesitation by Strother, 1991, to *W. acapulcensis* var. *hispida*) by Govindachari et al. (1956a). Wagner et al. (1986) identified apigenin from the same taxon. Farag et al. (1997) identified the anthochlors coreopsin, stillopsin, and sulfurein from extracts of *W. prostrata*. A recent study of *W. asperrima*, a toxic herb native to northern Australia, resulted in the identification of a new flavanone, 7,4'-dihydroxy-8,3'-dimethoxyflavanone, the corresponding chalcone, and its 4'-*O*-glucoside (Calanasan and MacLeod, 1998). (The toxic components of the plant are kaurene aminoglycosides.) A recent paper by Bellini et al. (1999) reported rutin from *Aspilia montevidensis*. Karis and Ryding (1994b) place this genus within *Wedelia*.

Wilkesia (Hele): *Wilkesia* consists of two species endemic to the Island of Kauai (Carr, 1985). One of them, *W. hobdyi*, is a threatened species. Both *W. hobdyi*, and the more common *W. gymnoxiphium*, have been examined for leaf surface flavonoids (Bohm and Fong, 1990), and shown to accumulate compounds with 6- and/or 8-oxygenation and moderate levels of *O*-methylation. Occurrence of these compounds is summarized in Table 11–17. The vacuolar flavonoid fraction of both species is dominated by quercetin 3-*O*-glucoside, 3-*O*-rutinoside, and patuletin 3-*O*-glucoside (Bohm, unpubl.).

Table 11–17. Exudate flavonoids from species of *Wilkesia*[1]

Compound	Species[2]	
	GYM	HOB
5,7,4'-Trihydroxy-3-methoxyflavone	+	–
5,3',4'-Trihydroxy-3,6,7,8-tetramethoxyflavone	–	+
5,7,4'-Trihydroxy-8-methoxyflavone	–	+
5,7,3',4'-Tetrahydroxy-8-methoxyflavone	–	+
5,4'-Dihydroxy-7,8-dimethoxyflavone	+	+
5,4'-Dihydroxy-6,7-dimethoxyflavone	+	–
5,3',4'-Trihydroxy-6,7-dimethoxyflavone	–	+
5,7,4'-Trihydroxy-6,8-dimethoxyflavone	+	+
5,4'-Dihydroxy-6,7,8-trimethoxyflavone	+	+
5,7,3',4'-Tetrahydroxy-6,8-dimethoxyflavone	+	+
5,3',4'-Trihydroxy-6,7,8-trimethoxyflavone	+	+

[1] Data from Bohm and Fong (1990).
[2] GYM = *W. gymnoxiphium*; HOB = *W. hobdyi*.

Wollastonia: *Wollastonia* is a monotypic genus of the Pacific region. Miles (1993) identified quercetin-3,3'-dimethyl ether, quercetin-7,3'-dimethyl ether, and the highly unusual 2,7,4'-trihydroxy-5,3'-dimethoxyisoflavone **[4–79]** from *W. biflora* (cited as *Wedelia biflora*; synonymy from Strother, 1991). Gardner (1976) stated that the flavonols of *W. biflora* were "virtually identical" to compounds present in the diploid species of *Lipochaeta*, which he showed to be glycosides of quercetin. (See discussion under that genus.)

Wyethia: *Wyethia* is a western North American genus of 14 species, flavonoid data for 10 of which are available. Flavanones, dihydroflavonols, flavones, isoflavones, and flavonols have been identified. Several exhibit 6-oxygenation and a modest level of *O*-methylation has been observed in many instances. Particularly noteworthy is the presence of compounds based upon 5,7,3',4'-tetrahydroxyisoflavone (orobol), namely orobol 7-methyl ether (santal), 3'-methyl ether, and 7,3'-dimethyl ether. One or more of these compounds have been reported from *W. angustifolia* (McCormick et al., 1986), *W. helenioides* (Bohlmann et al., 1981j; McCormick et al., 1986), *W. mollis* (Waddell et al., 1982; Bohm and Choy, 1987; Bohm et al., 1989a), and in the following four species of *Wyethia* sect. *Agnorhiza*: *W. invenusta, W. ovata, W. reticulata*, and *W. scabra* (McCormick et al., 1987). 6-Methoxyorobol 7-methyl ether was also reported as a constituent of *W. reticulata* by McCormick et al. (1987) in that paper. 6-*C*-Prenylorobol and 6-*C*-prenylorobol 3'-methyl ether were reported from *W. angustifolia* (McCormick et al., 1986), while 8-*C*-prenylorobol 3'-methyl ether was shown to be a component of *W. mollis* (Bohm et al., 1989a).

Naringenin 4'-methyl ether occurs in *W. angustifolia* and *W. helenioides* (McCormick et al., 1986), and dihydrokaempferol (aromadendrin) was reported from *W. elata* (McCormick et al., 1987). More complicated flavanones abound in the genus. Thus, *C*-prenylated naringenin (C-6 and C-8 isomers) and *C*-prenylated eriodictyol (C-6 and C-8 isomers) were identified from the following species: *W. arizonica* (Bohlmann et al., 1984h), *W. angustifolia* (McCormick et al., 1986), *W. helenioides* (Bohlmann et al., 1981j; McCormick et al., 1986; Bohm and Choy, 1987) and in *W. invenusta* and *W. scabra* (McCormick et al., 1987). 8-*C*-Prenyldihydroisorhamnetin was identified from *W. angustifolia* (McCormick et al., 1986) and from *W. mollis* (Bohm and Choy, 1987). Results of an unusual *C*-prenyl cyclization are seen in *W. angustifolia* (McCormick et al., 1986) from which the seven-membered ring dihydroöxepine derivatives of eriodictyol, dihydroquercetin **[4–44]**, and a chalcone have been isolated.

Leaf exudates of several *Wyethia* species also contain substantial quantities of *O*-methylated flavones and flavonols several of which exhibit sub-

Table 11–18. Flavones and flavonols from species of *Wyethia*

Species	Compound[1]	Reference
W. bolanderi	574'/3, 573/4', 57/34', 574'/36, 3564'/6, 57/364, 573'4'/3, 3574'/3', 573'4'/36, 574'/363'	McCormick et al. (1985b, 1987)
W. elata	574'/6, 3574'/6, 573'4'/36, 574'/363'	McCormick et al. (1987)
W. helenioides	573'4'/6, 573'4'/36, 574'/363'	McCormick et al. (1986)
W. invenusta	573'4'/3, 3574'/3', 573'4'/36, 3574'/63', 574'/363'	McCormick et al. (1987)
W. ovata	573'/4', 3574'/6	McCormick et al. (1987)
W. reticulata	573'4'/6, 573'4'/36	McCormick et al. (1987)
W. scabra	3574'/6	McCormick et al. (1987)

[1] Substitution: Hydroxyl positions/Methoxyl positions.

stitution at C-6. Several species are involved (McCormick et al., 1985b, 1986, 1987). See Table 11–18 for details. A study of the vacuolar flavonoids of *W. helenioides* showed the presence of eriodictyol 7-*O*-glucoside and kaempferol and quercetin 3-*O*-glycosides (Bohm and Choy, 1987).

Recently Weber (1998) has created a new genus, *Agnorhiza* (Jepson) W. A. Weber from species in *Wyethia* (*W. elata*, *W. ovata*, *W. reticulata*), *Balsamorhiza* (*B. bolanderi*), and *Helianthus* (*H. invenustus*). *Wyethia scabra* has also been moved into its own genus, *Scabrethia* W. A. Weber. Because neither comparative details nor arguments for these taxonomic alterations have yet been provided, it is impossilbe to evaluate their efficacy. We have chosen not to use these new concepts in this book pending availability of further information.

Xanthium: This small genus, two or three species, is native to the New World, but has become widely dispersed. Taylor and Wong (1965) identified 5,7,3',4'-tetrahydroxy-3,6-dimethoxyflavone from *X. strumarium* (cited as *X. pennsylvanicum*; synonymy from Gleason and Cronquist, 1991). Paschenko et al. (1966) reported 8-*C*-prenyl apigenin from *X. spinosum*. Wollenweber et al. (1997) have identified 6-methoxykaempferol and quercetagetin 3,6-dimethyl ether from a leaf rinse of *X. strumarium*.

Zaluzania: *Zaluzania* is a Mexican genus of 10 species eight of which were examined for their flavonoids by Olsen and Mabry (1980). The compounds identified were quercetin 3-*O*-mono- and diglycosides, quercetin-3,7-di-*O*-glycosides, patuletin 3-*O*-monoglycosides and a group of *O*-methylated flavonols. The latter group consisted of kaempferol and quercetin 3-methyl ethers, quercetin-3,7-dimethyl ether, 5,7,4'-trihydroxy-3,6-dimethoxyflavone, 5,7,3',4'-tetrahydroxy-3,6-dimethoxyflavone and 5,7,4'-trihydroxy-3,6,3'-trimethoxyflavone (jaceidin).

Zinnia: *Zinnia* is a genus of 22 American species. Common flavone, flavonol, and anthocyanin glycosides have been reported from *Z. elegans*, varieties of which are the zinnias of commerce (Willstätter and Bolton, 1916; Nakaoki, 1941; Dembinska-Migas et al., 1983; Yamaguchi et al., 1990). The anthochlor pair maritimein/marein has been reported from *Z. angustifolia* (cited as *Z. linearis*; synonymy from Torres, 1963; McVaugh, 1984) by Harborne et al. (1983). The seven species that comprise *Zinnia* section *Diplothrix* were subjected to a chromatographic study by Torres and Levin (1964) who reported two basic patterns but did not establish structures of the compounds involved.

CHAPTER 12

Flavonoids of Inuleae s.l.

According to the systematic review of Inuleae by Merxmüller et al. (1977), we are dealing with about 180 genera with some 2100 species. A quite different view of the tribe is held by Anderberg (1989, 1991a,b,c, 1994, 1996) who sees Inuleae *sensu stricto* as comprising 38 genera and about 480 species. He places the remaining genera in either Plucheeae (28 genera; 220 species) or Gnaphalieae (180 genera; 2000 species). The presentation of flavonoid data below follows the traditional view of Inuleae (as in Arriagada, 1998, who recognizes 245 genera and ca. 2600 species), but each genus is identified with a letter that designates the tribal placement according to Anderberg (1994): I = Inuleae, G = Gnaphalieae, and P = Plucheeae. In Chapter 19 we discuss whether flavonoid data offer support for recognition of one or several tribes. Flavonoid data have been reported for 43 genera. Ranges of occurrence and numbers of species are those given in Anderberg (1994) unless otherwise noted. Generic concepts follow those presented in Anderberg (1991a,b,c, 1994).

Achyrocline: (G) *Achyrocline* is a genus of about 25-30 species [15-20 in Dillon and Sagástegui (1991a)] that occur mainly in Central and South America with extensions into tropical Africa and Madagascar (Nesom, 1990c). Flavonoid data are available for several species. Naringenin and its 5-methyl ether occur in *A. flaccida* (Norbedo et al., 1982, 1984), while an unusual flavanone, 5,7,4'-trihydroxy-8,3'-dimethoxyflavanone, accumulates in *A. tomentosa* (Ferraro et al., 1985). 7-Hydroxy-3,5,4'-trimethoxyflavanone was identified in *A. alata* (Broussalis et al., 1993) where it occurs with apigenin, kaempferol, and quercetin 3-methyl ether, quercetin-7,3'-dimethyl ether (rhamnazin), and two B-ring deoxyflavones (see Table 12–1). Common flavone and flavonol derivatives occur in several species as well: luteolin 7-*O*-glucoside in *A. glumacea* (Saleh et al., 1988); luteolin 3'-methyl ether (chrysoeriol) and quercetin 4'-methyl ether (tamarixetin) in *A. flaccida* (Norbedo et al., 1984); glycosides of quercetin 3-methyl ether in *A. alata* (Bauer et al., 1987, 1989); luteolin, quercetin 3-, 3,7- and 7,3'-methyl ethers

Table 12–1. B-Ring unsubstituted flavonoids from species of *Achyrocline*

			Species of *Achyrocline*[1]		
Subn[2]	ALA[3]	BOG[4]	FLA[5]	SAT[6]	TOM[7]
57/8			+		
5/78		+			
-/578				+	
357/-				+	
57/3	+		+	+	
35/78	+				+
5/367				+	
5/378		+			
56/37	+				
57/38	+		+		
58/37			+	+	
7/358				+	
578/3			+		
-/3578				+	
35/678	+	+			

1 ALA = *A. alata*; BOG = *A. bogotensis*; FLA = *A. flacida*; SAT = *A. satureioides*; TOM = *A. tomentosa.*
2 Substitution: Hydroxyl positions/Methoxyl positions.
3 Bauer et al. (1985, 1989); Bohlmann et al. (1980b); Broussalis et al. (1993).
4 Torrenegra et al. (1982).
5 Norbedo et al. (1984).
6 Hänsel and Ohlendorf (1971); Ferraro et al. (1981); Hirschman (1984); Mesquita et al. (1986).
7 Ferraro et al. (1985).

and quercetagetin in *A. tomentosa* (Martino et al., 1988); apigenin, luteolin, and quercetin glycosides in *A. rufescens* (cited as *Gnaphalium rufescens*) (Torrenegra et al., 1987), and apigenin 5-methyl ether 7-epoxyprenyl ether in *A. flaccida* (Norbedo et al., 1984).

Several of the species for which data are available show the presence of at least one B-ring unsubstituted flavone or flavonol. Flavones with 5,7,8-substitutions and flavonols based on 3,5,7-, 3,5,6,7-, 3,5,7,8-, and 3,5,6,7,8-substitutions have been identified. Details of occurrence of these compounds in *Achyrocline* can be found in Table 12–1.

Actinobole (G): This genus comprises four species native to Australia. Jakupovic et al. (1988c) reported 5,4'-dihydroxy-7-methoxyflavanone (sakuranetin) from *A. uliginosum*.

Allagopappus (I): *Allagopappus* consists of two species endemic to the Canary Islands. Gonzalez et al. (1992a) reported naringenin, eriodictyol, kaempferol 3-methyl ether, and quercetin 3-methyl ether from *A. viscosissimus*, and in a later paper (Gonzalez et al., 1995), chrysoeriol, kaempferol and quercetin methyl ethers, naringenin and eriodictyol, and the 3,6-di and 3,6,4'-tri methyl ethers of 6-hydroxykaempferol from *A. dichotomus*. Differences were noted between the occurrence patterns in the two subspecies, *dichotomus* and *latifolius*.

Anaphalis (G): *Anaphalis* is a moderately large genus with 15-30 (possibly up to 60; Dillon and Sagástegui, 1991b), 60-100 (Merxmüller et al., 1977) or about 110 (Anderberg, 1994) species. Most are Asian but a few occur in North America. Ali et al. (1979) identified 5,7-dihydroxy-3,6,8-trimethoxyflavone and 5,7-dihydroxy-3,6,8,4'-tetramethoxyflavone from extracts of *A. araneosa*. B-Ring deoxyflavonoids were also reported from *A. velutina* by Ikramov et al. (1986) who isolated 3,5-dihydroxy-6,7,8-trimethoxyflavone and 5-hydroxy-3,6,7,8-tetramethoxyflavone. (We could find no record of this epithet.) Seven galangin (3,5,7-trihydroxyflavone) derivatives, 6-*C*-prenylpinocembrin, 2',4',6'-trihydroxy-3'-*C*-prenylchalcone and quercetin were identified as constituents of *A. margaritacea* by Wollenweber et al. (1993b). Kaempferol and quercetin 3-*O*-glucosides have been reported from *A. contorta* (Lin, 1993).

Antennaria (G): *Antennaria* consists of 40-50 holarctic species with a few known from the Andes (Merxmüller et al., 1977), or perhaps only a total of about 20 sexually reproducing species (Bayer, 1988). Anderberg (1994) suggests that there could be as many as several hundred depending on one's species concept. We have found flavonoid information for only one, however, *A. dioica*, from which 7- and 4'-*O*-glycosides of apigenin and luteolin have been obtained (Tira et al., 1969; Mericli, 1983).

Antiphiona (I): *Antiphiona* consists of two species native to Namibia. Naringenin has been reported from both, *A. fragrans* and *A. pinnatisecta*, by Zdero and Bohlmann (1989a).

Anvillea (I): This genus, which consists of two species, occurs from North Africa to Iran. Ulubelen et al. (1979), working with *A. garcinii*, isolated 6-methoxyapigenin (hispidulin), 6-methoxyluteolin (eupafolin) and a group of *O*-methylated flavonol derivatives: isorhamnetin 3-*O*-glycosides, quercetin-3',4'-dimethyl ether 3-*O*-rutinoside, 6-methoxykaempferol-3-*O*-glycosides and 6-methoxykaempferol 4'-methyl, 6,7-dimethyl, and 6,7,4'-trimethyl ethers. The latter three compounds occurred as 3-*O*-glycosides.

Asteriscus (I): *Asteriscus* is a small Mediterranean genus that has been well sampled for flavonoids. *Asteriscus graveolens* accumulates kaempferol and quercetin 3-*O*- and 7-*O*-monoglycosides, luteolin 7-*O*-glucoside, quercetin 4'-methyl ether (tamarixitin) 3-*O*-robinobioside and quercetin 3,4'-dimethyl and 3,6,3'-trimethyl ethers (Ahmed et al., 1991). A paper by Youssef et al. (1995) reported apigenin and luteolin and their 7-*O*-glucosides, luteolin 7,4'-di-*O*-glucoside, luteolin 7-methyl ether, and quercetin and its 3-methyl ether from *A. pygmaeus*. Ahmed et al. (1992) reported the following compounds from the glycosidic fraction of *A. spinosus* (cited as *Pallenis spinosa*; taxonomy after Anderberg, 1991c): tricin 5-*O*- and 7-*O*-glucosides, quercetin-3-*O*-galactoside, and patuletin 3-*O*-galactoside, 7-*O*-galactoside, and 3-*O*-rhamnosylgalactoside. The nonpolar fraction afforded kaempferol and quercetin 3-methyl ethers, patuletin, 4'-hydroxy-3,5,6,7,3'-pentamethoxyflavone, and 3,5,6,7,3',4'-hexamethoxyflavone. A recent survey by Wollenweber et al. (1997) showed the presence of quercetagetin 3,6-di-, 3,6,3'-tri-, and 3,7,3'-trimethyl ethers in a leaf surface extract of *A. sericeus*.

Bellida (G): *Bellida* is an Australian genus of one (Anderberg, 1994) or two (Merxmüller et al., 1977) species. *Bellida graminea* was shown to have luteolin and quercetin (Jakupovic et al., 1989a).

Belloa (G): This is a South American genus of nine (Anderberg, 1994) to 12 species (Anderberg, 1991c). The only report of flavonoids from the genus is the work of Bohlmann et al. (1990) who isolated 5-hydroxy-3,6,7,8-tetramethoxyflavone from *B. nivea* (reported as *Lucilia nivea*; synonymy from Anderberg, 1991c).

Blumea (I): *Blumea* consists of about 75 species according to Merxmüller et al. (1977) with occurrences in Africa, Asia, and Australia. Anderberg (1994) says that there are about 100 species in the genus but adds that many of them probably belong in Plucheeae. The most recent revision of the genus (Randeria, 1960) lists only 49 species. Flavonoids have been reported for four species. The earliest report was that of Bose et al. (1968) who isolated 5-hydroxy-3,6,7,3',4'-pentamethoxyflavone (artemetin) from *B. eriantha*. Artemetin and 5,3',4'-trihydroxy-3,6,7-trimethoxyflavone (chrysosplenol-D) were reported from *B. lacera* (Rao et al., 1977), and two taxifolin methyl ethers, (2*R*,3*R*)-dihydroquercetin-4'-methyl and 7,4'-dimethyl ethers, were isolated from *B. balsamifera* (Ruangrungsi et al., 1981). Kulkarni et al. (1987) reported 6-hydroxy-3,5,7,4'-tetramethoxyflavone and three flavonols supposedly with 2',5'-B-ring substitution from *B. malcolmii*. A reinterpretation of the spectral data led Markham (1989) to reassign structures to the trio of compounds. The revised structures are quercetagetin-3,6,7-trimethyl ether, quercetagetin-3,6,7,3'-tetramethyl, and quercetagetin-3,6,7,3',4'-pentamethyl ether. A recent paper by Barua and Sharma (1992) described the structural elucidation of (2*R*,3*R*)-3,5,2'-trihydroxy-7,5'-dimethoxyflavanone, which these workers isolated from *B. balsamifera*.

Bracteantha (G): *Bracteantha* is a genus of five species native to Australia. Formerly, these species were placed in *Helichrysum* (Anderberg, 1991b). Interest in the flavonoid chemistry of this genus goes back to the early 1960s when schools in Germany and the United States began to study the pigments in the well known and popular garden "everlastings," which were then referred to as *Helichrysum bracteatum*. To be consistent with other presentations in this book, we conform to the nomenclature in Anderberg (1991b, 1994). Among the earliest reports concerning *Bracteantha bracteata* was the establishment of the structure of "bracteatin" and its glucoside "bractein" by Hänsel et al. (1962). The structure of bracteatin was determined to be 4,6,3',4',5'-pentahydroxyaurone, and that of bractein to be the corresponding 4-*O*-glucoside (Hänsel et al., 1962, 1963; Hänsel and Langhammer, 1963). The corresponding chalcone, 2',4',6',3,4,5-hexahydroxychalcone 2'-*O*-glucoside, as well as 4,6,3',4'-tetrahydroxyaurone (aureusidin) were also identified as components of this plant (Rimpler et al., 1963; Rimpler and Hänsel, 1965; Krishnamoorthy and Seshadri, 1966; Kaufmann and El Baya, 1969). Another first report from this species, the identification of an 8-*C*-glucosylluteolin derivative, appeared at about the same time (Rimpler et al., 1963). The occurrence of the *C*-glycosylflavone isoörientin in *B. bracteata* was confirmed recently by Sivakumar et al. (1995). The paper by Rimpler et al. (1963) also documented the presence of cyanidin, as a glycoside, in this plant.

Several flavanones have been reported from members of this genus. One of the earlier reports described 5,7,3',4'-tetrahydroxyflavanone (eriodictyol) and 5,7,3',4',5'-pentahydroxyflavanone from *B. bracteata* (Rimpler et al., 1963). Forkmann (1983) reported the latter compound as a constituent of this species in his studies of flavonoid genetics. (Although the natural occurrence of flavanones is well established, it is prudent to remind the reader that the presence of a given flavanone may merely reflect the natural occurrence of the corresponding chalcone, a portion of which has undergone conversion to the cyclic form as a result of isolation procedures.) Eriodictyol and its 3'-methyl ether (homoeriodictyol) were reported as components of *B. viscosa* var. *bracteata* (Geissman et al., 1967). An assortment of flavanones was reported from *B. viscosa* by Wollenweber et al. (1993c): naringenin, 6-methoxynaringenin, 5,4'-dihydroxy-6,7-dimethoxyflavanone, eriodictyol, eriodictyol 3'-methyl ether, 6-methoxyeriodictyol, and 8-methoxyeriodictyol. A more recent survey of leaf surface flavonoids disclosed the presence of naringenin and its 7-methyl ether, eriodictyol and its 7-methyl ether, and eriodictyol 7,3'-dimethyl ether from *B. bracteata* (cited as *H. bracteatum*).

Geissman et al. (1967) reported 5,4'-dihydroxy-6,7-dimethoxyflavone (cirsimaritin) and 5,7,4'-trihydroxy-6,3'-dimethoxyflavone (jaceosidin) from *B. viscosa* var. *bracteata*. More recently, Wollenweber et al. (1993c) reported several flavone derivatives from this species: apigenin and its 7-methyl ether, scutellarein 6-methyl, 6,7- and 6,4'-dimethyl ethers, luteolin, its 7,3'-dimethyl ether, 6-methoxyluteolin, 5,7,4'-trihydroxy-6,3'-dimethoxyflavone (jaceosidin), and two compounds with 8-substitution, 5,4'-dihydroxy-6,7,8-trimethoxyflavone (xanthomicrol) and 5,7-dihydroxy-8,3',4',5'-tetramethoxyflavone. Other 8-substituted flavonoids, namely gossypetin 8-*O*-glucoside and 8-*O*-glucuronide have also been reported from this species (Gunasegaran et al., 1993). Included in the Gunasegaran et al. paper were other flavonoids including apigenin, glycosides of luteolin, chrysoeriol, hispidulin, scutellarein, and acylated glucosides of kaempferol and quercetin. Apigenin 7-methyl ether (genkwanin), apigenin 7,4'-dimethyl ether, luteolin, luteolin 3'-methyl ether (chrysoeriol), and luteolin 7,3'-dimethyl ether have been identified as components of a leaf rinse of *B. bracteata* (cited as *H. bracteatum*) by Wollenweber et al. (1997).

Cassinia (G): *Cassinia*, native to Australia and New Zealand, consists of perhaps 20 species. We have some flavonoid information on about half of them. 3,5-Dihydroxy-6,7,4'-trimethoxyflavone (mikanin) has been reported from *C. aculeata* (Zdero et al., 1990b) and *C. longifolia* (Zdero et al., 1987b), pinobanksin and its 3-acetate have been identified from *C. arcuata* (Zdero et al., 1991e), 5,7-dihydroxy-6,4'-dimethoxyflavone (pectolinarigenin) from *C. uncata* and quercetin from *C. subtropica* (Jakupovic et al., 1988c). Pinobanksin, its 3-acetate, pinocembrin, 3,5,7-trihydroxyflavone (galangin), 6-hydroxygalangin, and 6-methoxygalangin have been identified as components of *C. quinquefaria* (Wollenweber and Jay, 1988; Wollenweber et al., 1993c). Four of the five taxa that comprise the New Zealand endemic cassinias were shown by Reid and Bohm (1994) to exhibit species-specific arrays of both vacuolar and exudate flavonoids. The vacuolar components were known glycosides of kaempferol, quercetin, and eriodictyol; galangin was also present. The leaf exudate profile consisted of chalcones, dihydrochalcones, flavanones, and flavonols; several of the compounds lacked B-ring substitution.

Chiliadenus (I): *Chiliadenus* is a Mediterranean genus of nine species. El-Din et al. (1985) and Ahmed et al. (1988a, 1989a) have reported several *O*-

methylated flavonoids from *C. montanus* (cited as *Jasonia montana*): luteolin-7,3'-dimethyl ether, 6-hydroxykaempferol-3,6,7-trimethyl ether, quercetin-3,3'-dimethyl, 3,4'-dimethyl and 3,7,3'-trimethyl ethers, and quercetagetin 6-methyl, 3,6,7-trimethyl, 3,6,3'-trimethyl, 3,6,4'-trimethyl, 3,6,7,3'-tetramethyl, 3,6,7,4'-tetramethyl, 3,5,6,7,3'-pentamethyl and 3,5,6,7,3',4'-hexamethyl ethers. Quercetagetin-3,6,7-trimethyl ether (chrysosplenol D) was reported to be a constituent of *C. candicans* by Mossa et al. (1988) (cited as *Jasonia candicans*; synonymy after Anderberg, 1991a). Afifi (1991) reported kaempferol 3,7-dimethyl ether (kumatakenin), quercetin 3,3'-dimethyl ether, 5,4'-dihydroxy-6,7,8-trimethoxyflavone (xanthomicrol), and quercetagetin 3,6,3'-trimethyl ether (jaceidin) from *C. iphionoides* (cited as *Varthemia iphionoides*).

Glucosides of quercetin and patuletin were reported from *C. glutinosum* (cited as *Jasonia glutinosa*) by Villaescusa et al. (1992, 1995), and a recent report (Ahmed et al., 1994) described 6,4'-dihydroxy-3,5,7-trimethoxyflavone and 3'-hydroxy-3,5,7,4'-tetramethoxyflavone, along with well known *O*-methyl ethers of kaempferol and quercetagetin, from *C. candicans*. Two recent papers reported kaempferol and quercetin 3-*O*-glucuronides (Rubio et al., 1995) and a series of 3-*O*- and 7-*O*-monoglycosides of quercetin and patuletin from *C. glutinosa* (all cited as species of *Jasonia*).

Chrysocephalum (G): According to Anderberg (1994) *Chrysocephalum*, an Australian genus, consists of 10 species (in 1991c he listed seven). Zdero et al. (1992a) reported naringenin from *C. ambiguum*.

Dittrichia (I): Two species comprise this Mediterranean genus, *D. graveolens* and *D. viscosa*. Several reports have been published describing flavonoids from each of them, only two of which appeared under this generic name. Thus, *D. viscosa* yielded quercetin-7-methyl ether (rhamnetin) (Simöes and Nacimento, 1990), quercetin-3,3'-dimethyl ether, and (2*R*,3*R*)-dihydrokaempferol-7-methyl ether (Chiappini et al., 1982). All other reports of flavonoids from *Dittrichia*, which follow, appeared under *Inula*. 6-Hydroxyapigenin (scutellarein) and 6-hydroxyluteolin were identified from *D. graveolens* (Öksüz and Topçu, 1992), while *D. viscosa* afforded apigenin, apigenin 7-methyl ether (genkwanin), 6-methoxyapigenin (hispidulin), luteolin and 6-hydroxyluteolin (Grande et al., 1985; Wollenweber et al., 1991).

Flavanones of the genus consist of naringenin 7-methyl ether (sakuranetin) in *D. graveolens* (Öksüz and Topçu, 1992) and naringenin, naringenin 7-methyl ether (sakuranetin), and eriodictyol and its 7-methyl ether in *D. viscosa* (Grande et al., 1985; Ulubelen et al., 1987a; Wollenweber et al., 1991).

Dihydrokaempferol (aromadendrin) and dihydroquercetin (taxifolin), their 7-methyl ethers, and 3-acetates of both, have been reported from *D. graveolens* (Öksüz and Topçu, 1992) and from *D. viscosa* (Grande et al., 1985; Ulubelen et al., 1987a; Wollenweber et al., 1991). *Dittrichia viscosa* also afforded dihydroquercetin-4'-methyl ether 3-acetate (Bohlmann et al., 1977).

Dittrichia graveolens was shown to accumulate 6-hydroxyapigenin (scutellarein) and 6-hydroxyluteolin (Öksüz and Topçu, 1992); *D. viscosa* afforded apigenin 7-methyl ether (genkwanin), 6-methoxyapigenin (hispidulin), luteolin, and 6-hydroxyluteolin (Grande et al., 1985; Wollenweber et al., 1991).

Elytropappus (G): Eight species comprise this South African genus, only one of which has been examined for flavonoids. Quercetin, 5,7,4'-trihydroxy-6-methoxyflavone (hispidulin), 5,4'-dihydroxy-6,7-dimethoxyflavone (cirsimaritin), and 6-methoxyluteolin (eupafolin) were isolated from *E. rhinocerotis* by Proksch et al. (1982).

Epaltes (P): *Epaltes* consists of 14 (Anderberg, 1994) or 15 (Merxmüller et al. 1977) species distributed in warmer parts of both hemispheres. The sole source of flavonoid data appears to be the work of Nair et al. (1982) who found apigenin and luteolin 7-*O*-glucosides in *E. pygmaea*.

Erymophyllum (G): *Erymophyllum* consists of five species native to Australia. The only flavonoid information available comes from the work of Zdero et al. (1989b) who reported luteolin and chrysoeriol from *E. tenellum* (cited as *Helipterum tenellum*; synonymy from Anderberg, 1991c).

Filago (G): *Filago* consists of about 50 species widely distributed in Eurasia, North Africa, and in southwestern North America. N. Saleh et al. (1988) reported quercetin 3-*O*-glucoside and luteolin 7-*O*-glucoside from *F. desertorum* and *F. prolifera* but only the quercetin derivative from *F. mareotica*.

Geigeria (I): *Geigeria*, consisting of 27 species, occurs in the southern tropics and in South Africa. Coleman et al. (1984) reported 6-methoxyapigenin (hispidulin) and 6-methoxyluteolin (eupafolin) as constituents of *G. burkei* var. *zeyheri*.

Gnaphalium (G): *Gnaphalium* is a large, world-wide genus that has recently undergone taxonomic re-evaluation, primarily by Anderberg (1991c). Merxmüller et al. (1977) and Dillon and Sagástegui (1991b) earlier recognized some 150 species, but in the more recent view of Anderberg (1994) there are only about 50 species. Apparent flavonoid diversity in the genus has been lessened to a substantial degree by the movement of several species to *Pseudognaphalium* (see below).

Glycosides of apigenin, luteolin, and quercetin have been reported from *G. sylvaticum* (Konopleva et al., 1978). That paper also described identification of 5,7,4'- trihydroxy-3',5'-dimethoxyflavone (tricin) from *G. sylvaticum*, which appears to be the only B-ring trisubstituted flavonoid so far obtained from a member of the genus. Several 6-hydroxy and 6-methoxy compounds have been reported from *G. uliginosum* by Konopleva et al. (1979a,b): 6-hydroxyapigenin (scutellarein), 6-hydroxyluteolin, 6-methoxyluteolin (eupafolin) and jaceosidin. Eupafolin and jaceosidin both occurred as 7-*O*-(6"-*O*-caffeoyl)-glucosides in the plant.

Haeckeria (G): *Haeckeria* consists of four species native to mainland Australia. Recent work (Reid and Bohm, unpubl.) on *H. ozothamnoides* has shown the presence of 2',4',6',4-tetrahydroxychalcone, kaempferol and quercetin 3-methyl ethers, and quercetin 7-methyl ether (rhamnetin) in leaf exudate, and apigenin, luteolin, 3,5,7-trihydroxyflavone (galangin), and kaempferol and quercetin 3-*O*- and 7-*O*-glycosides from the vacuolar fraction.

Helichrysum (G): *Helichrysum*, with some 500 (Merxmüller et al., 1977) to 600 (Anderberg, 1994) species, is by far the largest genus of Inuleae, or whichever tribe is its rightful home. Species of *Helichrysum* occur in Eurasia, Africa, Madagascar, Australia, and New Zealand. Capitula of some species, especially *H. bracteatum* (see under *Bracteantha* above), the "everlastings," are dried and used as decorations. The genus displays a rich flavonoid chemistry including aurones, chalcones, dihydrochalcones, flavanones, a flavan, dihydroflavonols, flavones, flavonols, and anthocyanins, in various combinations. Noteworthy is the presence of numerous compounds lacking B-ring substitution.

The discovery of the aurone bracteatin and its glucoside bractein from plant material originally assigned to the genus *Helichrysum* was discussed above among entries under *Bracteantha*. A recent paper by Nair et al. (1989) reported both bractein and 4,6,3',4'-tetrahydroxyaurone 4-*O*-glucoside (cernuoside) from *H. buddleioides*.

Chalcone derivatives are widely distributed in the genus. They vary in structural complexity from 2',4',6'-trihydroxychalcone, and its methyl ethers, through derivatives of 2',4',6',4-tetrahydroxychalcone to a series of *O*- and *C*-prenylated compounds in some of which the prenyl functions have undergone cyclization with neighboring hydroxyls to form dimethyl chromane/chromene derivatives. Details of the occurrence of chalcones in *Helichrysum* can be found in Table 12–2.

Table 12–2. Chalcones from species of *Helichrysum*

Species	Compounds[1]	References
H. achyroclinoides	2'4'6'/3'-*C*-Pr,	
	2'6'4/4'	Randriaminahy et al. (1992)
H. acutatum	2'4'6'/-	Bohlmann and Abraham (1979c)
H. aphelexioides	2'4'/6',	
	2'4'/6'/3'-*C*-Pr,	
	2'6'/-/4'-*O*-Pr,	
	2'/6'/4'-*O*-Pr,	
	2'4'/-/5'-*C*-Pr(cy)	
	4'/2'/5'-*C*-Pr(cy)	Randriaminahy et al. (1992)
H. arenarium	2'4'6'4/-/2'-*O*-Glc	Vrkoc et al. (1959);
		Hänsel et al. (1960)
H. argyrolepis	2'4'6'/-/3'-*C*-Pr	Bohlmann et al. (1984c)
H. athrixiifolium	2'4'6'/-/3'-*C*-Pr,	
	2'6'/-/4'-*O*-Pr,	
	2'6'4/-/4'-*O*-Pr	Bohlmann and Ates (Gören) (1984)
H. bracteiferum	2'4'/6'	Randriaminahy et al. (1992)
H. cooperi	2'4'4/6'/4'-*O*-Glc	Wright (1976)
H. cymosum		
ssp. *calvum*	2'6'/4',	
	2'4'6'/-/3'-*C*-Pr,	
	2'6'/4'/3'-*C*-Pr,	
	2'6'/4'/3'-*C*-Pr(cy)	Bohlmann et al. (1979f)
H. forskahlii	2'6'/-/4'-*O*-Pr,	
	2'/6'/-/4'-*O*-Pr	Jakupovic et al. (1990b)
H. graveolens	2'4'6'4/2'-*O*-Glc	Çubukçu and Damatyan (1980, 1986)
H. heterolasium	2'4'4/6'	Bohlmann and Abraham (1979b)
H. kraussii	2'4'/-/3'-*C*-Pr(cy)	Jakupovic et al. (1989d)
H. odoratissimum	2'4'4/6'	Van Puyvelde et al. (1989a)
H. oreophilum	2'4'6'/-,	
	2'6'/4',	
	2'/4'6'	Jakupovic et al. (1986)
H. pallasii	2'4'6'4/2'-*O*-Glc	Çubukçu and Bingöl (1984)
H. retrorsum	2'4'6'/3'-*C*-Pr,	
	2'4'6'/3'-*C*-Ner,	
	2'6'/-/4'-*O*-Pr,	
H. rugulosum	2'6'/-/4'-*O*-Pr,	
	2'/6'/4'-*O*-Pr,	
	2'4'6'/-/3'-*C*-Pr,	
	2'6'/4'/3'-*C*-Pr	Bohlmann and Misra (1984)
H. sutherlandii	6'/2'3'4',	
	-/2'6'/3'4'-MDO	Bohlmann et al. (1978a)
H. tenuiculum	2'6'/4',	
	2'6'/4'/3'-*C*-Pr	Bohlmann et al. (1979f)

Table 12–2. (Continued)

Species	Compounds[1]	References
H. triplinerve	2'4'/6',	
	2'6'4/4'	Randriaminahy et al. (1992)
H. umbraculigerum	2'4'6'/-/3'-*C*-Ger	Bohlmann and Hoffmann (1979)

[1] Substitution: Hydroxyl positions/Methoxyl positions/Other substituents; Glc = glucose; Ger = geranyl; Pr = prenyl; Pr(cy) = cyclized prenyl; Ner = neryl; MDO = methylenedioxy.

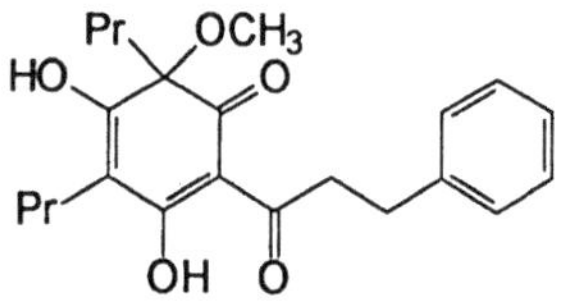

[12-1] Pr = Prenyl

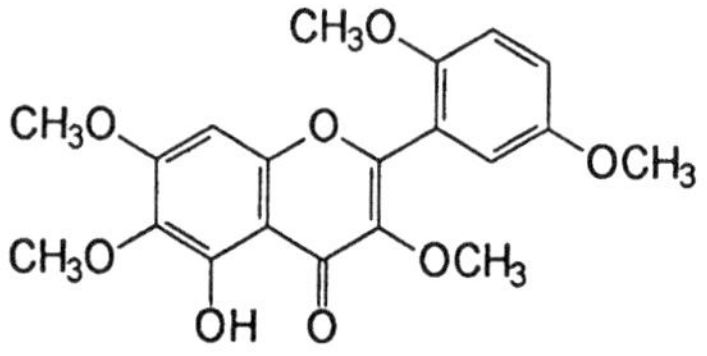

[12-2]

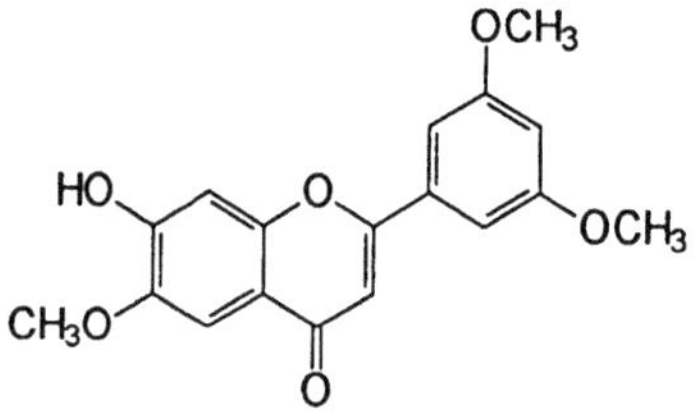

[12-3]

Fig. 12–1. Two substitutions at a single carbon result in the unusual A-ring seen in **[12–1]**. Compounds with unusual B-ring substitution patterns **[12–2, 12–3]** have been reported from *Iphiona grantioides*

Dihydrochalcones in the genus are perhaps even more structurally diverse than are the chalcones, although they appear not to be so widely distributed. The simplest member of this group is 2',4',6'-trihydroxydihydrochalcone, which was obtained from *H. tenuifolium* (Bohlmann and Abraham, 1979a). Other B-ring deoxydihydrochalcones from *Helichrysum* are 2',6'-dihydroxy-3',4'-methylenedioxydihydrochalcone from *H. sutherlandii* and its ß-hydroxy derivative from *H. mundtii* (Bohlmann et al., 1978a). *Helichrysum argyrolepis* (Bohlmann et al., 1984c), *H. forskahlii* (Jakupovic et al., 1990b), *H. cymosum* ssp. *calvum* and *H. tenuiculum* (Bohlmann et al., 1979f) accumulate dihydrochalcones that exhibit unusual A-ring structures. Owing to the double substitution at one of the ring carbons, thus preventing aromatization, these

Table 12–3. Dihydrochalcones from species of *Helichrysum*

Species	Compounds[1]	References
H. aphelexioides	A-Ring modified	Randriaminahy et al. (1992)
H. argyrolepis	A-Ring modified	Bohlmann et al. (1984c)
H. cymosum		
ssp. *calvum*	A-Ring modified	Bohlmann et al. (1979f)
H. forskahlii	A-Ring modified	Jakupovic et al. (1990b)
H. monticola	2'3'4'6'4/-5'-*C*-Ger	Bohlmann and Zdero (1980)
H. mundtii	ß/2'6'/3'4'-MDO	Bohlmann et al. (1978a)
H. splendidum	2'4'6'4/-	Bohlmann and Suwita (1979b)
H. sutherlandii	-/2'6'/3'4'-MDO	Bohlmann et al. (1978a)
H. tenuifolium	2'4'6'/-	Bohlmann and Abraham (1979a)
	A-Ring modified	Bohlmann et al. (1979f)

[1] Substitutions: Hydroxyl positions/Methoxyl positions/Other substituents; A-Ring modified = non-benzenoid ring; Ger = geranyl; MDO = methylenedioxy.

compounds are named as derivatives of cyclohexadienone. A typical member of this small group (Fig. 12–1) is 2-dihydrocinnamoyl-3,5-dihydroxy-6-methoxy-4,6-di-*C*-prenylcyclohexa-2,4-dienone **[12–1]** isolated from *H. forskahlii* by Jakupovic et al. (1990b). Other members of this group exhibit as many as three *C*-bound prenyl groups. In addition to *H. forskahlii*, these modified A-ring dihydrochalcones have been identified from the following species: *H. aphelexioides, H. argyrolepis, H. cymosum* ssp. *calvum*, and *H. tenuifolium*. References for these reports can be found in Table 12–3.

Returning to compounds with ordinary substitution patterns, we see that *H. splendidum* accumulates 2',4',6',4-tetrahydroxydihydrochalcone (Bohlmann and Suwita, 1979b) whereas *H. monticola* afforded two isomers of 2',3',4',6',4-pentahydroxy-5'-*C*-geranyldihydrochalcone. The isomeric forms involved orientation of groups in the geranyl group. Dihydrochalcone occurrences in the genus are summarized in Table 12–3.

The simplest flavanone, known from several species, is 5,7-dihydroxyflavanone (pinocembrin). Its 5-methyl ether is known from *H. herbaceum* (Bohlmann et al., 1979c). Several 8-substituted pinocembrin derivatives were described from *H. cymosum*: 8-*C*-prenylpinocembrin, 8-hydroxypinocembrin 7-methyl ether, 8-methoxypinocembrin and 8-methoxypinocembrin 7-prenyl ether (Jakupovic et al., 1989c). Higher levels of prenylation, involving both *O*- and *C*-substitution, are seen in a series of pinocembrin derivatives isolated from *H. rugulosum* (Bohlmann and Misra, 1984). 8-*C*-Geranylpinocembrin, along with 8-*C*-prenyl and 8-*C*-modified prenyl pinocembrin have been identified from *H. hyphocephalum* (Bohlmann and Abraham, 1979d). In

Table 12–4. Flavanones from species of *Helichrysum*

Species	Compounds[1]	References
H. acutatum	PC	Bohlmann and Abraham (1979c)
H. aphelexioides	PC	Randriaminahy et al. (1992)
H. arenarium	Nar	Vrkoc et al. (1975)
	Nar 5-*O*-Glc	Jerzmanowska and Grzybowska (1958); Hänsel and Heise (1959); Borkowski and Pasich (1961); Prokopenko et al. (1972); Zapesochnaya et al. (1972)
	Nar 5-*O*-DiGlc	Vrkoc et al. (1959)
H. athrixiifolium	PC 7-*O*-Pr,	
	Nar 7-*O*-Pr,	
	Nar 8-*C*-Pr	Bohlmann and Ates (Gören) (1984)
H. aurantiacum	Nar	Ovdienko et al. (1977)
H. bracteiferum	PC 7-Me	Randriaminahy et al. (1992)
H. callicomum	PC	Bohlmann and Abraham (1979b)
H. cymosum	PC,	
	PC 8-*C*-Pr,	
	5/8/7-*O*-Pr,	
	57/8,	
	58/7	Jakupovic et al. (1989a)
H. forskahlii	PC 7-*O*-Pr	Jakupovic et al. (1990b)
H. graveolens	Nar,	
	Nar 4'-*O*-Glc	Çubukçu and Damatyan (1980, 1986)
H. herbaceum	PC 5-*O*-Me	Bohlmann et al. (1979c)
H. hirtum	PC 7-*O*-Pr,	
	Nar 7-*O*Me,	
	Nar 7,4-Di*O*Me	Randriaminahy et al. (1992)
H. hyphocephalum	PC 6-*C*-Pr,	
	PC 8-*C*-Pr,	
	PC 8-*C*-Ger	Bohlmann and Abraham (1979d)
H. italicum	PC	Freitag (1979)
H. lepidissimum	PC 6-*C*(Odd),	
	PC 6-*C*(Odd) 8-*C*-Pr	Jakupovic et al. (1989a)
H. oreophilum	PC	Jakupovic et al. (1986)
H. pallasii	Nar, Nar 4'-O-Glc	Çubukçu and Bingöl (1984)
H. plicatum	Nar	Ovdienko et al. (1977)
H. plinthocalyx[2]	Nar	Ovdienko et al. (1977)
H. polyphyllum	Nar	Ovdienko et al. (1977)
	Nar 5-O-Glc	Zapesochnaya et al. (1972)
H. retrorsum	PC 7-*O*-Pr,	
	Nar 7-*O*-Pr	Randriaminahy et al. (1992)
H. rubicundum	Nar	Ovidenko et al. (1977)
H. rugulosum	PC 7-*O*-Pr,	
	PC 6,8-Di-*C*-Pr,	

Table 12–4. (Continued)

Species	Compounds[1]	References
	PC 7-*O*-Pr-8-*C*-Pr,	
	PC 6-*C*-Pr,	
H. tenuifolium	PC 5-*O*Me-7-*O*-Pr	Bohlmann and Misra (1984)
H. thapsus	PC 6-*C*-Pr	Bohlmann and Zdero (1983)
H. zeyheri	PC	Jakupovic et al. (1986)

[1] Substitution: Hydroxyl positions/Methoxyl positions; PC = pinocembrin; Nar = naringenin; Pr = prenyl; Ger = geranyl; Me = methyl; Glc = glucose; Odd = unusual structural feature.

[2] Not listed in Anderberg (1991c).

this same paper 5,7-dihydroxy-8-*C*-prenylflavan was also described from *H. hyphocephalum.*

Naringenin has been found in several species either as such, as the 5-*O*-glucoside or as the 4'-*O*-glucoside. 8-*C*-Prenylnaringenin and naringenin 7-prenyl ether were identified as components of *H. athrixiifolium*, where they were accompanied by several prenylated chalcones [Bohlmann and Ates (Gören), 1984]. Flavanone occurrences in the genus are summarized in Table 12–4.

Dihydroflavonols (3-hydroxyflavanones) have been reported from only a few species of *Helichrysum.* 3,5,7-Trihydroxyflavanone (pinobanksin) has been obtained from *H. tenuifolium* (Bohlmann and Abraham, 1979a), *H. lepidissimum* (Jakupovic et al., 1989c), and *H. platyterum* (Jakupovic et al., 1986). Two 3,5,7-trihydroxy-6-*C*-prenylflavanones differing in the stereochemistry at C-3, and (2*R*,3*S*)-3,5,7-trihydroxy-6-*C*-geranylflavanone have been identified as components of *H. thapsus* (Bohlmann and Zdero, 1983).

We have already noted the existence of B-ring deoxyflavonoids in the sections above involving chalcones, flavanones, and dihydroflavonols. B-Ring deoxyflavones (including flavonols) are also very common in *Helichrysum.* Compounds based on the following substitution patterns are known: 57, 567, 578, 5678, 357, 3567, 3578, and 35678. Over twenty-five different combinations of hydroxy and methoxy (including methylenedioxy) groups characterize the flavones known from *Helichrysum.* These range from the simplest, 5,7-dihydroxyflavone (chrysin), which occurs in *H. tenuifolium* (Bohlmann and Abraham, 1979a), through an assortment of *O*-methylated compounds, to 3,5,6,7,8-pentamethoxyflavone from *H. nitens* (Tomas-Barberan et al., 1988). The most frequently encountered compound, present in eight species, is 3,5-dihydroxy-6,7,8-trimethoxyflavone. Details of occurrence of B-ring deoxyflavonoids can be found in Table 12–5.

Table 12–5. B-Ring deoxyflavones and deoxyflavonols from species of *Helichrysum*

Species	Compounds[1]	References
H. arenarium	357/-	Vrkoc et al. (1975)
	35/678	Hänsel et al. (1967, 1981); Çubukçu (1982)
H. armenium	57/3	Çubukçu (1982)
H. aureum	357/-, 57/3, 35/7, 357/8, 57/38	Wollenweber et al. (1997)
H. bracteiferum	357/, 57/36, 57/38, 357/68, 57/368	Randriaminahy et al. (1992)
H. cephaloideum	5/3678	Jakupovic et al. (1986)
H. chrysargyrum	56/37	Bohlmann et al. (1979c)
H. decumbens	35/67, 35/678, 57368	Tomas-Lorente et al. (1989)
H. graveolens	57/3, 35/678	Hänsel and Çubukçu (1972); Çubukçu (1982); Çubukçu and Damatyan (1980, 1986)
H. herbaceum	-/57, 58/67, 8/567, -/5678	Bohlmann et al. (1979c)
H. heterolasium	57/36	Bohlmann and Abraham (1979b)
H. italicum	57/38	Opitz et al. (1971)
	35/78	Hänsel et al. (1980)
H. kraussii	35/678	Candy et al. (1975)
H. mimetes	58/67	Jakupovic et al. (1986)
H. mundtii	-/578, -/5/78-MDO	Bohlmann et al. (1978a, 1980f)
H. nitens	-/57, 5/67, -/567, 5/678, -/5678, -/357, -/3567, -/35678	Tomas-Barberan et al. (1988)
H. noeanum	35/678	Çubukçu (1982)
H. odoratissimum	35/678	Van Puyvelde et al. (1989)
H. orientale	35/678	Çubukçu (1976, 1982)
H. pagophilum	35/678	Bohlmann et al. (1980f)
H. pallasii		Çubukçu (1982); Çubukçu and Bingöl (1984)
H. picardii[2]	57/38	De La Puerta et al. (1990); Tomas-Lorente et al. (1991)
	57/3	Tomas-Lorente et al. (1991)
H. platypterum	57/3	Jakupovic et al. (1986)
H. sanguineum	57/38, 357/68	Çubukçu (1982); Mericli et al. (1984)
H. schimperi	5/7, 5/67, -/567, 5/78, 5/678, 5/367	Jakupovic et al. (1990b)
H. tenuifolium	57/-, 357/-, 57/3	Bohlmann and Abraham (1979a)

[1] Substitution: Hydroxyl positions/Methoxyl positions; MDO = methylenedioxy.
[2] Not listed in Anderberg (1991c).

Several species have the capacity for extra A-ring oxygenation and/or *O*-methylation. Quercetin 3-methyl ether has been identified in *H. kraussii* (Candy et al., 1975) and *H. odoratissimum* (Van Puyvelde et al., 1989). The only quercetagetin derivatives recorded are the 7-*O*-diglucoside from *H. stoechas* (Pinkas et al., 1973; Mericli et al., 1992), the 3,6,7,3',4'-pentamethyl ether (artemetin) from *H. chionosphaerum* (Bohlmann et al., 1980a), and the 3,6-dimethyl ether (axillarin) from *H. lecomtei*, a Madagascarian species that also accumulates quercetin and 5,3',4'-trihydroxy-3,6,7,8-tetramethoxyflavone (6-hydroxygossypetin-3,6,7,8-tetramethyl ether) (Randriaminahy et al., 1992). These workers also reported four 6- and/or 8-substituted flavones from *H. benthamii*: 5,4'-dihydroxy-6,7,8-trimethoxyflavone (xanthomicrol), 5,7,3',4'-tetrahydroxy-6-methoxyflavone (nepetin), 5,7,3',4'-tetrahydroxy-6,8-dimethoxyflavone, and 5,3',4'-trihydroxy-6,7,8-trimethoxyflavone (Randriaminahy et al., 1992). Other 8-oxygenated compounds described from the genus are 7-hydroxy-5,6,8,4'-tetramethoxyflavone from *H. herbaceum* (Bohlmann et al., 1979c), and 5,3',4'-trihydroxy-3,7,8-trimethoxyflavone (gossypetin 3,7,8-trimethyl ether) from *H. splendidum* (Bohlmann and Suwita, 1979b). Several common flavone and flavonol derivatives have been identified as components of leaf wash extracts of *H. aureum* (Wollenweber et al., 1997): luteolin 3'-methyl ether, kaempferol, quercetin 3-methyl ether, and quercetin 3,3'-dimethyl ether. *Helichrysum cassianum* yielded the new flavone 3,5-dihydroxy-6,7,8,4'-tetramethoxyflavone along with several known compounds including apigenin, hispidulin, cirsimaritin, nepetin, and quercetin (Wollenweber et al., 1998).

Species of *Helichrysum* also produce comparatively common apigenin, luteolin, kaempferol, and quercetin derivatives. Flavones occur as 7- or 4'-*O*-glucosides, flavonols as 3- or 7-*O*-glycosides. Flavonol 3-*O*-(*p*-coumaroylglucosides) have been reported from *H. kraussii* (Candy et al., 1975) and *H. orientale* (Çubukçu, 1976). A quercetin 7-*O*-triglycoside, which yielded arabinose, galactose and xylose upon acid hydrolysis, was encountered in *H. stoechas*. A summary of occurrences of simple flavones and flavonols in the genus is given in Table 12–6.

Pelargonidin and peonidin glucosides were reported from *H. sanguineum* by Mericli et al. (1984). More detailed information on these pigments in the genus appears to be lacking.

Ifloga (G): This genus consists of six species native to South Africa, the Canary Islands, Eurasia, Africa, and Madagascar. *Ifloga spicata* and *I. labillardieri* (epithet not in Anderberg) afforded quercetin 3-*O*-glucoside, quercetin 3,7-di-*O*-glucoside, and an acylated derivative of the latter (N. Saleh et al., 1988).

Table 12–6. Occurrences of simple flavones and flavonols from species of *Helichrysum*

Species	Flavonoids[1]				References
	Api	Lut	Kae	Que	
H. annuum[2]		+			Chakraborti et al. (1959-1960)
H. arenarium	+		+		Jerzmanowska and Grzybowska (1958, 1960)
	+		+		Vrkoc et al. (1959, 1975)
	+	+	+	+	Prokopenko et al. (1972)
H. aurantiacum	+	+	+		Ovdienko et al. (1977)
H. buddleioides		+			Nair et al. (1989)
H. cassianum	+			+	Wollenweber et al. (1998)
H. cooperi		+			Wright (1976)
	+				Bohlmann et al. (1978c)
H. graveolens	+	+	+		Ovdienko et al. (1977)
	+	+	+	+	Çubukçu and Damatyan (1980, 1986)
H. kraussii				+	Candy et al. (1975)
H. orientale			+	+	Çubukçu (1976)
H. pallasii	+	+		+	Çubukçu and Bingöl (1984)
H. pamphyticum			+		Sezik and Akdemir (1986)
H. plicatum	+	+	+		Ovdienko et al. (1977)
H. plinthocalyx[2]	+	+	+		Ovdienko et al. (1977)
H. polyphyllum	+	+	+		Ovdienko et al. (1977)
H. rubicundum	+	+	+		Ovdienko et al. (1977)
H. sanguineum			+	+	Mericli et al. (1984)
H. stoechas				+	Pinkas et al. (1973)
				+	Saleh et al. (1988)
		+	+	+	Mericli et al. (1992)

[1] Api= apigenin; Lut = luteolin; Kae = kaempferol; Que = quercetin.
[2] Name not listed in Anderberg (1991c).

Inula (I): Despite the large size of this genus, between 80 and 100 species, we have flavonoid information for only a few. *Inula britannica* afforded luteolin, 6-hydroxyluteolin, and 6-methoxyluteolin (nepetin), all as 7-*O*-glycosides (Öksüz and Topçu, 1987; Krolikowska and Wolbis, 1979, 1981). Chen et al. (1992) reported luteolin and Jeske et al. (1996) naringenin from *I. salsoloides*. The 1987 paper by Öksüz and Topçu included a report of quercetin 3-sulphate and 6-hydroxykaempferol 3-sulphate as constituents of *I. britannica*. Quercetin and quercetagetin 6-methyl ether (patuletin) were obtained from *I. britannica* (Dombrowicz and Greiner, 1968; Krolikowska and Wolbis, 1979, 1981; Öksüz and Topçu (1987), 6-hydroxykaempferol 3,7-dimethyl ether was isolated from *I. grandis* (Nikonova and Nikonov, 1975), and quercetin and its 3-methyl ether were obtained from *I. helenium* (Kowalewska and Lutomski, 1978).

A recent survey of leaf surface flavonoids of several asteraceous taxa (Wollenweber et al., 1997) included three species of *Inula*: *I. brittanica, I. germanica*, and *I. salicina*. The former afforded the most complex array of compounds consisting of apigenin, luteolin, various methyl ethers of 6-hydroxyapigenin and 6-hydroxyluteolin, and the 3,6,7-tri-, 3,6,3'-tri-, 3,6,7,3'-tetra-, and 3,6,7,3',4'-pentamethyl ethers of quercetagetin. From *I. germanica* these workers obtained the 3,6-di-, 3,7-di-, and 3,6,7-trimethyl ethers of quercetagetin, while the leaf wash of *I. salicina* afforded only 6-methoxyluteolin.

Iphiona (I): *Iphiona* consists of 11 species and ranges from northeastern Africa to central Asia. A detailed study of *I. scabra* (Ahmed and Mabry, 1987) and *I. mucronata* (Ahmed et al., 1988b) showed that the predominant compounds in both species were a series of quercetin and isorhamnetin mono-, di-, tri-, and tetrasulphates. 6-Methoxyapigenin (hispidulin) 7-sulphate was identified from *I. scabra*. Simple flavonol monoglycosides were also found in both species as were flavone and flavonol methyl ethers containing up to five methyl groups. Eriodictyol was isolated from *I. mucronata*.

Ahmad and Ismail (1991a,b) reported two of the most unusual flavonoids known from the family as a result of their work with *I. grantioides*. The structure of "grantioiden" was shown to be 5-hydroxy-3,6,7,2',5'-pentamethoxyflavone **[12–2]**. More unusual is "grantionin," 7-hydroxy-6,3',5'-trimethoxyflavone **[12–3]**. The presence of a 3',5'-substitution (no 4' substituent!) along with the 5-deoxy condition is unique, certainly within the family. These workers have also reported 5,7-dihydroxy-6,4'-dimethoxyflavone (pectolinaringenin) and 5-hydroxy-6,7,4'-trimethoxyflavone (salvigenin) from the same species (Ahmad and Ismail, 1993). More recently, another 5-deoxyflavonoid has been reported from the genus. El-Lakany et al. (1996) reported 3,7,3',4'-tetrahydroxy-6,5'-dimethoxyflavone, along with quercetagetin 3'-methyl ether, from *I. crithmoides*.

Laggera (P): *Laggera* consists of 17 species with a range extending from tropical Africa through Arabia to Asia. The only flavonoid information on the genus comes from a recent report by Li and Ding (1994) who found 5,4'-dihydroxy-3,6,7-trimethoxyflavone (penduletin), 5-hydroxy-3,6,7,4'-tetramethoxyflavone, and 5-hydroxy-3,6,7,3',4'-pentamethoxyflavone (artemetin) in *L. pterodonta*.

Leontopodium (G): *Leontopodium* is a genus of 60 species, distributed in the mountains of Europe, central and eastern Asia, and Japan. Only two species appear to have been analyzed for flavonoids, however. *Leontopodium alpinum*, the renowned Edelweiss, was shown to accumulate luteolin 7-*O*-glucoside and 4'-*O*-glucoside (Tira et al., 1970). Dashbalyn and Glyzin (1978) found apigenin and luteolin 7-*O*-glucosides in *L. ochroleucum*.

Leucogenes (G): *Leucogenes* is a New Zealand genus of four species. Recent studies of flavonoids of Southern Hemisphere Gnaphalieae (Reid and Bohm, unpubl.) included *L. grandiceps* and *L. leontopodium*, which had the following compounds in common: apigenin, luteolin, kaempferol and quercetin 3-*O*-glucosides, kaempferol and quercetin 3-*O*-rutinosides, and kaempferol 7-*O*-glucoside. The former also has quercetin 3-methyl ether; the latter 3,5,7-trihydroxyflavone (galangin), kaempferol 3-methyl ether, quercetin 7-methyl ether (rhamnetin), and a quercetin 3-*O*-diglucoside. The two recently described species, *L. neglecta* and *L. tarahaoa*, (Molloy, 1995) from the South Island entered the scene too late to be included in our study.

Metalasia (G): *Metalasia* consists of 52 species native to South Africa (Karis, 1989; Anderberg, 1994). Again, sampling has been meager with only *M. cymbifolia* having been examined (Zdero and Bohlmann, 1990c). The compounds identified, chalcones, dihydrochalcones, and a flavanone, are all prenylated and lack B-ring substitution. The flavanone identified is pinocembrin 7-prenyl ether. The two chalcones were shown to be 2',4',6'-trihydroxychalcone 4'-prenyl ether (this is the chalcone corresponding to the pinocembrin derivative) and 2',4',6'-trihydroxy-3'-*C*-prenylchalcone. Three dihydrochalcones were identified: 2',4',6'-trihydroxydihydrochalcone 4'-prenyl ether, 2',4'-dihydroxy-6'-methoxydihydrochalcone 4'-prenyl ether, and 2',4',6'-trihydroxy-3',5'-di-*C*-prenyldihydrochalcone. Also identified was "hemihumulone" an A-ring modified dihydrochalcone, 2-dihydrocinnamoyl-3,5-dihydroxy-6-methoxy-4,6-di-*C*-prenylcyclohexa-2,4-dienone **[12–1]**.

Myriocephalus (G): The only species of the eight species that comprise this Australian genus that has been studied for flavonoids is *M. guerinae* from which Jakupovic et al. (1991a) isolated 5-hydroxy-3,7,4'-trimethoxyflavone.

Ozothamnus (G): *Ozothamnus* is a moderately large genus, some 50 species, with representatives in Australia, New Zealand, and New Caledonia. Examination of several species of *Ozothamnus* (Reid and Bohm, unpubl.) has shown that the flavonoid glycoside fraction consists of kaempferol and quercetin 3-*O*-glucosides, kaempferol and quercetin 3-*O*-rutinosides, and kaempferol 7-*O*-rhamnoside. Aglycones identified include apigenin, luteolin, kaempferol and quercetin 3-methyl ethers, eriodictyol 7-methyl ether, 5,7-dihydroxyflavanone (pinocembrin), and 3,5,7-trihydroxyflavanone (pinobanksin). This latter compound had been reported, along with its 3-acetate, from *O. stirlingii* (cited as *Helichrysum stirlingii*) by Jakupovic et al. (1987b). Bohlmann and Suwita (1979c) reported 2',4-dihydroxy-6'-methoxy-3',4'-methylenedioxychalcone from *O. aggregatus* (cited as *Helichrysum glomeratum*).

Pechuel-Loeschea (I): This monotypic genus is known from southwestern South Africa to Zimbabwe. Zdero and Bohlmann (1989b) reported 5,4'-dihydroxy-6,7-dimethoxyflavone (cirsimaritin) from *P. leibnitziae*. Although traditionally included in *Pluchea* (see comments in Merxmüller et al., 1977), Anderberg (1991a) suggests that *Pechuel-Loeschea* is distinct and that the two genera "...belong to different groups" (p. 107).

Pluchea (P): *Pluchea* is a genus of about 80 species that occurs in warmer parts of both hemispheres. Flavonoid data are available for only a few, but show a diverse array of compound types. Polar flavonoids obtained from *P. dioscoridis* (Ahmed et al., 1987b) were identified as kaempferol and quercetin 3-*O*-glycosides, isorhamnetin 3-sulphate, kaempferol, quercetin and isorhamnetin-3,7-disulphates, and an apigenin 6,8-di-*C*-glucoside. *Pluchea dodoneaefolia* (reported as *Tessaria dodoneaefolia*; taxonomy after Robinson and Cuatrecasas, 1973) has been shown to accumulate naringenin and eriodictyol 7-methyl ethers (Guerreiro et al., 1973), naringenin 7,4'-dimethyl ether, eriodictyol, luteolin, and quercetin 3-methyl and 3,7-dimethyl ethers (Kavka et al., 1977). Dihydroquercetin 3-acetate has also been reported from *P. dodoneaefolia* (cited as *Tessaria dodoneaefolia*) by Kavka et al. (1977) and Nanayakkara et al. (1988).

Scholz et al. (1994) reported quercetin and isorhamnetin as components of *P. symphytifolia*. 5,7,3'-Trihydroxy-4'-methoxyflavanone (hesperetin) 7-*O*-rutinoside and dihydroquercetin (taxifolin) 3-*O*-arabinoside were identified in soils associated with *P. lanceolata* and discussed by the authors (Inderjit and Dakshini, 1991) as possible germination and growth inhibitors. Follow-up studies by these workers (Inderjit and Dakshini, 1994, 1995) showed the presence of several flavonoids, including 7,4'-dihydroxyisoflavone as the 7-*O*-glucoside in leachings from roots, and quercetin and its 7-*O*-rhamnoside in natural soils surrounding the roots of the plant.

All other flavonoids from species of *Pluchea* are methylated compounds many of which also exhibit substitution at C-6. Quercetin-3,3'-dimethyl ether was isolated from *P. sericea* (Romo de Vivar et al., 1982), 5,7,4'-trihydroxy-3,6-dimethoxyflavone from *P. chingoyo* (Chiang and Silva, 1978), and a series of flavonol methyl ethers of kaempferol, 6-hydroxykaempferol, quercetin and quercetagetin from *P. odorata* (Arriaga-Giner et al., 1983;

Wollenweber et al., 1985a). The only 8-substituted flavonol so far encountered in the genus is 5,7,3',4'-tetrahydroxy-3,6,8-trimethoxyflavone obtained by Martino et al. (1976) from *P. sagittalis*. *Pluchea absinthoides* (cited as *Tessaria absinthoides*; synonymy after Robinson and Cuatrecasas, 1973) afforded quercetagetin-3,6,7,4'-tetramethyl ether (casticin), 3,6,7,3'-tetramethyl ether (chrysosplenetin), and 3,6,7,3',4'-pentamethyl ether (artemetin) (Guerreiro et al., 1990).

Pseudognaphalium (G): Merxmüller et al. (1977) considered the genus to consist of 10 species, but Anderberg (1994) remodelled it to comprise about 80 (ca. 90 in Anderberg, 1991c). Dillon and Sagástegui (1991b) caution against accepting *Pseudognaphalium* as a distinct genus pending further in-depth studies on *Achyrocline*, *Gnaphalium*, and *Stenocline*. Much of the flavonoid chemistry of this genus comes about from the transfer of *Gnaphalium* species.

N. Saleh et al. (1988), in a study of Gnaphaliinae of Egypt, reported that *P. luteo-album* accumulated apigenin and luteolin 7-*O*-glucosides, kaempferol and quercetin 3-*O*- and 7-*O*-monoglycosides, and the two highly substituted compounds, 5,4'-dihydroxy-3,6,7,8-tetramethoxyflavone and 5,4'-dihydroxy-3,6,7,8,3'-pentamethoxyflavone. *Pseudognaphalium luteo-album* (reported as *Gnaphalium*) subjected to UV-B radiation exhibited increased amounts of the latter two compounds in the leaf surface exudate fraction but a decrease in the concentration of 5,7-dihydroxy-3,8-dimethoxyflavone (gnaphalin) (Cuadra et al., 1997). Resinous material from trichomes of *P. cheiranthifolium* afforded three B-ring deoxyflavonoids, 5,7-dihydroxy-3-methoxyflavone (galangin 3-methyl ether), 5,7-dihydroxy-3,6-dimethoxyflavone (3-methylalnusin), and 5,8-dihydroxy-3,6,7-trimethoxyflavone, while *P. heterotrichium* afforded only 5-hydroxy-3,7-dimethoxyflavone (3,7-dimethylgalangin) (Urzua et al., 1995). Other variants reported were a 5,7-dihydroxyflavanone (pinocembrin) derivative characterized by an unusual nine-carbon lactone unit at C-8, obtained from *P. obtusifolium* (Hänsel et al., 1970), and 5,7,8-trihydroxy-3-methoxyflavone derivatives esterified directly with 2-methylbutyric acid and (*Z*)-2-methyl-2-butenoic acids from *P. robustum* (Urzua and Cuadra, 1990). These are only a sampling of the B-ring deoxyflavonoids accumulated by members of *Pseudognaphalium*. Table 12–7 summarizes the occurrence of these compounds for 13 species.

Apigenin, luteolin, and quercetin, and glycosidic derivatives thereof have been reported in various combinations from *P. affine* (cited as *Gnaphalium affine*) (Aritomi et al., 1964; Aritomi and Kawasaki, 1974; Itakura et al., 1975) and from *P. luteo-album* (cited as *G. luteo-album*; Mericli, 1980), and *P. pellitum* (cited as *G. pellitum*; Torrenegra et al., 1978). The latter

Table 12–7. B-Ring deoxyflavonoids from species of *Pseudognaphalium*

Species	Compounds[1]	References
P. canescens[2]	57/36	Bohlmann and Ziesche (1980)
P. cheiranthifolium	57/3, 57/36, 58/367	Urzua et al. (1995)
P. domingense[3]	57/368, 58/367, 5678/3	Torrenegra et al. (1979, 1980)
P. gaudichaudianum	58/67	Guerreiro et al. (1982)
P. heterotrichium	5/37	Urzua et al. (1995)
P. lanuginosum	57/38	Torrenegra et al. (1979, 1980)
P. luteo-album	57/38, 357/68	Mericli (1980)
P. microcephalum	357/-, 357/6, 57/38, 35/678	Wollenweber et al. (1993b)
P. obtusifolium	5/378, 35/78, 357/68	Hänsel and Ohlendorf (1969); Ohlendorf et al. (1971)
	57/8	Hänsel et al. (1970)
P. pellitum	5/78	Escarria et al. (1977)
P. robustum	57/3, 57/38, 578/3, 5/378, 35/678	Urzua and Cuadra (1989, 1990)
P. stamineum[4]	35/678	Wollenweber et al. (1993b)
P. undulatum	5/3678	Bohlmann and Ziesche (1980)

1 Substitution: Hydroxyl positions/Methoxyl positions.
2 Cited as *P. wrightii*.
3 Cited as *P. elegans*.
4 Cited as *P. chilense*.

work also listed quercetin 7-methyl ether (rhamnetin) as a component of *P. pellitum*. Extra A-ring substitution is seen in two compounds from this genus, 6-methoxyapigenin (hispidulin) from *P. antenarioides* (Torrenegra et al., 1987), and 5,7,4'-trihydroxy-6,3'-dimethoxyflavone (jaceosidin) from *P. luteo-album* (Mericli, 1980).

2',4',4-Trihydroxy-6'-methoxychalcone 4'-*O*-glucoside (dehydro-*p*-asebotin) was identified as a component of *P. affine* (cited as *G. affine*) by Aritomi and Kawasaki (1974) and by Itakura et al. (1975), and from *P. luteo-album* (cited as *G. multiceps*; Jackson, 1895, referred this to *G. luteo-album*, which Anderberg, 1991b, moved into *Pseudognaphalium*) by Maruyama et al. (1974) and by Ahluwalia and Rani (1976), who also synthesized the compound.

Pterocaulon (P): *Pterocaulon*, a genus of 18 species, enjoys a wide distribution, with representatives in tropical America, Madagascar and Mauritius, tropical Asia, and Australia. Bohlmann et al. (1981a) identified the 7-prenyl ethers of both dihydrokaempferol (aromadendrin) and dihydroquercetin (taxifolin) from *P. virgatum*. More recently, the dihydroquercetin prenyl ether was reported from *P. alopecuroides* (Vilegas et al., 1995). *Pterocaulon virgatum* also yielded quercetin 3-*O*-glycosides, isorhamnetin, and quercetin-7,3'-dimethyl ether (rhamnazin) (Debenedetti et al., 1983). *Pterocaulon purpurescens*, afforded quercetin, isorhamnetin, and quercetagetin 3-methyl, 3,7-dimethyl, 3,3'-dimethyl, and 3,7,4'-trimethyl ethers (Debenedetti et al., 1987b). More recently, quercetin and isorhamnetin were reported from *P. polystachium* (Debenedetti et al., 1994). 5,7-Dihydroxyflavanone (pinocembrin) has been reported from *P. serratulum* (Macleod and Rasmussen (1999).

Pulicaria (I): *Pulicaria* is an Eurasian and African genus of about 80 species. The flavonoid profile of the genus seems to consist principally of dihydroflavonols and highly *O*-methylated flavonols. There is also a comment in the literature that *P. arabica* contains an unidentified flavonoid sulphate (El-Negoumy et al., 1982). Eriodictyol and its 7-methyl ether occur in *P. gnaphalodes* along with quercetin 3- and 3'-methyl ethers (Wollenweber and Rustaiyan, 1991). Eriodictyol 7-methyl ether also occurs in *P. undulata*, along with dihydrokaempferol 7-methyl ether (Metwally et al., 1986). A later study of *P. undulata* (Abdel-Mogib et al., 1989) resulted in identification of dihydrokaempferol and its 7-methyl ether, dihydroquercetin and its 7-methyl and 3,7-dimethyl ethers, kaempferol 3-methyl ether, and quercetin-3,7-dimethyl ether. Mansour et al. (1990) reported dihydroquercetin 7-methyl ether from *P. incisa* along with kaempferol and quercetin 3-*O*-galactosides and quercetin 3'-*O*-methyl ether (isorhamnetin).

Several major producers of *O*-methylated flavonols have been identified in *Pulicaria*: *P. arabica*, with quercetagetin 3,7-di-, 3,5,7-tri-, 3,6,7-tri-, 3,5,7,3'-tetra-, 3,5,6,7,3'-penta-, 3,6,7,3',4'-penta-, and 3,5,6,7,3',4'-hexamethyl ethers (El-Negoumy et al., 1982; Melek et al., 1988; Mossa et al., 1988); *P. dysenterica* with quercetagetin 3,7-di and 3,7,4'-trimethyl ethers, along with the 3-, 3,7-di-, and 3,6,7-trimethyl ethers of 6-hydroxykaempferol (Schulte et al., 1968; Pares et al., 1981); *P. incisa*, with quercetagetin 6,7-di-, 3,6,7-tri-, 3,6,3'-tri, and 3,6,7,3'-tetramethyl ethers (Mansour et al., 1990); *P. paludosa* with 5,7,4'-trihydroxy-3,6-dimethoxyflavone, 5,6-dihydroxy-7,4'-dimethoxyflavone, and 5,6,8-trihydroxy-7,4'-dimethoxyflavone (San Feliciano et al., 1989), and *P. somalensis* with quercetagetin 3,6,7-tri-, 3,6,7,3'-tetra-, and 3,6,7,4'-tetramethyl ethers (Al-Hazimi and Alkhathlan, 1993). Wollenweber et al. (1997) reported quercetin 3,7,3',4'-tetramethyl ether as a leaf surface component from *P. dysenterica.*

Pulicaria crispa was reported to accumulate quercetin by Sarg (1977). Rizk and Ismail (1982) and Rizk et al. (1993) reported glycosides of apigenin, quercetin, and rhamnetin and quercetin 3-methyl ether from this species as well, but recorded their plant material as *Francoeuria crispa*, which genus Anderberg (1991a) has submerged into *Pulicaria*.

Raoulia (G): *Raoulia* is a genus of 20 species native to New Zealand and New Guinea. Some of these species are referred to as "vegetable sheep" owing to their cushion growth habit. We have gathered flavonoid data for seven New Zealand species (nine taxa) of *Raoulia*: *R. australis, R. glabra, R. hookeri* (three vars.), *R. monroi, R. petriensis, R. subsericea*, and *R. tenuicaulis*. They are as alike as real sheep in a pen with invariant profiles consisting of quercetin and isorhamnetin 3-*O*-glucosides and quercetin 3-*O*-rutinoside (Reid and Bohm, 1995).

Relhania (G): This South African genus consists of 13 species. Tsichritzis and Jakupovic (1990) reported apigenin, luteolin, 5-hydroxy-6,7,8,4'-tetramethoxyflavone (gardenin-B), 5-hydroxy-6,7,8,3',4'-pentamethoxyflavone, and kaempferol from *R. corymbosa*, quercetin and dihydroquercetin (taxifolin) from *R. genistifolia*, and sakuranetin, homoeriodictyol, quercetin-3-methyl ether, and isorhamnetin from *R. squarrosa*.

Rhodanthe (G): *Rhodanthe* consists of 45 species that occur in Australia. Zdero et al. (1989b) reported 6-methoxyapigenin (hispidulin) from *R. propinquum* (cited as *Helipterum propinquum*; synonymy after Wilson, 1992) and *R. corymbiflorum* (as *Helipterum corymbiflorum*; synonymy after Wilson, 1992), and also 6-methoxyluteolin (eupafolin) from the latter.

Sphaeranthus (P): *Sphaeranthus* consists of 40 species that range from tropical Africa through southern Asia to Australia. Zdero et al. (1991b) obtained isorhamnetin, quercetin-3,3'-dimethyl and 3,3',4'-trimethyl ethers from *S. confertifolius*. Jakupovic and coworkers (1990a) reported "chrysosplenol" from *S. bullatus* and *S. suaveolens*. It was not clear from their description which of the chrysosplenols they meant.

Tessaria (P): According to Merxmüller et al. (1977) and Anderberg (1994), *Tessaria* consists of a single South American species. De Feo et al. (1990) reported naringenin, quercetin 3-*O*-glucoside, and quercetin 3-*O*-rutinoside from *T. integrifolia*. Two quercetagetin derivatives have also been identified, the 3,6,7,4'-tetramethyl and 3,6,7,3',4'-pentamethyl ethers (Guerreiro et al., 1990). These same workers (Guerreiro et al., 1990) reported 5,4'-dihydroxy-3,6,7-trimethoxyflavone (penduletin) and quercetagetin 3,6,7-trimethyl ether from a collection of *T. ambigua* from Entre Rios province, Argentina, while a collection from Catamarca province gave only 5-hydroxy-3,6,7,3',4'-pentamethoxyflavone.

CHAPTER 13

Flavonoids of Lactuceae

Bremer (1994) recognizes 98 genera and some 1550 species in his view of Lactuceae. Tomb (1977) listed only 70 genera but estimated the tribe as comprising about 2300 species. Our knowledge of the flavonoid chemistry of the tribe is based upon reports for 39 genera. Sampling is again a problem, with many of these genera being represented in our database by a single species. Somewhat more information is available for *Hieracium*, *Lactuca*, and *Sonchus*, which is not surprising in view of the size of these genera. However, some of the larger genera, *Crepis*, *Taraxacum*, and *Scorzonera* for example, have been very poorly sampled. These shortcomings are somewhat compensated for by the comparative simplicity of the flavonoid profile of the tribe. Details of flavonoid reports follow. Sizes of genera and distributional data are from Bremer (1994) unless otherwise noted.

Agoseris: *Agoseris* consists of about a dozen and a half species; two occur in Argentina and Chile, the rest in western North America. Harborne (1977b) reported quercetin 3-*O*-glycosides from *A. grandiflora* and *A. heterophylla* and apigenin and luteolin derivatives from *A. grandiflora*.

Andryala: *Andryala* is a Mediterranean genus of about 20 species only one of which appears to have been studied for flavonoids. Recio et al. (1993) reported apigenin 7-*O*-neohesperidoside, luteolin 7-*O*-glucoside, and the new luteolin 7-*O*-galactosylglucuronide from *A. ragusina*. More recently, Mañez et al. (1994) reported the same flavone glycosides as well as quercetin 3-*O*-glucoside from that taxon.

Babcockia: *Babcockia platylepis*, the only species in this genus, occurs on the Canary Islands. Its flavonoid profile was described by Mansour et al. (1983b) who reported apigenin and luteolin 7-*O*-mono- and diglycosides.

Catananche: This genus of five Mediterranean species is represented in the flavonoid literature by *C. caerulea*. Proliac (1972) and Proliac and Combier

(1974) reported luteolin 7-*O*-glucoside and apigenin 7-*O*-glucuronide, respectively. In a more recent paper, Proliac and Raynaud (1977) described the structure of a di-*C*-glycosylflavone, which they called "carlinoside," as a 6-*C*-glucosyl-8-*C*-arabinosylluteolin.

Chondrilla: *Chondrilla* is a temperate Eurasian genus comprising approximately 25 species. Quercetin 3-*O*-galactoside and luteolin 7-*O*-glycosides have been recently reported from *C. juncea* by Terencio et al. (1993) and by Mañez et al. (1994).

Cicerbita: The only flavonoid data for this north-temperate genus of about 35 species comes from the work of Rees and Harborne (1984) who reported kaempferol from *C. plumieri*.

Cichorium: *Cichorium* consists of six species of Europe, the Mediterranean Basin, and northern Africa. We have flavonoid information for four of them: *C. endivia* [including *C. pumilum* (Sell, 1976); Wagenitz and Bedarff (1989) treat the endives as *C. endivia* subsp. *divaricatum* (Schousb.) P. D. Sell], *C. intybus*, and *C. spinosum*. Rees and Harborne (1984) found luteolin 7-*O*-glucuronide, kaempferol 3-*O*-glucoside and 3-*O*-glucuronide, quercetin 3-*O*-galactoside and 3-*O*-glucuronide, and isorhamnetin 3-*O*-glucuronide. Their observations corroborated earlier reports of flavone and flavonol glycosides in *C. intybus* (Dem'yanenko and Dranik, 1973) and flavonol glycosides in *C. endivia* (Woeldecke and Herrmann, 1974; Saleh et al., 1975). Woeldecke and Herrmann (1974) also reported a malonyl derivative of kaempferol 3-*O*-glucoside from *C. endivia*. Bridle et al. (1984) identified cyanidin 3-*O*-(6"-*O*-malonylglucoside) from *C. intybus*.

Crepis: *Crepis* is a large genus with about 200 species with a wide distribution including Europe, temperate Asia, large parts of Africa, and in North and South America. We have flavonoid data for six species. *Crepis divaricata* and *C. pygmaea* both have luteolin and quercetin (Rees and Harborne, 1984). Mañez et al. (1992, 1994) reported luteolin 7-*O*-glucoside from *C. vesicaria* and *C. capillaris*. Luteolin 4'-*O*-glucoside also was found in the latter species. A key observation concerning flavonoids of this tribe was made by Harborne (1978) who reported luteolin, scutellarein, and isoëtin from *C. senecioides* and *C. tectorum*. Both scutellarein and isoëtin exhibit extra oxygenation (Fig. 13–1). Scutellarein **[13–1]** is the 6-hydroxy derivative of apigenin; it and its derivatives have a modest range of occurrence in Asteraceae. Isoëtin, 5,7,2',4',5'-pentahydroxyflavone **[13–2]**, on the other hand, is much more restricted in its occurrence having been recorded from several genera in Lactuceae as well as in a few species of *Artemisia* (see Chapter 7).

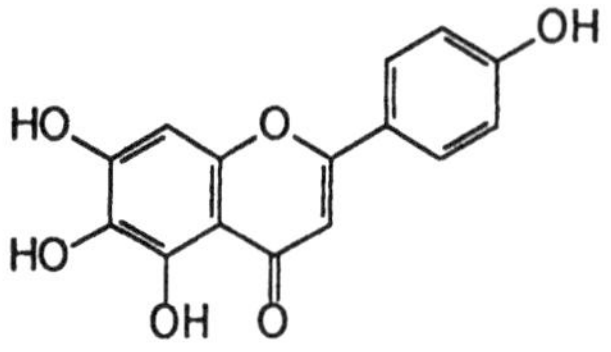

[13-1]

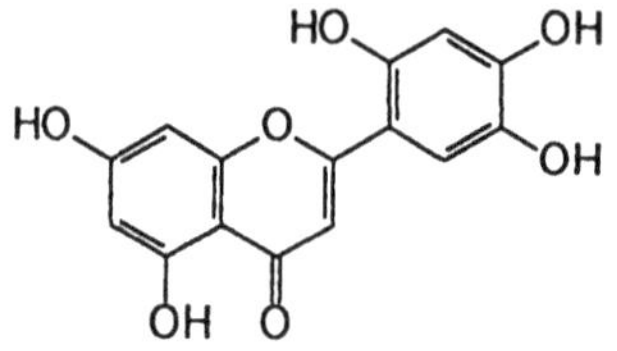

[13-2]

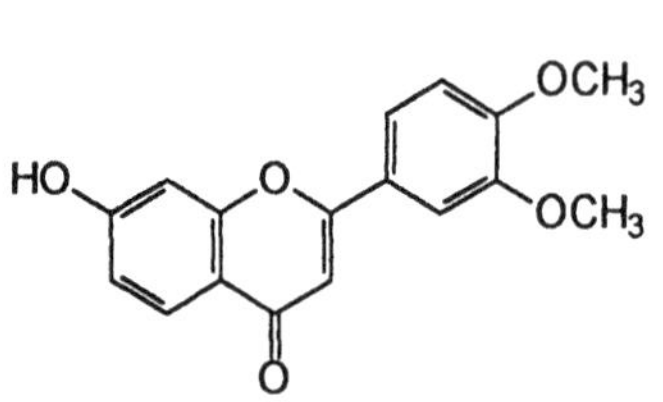

[13-3]

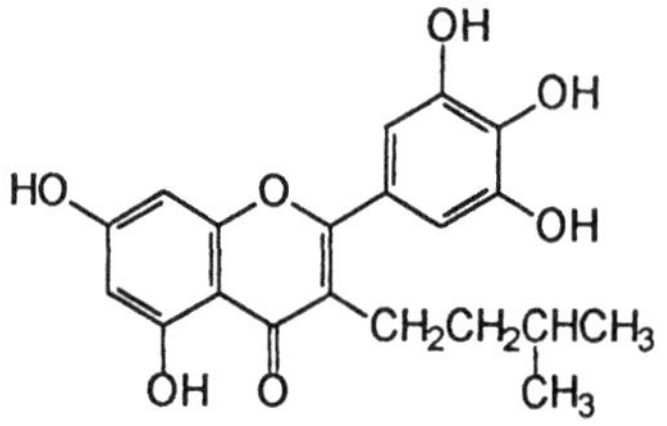

[13-4]

Fig. 13–1. Extra oxygenation in some members of Lactuceae results in the formation of scutellarein **[13–1]** and/or isoëtin **[13–2]**. Two unusual compounds identified from *Launaea asplenifolia* are the 5-deoxyflavone **[13–3]** and the *C*-alkylated tricetin derivative **[13–4]**

Dendroseris: All eleven species of this Juan Fernandez Islands endemic genus have been studied for flavonoids by Pacheco et al. (1991). Apigenin and luteolin 7-*O*-glycosides and quercetin 3-*O*-mono- and diglucosides were reported.

Embergeria: *Embergeria* consists of only one species, *E. grandifolia*, endemic to Chatham Island, New Zealand (Bremer, 1994; Given, 1996). Mansour et al. (1983b) reported the presence of luteolin 7-*O*-glucoside and 7-*O*-glucuronide from this taxon.

Hedypnois: This small genus of two species occurs on the Canary Islands, Madeira, and in the Mediterranean basin. *Hedypnois arenaria* and *H. cretica*, were both included in Harborne's (1978) survey; they contained luteolin, isoëtin, and possibly a trace of scutellarein.

Hieracium: *Hieracium* presents one of the most complex arrays of species within the family. Estimates of species number range from about 1000 macro-species (Tomb, 1977) to perhaps 5000 apomictic micro-species (Airy Shaw, 1983). Bremer (1994) suggests that the number of species may range from about 90 to more than 1000. The genus enjoys a cosmopolitan distribution in the temperate zone and in tropical mountains; it does not occur in Australasia.

Our picture of the flavonoids of this genus rests on data from about a dozen and a half species. Christy (1985) looked at the hybrid complex of eastern North America that consists of *H. gronovii, H. paniculatum, H. scabrum*, and *H. venosum*, and concluded that the flavonoid profiles, which were based on flavones (no details), are not sufficiently different to characterize these species. Guppy and Bohm (1976) reported a series of apigenin and luteolin 7-*O*- and 4'-*O*-glycosides plus chrysoeriol and quercetin 3-*O*-glucoside, in various combinations, from five taxa: *H. albiflorum, H. scouleri* var. *albertinum* (cited as *H. albertinum*), *H. scouleri* var. *griseum* (cited as *H. cynoglossum*), *H. triste* (cited as *H. gracile*), and *H. umbellatum*. Shelyuto et al. (1976a) reported apigenin, luteolin, and luteolin 7-*O*-glucoside from *H. umbellatum*. Petrovic et al. (1999b) reported the occurrence of luteolin 7-*O*-glucoside from *H. blecicii, H. coloriscapum, H. guentheri-beckii, H. gymnocephalum, H. naegelianum, H. rotundatum*, and *S. suborientum*. Luteolin 7-*O*-xylosyl(1→6)glucoside and 7-*O*-rhamnosyl(1→6)glucoside were identified from all but *H. rotundatum*. Apigenin 7-*O*-glucoside was also observed in five of those species (lacking in *H. naegelianum* and *H. rotundatum*). Several other unidentified apigenin, luteolin, and "flavonol" glycosides were reported as were caffeoylquinic acid derivatives.

Hieracium murorum subsp. *grandidens* var. *minoriceps* was shown to accumulate apigenin and luteolin in floral tissue and luteolin 7-*O*-glucoside and 7-*O*-diglucoside in leaves (Haag-Berrurier and Duquejois, 1969). *Hieracium pilosella* has been shown to accumulate luteolin and its 7-*O*-glucoside (Haag-Berrurier and Duquenois, 1962, 1963; Shelyuto et al., 1977; Harborne, 1978), apigenin and isorhamnetin (Shelyuto et al., 1977) and isoëtin (Shelyuto et al., 1977; Harborne, 1978). As part of his studies of flavonoid exudates of plants, Wollenweber (1984) reported naringenin 7-methyl ether, naringenin-7,4'-dimethyl ether, apigenin 4'-methyl ether (acacetin), apigenin itself, and apigenin-7,4'-dimethyl ether as components of the material excreted from leaves of *H. intybaceum*. Apigenin and its dimethyl ether were seen as minor components. *Hieracium amplexicaule* and *H. compositum* accumulate apigenin and luteolin 7-*O*-glucosides, luteolin 7-*O*-rhamnoside, and diosmetin (luteolin 4'-methyl ether) (Giner et al., 1992c; Mañez et al., 1994). It has recently been reported that *H. amplexicaule* accumulates apigenin 7,4'-dimethyl ether, naringenin 7-methyl ether, and naringenin 7,4'-dimethyl ether (Wollenweber et al., 1997). A recent report by Petrovic et al. (1999a) described identification of luteolin 7-*O*-a-L-rhamnosyl(1→6)-ß-D-glucoside and luteolin 7-*O*-ß-D-xylosyl(1→6)-ß-D-glucoside from *H. gymnocephalum*.

Hispidella: *Hispidella hispanica*, the sole member of this southwestern European genus, was shown to accumulate luteolin and isoëtin (Harborne, 1978).

Table 13–1. Flavones from species of *Hypochaeris*

Species	Flavone[1] Lut	Scu	Iso	References
H. achyrophorus		?	+	Harborne (1978);
			+	Fiasson et al. (1991)
H. cretensis	+			Fiasson et al. (1991)
H. glabra			+	Fiasson et al. (1991)
H. illyrica	+			Fiasson et al. (1991)
H. laevigata	+			Fiasson et al. (1991);
	+			Gluchoff-Fiasson et al. (1991)
H. maculata (2n)	+			Gluchoff-Fiasson et al. (1991)
H. maculata (4n)			+	Gluchoff-Fiasson et al. (1991)
H. radicata	+		+	Harborne (1978);
	+			Kaneta et al. (1978);
	+			Fiasson et al. (1991);
H. robertia			+	Fiasson et al. (1991)
H. uniflora	+	?	+	Harborne (1978)

[1] Lut = Luteolin; Scu = Scutellarein; Iso = Isoëtin.

Hyoseris: Five species comprise this genus which occurs in the Mediterranean area, southern Europe, North Africa, and Turkey. Rees and Harborne (1984) reported luteolin and quercetin from *H. radicata*, while Ghazy and Mahmoud (1987) reported apigenin and luteolin, both as aglycones and as the 7-*O*-glucosides, from *H. lucida*.

Hypochaeris: *Hypochaeris* consists of about 60 species distributed in South America, Eurasia, and the Mediterranean basin. Members of the genus so far examined accumulate various glycosides of luteolin, scutellarein, and isoëtin with reports of quercetin-3-*O*-glucoside from *H. radicata* (Rios et al., 1992; Giner et al., 1993a). A detailed analysis was recently published by Gluchoff-Fiasson et al. (1991) who reported the following glycosides of isoëtin from tetraploid *H. maculata*: 2'-*O*-xyloside, 2'-*O*-(4"-*O*-acetylxyloside), 7-*O*-glucoside-2'-*O*-xyloside, and 7-*O*-glucoside-2'-*O*-(4"-*O*-acetylxyloside). Chemotypes within *H. maculata* will be discussed in more detail in Chapter 24. The occurrence of flavones in the genus is summarized in Table 13–1.

Heywoodiella, with the single species *H. oligocephala*, was originally described as a monotypic genus restricted to the Canary Islands (Sventenius and Bramwell, 1971). More recent studies by Lack (1978), however, have referred it to *Hypochaeris* (*H. oligocephala*), and we follow this perspective, as did Bremer (1994). Two papers describing the flavonoid chemistry of this species (as *Heywoodiella*) have been published. In the first, Harborne (1978) reported luteolin and isoëtin. In the second paper, Harborne (1991) re-examined the spectral data for the isoëtin glycosides demonstrating that they were 7-*O*-glycosides and not 2'-*O*-glycosides as originally thought. The glycosides of isoëtin were, therefore, isoëtin 7-*O*-glucoside, 7-*O*-arabinoside and a 7-*O*-xylosylarabinosylglucoside.

Ixeris: *Ixeris* consists of about 20 species native to eastern and southeastern Asia. Nakaoki and Morita (1960b) reported luteolin 7-*O*-glucoside from *I. repens* [reported as *Lactuca repens*; taxonomy after Ohwi (1965)]. Ho et al. (1989) reported apigenin, its 7-*O*-glucoside, and luteolin from *I. laevigata*. An examination of *I. denticulata* f. *pinnatipartita* by Ma et al. (1999) revealed the presence of luteolin 7-O-glucoside and the methyl ester of luteolin 7-*O*-glucuronide.

Koelpinia: *Koelpinia* consists of five species that range from North Africa to eastern Asia. Our only flavonoid information is the finding of quercetin in *K. linearis* (Rees and Harborne, 1984).

Krigia: This North American genus consists of seven species, only one of which seems to have been examined for flavonoids. Harborne (1977b) reported the presence of apigenin and luteolin 7-*O*-glucosides and "quercetin 3-*O*-glycosides" in *K. virginica*.

Lactuca: *Lactuca* is a widely occurring genus consisting of about 75 species (Bremer, 1994) to 100 species (Vries and van Raamsdonk, 1994) mainly distributed in temperate Eurasia [17 species in Europe (Feráková, 1977)]; some species are also known from southern Africa and North America. *Lactuca sativa* is the common salad lettuce. Flavonoid data are available for 15 species including a number of cultivars of vegetable lettuce. Rees and Harborne (1984) reported flavone 7-*O*-glycosides to be present in all species tested (12) and flavonol 3-*O*-glycosides present in most (*L. acanthifolia, L. aculeata, L. altaica, L. graeca, L. perennis, L. saligna, L. sativa, L. serriola, L. tatarica, L. tenerrima*, and *L. viscosa*). Bilyk and Sapers (1985) reported the presence of kaempferol and quercetin in 13 varieties of lettuce. Rutin was reported by Lukovnikova (1965), while luteolin 7-*O*-glucuronide and quercetin 3-*O*-glucoside, 3-*O*-(malonyl)glucoside, and 3-*O*-glucuronide were recorded by Woeldecke and Herrmann (1974). *Lactuca indica* accumulates apigenin and luteolin 7-*O*-glucoside and quercetin 3-*O*-galactoside (Kaneta et al., 1978), while luteolin 7-methyl ether and 3'-methyl ether, plus apigenin and luteolin, have been reported from *L. saligna* (El-Din et al., 1987). *Lactuca viminea* accumulates apigenin, luteolin 7-*O*-

glucoside, luteolin 7-*O*-glucuronide methyl ester and quercetin and its 3-*O*-glucoside (Terencio et al., 1992). Apigenin and luteolin 7-*O*-glycosides and quercetin 3-*O*-glucoside have been reported from *L. tenerrima, L. viminea* and *L. virosa* in a recent paper by Mañez et al. (1994).

Launaea: The genus *Launaea* consists of 50 species that range from the Canary Islands to eastern Asia with representation in tropical and southern Africa. Flavonoids from about a dozen species have been reported. Flavones occur in all species tested; there are no reports of flavonols. Delphinidin and the *C*-glycosylflavone vitexin were reported from *L. aspleniifolia* (Gupta et al., 1985a,b). This species also afforded two unusual compounds, 7-hydroxy-3',4'-dimethoxyflavone **[13–3]** (Gupta et al., 1985b), and the *C*-alkylated 5,7,3',4',5'-trihydroxy-3-(3-methyl)-butyl flavone **[13–4]** (and its 5-*O*-neohesperidoside) (Gupta and Ahmed, 1985). Common apigenin and luteolin *O*-glycosides have been reported from several species: *L. aspleniifolia* (Gupta et al., 1985b), *L. arborescens, L. laniflora* (cited as *L. acanthoclada*), and *L. resedifolia* (Saleh et al., 1988; Giner et al., 1992b; Mañez et al., 1994), *L. nudicaulis* (Sarg et al., 1984), *L. spinosa* (Sarg et al., 1987), and *L. tenuiloba* (Abdel Salam et al., 1986). The most comprehensive listing of flavone glycosides for *Launaea* was that of Mansour et al. (1983a) who studied *L. capitata, L. cassiniana, L. nudicaulis, L. resedifolia, L. spinosa*, and *L. tenuiloba*. In addition to common apigenin and luteolin 7-*O*-glycosides, these workers reported luteolin 7,3'-di-*O*-glucoside, luteolin 7,4'-di-*O*-glucoside, and luteolin 7-*O*-gentiobioside-4'-*O*-glucoside from all of the listed species.

Leontodon: *Leontodon* consists of 50 species distributed in the Mediterranean basin, parts of Eurasia, and as far east as Iran. We have flavonoid data for five species: *L. crispus, L. hispidus, L. macrocanus, L. pyrenaicus* and *L. taraxacoides*. Harborne (1978) reported isoëtin for all five; apigenin, luteolin, and scutellarein were scattered. In a recent paper Rios et al. (1992) and Giner et al. (1993a) reported apigenin and luteolin and their glycosides as major components of *L. taraxacoides* along with lesser amounts of quercetin-3-*O*-glucoside.

Microseris: Sixteen species comprise this genus (Wallace and Jansen, 1990), which is principally western North American with one species in Chile, and one in Australia and New Zealand (Bremer, 1994). Published information reveals quercetin 3-*O*-glycosides, but neither apigenin nor luteolin, in *M. heterocarpa* and *M. linearifolia* (Harborne, 1977b).

Mycelis: *Mycelis* is a monotypic Eurasian genus with occasional populations in North Africa. Examination of *M. muralis* (cited as *Lactuca muralis*;

taxonomy after Sell, 1976) by Mañez et al. (1994) revealed the presence of apigenin and luteolin 7-*O*-glycosides and quercetin 3-*O*-glucoside.

Picris: *Picris* is a moderately large genus of about 40 species distributed in temperate Eurasia and in the Mediterranean area with representatives in Ethiopia. We have information for *P. hieracioides*, which contains luteolin and isoëtin, *P. hispanica*, which afforded only isoëtin (Harborne, 1978), and *P. echioides*, which contains luteolin 7-*O*-glucoside and quercetin 3-*O*-glucoside (Rios et al., 1992; Giner et al., 1993a).

Recently, Milovanovic and Djermanovic (1994) reported apigenin 7-*O*-glucoside, 6-hydroxykaempferol, and 4,6,7,4'-tetrahydroxyaurone, which they named "helmon," from a plant identified as *Helminthia echioides*. Neither Mabberley (1987) nor Bremer (1994) list this genus, but Bremer has *Helminthotheca* listed in Lactuceae where it is described as consisting of four species including the widely distributed weedy *H. echioides*. Prof. Milovanovic informed us (*in litt.*) that *Helminthia* was being used in the sense of Tutin et al. (1976), where it is indicated that the taxon was originally included in *Picris*. Since a record of flavonoids from *P. echioides* exists in the literature (this paragraph), we thought it expedient to include the more recent report here. This is an important report since it represents the first finding of an aurone in a member of Lactuceae (the chalcone coreopsin has been reported from *Pyrrhopappus*). While the report of 6-hydroxykaempferol also represents a first for the tribe, both flavonol and 6-hydroxyflavone (scutellarein) derivatives are known from several genera. This appears to be the first 6-hydroxyflavonol reported from the tribe, however.

Prenanthes: *Prenanthes* is a genus of 30 species with a very wide distribution, North America, the Canary Islands, tropical Africa, and tropical and temperate Asia. Our flavonoid information is based on the work of Fusiak and Schilling (1984) who found apigenin and luteolin glycosides in three species: *P. altissima, P. roanensis* and *P. serpentaria*. Luteolin 7,3'-di-*O*-glucoside and 7,4'-di-*O*-glucoside were found in some representatives of each species. These diglucosides were also seen in species of *Launaea* mentioned above.

Pyrrhopappus: *Pyrrhopappus* consists of three North American species (Turner and Kim, 1990; Bremer, 1994). Working with *P. carolinianus, P. grandiflorus, P. pauciflorus* (cited as *P. rothrockii*; taxonomy after Turner and Kim, 1990), and two varieties of *P. multicaulis*, Harborne (1977b) identified apigenin and luteolin 7-*O*-glucosides and coreopsin in the ligules.

Reichardia: This genus of eight (Gallego et al., 1980; Bremer, 1994) species is found in the Mediterranean basin. Harborne (1978) reported apigenin, luteolin, isoëtin, and possibly scutellarein in *R. picroides*. A series of apigenin and luteolin mono- and diglycosides has been reported from *R. picroides* and *R. tingitana* (Recio et al., 1992; Mañez et al., 1994).

Rhagadiolus: According to Meikle (1979) and Bremer (1994), *Rhagadiolus* consists of two species distributed in the Mediterranean region. Rees and Harborne (1984) reported quercetin from *R. stellatus*.

Rothmaleria: *Rothmaleria* is monotypic with *R. granatensis* native to Spain. Quercetin is the only flavonoid reported (Rees and Harborne, 1984).

Scariola: *Scariola* is listed as synonymous with *Lactuca* according to Airy Shaw (1983) but is not listed by Tomb (1977). Bremer (1994) says that the genus, which occurs in southern Europe, North Africa, southwestern and central Asia, consists of about 10 species. Rees and Harborne (1984) included *S. tetrantha* (Burtt and Davis) Sojak in their survey of Lactuceae. They reported the presence of luteolin and quercetin.

Scolymus: This Mediterranean genus consists of three species with flavonoid information available for two of them. Rees and Harborne (1984) reported kaempferol from *S. maculatus* and kaempferol and quercetin from *S. hispanicus*. An earlier paper by Romussi and Ciarallo (1978) reported the flavonoid composition of *S. hispanicus* to be kaempferol 3-*O*-galactoside and a kaempferol 3-*O*-rhamnosylgalactoside, quercetin 5-*O*-glucoside, and an apigenin 6,8-di-*C*-glucoside. More recently, Sanz and coworkers (1993) identified kaempferol 3-*O*-glucuronide and its 6"-methyl ester, quercetin 3-*O*-glucoside, quercetin 5-*O*-glucoside and quercetin 3-*O*-(2"-*O*-caffeoyl)-glucuronide from the same species. The presence of kaempferol and quercetin 3-*O*-glucuronides in this species was confirmed by Rubio-Garcia et al. (1994). Two recent papers (Rubio and Diaz, 1995; Rubio et al., 1995) described kaempferol, quercetin and isorhamnetin 3-*O*-glycosides from *S. hispanicus*.

Scorzonera: *Scorzonera* is a genus of 150 (Tomb, 1977) to about 175 (Bremer, 1994) species that occurs in the Mediterranean basin and from central Europe to central Asia. Rees and Harborne (1984) included nine species in their survey. Quercetin was recorded from each, while kaempferol, apigenin, and luteolin enjoyed only scattered occurrences. *Scorzonera columnae*, with apigenin, luteolin, and quercetin 3-*O*-(6"-*O*-caffeyl)-galactoside, was added to this list recently (Menichini et al., 1994).

Sonchus: *Sonchus*, with 60 species, occurs in Eurasia, the Mediterranean basin, tropical Africa, and on the Atlantic Islands. Unlike most genera of this size in the family, *Sonchus* has been very well sampled for flavonoids. Thus, Bramwell and Dakshini (1971) reported the presence of luteolin and/or its 7-*O*-glucoside in 20 of 21 species from the Canary Islands. Mansour et al. (1983b) examined thirty-three species from which they identified a variety of apigenin and luteolin 7-*O*-glycosides, some combination of which was present in each species. Other reports in the literature reporting flavone glycosides, plus the occasional flavonol, include the following: *S. asper* (Kaneta et al., 1978), *S. oleraceus* (Kaneta et al., 1978; Bondarenko et al., 1983), *S. pustulatus* (Rees and Harborne, 1984), and *S. maritimus*, *S. asper*, *S. oleraceous*, and *S. tenerrimus* (Giner et al., 1993b; Mañez et al., 1994). A variety of flavones, including isoëtin, and flavonols have been reported from *S. arvensis* by Bondarenko et al. (1973, 1974, 1976, 1978) and Qu et al. (1993, 1995).

Stephanomeria: *Stephanomeria*, native to western North America, consists of 24 species (Tomb, 1977; Stebbins, 1993). A study of the flavonoids of five annual species of *Stephanomeria* has been published (Bohm and Gottlieb, 1989). The taxa examined were *S. malheurensis, S. paniculata, S. diegensis, S. virgata*, and four subspecies of *S. exigua*. Apigenin, luteolin, kaempferol, quercetin, and eriodictyol glycosides were identified. Some variation in the occurrence of rutin and eriodictyol 7-*O*-glucoside was observed. Flavonoids of *S. exigua* subsp. *coronaria* and *S. malheurensis*, as they apply to the progenitor-derivative status of this pair, are discussed in Chapter 22 (within the section on the *Coreopsis neucensis-C. neucensoides* system).

Taeckholmia: *Taeckholmia*, which consists of seven species (Boulos, 1976), is endemic to the Canary Islands. Mansour et al. (1983b) included five species of *Taeckholmia* in their extensive study of members of the Lactuceae: *T. canariensis, T. capillaris, T. heterophylla, T. microcarpa*, and *T. regis-jubae*. These workers observed apigenin and luteolin 7-*O*-glycosides in various combinations in these taxa.

Taraxacum: According to Tomb (1977), this essentially northern temperate genus consists of 60 species. Bremer (1994) agrees with this number but points out that the number may be as high as 500 depending on species concept. This last number would recognize the hundreds of apomictic lines, or microspecies, that seem to characterize this taxon. Despite the apparent size of the genus and the conspicuousness of its most famous member, the dandelion, it is surprising that so few reports of flavonoids of *T. officinale*

Wigg. have appeared. The earliest report, apparently, was that of Hörhammer and Wagner (1962) who described apigenin and luteolin 7-*O*-glucosides from the species. Wolbis et al. (1993) reported luteolin 7-*O*- and 4'-*O*-glucosides, luteolin 7-*O*-rutinoside, quercetin 7-*O*-glucoside, and isorhamnetin 3-*O*-glucoside and 3,7-di-*O*-glucoside along with the aglycones, luteolin and quercetin. Kaneta et al. (1978) reported apigenin and luteolin 7-*O*-glucosides from *T. platycarpum*. Luteolin 7-*O*-glucoside and 7-*O*-rutinoside have been reported from *T. hallaisanensis* by Whang et al. (1994). The most recent report (Williams et al., 1996) describes the isolation of luteolin 7-*O*-glucoside, two luteolin 7-*O*-diglycosides, and chrysoeriol from various parts of the plant.

Tragopogon: According to Tomb (1977), *Tragopogon* consists of 50 species, whereas Bremer (1994) says there are about 110. Most of these are native to temperate Eurasia with one known from South Africa. Some of the ones introduced to North America have attracted a significant level of attention. Rees and Harborne (1984) reported quercetin in all five of the *Tragopogon* species examined: *T. crocifolius, T. hybridus, T. longirostris, T. porrifolius* and *T. pratensis*; apigenin and luteolin were also seen but they did not occur in all species. Apigenin, luteolin, and quercetin were reported from *T. orientalis* along with vitexin, orientin, and isoörientin (Smolarz and Krzaczek, 1988). The earliest study of flavonoids in the genus was the paper chromatographic study of *T. dubius, T. pratensis*, and *T. porrifolius* along with their tetraploid hybrids *T. mirus* and *T. miscellus* in eastern Washington (Brehm and Ownbey, 1965a). A detailed study of this system was done by Kroschewsky et al. (1969) who reported luteolin and quercetin glucosides and a suite of *C*-glucosylflavones.

A recent observation by Wollenweber et al. (1997) is of interest in that it involves an unusual exudate component. Normally, we would expect non-polar flavonoid derivatives to comprise the leaf surface flavonoid fraction, i.e., flavonoid methyl ethers. The paper noted described the identification of luteolin 7-*O*-glucoside and quercetin 3-*O*-glucoside as components of an acetone rinse of leaves of *T. pratensis*. Presumably, no non-polar components were found.

Urospermum: *Urospermum* consists of two species both of which are native to the Mediterranean basin. Quercetin and its 3-*O*-galactoside were reported from *U. picroides* (Metwally et al., 1973; Rees and Harborne, 1984) while luteolin, kaempferol, and quercetin glycosides were added to the list by Giner et al. (1992a, 1993a). *Urospermum dalechampii* accumulates luteolin 7-*O*- and 4'-*O*-glucosides and the 3-*O*-arabinoside, galactoside, and glucoside of quercetin (Rios et al., 1992). Marco et al. (1994)

reported naringenin, aromadendrin, and taxifolin 3-methyl ether from *U. dalechampii*.

Youngia: *Youngia* is a large genus, some 40 species, native to central, eastern, and tropical Asia. The only flavonoid information we have for this genus comes from the work of Kaneta et al. (1978) who reported luteolin 7-*O*-glucoside from *Y. denticulata*.

Chapter 14

Flavonoids of Mutisieae and Barnadesioideae

According to Cabrera (1977) Mutisieae, a largely South American tribe, comprise 89 genera with ca. 974 species. Bremer (1994) sees the tribe as comprising 76 genera with ca. 970 species, but he excludes the *Barnadesia*-group of genera. Mutisieae as presented by Cabrera (1977) were apportioned into four subtribes: Barnadesiinae, Gochnatiinae, Nassauviinae, and Mutisiinae. A recent major event disturbed this state of affairs. The finding of a 22kb inversion in the cpDNA genome of all tested members of Asteraceae, other than members of Barnadesiinae (Jansen and Palmer, 1987b, 1988), resulted in establishment of a new subfamily, Barnadesioideae, with Barnadesieae as its sole tribe (Bremer and Jansen, 1992). Although we accept this new treatment as reasonable (see Chapter 2), for convenience we present the flavonoid chemistry of Barnadesioideae in this chapter. Bremer (1994) now recognizes only two subtribes in Mutisieae: Mutisiinae (including Gochnatiinae) and Nassauviinae. We have information for members of 22 genera of Mutisieae and seven genera of Barnadesioideae. Numbers of species and distributional ranges come from Bremer (1994) unless stated otherwise.

Adenocaulon: This genus consists of five (Bittmann, 1990a, b; Bremer, 1994) species variously distributed in western North America, Chile, and eastern Asia (Mabberley, 1987; Bittmann, 1990a, b). Its distribution is quite broad for a genus of such small size, and its morphology makes tribal (and subtribal) placement uncertain (Hansen, 1991). Even the possession of cortical vascular bundles (Karrfalt, 1975) is unusual (although not unique) within the family. Recent macromolecular sequence investigations with the *ndh*F gene (Kim and Jansen, 1995b), however, reveal *Adenocaulon* to belong in Mutisieae near *Mutisia* (with some uncertainty; see Fig. 2–7). The only flavonoid chemistry of which we are aware is our finding (Bohm, unpubl.) of kaempferol and quercetin 3-*O*-mono- and 3-*O*-diglycosides in *A. bicolor*.

Aphyllocladus: *Aphyllocladus* consists of five species distributed in Bolivia, Chile, and Argentina. The only flavonoid data available appear to be the report of apigenin in *A. denticulatus* by Maldonado et al. (1988b).

Dicoma: There are about 65 species in *Dicoma* most of which are distributed in southern and tropical Africa with representatives in Madagascar. *Dicoma anomala* subsp. *anomala* was shown to accumulate scutellarein (5,6,7,4'-tetrahydroxyflavone) and cirsimaritin (5,4'-dihydroxy-6,7-dimethoxyflavone) along with sesquiterpene lactones (Zdero and Bohlmann, 1990a).

Dolichlasium: According to Cabrera (1977), Mabberley (1987), and Bremer (1994), *Dolichlasium* consists of a single species from the central Andes of Argentina. 5,7-Dihydroxyflavanone (pinocembrin) and naringenin have been reported from *D. lagascae* (Zdero et al., 1986b). Guerreiro (1989) reported 5,7,4'-trihydroxy-6,3'-dimethoxyflavone from material identified as *D. glandulifera*, but this is a synonym of *D. lagascae* according to Cabrera (1971).

Gerbera: *Gerbera* is a genus of about 30 species distributed in tropical Asia, Madagascar, tropical and southern Africa, and tropical South America (one species). Gu et al. (1987) reported luteolin 7-*O*-glucoside from *G. anandria*. Asen (1984) described a wide array of flavonoids from *G. jamesonii*, the so-called African or Transvaal daisy, including apigenin and luteolin 7-*O*-glucosides, their corresponding malonyl derivatives and 4'-*O*-glucosides, kaempferol and quercetin 3-*O*-glucosides and kaempferol and quercetin 4'-*O*-glucosides plus the corresponding malonyl derivatives of each. Pelargonidin and cyanidin 3-*O*-glucosides and their respective malonyl derivatives were identified as well.

Gochnatia: *Gochnatia*, with 68 species, is one of the largest genera of the tribe. Species occur in tropical Andes, Brazil, Mexico, southern United States, the West Indies, and southeastern Asia. We have information on flavonoids from six species. The first report involved the unusual compound (Fig. 14–1) 7-*O*-prenylaromadendrin **[14–1]**, which Bohlmann et al. (1983a) identified from *G. paniculata*. This was followed by a report of sakuranetin (naringenin 7-methyl ether) from *G. vernonioides* (Bohlmann et al., 1984f). Faini et al. (1984) reported kaempferol 3,7-dimethyl ether and the 3'-mono- and the 3,7- and 3,3'-dimethyl ethers of quercetin from *G. foliolosa* var. *fascicularis*. Garcia and Guerreiro (1988) reported 6-methoxyapigenin (hispidulin) and luteolin 7-methyl ether from *G. argentina* and eriodictyol, acacetin, quercetin 3,3'- and 3,4'-dimethyl ethers from *G. glutinosa*. Luteolin has been reported from *G. hypoleuca* by Maldonado et al. (1988a).

Gypothamnium: *Gypothamnium* is a monotypic genus endemic to Chile. Quercetin 3-*O*-mono- and diglycosides have been observed in extracts of *G. pinifolium* (Bohm and Stuessy, 1995).

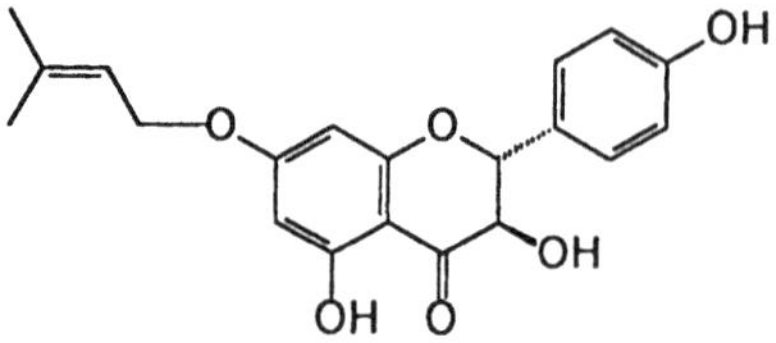

[14-1]

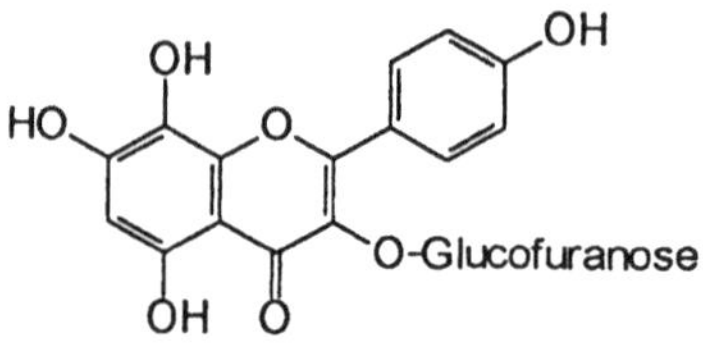

[14-2]

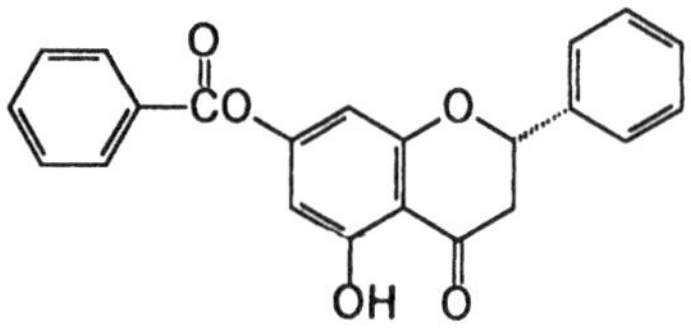

[14-3]

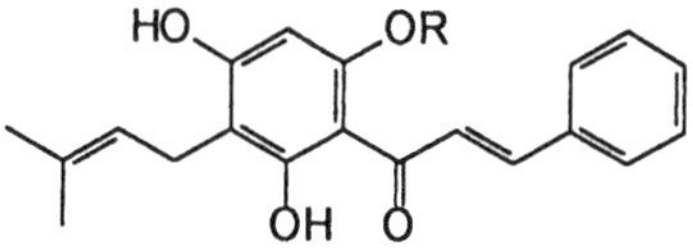

[14-4] R = H

[14-5] R = CH_3

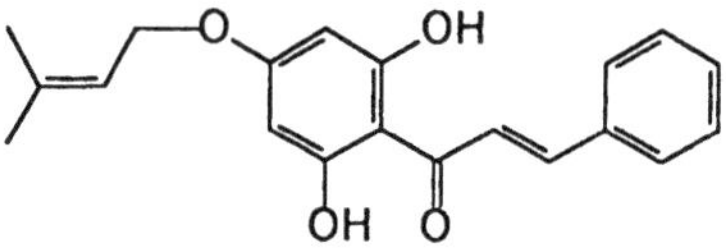

[14-6]

Fig. 14–1. New and unusual compounds reported from Mutisieae include the prenyl ethers **[14–1, 14–6]**, the *C*-prenyl chalcones **[14–4, 14–5]**, pinocembrin 7-benzoate **[14–3]**, and the 3-*O*-glucofuranoside of 8-hydroxykaempferol **[14–2]**

Jungia: The 26 species of *Jungia* occur in tropical America from Mexico to northern Argentina, Uruguay, and Brazil (Harling, 1995). Flavonoid information comes from two reports. Ybarra et al. (1992) described 5,7-dihydroxy-6,4'-dimethoxyflavone (pectolinarigenin) from *J. polita*. The second report involves the work of D'Agostino et al. (1995) who found the new flavonol derivative 8-hydroxykaempferol 3-*O*-ß-glucofuranoside **[14–2]** in *J. paniculata* along with the known kaempferol 3-*O*-(6"-galloylglucoside).

Leucheria: *Leucheria*, with 46 species (Crisci, 1976), is a moderately large genus within Mutisieae. It occurs in the Andes from southern Peru to Patagonia. The only flavonoid information appears to be the recent report of 5,7-dihydroxy-6,4'-dimethoxyflavone and 5,4'-dihydroxy-3,6,7,3'-tetramethoxyflavone from *L. runcinata* (Bittner et al., 1994).

Lophopappus: *Lophopappus* consists of six species that occur in the Andes in southern Peru, northern Chile, and northwestern Argentina. Hoeneisen et al. (1993) reported 5,7-dihydroxyflavanone (pinocembrin) and its 7-*O*-benzoate **[14–3]** from *L. tarapacanus*.

Mutisia: A large genus of 60 species, *Mutisia* occurs widely in the Andes of South America with a disjunct distribution in southeastern Brazil and adjacent Uruguay and northeastern Argentina (Cabrera, 1965). Daily et al. (1988) reported quercetin 3-*O*-glucuronide from *M. acuminata* var. *hirsuta*. Quercetin, isorhamnetin, and quercetin-7,3'-dimethyl ether (rhamnazin) were reported from *M. orbignyana* (Zdero et al., 1988b).

Nassauvia: *Nassauvia*, including *Calopappus*, consists of 39 species with a range that includes Argentina, Bolivia, and Chile. A study of the flavonoids of most of the species of *Nassauvia* (excluding *Calopappus*) showed the presence of flavone 7-*O*-glycosides, *C*-glycosylflavones, and flavonol 3-*O*-glycosides (Crawford, unpubl.). The flavones and *C*-glycosylflavones were based upon apigenin and luteolin; the flavonol fraction consisted of quercetin and isorhamnetin glycosides. Some unidentified compounds were also observed.

Perezia: Only one of the 32 known species of this South American genus has been studied for flavonoids. De Israilev and Gonzalez (1994) reported finding only kaempferol 3-*O*-glucoside and 3-*O*-rhamnoside in *P. scorzonera*, which is used as a medicinal plant by peoples in the high Andes.

Plazia: *Plazia* consists of three species and occurs from Peru to Argentina and Chile. *Plazia daphnoides* was shown by Zdero et al. (1988a) to accumulate apigenin 4'-methyl ether (acacetin), and 7-methyl ether (genkwanin), naringenin, and its two monomethyl ethers sakuranetin and isosakuranetin.

Pleiotaxis: *Pleiotaxis* is a tropical African genus of about 26 species. The only one for which we have flavonoid information is *P.* cf. *rugosa* from which Bohlmann and Zdero (1982) obtained three chalcones, 2',4',6'-trihydroxy-3'-*C*-prenylchalcone **[14–4]**, 2',4'-dihydroxy-6'-methoxy-3'-*C*-prenylchalcone **[14–5]**, and 2',4',6'-trihydroxychalcone 4'-prenyl ether **[14–6]**.

Pleocarphus: *Pleocarphus revolutus*, the sole member of this genus, was shown by Silva et al. (1977) to accumulate 5,6,4'-trihydroxy-7-methoxyflavone.

Proustia: *Proustia* consists of three species known from the Andes of Bolivia, Chile, and Argentina. Our only flavonoid information is the report of apigenin-7,4'-dimethyl ether, naringenin-7-methyl ether (sakuranetin), and naringenin 4'-methyl ether (isosakuranetin) from *P. cuneifolia* var. *cuneifolia* (Bittner et al., 1989).

Tarchonanthus: *Tarchonanthus* consists of two species from southern and tropical Africa. Mabberley (1987) lists *Tarchonanthus* as questionably in Inuleae, a position taken earlier by Bentham (1873), but it was excluded from that tribe by Merxmüller et al. (1977). The genus has also been suggested to belong to Anthemideae (Skvarla et al., 1977). Recent cpDNA restriction site data reveal an isolated position for *Tarchonanthus*, which has been joined with *Brachylaena* (also problematic in tribal disposition) into a possible new tribe, Tarchonantheae (Keeley and Jansen, 1991). We agree with Bremer (1996) that, although a tribal placement is clearly in doubt for these genera, more study is needed before a new tribe should be added to the family housing just these two genera. Furthermore, macromolecular evidence so far shows a close tie to Mutiseae (Keeley and Jansen, 1991; their Fig. 1). The only flavonoid information available appears to be the report of 5,7-dihydroxyflavanone from *T. trilobus* by Bohlmann and Suwita (1979a).

Triptilion: *Triptilion* consists of 12 species distributed in Argentina and Chile. Flavonoid data come from an examination of *T. spinosum* (Crawford, unpub.) which showed the presence of a flavone *O*-glycoside, two *C*-glycosylflavones, and two flavonols.

Trixis: This large genus, with about 50 species, occurs from southern United States, through Mexico to northern Argentina, northern Chile, and Uruguay. We have information for only *T. vauthieri* from which Bohlmann et al. (1981g) obtained kaempferol, kaempferol 7-methyl ether, and the 7-methyl ethers of naringenin, eriodictyol, aromadendrin, taxifolin, and apigenin.

Wunderlichia: Five species comprise this Brazilian genus, only one of which seems to have been examined for secondary compounds. Bohlmann et al. (1984c) reported 5-hydroxy-7,4'-dimethoxyflavone (acacetin 7-methyl ether) as a component of *W. mirabilis*.

Barnadesioideae

Arnaldoa: *Arnaldoa* consists of two species from the Peruvian Andes (Stuessy and Sagástegui, 1993). *Arnaldoa weberbaueri* (including *A. coccinosantha*; synonymy from Stuessy and Sagástegui, 1993) was shown to accumulate kaempferol and quercetin 3-*O*-mono- and 3-*O*-diglycosides (Bohm and Stuessy, 1995).

Barnadesia: *Barnadesia* consists of 18 species (Urtubey, 1999), all but one of which occur in the tropical Andes from Colombia to northwestern Argentina. One species occurs in eastern Brazil. The flavonoid profiles of *B. aculeata, B. arborea, B. dombeyana, B. hutchinsonianum, B. jelskii, B. kingii, B. lehmannii, B. parviflora,* and *B. spinosa* (some from Ecuador, some from Peru) consist of kaempferol and quercetin 3-*O*-mono- and diglycosides with the occasional appearance of eriodictyol 7-*O*-glucoside. *Barnadesia parviflora* also contains isorhamnetin 3-*O*-glucoside (Bohm and Stuessy, 1995). Recently, Mendiondo et al. (1997) reported kaempferol 3-*O*-glucoside and 3-*O*-rutinoside and quercetin 3-*O*-glucoside from *B. odorata.*

Chuquiraga: *Chuquiraga*, a genus of 21 (Bremer, 1994; Granda, 1997) species, occurs from Colombia to southern Argentina. *Chuquiraga jussieui* (Ecuador), *C. oblongifolia* and *C. weberbaueri* (both from Peru) accumulate 3-*O*-monoglycosides of kaempferol and quercetin (Bohm and Stuessy, 1995). *Chuquiraga calchaquina* and *C. longiflora* have been shown to accumulate kaempferol 3-*O*-glucoside and 3-*O*-rutinoside, to which profile the latter species adds quercetin 3-O-glucoside (Mendiondo et al., 1997).

Dasyphyllum: *Dasyphyllum*, the largest genus in the barnadesioid group, consists of 40 species. Members of the genus occur widely in South America. *Dasyphyllum argenteum* (Ecuador) and two species from Peru (*D. brevispinum* and *D. weberbaueri*) accumulate kaempferol and quercetin 3-*O*-mono- and diglycosides (Bohm and Stuessy, 1995). *Dasyphyllum argenteum* is a rare taxon, having previously gone uncollected since 1961 (Harling, 1991). It is also of interest because, despite its restricted geographical distribution, it displays a remarkable range of leaf variation (Stuessy, per. observ.). An isozyme study of this taxon would be most interesting. A recent report lists *D. spinescens* and *D. synacanthum* as possessing kaempferol and quercetin 3-*O*-glucosides and 3-*O*-rutinosides (Mendiondo et al., 1997).

Doniophyton: Two species of this genus are known from Chile and Argentina (Katinas and Stuessy, 1997). The flavonoid profile of *D. weddellii* (reported as *D. anomalum* (for new combination see Katinas and Stuessy, 1997) consists of kaempferol and quercetin 3-*O*-mono- and diglycosides (Bohm and Stuessy, 1995).

Fulcaldea: *Fulcaldea laurifolia*, the sole member of this genus, is native to northern Peru and southern Ecuador (Stahl et al., 1999). The flavonoid profile of this species consists of kaempferol and quercetin 3-*O*-mono- and 3-*O*-diglycosides (Bohm and Stuessy, 1995).

Schlechtendalia: This is a monotypic genus with its sole species *S. luzulaefolia*, occurring in northeastern Argentina, southeastern Brazil, and Uruguay. Flavonoids present in this species are kaempferol and quercetin 3-*O*-mono- and diglycosides (Bohm and Stuessy, 1995). This taxon has been suggested as the most basal genus of Barnadesioideae (Stuessy et al., 1996).

Chapter 15

Flavonoids of Senecioneae

Nordenstam (1977) listed 100 genera in his systematic review of Senecioneae, but Bremer's (1994) recent synthesis suggests 120. Even more recently, Nordenstam (1996) lists 123. The reason for the generic increase in the tribe stems mainly from segregates being pulled out of *Senecio* in a fashion similar to what has happened to *Vernonia*. The most active workers in this regard have been C. Jeffrey, B. Nordenstam, and H. Robinson. Once more, we follow these changes in this book so that the flavonoid data are available in the context of the most recent overview by Bremer (1994). The flavonoid literature includes references to several genera originally considered members of Senecioneae but now removed to Heliantheae, viz. *Amauria*, *Arnica*, *Eriophyllum*, *Flaveria*, *Hulsea*, *Pericome*, *Perityle*, and *Psathyrotes*. Of the genera remaining in Senecioneae we have some flavonoid information on 21.

Abrotanella: *Abrotanella* is a circum-Pacific genus consisting of 19 species (Swenson, 1995). Species occur in southeastern Australia, New Zealand, New Guinea, Argentina, including the Falkland Islands, and Chile, including the Juan Fernandez Islands. A recent study of seven species, representing the entire range of the genus, has shown the presence of quercetin 3-*O*-mono- and 3-*O*-diglycosides as the major components. Small amounts of kaempferol 3-*O*- glycosides were seen in a few samples and *O*-methylated flavonols were detected in one or two species (Swenson et al., 1999).

Acrisione: *Acrisione* is a small genus of two species confined to Chile. Jeffrey et al. (1977) and Nordenstam (1978) commented on the anomalous position of both within *Senecio*, which led eventually to formal segregation by Nordenstam (1985; see also for full synonymy for species). *Acrisione cymosa* (cited as *Senecio cymosa*) and *A. denticulata* (cited as *S. yegua*) were examined by Reyes et al. (1977) who reported "flavonol 3- and 3,7-diglycosides" from both.

Blennosperma: This genus consists of three species, one in Chile and two native to California. One of the Californian species, *B. nanum*, consists of two varieties. The flavonoid profiles of all three species were shown by Ornduff et al. (1973b) to be dominated by quercetin 3-*O*-glycosides. *Blennosperma nanum* var. *robustum* also possessed luteolin 7-*O*-glucoside.

Brachyglottis: *Brachyglottis* is a genus of some 30 or 31 species native to New Zealand, including the Chatham Islands, and Tasmania. Leaf exudate flavonoids of *B. cassinioides* identified were kaempferol 3-methyl ether and 8-hydroxykaempferol 3,8-dimethyl ether; vacuolar flavonoids were the 3-*O*-glucosides and 3-*O*-rutinosides of kaempferol and quercetin (Reid and Bohm, 1993). *Brachyglottis monroi* afforded kaempferol 3-methyl ether, naringenin, and 2',4',6',4-tetrahydroxychalcone from a leaf surface extract and quercetin 3-*O*-glucoside, quercetin 3-*O*-rutinoside, and isorhamnetin 3-*O*-glucoside were identified as vacuolar components (Reid and Bohm, unpubl.). *Brachyglottis greyii* and *B. repanda* gave no exudate flavonoids but their vacuolar components were identical to those of *B. monroi* (Reid and Bohm, unpubl.).

Cacalia: This genus consists of four species that occur in central and southern Europe. The only information for flavonoids of the genus appears to be the work of Schwendimann et al. (1974) who reported luteolin 7-*O*-glucoside and quercetin 3-*O*-glucoside in *C. alliariae* (reported as *Adenostyles alliariae*; synonymy from Jeffrey, 1992; generic synonymy also reiterated in Bremer, 1994). Murav'eva et al. (1974) reported quercetin in *Cacalia* sp. [cited as *Senecio platyphylloides*; this taxon is regarded by Jeffrey (1992) as part of the *Cacalia macrophylla* Bieb. complex that (p. 104) "...should be referred to an allied as yet undescribed genus"].

Crocidium: *Crocidium* consists of a single species, *C. multicaule*, that occurs from California north to southwestern British Columbia. The flavonoid profile of the species is dominated by quercetin 3-*O*-glycosides with luteolin 7-*O*-glucoside present in some California populations. Cyanidin and delphinidin 3-*O*-glycosides were detected in a specimen from each Oregon and British Columbia (Ornduff et al., 1973b).

Doronicum: *Doronicum* is a genus of about 40 species that occur in temperate Eurasia and North Africa (Bremer, 1994). The earliest flavonoid report we have on the genus is that of Alieva et al. (1979) who reported kaempferol and quercetin 3-*O*-glucosides, quercetin 3-*O*-rutinoside, and apigenin from *D. macrophyllum* and quercetin 3-*O*-glucoside from *D. oblongifolium*. Quercetin 3-*O*-(6"-*O*-acetylglucoside) was later shown to accumulate in *D.*

grandiflorum (Reynaud et al., 1985; Reynaud and Raynaud, 1986). The most interesting observation, however, was the report of derivatives of 8-hydroxyapigenin and 8-hydroxyluteolin in *D. grandiflorum* (Reynaud et al. 1983; Reynaud and Raynaud, 1984, 1986). Those workers identified 8-methoxyapigenin, 8-methoxyluteolin, and 8-hydroxyluteolin 8,3'-dimethyl ether.

Emilia: *Emilia* consists of about 90 species in tropical Africa and Asia. The only flavonoid report is that of Nair et al. (1982) who reported quercetin 3-*O*-galactoside from *E. sonchifolia*.

Euryops: *Euryops* consists of 97 species that occur in southern and eastern Africa north to Ethiopia and Arabia. Despite the size of the genus only one report of flavonoids seems to have appeared. Hafez et al. (1989) reported an acylated quercetin 3-*O*-rutinoside from *E. arabicus*.

Gynura: About 40 species comprise this genus, which occurs in tropical Africa and Asia, Madagascar, and northern Australia. A cyanidin tetraglucoside acylated with three molecules of caffeic acid and one of malonic acid was reported from *G. aurantiaca* cv. "Purple Passion" by Yoshitama et al. (1994).

Homogyne: The three species of *Homogyne* occur in central and southern Europe. The only flavonoid information is the presence of quercetin 3-*O*-rutinoside in *H. alpina* (Schwendimann et al., 1974).

Ischnea: *Ischnea* is a genus consisting of four species (Swenson, 1995) restricted to high elevations in New Guinea. *Ischnea elachoglossa* accumulates quercetin 3-*O*-glycosides (Swenson et al., 1999).

Kleinia: *Kleinia* is a genus of ca. 40 species with representatives in tropical Africa, the Middle East, Madagascar, and India. Two sets of workers reported kaempferol 7-*O*-rhamnoside and kaempferol 3,7-di-*O*-rhamnoside from *K. grandiflora* (Rao and Rao, 1972; Kotaiah et al., 1976). Several species of *Kleinia* (cited as species of *Senecio*; synonymy following Chater, 1976, and Jeffrey, 1986) were shown to accumulate kaempferol and/or quercetin (Glennie et al., 1971), viz. *K. ficoides* (as *S. ficoides*), *K. mandraliscae* (as *S. mandraliscae*), *K. squarrosa* (as *S. longiflorus* var. *violacea*) and *K. ueriifolia* (as *S. kleinia*). There was also a report of quercetin from *S. longiflorus* var. *madagascarensis* in the Glennie et al. (1971) paper. Jeffrey (1986) listed this taxon as *Kleinia* sp., but to our knowledge the formal combination has not yet been made.

Packera: *Packera* is a genus of about 65 species distributed in North America and into Siberia. This is another one of the generic segregates of *Senecio* now accepted by Jeffrey (1992) and Barkley et al. (1996). Only one species of the genus has been examined for flavonoids, *P. sanguisorbae* (cited as *Senecio sanguisorbae*; synonymy after Jeffrey, 1992), from which Betancourt and Proliac (1988) obtained isorhamnetin.

Pericallis: This genus consists of 15 species from Macaronesia and the Azores. Ashtakala et al. (1971) reported apigenin, cyanidin, and delphinidin from extracts of *P. cruenta* [= *Senecio cruentus*; synonymy from Nordenstam (1978); cited as *Cineraria cruenta*]. Among the most complex flavonoids reported from the family are anthocyanidin derivatives that Yoshitama and Abe (1977) and Yoshitama (1981) isolated from *P. cruenta* cv. "blue light" (cited as above). They showed that the pigment was a cyanidin 3,7,3'-tri-*O*-glucoside acylated with caffeic acid-4'-*O*-glucoside. Early work on "cinerarin," the pigment from the garden cineraria, showed the existence of a delphinidin derivative that involved glucose attachments at several positions and acylation with caffeic acid (Yoshitama and Hayashi, 1974; Yoshitama et al., 1975). A full structure for the compound, which also included a malonyl moiety, was reported by Goto and associates (1984): 3-*O*-(6-malonyl-ß-D-glucopyranosyl)-7-*O*-(6-*O*-(4-*O*-(6-*O*-caffeoyl-ß-D-glucopyranosyl)caffeoyl)-ß-D-glucopyranosyl)-3'-*O*-(6-*O*-caffeoyl-ß-D-glucopyranosyl)-delphinidin **[4–82]**. A recent paper by Terahara et al. (1993) described a similar, but somewhat simpler, pigment from a *P. cruenta* (cited as *Senecio cruentus*) red cultivar, namely cyanidin 3-*O*-(6"-malonylglucoside)-3'-*O*-(6"-*O*-caffeoyl)glucoside. Equally complex pigments based on pelargonidin have been reported from the cultivar "Jupiter Pink-White" by Toki et al. (1995).

Petasites: The northern circumpolar genus *Petasites* comprises 19 species, four of which have been examined for flavonoids. Kaempferol and quercetin as aglycones or common *O*-glycosides have been reported from *P. japonicus* (Kurihara and Kikuchi, 1972), *P. albus*, and *P. hybridus* [Schwendimann et al., 1974; Toropkina and Minina, 1976; reported as *P. georgicus*, now treated as *P. hybridus* subsp. *georgicus*, fide Toman (1972)]. Glyzin and Senov (1965) reported quercetin 3-*O*-galactoside and quercetin 3-*O*-rutinoside as constituents of *P. radiatus* (cited as *Nardosmia laevigata*; synonymy from Toman, 1972).

Psacalium: *Psacalium* comprises about 40 species (Robinson and Brettell, 1973d) that range from southwestern United States through Mexico to Guatemala. A recent report by Perez-Castorena et al. (1997a) described an acetylated kaempferol 3,7-di-*O*-rhamnoside from *P. megaphyllum*.

Robinsonia: *Robinsonia* is a genus of seven species (Pacheco et al., 1985; Sanders et al., 1987) restricted to the Juan Fernandez Islands. The species have *O*-glycosides of naringenin, eriodictyol, dihydroquercetin (taxifolin), apigenin, luteolin, kaempferol, and quercetin. A detailed discussion appears in Chapter 21.

Senecio: *Senecio* is an extremely large genus with about 1250 species. The group enjoys a cosmopolitan distribution but the greater concentrations of species occur in Latin America and southern Africa. Flavonoid data are available for over 50 species. Most species have been shown to accumulate various *O*-glycosides of common flavones and flavonols with several also testing positive for *C*-glycosylflavones (Glennie et al., 1971). Kaempferol and quercetin 3-*O*-glucosides also occur as acetate esters in *S. aureus* (Piegay et al., 1986). Direct sulphation of phenolic groups of isorhamnetin has been observed to occur in *S. gallicus*, *S. hoggariensis*, and *S. vulgaris* (Mansour and Saleh, 1981). A summary of occurrences of flavone and flavonol glycosides is given in Table 15–1. In several of the papers encountered in the literature, full chemical structural information was not given, which accounts for the abbreviated reports in the table. Apigenin, cyanidin, and delphinidin were reported from *S. cruenta* by Ashtakala et al. (1971) (cited as *Cineraria cruenta*; synonymy from Hansen and Sunding, 1979).

Naringenin was reported from *S. longilinguae* by Bohlmann and Zdero (1979a), which appears to be the only record of a flavanone in the genus. The only report of chalcones is the paper by D'Agostino et al. (1991b) in which they described okanin 4-methyl ether-3'-*O*-ß-D-(6"-*O*-acetyl)glucopyranoside, okanin 4'-*O*-ß-D-(3",4",6"-tri-*O*-acetyl) glucopyranoside and okanin 4-methyl ether 3'-*O*-ß-D-glucopyranoside from *S. pseudotites*. Okanin is 2,3,4,3',4'-pentahydroxychalcone. The first two of these compounds were newly described natural products.

Extra hydroxylation has not been detected in flavones or flavonols reported from members of *Senecio*. However, a modest level of *O*-methylation has been reported (beyond the presence of isorhamnetin). Kaempferol-3,7-dimethyl ether was reported from *S. adenotrichius* (Urzua et al., 1991) while the 3,4'-dimethyl ether was found in *S. rhyncholaenus* (Bohlmann et al., 1978b). Kaempferol-3,7,4'-trimethyl ether is known from *S. nebrodensis* where it occurs with quercetin-3,7,3',4'-tetramethyl ether (Barrero et al., 1988). Quercetin-3,7-dimethyl ether and 3,7,3'-trimethyl ether have been reported from *S. cachinalensis* (Dupre et al., 1991). Chrysoeriol 7-*O*-glucoside, along with apigenin 7-*O*-glucoside, were reported from *S. hollermayerii* (Sepulveda et al., 1994a). In a recent survey of asteraceous taxa for leaf exudate flavonoids, Wollenweber et al. (1997) obtained apigenin 7-methyl ether (genkwanin), luteolin 7,3'-dimethyl ether, kaempferol

Table 15–1. Common flavonoid *O*- and *C*-glycosides from species of *Senecio*

Species	Flavonoids[1] K	Q	IR	Others	References
S. acaulis		+			Glennie et al. (1971)
S. adenophylloides				Flav. aglycone	Pacheco et al. (1985)
S. adonidifolius	+	+		3,7-DiGlys	Ferry (1977)
S. aquaticus				Flol 3-, 3,7-Glys	Santavy et al. (1957)
S. azoides	+	+		Flavone *C*-Gly	Glennie et al. (1971)
S. angulatus				Apig 7-Gly, Flav *C*-Gly	Glennie et al. (1971)
S. argyreus				Flav & Flol 7-Gly	Pacheco et al. (1985)
S. aureus	+	+		Gly-acetates	Piegay et al. (1986)
S. bulbinefolius	+	+		Apig 7-Glc	Glennie et al. (1971)
S. canus		+	+		Bain and Denford (1975)
S. chordifolius	+	+		Flav *C*-Gly	Glennie et al. (1971)
S. cineraria		+	+		Cezard et al. (1984)
S. citriformis	+	+		Apig 7-Glc	Glennie et al. (1971)
S. crassiflorus		+		Chrysin, luteolin, tricin	Tettamanzi et al. (1994)
S. gallicus		+	+	3-Sulphation	Mansour and Saleh (1981)
S. glaber				Flol 3-Gly	Pacheco et al. (1985)
S. hallianus	+	+		Apig 7-Gly, Flav *C*-Gly	Glennie et al. (1971)
S. herreianus	+	+		Q-3-Me, Apig 7-Glc	Glennie et al. (1971)
S. hoggariensis		+	+	3-Sulphation	Mansour and Saleh (1981)
S. iosensis	+	+		Q-3-Me	Glennie et al. (1971)
S. jacobaea		+			Kowalska (1968);
	+	+			Ferry (1977)
S. kleiniiformis	+				Glennie et al. (1971)
S. macroglossus		+			Glennie et al. (1971)
S. magellanicus				Flav 7-Gly	Pacheco et al. (1985)
S. multilobatus	+	+	+		Bain and Denford (1985)
S. nemorensis		+			Nhia (1975)
S. neomexicanus		+	+		Bain and Denford (1985)
S. otites				Flol 3-, 7-, 3,7-Glys	Reyes et al. (1977)
S. ovoideus	+	+		Apig 7-Glc	Glennie et al. (1971)
S. pancicii		+			Dimitrova et al. (1981)
S. pauperculus	+	+		3,7-Gly	Bain and Denford (1985)
S. polyphyllus				Flol 3-Gly	Pacheco et al. (1985)
S. racemosus	+	+	+		Tanbug et al. (1997)
S. radicans	+	+		Q-3-Me, Flav *C*-Gly, Apig 7-Glc	Glennie et al. (1971)
S. rowleyanus	+	+			Glennie et al. (1971)
S. serpens	+	+		Flav *C*-Gly	Glennie et al. (1971)
S. streptanthifolius	+	+	+	3,7-Glys, Flavones	Bain and Denford (1985)

Table 15–1. (Continued)

Species	Flavonoids[1] K	Q	IR	Others	References
S. subdentatus		+	+	3,7-Glys	Chumbalov et al. (1975); Fadeeva and Nikischenko (1979)
S. subdiscoideus	+	+		3-DiGly, 7-Gly, 3,7-DiGly	Reyes et al. (1986)
S. subsessilis					
S. tamoides				Flav *C*-Gly	Glennie et al. (1971)
S. trifurcatus				Flol 7-Gly	Pacheco et al. (1985)
S. vira-vira		+	+		Jares and Pomilio (1987)
S. vulgaris			+	3,7-DiGly	Ferry (1977)
		+	+	3-Sulphation	Mansour and Saleh (1981)

[1] K = Kaempferol; Q = Quercetin; IR = Isorhamnetin; Flav = Flavone; Flol = Flavonol; Apig = Apigenin; Me = Methyl; Glc = Glucose; Gly = Glycoside.

3-, 3,7-di-, and 3,7,4'-trimethyl ethers, and quercetin 3,7,3'-trimethyl ether from an acetone wash of *S. viscosa*.

Traversia: This monotypic genus is native to New Zealand. *Traversia baccharoides* accumulates eriodictyol 7-methyl ether, quercetin 3'-methyl ether (isorhamnetin), quercetin 7-methyl ether (rhamnetin), and quercetin 7,3'-dimethyl ether (rhamnazin) (Kulanthaivel and Benn, 1986). These workers also found that quercetin occurred in this species in three acylated forms, the 3-, 3'-, and 4'-isobutyrates.

Tussilago: This genus from Europe, west-central Asia, and northern Africa, comprises the single species *T. farfara*. Common kaempferol and quercetin *O*-glycosides have been reported (Olechowska-Baranska and Lamer, 1962; Didry et al., 1980; Kikuchi and Mori, 1992).

CHAPTER 16

Flavonoids of Tageteae

Although Karis and Ryding (1994a) treat genera representing Tageteae as subtribe Pectidinae of resurrected tribe Helenieae (here treated within Heliantheae s.l., Chapter 11), we prefer to list it in this book as a distinct tribe following Strother (1977). Robinson (1981) elected to include these closely related genera as subtribe Pectidinae of tribe Heliantheae, including much of the former tribe Helenieae. However tribally positioned, Tageteae consist of 230-250 species distributed among 23 genera (Karis and Ryding, 1994a). Examination of the literature shows that the tribe has been poorly sampled for flavonoids. Not surprisingly, most interest in the flavonoids has been paid to *Tagetes*, which contains the cultivated marigolds. We present available flavonoid data after generic limits sketched by Strother (1986).

Chrysactinia: *Chrysactinia* consists of five species of shrubs native to southwestern North America. *Chrysactinia mexicana* accumulates glycosides of 6-hydroxykaempferol and 6-hydroxyquercetin (quercetagetin) (Harborne et al., 1991).

Dysodiopsis: *Dysodiopsis* has been resurrected by Strother (1986) to accommodate species referred to *Hymenatherum* sect. *Dysodiopsis*. Material of *D. tagetoides* has been examined for flavonoids (Bohm, unpubl., material provided by J. Strother). The major flavonoids present are glycosides of quercetin, quercetagetin (6-hydroxyquercetin), and patuletin (quercetagetin 6-methyl ether). Several unusual compounds require further study.

Dyssodia: The genus has recently been reduced in size to four species (Bremer, 1994) owing to inflation of subgeneric and some sectional taxa to generic rank (Strother, 1986). Earlier, Strother (1969) considered that there were 32 species native to southwestern North America. Material of *D. papposa* and *D. pinnata* has been examined, and both exhibit *O*-glycosides of quercetin and quercetagetin with a trace of a kaempferol derivative visible in *D. papposa* (Bohm, unpubl., material provided by J. Strother).

Pectis: About 100 species constitute this New World genus. *Pectis sessiliflora* was shown by De Israilev and Seeligmann (1990a) to accumulate 6-hydroxykaempferol 7-*O*-xylosylglucoside, quercetagetin 7-*O*-galactoside, patuletin, patuletin 3-*O*-glucoside, and patuletin 3-*O*-rhamnosylglucoside, which represents an array of compounds based upon 6-hydroxykaempferol and 6-hydroxyquercetin (quercetagetin) much akin to other members of this tribe (subtribe).

Porophyllum: *Porophyllum* consists of 28 (Bremer, 1994) species that occur in southwestern United States and Mexico, the Caribbean Islands, and in South America. From floral tissue of *Porophyllum lanceolatum*, Juarez et al. (1990) reported quercetin 7-*O*-glucoside, quercetin 3'-*O*-glucoside, quercetin 3-*O*-arabinosylrhamnoside, and 5,7,8,3'-tetrahydroxy-4'-methoxy 3'-*O*-glucoside. Van Baren et al. (1994) reported that *P. lanceolatum* (cited as *P. obscurum*; synonymy from Johnson, 1969) accumulates glycosides of quercetin (3-*O*-glucoside, 3-*O*-rhamnoside and 7-*O*-arabinoside), isorhamnetin (3-*O*-galactoside) and quercetagetin (7-*O*-glucoside). The floral flavonoids of *P. ruderale* were shown to be quercetin 3'-*O*-arabinoside, quercetin 7-*O*-glucoside, and gossypetin 4'-methyl ether 3-*O*-glucoside by Mendiondo et al. (1990b). In the same paper those authors reported quercetin 7-*O*-glucoside and the gossypetin derivative from *P. lanceolatum* (reported as *P. obscurum*, see comment on synonymy above).

Strotheria: *Strotheria gypsophylla*, the sole member of the genus, occurs in northern and central Mexico (Turner, 1972). The flavonoid profile of this species consists of *O*-glycosides of kaempferol, quercetin, quercetagetin, and patuletin (Bohm, unpubl., material provided by J. Strother). The array of compounds is complex and requires further study.

Tagetes: This genus consists of 55 species (Soule, 1996) occurring in warmer parts of the Americas and in Africa. The major flavonoids of *Tagetes* are flavonols with a preponderance of workers having reported quercetagetin and/or patuletin and their *O*-glycosides. The structure of quercetagetin was originally established by Perkin (1913) on material isolated from *T. patula*; Perkin referred to an even earlier work by Latour and Magnier de le Source (1877; Bull. Soc. Chim. 28: 337). Quercetin and *O*-glycosides of quercetin have been reported from several taxa, 6-hydroxykaempferol 7-*O*-glucoside from *T. erecta* (El-Emary and Ali, 1983), isorhamnetin *O*-diglycoside from *T. terniflora* (De Israilev and Seeligmann, 1983) and *T. elliptica* (D'Agostino et al., 1992), myricetin derivatives from *T. argentina*, *T. biflora*, *T. laxa*, *T. perezii* (De Israilev et al., 1991) and *T. elliptica* (D'Agostino et al., 1992), and luteolin from *T. patula* (Bhardwaj et al., 1980). *Tagetes patula* has also yielded 6-methoxykaempferol (Ivancheva and Zdravkova, 1993). De Israilev and Seeligmann (1990b) reported quercetagetin 3,6,3',4'-tetramethyl and 3,5,6,7,3'-pentamethyl ethers from *T. campanulata*. More recently, these workers (De Israilev and Seeligmann, 1994) reported kaempferol, quercetin, patuletin, and quercetagetin glycosides, quercetagetin 3,6,3',4'-tetramethyl ether, luteolin, and luteolin 7-*O*-glucoside from *T. rupestris*. Quercetin, patuletin, and quercetagetin, along with *O*-glycosides thereof, have been reported from *T. minuta* (Abdala and Seeligmann, 1995a) and *T. zipaquirensis* (Abdala and Seeligmann, 1995b).

Noteworthy in members of *Tagetes* is the capacity to oxygenate C-6 to produce quercetagetin in most cases, and 6-hydroxykaempferol in the case of *T. erecta* (El-Emary and Ali, 1983). Both the quercetagetin and myricetin oxygenation patterns occur (separately) in four species described by De Israilev et al. (1991). Glycosylation occurs commonly at position-7 of the flavonols with glycosylation at position-3 seen as well. Flavonol 5-*O*-glycosylation is seen in *T. argentina*, *T. biflora*, *T. laxa*, and *T. perezii* (De Israilev et al., 1991). *O*-Methylation is a prominent feature of the genus insofar as patuletin is widespread. 3-*O*-Methylation is seen in the four species just listed, to give quercetin 3-methyl ether, and in *T. patula* where quercetagetin 5-methyl ether was found (Bhardwaj et al., 1980). Flavonoids isolated from *Tagetes* species are summarized in Table 16–1.

Aerial parts of *Tagetes subulata* afforded quercetin 3-methyl ether, quercetagetin, quercetagetin 3,6,3',4'-tetramethyl ether (the authors said "quercetin" which cannot have a 6-methyl ether), and quercetagetin 3,5,6,7,3'-pentamethyl ether (Abdala and Seeligmann, 1997). In a later paper *Tagetes lucida* was described as accumulating isorhamnetin and its 7-*O*-glucoside, a quercetagetin 3-*O*-arabinosylgalactoside, and patuletin (Abdala, 1999).

Thymophylla: *Thymophylla* has been resurrected by Strother (1986) to accommodate species referred to *Dyssodia* sect. *Thymophylla*. Ruiz et al. (1992) reported apigenin, luteolin, luteolin 7-methyl ether, 6-methoxyluteolin (nepetin), and quercetagetin 7-*O*-glucoside from *Thymophylla belenidium* (cited as *Dyssodia pentachaeta* var. *belenidium*). Material of *T. setifolia* var. *setifolia* and *T. pentachaeta* var. *belenidium* has been examined for flavonoids. *O*-Glycosides of kaempferol (?), quercetin, quercetagetin, and

Table 16–1. Flavonoids from species of *Tagetes*

Species	Compounds[1]	References
T. argentina	K, Q, and M-3-*O*-Glc,	
	K, Q, M, Qt, and Pat-7-*O*-Glc,	
	Q 3-*O*-Rhm,	
	Q 5-*O*-Glc,	
	Q 3-*O*Me	De Israilev et al. (1991)
T. biflora	K, Q, and M3-*O*-Glc,	
	K, Q, M, Qt, Pat 7-*O*-Glc,	
	Q 3-*O*-Rhm,	
	Q 5-*O*-Glc,	
	Q-3-*O*Me	De Israilev et al. (1991)
T. campanulata	Qt-*O*Me ethers	De Israilev and Seeligmann (1990b)
T. dianthioflora	IR 3-*O*-Gal,	
	M 3-*O*-Glc,	
	Q 5-*O*-Glc,	
	Qt and Pat 7-*O*-Glc,	
	57/363'4'	Abdala and Seeligmann (1995c); Abdala (1997)
T. elliptica	Q 3-*O*-Gal, acetate, gallate	
	Q 3-*O*-Rhm,	
	Q 3-*O*-Gal-Rhm,	
	IR 3-*O*-Gal,	
	IR 3-*O*-Glc-Rhm,	
	M 3-*O*-Glc,	
	Rhamnetin	D'Agostino et al. (1992)
T. erecta	Kaempferitrin	Nakaoki and Morita (1956)
	Patuletin	Constantinescu et al. (1962)
	K 7-*O*-Rhm,	
	6-OH-K 7-*O*-Glc,	
	Kaempferitrin,	
	Qt 7-*O*-Glc	El-Emary and Ali (1983); Kaloshina and Mazulin (1983a)
	Quercetin	Tripathy and Gupta (1991)
T. laxa	K, Q, M 3-*O*-Glc,	
	K, Q, M, Qt and Pat 7-*O*-Glc,	
	Q 3-*O*-Rhm,	
	Q 5-*O*-Glc,	
	Q-3-OMe	De Israilev et al. (1991)
T. mandonii	Qt,	
	IR 3-*O*-Gal,	
	Pat 7-*O*-Glc	Abdala (1997)
	3563'4'/7 6-*O*-Glc,	
	3564'/73' 6-*O*-Glc	Senatore et al. (1999)

Table 16–1. (Continued)

Species	Compounds[1]	References
T. mendocina	Qt and Pat 3/7-*O*-Glc,	
	3/5673'4'	De Israilev and Seeligmann [1992 (1993)]
T. minuta	Q and Q 5-*O*-Glc	Abdala and Seeligmann (1995a)
	Qt 7-*O*-Glc	Ickes et al. (1973)
	Pat 7-*O*-Glc	Ickes et al. (1973); Abdala and Seeligmann (1995a)
	Qt and Pat 7-*O*-Glc	Jain (1977)
	Qt 3-*O*-Gal-Ara	Abdala and Seeligmann (1995a)
	Qt 7-*O*-Gal-Ara	De Israilev and Seeligmann (1983), Tereschuk et al. (1997)
T. multiflora	Q 5-*O*-Glc	De Israilev [1992 (1993)]
T. patula	Quercetagetin	Perkin (1913); Bannerjee and Seshadri (1956)
	Pat, Pat 7-*O*-Glc	Bannerjee and Seshadri (1956)
	Patuletin	Constantinescu et al. (1962); Tarpo (1967, 1968, 1969a)
	Quercetagetin	Tarpo (1969b, 1971)
	Luteolin,	
	Qt and Pat 7-*O*-Glc,	
	Qt-5-OMe	Bhardwaj et al. (1980)
	Qt 7-*O*-Glc	Kaloshina and Mazulin (1983a)
	IR 3-*O*-Gal	Tripathi et al. (1991)
	6-MeO-K	Ivancheva and Zdravkova (1993)
	Chry 7-*O*-Glc-Arab	Das et al. (1996)
T. perezi	K, Q and M 3-*O*-Glc,	
	K, Q, M, Qt, and Pat 7-*O*-Glc,	
	Q 3-*O*-Rhm,	
	Q 5-*O*-Glc,	Tripathi et al. (1991)
	Q-3-OMe	De Israilev et al. (1991)
T. pusilla	Qt 7-*O*-Gal-Ara	De Israilev and Seeligmann (1983)
T. riojana	Q 5-*O*-Glc and Qt deriv.	De Israilev [1992 (1993)]
T. rupestris	Lute 7-*O*-Glc,	
	K and Q 3-*O*-Glc,	
	Qt and Pat 7-*O*-Glc,	
	Q 3-*O*-Rhm,	
	57/363'4'	De Israilev and Seeligmann (1994)
T. terniflora	Q 3-*O*-Ara	
	Q 3-*O*-Gal	
	IR 3,7-Di-*O*-Gal	De Israilev and Seeligmann (1994)

[1] Substitution: Hydroxyl [positions/Methoxyl positions; K = kaempferol; Q = quercetin; IR = isorhamnetin; M = myricetin; Qt = quercetagetin; Pat = patuletin; Glc = glucose; Gal = galactose; Ara = arabinose; Rhm = rhamnose.

patuletin were detected in *T. pentachaeta*. Flavonoid profiles of both taxa are moderately complex and require further study (Bohm, unpubl., material from J. Strother).

Chapter 17

Flavonoids of Vernonieae and Liabeae

According to Jones (1977), Vernonieae consist of about 70 genera and something in excess of 1450 species. Bremer (1994) circumscribed the tribe to comprise 98 genera owing in part to new genera described by H. Robinson during the period 1987-1992 (See Bremer, 1994, for pertinent references), but his species count is around 1300. Robinson (1996a) includes 111 genera. Thirty-seven genera in Jones' (1977) treatment are monotypic; 46 are monotypic in Bremer's (1994). The subtribal classification of Vernonieae is at present in a state of flux (Bremer, 1994; Robinson, 1996a). Central to the problem is the large genus *Vernonia*, which is now undergoing increased taxonomic scrutiny. New genera will no doubt continue to be peeled off from the core taxa, especially in the Old World, whether rightly or wrongly remains to be seen. This tendency, in fact, has already been the main impetus for generic inflation since 1977 (70 genera in Jones, 1977, vs. 111 in Robinson, 1996a). Jeffrey (1988) has suggested some informal groupings in *Vernonia* of East Tropical Africa that could eventually be considered for generic status. For consistency we have chosen to follow the generic concepts developed by Robinson, which he summarized in 1996a (presented also by Bremer, 1994). Number of included species and distributional ranges are after Bremer (1994) unless stated otherwise. An important new synthesis of American Vernonieae, containing a valuable list of generic and specific names (with synonyms), recently has been published by Robinson (1999), but this paper was not available to us at the time our manuscript was prepared.

Flavonoid data for most of the genera will be presented in the same format as earlier chapters. However, we will alter the format for presenting information compiled by Harborne and Williams (1977) who, having surveyed a number of genera, presented only a summary statement about the observed flavonoids. Their data appear in Table 17–1, along with species numbers and ranges of occurrence taken from Bremer (1994) and Robinson (1983a, 1988a, 1992b, 1994, 1996a). In all, we have some flavonoid information for 35 genera.

Table 17–1. Flavonoids of genera of Vernonieae (from Harborne and Williams, 1977)

Genus	Number of sp.[1]	Number of sp. studied[2]	Distribution	Flavonoids[3] Flav.	Flavonoids[3] Flol.
Adenoön	1	1	Indo-Malaysia		+
Aedesia	3	1	Tropical West Africa	+	
Alcantara	1	1	Brazil	+	
Blanchetia	1	1	Brazil		+
Bolonosa[4]	1	1	Mexico	+	
Centauropsis	8	1	Madagascar	+	
Centratherum	3	1	Tropical So. America	+	
Corymbium[5]	17	1	South Africa	+	
Ekmania	1	1	Cuba		+
Elephantopus	2	2	Tropical	ND	
Eremanthus[6]	18	1	Brazil		+
Erlangea	6	1	Tropical Africa	+	
Ethulia	19	1	Africa; S.E. Asia	+	
Gutenbergia	25	1	Tropical Africa	+	
Herderia	1	1	Tropical Africa	+	
Lessingianthus[7]	100	2	Brazil		+
Lychnophora[8]	26	2	Brazil		+
Oliganthes	9	2	Madagascar	+	+
Pacourina	1	1	South America	+	+
Piptocarpha	45	2	Central & So. America	+	+
Piptocoma	3	2	West Indies		+
Piptolepis	9	1	Brazil		+
Pseudopiptocarpha[9]	2	1	Colombia	+	
Spiracantha	1	1	Central Am.; Colombia	ND	
Stilpnopappus	24	1	South America	+	+
Stokesia	1	1	S.E. United States	+	
Vernonia	500	12	America; Africa, S.E. Asia	+/–	+/–
Vernonanthura[10]	ca. 65	1	Central & So. America; West Indies	+	

1 Number of species according to Bremer (1994).
2 Number of species examined by Harborne and Williams (1977).
3 Flav. = Flavone; Flol. = Flavonol; ND = No flavonoids detected.
4 Not treated by Bremer (1994) but included by Robinson (1996a), but without subtribal referral.
5 Placed in Cichorioideae by Bremer (1994) but not assigned to a tribe.
6 Reported as *Vanillosmopsis erythropappa*; synonymy from MacLeish (1987).
7 Reported as *Vernonia rubricaulis* and *V. saltensis* = *L. rubricaulis* and *L. saltensis*, Robinson (1988a).
8 Includes one report in *Haplostephium* (*H. passerinia* var. *subulatum*); a generic synonym, *fide* Robinson (1983a) and Bremer (1994).
9 Reported as *Piptocarpha elaeagnoides*; = *Pseudopiptocarpha elaeagnoides*, Robinson (1994).
10 Reported as *Vernonia squamulosa*; = *Vernonanthura squamulosa*, Robinson (1992b).

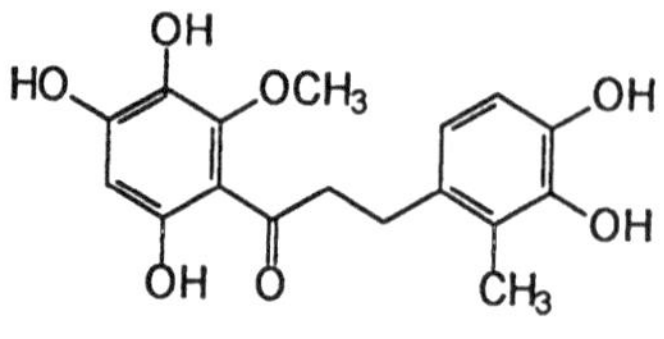

[17-1]

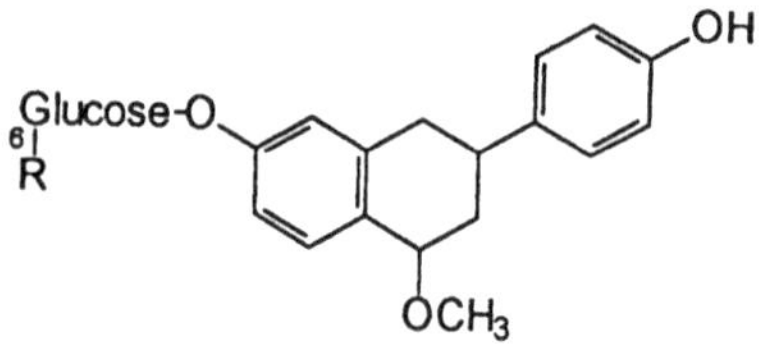

[17-2] R = H

[17-3] R = *p*-Hydroxybenzoyl

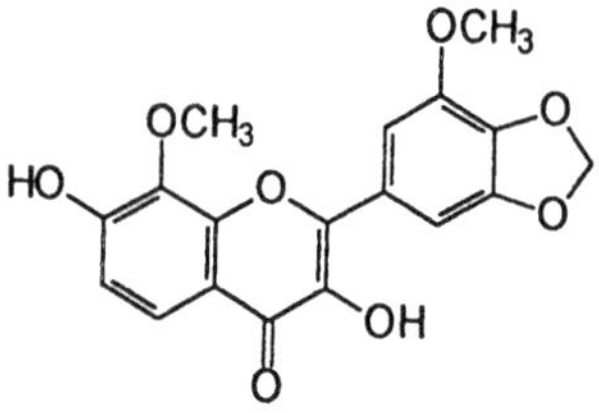

[17-4]

Fig. 17–1. Unusual flavonoids reported from members of Vernonieae include the *C*-methyl chalcone **[17–1]**, the two 4-methoxyflavan derivatives **[17–2, 17–3]**, and the 5-deoxyflavonol **[17–4]**

Baccharoides: *Baccharoides* is one of the new genera described or resurrected from within *Vernonia* by H. Robinson in a series of recent papers (1980b, 1988a,b,c, 1990a,b, 1992a,b, 1993a,b, 1994). It is circumscribed to contain approximately 30 species centered in tropical Africa (Bremer, 1994). *Baccharoides adoensis* (cited as *Vernonia adoensis*; synonymy from Robinson, 1990a) was reported to accumulate luteolin 7- and 3'-methyl ethers (Mabry et al., 1975). An unusual array of compounds was recently described from *Baccharoides anthelmintica* by Wu et al. (1991) (cited as *Vernonia anthelmintica*; synonymy from Robinson, 1988b). The identification of 2',4',6',4-tetrahydroxychalcone and the highly unusual 2',4',5',3,4,-pentahydroxy-6'-methoxy-2-*C*-methylchalcone (Fig. 17–1) **[17–1]** represent first reports of chalcones from Vernonieae. Also unusual are the other two compounds reported, 7,4'-dihydroxy-4-methoxyflavan 7-*O*-glucoside **[17–2]** and its 6"-*O*-*p*-hydroxybenzoyl derivative **[17–3]**. In view of these observations it would certainly seem that a detailed examination of other members of this genus would repay the efforts.

Centratherum: This small genus of tropical South America consists of three species. Harborne and Williams (1977) reported a positive "flavone" sighting for *C. phyllolaenum* in their survey. A recent paper by Yadev and Barsainya (1997) describes the isolation of a flavonol from *C. anthelminticum*, which they showed to be the novel 3,7-dihydroxy-8,5'-dimethoxy-3',4'-methylenedioxyflavone **[17–4]**.

Chrysolaena: Newly described by Robinson (1988b), this genus consists of nine species from central and southern South America. The only flavonoid information comes from the work of Kisiel (1975, 1976) who reported apigenin 7-methyl ether and luteolin 7,3'-dimethyl ether from *C. flexuosa* (cited as *Vernonia flexuosa*; taxonomy after Robinson, 1988b).

Critoniopsis: Re-established in 1980b by Robinson, this genus is now viewed by Bremer (1994) as containing 28 species concentrated in the northern Andes. Flavonoid reports involve *C. duncanii* and *C. tomentosa* (cited as *V. duncanii* and *V. paniculata*, respectively; taxonomy and nomenclature after Robinson, 1993a). Mabry et al. (1975) reported apigenin 7-methyl ether from the former and luteolin 7- and 3'-methyl ethers from the latter.

Cyanthillium: This genus contains about 25 species distributed in tropical regions of Africa and Asia and Madagascar. The only species examined for flavonoids appears to be *C. cinereum* (cited as *Vernonia cinerea*; taxonomy after Robinson, 1990a) from which Nair et al. (1982) obtained luteolin 4'-methyl ether (diosmetin).

Distephanus: Another genus nestled within the confines of *Vernonia* recently resurrected by Robinson and Kahn (1986) has been *Distephanus*. Bremer (1994) describes the genus as consisting of about 40 species centered in Madagascar but with extended distribution in tropical Africa and southeastern Asia. Rakotoarimanga et al. (1992) reported quercetin 3-methyl ether from *D. glutinosus* (cited as *Vernonia glutinosa*; taxonomy from Robinson and Kahn, 1986).

Eirmocephala: *Eirmocephala* consists of three South and Central American species. A recent paper described the isolation and identification of kaempferol 3,7-dimethyl ether from *E. megaphylla* (Borkosky et al., 1996). This taxon was previously considered as a species of *Vernonia*.

Elephantopus: *Elephantopus* is a tropical genus of about 30 species. Ghanin et al. (1963) reported luteolin 7-*O*-glucoside from *E. scaber*. Harborne and Williams (1977) failed to detect flavonoids in *E. carolinianus* and *E. mollis*.

Eremanthus: *Eremanthus* consists of 18 species native to Brazil. *Eremanthus glomerulatus* accumulates kaempferol 3,7-dimethyl ether (Bohlmann et al., 1980h). Vichnewski et al. (1989) stated that they isolated "...the flavone isorhamnetin..." from *E. seidelii* but gave the structure of chrysoeriol in their paper. Lunardello et al. (1995) reported naringenin from *E. mattogrossensis* and apigenin 7,4'-dimethyl ether and kaempferol 3,7,4'-trimethyl ether from *E. eriopus*. Harborne and Williams (1977) recorded "flavonols" from *E. erythropappus* (described as *Vanillosmopsis erythropappa*; see MacLeish, 1987, for synonomy). 5,7-Dihydroxy-3',4'-dimethoxyflavone (luteolin 3',4'-dimethyl ether) has been reported from *E. arboreus* by Vichnewski et al. (1999).

Ethulia: *Ethulia* is a genus of 19 species (Bremer, 1994; following Gilbert and Jeffrey, 1988) of Africa and southeastern Asia. Apigenin and luteolin have been reported from *E. vernonioides* (Schuster et al., 1993).

Gutenbergia: *Gutenbergia* consists of 25 species (Bremer, 1994, following Gilbert and Jeffrey, 1988) from tropical Africa. The only flavonoid information we have is the report by Mungarulire (1990) that *G. cordifolia* accumulates 5,7-dihydroxyflavone (chrysin).

Hesperomannia: This small genus consists of three species (Carlquist, 1957; Wagner et al., 1990) endemic to the Hawaiian Islands. *Hesperomannia* has traditionally been placed within Mutisieae (Carlquist, 1957; Bremer, 1994), but recent *ndh*F sequence studies have shown clear alliance with African species of *Vernonia* (Kim et al., 1996). *Hesperomannia arborescens* has a very simple flavonoid profile consisting of quercetin 3-*O*-monoglucoside (Bohm and Stuessy, 1995), 3-*O*-glucuronide (trace), 3-*O*-rutinoside, and eriodictyol 7-*O*-glucoside (Bohm, unpubl.).

Lepidaploa: Viewed as a relatively large genus of 116 species from Latin America and the West Indies, *Lepidaploa* was resurrected from within *Vernonia* by Robinson (1990b). 5,7,4'-Trihydroxy-3',5'-dimethoxyflavone (tricin) was obtained from *L. remotiflora* (cited as *Vernonia remotiflora*; taxonomy after Robinson, 1992a) by Jacobs et al. (1986a).

Lessingianthus: A genus of perhaps 100 species centered in Brazil, *Lessingianthus* was also plucked out of *Vernonia* by Robinson (1988a). Mabry et al. (1975) reported luteolin 7- and 3'-methyl ethers from *L. rubricaulis* (cited as *Vernonia rubricaulis*; taxonomy from Robinson, 1988a).

Lychnophora: *Lychnophora* is a Brazilian genus consisting of 26 species. In a paper published in 1976, Le Quesne et al. described five compounds from *L. ericoides* (cited as *L. affinis*, synonymy from Coile and Jones, 1981), the structures of three of which were subsequently corrected. The two correctly identified compounds from the original paper were luteolin 7,4'-dimethyl and 7,3',4'-trimethyl ethers. The corrected structures are quercetin 3,7,3'-trimethyl and 3,7,3',4'-tetramethyl ethers, and myricetin 3,7,4',5'-tetramethyl ether (Le Quesne et al., 1979). *Lychnophora phylicifolia* contains naringenin 7-methyl ether (sakuranetin) and eriodictyol 7,4'-dimethyl ether (Bohlmann et al., 1980j). Pinocembrin, pinobanksin, both of which are B-ring unsubstituted flavanones, and isorhamnetin have been reported from *L. diamantinana* (da Costa et al., 1993). Isorhamnetin 3-*O*-glucoside and acacetin were reported more recently from *L. rupestris* (Cunha et al., 1995). Harborne and Williams (1977) reported two species of *Lychnophora* as flavonol-positive (Table 17–1). Harborne and Williams (1977) recorded "flavonols" as being present in *L. passerina* (cited as *Haplostephium passerina* var. *sublatum*). *Haplostephium* has recently been treated as a synonym of *Lychnophora* by Robinson (1983a; followed by Bremer, 1994). Borella et al. (1998) have recently reported luteolin and quercetin from *L. pseudovillosissima* and pinobanksin from *L. ericoides*.

Piptocarpha: Other than *Vernonia*, *Piptocarpha* is one of the larger genera in the tribe with 45 species scattered about Central and South America. *Piptocarpha oblonga*, the only member for which flavonoid information is available, afforded naringenin 7-methyl and 7,4'-dimethyl ethers, and eriodictyol 7,3'-dimethyl ether (Bohlmann et al., 1980i).

Pseudopiptocarpha: This genus of only two species from Colombia has been so recently described (Robinson, 1994) from out of *Piptocarpha* and *Vernonia* that it does not appear in Bremer (1994). The only flavonoid data reported are "flavones" from *P. elaeagnoides* (cited as *Piptocarpha elaeagnoides*; transfer by Robinson, 1994) (Harborne and Williams, 1977).

Vernonanthura: Recently removed from *Vernonia* by Robinson (1992b), *Vernonanthura* includes about 65 species distributed in Latin America and the West Indies (Bremer, 1994). Several species have been examined for flavonoids: *Vernonanthura brasiliana* (cited as *Vernonia brasiliana*; all specific epithets remain unchanged; all transfers from Robinson, 1992b), *V. chamaedrys, V. liatroides*, *V. nudiflora*, *V. patens*, and *V. squamulosa*. *Veronanthura brasiliana* (Mabry et al., 1975) and *V. liatroides* (Rakotoarimanga et al., 1992) afforded identical profiles consisting of apigenin and luteolin 7- methyl ethers, luteolin 3'-methyl ether, and luteolin 7,3'-dimethyl ether. Catalan et al. (1988) reported luteolin 3'-methyl ether from *V. chamaedrys*, and Wagner et al. (1972k) reported quercetin 3-methyl ether from *V. patens*.

Vernonia: It is clear that the limits of variation within this genus have not been established with any certainty, although all authorities agree that *Ver-*

nonia is large with estimates ranging from about 500 (Bremer, 1994) to 1000 species (Jones, 1977; Keeley and Turner, 1990). Representatives occur widely in the Americas, Africa, and southeastern Asia. Much of what is known about flavonoids in the genus comes from three surveys, one by Mabry et al. (1975), involving 39 species, one by King and Jones (1982) that dealt with the eight species comprising *Vernonia* series *Flexuosae*, and one by Harborne and Williams (1977) that involved a large number of genera.

Mabry et al. (1975) identified 27 compounds, many of which were glycosides of apigenin, luteolin, kaempferol, and quercetin. The following *O*-methylated compounds were also obtained, although not all compounds occurred in all species: apigenin 7-methyl ether (genkwanin), luteolin 3'-methyl ether (chrysoeriol), luteolin 7-methyl ether, luteolin 7,3'-dimethyl ether (velutin), kaempferol 3-methyl ether, quercetin 7,3'-dimethyl ether (acylated), and quercetin 3,7,4'-trimethyl ether (acylated). Several other reports of *O*-methylated flavones and flavonols in *Vernonia* have appeared in the literature, an example of which is 5,4'-dihydroxy-6,7,3'-trimethoxyflavone (cirsilineol) from *V. thomsoniana* (Mungarulire, 1990). *Vernonia menthaefolia*, a Cuban endemic, was recently shown to accumulate apigenin 7-methyl ether (genkwanin) and luteolin 7,3'-dimethyl ether (velutin) (Budesinsky et al., 1994). See Table 17–2 for details of the occurrence of simple flavone and flavonol methyl ethers.

Eriodictyol 4'-methyl ether (hesperidin) was detected in *V. brevifolia* (Mabry et al., 1975). Other flavanone sightings involve eriodictyol in *V. hindei* and *V. syringifolia*, both reported by Zdero et al. (1991a).

The Harborne and Williams (1977) survey included several species of *Vernonia* some of which were reported to contain only flavones while others had only flavonols.

King and Jones (1982) discussed chemotaxonomic relationships within *Vernonia* series *Flexuosae* and reported *O*-glycosides of apigenin, luteolin, kaempferol, and quercetin and *C*-glycosyl derivatives of apigenin and luteolin. *O*-Methylated compounds were reported but structures were not established with certainty. The *C*-glycosylflavone isoörientin has been reported from *V. cinerea* (Gunasingh adnd Nagarajan, 1981).

O-Methylated flavones and flavonols have been reported from several other genera in the Vernonieae. Reference to Table 17–2 shows that some species of *Vernonia* make a series of apigenin and luteolin methyl ethers. *Vernonia amygdalina* and *V. brevifolia* accumulate quercetin 3-mono-, 3,3'-di- and 3,7,4'-trimethyl ethers (Rwangabo et al., 1986). *Vernonia nigritiana* accumulates quercetin 3-methyl ether along with quercetin 3-*O*-glucoside (Si et al., 1994). A recent paper by Igile et al. (1994) reported luteolin, its 7-*O*-glucoside, and 7-*O*-glucuronide as constituents of *V. amygdalina*. *Ver-*

Table 17–2. *O*-Methylated flavones and flavonols in *Vernonia*

Species	Compounds[1] A7	L7	L3'	L73'	Others	References
V. acaulis	+	+	+	+		Mabry et al. (1975)
V. alamanii	+	+	+	+		"
V. amygdalina					Q3, Q33'	Rwangabo et al. (1986)
V. arkansana	+	+	+	+		Mabry et al. (1975)
V. baldwinii	+	+	+	+		"
V. blodgettii		+	+	+		"
V. brevifolia	+			+	K3, Q73', Q374'	"
V. cinerea			+			Gunasingh and Nagarajan (1981)
V. divaricata	+			+		Mabry et al. (1975)
V. fasciculata	+	+	+	+		"
	+					Narain (1977)
V. gigantea	+	+	+	+		Mabry et al. (1975)
V. glauca	+	+	+	+		"
V. greggii	+	+	+	+		Rakotoarimanga et al. (1992)
V. incana		+	+	+		"
V. larsenii	+	+	+	+		"
V. lettermannii		+	+	+		"
V. lindheimeri	+	+	+	+		"
V. marginata	+	+	+	+		"
V. missurica	+	+	+	+		"
V. nitidula					Q3'	Bardón et al. (1988)
V. noveboracensis	+	+	+	+		Mabry et al. (1975)
V. pannosus[2]					K3, K37	Bohlmann et al. (1981c)
V. texana	+	+	+	+		Mabry et al. (1975)

[1] A = Apigenin; L = Luteolin; K = Kaempferol; Q = Quercetin; Numbers are locations of the methyl group locations.

[2] Reported as *Eremanthus mollis* (MacLeish, 1984).

nonia nigritiana afforded quercetin, its 3-*O*-glucoside and 3-methyl ether (Atindehou et al., 1993 [1994]). Another recent report described flavonoids from three Ethiopian species (Abegaz et al., 1994): *V. leopoldii* with luteolin and quercetin, *V. myriantha* with quercetin 3-methyl ether, and *V. filigera* with quercetin 3-methyl ether and apigenin. An East African species, *V. galamensis* ssp. *nairobiensis*, afforded three quercetin 3-*O*-glycosides, including an apiosylgalactoside, and a newly reported 3-*O*-apiosyl(1→2)galactoside of isorhamnetin (Miserez et al., 1996).

Liabeae

Liabeae are described by Bremer (1994) as one of the smallest and taxonomically best understood tribes of Asteraceae. It is unfortunate that so little is known about flavonoids in members of this tribe. We undertook a survey of a small number of species using herbarium specimens and thin layer chromatographic methods (Bohm and Stuessy, unpubl.). Liabeae consist of 14 genera distributed from Mexico through Central America southward to northern Argentina (Bremer, 1994). Most of the estimated 160 species occur in the Andes in Peru and Ecuador. Two-dimensional thin layer chromatograms of methanol extracts of fragments of herbarium specimens (from OS with permission) were examined for flavonoids in order to provide an idea of what compound types exist in the tribe (Bohm and Stuessy, unpubl.).

Liabum kingii H. Robinson from two sites in Ecuador gave identical chromatograms clearly showing the presence of kaempferol and quercetin 3-*O*-mono- and diglycosides as well as a spot that could be interpreted as arising from a quercetin methyl ether. *Liabum acaule* (H.B.K.) Less. [syn. *Chrysactinium acaule* (H.B.K.) Wedd.] from Ecuador appears to have quercetin 3-*O*-mono and diglycosides as well as a compound with chromatographic properties suggestive of a flavonol with additional oxygenation, possibly on the A-ring. Chromatograms of *L. asperifolium* Muschler [syn. *Paranephelius asperifolium* (Muschler) Robin. & Brettell] from Peru, and *L. volubile* (H.B.K.) Less. [syn. *Oligactis volubilis* (H.B.K.) Cassini] from Colombia each had a strong spot corresponding to quercetin.

In a recent paper, Juarez et al. (1995) reported quercetin, quercetin 3-*O*-glucoside, quercetin 3-methyl ether, and patuletin from *Liabum candidum* and *L. polymnioides*. These data, along with information on sesquiterpene lactones and pyrrolizidine alkaloids, as well as macromolecular observations, were taken to indicate a closer relationship between Liabeae and Vernonieae than between Liabeae and Senecioneae.

Returning to our brief survey (Bohm and Stuessy, unpubl.), we found that *Cacosmia rugosa* H.B.K., collected in Peru, gave spots corresponding in color and chromatographic properties to quercetin 3-*O*-mono- and diglycosides. *Philoglossa mimuloides* (Hieron.) H. Robin. & Cuatr., also from Peru, had a strong quercetin 3-*O*-monoglycoside spot and a smaller one for the corresponding diglycoside. The chromatogram of *Sinclairia moorei* (H. Robin. & Brettell) Robin. & Brettell, from Mexico, showed clear evidence for kaempferol and quercetin 3-*O*-monoglycosides, a small amount of a quercetin 3-*O*-diglycoside, and an eriodictyol 7-*O*-monoglycoside.

Section IV

Efficacy of Flavonoids at Different Taxonomic Levels

Section Four consists of six chapters in which we will present and discuss flavonoid data at various levels of the taxonomic hierarchy. Chapter 18 deals with subfamilial characteristics and features information on Barnadesioideae, newly elevated to subfamily status. Chapter 19 presents tribal flavonoid patterns. Since Bremer and colleagues (Bremer, 1994) view tribal boundaries somewhat differently from other contemporary workers, particularly with regard to delineation of Heliantheae and Helenieae (including Tageteae), comparison of flavonoid profiles becomes somewhat complex. Chapter 20, in which subtribal information is summarized, represents perhaps the least satisfactory story. It is at this level that sampling problems manifest themselves most conspicuously. Chapter 21 deals with relationships between and among various genera and, in contrast to disappointment at subtribal applications, shows in many instances the usefulness of flavonoids. Chapters 22 and 23, covering flavonoid distribution patterns at the specific and infraspecific levels, respectively, provide additional examples of the usefulness of these compounds.

Geography can play as important a role in understanding relationships among individuals of a species as it does between taxa. Rather than devote a chapter specifically to geographic patterns of flavonoid differentiation, we have incorporated relevant information into, mainly, Chapters 22 and 23. This allows for a more comprehensive and unbroken discussion of problems at intergeneric or interspecific levels. Flavonoid involvement in studies of island systems will be found, for example, under the genera themselves, *Dendroseris*, *Erigeron*, or *Robinsonia* for Juan Fernandez Islands, under *Scalesia* for Galapagos Islands, and under *Argyroxiphium*, *Dubautia*, or *Wilkesia* for the Hawaiian Islands.

A similar situation obtains for chromosome numbers. Those studies where both flavonoids and chromosome numbers are discussed will appear under the genera concerned. Both geographic and chromosomal information is amply available through the Subject Index to this book.

Chapter 18

Flavonoids at the Subfamilial Level

There has been ongoing debate over the years concerning tribal boundaries within Asteraceae and what the relationships among the tribes might be. Surprisingly, perhaps, much less attention has been given to the question of subfamilies. Any in-depth study of inter-tribal relationships, however, is bound to result in a grouping of tribes, a logical outcome of which would be recognition of these groups as subfamilies. Although a detailed discussion of these issues appears in Chapter Two, we herewith give a brief synopsis of recent efforts at classification to help the reader better follow correlations (or lack thereof) with the flavonoid data.

Bremer (1994) briefly reviews attempts to arrive at subfamily groupings in Asteraceae with the interesting observation that the issue really didn't emerge until quite recently. Early suggestions of subfamilial groups can be found in Cronquist (1955), Carlquist (1961), Poljakov (1967), and Robinson and Brettell (1973a). Two papers that appeared more recently addressed the subject directly. Carlquist (1976) concluded that there are two main groups of tribes, which he arranged in two subfamilies, Asteroideae, consisting of Astereae, Calenduleae, Anthemideae, Heliantheae, Inuleae, and Senecioneae; and Cichorioideae, consisting of Arctoteae, Cardueae, Cichorieae, Eupatorieae, Mutisieae, and Vernonieae. In the second paper, Wagenitz (1976) described a similar arrangement except that he placed Eupatorieae among the Asteroid tribes pointing out that Eupatorieae and Vernonieae, for example, have few things in common other than discoid capitula and flower color.

That was the state of affairs until quite recently. In 1987(b) Jansen and Palmer reported finding a 22 kb inversion in the chloroplast DNA of all members of Asteraceae except for members of the *Barnadesia*-group of genera, which was regarded, up to that time, as subtribe Barnadesiinae of Mutisieae. A cladistic analysis of morphological data supported the distinction (Bremer, 1987). In 1992 Bremer and Jansen elevated Barnadesiinae to subfamily status as Barnadesioideae. The family, as currently viewed, consists of three subfamilies: Asteroideae, Barnadesioideae, and Cichorioideae.

Table 18–1. Presence or absence of flavonoid classes and structural features in tribes and subfamilies of Asteraceae

Taxon	Gen[2]	Flavonoid classes and structural features[1]																		
		Flol	Flav	Anth	Fon	DiHF	DiHC	IsoF	CGly	OMe	6Ox	8Ox	68Ox	Bare	345	245	Other	SO_4	CAlk	Acyl
Asteroideae																				
Anthemideae	44	+	+	−	+	+	+	−	+	+	+	+	+	+	+	+	+	−	+	+
Astereae	48	+	+	+	+	+	+	−	+	+	+	+	+	+	+	+	+	−	−	−
Calenduleae	3	+	+	−	−	−	−	−	−	+	−	−	−	−	+	−	−	−	−	−
Eupatorieae	32	+	+	+	+	+	−	−	+	+	+	+	+	+	+	+	−	+	−	−
Heliantheae s.l.	110	+	+	+	+	+	−	+	+	+	+	+	+	+	+	−	+	+	+	+
Heliantheae s.s.	75	+	+	+	+	+	−	+	+	+	+	+	+	+	−	−	−	+	+	+
Helenieae (-Tag)	35	+	+	+	+	+	−	+	+	+	+	+	+	−	+	−	+	+	+	+
Helenieae (+Tag)	42	+	+	+	+	+	−	+	+	+	+	+	+	−	+	−	+	+	+	+
Inuleae s.l.	43	+	+	+	+	+	+	+	+	+	+	+	+	+	+	−	−	+	+	+
Inuleae s.s.	12	+	+	−	+	+	−	−	−	+	+	−	+	−	+	−	−	+	−	−
Gnaphalieae	25	+	+	+	+	+	+	−	−	+	+	+	+	+	+	−	−	−	+	+
Plucheeae	6	+	+	−	+	+	−	+	+	+	+	+	+	−	−	−	−	+	−	−
Senecioneae	21	+	+	+	+	+	−	−	−	+	−	+	−	−	+	−	−	−	−	+
Tageteae	7	+	+	−	−	−	−	−	−	+	+	−	−	−	+	−	−	−	−	−
Barnadesioideae																				
Barnadesieae	7	+	−	−	+	−	−	−	−	+	−	−	−	−	−	−	−	−	−	−
Cichorioideae																				
Cardueae	26	+	+	+	+	+	−	−	+	+	+	+	−	+	+	−	−	+	−	−
Lactuceae	39	+	+	+	+	+	−	−	+	+	+	−	−	−	+	+	−	−	−	−
Liabeae	4	+	−	−	+	−	−	−	−	+	+	−	−	−	−	−	−	−	−	−
Mutisieae	22	+	+	+	+	+	−	+	+	+	+	+	−	+	−	−	−	−	−	+
Vernonieae	35	+	+	+	+	−	−	−	+	+	+	−	−	+	+	−	−	−	+	−

Table 18–1. (Continued)

Taxon	Gen[2]	Flol	Flav	Anth	Fon	DiHF	DiHC	IsoF	CGly	OMe	6Ox	8Ox	68Ox	Bare	345	245	Other	SO_4	CAlk	Acyl
		Flavonoid classes and structural features[1]																		
Subfamily totals																				
Asteroideae	308	+	+	+	+	+	+	+	+	+	+	+	+	+	+	+	+	+	+	+
Barnadesioideae	7	+	–	–	+	–	–	–	–	+	–	–	–	–	–	–	–	–	–	–
Cichorioideae	126	+	+	+	+	+	–	+	+	+	+	+	–	+	+	+	–	+	–	+

[1] Flol = flavonol; Flav = flavone; Anth = anthochlor (chalcone and/or aurone); Fon = flavanone; DiHF = dihydroflavonols; DiHC = dihydrochalcone; IsoF = isoflavone; CGly = *C*-glycosylflavone; OMe = *O*-methylation; 6Ox = 6-oxygenation; 8Ox = 8-oxygenation; 68Ox = 6,8-dioxygenation; Bare = B-ring deoxyflavonoid; 345 = 3',4',5'-trioxygenation; 245 = 2',4',5'-trioxygenation (isoëtin type); Other = other B-ring oxygenation pattern; SO_4 = sulphate; CAlk = C-alkylation; Acyl = acylation of a phenolic group.

[2] Number of genera for which data are available.

Bremer appears to be comfortable with the notion that the first two are monophyletic groups. However, he has expressed reservations about Cichorioideae which presents problems that he feels are still unresolved (Bremer and Jansen, 1992). Nonetheless, he continues to use Cichorioideae in his treatment (Bremer, 1994). We shall follow this convention here. Even more recently, Bremer (1996; see our Fig. 2–4) added a fourth subfamily, although somewhat tentatively: Carduoideae to contain the single tribe Cardueae. This is a novel approach deriving from strict recognition of taxa based on cladogram topology. Further studies will test its suitability.

In this chapter we examine the flavonoid data to see if any characteristic structural features, or sets of features, exist at the subfamily level. To aid in the search for patterns, we have arranged the data in several ways. Table 18–1 simply shows whether the various structural features are present or absent in the various tribes with a summary entry for each of the three subfamilies. The first column gives the number of genera of each tribe for which relevant data are available. The next eight columns refer to the occurrence of flavonoid classes in the order: flavonols, flavones, anthochlors (aurones and/or chalcones), flavanones, dihydroflavonols, dihydrochalcones, isoflavones, and *C*-glycosylflavones. (Flavan derivatives, the biflavonoid, and 5-deoxyflavonoids, whose occurrences are very restricted, are not included.) The next 11 columns report occurrences of various structural elaborations in the order: *O*-methylation, 6-oxygenation, 8-oxygenation, 6,8-dioxygenation, lack of B-ring oxygenation, 3',4',5'-trioxygenation, 2',4',5'-trioxygenation, other B-ring oxygenation, sulphation, *C*-alkylation, and direct acylation of a phenolic function.

Note that the data for two of the larger tribes, Heliantheae and Inuleae, are presented in different ways in the tables. This facilitates comparisons between older views of the tribes and more recently published ideas (Bremer, 1994). Thus, in the case of Heliantheae we have entries for Heliantheae s.l., Heliantheae s.s., Helenieae (including Tageteae), Helenieae (excluding Tageteae), and Tageteae *per se*. For Inuleae, we have an entry for the Inuleae s.l., as well as separate entries for Inuleae s.s., Gnaphalieae, and Plucheeae.

While Table 18–1 adequately reports absences, it is misleading with regard to presences. Several of the structural features occur in only one, or at most, a small number of genera within some of the tribes. In fact, lack of these features would be the more commonly encountered character state. To minimize this problem, the data have been recast in Table 18–2 in terms of frequencies (%) of occurrence at the generic level of individual flavonoid classes and certain structural features within each tribe. Table 18–2 also presents summary values for the three subfamilies.

It is obvious that a comparison of the flavonoids of the three subfamilies involves comparing databases of grossly unequal size. Seven of the nine genera and nearly half of the species of Barnadesioideae have been examined for their flavonoids (Bohm and Stuessy, 1995; Mendiondo et al., 1997), whereas there are many genera of the other two subfamilies that have never been examined for flavonoids (or any chemical components for that matter). Despite having to work with an imperfect sampling of the family, we feel that the comparisons we make are valid since they are based on studies of species representing at least 120 genera of Cichorioideae and over 300 genera of Asteroideae. [These values reflect names used by Bremer and colleagues (in Bremer, 1994).]

Owing to wide variation in frequency of occurrence of flavonoids within tribes (e.g., Table 18–2), and owing in part to fairly large sample sizes (e.g., over 100 genera in Heliantheae s.l. alone), use of multivariate clustering and ordination techniques are judged inadequate with the presence/absence data. Such analyses, in fact, have been attempted by us and have yielded no profitable insights. A more fruitful approach is to inspect the data carefully and make judgements about their taxonomic import. We believe it is instructive to divide the data into two parts, first, variation in basic structural aspects (columns 2-9 in Tables 18–1 and 18–2), and second, variation in more minor structural features or "adornments" (columns 10-20 in Tables 18–1 and 18–2).

Barnadesioideae exhibit the simplest flavonoid profile of the three subfamilies with kaempferol and quercetin 3-*O*-mono- and 3-*O*-diglycosides present in all species tested. In addition to this basic profile, isorhamnetin was observed in one species of *Barnadesia* and traces of eriodictyol in three others. Features otherwise common in the Asteraceae, but not seen in Barnadesioideae, include extra oxygenation in the A-ring, B-ring substitution patterns other than the kaempferol and quercetin types, all other flavonoid classes, and such modifications as sulphation, *O*-, or *C*-alkylation. Isorhamnetin is widespread in the family so its occurrence in *Barnadesia* is not particularly unusual. Higher levels of *O*-methylation, however, are very common in the other subfamilies, particularly in the case of some genera in Asteroideae. Levels of *O*-methylation in each tribe are summarized in Table 18–3. The absence of other *O*-methylated products in Barnadesioideae can be taken to indicate a lack of enzymatic specialization seen in other taxa. The presence of trace amounts of the flavanone eriodictyol should, likewise, not be greeted with surprise. The common flavanones, naringenin and eriodictyol, which are the presumed intermediates in the biosynthesis of all flavonoids with the common patterns of B-ring oxygenation, would be expected to be present in most species, albeit in low concentrations.

By comparison, both of the other two subfamilies offer more complicated arrays of pigments, with Asteroideae being the more complex of the two. Quantitatively, some interesting differences exist. Flavones and fla-

Table 18–2. Frequencies of occurrence of flavonoid classes and structural features among genera surveyed in tribes and subfamilies of Asteraceae

		Flavonoid classes and structural feature[1]																		
Taxon	Gen[2]	Flol	Flav	Anth	Fon	DiHF	DiHC	IsoF	CGly	OMe	6Ox	8Ox	68Ox	Bare	345	245	Other	SO_4	CAlk	Acyl
Asteroideae																				
Anthemideae	44	0.77	0.68	—	0.18	0.02	0.05	—	0.16	0.64	0.73	0.09	0.05	0.09	0.09	0.02	0.02	—	0.02	0.02
Astereae	48	0.83	0.52	0.04	0.20	0.04	0.02	—	0.14	0.63	0.41	0.09	0.20	0.14	0.19	0.04	0.02	—	—	—
Calenduleae	3	0.67	0.33	—	—	—	—	—	—	0.33	—	—	—	—	0.33	—	—	—	—	—
Eupatorieae	32	0.69	0.69	0.06	0.28	0.03	—	—	0.03	0.78	0.69	0.06	0.09	0.09	0.13	0.09	—	0.09	—	—
Heliantheae s.l.	110	0.67	0.47	0.30	0.23	0.06	—	0.05	0.05	0.63	0.46	0.05	0.17	0.03	0.01	—	0.01	0.05	0.05	0.03
Heliantheae s.s.	75	0.61	0.47	0.41	0.21	0.04	—	0.05	0.05	0.52	0.36	0.04	0.17	0.04	—	—	—	0.01	0.05	0.01
Helenieae (-Tag)	35	0.80	0.49	0.06	0.26	0.11	—	0.03	0.09	0.86	0.69	0.09	0.17	—	0.03	—	0.03	0.11	0.03	0.06
Helenieae (+Tag)	42	0.83	0.45	0.05	0.21	0.10	—	0.02	0.07	0.88	0.74	0.07	0.14	—	0.02	—	0.02	0.10	0.02	0.05
Inuleae s.l.	43	0.72	0.60	0.16	0.35	0.16	0.05	0.02	0.02	0.72	0.49	0.14	0.23	0.16	0.07	—	—	0.07	0.09	0.02
Inuleae s.s.	12	0.75	0.67	—	0.42	0.33	—	—	—	0.83	0.83	—	0.08	—	0.17	—	—	0.17	—	—
Gnaphalieae	25	0.68	0.64	0.28	0.36	0.04	0.08	—	—	0.72	0.32	0.20	0.32	0.28	0.04	—	—	—	0.16	0.04
Plucheeae	6	0.83	0.33	—	0.17	0.33	—	0.17	0.17	0.50	0.50	0.17	0.17	—	—	—	—	0.17	—	—
Senecioneae	21	0.95	0.33	0.10	0.14	0.05	—	—	—	0.24	—	0.10	—	—	0.14	—	—	—	—	0.05
Tageteae	7	1.0	0.29	—	—	—	—	—	—	1.0	1.0	—	—	—	—	—	—	—	—	—
Barnadesioideae																				
Barnadesieae	7	1.0	—	—	0.14	—	—	—	—	0.14	—	—	—	—	—	—	—	—	—	—
Cichorioideae																				
Cardueae	26	0.73	0.77	0.12	0.16	0.04	—	—	0.24	0.60	0.40	0.12	—	0.08	0.08	—	—	0.04	—	—
Lactuceae	39	0.67	0.85	0.03	0.08	0.03	—	—	0.08	0.15	0.13	—	—	—	0.03	0.23	—	—	—	—
Liabeae	4	1.0	—	—	0.25	—	—	—	—	0.25	0.25	—	—	—	—	—	—	—	—	—
Mutisieae	22	0.55	0.64	0.05	0.32	0.10	—	0.05	0.10	0.55	0.27	0.05	—	0.10	—	—	—	—	—	0.06
Vernonieae[3]	35/15	0.46	0.71	0.13	0.27	—	—	—	0.07	0.80	0.07	—	—	0.13	0.20	—	—	—	0.13	—

Table 18–2. (Continued)

		Flavonoid classes and structural feature[1]																		
Taxon	Gen[2]	Flol	Flav	Anth	Fon	DiHF	DiHC	IsoF	CGly	OMe	6Ox	8Ox	68Ox	Bare	345	245	Other	SO_4	CAlk	Acyl
Subfamily totals																				
Asteroideae	308	0.75	0.54	0.15	0.20	0.06	0.02	0.02	0.05	0.68	0.50	0.09	0.14	0.07	0.08	0.02	0.01	0.04	0.03	0.02
Barnadesioideae	7	1.0	—	—	0.14	—	—	—	—	0.14	—	—	—	—	—	—	—	—	—	—
Cichorioideae[4]	126/101	0.62	0.74	0.07	0.17	0.04	—	0.01	0.12	0.42	0.22	0.04	—	0.08	0.06	0.09	—	0.01	0.02	0.01

[1] Flol = flavonol; Flav = flavone; Anth = anthochlor (chalcone and/or aurone); Fon = flavanone; DiHF = dihydroflavonols; DiHC = dihydrochalcone; IsoF = isoflavone; CGly = *C*-glycosylflavone; OMe = *O*-methylation; 6Ox = 6-oxygenation; 8Ox = 8-oxygenation; 68Ox = 6,8-dioxygenation; Bare = B-ring deoxyflavonoid; 345 = 3',4',5'-trioxygenation; 245 = 2',4',5'-trioxygenation (isoëtin type); Other = other B-ring oxygenation pattern; SO_4 = sulphate; CAlk = C-alkylation; Acyl = acylation of a phenolic group.

[2] Number of genera for which data are available.

[3] Flavonol and flavone values for Vernonieae come from all sources (35 genera) but primarily from the survey of Harborne and Williams, 1977. Values for the remaining flavonoid classes and structural features come from 15 genera described by other authors. See Chapter 17 for details.

[4] The Vernonieae dilemma is also reflected in the subfamily values for Cichorioideae.

Table 18–3. Levels of flavonoid *O*-methylation in tribes of Asteraceae

		Number of genera having given level of *O*-methylation								
	N[1]	0	1	2	3	4	5	6	7	8
Asteroideae										
Anthemideae	42	33	25	14	16	10	11	—	—	—
Astereae	48	35	23	20	19	21	10	4	—	—
Calenduleae	3	3	1	—	—	—	—	—	—	—
Eupatorieae	32	15	23	13	17	11	8	3	3	1
Heliantheae s.l.	91	68	60	49	42	19	6	—	—	—
Heliantheae s.s.	58	46	36	28	25	11	4	—	—	—
Helenieae	33	22	24	21	17	8	2	—	—	—
Inuleae s.l.	43	32	31	22	19	15	11	3	—	—
Inuleae s.s.	12	6	10	10	8	4	5	3	—	—
Plucheeae	6	4	4	4	4	3	3	—	—	—
Gnaphalieae	25	22	17	8	7	8	3	—	—	—
Senecioneae	21	20	5	4	1	1	—	—	—	—
Tageteae	7	7	5	—	—	1	1	—	—	—
Barnadesioideae										
Barnadesieae	7	7	1	—	—	—	—	—	—	—
Cichorioideae										
Cardueae	25	25	20	12	3	—	—	—	—	—
Lactuceae	39	39	7	2	—	—	—	—	—	—
Liabeae	4	4	1	—	—	—	—	—	—	—
Mutisieae	22	17	9	9	—	1	—	—	—	—
Vernonieae	15	8	10	9	3	1	—	—	—	—

[1] Number of genera for which necessary information is available.

vonols occur in both subfamilies but differences between the two taxa are marked. In Asteroideae flavonols (frequency value 0.75) predominate over flavones (0.54), in contrast to Cichorioideae where flavonols (0.62) have been less often reported than flavones (0.74). It should be noted here that flavonol frequencies in Asteroid tribes are consistently higher than flavone frequencies, and that in Cichorioideae the reverse is true. Frequency values for anthochlors, flavanones, and dihydroflavonols are not radically different for the two subfamilies. Dihydrochalcones occur in members of Asteroideae but not in the other two subfamilies.

Differences are also seen in frequency data for the features involving elaborations of the fundamental flavonoid nucleus (columns 10 to 20 of Tables 18–1 and 18–2). A significant difference between Asteroideae and Cichorioideae lies in the two groups' relative capacities to produce flavonoids with oxygenation at C-6. Half of the tested genera of Asteroideae have

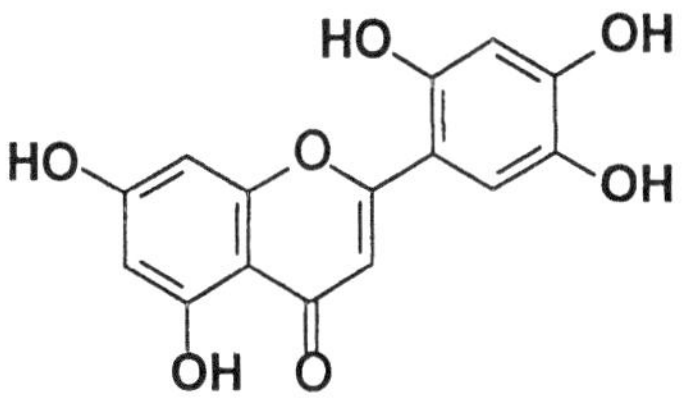

[18-1]

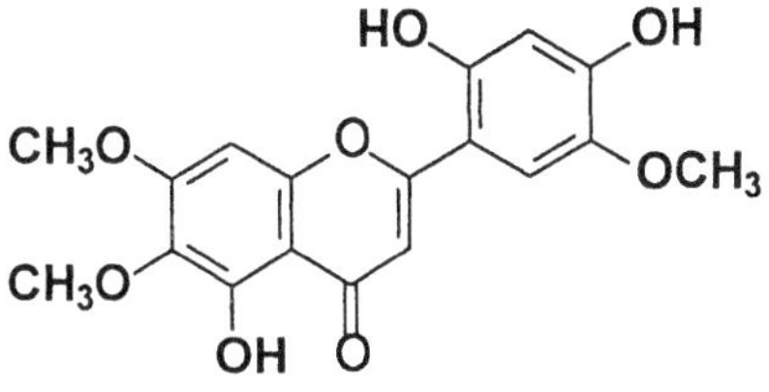

[18-2]

Fig. 18–1. B-Ring oxygenated flavonoids in Cichorioideae (isoëtin) **[18–1]** and Asteroideae (arcapillin) **[18–2]**

flavonoids with an oxygen substituent, either hydroxyl or methoxyl, at C-6, in contrast to the much lower value of 22% in Cichorioideae. Flavonoids with oxygenation at C-8 and those with oxygenation at both C-6 and C-8 also occur with a higher frequency in Asteroideae than in Cichorioideae.

Members of Asteroideae and Cichorioideae have the capacity to make flavonoids with B-ring oxygenation patterns other than those based on 4'- or 3',4' substitutions. This information appears in the tables under four headings: B-ring deoxy, 3',4',5'-B-ring, 2',4',5'-B-ring, and Other B-ring. No such patterns have been reported to occur in Barnadesioideae. Cichorioideae have genera with representatives in the first three categories (no Cichorioideae have entries in the "Other B-ring" category). The most striking feature of Cichorioideae in this regard is the occurrence of isoëtin and its glycosides in nine genera. Isoëtin is 5,7,2',4',5'-pentahydroxyflavone **[18–1]**. Flavonoids with this B-ring oxygenation pattern are also known from members of Astereae, Anthemideae, and Eupatorieae, but in these genera the compounds in question are *O*-methylated, such as "arcapillin" **[18–2]** from *Artemisia capillaris* (Kiso et al., 1984). Other B-ring substitution patterns are confined to Asteroideae. Similarly, possession of sulphates and phenolic acylation are both weak tendencies in Asteroideae and are nearly absent from Cichorioideae.

Other than absence of most flavonoids from Barnadesioideae, and tendencies of more complex flavonoid chemistry in Asteroideae, no specific positive correlations of major structural classes of compounds in Asteraceae (columns 2 to 9 in Table 18–1) occur with morphologically delimited subfamilies. As flavonols occur in all tribes and subfamilies, this is not a useful diagnostic taxonomic character within the family at this higher level, and hence can be ignored. Flavones occur in all tribes in all subfamilies except Barnadesioideae. Because Barnadesioideae is defined primarily by pollen features (Stuessy et al., 1996) and absence of a chloroplast DNA inversion (Jansen and Palmer, 1987b), and because it is conspicuously depauperate in flavonoid constituents (Bohm and Stuessy, 1995), it is not meaningful to attempt a flavonoid characterization of this taxon. The only other flavonoid class represented are the flavanones, which compounds occur throughout the other two subfamilies. Because flavanones occur in all other tribes of the family, therefore, neither is this a useful character for testing subfamilial structure of the family. None of the other structural classes, i.e., anthochlors, flavanones, dihydroflavonols, dihydrochalcones, or isoflavones, show correlations with subfamilial delimitation either. The patterns are of nearly pervasive occurrence (flavanones to dihydroflavonols) or nearly complete absence (dihydrochalcones and isoflavones).

Chapter 19

Flavonoids at the Tribal Level

It is the purpose of the present chapter to look at flavonoid occurrence patterns at the tribal level to see if any characteristic flavonoid feature or collection of features emerges. Since several major reassessments of tribal boundaries have appeared recently, we are presented with opportunities to examine the usefulness of flavonoids in arguing for or against the new alignments. This situation is particularly well represented by the recent work of Bremer and his associates (in Bremer, 1994) who discuss historical views of tribal boundaries and offer some newer views, particularly with regard to Inuleae s.l. and Heliantheae s.l. In addition to the tribes currently recognized by Bremer and colleagues (Bremer, 1994), some smaller tribes will be commented upon toward the end of this chapter, including Coreopsideae, Eremothamneae, Flaverieae, Madieae, Nassauvieae, and Tarchonantheae.

Several attempts to apply macromolecular data to discern relationships within the family have been made, with varying degrees of success. Tribal relationships were discussed in terms of plastocyanin amino acid sequence data (Haslett et al., 1977; Boulter et al., 1978). Results from those studies, which involved 22 species representing eight tribes, were only in weak agreement with relationships based upon morphological features (Boulter et al., 1978). Recent macromolecular studies of Asteraceae (Kim et al., 1992, and citations therein) using *rbc*L sequence data have greatly enhanced our understanding of relationships among the tribes, especially in view of the reassessment of subfamilies based on the findings of Jansen and Palmer (1987a,b, and subsequent work).

We are primarily concerned with micromolecular data in the present work, of course, so a few comments on applications involving compounds other than flavonoids are in order. Acetylenic compounds (or polyynes, as they are also known; Fig. 19–1), for example [**19–1**, terthienyl], have attracted a great deal of attention. Much of the early work on the natural occurrence of these compounds in plants was reviewed by Bohlmann et al. (1973). A series of reviews dealing with acetylenes in tribes of Asteraceae appeared in the early 1990s: Cynareae (Christensen and Lam, 1990), Heliantheae (Christensen and Lam, 1991a), Astereae (Christensen and Lam,

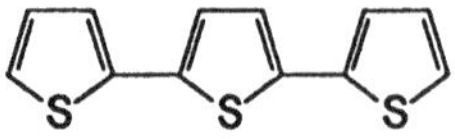

[19-1]

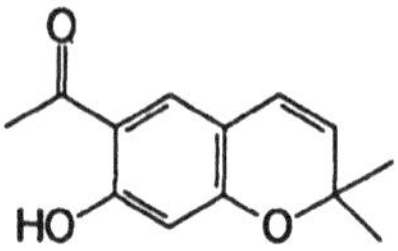

[19-2]

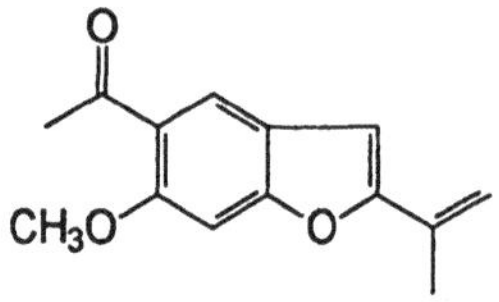

[19-3]

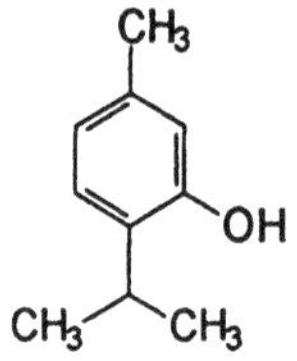

[19-4]

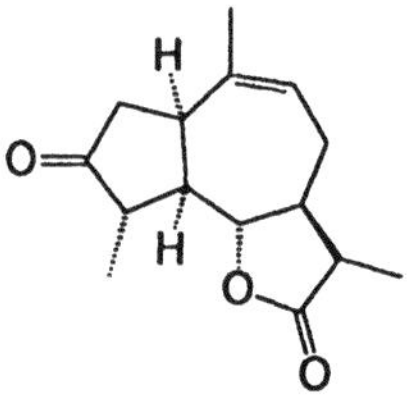

[19-5]

Fig. 19–1. Other secondary plant products useful taxonomically within Asteraceae. Acetylenic compounds (polyynes) **[19–1]**; chromenes (prenylated acetophenones) **[19–2]**; benzofurans **[19–3]**; monoterpenoids (thymol) **[19–4]**; sesquiterpene lactones **[19–5]**

1991b), and Anthemideae (Christensen, 1992). The authors of the tribal surveys point out that, although many acetylenic compounds are widespread in the family, some compounds appear to be restricted to individual tribes. The patterns of occurrence of prenylated acetophenones, i.e., chromenes **[19–2]** and benzofurans **[19–3]**, within the family have also been addressed (Proksch and Rodriguez, 1983; Proksch, 1985). More recently, Proksch and Kunze (1996) discussed the application of these compounds to relationships at the tribal, inter- and intra-generic levels. Of particular relevance to us at this point is the observation that chromene and benzofuran derivatives are particularly abundant in Heliantheae and Eupatorieae adding support to the view that these tribes are closely related to each other. Further support for this relationship comes from the number and variety of derivatives of thymol **[19–4]** elaborated by members of these two tribes (Zdero and Bohlmann, 1990d). The occurrence of sesquiterpene lactones, e.g. **[19–5]**, in the family has been discussed in detail (Seaman, 1982; Seaman and Funk, 1983). While these works are all rich sources of information, a major problem in dealing with any of these compound types, as in the case of flavonoids to be sure, is the lack of a reasonably complete distributional database for any given set of taxa under consideration. The ideal solution to this problem would involve a thorough study of all of these compound types, where appropriate, for whatever taxa are under examination. Rarely, if ever, has this approach been taken.

Some efforts have been made in that direction, however. The most noteworthy in this regard is the voluminous work of Prof. F. Bohlmann and his many colleagues and students at Berlin. The question of chemical relationships among tribes of Asteraceae was a long standing objective of their research program. A recent review article (Zdero and Bohlmann, 1990d) presented a condensed view of relationships among the tribes based upon their data. Playing prominent roles in this work have been sesquiterpene derivatives, acetylenic compounds, thymol derivatives, and, to a somewhat lesser extent, alkaloids. Of more than passing interest to us in the present endeavor is a statement in that review (Zdero and Bohlmann, 1990d) that effectively eliminated flavonoids from consideration! They wrote (p. 7), "Of course also flavanoids (sic) are widespread in the family but they are distributed over all tribes." Wollenweber and Valant-Vetschera (1996) also commented on the usefulness of flavonoid data but were more positive in their appraisal concluding that exudate flavonoids of the family "...appear not to be suited for chemotaxonomic purposes at the tribal or generic level, whereas on the specific level there are good examples for their usefulness in taxonomic studies." This is a reasonable conclusion based on their long-standing interest and contributions in this area. The following view of the subject is much less easily accepted.

In a recent analysis of flavonoid distribution in Asteraceae, Seeligmann (1996, p. 159) concludes "...that flavonoids should not be used in the Compositae as intertribal evolutionary markers". Whereas the conclusions of Wollenweber and Valant-Vetschera (1996), cited in the preceding paragraph, were based upon extensive contributions to and reference to the literature, Seeligmann's view was based upon his almost total reliance on the four volumes of *Flavonoids – Advances in Research* as his source of data. Significant literature that has appeared during the last few years was overlooked in the preparation of that treatment. For example, he states that the flavonoid chemistry of Mutisieae is poorly understood (no argument there) with information being available for only three (of 89) genera. Our survey of the literature revealed information on 18 genera. More troublesome, however, is the opinion expressed (p. 165) that, "...so far, the flavonoids have failed to pave the road which might lead us to the origins of our family" (emphasis ours). Several recent studies, including flavonoid analyses, have clearly implicated Calyceraceae as sister group to Asteraceae, and

Barnadesioideae as the subfamily of Asteraceae most likely basal to the rest of the family. Flavonoid profiles allow one to go even further in speculating which genera might have been involved in the evolutionary events leading to establishment of Asteraceae. In this respect, flavonoid data have clearly been useful.

It is obvious that clear-cut differences among the tribes do not exist, and therefore a flavonoid profile could not be used as a precise diagnostic tool for taxonomic purposes. Indeed, it would be nice if that were the case, but it is naïve to expect that any single character, micromolecular, macromolecular, morphological, or whatever, would assort so conveniently. It is safe to say, in fact, that many features of the family, taken individually, would fail to show clear-cut differences among all tribes. The family has undergone a remarkable amount of diversification in its history, during which time morphological convergence has certainly occurred repeatedly. There is no doubt that convergence in establishment of flavonoid profiles has also occurred. In fact, this would be anticipated given the limited number of flavonoid structural classes that are possible. Thus, we agree in principal with those workers who have concluded that flavonoids are not useful in explicitly defining tribes. However, we disagree with the idea that flavonoids lack value in examining evolutionary processes. In evolutionary studies, one rarely has the good fortune to deal with absolutes; one has to be satisfied with detecting trends. As the reader has already seen in our discussion of trends at the subfamily level, and will see again in the tribal discussions below, flavonoids should not be dismissed so easily!

A more analytical, and in our view, reasonable approach to the study of flavonoids as an aid to understanding tribal relationships comes from the work of Prof. O. Gottlieb. It is his contention that the oxidation levels and structural specializations of sesquiterpene lactones afford highly useful indicators of evolutionary events in the family. He and his coworkers also believe that flavonoid structures are inherently useful for the same purpose. A major discussion of the application of sesquiterpene lactone and flavonoid structures to the question of tribal relationships in the Asteraceae appeared in 1987 (Emerenciano et al., 1987). In this approach "Evolutionary Advancement Indices" were calculated for each of three structural features: (1) occurrence of flavonoid glycosylation; (2) degree of flavonoid *O*-methylation; and (3) the degree of "protection" of phenolic hydroxyl groups, which is a measure of the ratio of protected groups to total hydroxyl groups in the molecule.

The indices calculated in the Gottlieb treatment would seem to be subject to the same sources of error that were discussed above, however. For example, in the calculation of the *O*-methylation index, it is difficult to see how one can get accurate values when so many studies have focused on external (glandular), lipophilic compounds without equal attention having been given to the vacuolar, glycosylated compounds. It is also difficult to assess the significance of some of these data because many taxa produce only glycosylated flavonoids. With these caveats expressed, it is nonetheless interesting to see that certain trends do indeed appear to emerge from the treatments. Although the authors' graphic presentations of the results are difficult to interpret, it is stated in the discussion that the tribes Vernonieae, Cardueae, Mutisieae, and Lactuceae exhibit flavonoids with smaller variation in oxidation in both A and B rings. Similarly, Heliantheae s.l., Inuleae s.l., Eupatorieae, and Astereae contain compounds with relatively greater variation level of oxidation in both rings. These workers found Anthemideae to lie in a somewhat intermediate position between the two assemblages just listed, and took that as support for Wagenitz's (1976) placement of Anthemideae in his Asteroideae Group 2, but with the reservation that pollen structure of Anthemideae is not consistent with this group of tribes.

Subfamily Asteroideae

Tribe Anthemideae

Detailed flavonoid information on Anthemideae appears in Chapter 7 (see also Greger, 1977). Owing to size, floristic conspicuousness, and long-time use by people, certain genera have been studied in much greater detail than others, prominent among which are *Achillea*, *Artemisia*, and *Tanacetum*. It is not surprising, then, that the flavonoid chemistry of these genera appears to be richer than many of the other genera that comprise the tribe. Most of the problem lies simply in lack of detailed studies of the majority of genera.

Anthemideae display a moderately complex flavonoid chemistry. Flavones and flavonols are about equally represented in the larger genera. Reference to Table 18-1 shows that anthochlors (aurones and chalcones) and isoflavones have not been reported from the tribe and that dihydrochalcones, flavanones, dihydroflavonols, and *C*-glycosylflavones are present in only a few members. Of the structural modifications reported, only flavonoid sulphates are missing. Frequencies of some of these modifications, however, are quite low (Table 18–2).

Additional insights into the biosynthetic capacities of Anthemideae appear in Table 19–1. This table shows the number of oxygenation patterns elaborated by members of the tribe (for which appropriate data are available), as well as the level of *O*-methylation achieved. This table, along with equivalent ones below for Astereae and Eupatorieae, record the obvious variable biosynthetic capacities of the genera that comprise these tribes. As

Table 19–1. Oxygenation patterns, including combinations, and levels of *O*-methylation in Anthemideae

Genus	Oxygenation Pattern[1]												Me[2]
	K	Q	M	K6	Q6	M6	K8	Q8	M8	K68	Q68	Other	
Achillea	+	+	–	+	+	–	+	–	–	+	–	BN	0–5
Ajania	+	+	–	–	+	–	–	–	–	–	–	–	0,2,3,5
Anacyclus	–	+	–	–	–	–	–	–	–	–	–	–	0,1
Anthemis	+	+	–	+	+	–	–	–	–	–	–	–	0–3
Arctanthemum	+	–	–	–	–	–	–	–	–	–	–	–	1
Argyranthemum	?[3]	?	–	–	–	–	–	–	–	–	–	–	?
Artemisia	+	+	+	+	+	+	–	–	–	+	–	+[4]	0–5 (&BN)
Chamaemelum	+	–	–	–	–	–	–	–	–	–	–	–	0
Chrysanthemum	+	+	–	–	+	–	–	+	–	–	–	–	0,1
Cladanthus	–	+	–	–	+	–	–	–	–	–	–	–	0,5
Coleostephus	+	+	–	–	+	–	–	–	–	–	–	–	0,1
Cotula	+	+	–	–	+	–	–	–	–	–	–	–	0,1,3
Dendranthema	+	+	–	+	+	–	–	–	–	–	–	BN	0,1,3
Endopappus	+	+	–	+	–	–	–	–	–	–	–	–	0,1
Eriocephalus	+	+	–	+	+	–	–	–	–	–	–	–	0–4
Filifolium	+	+	–	+	–	–	–	–	–	–	–	2'5'	0,1
Glossopappus	–	+	–	–	+	–	–	–	–	–	–	–	0
Gymnopentzia	+	–	–	–	–	–	–	–	–	–	–	–	1
Heliocauta	–	+	+	–	–	–	–	–	–	–	–	–	0
Heteranthemis	–	+	–	–	+	–	–	–	–	–	–	–	1
Hippolyta	–	+	–	–	+	–	–	–	–	–	–	–	5
Inulanthera	+	–	–	–	–	–	–	–	–	–	–	–	0
Ismelia	–	+	–	–	+	–	–	–	–	–	–	–	0,1
Lepidophorum	+	+	–	–	+	–	–	–	–	–	–	–	0–2
Leptinella	–	+	–	–	–	–	–	–	–	–	–	–	1
Leucanthemopsis	–	+	–	–	+	–	–	+	–	–	–	–	0
Leucanthemum	+	+	–	–	+	–	–	–	–	–	–	–	0,1
Leucocyclus	+	+	–	+	+	–	–	–	–	–	–	–	0–5
Lonas	+	+	–	–	–	–	–	–	–	–	–	–	0
Matricaria	+	+	–	–	+	–	–	–	–	–	–	–	0–4
Nivellea	+	+	–	–	–	–	–	–	–	–	–	–	0
Oncosiphon	–	+	–	+	+	–	–	–	–	–	–	–	0,2–5
Otanthus	+	+	+	–	+	+	–	–	–	–	–	–	0,1,3–5
Pentzia	+	+	–	+	+	–	–	–	–	–	–	–	0,2–5
Plagius	?[3]	?	–	–	–	–	–	–	–	–	–	–	?
Rennera	–	+	–	–	+	–	–	–	–	–	–	–	4
Santolina	+	+	–	+	+	–	–	–	–	–	–	–	0–2
Schistostephium	+	–	–	+	–	–	–	–	–	–	–	–	3
Seriphidium	+	+	+	+	+	+	–	+	–	–	–	–	0–4
Sphaeromeria	–	+	–	–	+	–	–	–	–	–	–	–	5

Table 19–1. (Continued)

	Oxygenation Pattern[1]												
Genus	K	Q	M	K6	Q6	M6	K8	Q8	M8	K68	Q68	Other	Me[2]
Stilpnolepis	–	–	+	–	+	+	–	–	–	–	–	–	2,3
Tanacetum	+	+	–	+	+	–	–	–	–	–	–	–	0–5
Tripleurospermum	+	+	–	–	+	–	–	–	–	–	–	–	0,1
Ursinia	+	–	–	+	–	–	–	–	–	–	–	–	3

[1] K = 4'-hydroxylation; Q = 3',4'-hydroxylation; M = 3',4',5'-trihydroxylation; K6 = 4'-hydroxylation with extra oxygenation at C-6; Q6 = as for K6 but with 3',4'-oxygenation; M6 as for K6 but with 3',4',5'-trioxygenation; K8, Q8, M8, K68, Q68 = corresponding combinations with 8- and, 6,8-oxygenation.

[2] Me = levels of flavonoid *O*-methylation.

[3] ? = paper reported only "flavonols."

[4] Other compounds included B-ring deoxy (BN), those with 2',4',5'-trioxygenation, and compounds with 5,8,4'- and 5,6,4'-substitutions.

is often the case in wide-ranging comparative studies of this sort, unusual distinguishing features rarely occur in all of members of the group.

The common occurrence in the family of flavonoids with the apigenin or kaempferol (4') and luteolin or quercetin (3',4') oxygenation patterns provides little information on relationships. Of more limited occurrence, and therefore possibly more likely to be of value to us here, is the presence of flavonoids with additional oxygenation in the A-ring, features that are identified in Table 18–2 as 6-oxygenation, 8-oxygenation, and 6,8-dioxygenation. It is also obvious from Table 19–1 that flavonoids with extra oxygenation at C-6 combined with a 3',4'-dioxygenated B-ring are the predominant type in the tribe. Other oxygenations are known but, other than in the case of *Artemisia*, they are scattered in several other genera. The flavonoid profile of *Artemisia* is one of the richest in Anthemideae, when one considers both the nature of oxygenation patterns and level of *O*-methylation.

Tribe Astereae

Detailed information on members of Astereae was presented in Chapter 8. Flavones are somewhat less well represented in Table 18–2 than are flavonols, 0.52 vs. 0.83. Anthemideae and Astereae exhibit the same frequency of occurrence of 8-oxygenated flavonoids (0.09), whereas Astereae have four times the level of occurrence of 6,8-dioxygenated flavonoids compared to Anthemideae (0.20 vs. 0.05).

Table 19–2. Oxygenation patterns, including combinations, and levels of *O*-methylation in Astereae

Genus	Oxygenation pattern[1]												
	K	Q	M	K6	Q6	M6	K8	Q8	M8	K68	Q68	Other[2]	O-Me[3]
Amphiachyris	–	+	+	–	–	–	–	–	–	–	–	M68	2,4–6
Aster	+	+	+	–	+	–	+	+	+	–	+	BN	0,4,5
Baccharis	+	+	+	+	+	+	+	–	–	+	+	BN	0–5
Bellis	+	–	–	–	–	–	–	–	–	–	–	–	0
Callistephus	+	–	–	–	–	–	–	–	–	–	–	–	0
Centipida	–	–	+	–	–	+	–	–	–	–	–	–	5,6
Ceruana	+	–	–	–	–	–	–	–	–	–	–	–	0
Chrysopis	–	+	–	–	–	–	–	–	–	–	–	–	0
Chrysothamnus	+	+	–	–	–	–	–	–	–	–	+	–	0–4
Conyza	+	+	+	+	–	–	+	+	+	–	–	–	0–5
Conyzanthus	+	–	–	–	–	–	–	–	–	–	–	–	0
Dichrocephala	+	+	–	–	–	–	–	–	–	–	–	–	0
Diplostephium	+	+	–	–	–	–	–	–	–	–	–	–	0–2
Egletes	–	+	–	–	–	–	–	+	–	–	–	–	4
Ericameria	+	+	–	–	+	–	–	–	–	–	–	BN	0–3
Erigeron	+	+	–	–	+	–	–	–	–	–	–	–	0,1
Euthamia	+	–	–	–	–	–	–	–	–	–	–	–	0
Felicia	+	–	+	–	–	–	–	–	–	–	–	–	1
Grangea	–	+	–	–	–	–	–	–	–	–	–	–	0
Grindelia	+	+	–	+	+	–	–	–	–	–	–	–	0–4
Gutierrezia	+	+	+	+	+	+	+	+	+	+	+	2'4'5'	0–6
Gymnosperma	+	+	+	–	–	+	–	–	–	+	+	2'4'5'	0,2–6
Haplopappus	+	+	+	–	+	–	–	–	–	–	–	BN	0–4
Hazardia	+	+	+	+	+	–	–	–	–	–	–	–	0–4
Heteromma	–	+	+	–	–	–	–	+	+	–	–	–	3–5
Heteropappus	–	+	–	–	–	–	–	–	–	–	+	–	0,1,5
Heterothalamus	–	–	–	–	–	–	–	–	–	–	–	BN	0,1
Heterotheca	+	+	–	+	+	–	–	–	–	–	–	–	0–4
Isocoma	+	+	–	+	–	–	–	–	–	–	–	–	0–4
Kalimeris	+	+	–	–	–	–	–	–	–	–	–	–	0,1
Lepidophyllum	–	+	–	–	+	–	–	–	–	–	–	–	4
Machaeranthera	–	+	–	–	+	–	–	–	–	–	–	–	4
Microglossa	–	+	–	–	–	–	–	–	–	–	+	–	4
Nidorella	–	+	–	–	–	–	–	–	–	–	–	–	3
Olearia	+	+	–	+	+	–	–	+	–	+	–	–	2,3
Parastrephia	+	+	–	+	–	–	–	+	–	+	–	–	1,2,4
Prionopsis	+	+	–	+	–	–	–	–	–	–	–	–	0–2
Psiadia	+	+	+	+	–	–	+	+	–	+	–	2'3'4'5'	0–5
Pteronia	+	+	–	–	+	–	–	–	–	+	+	–	0,2–4
Rigiopappus	+	+	–	–	–	–	–	–	–	–	–	–	0,1
Solidago	+	+	–	–	–	–	–	–	–	–	–	–	0,1

Table 19–2. (Continued)

Genus	Oxygenation pattern[1]												
	K	Q	M	K6	Q6	M6	K8	Q8	M8	K68	Q68	Other[2]	O-Me[3]
Stephanodoria	+	+	–	–	–	–	–	–	–	–	–	–	0,1
Tracyina	+	+	–	–	–	–	–	–	–	–	–	–	0
Trimorpha	+	+	–	–	–	–	–	–	–	–	–	–	0
Vanclevea	+	+	–	–	–	–	–	–	–	–	–	–	2–4
Xanthisma	–	+	–	–	–	–	–	–	–	–	–	–	0
Xanthocephalum	+	+	+	–	–	–	–	–	–	–	–	–	0–3
Xylothamnia	+	–	–	–	–	–	–	–	–	–	–	–	3

1 K = kaempferol-type B-ring (4'); Q = quercetin-type B-ring (3',4'); M = myricetin-type B-ring (3',4',5'); K6 = kaempferol-type B-ring plus 6-oxygenation; etc.
2 BN = B-ring deoxyflavonoid.
3 Levels of *O*-methylation observed in the genus.

Astereae exhibit one of the highest levels of oxygenation pattern diversity in the family owing to the extraordinary flavonoid richness of *Gutierrezia*, with 15 different oxygenations, and *Gymnosperma* with 10. With representatives having six *O*-methyl groups these two genera are among the highest in that category as well. These data are presented in Table 19–2. A comparison of the levels of *O*-methylation in tribes of Asteraceae appeared as Table 18–3. The complexity of oxygenation patterns in *Gutierrezia* and *Gymnosperma* are too unwieldy to be accommodated easily in the table. Both of these genera have members that accumulate an array of flavonoids based on 2',4',5'-oxygenation in combination with extra A-ring oxygenation. This pattern is seen elsewhere in the family only in *Ageratum* and *Brickellia*, both in Eupatorieae.

Tribe Calenduleae

As pointed out in Chapter 9, the amount of flavonoid information for this tribe is too small to allow any serious discussion of relationships. *Dimorphotheca sinuata* has the capacity to make flavonoids with the 3',4',5'-trioxygenation pattern (delphinidin). It would be of critical interest to explore this tribe in more detail, especially with regard to its capacity to accumulate flavonoids with the 3',4',5'-oxygenation pattern.

Tribe Eupatorieae

Flavonoid profiles of Eupatorieae were described in Chapter 10. As mentioned above, a few members of the tribe have the capacity to produce some of the most highly substituted flavonoids seen in the family in terms of both oxygenation patterns and levels of *O*-methylation. Data for Eupatorieae can be found in Table 19–3. Eupatorieae exhibit a level of 6-oxygenation (0.69), which approaches that of Anthemideae (0.73) (Table 18–2). The tribe is distinguished by having genera with one of the highest levels of *O*-methylation seen in the family. The richness of *O*-methylated derivatives can be appreciated by examining the right hand column of Table 19–3 where one sees several genera with penta-, hexa-, and heptamethoxy flavonoids and one, *Ageratum*, with an *O*-methylation level of eight.

Tribes Heliantheae and Helenieae

Three major treatments of Heliantheae s.l. have appeared in recent years, the presentation that appeared in *The Biology and Chemistry of the Compositae* by Stuessy (1977), the re-evaluation of the tribe by Robinson (1981), and the recent cladistic treatment by Karis and Ryding (1994b). The first two treatments deal with attempts to delineate subtribes and will be discussed in the next chapter. The Karis and Ryding (1994b) treatment, however, has particular relevance in the present context, because it resulted in re-instatement of Helenieae, a tribe whose "passing" was documented in the 1977 symposium volumes. Karis and Ryding (1994b) noted from their cladistic analysis that, whereas Heliantheae s.s. appear to be holophyletic, Helenieae are paraphyletic (Karis, 1993b). Further complicating the picture is the inclusion of *Tagetes* and related genera within the newly resurrected Helenieae. These genera have in the past been accorded their own tribe, Tageteae, which enjoys its own chapter in this book (Chapter 16). We are in a position to discuss these re-assortments of genera in terms of flavonoids since information is available for over a hundred genera of Heliantheae s.l., as well as for seven genera of Tageteae.

Karis (1993b) included several species from Eupatorieae in his cladistic analyses, which led to the conclusion that Eupatorieae may be the sister group to Heliantheae and Helenieae. Eupatorieae are also well represented in the flavonoid literature, as discussed in Chapter 10, and earlier in this chapter. The availability of these data should enable us to compare the entire assemblage of taxa that have been brought into play through cladistic treatments. Details of the flavonoid arrays in Heliantheae and Helenieae were presented in Chapter 11. Reference to Table 18–2 shows a predomi-

Table 19–3. Oxygenation patterns, including combinations, and levels of *O*-methylation in Eupatorieae

Genus	Oxygenation pattern[1] K	Q	M	K6	Q6	M6	Other[2]	*O*-Me[3]
Acanthostyles	+	+	–	+	+	–	2'4'5'	1, 3–5
Acritopappus	+	–	–	–	–	–		1
Ageratina	+	+	+	+	+	–	K8, Q8	0–4
Ageratum	+	+	+	–	+	+	M8, M68, 2'3'4'5'	0, 4–8
Aristeguietia	+	–	–	+	–	–		2, 3
Austroeupatorium	–	+	–	–	+	–		1, 2
Bejaranoa	+	–	–	+	–	–		1
Brickellia	+	+	–	+	+	–	2'4'5', K68, 2'3'4'5', BN	0–5
Brickelliastrum	+	+	–	+	–	–		0, 1, 3
Campuloclinum	+	+	–	–	–	–		1
Carphochaeta	+	+	+	+	+	+		1–4
Chromolaena	+	+	+	+	+	+	K8, Q8, BN	0–5
Conoclinium	–	+	+	–	+	–	Q68	3, 5–7
Critonia	+	+	–	–	–	–		0
Decachaeta	+	+	–	+	+	–		1–3
Disnaphia	–	+	–	–	–	–		0
Eupatorium	+	+	+	+	+	–	Q68, BN	0–7
Fleischmannia	–	+	–	–	–	–		3
Flyriella	–	+	–	–	–	–		0–2
Goyazianthus	+	–	–	–	–	–		0
Graziela	+	+	–	+	+	–		1
Lasiolaeana	–	+	–	–	+	–		4
Liatris	+	+	+	+	+	–		0, 1, 3, 4
Lourteigia	+	+	–	+	+	–		1
Mikania	+	+	–	–	+	–		0, 1, 3
Nothobaccharis	–	+	–	–	+	–		1, 3
Ophryosporus	+	–	–	–	–	–		1, 2
Piptothrix	+	+	–	+	+	–		0–3
Stevia	+	+	–	+	+	–	K68	0–5
Steviopsis	+	+	–	+	+	–		0–2
Symphyopappus	+	+	–	–	+	–		1, 5
Urolepis	–	+	–	–	+	–		3, 4

1 K = Kaempferol-type B-ring (4'); Q = Quercetin-type B-ring (3',4'); M = Myricetin-type B-ring (3',4',5'); K6 = Kaempferol-type B-ring plus 6-oxygenation; etc.

2 BN = B-ring deoxyflavonoid.

3 Levels of *O*-methylation in the genus.

nance of flavonols over flavones, regardless of whether one considers the more broadly circumscribed Heliantheae s.l., or two smaller, more restricted tribes, Heliantheae s.s. and Helenieae. Inclusion of Tageteae serves to increase the flavonol frequency value (to 0.83) since these compounds are the predominant flavonoid class in this small group.

Isoflavonoid derivatives have been reported in Asteraceae from members of seven genera, two of which belong to Helenieae, three to Heliantheae s.s., and two to Mutisieae (see Chapter 4 for details). Another flavonoid class, the anthochlors, although present in a few other groups within the family, makes a much greater contribution to the tribal chemistry of Heliantheae s.l. than they do to any of the other tribes. Anthochlors also make a greater contribution to the flavonoid chemistry of Heliantheae s.s. than they do to that of Helenieae.

6-Oxygenated flavonoids make a significantly greater contribution to the profile of Helenieae (0.69) than they do to Heliantheae s.s. (0.36). Although of lesser frequency, 8-oxygenated compounds are also more often found in members of Helenieae (0.09 compared to 0.04). Flavonoids with 6,8-dioxygenation are present in both tribes at about the same frequency. Levels of *O*-methylation are approximately equal in the two as well (Table 18–3). In order to reduce reader impatience, no detailed table(s) of oxygenation patterns and *O*-methylation levels have been prepared for Helenieae and Heliantheae. To show their similarities, however, as well as to provide an opportunity for comparing these structural features among tribes, we have summarized *O*-methylation level and frequency data for all tribes in Table 18–3.

Although 6-oxygenated flavonoids occur in many genera of Heliantheae and Helenieae, they do not occur as widely in these tribes as they do in Anthemideae, Astereae, and Eupatorieae. 8-Oxygenated flavonoids occur in these two tribes less frequently than in the others with the 8,4'-combination having been reported from members of five genera. Flavonoids with the 8,3',4'-combination occur in species of six genera, those with the 6,8,4'-combination have been reported from 12 genera, and those with the 6,8,3',4'-pattern from seven genera. The 8,3',4',5'-combination was observed in one genus. The highest level of *O*-methylation achieved in Heliantheae s.l. is five, that value having been observed in species representing six genera (four in Heliantheae s.s. and two in Helenieae).

The suggestion has been made that Eupatorieae is the sister group to the Heliantheae s.s./Helenieae clade. With the exception of four or five genera, Eupatorieae exhibit a somewhat higher degree of flavonoid structural consistency than does Heliantheae s.l., and have a richer array of highly substituted compounds. An interesting observation, but one that may be compromised by sampling error, involves the apparent lower frequency of occurrence of flavonoids that lack *O*-methylation in Eupatorieae and Helenieae compared to Heliantheae s.s. (see Table 18–3).

Tribe Inuleae

The flavonoid chemistry of Inuleae s.l. was discussed in Chapter 12 with individual genera identified as belonging to Inuleae s.s., Gnaphalieae, or Plucheeae. In the broader view, Inuleae has an overall flavonoid chemistry similar to what was seen in Anthemideae, Astereae, and Heliantheae s.l. in that its members accumulate more flavonols than flavones, have a high frequency of compounds with *O*-methyl groups, and a modest level of extra oxygenation on the A-ring. Inuleae s.l. is characterized by a tendency to accumulate flavanones (0.35), a moderately high frequency of 6,8-dioxygenated flavonoids (0.23), and the highest frequency of flavonoids with unsubstituted B-rings (0.16).

The frequencies of some of these features take on larger significance, however, when one considers their occurrence patterns in the three segregate tribes, Inuleae s.s., Gnaphalieae, and Plucheeae (Anderberg, 1994, and citations therein). Anthochlors, dihydrochalcones, and flavonoids lacking B-ring oxygenation are known only from genera assigned to Gnaphalieae. Conversely, the frequency of dihydroflavonol accumulation by members of Gnaphalieae (0.04) is significantly lower than the values in the other two tribes, 0.33 for both Inuleae s.s. and Plucheeae. Examination of Table 18–3 shows that the array of *O*-methylation in Inuleae s.l. is similar to that in Heliantheae/Helenieae, with the exception that flavonoids with six *O*-methyl groups have been described from some members of three genera of Inuleae s.s. [*Asteriscus* (cited as *Pallenis*), *Chiliadenus*, and *Pulicaria*]. In short, flavonoid distribution patterns tend to support the division of Inuleae s.l. into the three smaller tribes, or at least into discrete units within Inuleae s.l.

Tribe Senecioneae

Senecioneae produce one of the simplest arrays of flavonoids in the subfamily. Flavonoids with the 4'- and 3',4'-oxygenation patterns predominate with only two genera elaborating compounds with 3',4',5'-oxygenation and two having any 8-oxygenated compounds. No 6-oxygenated flavonoids have so far been identified from any member of the tribe. The majority of compounds identified lack *O*-methylation; the level of *O*-methylation is most often zero, and the highest level seen is four (in *Senecio*).

Tribe Tageteae

The flavonoid profile of Tageteae is dominated by quercetagetin derivatives. In most instances, although the sampling of the tribe has been comparatively poor, quercetagetin and patuletin glycosides are the major compounds. There have been reports of higher levels of *O*-methylation (four and five) in species of *Tagetes*, however. This feature would likely be seen more often if a wider sampling of the tribe were done. The occurrence of tetra- and penta methyl ethers ties in well with the general tendency for higher levels of *O*-methylation in the subfamily. Simpler compounds, common flavones and flavonols, have also been reported.

Subfamily Cichorioideae

Tribe Cardueae

The flavonoid chemistry of Cardueae was described in detail in Chapter 9 where information for 26 genera appeared. The flavonoid profiles of the majority of these are dominated by simple flavones and flavonols, while some species of five genera (*Atractylis*, *Carlina*, *Centaurea*, *Notobasis*, *Staehelina*) produce *C*-glycosylflavones. About half of the genera have representatives with two *O*-methyl groups (Table 18–3) with higher levels of *O*-methylation being restricted to three genera, *Centaurea*, *Cirsium*, and *Picnomon*. Members of two of these genera, *Centaurea* and *Picnomon*, along with *Carthamus*, also accumulate flavonoids with extra oxygenation at C-8. 6-Oxygenation is more common in the tribe, occurring in members of eight genera. B-Ring deoxyflavonoids have been identified in species of *Carthamus* and *Centaurea*. B-Ring deoxyflavonoids have also been found in species belonging to Mutisieae. It is interesting to note that the frequency of occurrence of B-ring deoxyflavonoids is about the same in Cichorioideae and Asteroideae, 0.08 and 0.07, respectively); they were not observed in any member of Barnadesioideae.

In a recent review of the distribution of anthocyanins in the family, Harborne (1996) pointed out an interesting tribal difference. Acylated anthocyanins from members of Cardueae (he uses Cynareae) involve succinic acid, whereas in other tribes malonic acid is the acylating acid. An unusual flavonoid type, the flavonolignans, occur in *Silybum* and *Onopordum*, both members of subtribe Carduinae. A wider search for these unusual compounds in the tribe could well repay the effort.

Tribe Lactuceae

Flavonoid information for Lactuceae was presented in detail in Chapter 13. The flavonoid profile of Lactuceae has a predominance of flavones over flavonols. The vast majority of genera have only 4' and/or 3',4'-oxygenation. A species of *Launaea* has been reported to have delphinidin so the capacity for producing the 3',4',5'-oxygenation pattern exists. 6-Oxygenation has been reported from only three of the 39 genera for which data are available. The highest level of *O*-methylation achieved by any member of the tribe is two, that having been found in species of *Hieracium* and *Launaea*. Lactuceae are unique in the subfamily in having species (representing nine genera) with the capacity to make 5,7,2',4',5'-pentahydroxyflavone (isoëtin) **[18–1]** derivatives. The tribe is one of the more conservative and internally consistent flavonoid-producing taxa in the family, with the obvious exception of Barnadesioideae. Sampling has been comparatively good so this generalization seems a reasonable one.

Tribe Mutisieae

Details of the flavonoid constituents of Mutisieae appeared in Chapter 14. Despite the comparatively poor sampling of the tribe for flavonoids (less than a fifth of known genera and often only a single species from a genus), Mutisieae present quite a varied chemistry. Flavonols and flavones are present in similar proportions, 0.55 for the former, 0.64 for the latter. Flavanones are comparatively well represented. 6-Oxygenated flavonoids are known from representatives of four genera. The highest level of *O*-methylation seen in the tribe is four as seen in one species of *Leucheria*; the highest otherwise is two, seen in representatives of several other genera.

Tribe Vernonieae

Flavonoids of Vernonieae were described in Chapter 17 where at least some information is available for nearly three dozen genera. Whereas *Vernonia* has been thoroughly studied (Mabry et al., 1975; King and Jones, 1982), information for most of the other genera comes from a survey (Harborne and Williams, 1977) that reported only that flavones or flavonols were present. It is difficult to know precisely what compounds were reported although it is implied that only common compounds are involved. Information on the occurrence of flavonoids with other structural features, e.g., flavanones, *O*-methylation, 8-oxygenation, 3',4',5'-oxygenation, and *C*-glycosylation, comes from more detailed work on 15 genera. As reports dealing with individual members of various genera have appeared, it has become clear that a richer flavonoid chemistry exists in the tribe than may have been indicated by the earlier surveys. Oxygenation patterns are simple, flavonoids with the 3',4',5'-trioxygenation pattern have been reported from only two genera (*Lychnophora* and *Vernonia*). The report of an 8-oxygenated flavonoid in *Lychnophora affinis* (Le Quesne et al., 1976) has been withdrawn (Le Quesne et al., 1979). A single report of a 6-oxygenated flavonoid (5,4'-dihydroxy-6,7,3'-trimethoxyflavone, cirsilineol) came from the work of Mungarulire (1990) on *Vernonia thomsoniana*.

Tribe Liabeae

What limited information available for this small tribe also appeared in Chapter 17. Chromatographic examination of representatives of six genera (Bohm and Stuessy, unpubl.) clearly showed the presence of eriodictyol, kaempferol, and quercetin *O*-glycosides. Other, more detailed, work confirmed the presence of kaempferol and quercetin and demonstrated quercetin 3-methyl ether and patuletin as components of a species of *Liabum*. At the very least, then, this tribe has the capacity for extra oxygenation at C-6 and *O*-methylation. Additional information is sorely needed.

Subfamily Barnadesioideae

Tribe Barnadesieae

Little more needs to be said about the Barnadesiinae since its flavonoid chemistry, described in detail in Chapter 14, was commented upon in the discussion of subfamilies in Chapter 18. The uniformity of flavonoid profile observed in the tribe, which consists of kaempferol and quercetin glycosides with no flavones, precludes any attempts to apply pigment chemistry to relationships within the tribe as indicated by cladograms in Bremer (1994) and Stuessy et al. (1996). It may be useful to note that we have sampled species from each of the clades represented. More recently published data from Mendiondo et al. (1997) corroborate our earlier findings.

Other tribes

Disagreement in generic affinities has resulted in the description of a number of tribes in addition to those that are recognized by Bremer (1994). Bremer (1994, p. 22) lists several and suggests that they are maintained as subtribes within "larger, insufficiently known complexes...needing further study." A few comments on the flavonoids of members of these tribes and suggestions as to the need for additional sampling would seem in order.

Tribe Coreopsideae

The genera that comprise Coreopsideae are those that reside, quite comfortably according to Bremer (1994), within subtribe Coreopsidinae in his treatment. The single most noteworthy flavonoid characteristic by which this group can be distinguished is the high frequency of occurrence of anthochlors (aurone and chalcone derivatives), 0.41 in Heliantheae s.s., 0.30 in Heliantheae s.l. (The next highest value is 0.28 for Gnaphalieae.) Aspects of this tribal/subtribal flavonoid trait were discussed in some detail by Crawford and Stuessy (1981).

Tribe Echinopeae

A new classification of Cardueae s.l. (Petit et al., 1996) consists of tribe Echinopeae and tribe Cardueae (which consists of subtribes Carlininae and Carduinae). Flavonoid data are not helpful in discussions of this system.

Tribe Eremothamneae

Eremothamnus consists of a single Namibian species (Bremer, 1994). It has been variously dealt with, including having been accorded tribal status (see Bremer 1994, p. 66). According to Bremer (1994) it currently resides as an unassigned genus in Cichorioideae. Unfortunately, no flavonoid data appear to be available for this enigmatic taxon.

Tribe Flaverieae

Of seven genera included in Bremer's (1994) subtribe Flaveriinae of Helenieae, *Clappia*, *Flaveria*, *Haploëstes*, *Jaumea*, *Pseudoclappia*, *Sartwellia* and *Varilla*, we have found flavonoid information for only the first three, and much of this is incomplete. *Flaveria* has been the most thoroughly studied, presumably because of its accumulation of highly sulphated flavonoids (Barron et al., 1988). *Haploëstes* is also credited as accumulating a similar array of flavonols, some with extra oxygenation in the A-ring and some with sulphate groups attached (Al-Khubaizi, 1977). The only information on *Jaumea* deals with anthochlor occurrence in the genus (Crawford and Stuessy, 1981). Sulphated flavonoids, at least at the level exhibited by species of *Flaveria*, are sufficiently uncommon in the family that a detailed search for them in other putative members of this group of genera might prove useful in assessing relationships.

Tribe Madiinae

Madiinae consist of those genera treated as subtribe Madiinae of Helenieae *sensu* Bremer. The group consists of the so-called tarweeds and enjoys an interesting distribution in western North America, principally California, but also present in Mexico, Chile, and Hawaii. Flavonoid profiles of several genera have been established (Crins and Bohm, 1990). On the basis of the information currently available it would appear that the flavonoid profiles of these genera are of more significance in investigating interspecific and intergeneric relationships (e.g., see Chapters 21 and 22).

Tribe Nassauvieae

Nassauvieae consists of those genera currently seen as comprising subtribe Nassauviinae of Mutisieae (Bremer, 1994). We have flavonoid data for seven genera, eight counting the anomalous *Adenocaulon*, recognized as constituting this assemblage. No single flavonoid factor unites the group, although it is difficult to assess the significance of the data since much of the information came via volunteer compounds in studies of sesquiterpenes. As a result, we have very little information concerning vacuolar constituents. *Adenocaulon* (*A. bicolor*) afforded flavonol glycosides (Bohm, unpubl.), while the other genera yielded a diverse array of flavones, flavonols, and flavanones. Some had no B-ring substituents, and extra oxygenation at *C*-6 was common. A study of polar flavonoid derivatives is needed to remedy this all too common situation.

Tribe Tageteae

Many workers have viewed Tageteae as a legitimate tribe, a practice we follow in our presentation of data (Chapter 16). Bremer (1994) considers this assemblage of genera to constitute subtribe Pectidinae in his Helenieae. Flavonoids from several genera, including several species of *Tagetes*, suggest a fairly uniform group. A few flavones have been reported, but all taxa studied accumulate flavonols with extra oxygenation at *C*-6 (quercetagetin seems to predominate) and only a low level of *O*-methylation.

Tribe Tarchonantheae

Keeley and Jansen (1991) pointed out that the three genera thought to comprise this tribe, *Brachylaena*, *Pluchea*, and *Tarchonanthus*, had been placed together in Vernonieae by Cassini (1828), but have been treated in widely different ways by subsequent workers. Prior to their work (Keeley and Jansen, 1991), *Brachylaena* had been placed in six tribes, *Pluchea* in four, and *Tarchonanthus* in six. Analysis of chloroplast DNA restriction site data clearly established that *Brachylaena* and *Tarchonanthus* are more closely related to each other than they are to any other of the 64 genera (representing 15 other tribes) tested. To accommodate the results, these workers described the new tribe Tarchonantheae (Cassini) Keeley and Jansen. *Pluchea* fell more or less within normal bounds for Inuleae in all of their trees.

Sesquiterpene lactone chemistry of these genera provided some support for a placement near Mutisieae, but the data were not sufficient to rule out separate tribal status (Bohlmann and Jakupovic, 1979, 1990; Zdero and Bohlmann, 1987). It would be of some interest, of course, to see if flavonoid data lend any further support for the close association between *Brachylaena* and *Tarchonanthus* as indicated by the cpDNA data. Regrettably, this isn't possible. Bohlmann and Suwita (1979a) isolated 5,7-dihydroxyflavanone (pinocembrin) from *Tarchonanthus trilobus*, but to the best of our knowledge no flavonoid information is available for the other species of *Tarchonanthus* or any of the species of *Brachaelaena* [ca. 15 according to Bremer (1994)]. An added incentive to pursuing flavonoid studies of these genera is the mainland-island distributions of the two genera; *Tarchonanthus* is South African while *Brachylaena* occurs on the African mainland as well as on Madagascar and related islands. Keeley and Jansen (1991) suggested that these taxa may have been isolated from the main line (or lines) of evolution within Asteraceae for a considerable period of time.

Chapter 20

Flavonoids at the Subtribal Level

In this chapter we examine flavonoid occurrence patterns at the subtribal level. Two problems plague us in our efforts to find patterns of flavonoids at this level. The first has been with us from the outset, but it appears to manifest itself most acutely in examining subtribes, namely, sampling error. There are vast gaps in our knowledge of flavonoid occurrence. All too often, for example, one encounters clades [in the treatments in Bremer (1994)] where none of the component genera has been examined chemically. The second problem arises from the smaller range of structural variation that may exist within any subtribe; there are simply fewer variables available for comparison purposes. The combination of fewer OTUs and a reduced range of structural variation precludes the use of sophisticated numerical methods. The discussions below, therefore, will be based on visual comparisons of flavonoid profiles presented in a series of tables. In a few instances we use frequency values when warranted by sample size.

We have noted in the data chapters (7-17) that certain genera are not recognized by Bremer and colleagues (Bremer, 1994), and that others have been placed in tribes (and hence subtribes) different from those in which they appeared in the 1977 Compositae symposium volumes. When appropriate, comments on these realignments will be included. The order of presentation of tribes will follow the same order as in Chapters 7 to 17.

Tribe Anthemideae

Detailed information on 44 genera belonging to this tribe appeared in Chapter 7. Before looking at subtribal flavonoid chemistry, however, we will address two problematic genera, *Pyrethrum* and *Centipeda*. The situation with regard to *Pyrethrum* appears to be simply a case of synonymy. Bremer and Humphries (1993) list *Pyrethrum* Zinn as a synonym of *Tanacetum*. Information for two species identified as *P. corymbosum* and *P. roseum* has appeared in the literature (Zemtsova and Molchanova, 1978, 1979a). In line with the new monographic view, we have included these data in *Tanacetum*. A more complicated situation exists with regard to *Centipeda*. Although *Centipeda* is included by Heywood and Humphries (1977) in their list of genera in Anthemideae, they mark it as a problematic member quoting Skvarla et al. (1977) who place it in Astereae on the basis of pollen characteristics. We treat *Centipeda* as a member of Astereae.

Most of the reports of flavonoids from members of Anthemideae involve, not surprisingly, three of the largest and most conspicuous genera, *Achillea*, *Artemisia*, and *Tanacetum*. Sampling is further skewed by the enormous amount of attention paid to *Artemisia*. Sampling in the tribe has otherwise not been very thorough so that comments on relationships must be taken as speculative. As Heywood and Humphries (1977) pointed out, Anthemideae have traditionally been divided into two main groups, Anthemidineae Dumort. and Chrysantheminae Less., a division based upon whether or not the receptacles have scales. Heywood and Humphries (1977) go on to discuss a number of generic assemblages: the *Chrysanthemum* complex, the *Artemisia* group, the *Anthemis* assemblage, *Achillea*, the African complex, and the "Cotuleae." Taking a more formal approach, Bremer and Humphries (1993) applied cladistic techniques and observed 12 more or less clear-cut groups of genera. The subtribes resulting from this treatment, sorted with respect to geography, are as follows: the mostly Asian group, which consists of Artemisiinae Less., Cancriniinae K. Bremer & Humphries, Handeliinae K. Bremer & Humphries, and Tanacetinae K. Bremer & Humphries; the African group, which consists of Gonosperminae K. Bremer & Humphries, Thaminophyllinae K. Bremer & Humphries, and Ursiniinae K. Bremer & Humphries; and the mainly Mediterranean group (except for Chrysantheminae, which are Macaronesian with some South African members), Achilleinae K. Bremer & Humphries, Anthemidinae Less., Chrysantheminae Less., Leucantheminae K. Bremer & Humphries, and Matricariinae K. Bremer & Humphries (includes the *Cotula* group with members in Australia and South America).

Flavonoids from members of some of these subtribes are presented in Table 20–1 along with information on the number of genera recognized by Bremer (1994) and the number of genera for which we have flavonoid data. The subtribes have been arranged into the three geographic groups discussed by Bremer. The first thing that is evident is that we have no information for four subtribes Cancriniinae, Handeliinae, Gonosperminae, and Thaminophyllinae. This effectively limits comparisons. Despite this unfortunate state of affairs, some observations can be made. An obvious one is the apparent absence of flavonols and only a single report of a flavone in members of Ursiniinae. *Athanasia calva* was shown by Zdero et al. (1991f) to accumulate naringenin and aromadendrin (dihydrokaempferol),

Table 20–1. Distribution of flavonoids in subtribes of Anthemideae *sensu* Bremer

Geographical group/ Subtribe/Genus	Flavonoid class and structural feature[1] Fol	Flav	Fon	CGly	OMe	6Ox	8Ox	68Ox	Bare	345	Other[2]
"Asian group"											
Artemisiinae (8/19)[3]											
Ajania	+	+	–	+	5	+	–	–	–	+	
Arctanthemum	–	–	+	–	0	–	–	–	–	–	
Artemisia	+	+	+	+	5	+	+	+	+	+	Other B-rings
Dendranthema	+	+	–	–	3	+	–	–	+	–	Cyanidin
Filifolium	–	–	+	–	1	–	–	–	–	–	
Seriphidium	+	+	+	+	4	+	+	–	–	+	
Sphaeromeria	+	–	–	–	5	+	–	–	–	–	
Stilpnolepis	–	+	–	–	3	+	–	–	–	+	
Cancriniinae (0/6)											
Handeliinae (0/5)											
Tanacetinae (3/8)											
Heliocauta	+	+	–	–	0	+	–	–	–	–	
Hippolytica	+	–	–	–	5	+	–	–	–	–	
Tanacetum	+	+	–	+	5	+	–	–	–	–	
"African group"											
Gonosperminae (0/3)											
Thaminophyllinae (0/5)											
Ursiniinae (3/7)											
Gymnopentzia	–	–	+	–	1	–	–	–	–	–	
Inulanthera	–	–	+	–	0	–	–	–	–	–	DiHF
Ursinia	–	+	–	–	3	+	–	–	–	–	
"Mediterranean group"											
Achilleinae (7/9)											
Achillea	+	+	–	+	5	+	–	+	+	–	
Anacyclus	+	+	–	–	0	–	–	–	–	–	
Chamaemelum	–	+	–	–	0	–	–	–	–	–	
Cladanthus	+	+	–	–	5	+	–	–	–	–	
Leucocyclus	+	+	–	+	5	+	–	–	–	–	
Otanthus	+	+	–	+	5	+	–	–	–	–	
Santolina	+	+	–	–	2	+	–	–	–	–	
Anthemidinae (1/2)											
Anthemis	+	+	–	–	3	+	–	–	–	–	
Chrysantheminae (4/4)											
Argyranthemum	+	–	–	–	0	–	–	–	–	–	
Chrysanthemum	+	+	–	–	1	+	+	–	–	–	Cyanidin
Heteranthemis	+	–	–	–	1	+	–	–	–	–	
Ismelia	+	–	–	–	1	+	–	–	–	–	
Leucantheminae (7/16)											
Coleostephus	+	–	–	–	1	+	–	–	–	–	

Table 20–1. (Continued)

Geographical group/	Flavonoid class and structural feature[1]										
Subtribe/Genus	Fol	Flav	Fon	CGly	OMe	6Ox	8Ox	68Ox	Bare	345	Other[2]
Glossopappus	+	+	–	–	0	+	–	–	–	–	
Lepidophorum	+	–	–	–	2	+	–	–	–	–	
Leucanthemopsis	+	+	–	–	0	+	+	–	–	–	
Leucanthemum	+	+	–	–	1	+	–	–	–	–	
Nivellea	+	+	–	–	0	–	–	–	–	–	
Plagius											
Matricariinae (11/25)											
Cotula	+	+	–	–	3	+	–	–	–	–	
Endopappus	+	–	–	–	1	+	–	–	–	–	
Eriocephalus	+	+	+	4	+	–	–	–	–		
Leptinella	+	–	–	–	1	–	–	–	–	–	
Lonas	–	+	–	–	0	–	–	–	–	–	
Matricaria	+	+	–	–	3	+	–	–	–	–	
Oncosiphon	+	+	+	–	5	+	–	–	–	–	
Pentzia	+	+	+	–	5	+	–	–	–	–	
Rennera	+	–	–	–	4	+	–	–	–	–	
Schistostephium	–	+	–	–	3	+	–	–	–	–	
Tripleurospermum	–	+	–	–	0	–	–	–	–	–	

1 Fol = flavonols; Flav = flavones; Fon = flavanones; CGly = *C*-glycosylflavones; OMe = highest level of *O*-methylation observed; 6Ox = 6-oxygenation; 8Ox = 8-oxygenation; 68Ox = 6,8-dioxygenation; Bare = B-ring deoxyflavonoids; 345 = 3',4',5'-trioxygenation (myricetin-type B-ring).

2 *Artemisia* has flavonoids with 2',4'-dioxygenated, 2',5'-dioxygenated, and 2',4',5'-trioxygenated flavonoids. DiHF = Dihydroflavonol.

3 Fraction of genera for which flavonoid data are available.

Gymnopentzia bifurcata afforded naringenin 7-methyl ether (sakuranetin) and naringenin 4'-methyl ether (isosakuranetin) (Bohlmann and Zdero, 1978), while *Ursinia nana* ssp. *leptophylla* gave 5-hydroxy-6,7,4'-trimethoxyflavone (Bohlmann and Gupta, 1982). It is tempting to offer this unusual array of flavonols and flavones in support of the individuality of Ursiniinae. *Ursinia* itself has been considered difficult to accommodate in Anthemideae (Heywood and Humphries, 1977, p. 855). Robinson and Brettell (1973c) went so far as to place it in a new tribe, the Ursineae. Bohlmann et al. (1973) presented acetylene occurrence data that could be taken as support for relationship between *Ursinia* and other Anthemideae. A more complete flavonoid profile for *Ursinia*, as well as for the other members of Ursiniinae, is essential before the significance of the reported differences can be assessed. In all three cases, *Athanasia*, *Gymnopentzia*, and *Ursinia*, the primary goals of the chemical studies centered on nonpolar compounds and not on the flavonoids. The flavonoids that did emerge from these studies did so as a result of characteristic solvent properties of the compounds and not through any directed efforts of the workers. Searches for nonpolar components would, obviously, tend to overlook polar constituents of the plant with the result that compounds in the vacuole, i.e., the flavonol and flavone glycosides, would be lost. As the reader will recall, formation of flavonols requires two enzymatic steps beyond flavanones, one to convert the flavanone to a 3-hydroxyflavanone (a dihydroflavonol) and a second to convert that product to the flavonol. The identification of dihydrokaempferol from *Athanasia* indicates that the first of those two enzymes, at least, is operative in that genus. A critical analysis of the flavonoid fraction of representatives of these genera is needed before it can be stated with certainty that flavonols are absent. If they truly are not formed in members of Ursiniinae, or *Ursinia* alone, then the chemotaxonomic significance of the flavonoid data would be established.

Another observation that might be of significance is the limited occurrence of *C*-glycosylflavones. These compounds occur in representatives of two genera of Artemisiinae, *Ajania* and *Artemisia*, and three genera of Achilleinae, *Achillea*, *Leucocyclus*, and *Otanthus*. Again, several of the studies of these genera involved nonpolar plant constituents; *C*-glycosylflavones, if present, would not have been detected.

With the removal of Ursiniinae and the *C*-glycosylflavone-bearing groups from the list of taxa studied, the remaining genera and subtribes are remarkably homogeneous. The only biasing factor in this is *Artemisia* from whose members an extremely rich array of flavonoids has been described. The maximum level of *O*-methylation, five, is seen in members of two subtribes: Artemisiinae, where two genera, *Ajania* and *Artemisia*, are so blessed, and Achilleinae, where some species of three genera, *Achillea, Leucocyclus*, and *Otanthus*, are likewise favored with flavonoids having four or five *O*-methyl groups.

It might also be worth pointing out that the presence of flavonoids with 2'-oxygenation is restricted to two genera, *Artemisia* (2',4',5' pattern) and *Tanacetum* (2',5' pattern), of Artemisiinae and Tanacetinae, respectively. These two are part of the Asian group of subgenera, according to Bremer and Humphries (1993). Unfortunately, again, poor sampling of these two subtribes and total lack of information on the other two, Cancriniinae and Handeliinae, prevent assessment of the significance of the presence of this unusual oxygenation pattern.

The question naturally arises as to whether the flavonoid occurrences afford support for any other subtribal treatment. With the exceptions already discussed, Ursiniinae and the *C*-glycosylflavone-containing taxa, there remains little structure in the data set. It would be reasonable to defer detailed comments on this subject until a better representation of subtribes has been achieved. See Francisco-Ortega et al. (1997) for a summary of macromolecular data that bear on relationships within *Argyranthemum* and among selected genera of the "Mediterranean group."

Tribe Astereae

Detailed information on Astereae appeared in Chapter 8. Early treatments of Astereae were presented by Cassini (1819), Lessing (1832), and Candolle (1838). Bremer (1994) did not discuss these treatments except to point out that they are artificial assemblages based on homoplasious characters. Two more treatments, those of Bentham (1873) and Hoffmann (1890), were taken more seriously by Bremer. These two systems differed little other than in the names given to the six subtribes, which, according to Hoffmann (1890), were: Asterinae Dum., Baccharidinae Less., Bellidinae Benth., Conyzinae Benth., Grangeinae Benth., and Solidagininae O. Hoffm. Subsequently, Hinterhuberinae Cuatrec. was established to accomodate the unusual genus *Hinterhubera*, to which were added three more genera (Cuatrecasas, 1969, 1986). We have failed to find any flavonoid information for members of Hinterhuberinae.

Zhang and Bremer (1993) performed a cladistic analysis on members of 24 genera representing all of these groups. Their analysis showed that the Hinterhuberinae, Bellidinae, Baccharidinae, and Conyzinae are nested within the Asterinae. The other two subtribes, Grangeinae and Solidagininae, survived the cladistic evaluation and have been retained. We have flavonoid data for members of the three subtribes recognized by Bremer (1994), although we have information for only two genera of Grangeinae.

Member genera of Astereae have also been arranged into two subtribes according to flower color. Homochrominae Benth. accommodates those genera whose flowers are uniformly yellow, while Heterochrominae Benth. consists of genera with variously colored ray florets, white, blue, violet, or purple, in contrast to yellow disc florets. This feature, which is generally taken to be a technical one (Bremer, 1994, p. 381), is based upon the presence of anthocyanins in the ray florets, which certainly demands that it be mentioned in the present context! Although this feature is not consistently present in members of Heterochrominae (anthocyanin biosynthesis is notoriously fickle), the correlation with Bremer's (1994) circumscription of Astereae is certainly worth mentioning. Homochromous taxa fall within the limits of Bremer's view of Solidagininae while heterochromous taxa are to

Table 20–2. Frequency of occurrence of flavonoid classes and structural features of subtribes of Astereae *sensu* Bremer (1994). (Details in Table 20-3.)

		Flavonoid class and structural feature[1]									
Subtribe	No.[2]	Fol	Flav	Fon	CGly	OMe	6Ox	8Ox	68Ox	Bare	345
Solidagininae	19	0.89	0.63	0.26	0.21	0.89	0.63	0.11	0.32	0.05	0.16
Asterinae	21	0.86	0.62	0.19	0.10	0.76	0.38	0.19	0.33	0.14	0.19
Grangeinae	3	1.0	—	—	—	—	—	—	—	—	—

[1] Fol = flavonols; Flav = flavones; Fon = flavanones; CGly = C- Glycosylflavones; OMe = O-methylation; 6Ox = 6-oxygenation; 8Ox = 8-oxygenation; 68Ox = 68-oxygenation; 345 = 3',4',5'-trioxigenation (myricetin-type B-ring); Bare = B-ring deoxyflavonoids; 245 = 2'4'5'-trioxygenation (isoëtin-type B-ring).

[2] No. = Number of genera for which data are available.

be found in members of genera comprising his Asterinae. In summarizing this situation, Bremer (1994, p. 382) states that, "...on the whole, the old homochromous/heterochromous subdivision of the Astereae did not turn out to be such a bad disposition after all."

We are fortunate to have information on most of the subtribes involved (*Hinterhubera* group is missing). It is especially rewarding that we have information on three genera of Grangeinae, the monotypic *Ceruana*, one of the species (of 10) of *Grangea*, and one of *Dichrocephala* (of four). *Ceruana pratensis* afforded kaempferol 3-*O*-rutinoside, while *Grangea maderaspatana* gave quercetin 3-*O*-gentiobioside and quercetin 3-*O*-gentiobioside-7-*O*-glucoside (El-Karemy et al., 1987). Khan et al. (1993) reported kaempferol and quercetin 3-*O*-rutinosides from *Dichrocephala bicolor*. This is a very simple profile compared to the overall profiles of the other two subtribes (*sensu* Bremer) where several different flavonoid classes and a significant amount of structural elaboration have been observed. This simple profile is all the more significant when one considers that one third (three of nine) of the genera of this subtribe have been examined. Generic sampling in the other two subtribes was 17/48 for Solidagininae and 20/117 for Asterinae.

Table 20–2 presents frequency values for the various structural features that have been identified in members of the three subtribes. The format is the same as that in the major frequency table, Table 18–2, except that we have simplified presentation here by removing features that are absent in all members of the three subtribes in question or present in only one or a few genera. These low frequency features are identified in the "Other" category in Table 20–3. Table 20–3 shows the distribution of flavonoid structural features by genus according to two systems. There is general agree-

Table 20–3. Distribution of flavonoid classes and structural features within subtribes of Astereae sensu Bremer (Solidagininae, Asterinae, Grangeinae)

Subtribe/Genus	Flavonoid classes and structural features[1]										
	Fol	Flav	Fon	CGly	OMe	6Ox	8Ox	68Ox	Bare	345	Other[2]
Solidagininae											
Amphiachyris	+	–	–	–	6	–	–	+	–	+	SO_4
Chrysothamnus	+	+	–	–	4	+	–	+	–	–	
Ericameria	+	+	–	–	3	+	–	–	+	–	
Euthamia	–	–	–	+	0	–	–	–	–	–	
Grindelia	+	+	–	–	4	+	–	–	–	–	DiHF
Gutierrezia	+	+	+	–	7	+	+	+	–	+	245, DiHF
Gymnosperma	+	+	–	+	6	+	–	+	–	+	245
Haplopappus	+	+	+	–	4	+	–	+	–	–	
Hazardia	+	+	+	–	4	+	–	–	–	–	
Heterotheca	+	+	+	–	4	+	–	–	–	–	
Isocoma	+	+	+	–	4	+	–	–	–	–	
Machaeranthera	–	+	–	–	4	+	–	–	–	–	
Prionopsis	+	–	–	–	2	+	–	–	–	–	
Pteronia	+	+	–	–	4	+	–	+	–	–	
Solidago	+	–	–	+	1	–	–	–	–	–	
Stephanodoria	+	–	–	+	1	–	–	–	–	–	
Vanclevea	+	–	–	–	4	–	–	–	–	–	
Xanthisma	+	+	–	–	0	–	–	–	–	–	
Xanthocephalum	+	–	–	–	3	–	+	–	–	–	
Asterinae											
Aster	+	+	–	–	5	+	–	+	+	+	
Callistephus	–	+	–	–	0	–	–	–	–	–	Chalcone
Erigeron	+	+	–	–	1	+	–	–	–	–	
Heteropappus	+	+	–	–	5	–	–	+	–	–	
Kalimeris	+	+	–	–	1	–	–	–	–	–	
Lepidophyllum	+	–	–	–	4	+	–	–	–	–	
Olearia	+	+	+	–	3	+	+	+	–	–	
Rigiopappus	+	–	–	–	1	–	–	–	–	–	
Tracyina	+	–	–	–	0	–	–	–	–	–	
Trimorpha	+	+	–	–	0	–	–	–	–	–	
Baccharidinae Less.											
Baccharis	+	+	+	–	5	+	–	+	+	+	DiHchalcone
Heterothalamus	+	–	+	–	1	–	–	–	+	–	
Parastrephia	+	+	–	–	4	+	–	+			
Bellidinae Benth.											
Bellis	–	+	–	–	0	–	–	–	–	–	
Egletes	+	–	–	–	4	–	+	–	–	–	
Conyzinae Benth.											
Conyza	+	+	+	+	5	+	–	–	–	+	
Conyzanthus	+	–	–	–	0	–	–	–	–	–	

Table 20–3. (Continued)

Subtribe/Genus	Flavonoid classes and structural features[1]										
	Fol	Flav	Fon	CGly	OMe	6Ox	8Ox	68Ox	Bare	345	Other[2]
Heteromma	+	–	–	–	5	–	+	–	–	–	
Microglossa	–	+	–	–	4	–	–	+	–	–	Aurone
Nidorella	+	–	–	–	3	–	–	–	–	–	
Psiadia	+	+	–	+	5	+	+	+	–	+	2345, M68
Grangeinae Benth.											
Ceruana	+	–	–	–	0	–	–	–	–	–	
Dichrocephala	+	–	–	–	0	–	–	–	–	–	
Grangea	+	–	–	–	0	–	–	–	–	–	

1 Fol = flavonols; Flav = flavones; Fon = flavanones; CGly = *C*-glycosylflavones; OMe = highest level of *O*-methylation; 6Ox = 6-oxygenation; 8Ox = 8-oxygenation; 68Ox = 6,8-dioxygenation; Bare = B-ring deoxyflavonoids; 345 = 3',4',5'-trioxygenation (myricetin-type B-ring).

2 SO_4 = sulphate group; DiHF = dihydroflavonol; 245 = 2',4',5'-trioxygenation (isoëtin-type B-ring); 2345 = 2',3',4',5'-tetraoxygenation; M68 = myricetin-type B-ring plus 6,8-dioxygenation.

ment as to the generic composition of Solidagininae. Genera of Asterinae are listed according to the subtribes as they appeared (mostly) in Hoffmann (1890). For convenience, the same abbreviations for flavonoid classes and structural features have been used in these tables as were used in Table 18-2.

Several distinctions between Solidagininae and Asterinae can be seen, the most pronounced of which is the occurrence of *C*-glycosylflavones. Although these compounds have been reported from members of six genera from Astereae, their occurrence is limited to members of Solidagininae. Other differences, with Solidagininae being the richer, involve higher frequencies of flavanones, dihydroflavonols, presence of 6-oxygenated flavonoids, and flavonoids with 3',4',5'-oxygenation, and with 2',4',5'-oxygenation. There is a higher frequency of flavonoids with 8-oxygenation in Asterinae than in Solidagininae. It is interesting that the occurrence of flavonoids with 6,8-dioxygenation is about the same in the two subtribes. Differences in frequencies of occurrence of other features are probably not sufficiently different to be of much significance. A larger sampling might be useful in this regard.

Tribe Cardueae

Chapter 9 describes the flavonoid chemistry of 26 genera of Cardueae. It is the purpose of this section to see how well the chemical features fit the systematic treatments of this taxon. Dittrich (1977) saw Cynaroideae (Carduoideae) as consisting of three tribes, Echinopeae Cass., Carlineae Cass., and Cardueae Cass. Members of Cardueae were further segregated into subtribes Carduinae Dumort. and Centaureinae Dumort. Bremer (1994) sees Cardueae as comprising four subtribes, Carduinae Dumort., Centaureinae Dumort., Carlininae Dumort., and Echinopsidinae Dumort. Petit et al. (1996) view the tribe as having only two subtribes: Carlininae and Carduinae (including Centaureinae). They also treat the *Echinops* group as a distinct tribe, Echinopeae. Cynaroideae were not treated in our chapter dealing with subfamilies nor were the tribes *sensu* Dittrich treated in the tribal chapter. A comparison of these two treatments shows that Echinopeae and Carlineae of Dittrich are equivalent to subtribes Echinopsidinae and Carlininae of Bremer, while Carduinae and Centaureinae of the two treatments are identical. The question to be addressed is whether any flavonoid differences among these groups exist. The occurrence of flavonoid structural features is summarized in Table 20–4.

Echinopsidinae consist of two genera, *Echinops*, for which we have information on three species (of ca. 120), and the monotypic *Acantholepis* for which we have no information. For what it is worth, we can state that the flavonoid profile of this subtribe is a simple one consisting of simple flavone glycosides and quercetin. Four genera of the 13 that comprise Carlininae have been sampled. Three of the four afforded *C*-glycosylflavones, while this class of compound was not seen in *Echinops* and has been reported from only one genus from each of the other two subtribes, *Notobasis* (one species tested) of Carduinae and *Centaurea* (reported from eight species) of Centaureinae.

Carduinae and Centaureinae show comparable levels of structural variation with several combinations of features present in genera of each subtribe. Both have members with a modest level of *O*-methylation with the edge going to Centaureinae because of the compounds in *Aegialophila* and *Centaurea*, both with *O*-methylation level of four. It might be noted here that Bremer (1994) considers *Aegialophila* to be synonymous with *Centaurea*. The frequency of both 6- and 8-oxygenation appears to be comparable in the two subtribes. The only two genera from which flavonolignans have been obtained, *Onopordum* and *Silybum*, are members of Carduinae. B-Ring deoxyflavonoids have been reported to occur only in two genera of Centaureinae, *Carthamus* and *Centaurea*, while flavonoids with the 3',4',5'-oxygenation pattern have been reported only from members of Carduinae, namely, *Carduus* and *Silybum*. Two conclusions arise from all this. First,

Table 20–4. Distribution of flavonoid classes and structural features in subtribes of Cardueae

	Flavonoid class and structural feature[1]							
Subtribe/Genus	Fol	Flav	Fon	CGly	OMe	6Ox	8Ox	Other[2]
Echinopsidinae								
Echinops	+	+	–	–	1	–	–	
Carlininae								
Atractylis	–	–	–	+	0	–	–	
Carlina	–	+	–	+	0	–	–	
Staehelina	–	+	–	+	1	+	–	
Xeranthemum	+	+	–	–	0	–	–	
Carduinae								
Arctium	+	–	–	–	0	–	–	
Carduus	+	+	–	+	2	+	–	345
Cirsium	+	+	–	–	2	+	–	
Cousinia	–	+	–	–	2	+	–	
Cynara	+	+	–	–	1	–	–	Aurone
Notobasis	+	+	–	+	1	–	–	
Onopordum	+	+	–	–	1	+	–	Flavonolignan
Picnomon	+	+	–	–	3	+	+	
Saussurea	+	+	–	–	2	+	–	
Silybum	+	+	+	–	2	–	–	345, Flavonolignan, SO_4
Centaureinae								
Acroptilon	–	+	–	–	0	–	–	
Carthamus	+	+	+	+	2	+	+	Bare, chalcones
Centaurea	+	+	+	+	4	+	+	Bare, 68Ox, SO_4
Cheirolophus	+	+	–	–	2	+	–	
Cnicus	+	+	–	–	1	–	–	
Leuzea	+	–	–	–	2	–	–	
Serratula	+	+	–	–	1	–	–	
Stemmacantha[3]	+	–	+	–	2	–	–	
Tricholepis	+	–	–	–	0	–	–	
Volutaria	–	+	–	–	2	+	–	

1 Fol = flavonols; Flav = flavones; Fon = flavanones; CGly = *C*-glycosylflavones; OMe = highest level of *O*-methylation; 6Ox = 6-oxygenation; 8Ox = 8-oxygenation.

2 345 = 3',4',5'-trioxygenation (myricetin-type B-ring); SO_4 = sulphate; Bare = B-ring deoxyflavonoids; 68Ox = 6,8-oxygenation.

3 Reported as *Rhaponticum*; see text for discussion.

the presence of *C*-glycosylflavones in three-quarters of the genera tested along with a lower frequency of flavonols and a much lower tendency to accumulate *O*-methylated flavonols set Carlininae off from the other subtribes. The second observation we can make based on the flavonoid data is that subtribes Carduinae and Centaureinae are sufficiently similar in flavonoid composition that it could be argued that they represent the most closely related pair of subtribes in Cardueae. This would be in line with the earlier view that these were, in fact, subtribes within Cardueae while the remainder of genera involved were contained in Carlineae and Echinopeae.

The most recent treatment of Cardueae that we have is a cladistic analysis of 75 morphological features of 45 genera by Petit (1999). He concluded that the traditional Carlininae (sensu Dittrich, 1977) is a paraphyletic group and that the patterns of variation observed were best accommodated by recognizing two tribes, Echinopeae, represented in his study by *Acantholepis*, *Cardopatium*, *Cousiniopsis*, and *Echinops*, and Cardueae. The latter tribe consists of subtribe Carlininae, represented by *Atractylis*, *Atractylodes*, *Carlina*, *Chamaeleon*, and *Carlina*, and subtribe Carduinae, which consists of the remaining genera assorted into three series. Re-examination of the flavonoid data in light of this new arrangement provides little or no additional insights into relationships among and between these taxa. That statement has to be tempered with the usual comment concerning sampling error. Only one of the four genera in tribe Echinopeae (*Echinops*) appears to have been examined for flavonoids, but at least there would be the simple flavone profile characterizing *Echinops* to serve as a comparator should analysis of the other genera be undertaken. We have found flavonoid information for only two of the five genera that now sit in subtribe Carlininae, *Atractylis* and *Carlina*. *C*-Glycosylflavones have been reported from members of both genera suggesting that an examination of the others might provide useful information. This is borne out in work referred to by Petit (1999), of which we were unaware. Thus, a study of *Carlina* and *Chamaeleon* showed the presence of the *C*-glycosylflavones isovitexin, shaftoside, isoshaftoside, orientin, and unidentified di-*C*-glycosyl derivatives of apigenin and luteolin variously distributed in members of the two genera (Meusel and Kästner, 1990). Petit concluded that the chemical data were in agreement with the overall homogeneity of the subtribe.

The flavonoid situation with regard to subtribe Carduinae, however, is a good deal more chaotic. Although some flavonoid data are available for about 20 genera that Petit (1999) included in his study, the arrays run the gamut from the simplest to the most complex observed in the tribe. Compounding the problem is the fact that several of the genera have been extensively examined, e.g., *Centaurea*, *Cirsium*, and *Silybum*, owing to either their conspicuousness in Nature or their physiological activities, whereas many others have been little more than parts of broad surveys. In a word, flavonoids do not appear to be of any help in assessing relationships at this level. It would seem more reasonable to expect these chemicals to be useful in studies of small groups of more or less closely related genera as will be clear from a reading of Chapter 21.

Tribe Eupatorieae

Detailed notes on flavonoids from members of 32 genera in Eupatorieae can be found in Chapter 10. This is not a particularly good sampling, since some 170 genera are thought to exist (Bremer et al., 1994). Also, the number of genera sampled within some of the larger subtribes is not large. Nonetheless, flavonoid structural variation is extensive, ranging from genera that exhibit profiles with simple flavones and flavonols to others with some of the most complex substitution patterns seen in the family. Noteworthy among the latter are *Ageratum, Brickellia*, and *Eupatorium.*

The treatment of Eupatorieae by Robinson and King (1977) involved several groups of genera but no formal subtribal classification was proposed. A more recent treatment of the tribe (King and Robinson, 1987) resulted in recognition of 18 subtribes. Bremer et al. (1994) commented on the basically phenetic nature of the King and Robinson (1987) treatment and those authors' view of reticulate evolutionary relationships within the tribe. Bremer et al. (1994) agree with the splitting off of some smaller genera from the large genus *Eupatorium*. In our description of the flavonoids of the tribe (Chapter 10) we also followed the suggested changes in taxonomy and nomenclature.

Bremer and coworkers (1994) described Eupatorieae in terms of 16 subtribes based on a cladistic analysis of representatives of 30 genera, which included the type genus of each subtribe. We have assembled available flavonoid data for members of 12 of these subtribes, although in several instances we have managed to find information for only one genus from the subtribe in question. Table 20–5 summarizes the occurrence of several structural features of the flavonoids of this tribe. Only those compound types or structural features that occur more or less consistently in members of the tribe are included. Very little, if any, pattern can be seen in these data. Flavonols, flavones, and flavonoids with 6-oxygenation are fairly evenly distributed among the subtribes. *O*-Methylation levels are reasonably high in most subtribes.

All subtribes, where there are data for two or more genera, show a good deal of infra-group variation. We do not see any correlation between flavonoid patterns and the degree of presumed advancement indicated by the placement of subtribes in the cladograms presented by Bremer et al. (1994). One finds as complex a pattern of flavonoid substitutions in the Ageratinae Less., which is segregated as a relatively early clade, as in Alomiinae Less.,

Table 20–5. Distribution of flavonoid classes and structural features in subtribes of Eupatorieae according to Bremer (1994)

Subtribe/Genus[1]	Flavonoid class and structural feature[2]								
	Fol	Flav	OMe	6Ox	8Ox	68Ox	345	245	Other[3]
Hofmeisteriinae (0/1)									
Oxylobinae (2/9)									
Ageratina	+	+	4	+	+	–	–	–	*C*-Gly, DiHF
Piptothrix	+	–	2	+	–	–	–	–	
Mikaniinae (1/1)									
Mikania	+	+	1	+	–	–	–	–	Fon
Trichocoroninae (0/3)									
Adenostemmatinae (0/3)									
Ageratinae (3/22)									
Acritopappus	+	–	1	–	–	–	–	–	
Ageratum	+	+	8	+	+	+	+	+	2345
Carphochaete	–	+	3	+	–	–	+	–	
Stevia	+	+	5	+	–	+	–	–	
Fleishmanniinae (1/2)									
Fleischmannia	–	+	3	–	–	–	–	–	
Praxelinae (1/7)									
Chromolaena	+	+	5	+	+	–	+	–	Bare, Fon, Chal
Gyptidinae (6/27)									
Bejaranoa	–	+	1	+	–	–	–	–	
Campuloclinum	+	–	1	–	–	–	–	–	DiHF
Conoclinum	–	+	7	+	–	+	–	–	
Lasiolaena	–	+	4	+	–	–	–	–	
Lourtegia	+	+	1	+	–	–	–	–	
Urolepis	+	–	4	+	–	–	–	–	
Liatrinae (1/6)									
Liatris	+	+	4	+	–	–	–	–	
Eupatoriinae (2/4)									
Austroeupatorium	–	+	2	+	–	–	–	–	
Eupatorium	+	+	6	+	–	+	+	–	Bare, *C*-Gly, Fon
Disynaphiinae (4/6)									
Acanthostyles	+	–	5	+	–	–	–	+	
Disynaphia	+	–	0	–	–	–	–	–	
Graziela	+	–	1	+	–	–	–	–	
Symphyopappus	+	+	5	+	–	–	–	–	
Ayapaninae (0/12)									
Alomiinae (5/20)									
Brickellia	+	+	5	+	–	–	–	+	Bare, SO_4
Brickelliastrum	+	+	3	+	–	–	–	–	
Flyriella	+	–	2	–	–	–	–	–	
Goyazianthus	–	+	0	–	–	–	–	–	Fon
Steviopsis	+	+	2	+	–	–	–	–	

Table 20–5. (Continued)

Subtribe/Genus[1]	Fol	Flav	OMe	6Ox	8Ox	68Ox	345	245	Other[3]
	Flavonoid class and structural feature[2]								
Critoniinae (4/39)									
Aristeguietia	–	+	3	+	–	–	–	–	
Critonia	+	–	0	–	–	–	–	–	
Nothobaccharis	+	–	3	+	–	–	–	–	SO_4
Ophryosporus	–	+	2	–	–	–	–	–	Fon
Hebecliniinae (1/8)									
Decachaeta	+	+	3	+	–	–	–	–	

1 The fraction in parentheses represents the number of genera in each subtribe for which flavonoid information is available.

2 Fol = flavonol; Flav = flavone; OMe = highest level of *O*-methylation; 6Ox = 6-oxygenation; 8Ox = 8-oxygenation; 68Ox = 6,8-oxygenation; 345 = 3',4',5'-oxygenation (myricetin-type B-ring; 245 = 2',4',5'-oxygenation (isoëtin-type B-ring).

3 Bare = B-ring deoxyflavonoid; 2345 = 2',3',4',5'-oxygenation; SO_4 = sulphate; Fon = flavanone; Chal = chalcone; DiHF = dihydroflavonol.

which is a later one. It is interesting to note, however, that in Critoniinae sensu King and Robinson and Hebecliniinae sensu King and Robinson, among the last clades to be separated out, the flavonoid patterns are actually somewhat less complex than in the subtribes that branch earlier. It is difficult to assess whether this is a real phenomenon or simply an artefact caused by sampling error. Without further sampling, it seems reasonable to conclude that flavonoid occurrences do not provide support for subtribal delineation within Eupatorieae.

Tribe Helenieae

Flavonoid information for members of Helenieae *sensu* Karis and Ryding appeared in Chapter 11, where the various genera were presented under the broader concept of the Heliantheae. Those genera considered to comprise Helenieae (Karis and Ryding, 1994a) are identified by the symbol (H) in Chapter 11. The appearance of Helenieae in these discussions represents a resurrection of sorts. The analysis of Helenieae by Turner and Powell (1977) involved dismemberment of the tribe, accompanied by speculation as to where the various genera should best be located. Remnants of the tribe came to rest, not always to unanimous approval, in various other tribes. A cladistic analysis by Karis (1993b), however, found evidence to suggest

Table 20–6. Distribution of flavonoid classes and structural features in subtribes of Helenieae *sensu* Karis and Ryding (1994a)

	Flavonoid class and structural feature[1]							
Subtribe/Genus	Fol	Flav	Fon	CGly	OMe	6Ox	68Ox	Other[2]
Chaenactidinae								
Arnica	+	+	+	–	3	+	–	345
Bahia	+	–	–	–	5	+	–	
Hulsea	+	+	–	–	1	–	–	
Picradeniopsis	+	–	–	–	3	+	–	
Schkuhria	–	+	–	–	2	+	–	
Venegasia				No Data			(–)	antho
Flaveriinae								
Flaveria	+	–	–	–	3	+	–	SO_4
Haploëstes	+	–	–	–	2	+	–	SO_4
Jaumea				No Data			(+)	antho
Sartwellia	+	–	–	–	2	+	–	SO_4
Pectidinae								
Chrysactinia	+	–	–	–	0	+	–	
Dysodiopsis	+	–	–	–	1	+	–	
Dyssodia	+	–	–	–	0	+	–	
Porophyllum	+	–	–	–	1	+	–	
Strotheria	+	–	–	–	1	+	–	
Tagetes	+	+	–	–	4	+	–	
Thymophylla	+	+	–	–	1	+	–	
Hymenopappinae								
Galeana	+	–	–	–	3	–	+	Direct acyl.
Baeriinae								
Eriophyllum	+	–	–	–	1	+	–	
Lasthenia	+	+	+	–	1	+	–	SO_4
Gaillardiinae								
Amblyolepis	+	–	–	–	1	–	–	
Baileya	+	–	–	–	2	–	–	Isoflavone
Balduina	+	+	–	+	0	–	–	
Dugaldia	+	+	–	–	2	+	–	
Gaillardia	–	+	+	+	0	–	–	Isoflavone
Helenium	–	+	–	+	1	+	–	
Hymenoxys	+	+	–	–	4	+	+	DiHF
Marshallia	+	–	+	–	0	–	–	DiHF, *C*-Alk, Bare
Plummera	+	+	–	–	2	+	–	
Psathyrotes	–	+	–	–	3	+	–	
Peritylinae								
Amauria	+	–	–	–	4	+	–	
Pericome	+	–	–	–	4	+	–	
Perityle	+	–	+	–	4	+	–	

Table 20–6. (Continued)

Subtribe/Genus	Fol	Flav	Fon	CGly	OMe	6Ox	68Ox	Other[2]
	Flavonoid class and structural feature[1]							
Madiinae								
Argyroxiphium								
Calycadenia	–	+	–	–	3	–	+	8Ox
Dubautia	+	–	+	–	1	+	–	
Hemizonia	+	+	+	–	3	+	+	
Holocarpha	+	+	+	–	2	+	–	DiHF
Lagophylla	+	+	–	–	4	+	–	
Layia	+	+	–	–	3	+	–	
Madia	+	–	+	–	3	+	–	
Wilkesia	+	+	–	–	4	+	+	8Ox

[1] Fol = flavonols; Flav = flavones; Fon = flavanones; CGly = *C*-glycosylflavones; OMe = highest level of *O*-methylation; 6Ox = 6-oxygenation; 68Ox = 6,8-oxygenation.

[2] 345 = 3',4',5'-oxygenation (myricetin-type B-ring); antho = anthochlors; SO_4 = sulphate; DiHF = dihydroflavonol; 8Ox = 8-oxygenation; Direct acyl. = direct acylation of a phenolic group; Bare = B-ring deoxyflavonoid.

that Helenieae could stand again as a tribe in its own right. This analysis also resulted in recognition of several subtribes. One of the subtribes that emerged was Pectidinae Less., which appeared in earlier treatments as tribe Tageteae. The flavonoids of Tageteae appear in our Chapter 16.

The distribution of flavonoid structural features in the subtribes of Karis and Ryding (1994a) is shown in Table 20–6. We have located information for at least one member of each subtribe and in some cases, e.g., Gaillardiinae Less. and Madiinae Benth., sampling has been comparatively good. A lack of immediately obvious differences between the subtribes is evident. There are few unique features evident in the array. One feature of note, though, is the presence of *C*-glycosylflavones in three genera of Gaillardiinae. The usual problem here is whether the compounds are truly missing from the other genera or whether they simply haven't been seen or looked for. It should not be difficult to remedy this situation.

The presence of sulphated flavonoids in members of three genera of Flaveriinae Less., *Flaveria, Haploëstes*, and *Sartwellia*, is of some interest. The other member of this subtribe that appears in the flavonoid literature is *Jaumea*, about which we know only that it accumulates anthochlors (Crawford and Stuessy, 1981). Unfortunately, there is no information concerning the flavonoid sulphating capacities of members of this genus or others in the subtribe. The profiles of *Flaveria, Haploëstes*, and *Sartwellia* are very similar. They share the capacity to accumulate flavonols, appear to lack flavones, have the capacity to perform 6-oxygenation, and achieve the same level of *O*-methylation. Two of them accumulate flavonols with the 3',4',5'-oxygenation pattern as well. This flavonoid profile clearly distinguishes Flaveriinae from other members of Helenieae.

Madiinae have been subjected to intensive investigation owing in large measure to their disjunct distribution. Three of the genera, *Argyroxiphium*, *Dubautia*, and *Wilkesia*, are Hawaiian endemic genera, while the rest are primarily western North American. Flavonoid studies are not complete for the subtribe, but it is interesting to note the differences among Hawaiian genera and the similarity, for example, between *Wilkesia* (Bohm and Fong, 1990) and the mainland genera *Calycadenia* and *Hemizonia* (Bohm et al., 1992; Proksch et al., 1984; Tanowitz et al., 1987). Flavonoid information would not support, at least at this time, a grouping of genera within Madiinae. A detailed review of this subtribe has been assembled by Carlquist et al. (2001).

Tribe Heliantheae

Heliantheae present an interesting challenge to any attempt to correlate flavonoid occurrence with subtribal delineation (Table 20–7). First, Karis and Ryding (1994b) see Heliantheae from a significantly different perspective than do some other current workers. Flavonoid differences between Heliantheae and Helenieae, both *sensu* Karis and Ryding, were discussed in Chapter 19. Two major efforts to sort genera into subtribes have been described, one by Stuessy (1977) and one by Robinson (1981). Comparisons of these systems is complicated by the inclusion by Robinson (1981) of several genera not treated by Stuessy (1977). Robinson (1996b) has also added additional comments on his 1981 system, but without a comprehensive revision. A recent study of subtribe Ecliptinae using cpDNA restriction site data treats many of the issues surrounding the three treatments of Heliantheae (Panero et al., 1999). Efforts to search out correlations with flavonoid occurrences suffer from the absence of flavonoid data for about half of the genera studied by Panero and his associates. What information is available will be discussed in the following paragraphs.

We will start our discussion with a look at Coreopsidinae Less., one of the better sampled subtribes with regard to flavonoids, particularly anthochlor derivatives. Our understanding of the chemistry of this group of genera comes from a collection of very detailed studies of *Bidens, Coreopsis, Cosmos*, and *Dahlia*, coupled with an extensive survey of genera for anthochlors (Crawford and Stuessy, 1981). Table 20–8 lists the genera for which we

Table 20–7. Distribution of flavonoid classes and structural features in subtribes of Heliantheae *sensu* Karis and Ryding (1994b)

	Flavonoid class and structural feature[1]							
Subtribe/Genus	Fol	Flav	Anth	Fon	OMe	6Ox	68Ox	Other[2]
Unassigned								
Clibadium	+	–	+	–	1	+	–	
Desmanthodium	+	–	–	–	3	+	–	
Eclipta	–	+	–	–	1	–	–	Isoflavonoid
Guardiola								C&S survey
Ichthyothere	+	–	–	–	1	–	–	DiHF
Montanoa	+	+	–	–	4	+	+	
Rudbeckiinae								
Echinacea	+	+	–	–	0	–	–	
Ratibida	–	–	–	+	0	–	–	
Rudbeckia	+	+	–	–	3	+	–	
Zinniinae								
Acmella[4]	+	+	–	–	0	–	–	
Heliopsis	+	–	–	–	5	–	+	
Zinnia	+	+	+	–	0	–	–	
Verbesininae								
Balsamorhiza	+	–	–	–	3	+	–	
Encelia	+	+	–	+	4	–	+	8Ox
Enceliopsis	+	–	–	+	3	–	–	8Ox
Espeletia	+	+	–	–	0	–	–	
Flourensia	+	+	+	+	3	+	–	C-Alk, Acyl.
Geraea	+	+	–	–	3	–	+	DiHF
Lipochaeta	+	+	–	–	0	–	–	
Perymenium	–	–	–	+	2	–	–	
Phoebanthus	–	+	–	–	4	+	+	
Steiractinia	–	–	–	+	0	–	–	
Verbesina	+	–	–	–	1	–	–	
Wedelia	–	+	–	–	1	–	–	Isoflavonoid
Wollastonia	+	–	–	–	2	–	–	Isoflavonoid
Wyethia	+	+	–	+	3	+	–	Isoflavonoid, C-Alk
Zaluzania	+	–	–	–	3	+	–	
Helianthinae								
Helianthus	+	+	+	+	4	+	+	
Heliomeris	+	–	+	–	0	–	–	
Iostephane	+	–	+	–	0	–	–	
Lagascea	+	–	–	–	2	+	–	
Simsia	+	+	+	–	3	–	+	
Scalesia	–	–	–	+	1	+	–	DiHF
Tithonia	–	+	–	–	3	–	+	
Viguiera	+	+	+	+	4	+	+	

Table 20–7. (Continued)

Subtribe/Genus	Flavonoid class and structural feature[1]							
	Fol	Flav	Anth	Fon	OMe	6Ox	68Ox	Other[2]
Galinsoginae								
Galinsoga	–	+	–	–	0	–	–	
Melampodiinae								
Acanthospermum	+	+	–	–	3	+	+	
Calea	–	+	–	–	2	–	–	
Guizotia			+					C&S survey
Unxia	–	+	–	–	2	–	–	
Lecocarpus	+	–	–	–	4	+	–	
Melampodium	+	+	–	–	5	+	–	
Neurolaena	+	+	–	–	3	+	–	SO_4
Smallanthus[3]	+	–	–	+	3	+	–	
Staurochlamys			–					C&S survey
Tetragonotheca	+	–	–	–	3	+	–	
Coreopsidinae								
Bidens	+	+	+	+	4	+	+	
Chrysanthellum	+	+	+	+	0	–	–	
Coreocarpus			+					C&S survey
Coreopsis	+	+	+	+	1	–	–	*C*-Gly, 8Ox
Cosmos	+	+	+	–	1	–	–	
Dahlia	+	+	+	+	2	+	–	*C*-Gly
Dicranocarpus			+					C&S survey
Ericentrodea			+					C&S survey
Glossocarda			+					C&S survey
Goldmanella			+					C&S survey
Helianthopsis			+					C&S survey
Henricksonia			+					C&S survey
Heterosperma			+					C&S survey
Hidalgoa			+					C&S survey
Isostigma			+					C&S survey
Megalodonta	–	–	+	–	1	–	–	
Moonia			+					C&S survey
Narvalina			+					C&S survey
Oparanthus			+					C&S survey
Petrobium			+					C&S survey
Selleophytum[5]			+					C&S survey
Thelesperma	–	+	+	–	0	–	–	
Engelmanniinae								
Berlandiera	+	+	–	–	1	+	–	
Ambrosiinae								
Ambrosia	+	+	+	–	4	+	+	Bare
Hymenoclea	–	–	–	+	0	–	–	DiHF
Iva	+	+	–	–	3	+	+	

Table 20–7. (Continued)

Subtribe/Genus	Flavonoid class and structural feature[1]							
	Fol	Flav	Anth	Fon	OMe	6Ox	68Ox	Other[2]
Parthenium	+	+	–	–	5	+	–	
Xanthium	+	+	–	–	2	+	–	*C*-Alkyl
Pinillosiinae	No flavonoid data							

[1] Fol = flavonols; Flav = flavones; Anth = anthochlors (chalcones and/or aurones); Fon = flavanones; OMe = highest level of *O*-methylation; 6Ox = 6-oxygenation; 68Ox = 6,8-oxygenation.

[2] DiHF = dihydroflavonol; 8Ox = 8-oxygenation; Bare = B-ring deoxyflavonoids; *C*-alkyl = *C*-alkylation; *C*-Gly = *C*-glycosylflavones; Acyl. = direct phenol acylation; SO_4 = sulphate; C&S survey = anthochlor survey of Crawford and Stuessy (1981).

[3] Reported as *Polymnia*; see text for discussion.

[4] Reported as *Spilanthes*; see text for discussion.

[5] Currently considered as *Coreopsis*.

have information on anthochlor occurrence, arranged according to Stuessy's (1977) system of groups within the subtribe, and then shows how the taxa are treated by Robinson (1981). All genera that were shown not to accumulate anthochlors (Crawford and Stuessy, 1981), *Guizotia, Guardiola, Staurochlamys, Jaumea*, and *Venegasia*, are seen by Robinson (1981) as belonging to subtribes other than Coreopsidinae: Milleriinae Benth. & Hook., Guardiolinae H. Robinson, Neurolaeninae (Rydb.) Stuessy, Turner & Powell, Jaumeinae Benth. & Hook., and Chaenactidinae Rydb., respectively. Thus, Robinson's view of Coreopsidinae involves only those genera that have been shown to be anthochlor positive. This level of flavonoid homogeneity is unusual.

Anthochlors are not restricted to Coreopsidinae, however. They have been reported from members of *Clibadium*, which was not assigned to a subtribe by Robinson (1981) but belongs in Milleriinae according to Stuessy (1977), *Zinnia* (Zinniinae Benth.), *Flourensia* (Verbesininae Benth.), and four genera of Helianthinae Dumort. (*Helianthus, Iostephane, Simsia*, and *Viguiera*). Only in the case of Helianthinae might the presence of anthochlors be considered characteristic; the other reports are likely to be random sightings. Although the presence of anthochlor derivatives in four of the six genera tested from Helianthinae does not correlate well with other flavonoids present in these taxa, one could ask whether their higher frequency of occurrence in the subtribe in general might suggest a relationship with Coreopsidinae. The strict consensus tree obtained using morphological data (Karis and Ryding, 1994b, p. 567), however, does not suggest such a relationship.

Table 20–8. Anthochlor occurrence in Coreopsidinae groups as seen by Stuessy (1977) compared to treatment by Robinson (1981)

Stuessy Group/Genus	Anthochlor(s) Present	Robinson's Coreop.	Comment[1]
Group 1			
Bidens	+	Yes	
Coreopsis	+	Yes	
Cosmos	+	Yes	
Dahlia	+	Yes	
Ericentrodea	+	Yes	
Guizotia	–	No	Milleriinae (R)
Hidalgoa	+	Yes	
Megalodonta	+	Yes	
Microlecane	+	Yes	
Moonia	+	Yes	
Narvalina	+	Yes	
Selleophytum	+	Yes	syn. *Coreopsis* (R)
Thelesperma	+	Yes	
Group 2			
Coreocarpus	+	Yes	
Dicranocarpus	+	Yes	
Glossocardia	+	Yes	
Guardiola	–	No	Guardiolinae (R)
Heterosperma	+	Yes	
Group 3			
Chrysanthellum	+	Yes	
Eryngiophyllum	+	Yes	syn. *Chrysanthellum* (K&R)
Glossogyne	+	Yes	syn. *Glossocardia* (K&R)
Guerreroia	+	Yes	syn. *Glossocardia* (K&R)
Isostigma	+	Yes	
Group 4			
Oparanthus	+	Yes	
Petrobium	+	Yes	
Group 5			
Staurochlamys	–	No	Neurolaeninae (R)
Group 6			
Goldmanella	+	Yes	
Group 7			
Jaumea	–	No	Jaumeinae (R)
Venegasia	–	No	Chaenactidinae (R)[2]
Other			
Henricksonia	+	Yes	Not in Stuessy (1977)

1 R = According to Robinson (1981); K&R = According to Karis and Ryding (1994b).
2 Karis and Ryding (1994a) place *Venegasia* in Helenieae.

It is of interest to notice, in the following discussion of Melampodiinae, the number of other subtribes that enter the discussion, along with all or some of their respective genera. This flavonoid domino effect is not unique to Melampodiinae, as we shall see below in the comparison of Robinson's Ecliptinae and Stuessy's Helianthinae. This phenomenon reflects two fundamental aspects of the plant groups under consideration: (1) their underlying morphological similarity, and (2) their similar flavonoid biosynthetic capacities. In order to resolve problems encountered in both of these areas, further studies at the developmental level, for problems arising in the morphological realm, and macromolecular studies involving comparative enzymological, and possibly DNA sequence work as well, for difficulties in resolving problems at the micromolecular level, are clearly needed.

Three different views of the Melampodiinae can be compared using available flavonoid data. Table 20–9 summarizes these data and includes information on members of Milleriinae, Galinsoginae Benth., and Neurolaeninae, all of whose involvement may become clear below. The four genera for which flavonoid data are available and that fall within Stuessy's (1977) view of Melampodiinae are *Acanthospermum*, *Lecocarpus*, *Melampodium*, and *Polymnia*. Robinson (1981) placed *Polymnia* in its own subtribe, Polymniinae H. Robinson, but added *Ichthyothere* to his version of Melampodiinae. Only limited information is available for *Polymnia*: 6-oxygenated flavonols with *O*-methylation were obtained, one from each of two species (Herz and Bhat, 1970; Beutler et al., 1993).

The flavonoids of *Ichthyothere* (Bohm and Stuessy, 1982) lack 6-oxygenation and the higher levels of *O*-methylation that distinguish other members of the subtribe. Stuessy (1977) viewed *Ichthyothere* as part of his Milleriinae along with *Clibadium* and *Desmanthodium*. Detailed flavonoid studies of the latter two genera (Bohm and Stuessy, 1981a, 1981b, 1985; Bohm et al., 1983) showed the presence of an array of flavonol derivatives having both 6-oxygenation and moderate levels of *O*-methylation in both. The flavonoid profile of *Ichthyothere* no more supports its placement in Milleriinae than it does putting it in Melampodiinae. Karis and Ryding (1994b) would seem to agree insofar as *Ichthyothere* is one of the genera not assigned to a subtribe in their treatment of Heliantheae.

Karis and Ryding's (1994b) view of Melampodiinae is also different from either the Stuessy or Robinson view of the subtribe. It is difficult to judge the significance of the flavonoid profile of two of their added genera, *Guizotia* and *Staurochlamys*, however, since the only information available for them is their apparent lack of anthochlors (Crawford and Stuessy, 1981). For what it's worth, this absence is consistent with other genera under discussion here (except for a report of anthochlors from a single population of *Clibadium trianae*; Bohm et al., 1983).

Table 20–9. Comparisons of different views of Melampodiinae and related groups

Subtribe/Genus[2]	Flavonoid feature[1] Fol	Flav	Antho	OMe	6Ox	68Ox	Comment
Melampodiinae (ST)							
Acanthospermum	+	+	–	3	+	+	
Espeletia	+	+	–	0	–	–	
Lecocarpus	+	–	–	3	+	–	
Melampodium	+	+	–	5	+	–	
Polymnia	+	–	–	5	+	–	Polymniinae (R)
Melampodiinae (R)							
Acanthospermum	+	+	–	3	+	+	
Ichthyothere	+	–	–	1	–	–	Unassigned (K&R)
Lecocarpus	+	–	–	3	+	–	
Melampodium	+	+	–	5	+	–	
Melampodiinae (K&R)							
Acanthospermum	+	+	–	3	+	+	
Calea	–	+	–	2	–	–	
Guizotia			–				C&S survey
Lecocarpus	+	–	–	3	+	–	
Melampodium	+	+	–	5	+	–	
Neurolaena	+	+	–	3	+	–	Sulphate
Polymnia	+	–	–	5	+	–	
Staurochlamys			–				C&S survey
Tetragonotheca	+	–	–	3	+	–	
Milleriinae (ST)							
Clibadium	+	–	+	2	+	–	Clibadiinae (R)
Desmanthodium	+	–	–	3	+	–	Desmanthodiinae (R)
Ichthyothere	+	–	–	1	–	–	Unassigned (K&R)
Milleriinae (R)							
Guizotia			–				C&S survey
Espeletiinae (R)							
Espeletia	+	+	–	0	–	–	
Galinsogiinae (ST)							
Calea	–	+	–	2	–	–	
Galinsoga	–	+	–	0	–	–	
Spilanthes	+	+	–	0	–	–	Zinniinae (K&R); Ecliptinae (R)
Galinsogiinae (R)							
Galinsoga	–	+	–	0	–	–	
Tetragonotheca	+	–	–	3	+	–	
Neurolaeninae (ST)							
Neurolaena	+	+	–	3	+	–	
Zaluzania	+	–	–	3	+	–	Verbesininae (K&R); Zaluzaniinae (R)
Neurolaeninae (R)							
Calea	–	+	–	2	–	–	

Table 20–9. (Continued)

Subtribe/Genus[2]	Fol	Flav	Antho	OMe	6Ox	68Ox	Comment
Neurolaena	+	+	–	3	+	–	
Staurochlamys			–				C&S survey

[1] Fol = flavonols; Flav = flavones; Antho = anthochlors; OMe = highest level of *O*-methylation; 6Ox = 6-oxygenation; 68Ox = 6, 8-dioxygenation.

[2] C&S = Crawford and Stuessy (1981); K&R = Karis and Ryding (1994b); (R) = Robinson (1981); ST = Stuessy (1977).

Studies of several species of *Calea* have consistently shown that members of this genus produce only apigenin and its mono- and dimethyl ethers (see Ortega et al., 1989 for citations). Lacking both flavonols and 6-oxygenation *Calea* does not fit comfortably in Melampodiinae. *Calea* is treated differently by Stuessy (1977) and by Robinson (1981). Stuessy (1977) places *Calea* in Galinsoginae Benth. along with *Galinsoga* and *Spilanthes*. *Galinsoga parviflora* was shown to accumulate simple glucosides of apigenin and luteolin (Plekhanova et al., 1977), which fits with the flavone-accumulating capacity of *Calea*.

Returning to Karis and Ryding's (1994b) view of Melampodiinae, we next note their inclusion of *Neurolaena* in this subtribe. Several detailed flavonoid studies of *Neurolaena* have been described (Ulubelen et al., 1980a, 1981; Ulubelen and Mabry, 1981b; Kerr et al., 1981). While flavonols, including those with 6-oxygenation and *O*-methylation, predominate in the flavonoid profiles, simple flavones have also been reported. A striking difference, however, lies in the observation that sulphated flavonols occur in members of the genus. Robinson (1981) places *Neurolaena* in Neurolaeninae along with *Calea* and *Staurochlamys*. Flavonoids suggest that this might not be a comfortable association since *Calea* produces flavones but not flavonols, while *Neurolaena* is very much a flavonol specialist. Again, the limited information available for *Staurochlamys* precludes further comment.

Stuessy (1977) placed *Neurolaena* in Neurolaeninae, which was newly defined at that time. He included *Zaluzania* for which we do have flavonoid information thanks to the efforts of Olsen and Mabry (1980) who studied all nine species. Flavonols, some of which were shown to be *O*-methylated and have 6-oxygenation, were present in some of the species, simpler flavonol glycosides in others. Neither flavones nor sulphated flavonoids were reported. Robinson (1981) placed *Zaluzania* in Zaluzaniinae H. Robinson; Karis and Ryding (1994b) in Verbesininae Benth. No other genera are available for comparison.

We encountered Galinsoginae above when we considered Stuessy's (1977) placement of *Calea*, which was accompanied in that subtribe by *Galinsoga* and *Spilanthes*. Robinson's (1981) view of Galinsoginae has *Galinsoga*, of course, as well as *Tetragonotheca*, which we have already seen as a member of Karis and Ryding's (1994b) Melampodiinae. *Galinsoga parviflora*, as we saw, exhibited only simple flavones, whereas *Tetragonotheca* was shown by Bacon et al. (1978) to elaborate a variety of flavonols that exhibited both 6-oxygenation and *O*-methylation. All members of the genus were examined and no flavones were found.

Our next task involves a search for patterns that may exist in the distribution of flavonoids in the genera treated variously as Helianthinae, Ecliptinae, or Verbesininae (and others, for that matter) in the treatments of Stuessy (1977), Robinson (1981), and Karis and Ryding (1994b). (Also refer to Panero et al., 1999). The first problem that arises deals with the fact that these three authorities disagree significantly as to the component genera of the subtribes. An effort has been made to list the genera involved in such a way that comparison of the flavonoid profiles of the subtribes is somewhat easier than trying to single out unique combinations in the master data table (Table 20–7). Thus, Table 20–10 first lists the genera that comprise Helianthinae in all three systems. The first five genera listed, *Helianthus, Iostephane, Simsia, Tithonia*, and *Viguiera*, are consensus members of Helianthinae, while the remaining members of the subtribe *sensu* Stuessy (1977) are presented quite differently in the other two treatments. Most of the remaining genera in the Stuessy Helianthinae belong to Robinson's (1981) Ecliptinae. The five genera not assigned to Ecliptinae fall into four subtribes. *Ratibida* and *Rudbeckia*, seen as members of Helianthinae by Stuessy, are placed in Rudbeckiinae H. Robinson in the other two treatments. *Lagascea* is seen as a member of Verbesininae by Stuessy, but is included in Helianthinae by the other workers. *Tetragonotheca* and *Zaluzania*, whose flavonoids were mentioned above, are variously positioned in the other two systems, with *Tetragonotheca* residing within Helianthinae according to Stuessy.

Despite the apparent richness of flavonoid chemical data in the literature, the actual usefulness of the information in discerning differences among the various subtribes is disappointing. Very few clear-cut distributional differences are evident, and the ones that do appear often involve groups of genera that have not been well sampled. Conclusions drawn from these observations do not stand on firm ground. One interesting correlation that appears to be on firmer ground than most involves the presence of anthochlor derivatives in four of the five genera that are consensus members of Helianthinae. *Helianthus, Iostephane, Simsia, Tithonia*, and *Viguiera* have been studied for a sufficiently long period of time by several groups of

Table 20–10. Comparison of different taxonomic treatments of Helianthinae, Ecliptinae, Verbesininae, and related subtribes, and flavonoid structural features found within included genera

	Taxonomic Treatment[1]			Flavonoid structural feature[2]							
Genus	ST	R	K&R	Fol	Flav	Antho	OMe	6Ox	8Ox	68Ox	Comment
Helianthus	Hel	Hel	Hel	+	+	+	4	+	–	+	
Iostephane	Hel	Hel	Hel	+	–	+	0	–	–	–	
Simsia	Hel	Hel	Hel	+	+	+	0	–	–	–	
Tithonia	Hel	Hel	Hel	–	+	–	4	–	–	+	
Viguiera	Hel	Hel	Hel	+	+	+	4	+	–	+	
Balsamorhiza	Hel	Ecl	Ver	+	–	–	3	+	–	–	
Echinacea	Hel	Ecl	Rud	+	+	–	0	–	–	–	Cyanidin
Encelia	Hel	Ecl	Ver	+	+	–	4	+	+	+	Bare
Enceliopsis	Hel	Ecl	Ver	+	–	–	3	–	+	–	
Flourensia	Hel	Ecl	Ver	+	+	+	3	+	–	–	*C*-Gly, Bare
Geraea	Hel	Ecl	Ver	+	+	–	3	–	+	–	
Phoebanthus	Hel	Ecl	Ver	–	+	–	4	+	–	+	
Wyethia	Hel	Ecl	Ver	+	+	–	2	+	–	–	Isoflavone
Eclipta	Ecl	Ecl	None	–	+	–	4	+	–	+	Isoflavone
Heliopsis	Zin	Ecl	Zin	+	–	–	5	+	–	+	
Lipochaeta	Ecl	Ecl	Ver	+	+	–	0	–	–	–	
Perymenium	Ver	Ecl	Ver	–	–	–	2	–	–	–	Flavanone
Spilanthes	Gal	Ecl	Zin	+	+	–	0	–	–	–	
Steiractinia	Ver	Ecl	Ver	–	–	–	0	–	–	–	Flavanone
Verbesinia	Ver	Ecl	Ver	+	–	–	1	–	–	–	
Wedelia	Ecl	Ecl	Ver	+	+	–	2	–	–	–	Isoflavone
Zinnia	Zin	Ecl	Zin	+	+	+	0	–	–	–	
Berlandiera	Eng	Ecl	Eng	+	+	–	1	+	–	–	
Tetragonotheca	Hel	Gal	Mel	+	–	–	3	+	–	–	
Ratibida	Hel	Rud	Rud	–	–	–	0	–	–	–	Bare
Rudbeckia	Hel	Rud	Rud	+	–	–	3	+	–	–	
Lagascea	Ver	Hel	Hel	+	–	–	2	+	–	–	
Zaluzania	Neu	Zal	Ver	+	–	–	3	+	–	–	

[1] ST = Stuessy (1977); R = Robinson (1981); K&R = Karis and Ryding (1994b); Hel = Helianthinae; Ecl = Ecliptinae; Ver = Verbesininae; Eng = Englemanniinae; Zin = Zinninae; Gal = Galinsoginae; Mel = Melampodiinae; Rud = Rudbeckiinae; Neu = Neurolaeninae; None = not assigned to a subtribe by Karis and Ryding (1994b).

[2] Fol = flavonols; Flav = flavones; Antho = anthochlors; OMe = highest level of *O*-methylation; 6Ox = 6-oxygenation; 8Ox = 8-oxygenation; 68Ox = 6,8-dioxygenation; Bare = B-ring deoxyflavonoid; *C*-Gly = *C*-glycosylflavone.

workers that one is reasonably confident in stating that the capacity to accumulate anthochlors is a biochemical characteristic of the group.

An interesting observation, but one that requires a good deal of follow-up work, involves the presence of isoflavones in members of this tribe. Isoflavones have been identified from members of *Eclipta*, *Wedelia*, and *Wyethia*. Stuessy (1977) sees the first two of these as belonging to Ecliptinae. Robinson (1981) includes all three in his view of Ecliptinae. One might suggest that a detailed survey of Ecliptinae might find these compounds more widely distributed than we currently are aware. These three genera are widely separated in the cpDNA array of Panero et al. (1999). Since this class of flavonoid is known only from a few genera in Heliantheae s.l., additional information on the extent of their occurrence might provide interesting evolutionary insights.

The agreement between the various workers in placing *Ratibida* and *Rudbeckia* in the same group, Group 2 of Helianthinae according to Stuessy, and Rudbeckiinae according to the other authorities, offers an interesting opportunity to look for chemical similarities. However, there are significant gaps in our knowledge. Studies of six species (see Cisowski et al., 1993 for citations) have provided us a reliable view of *Rudbeckia*'s capacity to accumulate both simple flavonol glycosides as well as more complex ones with both 6-oxygenation and modest levels of *O*-methylation. Unfortunately, the only information available for *Ratibida* is the report of 5,7-dihydroxyflavanone (pinocembrin) from a single species (Ellmauerer et al., 1987). Since *Ratibida* consists of seven species all of which occur in North America, it should not be a difficult task to get a better idea of its flavonoids.

Although again based upon very limited sampling, another tempting observation involves the reports of 8-oxygenated flavonoids in three genera of Karis and Ryding's (1994b) Verbesininae, namely, *Encelia, Enceliopsis*, and *Geraea*. The number rises to four with the inclusion of *Phoebanthus*, both species of which have been shown to possess a 6,8-dioxygenated flavonoid (Schilling and Rieseberg, 1985). 6,8-Dioxygenated flavonoids occur in a few other genera that don't belong to Verbesininae, but the 8-oxygenation correlation holds.

A cladistic analysis of Ambrosiinae Less. indicated that the subtribe is a well supported monophyletic group (Karis, 1993b). Because that analysis was based on only three genera [eight are listed by Bremer (1994)], a larger sampling of the subtribe was subjected to cladistic analysis more recently using morphological and cytological data (Karis, 1995). Reasonably detailed flavonoid information is available for members of four genera, *Ambrosia*, *Iva*, *Parthenium*, and *Xanthium*. Only two incomplete studies of flavonoids of *Hymenoclea* have been described, hence its involvement in discussion of the subtribe is much curtailed. In general terms, however, the first four genera listed have much in common in that they all accumulate flavones and flavonols, they have moderate to high levels of *O*-methylation, and all have 6-oxygenated flavonoids. *Ambrosia* is the most structurally diverse in that it also has taxa with 8-oxygenated and 6,8-oxygenated flavonoids as well as a B-ring deoxyflavonoid (Wollenweber et al., 1987a). A closer look at the results of the cladistic analyses cited above, however, reveals that all is not quiet with regard to generic identities within the genera comprising this subtribe. Details of these generic complexities will be discussed in the next chapter.

Tribe Inuleae

Detailed information on flavonoids of Inuleae, taken in the broad sense, appeared in Chapter 12. Genera were marked to indicate their membership in Inuleae s.s., Gnaphalieae, or Plucheeae, according to recent treatments summarized by Anderberg (1994). In the present chapter we will look for patterns in flavonoid distribution in the subtribes of Gnaphalieae, as currently viewed by Anderberg (1994): Angianthinae Benth., Cassiniinae Anderb., Gnaphaliinae Less., Loricariinae Anderb. (for which we have no chemical data), and Relhaniinae Less. Merxmüller et al. (1977) discussed Inuleae s.l. in terms of two main groups, Inulinae *sensu amplo* and Gnaphaliinae *sensu amplo*. A third, smaller, group, Athrixiinae *sensu amplo*, accommodated those genera whose placement in the larger two would disrupt their delineation. Table 20–11 summarizes the occurrence of flavonoids in Gnaphalieae following the subtribal delineation of Anderberg (1994), along with corresponding information for Inuleae s.s. and Plucheeae. The placement of the various genera in the treatment by Merxmüller et al. (1977) is also indicated in the table.

There is strong correspondence between Inuleae s.s. of Anderberg (1994) and the Inulinae *sensu amplo* of Merxmüller et al. (1977). Genera seen as comprising Angianthinae, Cassiniinae, and Gnaphaliinae by Anderberg (1994) are all members of Gnaphaliinae *sensu amplo* in the Merxmüller et al. (1977) scheme. Two of the three members of Relhaniinae in the Anderberg treatment, *Elytropappus* and *Relhania*, appear in the Merxmüller et al. system as belonging to Athrixiinae *sensu amplo*. The third genus of this group for which we have some flavonoid information is *Metalasia* which Merxmüller et al. (1977) mark "?II" in their notation suggesting that its position in their Gnaphaliinae is uncertain.

In general, the flavonoid profiles of *Elytropappus* and *Relhania* do not distinguish them from most other members of tribes or subtribes under consideration. *Metalasia*, on the other hand, has exhibited the capacity to accu-

Table 20–11. Flavonoid structural features of subtribes of Gnaphalieae and tribes Inuleae and Plucheeae, all *sensu* Anderberg (1994), compared to the system of Merxmüller et al. (1977)

		Flavonoid structural feature[1]										
Tribe/Subtribe/Genus	MX[2]	Fol	Flav	Antho	Fon	OMe	6Ox	8Ox	68Ox	Bare	345	Comment[3]
Gnaphalieae												
Angianthinae												
Actinobole	II	–	–	–	+	1	–	–	–	–	–	
Bellida	I	+	+	–	–	0	–	–	–	–	–	
Bracteantha	*	–	+	–	+	4	+	+	+	–	+	
Chrysocephalum	*	–	–	–	+	0	–	–	–	–	–	
Myriocephalus	II	+	–	–	–	3	–	–	–	–	–	
Cassiniinae												
Anaphalis	II	+	–	+	–	4	–	–	+	+	–	
Antennaria	II	–	+	–	–	0	–	–	–	–	–	
Cassinia	II	+	+	+	+	3	+	–	–	+	–	
Haekeria	*	+	+	+	–	1	–	–	–	+	–	
Ozothamnus	*	+	+	+	–	1	+	–	–	+	–	
Raoulia	II	+	–	–	–	1	–	–	–	–	–	
Gnaphaliinae												
Achyrocline	II	+	+	+	+	4	+	+	+	+	–	
Filago	II	+	+	–	–	0	–	–	–	–	–	
Gnaphalium	II	+	+	+	+	4	+	+	+	+	+	
Helichrysum	II	+	+	+	+	5	+	+	+	+	+	*C*-Glycosylflavone
Ifloga	II	+	–	–	–	0	–	–	–	–	–	
Leucogenes	II	+	+	–	–	1	+	–	–	+	–	
Leontopodium	II	–	+	–	–	0	–	–	–	–	–	
Lucilia	II	+	–	–	–	4	–	–	+	+	–	
Pseudognaphalium	II	+	+	–	–	5	–	–	+	–	–	

Table 20–11. (Continued)

Tribe/Subtribe/Genus	MX[2]	Flavonoid structural feature[1] Fol	Flav	Antho	Fon	OMe	6Ox	8Ox	68Ox	Bare	345	Comment[3]
Syncarpha	II	–	+	–	–	1	+	–	–	–	–	syn. *Helipterum*
Relhaniinae												
Elytropappus	III	+	+	–	–	2	+	–	–	–	–	
Metalasia	?II	–	–	+	+	1	–	–	–	+	–	Odd A-ring (see text)
Relhania	III	+	+	–	+	5	–	–	+	–	–	Dihydroflavonol
Inuleae *s.s.*												
Allagopappus	I	+	+	–	+	3	+	–	–	–	–	
Antiphionia	I	–	–	–	+	0	–	–	–	–	–	
Anvillea	I	+	+	–	–	3	+	–	–	–	–	
Asteriscus	I	+	+	–	–	6	+	–	–	–	+	
Blumea	I	+	–	–	+	5	+	–	–	–	–	Dihydroflavonol; B-2'5'
Dittrichia	I	+	–	–	–	2	–	–	–	–	–	Dihydroflavonol
Geigeria	I	–	+	–	–	1	+	–	–	–	–	
Inula	I	+	+	–	+	5	+	–	–	–	–	Dihydroflavonol; B-2'5'; sulphate
Iphiona	I	+	+	–	+	5	+	–	–	–	–	Sulphate
Jasonia	I	+	+	–	+	6	+	–	–	–	–	
Pechuel–Loeschea	I	–	+	–	–	2	+	–	–	–	–	
Pulicaria	I	+	+	–	+	6	+	–	–	–	–	Dihydroflavonol
Varthemia	I	+	+	–	–	3	+	–	+	–	–	
Plucheeae												
Epaltes	I	–	+	–	–	0	–	–	–	–	–	
Pluchea	I	+	–	–	–	5	+	+	+	–	–	*C*-Glycosylflavone; sulphate
Pterocaulon	I	+	–	–	–	3	+	–	–	–	–	Dihydroflavonol
Sphaeranthus	I	+	–	–	–	3	+	–	–	–	–	
Tessaria	I	+	+	–	+	5	+	–	–	–	–	Dihydroflavonol

Table 20–11. (Continued)

1 Fol = flavonols; Flav = flavones; Antho = anthochlors; Fon = flavanones; OMe = highest level of *O*-methylation; 6Ox = 6-oxygenation; 8Ox = 8-oxygenation; 68Ox = 6,8-dioxygeantion; Bare = B-ring deoxyflavonoids; 345 = 3',4',5'-trioxygenation.

2 Placement according to Merxmüller et al. (1977) where I = Inuleae *sensu amplo*; II = Gnaphaliinae *sensu amplo*; III = Athrixiinae *sensu amplo*; an asterix indicates that the genus was not listed by those authors.

3 B-2'5' = 2',5'-dioxygenation.

mulate flavonoids that lack B-ring oxygenation, which feature occurs with some consistency in Cassiniinae and Gnaphaliinae. *Metalasia* is also characterized by the presence of a dihydrochalcone derivative with an unusual hepta-substituted A-ring. Most compounds identified from *M. cymbifolia* exhibit both *C*- and *O*-prenylation (Zdero and Bohlmann, 1990c). Unfortunately, we do not have any information on the vacuolar flavonoids of this genus, which prevents us from commenting on the significance of the apparent lack of both flavonols and flavones in the taxon.

The presence of B-ring deoxyflavonoids in *Metalasia* is interesting. This structural feature has been observed in flavonoids from members of four of six genera of Cassiniinae studied and in five of 10 genera of Gnaphaliinae. It is a very commonly encountered feature in species of *Achyrocline, Gnaphalium*, and *Helichrysum*, all members of Gnaphaliinae. B-Ring deoxyflavonoids have not been reported from any member of Angianthinae (no information available for Loricariinae). Their presence in *Metalasia* is interesting in that it suggests, as usual, that a further search for these compounds in other members of the *Relhania* group might prove rewarding.

Within Asteroideae the highest frequency of occurrence of flavanones is seen in Inuleae s.l. (0.35). Flavanones have been identified as components of all tribes in Asteraceae except Tageteae. In the context of the present chapter, however, where we are looking at the more dissected view of the everlastings, Inuleae s.s. can be seen to have 42% of genera that accumulate flavanones. Within Gnaphalieae, flavanones have been reported from members of all subtribes for which data are available, but the frequencies of occurrence are somewhat lower than in Inuleae. It is also worth noting that sulphated flavonoids and dihydroflavonols, shown in the Comment section of Table 20–11, have been reported from several members of Inuleae s.s. and Plucheeae but only from one representative of Gnaphalieae, *Relhania genistifolia* (Tsichritzis and Jakupovic, 1990) of subtribe Relhaniinae.

Anthochlors, mostly chalcones, but aurones as well, have been reported from one of three genera in Relhaniinae, three of 10 genera of Gnaphaliinae, and four of six genera from Cassiniinae. There have been no reports from Angianthinae, and, as already noted, we have no information at all about Loricariinae. It would be of interest to establish the Angianthinae as a non-accumulating subtribe. Sampling has been sufficient, however, to indicate that Gnaphalieae can be credited as being an accumulating taxon as opposed to Inuleae and Plucheeae, which appear not to be.

While flavonoids with 6-oxygenation are fairly common in Asteraceae, there are nonetheless some distinctions within Inuleae that can be made on the basis of presence or absence of this character. The frequency of occurrence of 6-oxygenated flavonoids in Angianthinae is 0.20, in Cassiniinae 0.40, in Gnaphaliinae 0.50, and in Relhaniinae 0.33. The frequency of occurrence of these compounds in Inuleae s.s. is 0.83. Although this does not count when comparing subtribes in the Anderberg scheme (1994), it certainly offers support for the Merxmüller et al. (1977) view of subgroups within Inuleae s.l. Perhaps adding to the distinctiveness of subtribes in the latter sense is the observation that flavonoids with 8-oxygenation, but lacking 6-oxygenation, are known from members of Gnaphalieae (four genera) but have not been reported from any member of Inuleae s.s. This distinction has been emphasized by Bayer and Starr (1998) with data from *trn*L intron and *trn*L/*trn*F intergenic spacer regions of cpDNA, in which Gnaphalieae are quite distinct from Inuleae s.s. and Plucheeae.

A recent study of ITS sequences in Inuleae s.s. by Eldenäs et al. (1998) resulted in two conclusions: (1) the position of *Anisopappus* as sister group to the remainder of the tribe is well supported; and (2) the subdivision of the tribe into paleate and epaleate grade groups is not supported. We have found no flavonoid information for *Anisopappus* and for over half of the genera examined by those workers. Additional comment would not be fruitful.

Tribe Lactuceae

Flavonoids of Lactuceae were described in Chapter 13 where we presented information for 39 genera. Three treatments of Lactuceae provide us with an interesting opportunity to examine the potential usefulness of the flavonoid data. The three systems are those of Stebbins (1953), Jeffrey (1966), and the recent cladistic study of Bremer (1994). Stebbins' treatment employed chromosomal, pollen, and geographic distributional information, to which Jeffrey added measurements of microcharacters. Bremer's (1994) treatment is based on a variety of morphological features as well as chromosome number. The only chemical character employed in Bremer's analysis would be that inferred from the use of floret color, although it would be difficult to assign any chemical structure to the character states since he included white florets along with pink, purple, violet, and blue ones. The other character state was "yellow or cream."

Lactuceae exhibit one of the simplest and most uniform flavonoid profiles seen in the family. With few exceptions, members of the tribe accumulate glycosides of apigenin, luteolin, kaempferol, and quercetin, or their simple *O*-methyl ethers. Flavones occur in 85% of the genera for which data are available with flavonols having been reported in only 67%. The tribe also exhibits one of the lowest levels of *O*-methylation seen in the family, at least in groups that have been comparably sampled. Other tribes with similar low levels of *O*-methylation are Vernonieae and Barnadesieae.

Extra A-ring oxygenation is seen, in the form of 6-hydroxyapigenin (scutellarein), in species of *Hedypnois, Hypochaeris*, and *Reichardia*. Species of *Cichorium, Hieracium, Lactuca*, and *Sonchus* have the capacity for *O*-methylation, but the level achieved rarely exceeds the formation of chrysoeriol or isorhamnetin (both 3'-*O*-methyl ethers). Only in the case of *Hieracium* and *Launaea* does one see dimethyl ethers of flavones and/or flavanones formed. In the case of *Launaea* (*L. asplenifolia*) two unusual compounds have been described, 7-hydroxy-3',4'-dimethoxyflavone, one of the few 5-deoxyflavonoids known from the family, and 3-*C*-(3-methylbutyl)-5,7,3',4',5'-pentahydroxyflavone (Gupta and Ahmed, 1985; Gupta et al., 1985a,b). In the second paper, these workers also reported the presence of delphinidin, which also possesses the 3',4',5'-hydroxylation pattern. The chalcone coreopsin has been reported from *Pyrrhopappus*. *C*-Glycosylflavones have been reported from members of *Catananche, Launaea, Scolymus*, and *Tragopogon*.

Flavonoid occurrences in the tribe, presented according to the subtribal view of Bremer (1994), are summarized in Table 20–12. The most unusual flavonoid in the tribe is 5,7,2',4',5'-pentahydroxyflavone (isoëtin), which has been identified in species from the following genera: *Crepis, Hedypnois, Heywoodiella* (referred to *Hypochaeris* by Lack, 1978), *Hieracium, Hispidella, Hypochaeris, Leontodon, Picris, Reichardia*, and *Sonchus*. If we look at the occurrence of isoëtin within the tribe, we find that five of these genera (*Hedypnois, Heywoodiella, Hypochaeris, Leontodon and Picris*) are placed in Hypochaeridinae Less. by Bremer (1994). Stebbins (1953) placed these genera (except *Heywoodiella*, which hadn't been described at that time) in Leontodontinae O. Hoffm. These genera were seen by Jeffrey (1966) as belonging to his *Hypochaeris* group-subgroup *Hypochaeris*. Bremer placed *Andryala, Hieracium*, and *Hispidella* in subtribe Hieraciinae Dumort. *Hieracium* and *Hispidella* accumulate isoëtin, *Andryala* apparently not. Stebbins (1953) placed these two genera in different subtribes, *Hispidella* in Cichoriinae and *Hieracium* in Crepidinae Dumort. Jeffrey (1966) has them together in his group *Tolpis*-subgroup *Tolpis*. *Reichardia* and *Sonchus* are also isoëtin-positive. They appear together in Sonchinae K. Bremer according to Bremer. Stebbins has them together in Crepidinae (with *Hieracium* and *Crepis*) while Jeffrey has them in different groups.

In a recent paper, Kim et al. (1999a) described a test of a non-coding region (*psb*A-*trn*H^{GUG}) of cpDNA as a source of phylogenetic information for the subtribe Sonchinae. They found that this region does not evolve fast enough to help in the resolution of closely related insular endemics of Asteraceae.

Bremer's (1994) strict consensus tree of equally parsimonious cladograms (his Figure 9–2, page 171) shows a clustering of *Hedypnois*,

Table 20–12. Flavonoid occurrences in Lactuceae *sensu* Bremer (1994)

	Flavonoid structural feature[1]						
Subtribe/Genus	Fol	Flav	CGly	OMe	6Ox	245	Comment
Unassigned							
Cichorium	+	+	–	1	–	–	
Scolymus	+	–	+	0	–	–	
Catananchinae							
Catananche	–	+	+	0	–	–	
Rothmaleria	+	–	–	0	–	–	
Crepidinae							
Chondrilla	+	+	–	0	–	–	
Crepis	+	+	–	0	+	+	
Ixeris	–	+	–	0	–	–	
Taraxacum	+	+	–	1	–	–	
Youngia	–	+	–	0	–	–	
Lactucinae							
Cicerbita	+	–	–	0	–	–	
Lactuca	+	+	–	1	–	–	
Prenanthes	–	+	–	0	–	–	
Scariola	+	+	–	0	–	–	
Sonchinae							
Babcockia	–	+	–	0	–	–	
Embergeria	–	+	–	0	–	–	
Launaea	–	+	+	2	–	–	
Reichardia	–	+	–	0	+	+	3',4',5'-B-ring
Sonchus	+	+	–	1	–	+	
Taeckholmia	–	+	–	0	–	–	
Dendroseridinae							
Dendroseris	+	+	–	0	–	–	
Microseridinae							
Agoseris	+	+	–	0	–	–	
Krigia	+	+	–	0	–	–	
Microseris	+	–	–	0	–	–	
Pyrrhopappus	–	+	–	0	–	–	Chalcone
Stephanomeriinae							
Stephanomeria	+	+	–	0	–	–	Flavanone
Hieraciinae							
Andryala	+	+	–	0	–	–	
Hieracium	+	+	–	2	–	+	
Hispidella	–	+	–	0	–	+	
Hypochaeridinae							
Hedypnois	–	+	–	0	+	+	
Heywoodiella[2]	–	+	–	0	–	+	
Hyoseris	+	+	–	0	–	–	
Hypochaeris	–	+	–	0	+?	+	

Table 20–12. (Continued)

Subtribe/Genus	Fol	Flav	CGly	OMe	6Ox	245	Comment
	Flavonoid structural feature[1]						
Leontodon	+	+	–	0	+?	+	
Picris	+	+	–	0	–	+	
Rhagadiolus	+	–	–	0	–	–	
Urospermum	+	+	–	0	–	–	
Scorzonerinae							
Koelpinia	+	–	–	0	–	–	
Scorzonera	+	+	–	0	–	–	
Tragopogon	+	+	+	0	–	–	

[1] Fol = flavonols; Flav = flavones; CGly = *C*-glycosylflavones; OMe = highest level of *O*-methylation; 6Ox = 6-oxygenation; 245 = 2',4',5'-trihydroxylation (isoëtin).

[2] Referred to *Hypochaeris* by Lack (1978).

Leontodon, and *Hypochaeris* all of which are isoëtin-producers. *Hieracium* (and presumptively *Hispidella*), not cleanly resolved from this trio of genera in his tree, also accumulates isoëtin. It would seem safe to conclude that production of isoëtin is a fairly reliable indicator of relationships among this group of genera. Although its occurrence is even more limited than isoëtin, the presence of 6-hydroxyapigenin (scutellarein) parallels the occurrence of isoëtin in *Hedypnois, Hypochaeris*, and *Leontodon* in Hypochaeridinae. It is interesting to note that scutellarein and isoëtin occur together in *Crepis* and in *Reichardia*. The distribution of apigenin, luteolin, and quercetin glycosides in members of Crepidinae was discussed by Mañez et al. (1994).

C-Glycosylflavones occur in members of four different subtribes regardless of which treatment is considered and provide no additional insights.

The usual complaint can be registered that sampling could be improved in terms of both more genera and more species within the genera in question. It seems reasonable to suggest that isoëtin, with its unique structure, could be exploited to a greater extent. It is entirely possible that inexperienced workers may have overlooked isoëtin in their studies. Flavonoids with the isoëtin B-ring oxygenation pattern are known from other groups, e.g. *Artemisia*, but all but one of these compounds carry several *O*-methyl groups. It would be interesting to compare the enzymes responsible for generating the B-ring oxygenation pattern in these very different groups. Also, is there any enzymatic connection between scutellarein and isoëtin formation?

We conclude discussion on subdivisions within Lactuceae by briefly mentioning the study of Whitton et al. (1995) on cpDNA restriction site variation within the tribe. Little can be made of these new relationships, however, because, once again, two problems exist with regard to comparing the results of that study with flavonoid profiles: (1) absence of information in the flavonoid literature on many of the genera included in the DNA analysis; and (2) the overall homogeneity of flavonoid profiles that have been published.

Tribe Mutisieae

Flavonoids of Mutisieae were discussed in Chapter 14. Cabrera (1977) discussed Mutisieae in terms of four subtribes: Barnadesiinae Benth. & Hook., Gochnatiinae Benth. & Hook., Mutisiinae Less., and Nassauviinae Less. Barnadesiinae have now become Barnadesioideae and need not be extensively commented upon here. Bremer (1994) discussed the Karis et al. (1992) cladistic treatment of Mutisieae and related groups which showed, in general, that Gochnatiinae were not a holophyletic group, and that the tribe could more reliably be treated as consisting of two subtribes, Mutisiinae and Nassauviinae. The composition of Nassauviinae did not change while those genera originally assigned to Gochnatiinae found a new home in Mutisiinae.

It should come as no surprise that representation in the flavonoid literature for Mutisieae is poor. We have flavonoid information for only three of 52 genera listed by Bremer (1994) for Mutisiinae. Sampling of Nassauviinae is better, nine of 24 genera, but still not particularly good. Sampling at the species level has not been exhaustive, or even reasonably detailed; studies of most of the members of the tribe have not been carried out. With those caveats in mind we can look in Table 20–13 where the available information is summarized. Genera in Mutisiinae have been arranged with *Gerbera*, *Mutisia*, and *Tarchonanthus* listed first followed by five genera that were originally placed in Gochnatiinae. The search for patterns in this depauperate data set is disappointing. The structural features reported are more or less equally spread between the two subtribes. Clearly, no meaningful correlations are to be found. It is interesting to note that the only difference that would have been argued as significant involved the consistent occurrence of kaempferol and quercetin 3-*O*-mono- and 3-*O*-diglycosides in members of the former subtribe Barnadesiinae. This simple but consistent profile, along with a number of other features, was taken as evidence supporting the macromolecular-based elevation of Barnadesiinae to the status of subfamily (Bohm and Stuessy, 1995).

Table 20–13. Flavonoids of subtribes of Mutisieae *sensu* Bremer (1994)

Subtribe/Genus	Flavonoiod structural feature[1] Fol	Flav	Fon	OMe	6Ox	Bare	Comment
Mutisiinae							
Gerbera	+	+	–	0	–	–	Cyanidin; pelargonidin
Mutisia	+	–	–	2	–	–	
Tarchonanthus	–	–	+	0	–	+	
Gochnatia (G)[2]	+	+	+	2	+	–	Dihydroflavonol
Gypothamnium (G)	+	–	–	0	–	–	
Hesperomannia (G)	+	–	+	0	–	–	
Plazia (G)	–	+	+	1	–	–	
Pleiotaxis (G)	–	–	–	1	–	1	Chalcones
Nassauviinae							
Adenocaulon	+	–	–	0	–	–	
Dolichlasiuim	–	+	+	2	+	+	
Jungia	–	+	–	2	+	–	
Leucheria	+	+	–	4	+	–	
Lophopappus	–	–	+	0	–	+	
Perezia	+	–	–	0	–	–	
Pleocarphus	–	+	–	1	+	–	
Proustia	–	+	+	1	–	–	
Trixis	+	+	+	1	–	–	Dihydroflavonol

[1] Fol = flavonols; Flav = flavones; Fon = flavanones; OMe = highest level of *O*-methylation; 6Ox = 6-oxygenation; Bare = B-ring deoxyflavonoids.
[2] G = Gochnatiinae *sensu* Cabrera (1977).

Tribe Senecioneae

The flavonoids of Senecioneae were described in detail in Chapter 15. Senecioneae were discussed in the earlier Reading Compositae Symposium by Nordenstam (1977). His view of the tribe involved two subtribes, Senecioninae Dumort. with 96 genera, and Blennospermatinae Rydb. with four small genera "with obscure affinities." Jeffrey and Chen (1984) treated Senecioneae as containing three subtribes: Tussilagininae (equating to the "Cacalioid" group of Nordenstam, 1977); Senecioninae; and a new subtribe Tephroseridinae C. Jeffrey & Y. L. Chen, containing the three genera *Nemosenecio*, *Sinosenecio*, and *Tephroseris*. They also excluded four genera making up Blennospermatinae (i.e., *Abrotanella, Blennosperma, Crocidium,* and *Ischnea*) from the tribe, but failed to suggest an alternative placement. More recently, Jeffrey (1992) accepted Blennospermatinae as sister group to the rest of the tribe, which he now lumped together into just one subtribe, Senecioninae [essentially the position of Nordenstam (1977)]. The recent discussion of Senecioneae by Bremer (1994) involves three subtribes, Blennospermatinae with four genera in agreement with Nordenstam, Senecioninae with 68 genera, and Tussilagininae Dumort. with 48 genera. We have flavonoid information for all four genera of Blennospermatinae and 17 representing Senecionineae and Tussilagininae, all in all a rather dismal sampling of a large and conspicuous tribe. The data appear in Table 20–14.

The reader must be growing accustomed to, or more likely, tired of, hearing that the sampling of this or that group has been too limited to afford much in the way of useful information. Such would appear to be the case again with Senecioneae. Many of the genera are monotypic and/or difficult of access. However, even some of the larger genera have not been well sampled, e.g., 1/90 for *Emilia*, 1/97 for *Euryops*, 1/40 for *Kleinia*, etc.

Table 20–14. Flavonoids of subtribes of Senecioneae sensu Bremer (1994)

	Flavonoid type[1]				
Subtribe/Genus	Fol	Flav	Fon	OMe[2]	Comment
Blennospermatinae					
Abrotanella	+	–	–	1	Delphinidin
Blennosperma	+	+	–	0	
Crocidium	+	+	–	0	Cyanidin, delphinidin
Ischnea	+	–	–	0	
Senecioninae					
Cacalia	+	+	–	0	
Cineraria	–	+	–	0	Cyanidin, delphinidin
Emilia	+	–	–	0	
Euryops	+	–	–	0	
Kleinia	+	–	–	0	
Robinsonia	+	+	+	0	Dihydroflavonol
Senecio	+	+	+	4	Anthochlors, sulphate, delphinidin
Tussilagininae					
Brachyglottis	+	–	+	2	Anthochlor, 8-oxygenation
Doronicum	+	+	–	2	
Homogyne	+	–	–	0	
Petasites	+	–	–	0	
Traversia	+	–	+	2	
Tussilago	+	–	–	0	

[1] Fol = flavonols; Flav = flavones; Fon = flavanones.
[2] OMe = highest level of *O*-methylation.

No conclusions, and few speculations, seem justified. Perhaps one could suggest a relationship between Blennospermatinae and some element of Senecioninae on the basis of combined flavonol-flavone presence with no apparent *O*-methylation. Luteolin 7-*O*-glucoside has been reported from *Blennosperma* (all three species) and *Crocidium* (monotypic) (Ornduff et al., 1973b). More recently Swenson et al. (1999) reported a mix of kaempferol and quercetin 3-*O*-glycosides and delphinidin 3-*O*-glucoside from seven species of *Abrotanella* and one species of *Ischnea*. Flavones were not observed in the two latter genera.

Some unusual combinations of compounds make appraisal difficult. *Senecio* has been fairly well studied, so it is not altogether odd that it has been shown to be one of the richer sources of flavonoids in the tribe. The situation with *Brachyglottis* is unusual in that 8-oxygenated compounds were obtained from exudates of *B. cassinioides* glands, structures not universal in the genus (Reid and Bohm, 1993). The esterified quercetin derivatives of *Traversia baccharoides* (Kulanthaivel and Benn, 1986) are unique in the tribe, and since *Traversia* is a monotypic genus, the significance of these compounds is difficult, if not outright impossible, to determine.

Tribe Vernonieae

Jones (1977) concluded that the two subtribes into which Bentham (1873) and Hoffmann (1890) had apportioned the genera of Vernonieae were not natural groups and proceeded to discuss the tribe as an assemblage of genera, a treatment that he clearly recognized as not without problems. Bremer (1994) discussed the tribe in terms of six subtribes. The flavonoid chemistry, presented in Chapter 17, is quite homogeneous and may better be related to geographic origins of the species than to any subtribal classification (Harborne and Williams, 1977).

A summing up

Although some interesting correlations between flavonoid profiles and subtribal relationships have emerged, the overall conclusion here is that flavonoids are not at their most informative at this level of the taxonomic hierarchy. This should not be too surprising in some cases, Lactuceae for example, in view of the homogeneity of flavonoids across the tribe. A problem that seems to have manifested itself to a greater extent at this level than in any other is sampling error. This is particularly vexing in view of the very wide sampling of tribes and subtribes in current macromolecular studies. Unless a major change in focus occurs, and this seems highly unlikely to us, these holes in the secondary compound literature are likely to remain. Nevertheless, filling in the odd gap here and there within some of these subtribes (and other taxa) could still provide useful information.

CHAPTER 21

Flavonoids at the Generic Level

In this chapter we present examples of the application of flavonoid distribution patterns to studying relationships among genera that are, or have been, the topic of debate in the literature. One of the guides used for selecting examples to discuss in this context has been the taxonomic treatments of the various tribes presented by Bremer and his colleagues (Bremer, 1994). The submergence of some genera into others according to their views will be examined in terms of whether flavonoid profiles agree or disagree with such treatments. Similarly, the breaking up of larger genera into smaller groups offers other opportunities to apply flavonoid chemical data. With regard to this latter point, and specifically to the removal of some species of *Artemisia* to *Seriphidium*, flavonoid data are not necessarily useful. In this chapter we will not usually concern ourselves with the patterns of flavonoid variation among species. Some of these problems will be discussed in Chapter 22.

Argyroxiphium-Wilkesia-Dubautia (Madiinae)

The Hawaiian silversword alliance consists of three genera, *Argyroxiphium* (the silverswords themselves), *Wilkesia*, and *Dubautia*, with five, two, and 21 species, respectively. The group has received an exceptionally high degree of botanical attention owing to the stunning appearance of the silverswords themselves, as well as that the group is one of the most thoroughly studied examples of adaptive radiation in the plant kingdom (Carlquist, 1974, 1980; Carr, 1985; Baldwin, 1992, 1996, 1998; Baldwin et al., 1990, 1991; Carlquist et al., 2001). Some members of each genus have been subjected to flavonoid analysis. Nonpolar flavonoid aglycones of *Argyroxiphium sandwicensis* subsp. *sandwicensis* and both *Wilkesia* species have been identified (Bohm and Fong, 1990). Identified from the former were 5,4'-dihydroxy-7,8-dimethoxyflavone and 5,4'-dihydroxy-6,7,8-trimethoxyflavone, which represents a subset of the profile of *Wilkesia*. Additional spots were observed on chromatograms but very limited material precluded further work. Profiles comprising 6,8-dioxygenated flavonoids have been seen in two mainland tarweed genera, *Calycadenia* (Emerson et al., 1986) and *Hemizonia* (Proksch et al., 1984; Tanowitz et al., 1987).

The principal flavonoid study of *Dubautia*, described in Chapter 25, involved naturally occurring hybrids between *D. scabra* subsp. *scabra* and *D. ciliolata* subsp. *glutinosa* (Crins et al., 1988a). However, a chromatographic survey of some dozen and a half taxa showed the presence of nonpolar flavonoid aglycones in extracts of other glandular species (sect. *Railliardia*) and flavonoid glycosides (presumably kaempferol and quercetin) from members of all three sections (*Venoso-reticulatae*, *Dubautia*, and *Railliardia*) (Crins and Bohm, unpubl.).

Artemisia-Seriphidium

There appears to be increasing acceptance of the recognition of the species of *Artemisia* that constituted subgenus *Seriphidium* as the genus *Seriphidium*, as originally suggested by Poljakov (1967). Limited sampling for ITS and cpDNA analyses, however, do not support this perspective (Kornkven et al., 1998). As pointed out by Bremer and Humphries (1993), who also review the taxonomic history of the groups, species of *Seriphidium* are characterized by discoid, homogamous capitula in contrast to the disciform, heterogamous capitula of *Artemisia*. With flavonoid data for about two dozen species of *Seriphidium* and nearly 50 species of *Artemisia*, give or take a few synonyms, we ought to have sufficient information to detect differences between the two genera. We use "ought to" because once again we encounter a major problem in comparative flavonoid studies: although there is a considerable amount of information available for these genera, it was not necessarily obtained in a systematic fashion. It is not possible to judge the significance of the pigment profile of a species if only its nonpolar (often called exudate) components have been reported, either as a result of dedicated studies of those compounds, or as volunteer compounds from studies of, say, sesquiterpene lactones. A case in point would be the occurrence of *C*-glycosylflavones, which would not emerge in studies of nonpolar constituents. With this pervasive problem in mind, let's turn our attention to the data at hand.

The flavonoid chemistry of *Artemisia*, taken in the broad sense for the moment, reflects the richness and variety of the family itself. Common flavones and flavonols have been reported from almost every species specifically examined for polar compounds. Extra A-ring oxygenation at C-6 is common, less so at C-8. High levels of *O*-methylation are seen with both flavones and flavonols in many species. Common *O*-glycosides of many compounds have been reported as have mono- and di-*C*-glycosylflavones from a number of species. Chalcones, dihydrochalcones, flavanones, dihydroflavonols, and occasional B-ring deoxyflavonoids have been reported, but no apparent pattern emerges. Identification of flavonoids with higher levels of B-ring oxygenation has been reported, in particular the 3',4',5'- and the somewhat more unusual 2',4',5'-patterns. A comparison of the occurrences of flavonoids with these oxygenation patterns shows that the 3',4',5'-pattern occurs in about 24% of *Artemisia* species and in about 29% of *Seriphidium* species. Five species of *Artemisia* have flavonoids with 2',4',5'-oxygenation whereas this pattern has not been observed in *Seriphidium* to date. *C*-Glycosylflavones are known from three species of each genus. A few rare or unusual compounds have been identified but are of no value in helping to define generic limits. All in all, flavonoid chemistry does not provide, at least in its present incomplete form, any clear support for recognition of *Seriphidium* as a genus in its own right. This is essentially the same conclusion drawn by Belenovskaja (1996) who summarized a large body of data in her contribution to the 1994 Kew symposium. Her observations were based on a study of 130 species of *Artemisia*. Flavonoid patterns do support the recognition of three subgenera: *Artemisia*, *Dracunculus*, and *Seriphidium*. Further comments on infrageneric groups of species within *Artemisia* s.l. can be found in the next chapter.

Asanthus-Brickellia

Two species identified as *Asanthus solidaginifolia* and *A. thrysiflora* were analyzed for their flavonoid components by Yu et al. (1986b) as part of the examination of chemistry of Eupatorieae by T. J. Mabry and colleagues at the University of Texas. *Asanthus* is a segregate from *Brickellia*, but the picture is slightly more complicated since Bremer (1994) places *Asanthus* under *Steviopsis*. Whatever the genus is called, its flavonoid chemistry is instructive *vis-a-vis* relationships with other members of the tribe. The Yu et al. study showed that *Asanthus* accumulates common 3-*O*-glycosides of kaempferol, quercetin, and isorhamnetin, quercetin 3-methyl ether 7-*O*-glucoside, 6-methoxyapigenin and its 4'-methyl ether and 6-methoxyluteolin and its 4'-methyl ether. Those workers concluded that the presence of flavonoids that lacked 6-methoxylation argued for generic status since they had shown that a variety of flavonoids "all with 6-methoxylation" were "characteristic of the main evolutionary line in *Brickellia*." This statement is not strictly correct. Although 6-methoxylated flavonoids certainly dominate in *Brickellia*, flavonoids lacking this feature are by no means universally absent. Examples include reports of luteolin 4'-methyl ether from *B.*

laciniata (Timmermann et al., 1979), isorhamnetin 3-*O*-galactoside from *B. chlorolepis*, isorhamnetin and quercetin 3,7-dimethyl ether from *B. glutinosa* (Goodwin et al., 1984), isorhamnetin 3-*O*-glucoside from *B. arguta* var. *odontolepis* (Rosler et al., 1984), and quercetin 3-methyl ether from *B. monocephala* (Norris and Mabry, 1985). The flavonoid chemistry alone would not provide particularly strong evidence for segregation of *Asanthus*; presumably other factors were taken into account as well.

In passing, it should be noted that those workers (Yu et al., 1986b) concluded that the very similar flavonoid chemistries of *A. solidaginifolia* and *A. thrysiflora* support the view of McVaugh (1984) that these two taxa should be treated as a single taxonomic entity.

Asteriscus-Pallenis

According to Merxmüller et al. (1977), *Asteriscus* consists of three species and *Pallenis* one. Both genera are native to the Mediterranean area. Anderberg (1994) disagrees with the Merxmüller et al. taxonomy and places *Pallenis* in synonymy with *Asteriscus*. Two published flavonoid studies allow us some comment on this system. Ahmed et al. (1991) reported luteolin 7-*O*-glucoside, the 3-*O*-glucoside, 3-*O*-galactoside, and 7-*O*-glucoside of kaempferol and quercetin, quercetin 4'-methyl ether 3-*O*-robinobioside, and the 3,4'-dimethyl and 3,6,3'-trimethyl ethers of quercetagetin from *Asteriscus graveolens*. The flavonoid profile of *Pallenis spinosa*, according to Ahmed et al. (1992), consists of 3-*O*-glycosides of quercetin and patuletin, kaempferol and quercetin 3-methyl ethers, quercetagetin 3,5,6,7,3'-penta- and 3,5,6,7,3',4'-hexamethyl ethers, and tricin 5-*O*- and 7-*O*-glucosides. The two taxa can be distinguished on the basis of the presence of 3',4',5'-oxygenation and a much higher level of *O*-methylation in *Pallenis*. It would, of course, be necessary to look at all three species that comprise Anderberg's *Asteriscus* before challenging this generic concept on the grounds of flavonoid chemistry.

Balsamorhiza-Wyethia

Several features of the flavonoid chemistry of *Wyethia* set the genus off markedly from most, if not all, others in the tribe. Although not all species appear to accumulate the unusual compounds, the summed profiles of *Wyethia* and *Balsamorhiza* show distinct differences from one another. Eight of the 11 species of *Wyethia*, for which some flavonoid data are available (14 species are known), exhibit derivatives of 5,7,3',4'-tetrahydroxyisoflavone (orobol), seven species accumulate flavanones that bear a *C*-prenyl function, and one, *W. angustifolia*, accumulates two *C*-prenylated isoflavone derivatives, among other compounds (McCormick et al., 1986; Bohm and Choy, 1987 and citations therein). Two species of *Wyethia*, *W. bolanderi* and *W. angustifolia*, accumulate *C*-prenyl flavonoids in which the prenyl group has cyclized to form a seven-membered heterocyclic ring (the dihydroöxepine ring system) instead of the normal six membered chromene type. *Wyethia* is further distinguished by the presence of naringenin and eriodictyol methyl ethers. Flavanones have not been detected in any *Balsamorhiza* species to date.

None of the five species of *Balsamorhiza* studied to date yielded either isoflavones or *C*-prenylated flavonoids. With one exception, the flavonoid profiles of this genus consist of methyl ethers of kaempferol, 6-hydroxykaempferol, quercetin, or quercetagetin in various combinations. The sole exception was the identification of 6-methoxyapigenin from one species. Flavonoid occurrences are summarized in Table 21–1. Although fine structure is lost in this presentation (intersectional differences are discussed in Chapter 22), differences in generic biosynthetic capacities are obvious.

These two taxa are generally accepted as good genera in their own right although Cronquist, writing in Hitchcock et al. (1955), said (Vol. 5, p. 332) that it would "...not be unnatural to treat the two as subgenera." More recently he (Cronquist, 1994) continued to treat them as distinct. The two genera can be distinguished on the basis of differences in scapiform stems

Table 21–1. Flavonoid structural features in sections of *Wyethia* and *Balsamorhiza*

	Wyethia[1]			*Balsamorhiza*[2]	
Structural feature	AG	AL	EUW	AR	EUB
Isoflavones	+	+	+	–	–
C-Prenylation	+	+	+	–	–
Dihydroöxepines	–	–	+	–	–
Flavanones	–	+	+	–	–
Flavone *O*-methylation	+	+	–	+	–
Kaempferol *O*-methylation	+	+	–	+	–
6-Hydroxykaempferol *O*-methylation	+	–	–	+	–
Quercetin *O*-methylation	+	–	–	+	+
Quercetagetin *O*-methylation	+	+	–	+	+

[1] Sections of *Wyethia*: AG = *Agnorhiza*; AL = *Alarconia*; EUW = *Euwyethia* (= *Wyethia*).
[2] Sections of *Balsamorhiza*: AR = *Artorhiza*; EUB = *Eubalsamorhiza* (= *Balsamorhiza*).

and the nature of cauline leaves, which, when present in *Balsamorhiza*, tend to be opposite; they are alternate in *Wyethia* (Weber, 1946). There is also indication that hybridization between members of the two groups occurs naturally (Hitchcock, pers. comm.), and it is well known that hybridization occurs within them (Ownbey and Weber, 1943). Flavonoid studies of authenticated intergeneric hybrids would be of value and a more thorough sampling of *Balsamorhiza* is needed. On the basis of the data on hand, however, one is justified in arguing for separate status for the two groups of species. Whether this should continue at the generic level or whether the groups should be treated as subgenera requires additional work. Macromolecular data would be a welcome addition.

A recent reassignment of several species of *Wyethia* along with one each of *Balsamorhiza* and *Helianthus* would result in two new genera, *Agnoriza* and *Scabrethia* according to Weber (1998). See Chapter 11 (p. 302) for details.

Bidens-Coreopsis

According to the presentation of Heliantheae at the Reading conference, *Bidens* and *Coreopsis* appear comfortably situated in Coreopsidinae (Stuessy, 1977). *Bidens* enjoys a world-wide distribution and is a comparatively large genus with some 230 species. *Coreopsis*, with 114 species, is a New World and African taxon. Their position together in Coreopsidinae seems to be generally agreed upon (Stuessy, 1977; Robinson, 1981; Bremer, 1994). Other opinions do exist, however. As pointed out by Bremer (1994), Mesfin (1984a,b, 1986) has suggested that the two genera be united because the differences in cypsela morphology, by which the two are usually distinguished, break down upon closer study; ITS data reinforce this view (Kim et al., 1999b). Bremer (1994) chose not to take *Bidens* apart, but does state that the genus in the strict sense is probably paraphyletic. The reader is referred to recent studies of comparative capitular morphology, anatomy, palynology, and ITS sequences in which the taxonomic problems surrounding these genera are reviewed (Mesfin et al., 1995; Kim et al., 1999b).

Owing to a quite sizeable flavonoid literature involving these two genera, it ought to be possible to see if any differences in their biosynthetic capabilities exist. Superficially at least, *Bidens* and *Coreopsis* are very similar in their capacity to elaborate most of the common chalcones and aurones, the anthochlors of chemosystematic fame. They also both produce the common flavones apigenin and luteolin, and the common flavonols, kaempferol, quercetin, and in one instance isorhamnetin (Crawford, 1981).

Unusual B-ring hydroxylation patterns are rare, 5,7,2',5'-tetrahydroxyflavone from *B. parviflora* (De Tommasi et al., 1992) being the only example of which we are aware. Extra oxygenation on the A-ring, however, is a common feature of flavonoids from both genera with derivatives of 2',3',4',3,4-pentahydroxychalcone (okanin) and 6,7,3',4'-tetrahydroxyaurone (maritimetin) widely distributed. It should also be noted that these compounds are based on the resorcinol or 5-deoxy-type A-ring. There appears to be a difference in production of extra-oxygenated flavones and flavonols, however. Whereas 6-hydroxyflavone derivatives have been reported from *Coreopsis nuecensoides* and *C. neucensis* (Smith and Crawford, 1981), *C. mutica* (Crawford, 1970), and *C. bigelovii* (Nicholls and Bohm, 1979), 6-oxygenated flavones have not apparently been seen in *Bidens*. 6-Oxygenated flavonol derivatives, on the other hand, are known from members of both genera.

Although *O*-methylated compounds are common in members of both genera, the levels of methylation in representatives from *Bidens* are higher than those seen in species of *Coreopsis*. Thus, 2',4',3,4-tetrahydroxy-3'-methoxychalcone (lanceoletin) and the related 6,3',4'-trihydroxy-7-methoxyaurone (leptosidin) have been isolated from species of *Coreopsis*, while di-, tri-, and tetra-*O*-methylated chalcones have been reported from several *Bidens* species (Hart, 1979; McCormick et al., 1984; Ballard, 1986).

The most striking difference between the two genera, however, may lie in the elaboration of complex acylated aurone and chalcone glycosides in *Bidens* (Hoffmann and Hölzl, 1988a,b,c, 1989a,b). Typical of the chalcone derivatives reported by these workers are okanin 4-*O*-(6"-acetyl-2"-caffeoyl)glucoside, 4'-*O*-(2",4"-diacetyl-6"-p-coumaroyl)glucoside, and 4-*O*-methylokanin-4'-*O*-(6"-acetyl-2"-caffeoyl)glucoside. Among aurone derivatives identified by these workers are maritimein 4",6"-di- and 2",4",6"-triacetates. The only acylated derivatives of this general sort from a species of *Coreopsis* was the report of caffeoyl derivatives of sulfuretin 6-*O*-di- and triglucosides in *C. mutica* by Julian and Crawford (1972). We are not aware of any published efforts to look specifically for highly acylated flavonoid glycosides in *Coreopsis*. The absence of such compounds in *Coreopsis* would provide an interesting difference between the two genera. Such a search could prove interesting.

Bidens-Megalodonta

Megalodonta consists of the single species, *M. beckii* (Torr.) Greene, distributed widely in the glaciated parts of North America. It is an aquatic with an unusual achene morphology. Roberts (1980) examined the flavonoid profile of fresh material from Michigan and samples from other sites in the species' range by means of comparative chromatography. The pigment array of *M. beckii* is very similar to that expected from *Bidens*: glycosides of the chalcones butein and okanin, and of the aurones sulfuretin and maritimetin along with glycosides of quercetin and cyanidin. It also afforded the otherwise rare 2',4',3-trihydroxy-4-methoxychalcone 4'-*O*-glucoside. This compound occurs otherwise in members of *Bidens* sections *Platycarpaea* and *Heterodonta*, which occurrence suggests a closer relationship with these groups, or, more likely, a common ancestor, than with other groups within *Bidens*. Giannasi (1975b) reported the aglycone from *Dahlia*. One further note of interest from Roberts' (1980) report is the lack of variation among populations collected from British Columbia, Ontario, New York, and Michigan.

Blennosperma-Crocidium

These two genera have been the subject of a great deal more debate than their small sizes would seem to warrant. *Crocidium* occurs from southern British Columbia to central California and consists of one species, *C. multicaule*. *Blennosperma* consists of three species, *B. bakeri* and *B. nanum* (two varieties) from California and *B. chilense* from Chile. Ornduff et al. (1973b) summarized the various opinions on relationships and tribal affinities of the two genera, and described their flavonoid chemistries. All specimens of all four species accumulated quercetin 3-*O*-galactoside and 3-*O*-glucoside (an apparently unusual isomer). Of more scattered occurrence but, still present in all four species, was a quercetin-3-*O*-"polyglucoside." Luteolin 7-*O*-glucoside occurred in several populations of *C. multicaule* from California and in *B. nanum* var. *robustum* from Marin County, California. Cyanidin and delphinidin glycosides were observed in material of *C. multicaule* from Oregon and from British Columbia. The authors were unable to confirm the suggestion made by Skvarla and Turner (1966) that these genera may contain isoflavones. The similarity of flavonoid profiles, along with other data summarized by Ornduff et al. (1973b), led to the suggestion that the two genera are closely related, and that they ought to be placed in Senecioneae. Bremer (1994) concurs.

Centaurea-Aegialophia-Cnicus-Grossheimia

Aegialophia, *Cnicus*, and *Grossheimia*, all of which appear in the flavonoid literature, have been submerged in *Centaurea* by Bremer (1994). Wagenitz and Hellwig (1996, p. 496) agree that *Cnicus* "...is part of the *Jacea* group of *Centaurea*" but do not comment on the other two segregates. It is convenient to include all of these in this section in an effort to see what flavonoid chemistry has to offer in supporting or refuting this taxonomic view.

El-Masry et al. (1980, 1981) described the flavonoid chemistry of a plant identified as *Aegialophia pumila* (Jusl.) Boiss. The profile of this species was shown to consist solely of flavones based on 6-hydroxyapigenin and 6-hydroxyluteolin, which occurred as a variety of methyl ethers, and sudachitin, which is 5,7,4'-trihydroxy-6,8,3'-trimethoxyflavone. The flavonoid profile of *Centaurea* is a complex one consisting of a multitude of flavone and flavonol derivatives, many of which have 6-oxygenation and moderate levels of *O*-methylation. B-Ring deoxyflavonoids have been reported from several species, *C*-glycosylflavones are known, and there have been reports of flavonoids with 8-oxygenation (El-Emary et al., 1983; Sepulveda et al., 1994b). There appear to have been no flavonoids with 6,8-oxygenation found. The differences in flavonoid profiles could certainly be used to argue for recognition of *Aegialophia* if morphological and other features were thought to warrant it.

Although only one remaining species of *Cnicus* is listed by Dittrich (1977) and by Mabberley (1987) from originally more than 200 published epithets (Index Kewensis), there are flavonoid reports for two species in the literature, *C. benedictus*, the so-called blessed thistle (Sendra and Walczak, 1967), and *C. wallichii* (Singh and Pandey, 1986; Singh et al., 1988). *Cnicus benedictus* accumulates apigenin 7-*O*-glucoside, a luteolin diglycoside, and kaempferol 3-*O*-glucoside, while *C. wallichii* afforded luteolin 7-*O*-glucoside and isorhamnetin 7-*O*-glucoside. The compounds described in these three papers fall well within the limits of flavonoid structural variation seen in *Centaurea* and, thus, would not support an argument for separate recognition of *Cnicus*.

The situation for *Grossheimia* is much the same. Monea (1986) reported naringenin and 6-hydroxyapigenin (scutellarein) from *G. macrocephala*. The occurrence of these compounds in several species of *Centaurea* could be used to argue for submergence of *Grossheimia* in the former.

Clibadium-Desmanthodium-Ichthyothere

In his treatment of Heliantheae for the Reading Congress, Stuessy (1977) included *Clibadium, Desmanthodium*, and *Ichthyothere* in Group 3 of his Milleriinae. Robinson's (1981) view of this group of genera differs significantly. He placed *Clibadium* and *Desmanthodium* in their own subtribes and placed *Ichthyothere* in Melampodiinae. Bremer (1994) lists these three among others that are not assigned to a subtribe. Flavonoid profiles of many species of *Clibadium* (Bohm and Stuessy, 1981a, 1985; Bohm et al., 1983), three species of *Desmanthodium* (Bohm and Stuessy, 1981b), and *Ichthyothere terminalis* (Bohm and Stuessy, 1982) have provided some information for some generalizations about relationships to be made. *Clibadium* and *Desmanthodium* have the capacity to hydroxylate flavonols at C-6 leading to quercetagetin derivatives; *I. terminalis* lacked 6-oxygenated flavonoids. In contrast, *I. terminalis* had flavonol 3-*O*-triglycosides and dihydroquercetin. *O*-Methylated flavonoids were observed in all three genera but *Ichthyothere* had only quercetin 3-methyl ether. *O*-Methylation at C-3 was seen in *Clibadium* and *Desmanthodium* but always in combination, e.g. 3,3'-dimethyl.

With the usual caveat concerning the poor sampling, particularly of *Ichthyothere*, flavonoid similarities between *Clibadium* and *Desmanthodium* could be used to argue for a closer relationship between them than between either of them and *Ichthyothere*.

Dendroseris

Dendroseris consists of 11 species all of which are endemic to the Juan Fernandez Islands of Chile. Eight of the species occur only on Masatierra (the older island), the other three only on Masafuera. The flavonoid chemistry has been thoroughly studied by Pacheco et al. (1991) who found the generic profile to consist of a series of apigenin and luteolin 7-*O*-glycosides and two quercetin glycosides. The subgeneric flavonoid profiles will be discussed in Chapter 22. The authors surveyed several likely relatives from mainland South America. Flavonoid profiles observed in these surveys suggested species of *Hieracium* or *Hypochaeris* as possible ancestors with the choice going to a species of the former. Apigenin and luteolin glycosides were identified in species from Chile. The absence of quercetin glycosides was not considered troublesome, since these compounds are otherwise well known from *Hieracium*. Recent DNA investigations (ITS sequences) (Crawford et al., 1992b; Sang et al.,1994; Kim et al., 1995, 1996b), however, suggest a closer tie to genera of Sonchinae, such as *Sonchus, Sventenia* or *Embergeria*. Non-coding regions of cpDNA (*psb*A-*trn*H) were not able to resolve further these intergeneric relationships (Kim et al., 1999a).

Erigeron

Erigeron contributes six endemic species to the flora of the Juan Fernandez Islands. One of these occurs on both islands while the remaining five occur only on the younger island, Masafuera. Valdebenito et al. (1992b) described the flavonoid profiles of all six species plus five more from the mainland. The overall profiles were quite similar with glycosides of apigenin, luteolin, 6-*C*-glucosylapigenin, quercetin, and quercetagetin present in various combinations although not all taxa, nor all populations, exhibited all compounds. Those authors concluded that the ancestor of the island species likely came from mainland South America. It is of note that four of the five mainland species examined produced the 6-*C*-glucosylflavones, features apparently not known otherwise in the *Erigeron* literature. A detailed discussion of the relationships among these species is found in Chapter 22.

Eupatorium fragments

One of the outcomes of the extensive study of Eupatorieae by King and Robinson (1987) was the reclassification of the once very large genus *Eupatorium* into a number of smaller genera while retaining the parent genus. Table 10–1 lists the name changes relevant to the flavonoid literature. Some former species of *Eupatorium* then found themselves in *Acanthostyles, Ageratina, Aristeguietia, Austroeupatorium, Campuloclinium, Chromolaena, Conoclinium, Disynaphia*, and *Piptothrix*, among others. These particular genera are listed because of the availability of flavonoid data for them which allows a comparison with the flavonoid profile of *Eupatorium* s.s. Table 21–2 lists *Eupatorium* s.s. and the fragment genera along with their oxygenation patterns and highest level of *O*-methylation. Excepting *Conoclinium* for the moment, it appears that the genera that resulted from the splitting range from somewhat complex to a good deal simpler with regard to oxygenation pattern than *Eupatorium*. Likewise, their maximum levels of *O*-methylation are consistently lower.

Conoclinium lacks simple 4'- and 3',4'-flavonoids but does have the 6,8,3',4'-combination. Compounds with the 6,3',4',5'-combination are unique to *Conoclinium* within this group of genera. *Conoclinium* shares with *Eupatorium* the capacity to accumulate flavonoids with the 6,8,3',4',5'-combination and the maximum level of *O*-methylation in both is seven.

Table 21–2. Comparison of oxygenation patterns and levels of *O*-methylation of *Eupatorium* s.s. and genera removed from *Eupatorium* s.l. by King and Robinson (1987)

	Flavonoid oxygenation pattern						
A-Ring oxygens	57			567			
B-Ring oxygens	4'	3'4'	3'4'5'	4'	3'4'	Other patterns	OMe[1]
Genus							
Eupatorium	+	+	+	+	+	5784', 56783'4', 57683'4'5'	7
Acanthostyles	+	+		+	+		3
Ageratina	+	+		+	+		4
Aristeguietia	+			+			3
Austroeupatorium		+			+		2
Campuloclinium	+						1
Chromolaena	+	+			+	5783'4'	5
Conoclinium			+			56783'4', 5673'4'5', 5783'4'5'	7
Disynaphia		+					0
Piptothrix	+	+		+	+		2
Brickellia	+	+		+	+	5672'4'5', 5672'3'4'5'	5
Nothobaccharis		+			+		3

[1] Highest level of *O*-methylation.

Although comparisons of total flavonoid profiles of these two genera show differences, their overall similarity in accumulating compounds with complex oxygenation patterns suggests that they are quite closely related. This is further supported by the presence of flavonoids with the methylenedioxy function in both genera, a structural feature seen otherwise in the tribe in *Ageratum*.

Table 21–2 also contains the oxygenation patterns and *O*-methylation values for *Brickellia* and *Nothobaccharis*. As in the cases just mentioned, the flavonoid profile of *Nothobaccharis* is quite simple compared to that of *Brickellia* from which it was split. 4'-Oxygenated flavonoids, present in *Brickellia*, are missing from *Nothobaccharis* and the latter has apparently lost the capacity to make flavonoids with 2',4',5'- and 2',3',4',5'-oxygenations. There has also been a reduction in *O*-methylation value. Both genera accumulate sulphated flavonoids which is a feature seen otherwise only in *Decachaeta* within Eupatorieae.

Flyriella

The *Flyriella* story begins, essentially, with the removal of *Eupatorium parryi* from *Eupatorium* and relocation in *Brickellia* as *B. shinneri* nom. nov. (Flyr, 1968). Although the move was predicated upon the taxon's possession of brickellioid features, its 5-ribbed achenes did not rest well with the remainder of *Brickellia*, all of which, save one, have achenes with 10 ribs. The other exception was *B. fendleri*, which was soon to become the monotypic genus *Brickelliastrum* (King and Robinson, 1972a). In that same year King and Robinson (1972b) established the genus *Flyriella* with five species (including *Brickellia shineri*, properly called in this genus *F. parryi*). A sixth species was to follow a few years later (King and Robinson, 1982). A more recent examination of the genus by Baker and Turner (1986) place the number of species at four: *F. harrimanii, F. leonensis, F. parryi* (which also includes *F. chrysostyla* and *F. sphenopoda* of King and Robinson's taxonomy) and *F. stanfordii*.

Before commenting on the flavonoid chemistry of these taxa it is useful to point out that chromosomal data are significant in this rearranged affiliation. *Eupatorium* has a base number of $x = 10$ as does *Flyriella*; *Brickellia* has a base number of $x = 9$. King and Robinson (1977) place *Flyriella* with *Alomia*, which has $x = 10$, together in subtribe Alomiinae.

Mabry et al. (1981) reported flavonoids from three species of *Flyriella* (they named four but one of these, *F. sphenopoda*, had been submerged in *F. parryi* in the Baker and Turner work mentioned above). The flavonoid profiles of these taxa were based on quercetin 3-*O*-arabinoside, quercetin 3-*O*-glucoside, quercetin 4'-methyl ether (tamarixetin) 3-*O*-rhamnoglucoside, and quercetin 7,4'-dimethyl ether (ombuin) 3-*O*-rhamnoglucoside. The four species studied exhibited different combinations of these compounds but the significant feature of the study was the simplicity of the overall flavonoid chemistry of *Flyriella* compared to that of *Brickellia*. *Brickellia* has an extensive array of flavonoid compounds with 6-oxygenation, 6-oxygenation combined with 2',4',5'- and 2',3',4',5'-B-rings, a diverse array of flavones as well as flavonols, several flavanones, and sulphated compounds in some species (see Chapter 10 for details). Mabry et al. (1981) stated that more than 70 species of *Brickellia* exhibited complex chromatographic profiles. They also stated that an examination of *Alomia alata* showed a closer chemical affinity to *Flyriella* than to *Brickellia* (we are unaware, however, of published details on *Alomia* flavonoids).

To complete this story, it is relevant to comment on the flavonoid chemistry of *Brickelliastrum fendleri*, which has been described by Bulinska-Radomska et al. (1985b). The comparatively simple pigment profile comprises quercetin 3-*O*-glycosides, isorhamnetin 3-*O*-glucoside, and 5,4'-dihydroxy-

3,6,7-trimethoxyflavone. The presence of a flavonol with 6-oxygenation and three *O*-methyl groups suggests a biochemical affinity with *Brickellia.*

Geraea-Encelia-Enceliopsis

An interesting observation was made in a study of simple phenolic compounds in both species of *Geraea* (Proksch et al., 1986). Differences between the two species would normally qualify for inclusion in Chapter 22, but since the information has ramifications with regard to relationships among *Geraea*, *Encelia*, *Enceliopsis*, and possibly *Flourensia*, the story is best told here. *Geraea*, the three other genera named, and *Phoebanthus*, were considered by Robinson (1981) to be closely related. Bremer (1994) places all five of these genera, along with several others, in subtribe Verbesininae. Before commenting on the work by Proksch and his colleagues, it is appropriate to recognize the finding of Harborne and his coworkers (Harborne and Smith, 1978; Harborne et al., 1978) that floral tissue of both species of *Geraea* yielded 3,5,7,3',4'-pentahydroxy-8-methoxyflavone 3-*O*-galactoside. Work by the Proksch group (Proksch et al., 1987) confirmed the presence of 8-methoxy derivatives in *G. canescens*. It was a paper published the previous year, however, (Proksch et al., 1986) that described the group's most interesting observations, at least in the present context. In that work individual plant parts were analyzed with surprising outcome. The leaf resin of *G. viscida* contained no benzofurans or benzopyrans and only *O*-methylated derivatives of the common flavonols kaempferol and quercetin, while the resin of *G. canescens* afforded *O*-methylated derivatives of 8-oxygenated flavonols (herbacetin and gossypetin), an 8-oxygenated flavone derivative, and a complex array of benzofurans.

The 1987 paper by Proksch et al. also contained a description of flavonoids and benzofurans in leaves of two varieties of *Enceliopsis nudicaulis*. Ignoring differences between the two varieties for the time being (see Chapter 23), we see that this species accumulates an array of 8-oxygenated flavonols and dihydroflavonols and benzofurans that approaches the profile of such compounds seen in *Geraea canescens* closer than to any of the other taxa involved in this group including *G. viscida*. A study of most of the species of the related genus *Encelia* by Proksch et al. (1988) showed that several exhibited flavonoid profiles dominated by 8-oxygenated flavonols. Flavanones oxygenated at C-8 were identified in one species of *Encelia* as well (*E. frutescens* var. *resinosa*). *Encelia* is also known to be a major producer of benzofurans and benzopyrans [see Proksch et al. (1987) for leading references]. These various results leave the relationships between the two species of *Geraea* and among *Geraea* and the other genera mentioned unclear. The possibility that the chemical differences between the two species of *Geraea* may be a reflection of their different habitats was mentioned by Proksch et al. (1986) who pointed out that *G. viscida* occurs in semi-arid chaparral while *G. canescens* is found in desert settings. Those authors, however, offered no discussion of how the different habitats may have influenced establishment or maintenance of the different chemistries.

One last issue remains and that is the status of *Flourensia* in all of this. *Flourensia* has enjoyed a modest degree of attention to the extent that we have found information for eight species [see Stuppner and Müller (1994) and Wollenweber and Yatskievych (1985) for leading references]. Although *Flourensia* accumulates benzofurans and benzopyrans [thesis of M. Aregullin cited by Proksch et al. (1986)], its flavonoid profile is quite different from those we have just been talking about. For one thing, no 8-oxygenated flavonoids have been reported, although prenylation at C-8 is known in compounds from two species. *C*-Glycosylflavones have been reported from one species, and B-ring deoxyflavonoids from five (see Chapter 11 for citation details). Flavonoid chemistry of *Flourensia* clearly sets it off from the others in this group. Dillon (1984) emphasized the morphological distinctiveness of *Flourensia*, but added that (p. 5) "the only genus that could be considered related is *Encelia*."

Gymnosperma-Gutierrezia

Gymnosperma glutinosum, the sole member of the genus, occurs from Central America to the southwestern United States. It has attracted attention from a practical point of view because of its toxicity to range animals, and also because of its apparent relationship to woody species of *Gutierrezia* (Yu et al., 1988). The availability of detailed information on the flavonoids of *Gutierrezia* (Fang et al., 1986b,c,d and citations therein) provide an excellent opportunity to compare the two genera. The flavonoid profile of *Gutierrezia* is among the most complex in the family. Both flavones and flavonols are known with extra oxygenation at C-6, C-8, and at both C-6 and C-8, all in various combinations with B-rings having 4'-, 3',4'-, 3',4',5'-, or 2',4',5'-substitution. Levels of *O*-methylation vary from zero to as high as six. Representing the extreme end of all this is 5,7-dihydroxy-3,6,8,3',4',5'-hexamethoxyflavone, one of 51 (!) compounds identified from *G. microcephala* (Fang et al., 1986b).

A study of flavonoids in *Gymnosperma glutinosum* (Yu et al., 1988) resulted in identification of 21 flavonoids including seven new ones. The close similarity of the *Gymnosperma* compounds with flavonoids of *Gutierrezia* is evident in the presence of flavones and flavonols with the same oxygenation patterns as those seen in the latter taxon including those with 2',4',5'-oxygenation. The most heavily substituted compound from *Gymnosperma* was 5,7-dihydroxy-3,6,8,2',4',5'-hexamethoxyflavone. The richness of flavonoid chemistries of these taxa provide strong support for suggesting a close relationship between them.

Haplopappus

Haplopappus has had anything but a peaceful taxonomic existence. The extensive study of Hall (1928) saw the reduction of a number of small genera to sectional status within his very broadly conceived genus (ca. 160 species). That treatment has been the subject of a good deal of reappraisal to the extent that, at the present time, *Haplopappus* is considered to be solely a South American genus of some 70 species centered in Chile (Bremer, 1994). Accompanying the recognition of "true" *Haplopappus* as a much more circumscribed genus was the return to generic status of many of the smaller taxa. One of the more striking differences among the genera that comprised *Haplopappus*, in the sense of Hall, was the divergent chromosome numbers with base numbers of n = 2, 3, 4, 5, 6, and 9. Flavonoid data proved to be an equally powerful source of information (Clark et al., 1980). That study provided information of the occurrence of pigments in *Asiris, Ericameria*, the "phyllocephalus group" of *Haplopappus*, *Hazardia* (three sections), *Machaeranthera* (three sections), *Macronema, Prionopsis, Pyrrocoma*, *Stenotus*, and *Tonestus*. Owing to the amount of information presented in the Clark et al. paper, it is most convenient to reproduce their flavonoid data in a summarized form along with chromosome numbers and growth habit (Table 21–3). Not included in the table are other sources of variation among this group of genera including the existence of polyploidy and the nature of the sugars involved in the flavonoid glycosides. Polyploidy is known from two of the five taxa with n = 9 (*Ericameria* and *Asiris*) and one each with n = 6 (*Pyrrocoma*), n = 4 (*Machaeranthera* sect. *Blepharodon*), and in *Machaeranthera* sect. *Sideranthus* which has species with chromosome numbers of n = 2-5. Variation in the array of flavones and flavonols includes scattered occurrences of methyl ethers, but the presence of methyl ethers in one structural class does not necessarily indicate them in the other. The variation in number of glycosylating sugars (one to four) shows no correlation with any of the other features of the group.

Clark et al. (1980) concluded that the ancestral stock was a woody shrub, it had n = 9 and exhibited polyploidy, and had a flavonoid profile consisting of flavones and flavonols, some of which were *O*-methylated, and *C*-glycosylflavones. Evolution proceeded, in general, with reduction in chro-

Table 21–3. Flavonoids, chromosome numbers (n), and growth habit of genera formerly considered part of *Haplopappus*[1]

Genus	n	Habit[3]	Flavonoid class[2]		
			Flav	Fol	CGly
Asiris	9	WS	+	+	+
Ericameria	9	WS	+	+	+
Macronema	9	WS	–	+	–
Stenotus	9	WS	–	+	+
Tonestus	9	HR	–	+	–
Hazardia					
Sect. *Hazardia*	5,6	HW	+	+	+
Sect. *Bracteofolia*	5	WS	+	+	+
Sect. *Machaerantheroides*	4	HW	+	+	+
Prionopsis	6	HP	+	+	–
Pyrrocoma	6	HP	+	+	–
Phyllocephalus group	6	HP	+	–	–
Machaeranthera					
Sect. *Sideranthus*	2-5	HP	+	+	–
Sect. *Blepharodon*	4	HP	–	+	–
Sect. *Havardi*	4	HP	–	+	–
Hypothetical ancestor	9	WS	+	+	+

1 Modified from Clark et al. (1980).
2 Flav = flavone; Fol = flavonol; CGly = *C*-glycosylflavone.
3 WS = woody shrub; HR = herbaceous perennial from rootstock; HW = herbaceous perennial from woody root crown; HP = herbaceous perennial from taproot.

mosome number, simplification of flavonoid profile, and a change away from woodiness, although not all of these changes occur in concert. Overall, the data provide excellent agreement for the view that *Haplopappus* s.l. is polyphyletic. What remains of *Haplopappus* is discussed in the following section.

Hazardia-Haplopappus

Bremer (1994) refers to the "*Haplopappus* group" as consisting of *Haplopappus*, *Hazardia*, and the South American *Llerasia*. We have no flavonoid data for the latter, but we can make some comparisons between the pigment profiles of the other two, which have been studied in some detail. In fact, flavonoids of all 13 species of *Hazardia* (Clark and Mabry, 1978; Clark and Wollenweber, 1985) have been described, and we have accumulated information on 16 (of ca. 70) species of *Haplopappus* (see entry in Chapter 8 for full details). The earlier *Hazardia* paper reported the presence, in various combinations, of apigenin, luteolin, kaempferol, quercetin, and isorhamnetin *O*-glycosides, *C*-glycosylflavones based on apigenin, and four unidentified compounds identified only as flavonol methyl ethers. The more recent report described no less than 33 compounds isolated from the resin of *H. squarrosa* var. *grindelioides*. Methyl ethers of kaempferol, quercetin, 6-hydroxykaempferol, quercetagetin, several flavones, naringenin, and other flavanones were identified. It is highly likely that the flavonol methyl ethers reported in the larger survey are members of this group.

There are extensive reports of the flavonoid chemistry of *Haplopappus* in the literature so only representative accounts need be mentioned here. Common flavonols have been identified from several species, *C*-glycosylflavones are known (Ulubelen and Mabry, 1981a; Ates et al., 1982), an array of naringenin and eriodictyol methyl ethers has been identified (Maldonado et al., 1993; Nuñez-Alarcón et al., 1993), and flavones and flavonols bearing extra oxygenation at C-6, and a modest level of *O*-methylation, have been reported (Öksüz et al., 1981; Ulubelen et al., 1982; Maldonado et al., 1993). The overall flavonoid similarities of the two genera are striking and could certainly be taken as evidence of a close relationship. One additional comment should be added at this point, namely, that *Hazardia squarrosa* yielded a flavanone with the 3',4',5'-trioxygenation pattern (Clark and Wollenweber, 1985), while *Haplopappus remyanus* was shown to accumulate two compounds that lacked B-ring substitution, 5-hydroxy-7-methoxyflavanone and 3,5-dihydroxy-7-methoxyflavanone (Zdero et al., 1991d). Whether these compounds reflect (systematically) significant features or whether this is the sort of structural variation that one finds in extended surveys of large genera, of course, can not easily be judged.

Helenium-Dugaldia

Dugaldia was established in 1828 by Cassini but few workers followed his taxonomy. Most workers saw species of *Dugaldia* placed in *Helenium* (see Bierner 1974 for a detailed history of the taxonomic changes involved). In that paper, Bierner states his view that the differences between these two genera, although not great, exceed the differences that distinguish sections within *Helenium*. Bierner's (1974) study included an examination of the flavonoids of all three species of *Dugaldia*: *D. pinetorum, D. integrifolium*, and *D. hoopesii*.

Several studies have shown that the flavonoid profile of *Helenium* consists of flavones (Carman et al., 1972; Maldonado et al., 1986) and *C*-

glycosylflavones (Wagner et al., 1972c,g). The profile of *Dugaldia* consists of common kaempferol and quercetin derivatives (Wagner et al., 1971c; Bohlmann et al., 1985d), 6-methoxyluteolin (Bohlmann et al., 1985d), and 5,7-dihydroxy-6,4'-dimethoxyflavone (pectolinarigenin) (Ozawa et al., 1983; Bierner, 1994a). No *C*-glycosylflavones have been reported. Combination of the flavonoid profile with several clear-cut morphological differences argues for recognition of *Dugaldia* as a genus in its own right. Karis and Ryding (1994a) agree. In a re-evaluation of summed chemical and morphological data, however, Bierner (1994b) has concluded that *Dugaldia* (and *Plummera*, see below) should be submerged in *Hymenoxys* and treated there as a subgenus.

Hymenoxys-Plummera

Plummera differs from *Hymenoxys* in the presence of sterile disc florets in the former as opposed to fertile ones in the latter (Bierner, 1978). *Plummera* is recognized by Karis and Ryding (1994a), but there are some authors (Turner et al., 1973; Bierner and Jansen, 1998; Wagner et al., 1999) who have argued that the differences are too slight to warrant such a treatment. Bierner (1978) studied the flavonoids of both species, *P. floribunda* and *P. ambigens*, in a attempt to gather additional information in order to arrive at a better understanding of the relationships of these genera. Bierner (1978) reported quercetin 3-*O*-glucoside and an acetylated derivative thereof, 5,7-dihydroxy-6,4'-dimethoxyflavone, and 5,7,4'-trihydroxy-6,3'-dimethoxyflavone from both species but failed to find kaempferol 3-*O*-glucoside in *P. floribunda* or patuletin derivatives that had been reported from *P. ambigens* (Wagner et al., 1972c,i). The flavonoid profile of *Hymenoxys*, for comparison, consists of kaempferol, quercetin, and isorhamnetin glycosides (Wagner and Iyengar, 1972), 6-oxygenated flavones (Herz et al., 1970a,b; Carman et al., 1972; Gao et al., 1990), 6-oxygenated flavonols (Herz et al., 1970a), and 6,8-dioxygenated flavones (Herz et al., 1970a; Dominguez et al., 1980; Gao et al., 1990). The capacity to make 8-oxygenated flavonoids appears to be a feature of *Hymenoxys*; such compounds have not been reported from *Plummera*. Bierner (1978) took these data to support the close relationship of *Plummera* to *Hymenoxys* (excluding *Tetraneuris*, which see below). As noted above in the discussion of *Dugaldia*, Bierner (1994b) has concluded that *Plummera* should be submerged in *Hymenoxys* and treated therein as a subgenus.

Hymenoxys-Tetraneuris

According to some authorities (Nelson, 1899; Robinson, 1981) the genus *Tetraneuris* should be recognized in its own right. This position was not accepted by Karis and Ryding (1994a) in their recent treatment of Helenieae where *Tetraneuris* was included as a synonym under *Hymenoxys*. Several studies provide us with a reasonable idea of the flavonoid biosynthetic capacities of both *Hymenoxys* and *Tetraneuris*, accepting for the moment that the latter is a good genus. Common glycosides of kaempferol, quercetin, and isorhamnetin were identified from a number of *Hymenoxys* species by Wagner and Iyengar (1972) while the occurrence of 6-methoxyapigenin (hispidulin) in several species has been documented (Herz et al., 1970b; Carman et al., 1972). *Hymenoxys acaulis* and *H. subintegra* afforded kaempferol 3,3'-dimethyl ether and 5,7-dihydroxy-6,4'-dimethoxyflavone (pectolinaringenin), respectively (Herz et al., 1970a). Several 6-methoxydihydroflavonols along with hispidulin and 5,7-dihydroxy-6,8,4'-trimethoxyflavone were reported from *H. turneri* by Gao et al. (1990). Much of the flavonoid chemistry of *Tetraneuris* has appeared under the name of *Hymenoxys scaposa* (Thomas and Mabry, 1967, 1968a,b; Ferracini et al., 1994) or *H. linearifolia* (Dominguez et al., 1980). The striking feature of these reports is the frequency of occurrence of flavonoids that carry a methoxy group at C-8, i.e., 5,7-dihydroxy-6,8,3',4'-tetramethoxyflavone (hymenoxin), 5,7-dihydroxy-6,8,4'-trimethoxyflavone (nevadensin), and 5,7,4'-trihydroxy-6,8,3'-trimethoxyflavone (sudachitin), its 3'-desmethoxy relative, and 5,4'-dihydroxy-6,7,8-trimethoxyflavone (xanthomicrol). Ferracini et al. (1994) focused attention on the difference between *Hymenoxys*, with 6-methoxy substitution, and *Tetraneuris*, with both 6 and 8-substitution. In view of the finding of 5,7-dihydroxy-6,8,4'-trimethoxyflavone in *Hymenoxys turneri* by Gao et al. (1990), however, the distinction between the two genera based on flavonoid chemistry is weakened.

Before leaving the Gaillardiinae one further genus deserves mention. *Psathyrotes annua*, one of three species in the genus, was reported to yield 5-hydroxy-6,7,4'-trimethoxyflavone (salvigenin) by Finnegan et al. (1972). The presence of extra oxygenation at C-6 and a modest level of *O*-methylation suggests that a dedicated study of flavonoids of the genus might well reveal useful taxonomic markers.

Helianthus-Viguiera-Simsia-Tithonia

The four genera to be discussed in this section are closely related and have had sufficient attention paid to their flavonoid chemistry to enable a reasonable assessment to be made of the usefulness of these data in understanding relationships. An important analysis of these genera, along with *Helianthopsis* and *Heliomeris*, using chloroplast DNA restriction fragment analysis was described by Schilling and Jansen (1989) and Schilling and Panero (1996). Unfortunately, there appears to be no information on flavonoids of *Heliomeris* and only a statement that *Helianthopsis* is anthochlor-positive (Schilling and Spooner, 1988).

The suggestion has been made that some member of *Viguiera* section *Maculatae* (series *Maculatae* according to Blake) may have served as the progenitor of *Helianthus* (Blake, 1918; Heiser et al., 1969). Recent cpDNA studies (Schilling and Panero, 1996) and cladistic evaluation of the data, however, show the two taxa as sister groups. This issue was addressed briefly by Schilling et al. (1988) in their study of the flavonoids of section *Maculatae*. Species in that section accumulate, among other flavonoids, 6-hydroxyluteolin, isoliquiritigenin 4'-*O*-glucoside, and coreopsin 3-methyl ether, none of which is known from *Helianthus*. Several species of *Helianthus*, on the other hand, accumulate flavones with extra oxygenation at positions C-6 and C-8 (often at both) along with a moderate level of *O*-methylation (see Table 11-11 for full details). Commonly encountered flavones in *Helianthus* are 5,7-dihydroxy-6,8,4'-trimethoxyflavone (nevadensin) and 5,7-dihydroxy-6,8,3',4'-tetramethoxyflavone (hymenoxin). Since nevadensin was isolated from both *Viguiera bicolor* and *V. greggii* and hymenoxin was obtained from the latter, perhaps it is appropriate to suggest that a link exists between *Helianthus* and a member of section *Breviflorae* to which these two species of *Viguiera* belong. The cpDNA data (Schilling and Panero, 1996) suggest a near basal position of the latter within Helianthinae, some distance removed from the more highly derived *Helianthus*.

Detailed flavonoid analyses of both *Simsia* and *Tithonia* have been described, the former by Schilling and Spooner (1988), the latter by La Duke (1982a). The *Simsia* study, focusing on floral flavonoids, detailed the occurrence of glycosides of apigenin, luteolin, and quercetin as well as the coreopsin/sulfurein and marein/maritimein anthochlor pairs (for details of interspecific patterns see Chapter 22). Information on leaf constituents is more limited but the report of nevadensin in *S. cronquistii* (Maldonado et al., 1992a) is significant because of the common occurrence of this compound in *Helianthus*, as discussed above. The flavonoid profiles observed in *Tithonia* are complex and will be discussed in some detail in Chapter 22. Suffice it to say that the overall pattern of compounds, with flavones, flavonols, and a variety of anthochlors (La Duke, 1982a), shows strong relationships with the patterns reported for some species of *Simsia*, *Viguiera*, and *Helianthus* all clearly belonging in monophyletic Helianthinae (Schilling and Panero, 1996). As well, nevadensin, hymenoxin, and acerosin (5,7,3'-trihydroxy-6,8,4'-trimethoxyflavone) have been reported from species of *Tithonia* by Perez et al. (1988, 1992).

Iostephane-Viguiera

Iostephane is a Mexican genus consisting of four species (Strother, 1999). *Viguiera* is by far the larger of the two with some 180 species with occurrences reported from both North, Central, and South America. Robinson (1981) placed *Iostephane* in his Helianthinae along with such genera as *Helianthus, Heliomeris, Pappobolus, Simsia, Tithonia*, and *Viguiera*. A detailed examination of the relationships among this set of genera was undertaken by Schilling and Panero (1991, 1996) who employed morphology, flavonoid chemistry, and restriction fragment analysis of chloroplast and nuclear DNA.

The flavonoid profiles of the four species of *Iostephane* were essentially superimposable in terms of overall composition, although there were differences in localization of the pigments. The generic profile, then, can be defined as consisting of quercetin 3-*O*-glucoside and 3-*O*-glucuronide and the anthochlor pair coreopsin/sulfuretin. The authors seemed content in using the presence of anthochlors in this genus to place it among anthochlor-positive genera. Other features of the group of the putatively related genera are of interest, however. The overall flavonoid profile of *Iostephane* is simpler than that seen in any of the genera named above for which flavonoid data are available. They all have compounds with 6,8-oxygenation and a modest level of *O*-methylation. Some members of *Helianthus*, *Tithonia*, and *Viguiera* accumulate 5,7-dihydroxy-6,8,3',4'-tetramethoxyflavone. At least one member of *Simsia* (*S. cronquistii*) accumulates 5,7-dihydroxy-6,8,4'-trimethoxyflavone (Maldonado et al., 1992a). *Helianthus* and *Tithonia* have at least one member each that accumulates a 5-deoxyflavonoid, 3'-hydroxy-7,4'-dimethoxyflavone in the case of *Tithonia* (Correa and Cervera, 1971) and 3,7,3',4'-tetrahydroxyflavone (5-deoxyquercetin) in the case of *Helianthus* (Harborne and Smith, 1978). In light of this information it seems reasonable to conclude that the only thing that *Iostephane* and these other genera have in common, within the flavonoid realm, is the capacity to accumulate anthochlors.

Iva-Ambrosia-Parthenium-Hymenoclea-Xanthium (Ambrosiinae)

Difficulties with *Iva* will be encountered in Chapter 22. Intimately associated with the circumscription of *Iva* is the relationship of the genus, in either the broader or the more limited context, with several others. Although the sampling of *Iva* and its relatives (in Ambrosiinae) has not been particularly good, one or two flavonoid features might be of interest. A broadly based study of Ambrosiinae and Engelmanniinae, including comments on sesquiterpenes but not flavonoids, was described by Bolick (1983).

The members of Ambrosiinae for which we have flavonoid information are *Ambrosia, Hymenoclea, Iva* s.l., *Parthenium*, and *Xanthium*. It is difficult to come to any conclusion about relationships involving *Hymenoclea* and *Xanthium* since only flavanone derivative has been obtained from the former and only a single quercetagetin derivative and 8-*C*-prenylapigenin from the latter. *Ambrosia*, the best studied in this little group of genera, has a rich flavonoid profile consisting of simple and 6-oxygenated flavones and flavonols, some flavanones, and, likely most significant in the present context, 6,8-dioxygenated flavones and flavonols. 6,8-Dioxygenated flavonoids have been reported from only one other source in this subtribe, *Iva acerosa* and *I. nevadensis*, both members of *Iva* sect. *Cyclachaena*. If one subscribes to the narrower view of *Iva*, where *I. acerosa* becomes *Oxytenia acerosa* and *I. nevadensis* becomes *Chorisiva nevadensis*, [see Bolick (1983) for details], then it is these taxa that share the 6,8-dioxygenated flavonoid feature with *Ambrosia. Parthenium* has also been studied intensively for its flavonoids but no 6,8-dioxygenated compounds have been reported. Quercetagetin methyl ethers have been reported from all of the genera discussed here except for *Hymenoclea*. Three of the genera from Ambrosiinae mentioned above were part of a cladistic morphological study of Heliantheae performed by Karis (1993b) and reproduced by Karis and Ryding (1994b). *Ambrosia* and *Iva* emerged as sister groups with *Parthenium* as sister group to them. In this instance the limited flavonoid chemistry can be seen as corroborating the relationships that arose from morphological criteria. The reader may wish to consult Bolick's papers (1983, 1985) for comments on the correspondence of sesquiterpene chemistry with taxonomic position of these taxa as well.

Leucogenes-Leontopodium

Leucogenes, the so-called New Zealand Edelweiss, consists of four alpine species, one native to the North Island, the other three [two described only recently by Molloy (1995)] to the South Island with the range of *L. grandiceps* extending to Stewart Island. According to some (Anderberg, 1991b, 1994), *Leucogenes* has as its nearest relative the true Edelweiss of the European Alps, *Leontopodium*. The flavonoid chemistries of the two genera are quite different with the compounds identified from *Leontopodium* apparently limited to apigenin and luteolin glucosides (Tira et al., 1970; Dashbalyn and Glyzin, 1978), whereas members of the New Zealand genus studied, *L. leontopodium* and *L. grandiceps*, accumulate, in different combinations, apigenin and luteolin, kaempferol and quercetin 3-*O*-glycosides, kaempferol and quercetin 3-methyl ethers, quercetin 7-methyl ether, and, in the case of *L. leontopodium*, 3,5,7-trihydroxyflavone (galangin). The chemistries of these two genera do not support a very close relationship between them.

Lipochaeta-Wedelia

One of the earliest studies of a Hawaiian endemic genus in which flavonoids played a part involved *Lipochaeta* (Gardner, 1976, 1979). *Lipochaeta* consists of 20 (Wagner et al., 1990; Bremer, 1994) or 23 (Rabakonandrianina and Carr, 1981) species divided into two sections. The sections differ in chromosome number, one is diploid, the other tetraploid, and in flavonoid chemistry. The diploids accumulate only flavonols whereas the tetraploids accumulate both flavones and flavonols (Gardner, 1976, 1977). Suggestions on how this system may have become established will be discussed in some detail in Chapter 22.

Here, the task is to introduce *Wedelia*, a genus that will continue to play a key role in understanding the evolution of this Hawaiian floristic element. *Wedelia biflora*, considered to be the species most closely related to *Lipochaeta*, was shown by Gardner (1976) to have an array of quercetin glycosides "virtually identical to that seen in the diploid species of *Lipochaeta*." A recent report by Miles (1993), however, suggests that the flavonoid chemistry of *W. biflora* is a good deal more complex than that reported by Gardner. This worker identified the 3,3'- and 7,3'-dimethyl ethers of quercetin, and the very unusual 2,7,4'-trihydroxy-5,3'-dimethoxyisoflavone from leaves of *W. biflora*. The observation of an isoflavone derivative, a flavonoid type otherwise quite rare in Asteraceae, in *W. biflora* is apparently not an isolated event. Govindachari et al. (1956a) showed that the structure of wedelolactone, obtained from *W. calendulacea*, is based on the isoflavone substitution pattern. A recent report of apigenin from *W. calendulacea* (Wagner et al., 1986) clearly indicates that this genus possesses the capacity to produce an array of flavonoid structural types. Detailed studies of additional species of *Wedelia* would be necessary to evaluate the significance of flavonoid data in the *Lipochaeta-Wedelia* question.

The status of these genera is unresolved. Wagner et al. (1990) consider *Lipochaeta* to be biphyletic and suggest that the problem will only be resolved when *Wedelia* and related genera are subjected to detailed scrutiny. They go on to suggest two possibilities: (1) treat the sections of *Lipochaeta* as sections of a larger *Wedelia*; or (2) group all of the Hawaiian species with *Wedelia biflora* and include them all within *Wollastonia*, which has been discussed and recommended by Fosberg and Sachet (1980). The part flavonoids play in this story, other than to help distinguish apparent sections within *Lipochaeta*, remains to be seen.

Matricaria-Tripleurospermum

Nomenclatural problems involving these genera were presented briefly in Chapter 7. We will follow Greger (1975) in the following. *Matricaria* and *Tripleurospermum*, components of Matricariinae (Bremer, 1994), were examined for flavonoids by Greger (1975) as part of his long-time interest in problems within Anthemideae. Twenty-five named and two other "species" of *Tripleurospermum* and three of *Matricaria* were shown to have profiles composed of various combinations of apigenin, luteolin, 6-hydroxyluteolin, chrysoeriol, quercetin, and isorhamnetin. A further distinguishing feature involves the position of glycosylation, always position-7 in the flavonols, but either position-7 or 4' in the flavones. Although the amount of information amassed by Greger is quite large and there is some variation within and among taxa, it is quite clear that the overall *Tripleurospermum* profile consists of quercetin, isorhamnetin, and luteolin 7-*O*-glucosides as major components along with lesser amounts of apigenin and chrysoeriol 7-*O*-glucosides. Greger's Group V, which consists of *T. microcephalum*, *T. disciforme*, and the two incompletely identified taxa, make the same compounds as seen in other members of the genus but add apigenin and luteolin 4'-*O*-glucosides as major components.

Three species of *Matricaria* were examined, *M. chamomilla*, *M. aurea*, and *M. discoidea*. All three had, to greater or lesser degrees, flavonol 7-*O*-glucosides along with luteolin 7-*O*-glucoside as a major component. Distinguishing *M. chamomilla* and *M. aurea* from all other species in the study, including *M. discoidea*, was the presence of major quantities of 6-hydroxyluteolin 7-*O*-glucoside and luteolin 7-*O*-rhamnoglucoside. *Matricaria discoidea* exhibited the simplest flavonoid profile seen in the study with only luteolin 7-*O*-glucoside as its main flavonoid along with a smaller amount of quercetin 7-*O*-glucoside. The simplification of flavonoid profiles seen in *Matricaria* was taken by Greger (1975) as an indication of advancement respective to *Tripleurospermum*. This agrees with the polyacetylenic chemistry of the two genera as well as their geography.

More recent studies by Exner et al. (1980, 1981) have shown the existence of several *O*-methylated derivatives of 6-hydroxykaempferol and quercetagetin in the non-polar extracts of the herbal preparation "Matricariae Flos." Other studies resulted in isolation of 5,4'-dihydroxy-3,6,7,3'-tetramethoxyflavone (chrysosplenetin) from *M. chamomilla* var. *recutita* (Mericli and Korkmaz, 1992). Other papers dealing with *M. chamomilla* reported simple glycosides or their acylated derivatives (see Chapter 7 for details). These data do not materially alter Greger's (1975) conclusions. 5,4'-Dihydroxy-6,7,3'-trimethoxyflavone, 5-hydroxy-6,7,3'4',-tetramethoxyflavone, and 5-hydroxy-3,6,7,3',4'-pentamethoxyflavone (artemetin) were obtained from *M. subulosa* by Zdero and Bohlmann (1990a).

Phoebanthus-Helianthus

Phoebanthus is a genus of two species distributed in the southeastern United States. Although recognized as a genus in its own right (Bremer, 1994), it has been suggested that it might be combined with *Helianthus*, with which it shares many features (Heiser et al., 1969). A study of the flavonoids of both species showed that the diploid *P. tenuifolius* and the tetraploid *P. grandiflorus* afforded identical profiles that consisted of 5,7,3',4'-tetrahydroxy-6-methoxyflavone (nepetin), 5,7,4'-trihydroxy-6,3'-dimethoxyflavone (jaceosidin), and 5,7-dihydroxy-6,8,3',4'-tetramethoxyflavone (hymenoxin). These data were taken as possible evidence for merging the two genera (under *Helianthus*). One further item from this study might be mentioned. The flavonoid data did not help to clarify the possible origin of the tetraploid species by autoploidy. A second possibility, that the tetraploid might have arisen through crossing of the diploid with *Helianthus angustifolius*, which bears similarity both in morphology and geography with the diploid, was not discounted by the flavonoid data insofar as the compounds identified are common components of *Helianthus* species.

Robinsonia

Robinsonia is the second largest endemic asteraceous genus on the Juan Fernandez Islands. A detailed study of all seven species, including *R. berteroi*, which was previously considered to constitute the monotypic genus *Rhetinodendron*, was described by Pacheco et al. (1985). The flavonoid profile of the genus consisted of glycosides of apigenin, luteolin, kaempferol, quercetin, isorhamnetin, naringenin, eriodictyol, and dihydroquercetin (taxifolin). Details of relationships among these species will be found in Chapter 22. For the purpose of the present chapter, however, we need only point out that ancestral relationships with a species of *Senecio*, from the mainland but otherwise unidentified, are supported by flavonoid as well as other sources of information.

Tracyina-Rigiopappus

Tracyina and *Rigiopappus* are both monotypic genera. *Tracyina rostrata* is a rare taxon from northern California, whereas *Rigiopappus leptocladus* occurs widely in the western United States. Morphological features of the two genera have been discussed by several authors (Blake, 1937; Robinson and Brettell, 1973b; Van Horn, 1973) and their identical chromosome number of $2n = 18$ has been recorded (Raven and Kyhos, 1961). Possible relationships and tribal disposition of the two genera were reviewed by Ornduff and Bohm (1975) who also described their flavonoid chemistry. Both taxa accumulate kaempferol and quercetin 3-*O*-glucosides and 3-*O*-rutinosides to which *Rigiopappus* adds the equivalent isorhamnetin glycosides. Successful artificial crossing of the two offer further indication of close relationship. The hybrids displayed intermediate morphological characters, the most striking of which was the number of pappus bristles. *Tracyina* has a pappus consisting of 24-40 capillary bristles whereas *Rigiopappus* has 3-5 awn-like scales. Hybrids exhibited 10-15 pappus bristles of intermediate width. The flavonoid profile of the hybrids consisted of the summed chemistry of the parents. The possibility of linking these taxa even closer as members of a single genus was mentioned, but it was concluded that the differences in pappus, achene, and capitulum structures were such that the genera should be maintained, and that they should be placed side by side in Astereae.

A summing up

The usefulness of flavonoids as indicators of intergeneric relationships (or lack thereof) as described in this chapter stands in contrast to their poor performance in dealing with subtribal problems seen in Chapter 20. Examples in this chapter included support for a close relationship between *Gymnosperma* and *Gutierrezia* based upon their mutual possession of rich and complex arrays of *O*-methylated flavonols, whereas the simple profile of *Flyriella*, in addition to chromosome data, was used as an argument for distinguishing it from the flavonoid-rich genus *Brickellia*. The similarity of the pigment profiles of *Phoebanthus* and *Helianthus* could be used to support submerging the former into the latter. Similarly, the profiles of *Cnicus* and *Grossheimia* do not help to distinguish them from *Centaurea*, whereas the profile of *Aegialophia* is such that the opposite argument can be made. The lumping together of *Clibadium*, *Desmanthodium*, and *Ichthyothere* did not receive as much support from their respective pigment profiles as did separating the first two genera from the third. It seems somewhat less likely that flavonoid data would support placement of each of these in different subtribes, however. Flavonoids have also been a useful source of data in efforts to establish relationships of insular genera. For example, flavonoids have been of value in helping to identify the ancestor of the Juan Fernandez Islands endemic genus *Robinsonia* among mainland South American *Senecio*. It is clear from these several examples, as well as the many others discussed in the chapter, that flavonoids can be expected to continue to provide useful information at this level of the taxonomic hierarchy.

CHAPTER 22

Flavonoids at the Specific Level

In this chapter we present summaries of the application of flavonoid chemistry to relationships among species within various genera. Some of the situations to be discussed involve extensively studied and well known genera, e.g., *Achillea, Artemisia, Bidens, Coreopsis, Dahlia,* and *Parthenium.* Other examples involve much smaller groups of species as well as genera that have not attracted the same level of attention. Although there may be taxonomic difficulties among some of the genera dealt with (see Chapter 21), our intent in this chapter is to look at the way in which flavonoid data have been utilized in sorting out relationships among species. In this regard, we will be dealing with sectional or sometimes less formal groupings of species. Insofar as the amount of information available for some of these genera is very large, we have selected examples that serve to exemplify the usefulness of flavonoids. Detailed comments on all situations where flavonoids have been employed would be unwieldy and highly repetitive.

Achillea

Taken as a whole, the accumulated flavonoid database for *Achillea* is dauntingly large and complex. Flavonoids have been, and will continue to be, a useful source of information for studying problems within this large genus. Despite the size and richness of the database, however, the usefulness of these chemical characters cannot be guaranteed. In the case of sections *Millefolium*, *Ptarmica*, and *Santolinoides*, Valant-Vetschera and Wollenweber (1986) observed differences in exudate composition among several species, whereas within the large section *Achillea* significant qualitative differences were not seen (Valant-Vetschera and Wollenweber, 1994) despite clear morphological differences. In contrast, species within section *Babounya* exhibited specific flavonoid profiles, although morphological differences are not so clear-cut within this group.

An interesting situation was described in a recent note concerning exudate flavonoids of *A. clusiana, A. multifida*, and *A. nana* (section *Ptarmica*)

(Valant-Vetschera and Wollenweber, 1996b) and discussed further in a paper describing exudate flavonoids of *Achillea* subsection *Filipendulinae* (Valant-Vetschera and Wollenweber, 1996a). *Achillea multifida* exudate contained two flavonoids, 6-hydroxykaempferol-3,6,4'-trimethyl ether and quercetagetin-3,6,4'-trimethyl ether (centaureidin). Relevant to the issue was discovery that *A. clusiana* and *A. nana* accumulated 3'-*O*-methyl derivatives, quercetagetin-3,6,3'-trimethyl ether (jaceidin) in the case of the former, quercetagetin-3,6,7,3'-tetramethyl ether (chrysosplenetin) in both, and 3,6,7,3',4'-pentamethyl ether (artemetin) in the latter. The implications of this apparently simple difference in B-ring *O*-methylation are not trivial; they are representative of the flavonoid structural changes that distinguish species (or other taxa). Since 4'-*O*-methylation is apparently the norm in this genus, what has happened to the *O*-methyltransferase system that resulted in the change from the 4'- to 3'-position? Sequence studies of sulphation enzymes associated with a change from 3-*O*-sulphation to 3'-*O*-sulphation in *Flaveria* have been documented (Varin et al., 1992). It would be of interest to see if a similar situation exists in the present system. The change of position of sulphation, from 3 to 3', involves a fairly large change in flavonoid structural environment whereas the site change in *Achillea* is presumably more subtle. An understanding of changes of this sort is crucial to addressing the broader issue of flavonoid evolution. Only when a sequence database for enzymes of flavonoid biosynthesis has been established will we be able to apply the tools of phylogenetic analysis that have been so successfully applied to other systems. A second question, and one even more difficult to answer, involves attempting to find specific requirements in the biology of these species that are better met by a change in flavonoid structure.

Anacyclus

The 13 taxa that comprise *Anacyclus* were shown by Greger (1978) to accumulate a variety of flavone and flavonol glycosides based on quercetin, isorhamnetin, luteolin and luteolin 4'-methyl ether (diosmetin). Luteolin glycosides were essentially ubiquitous, while trace quantities of apigenin and kaempferol glycosides were of scattered occurrence. Five groups could be distinguished within the genus on the basis of the flavonol glycosylation and the presence or absence of diosmetin. Group 1, consisting of *A. depressus, A. pyrethrum*, and *A. officinarum*, had profiles dominated by quercetin and isorhamnetin 5-*O*- and 7-*O*-glucosides and diosmetin 7-*O*-glycosides. Within the group, a small amount of quercetin 3-*O*-diglycoside was detected in *A. pyrethrum*, and *A. officinarum* lacked flavonol 7-*O*-glycosides. It should be emphasized that this set of species was unique in its production of diosmetin.

Group 2 consisted of *A. maroccanus, A. radiatus*, and a cultivar identified as "purpurescens." There was a trace of the quercetin 3-*O*-glycosides in *A. radiatus* while all three taxa accumulated major amounts of the quercetin and isorhamnetin 5-*O*-glycosides. Group 3 consists of *A. monanthos, A. linearilobus*, and *A. nigellifolius* and is characterized by accumulating the quercetin 3-*O*-glycosides, and in the case of *A. monanthos*, quercetin and isorhamnetin 7-*O*-glycosides as well. *Anacyclus linearilobus* also showed a small amount of quercetin 5-*O*-glucoside and *A. nigellifolius* was the only taxon in the genus that lacked luteolin glycosides.

Group 4 consists solely of *A. atlanticus* which is characterized by a flavonoid profile consisting only of quercetin 3-*O*-glycosides and luteolin glycosides. Except for a small amount of quercetin 5-*O*-glucoside in a cultivar called "coronatus," small amounts of the quercetin 3-*O*-glycosides and 7-*O*-glucoside in *A. velutinus*, and uncertain quercetin 7-*O*- and 5-*O*-glycoside sightings in *A. clavatus*, the set of three species that comprise Group 5 is consistent only with regard to the uniform occurrence of the luteolin glycosides.

The overall evolutionary significance of these flavonoid arrays is obscure, but Greger (1978) does point out the possibility that accumulation of quercetin 3-*O*-glycosides and luteolin 7-*O*-glycosides may represent the ancestral profile. Divergence in glycosylation has occurred, then, as a more recent development. A detailed study of glycosylation enzymes in this system would be of interest since the three most often glycosylated positions in flavonols are involved. The limited occurrence of isorhamnetin and diosmetin is also of interest from a comparative enzymological perspective. It would seem that this system might be ideal for comparative macromolecular studies, especially in that cultivation of these species is straightforward.

Arnica

One of the most thoroughly studied genera in northern North America has been, and is likely to continue to be, *Arnica*. The problems encountered in this mainly boreal and montane genus range through the themes of this and the next few chapters. We start with a study of *A. fulgens* and *A. sororia*, two species native to a wide area in western and northwestern North America. *Arnica fulgens* is the more widespread with its range extending from British Columbia to Manitoba, south to Colorado, and westward to northern California. *Arnica sororia* occupies a similar range but does not extend as far east nor as far south. As Downie and Denford (1987) pointed out in their detailed study of this species pair, the occurrence of these two species as components of prairie and grassland floras is unique in the genus.

The difficulties surrounding these species arise from their similar morphologies, extensive range overlap, and confusion in the taxonomic literature regarding correct names. The potential for hybridization was also noted (Downie and Denford, 1987). The biosystematic study described in that paper included morphometric analysis involving both floral and vegetative features, reproductive biology, ecological considerations, and foliar flavonoid chemistry. The descriptive data along with failure in attempts to effect hybridization led those authors to conclude that each was a species in its own right. The flavonoid profiles, although comparatively simple, aided in that decision. All twelve specimens of *A. fulgens* gave the same profile which consisted of apigenin, luteolin 7-*O*-glucoside, 6-methoxyluteolin 7-*O*-glucoside, kaempferol 3-*O*-glucoside, and quercetin 3-*O*-galactoside and 3-*O*-diglucoside. Eleven of the 13 specimens of *A. sororia* examined gave a profile that consisted of the listed compounds less apigenin and the methoxyluteolin derivative. Two specimens of *A. sororia*, however, exhibited a profile identical to that of *A. fulgens*. These two populations occur among other four-flavonoid race populations in southeastern Alberta (near Medicine Hat). Although Downie and Denford (1987) considered the existence of these unusual forms as significant, they did not speculate but suggested only that further chemical work would be necessary before phylogenies can be resolved.

One of the most complicated and evolutionarily interesting groups of species in the genus constitutes the subgenus *Austromontana*. According to a reassessment of the group, Wolf (1981) sees the subgenus as consisting of nine species with no infraspecific taxa. Morphological, cytological, ecological, and phytochemical information suggest that the widespread *A. cordifolia* and the closely related *A. latifolia* are ancestral to the group. Radiation from these has produced several highly specialized, rare, endemic species mainly centered in southern Oregon and northern California in the old and diverse Klamath Mountains. Wolf and Denford (1984b) studied the flavonoid profiles of all of these taxa, eight of which are discussed here. Since *A. gracilis* is considered to be a hybrid of *A. cordifolia* and *A. latifolia*, its story will be presented in Chapter 25.

The total flavonoid profile for the subgenus consists of 22 compounds, equally distributed between glycosides and aglycones. Quercetin 3-*O*-gentiobioside and a 3-*O*-diglucoside were nearly ubiquitous and, significantly, have not been found in members of other subgenera of *Arnica*. Kaempferol and quercetin 3-*O*-glucosides were also widely distributed in the species as was 6-methoxyapigenin (hispidulin). Several other *O*-methylated flavones and flavonols were identified and will be introduced in the following discussion. The two main members of the subgenus, *A. cordifolia* and *A. latifolia*, enjoy the widest ranges, the former ranging from the Yukon south to New Mexico and central California with disjunct populations in Ontario and Michigan, the latter from Alaska through Colorado and northern California. *Arnica cordifolia* exhibited a profile consisting of 10 compounds but there was a good deal of variation among the 33 populations visited. The presence of the kaempferol and quercetin glycosides, scutellarein, 6-methoxykaempferol 7-*O*-glucoside, and luteolin 7-*O*-glucoside were assumed to be the ancestral profile. Two components, apigenin 7-methyl ether (genkwanin) and quercetin 3-methyl ether, were seen only in populations of this species in the Klamath region. It was suggested that these two compounds were of recent origin, but it is not clear from their discussion how this conclusion was reached.

The flavonoid profile of *A. latifolia* consists of six glycosides and two aglycones. Notable differences from *A. cordifolia* are the presence of kaempferol 3-*O*-galactoside and the replacement of 6-methoxykaempferol by 6-methoxyquercetin (patuletin). Again, considerble variation was seen in the number of compounds per population. Patuletin appears to be restricted to tetraploid, therefore recently derived, populations in the Klamath area.

Arnica discoidea, which occurs in the California Coast Ranges, in the foothills of the Sierra Nevada, and northward in the foothills of the Cascade Mountains, has a moderately complex flavonoid profile. In terms of chromosome number, this species is also diverse consisting of diploid, triploid, and tetraploid races. Diploids are largely restricted to the Klamath area whereas the polyploids occur at the limits of the range of the species. The flavonoid profile of this species is much like that of *A. cordifolia* from which it is thought to have arisen. The differences between the two lie in the presence of 5,7-dihydroxy-6,4'-dimethoxyflavone (pectolinarigenin), restricted to the Klamath region, and 5,4'-dihydroxy-6,7-dimethoxyflavone (cirsimaritin). Considerable variation is again evident with the diploid Klamath populations exhibiting the greatest diversity of profiles and being most similar to *A. cordifolia*. Polyploid populations, by contrast, have reduced profiles, especially with regard to glycoside composition. These observations led to the conclusion that *A. discoidea* arose from a diploid population of *A. cordifolia* in the Klamath region and that the other populations of *A. discoidea* arose through migration away from the region, experiencing ploidy change and flavonoid profile reduction in the process.

The next species to be considered, *A. spathulata*, is a serpentine endemic of the Klamath region. The flavonoid profile of this species is a subset of the *A. discoidea* profile, which suggests that it arose from that taxon.

In particular, the presence of pectolinarigenin, which occurs in Klamath populations of *A. discoidea*, and genkwanin argue for a close relationship between these two taxa. Tetraploid populations of *A. spathulata* have reduced flavonoid profiles.

Arnica venosa, known only from dry hills in western Shasta County and adjacent eastern Trinity County in California, was described by Wolf and Denford (1984b) as the most geographically restricted species in the genus. Its flavonoid profile, which includes genkwanin, is a subset of the profile of *A. discoidea* and serves as additional evidence for a close relationship between the two. Occasional natural hybrids between *A. venosa* and *A. discoidea* are known.

Another comparatively rare species from the Klamath area is the diploid, serpentine endemic *A. cernua*. Only three compounds comprise the flavonoid profile of this species: kaempferol 3-*O*-glucoside and the two quercetin 3-*O*-diglycosides. This is the sort of extreme reduction in flavonoids that has been observed in other narrow edaphic endemics. An origin from *A. cordifolia* was suggested but whether that population was already depauperate in flavonoids or whether the reduction occurred subsequent to morphological and ecological differentiation can not be determined. However, the authors (Wolf and Denford, 1984b) report having found a population of diploid *A. cordifolia* that had only two flavonoids.

Arnica nevadensis occurs at higher elevations in the Sierra Nevada and extends northward to the Cascade Range and Olympic Mountains of Washington. The species is an apomictic tetraploid with morphological features that suggest a relationship with *A. cordifolia*. The presence of the two quercetin 3-*O*-diglycosides, characteristic compounds in the *A. cordifolia* profile, provides additional evidence for such a relationship. Although the presence of these compounds clearly places *A. nevadensis* in the subgenus, certain features of its pappus suggest that another species besides *A. cordifolia* has been involved. Luteolin 4'-methyl ether (diosmetin) and 6-methoxyluteolin (nepetin) are unique in *A. nevadensis*, but it was not established whether these compounds are diagnostic of the species. The need for further sampling was stressed.

The final species of the subgenus is the rare, diploid *A. viscosa*. This taxon, known from only seven localities, is the rarest species in the genus. It is restricted to young (<14,000 yr) volcanic substrates at high elevations, largely in the Klamath Mountains. The species is, thus, considered to be of comparatively recent origin which, Wolf and Denford (1984b) believe, is reflected by its having the most complex flavonoid profile in the subgenus (seven glycosides and seven aglycones). The presence of the diagnostic patuletin 3-*O*-glucoside in common with *A. latifolia* suggests a relationship with that widespread species. Successful artificial hybridization between these two species (Straley, 1980) provides additional support for a close relationship. This is all the more meaningful in view of the very limited success in achieving successful crosses between other combinations in the subtribe.

Much of the uncertainty associated with applying flavonoid data to complex problems of this sort arises from variation in pigment profiles, both within and among populations. In order to determine whether the observed variation is real, and not an artifact of sampling, it is necessary to conduct extensive sampling. This often leads to an array of data that is beyond the human capacity to analyze by simple visual examination. Multivariate analysis using high speed computational technology eliminates this problem. This approach was used by Wolf and Whitkus (1987) in a statistical analysis of flavonoid profiles of 113 populations representing all nine species of *Arnica* subgenus *Austromontana.* To a very substantial degree the numerical analyses confirmed the evolutionary hypotheses erected by Wolf and Denford (1984b). Thus, the likely ancestral species is *A. cordifolia* whose range of variation encompasses most of the species in the subgenus. Extensive differentiation in the Klamath region has occurred to produce a complex of ecologically specialized species some of which are of comparatively recent origin. Speculation in the earlier paper on the origin of *A. nevadensis* was based on limited sampling. Further collection of this species confirmed the original idea that *A. nevadensis* and *A. cordifolia* are closely related.

The foregoing detailed presentation of the work of Wolf, Denford, and Whitkus serves to point out the potential importance of flavonoid profile analysis in broadly based biosystematic research. The combination of chemistry with studies of morphology, ecological (especially edaphic) features, and chromosome numbers of the group has led to several eminently testable hypotheses concerning the evolutionary relationships among members of the subgenus. This group of taxa would seem an ideal subject for application of allozyme analyses and possibly other macromolecular techniques.

Artemisia

Although not the largest genus in the family, *Artemisia* is probably among the most widespread and is one of the most conspicuous elements in Northern Hemisphere asteraceous floras. A few species can be found in South America, in Africa south of the Sahara desert, and in the Hawaiian Islands (Bremer, 1994). We have managed to accumulate flavonoid information for more than 70 species of *Artemisia* taken in the broader context. This sample contains representatives of most of the infrageneric groups that have been postulated to account for the range of variation seen in the genus. A discussion of flavonoids as they apply to the segregate genus *Seriphidium* appeared in Chapter 21.

The species that comprise *Artemisia* have, until fairly recently, been sorted into various combinations, compressions, or rearrangements of five subdivisions: sect. *Abrotanum* Bess. (= *Artemisia* L.), sect. *Absinthium* DC., sect. *Dracunculus* (Bess.) Rydb., sect. *Seriphidium* (Bess.) Rouy, and sect. *Tridentatae* Rydb. More recent treatments have shown the original groupings to be unsatisfactory [see Bremer and Humphries (1993) for the taxonomic history of the groups], the result of which is recognition of two genera, *Artemisia* and *Seriphidium* (but see Kornkven et al., 1998). We have already seen that flavonoids do not help to distinguish these two genera. In fact, it appears that flavonoids and flavonoid classes are more or less randomly scattered throughout both groups of species. The only flavonoid feature that appears to be limited in its occurrence is the 2',4',5'-oxygenation pattern, which occurs in five species of *Artemisia* s.s. but which has apparently not been observed in any member of *Seriphidium*.

An observation on flavonoids of sect. *Tridentatae* would be in order at this point. The North American members of *Seriphidium* have been considered by some workers as sect./subgen. *Tridentatae* Rydb. This situation was reviewed by McArthur and Plummer (1978) who were concerned with the biogeography and management of *Artemisia tridentata* and related species in western North America. A more recent study discussed cytogeography and chromosome evolution of subgenus *Tridentatae* (McArthur and Sanderson, 1999). These workers felt that *Tridentatae* and *Seriphidium* should be considered as separate groups within the genus. Although we are again forced to work with frequencies of occurrence rather than absolute presence/absence data, it is interesting to note that the frequency values for two flavonoid derivatives change with the removal of the *Tridentatae* group from the *Seriphidium* group. Flavonoids with the 3',4',5'-oxygenation pattern and *C*-glycosylflavones do not appear to occur in the *Tridentatae* group. Thus the frequency of the former goes from 7/24 (ca. 29%) to 7/17 (ca. 41%), while the latter goes from 3/24 (12.5%) to 3/17 (ca. 18%). These results suggest that a closer look at these groups of species might yield more definitive results.

A further indication of the apparent lack of defining flavonoid features in *Artemisia* s.l. is provided by using Stangl and Greger's (1980) study of monoterpenes in *Artemisia* as a model. They discussed their data in terms of three subgenera, subgen. *Artemisia* with sections *Artemisia*, *Abrotanum*, and *Absinthium*, subgen. *Seriphidium*, including sect. *Tridentatae*, and subgen. *Dracunculus*. Simply put, flavonoid data showed no pattern of occurrence. This is in contrast to the monoterpene data which did show patterns within the genus.

The English abstract of the Stangl and Greger paper (1980) states that cineol and thujane and camphane derivatives together with small amounts of pinanes, monocyclic and acyclic monoterpenes apparently characterize sect. *Artemisia* and sect. *Abrotanum* (in part), and probably also the North American sect. *Tridentatae*. Subgenus *Dracunculus* and the series *Laciniatae* and *Latifoliae* (no flavonoid data for their species) of sect. *Abrotanum* are clearly distinguished by only small amounts of pinanes and monocyclic and acyclic monoterpenes together with a nearly complete reduction of all major components, e.g., cineol, thujane, and camphane derivatives. Other classes of secondary metabolites also appear to occur in *Artemisia* along certain taxonomic lines (Greger, 1982). In that paper Greger reminds us that most of the chemical studies of *Artemisia* have involved analyses of aerial parts for sesquiterpene derivatives and flavonoids. However, root components including polyacetylenes, coumarin-sesquiterpene ethers, and sesamin-type lignans, appear to be potentially useful as taxonomic indicators as well. Although flavonoid occurrence appears not to follow taxonomic lines, compared to these other compound types, it would be of considerable interest to select an appropriate group of species from *Artemisia* and from *Seriphidium* and perform exhaustive (preferably quantitative) studies of all of these secondary compounds. Although such an approach would be labor-intensive, it would be a significant step in establishing whether any or all of these compounds are truly useful in systematic work.

Coreopsis nuecensis and *C. nuecensoides*

Coreopsis nuecensis, as formerly constituted, comprised two races, one coastal with $n = 9$ and 10, and one that grew in more inland sites with $n = 6$, 7, and 8. These chromosomal differences, differences with regard to pubescence of inner involucral bracts and stems, and the existence of a breeding barrier between the two types, both in the glasshouse and in nature, led Smith (1974) to recognize the coastal form as *C. nuecensis* and consider the inland form to be the new species *C. nuecensoides*.

An examination of the flavonoids of the two species disclosed much greater differences between them than might have been anticipated from their morphological similarity (Smith and Crawford, 1981). The flavonoid profile of *C. nuecensoides* was shown to consist of apigenin and luteolin 7-*O*-glucosides, quercetin 3-*O*-glucoside and 3-*O*-rutinoside, quercetagetin 7-*O*-glucoside, 6-hydroxyluteolin 7-*O*-glucoside (variable among populations), and the marein-maritimein and lanceolin-leptosin anthochlor pairs.

Coreopsis nuecensis lacked apigenin 7-*O*-glucoside and quercetin 3-*O*-rutinoside but had two "new" compounds, 6-methoxyluteolin 7-*O*-glucoside and 6-methoxyquercetin (patuletin) 7-*O*-glucoside (variable). The authors suggested two explanations for these results. The chemical differences may have occurred as a result of divergence of structural genes involved in flavonoid biosynthesis, or differences might have arisen through chromosomal re-patterning involving structural or regulatory genes associated with aneuploidy.

It is possible to be a bit more specific about changes that may have occurred. It will be noticed that the capacity for extra oxygenation of the A-ring at C-6 already exists in *C. nuecensoides*, as indicated by the presence of quercetagetin in that species. Also present in the parent species is the capacity to make *O*-methylated flavonoids, evidenced by the formation of *O*-methylated anthochlors (lanceolin-leptosin). The new species shows two "new" compounds, 6-methoxyluteolin, which requires oxygenation at C-6, and the *O*-methyl ethers of the newly formed 6-hydroxyluteolin and quercetagetin. These flavonoids are new only in the sense that they have arisen by novel combinations of substrates and enzyme systems already present in the progenitor species.

A recent study of the flavonoids of *Stephanomeria* (Lactuceae) also involved a progenitor-derivative species pair (Bohm and Gottlieb, 1989). The flavonoid chemistry of several annual members of this western North American genus was established in an effort to see if any pigment differences could be detected in the species pair *S. exigua* subsp. *coronaria* and its derivative *S. malheurensis*. Morphological and other differences between these species include root/shoot ratios, duration of rosette growth, branch structure, number of flower heads, and seed weight (Gottlieb and Bennett, 1983). In contrast to the example from *Coreopsis*, the flavonoid profiles of the two *Stephanomeria* species were identical. Since the two species, which grow closely sympatric in eastern Oregon, are identical in their pigment profiles, it might be suggested that there has been no change in the selective value of the flavonoid profile. Insofar as the "new" *Stephanomeria* species is thought to be of comparatively recent origin, there simply may not have been time for changes in flavonoid genes to manifest themselves. In the case of the *Coreopsis* pair, where major chromosomal restructuring events occurred, time was not discussed as a contributing factor.

We turn our attention now to the broader issue of how flavonoids sort themselves out along sectional lines in *Coreopsis*. Detailed surveys of flavonoid profiles in both foliar and floral tissue of the entire genus in North America have been determined (Crawford and Smith, 1983a,b, and citations therein). They commented in that paper that the distribution of flavones and flavonols, although useful at lower taxonomic levels, was too

Table 22–1. Flavonoid profiles, areas of origin, and chromosome numbers (n) for sections of North American *Coreopsis*

			Flavonoids[1,2]										
			Foliar						Floral				
Section	Number	Area	1	2	3	4	5	6	7	8	9	10	11
Pseudo–Agarista	13	Mexico	+	+	+	–	–	–	+	+	–	–	–
Electra	14	Mexico	+	+	+	–	–	+	+	–	–	–	–
Anathysana	12	Mexico	+	+	+	–	–	+	+	–	–	–	–
Tuckermannia	12	California	+	+	+	–	–	–	+	+	–	–	–
Pugiopappus	12	California	+	+	+	–	–	+	+	+	v	–	–
Euleptosyne	12	California	+	–	+	–	–	+	+	+	v	–	–
Silphidium	13	SE U.S.A.	+	–	–	–	–	–	+	+	–	+	v
Palmatae	13	SE U.S.A.	+	+	–	+	+	–	v	+	–	+	v
Coreopsis	Var[3]	SE U.S.A.	+	+	+	+	+	–	v	+	–	+	–
Calliopsis	12	SE U.S.A.	+	–	–	+	+	–	+	+	–	v	–
Eublepharis	12	SE U.S.A.	+	+	+	+	+	–	+	+	-	v	v

[1] 1 = flavones; 2 = flavonols; 3 = 6-hydroxyflavones; 4 = 6-hydroxyflavonols; 5 = 6-methoxyflavones; 6 = *C*-glycosylflavones; 7 = butein derivatives; 8 = okanin derivatives; 9 = stillopsidin derivatives; 10 = lanceolatin derivatives; 11 = isoliquiritigenin derivatives.

[2] v = compound type variable within this section.

[3] Variable with n = 6, 7, 9, 10, and 13.

variable to be of use at the sectional level. However, the occurrence patterns of the floral chalcones correspond well with evolutionary relationships arrived at from other data sets. Chalcones with the following substitution patterns were involved: 2',4',4-trihydroxy- (isoliquiritigenin), 2',4',3,4-tetrahydroxy- (butein), 2',3',4',3,4-pentahydroxy- (okanin), 2',4',5',3,4-pentahydroxy- (stillopsidin), and 2',4',3,4,-tetrahydroxy-3'-methoxy- (lanceolatin).

In Table 22–1 we have summarized the occurrence data for floral anthochlors, foliar flavonoids, area of occurrence, and chromosome number for the sections of North American *Coreopsis*. (Foliar flavonoids are included despite those authors' concerns about excessive variation.) The simple pigment profile of section *Electra*, whose floral anthochlor profile is based solely on butein, accords well with the primitive position of the section. The presence of butein derivatives as the only chalcone type in flowers of section *Anathysana* supports its close affinity with section *Electra*. Sections *Eublepharis* and *Calliopsis* were considered by Smith (1975) to be closely related and highly advanced with both sections *Coreopsis* and *Palmatae* occupying branches of their own. The pigment chemistry of section *Eublepharis*, with derivatives of all four chalcones, is the most complex seen in the genus. The pigments of section *Calliopsis*, based on butein and okanin as major components, and lanceolatin as a minor contributor, are the same as those of section *Coreopsis* except that in the latter butein derivatives are only minor components while lanceolatin derivatives make a larger contribution. Section *Palmatae* has all four floral chalcones, as in the case of section *Eublepharis*, except that relative contributions of the aglycones is different. It is also interesting that the two closely related mainly Californian sections, *Leptosyne* (*Euleptosyne* in earlier papers) and *Pugiopappus*, have similar arrays of chalcones based on butein and okanin with lesser amounts of the otherwise rare stillopsidin.

A more recent study of 44 North American species of *Coreopsis* involved a cladistic analysis that utilized 35 characters (Jansen et al., 1987c). Two lineages emerged, one that comprised the six essentially Californian and Mexican sections, and one that consisted of the sections limited primarily to southern and southeastern United States. The authors presented cladograms on which the occurrence of flavonoid class data had been superimposed in one case, and floral chalcones in another. Flavonoid data agree well with the phylogeny obtained from morphological data alone. The sections judged to be advanced on the basis of earlier studies emerge as a major clade, along with section *Silphidium*, which in the earlier analyses had been allied with more primitive sections and not with the southern and southeastern sections. Flavonoid chemistry of section *Silphidium* does not support its association with this group, considering that it accumulates only flavone derivatives. Furthermore, its chalcone profile, based on butein and okanin, is among the simpler ones seen. Another difference involves the emergence of section *Anathysana* as most closely connected to the *Leptosyne-Pugiopappus-Tuckermannia* clade despite its simpler chalcone profile. However, members of *Anathysana* possesses *C*-glycosylflavones in common with sections *Leptosyne* and *Pugiopappus*. *C*-Glycosylflavones are also known from section *Electra*, however, with which *Anathysana* was more closely related in the earlier studies. Despite these apparent anomalies, the flavonoid data accord very well with the relationships indicated by the other data sets.

Further support for several aspects of the sectional phylogeny came from a study of plastid phosphoglucoisomerase isozymes (Crawford et al., 1990). Species from nine of the 11 sections of *Coreopsis* had two PGI isozymes, as did members of four other genera, *Bidens*, *Coreocarpus*, *Cosmos*, and *Thelesperma*. Since the normal number of PGI isozymes in diploid plants is one, there must have been a gene duplication in some ancestor common to these five genera. Members of the closely related sections *Electra* and *Anathysana* had only a single PSI isozyme which suggested to those workers that a common ancestor had likely undergone silencing of one of the two copies of PGI. The PGI data are in agreement with morphological, chromosomal, and flavonoid information and support those workers' view that the two sections comprise a monophyletic unit. DNA information on these taxa is also available (Kim et al., 1999b).

Dahlia

An exhaustive study of flavonoids in *Dahlia* (Giannasi, 1975a,b) disclosed significant differences in flavonoid profiles among the three sections of the genus. Species comprising sections *Dahlia* and *Pseudodendron* had flavonoid profiles based on flavone and flavonol glycosides, compound types totally lacking in members of section *Entemophyllon*. Species belonging to section *Entemophyllon*, in contrast, are characterized by the consistent presence of flavones bearing a 6-methoxy group.

This flavonoid distinction was shown to hold true in the case of *D. merckii* which some botanists had placed in section *Entomophyllon* on the basis of its possessing a few of the morphological features of that section. Most others, however, considered the species best allied with section *Dahlia*, a position supported by its flavonoid chemistry.

Ericameria

An examination of leaf and stem exudate from *Ericameria laricifolia* revealed a rich array of flavonoids including luteolin, 6-methoxyluteolin, and *O*-methylated derivatives of apigenin, kaempferol, and quercetin (Clark and Wollenweber, 1984). Eight of the 19 compounds reported in that study had been reported from *E. diffusa* earlier by Urbatsch et al. (1976). The possibility that such an overlap might suggest a close relationship between the two species was tempered by the additional observation that two populations of *E. diffusa* shared only three of the compounds found in *E. laricifolia*. This situation was used as an example of a system in need of further study with special emphasis to be given to the extent of variation among populations. Clark and Wollenweber (1984) noted one other feature about the flavonoids of this system that has wider applicability. They found a greater diversity of structure in the exudate flavonoid profile than in the internal or vacuolar flavonoid glycosides profile. They recommended that future work on these genera should focus on the exudate chemistry rather than on the more polar components. This may well be the case in the present situation, but it is always wise to determine whether this is true in all situations.

Erigeron (Juan Fernandez Islands)

The Juan Fernandez Islands, which lie about 600 km off the coast of Chile, offer an ideal situation in which to study evolutionary relationships of an island flora. The archipelago consists of only two islands of significant size, they are volcanic in origin and have been reliably dated (Stuessy et al., 1984a), they feature a flora that has interesting levels of endemism at both the generic and specific levels, the proximity of the islands to the South American continent clearly establishes it as the most likely source for propagules, and taxa related to the island endemics are known from the mainland. What follows immediately is a summary of studies of *Erigeron*, with six species, all island endemics. Two other genera (*Dendroseris* and *Robinsonia*), both endemic, will be discussed later in this chapter.

By way of introduction, the archipelago, which lies at about 80° W and 33° S, consists of Masatierra, the older island (3.7-4.4 M.Y.), and nearer of the two islands to the mainland, and the further away Masafuera (dated at 1-2.4 M.Y.). A small island, Santa Clara, which was at one time probably a part of its larger neighbor (Stuessy et al., 1984b), lies off the southwestern corner of Masatierra. The larger islands are separated by a distance of 181 km.

Erigeron contributes six species to the endemic flora of the islands: *E. fernandezianus, E. ingae, E. luteoviridis, E. rupicola, E. stuessyi*, and *E. turricola*. *Erigeron fernandezianus* can be found on both islands while the other five species occur only on the younger Masafuera. Valdebenito et al. (1992b) described the flavonoid profiles of all six species plus five more from the mainland. The overall profiles were quite similar with glycosides of apigenin, luteolin, 6-*C*-glucosylacacetin, quercetin, and quercetagetin present in various combinations although not all taxa, nor all populations, exhibited all compounds. Owing to the level of variation in the pigment profiles the authors placed little emphasis on this set of traits, except to state that the data agreed in general with relationships based upon morphological data.

A combination of morphological data and the unique occurrence of five of the species on Masafuera led the authors to conclude that colonization of the younger island had occurred first, followed by adaptive radiation and subsequent migration of *E. fernandezianus* to Masatierra. Flavonoid data point to an interesting problem, however. Luteolin 7-*O*-diglucoside and 6-*C*-glucosylacacetin 7-*O*-diglucoside were reported from collections of the mainland species examined, except for *E. campanensis* which exhibited the most depauperate flavonoid pattern in the group, and in the island endemics *E. ingae*, *E. luteoviridis*, and *E. turricola*. The compounds were found neither in *E. rupicola* nor, interestingly, in populations of *E. fernandezianus* from Masafuera. It was suggested that the most parsimonious explanation for these flavonoid (not morphological) results would involve colonization of Masatierra (the older, closer island) first, followed by dispersal to Masafuera and subsequent speciation. A few populations of *E. fernandezianus* lack one or both of these compounds and could have provided the propagules from which the Masatierra populations arose. This would, however, require that *E. fernandezianus* be the most primitive species of the endemics, which is not supported by morphological data.

A closer look at the two flavonoids themselves can provide an idea of the genetic divergence between the two chemotypes. Three biosynthetic steps distinguish the chemotypes: (1) *C*-glucosylation; (2) placement of the outer glucose to form the diglucosides; and (3) transfer of a methyl group to the 4'-position of an appropriate precursor to yield the acacetin (apigenin 4'-methyl ether) derivative. Establishment of the 7-*O*-diglucosides from the respective monoglucosides likely involves the same glucosyltransferase. 4'-*O*-Methylation occurs nowhere else in the species under study, although 6-*O*-methylation does. *O*-Methyltransferases are generally thought to be fairly selective in their position of action, so it seems reasonable to assume that the 4'-*O*-methyltransferase necessary for the formation of the acacetin derivative is unique to this group of species. Thus, three enzymes are required. If the direction of movement of the colonizer was from mainland to the younger island (Masafuera) and then to the older one, one must give some consideration to the "disappearance" of these enzymes and their reappearance at a later time. This seems like a drastic mechanism. Could it be that the genes responsible for these enzyme systems were not lost but were merely silenced during residence on the younger island? Perhaps in moving to the older island some ecological factor enabled the genes to regain function. This is the sort of situation where a detailed molecular biological study of the enzymes and their control might be helpful. It would also be important to learn if, or what, environmental factors might be involved.

Dendroseris

The Juan Fernandez Islands endemic genus *Dendroseris* consists of 11 species, which have been placed in three subgenera (Pacheco et al., 1991; Crawford et al., 1998). Eight species are restricted to Masatierra; the remaining three, one from each subgenus, occur only on Masafuera. The flavonoid profiles of these taxa comprise a series of 7-*O*-glycosides of apigenin and luteolin and two quercetin glycosides. Subgenera *Dendroseris* and *Rea* (each with four species) exhibited similar profiles having some representatives from each of the classes but could still be distinguished on the basis of individual compounds. Subgenus *Phoenicoseris*, with the remaining three species, lacked quercetin derivatives and exhibited an apigenin glycoside array easily distinguishable from the species in the other subgenera. The flavonoid profiles of species within subgenera *Dendroseris* and *Phoenicoseris* are reasonably uniform whereas there are larger differences among the profiles of the four species in subgenus *Rea*. The among-species variation in subgenus *Rea* is in accord with its paraphyletic status as judged from morphological data (Pacheco et al., 1991). Recent studies of restriction site variation in cpDNA (Crawford et al., 1992b) and sequences of the ITS regions of nuclear ribosomal DNA (Sang et al., 1994) did not resolve subgeneric inter-relationships, which is in line with the idea that rapid speciation within this genus has occurred on the islands.

With regard to the usefulness in interpreting species relationships, we once again find chemical and morphological data sets somewhat at odds. For example, *D. gigantea* from Masafuera has the most reduced flavonoid profile, not only in the genus, but in subgenus *Rea* as well. At face value one might expect that this reduction in flavonoid biosynthetic capacity might reflect similar significant changes in morphology. Such does not appear to be the case since morphologically, *D. gigantea* is no more divergent from the other species than most other species are from each other. Further speculation is not warranted at this time.

Gutierrezia

Some of the most spectacular arrays of flavonoids yet to be described come from species of *Gutierrezia*. Examples come from studies of *G. grandis* (Fang et al., 1985a) and *G. microcephala* (Fang et al., 1986b), from which, literally, dozens of compounds were obtained. Many of these compounds had 6,8-dioxygenation and the level of *O*-methylation was quite high, often reaching five or six groups. It appears that 6,8-dioxygenated flavonoids are characteristic of the woody species of this genus. A study of *G. alamanii* var. *megacephala*, one of the herbaceous perennial species, was described by Lenherr et al. (1986). They identified eight flavonoids: kaempferol, quercetin, quercetin 3-*O*-rhamnoside, rutin, quercetin 3,7-dimethyl and 3,7,4'-trimethyl ethers, 5,7-dihydroxy-3,3',4',5'-tetramethoxyflavone, and 5,3',5'-trihydroxy-3,7,4'-trimethoxyflavone. Those authors pointed out the absence of A-ring oxygenated flavonoids in this species and suggested that this feature could be used to distinguish between woody and non-woody taxa in the genus.

Hazardia

Hazardia, distributed in southwestern United States and adjacent Mexico, consists of 13 species arranged in three sections, sect. *Hazardia* with eight species, sect. *Bracteofolia* with three, and sect. *Machaerantheroides*, which consists of only one species, *H. whitneyi* with two varieties. Flavonoids of all thirteen species, including both varieties of *H. squarrosa* and *H. whitneyi*, have been determined (Clark and Mabry, 1978). The chemical data, coupled with morphological and chromosomal observations, show a high degree of correlation with, and thus support for, the infrageneric taxonomy. *Hazardia whitneyi* produces only flavonols, has $n = 4$, is ecologically as well as geographically separated from the other sections, and is the only species in the genus with herbaceous stems. The two varieties, *whitneyi* and *grindelioides*, exhibited identical flavonoid profiles.

The three species that comprise sect. *Bracteofolia* produce flavonols as their major flavonoid type but also have apigenin and luteolin *O*-glycosides as well as *C*-glycosylflavones. Members of this section have $n = 5$. Although there is some overlap in flavonoid profiles between sections *Bracteofolia* and *Hazardia*, the combination of morphological and chemical characters clearly distinguish section *Bracteofolia* from the others. The third group, section *Hazardia*, exhibits a flavonoid profile that consists of flavone *O*- and *C*-glycosides as major components with somewhat lesser amounts of flavonols. (Not all taxa have all compounds.) *Hazardia brickelliodes*, although maintained in section *Hazardia* on morphological grounds, has $n = 5$ and a flavonoid profile that again has flavonols as the major components, although both flavone *O*- and *C*-glycosides are also present.

Clark and Mabry (1978) recorded the presence of four flavonol methyl ethers, without defining their structures, in some members of each of the three sections of *Hazardia*. A later paper by Clark and Wollenweber (1985) reported identification of no less than 33 flavonoid aglycones from *H. squarrosa* var. *grindelioides*. An additional 13 suspected flavonoids were also observed on chromatograms. A comparison of the flavonoids of *Hazardia* with those of *Haplopappus* can be found in Chapter 21.

Helianthus

The sunflowers have commanded a considerable amount of attention because of their commercial importance as well as their conspicuousness, in terms of both size and abundance. The outcome of this attention has been an increasing understanding of the biology of the main participants, principally *H. annuus* including its many agriculturally important variants. *Helianthus annuus*, the true sunflower, will reappear in Chapter 23 where profiles of infraspecific variation will be discussed and in Chapter 24 where we will address variation at the population level. In this chapter we will look at work that has been done to understand relationships within and among groups of species. Prominent in this area has been the work of Ed Schilling at the University of Tennessee and some of his colleagues.

The first group to come under scrutiny was *Helianthus* series *Corona-solis* which consists of 15 species and includes diploids, tetraploids, and hexaploids (Schilling and Mabry, 1981). The workers hoped that flavonoid profiles would help in sorting out species boundaries, somewhat blurred by natural hybridization, and to aid in unravelling the ancestry of the polyploids, especially *H. tuberosus*, the Jerusalem artichoke. In general, the profile of the entire group was remarkably consistent with only a few exceptional cases. 6-Methoxyapigenin, 6-methoxyluteolin, and 5,7-dihydroxy-6,4'-dimethoxyflavone (pectolinarigenin) were almost universally present. Kaempferol was present in all taxa but its glycosidic form varied. Quercetin 3-*O*-glucoside was more scattered with a higher frequency of occurrence in diploid species. Quercetin 7-*O*-glucoside was restricted to *H. decapetalus* (4*n*) which species, as we will shortly see, has one of the more unusual profiles in the series. Four species yielded leaf anthochlors, the aurone derivatives sulfuretin 6-*O*-glucoside and sulfuretin 6-*O*-galactoside, and the chalcone coreopsin. Only one of these, *H. maximilianii*, is diploid, while the other three are polyploids, *H. decapetalus* (4*n*), *H. strumosus* (4*n*, 6*n*), and *H. tuberosus* (6*n*). This naturally led to the suggestion that *H. maximilianii* may have been involved in the ancestry of the polyploids. A closer examination of the total flavonoid complement of the polyploids indicates that the situation may not be so straightforward, at least for some of the polyploids. For example, if *H. maximilianii*, or a related ancestor, did contribute its genome in the formation of *H. decapetalus*, the capacity to make 6-oxygenated flavones, and methylate them, has been lost. *Helianthus decapetalus* has also gained the capacity to accumulate quercetin derivatives relative to the profile of the supposed diploid progenitor. In the case of *H. strumosus*, which consists of both tetraploids and hexaploids, involvement of *H. maximilianii* is more likely, in terms of flavonoids, owing to their very similar profiles. The situation with regard to the third polyploid with leaf anthochlors, *H. tuberosus*, is similar to that at which we just looked. The flavonoid profiles of the diploid and the hexaploid *H. tuberosus* are very similar and require only that the latter gain the capacity to accumulate kaempferol 3-*O*-glucuronide. Thus, in two of the polyploid species flavonoid profiles could be mustered as support for involvement of the diploid *H. maximilianii*. The authors of that work (Schilling and Mabry, 1981), however, suggest that "a more plausible explanation may lie with loss of tissue specificity of flavonoid production..." Without additional information, the question remains open.

The second paper in this series concerned flavonoids of *Helianthus* series *Angustifolii* (Schilling, 1983). Leaf and ray florets were examined separately. Variation was observed in the occurrence of flavone aglycones and anthochlors in leaves of all seven species, and flavone aglycones (e.g., hispidulin and nevadensin), anthochlors, and quercetin 7-*O*-glucoside in ray florets of five species (one not available and one lacks ray florets). Flavonol glycosides were present in leaves of all seven species, while quercetin 3-*O*-glucoside was present in all ray florets of the five tested species. It is very interesting that the two species that lacked anthochlors in the ray florets, *H. carnosus* and *H. floridanus*, had quercetin 7-*O*-glucoside instead. Cladistic analysis of the flavonoid data set was performed using series *Corona-solis* as the outgroup. The most significant feature emerging from this analysis was the highly branching nature of the relationships which was taken to indicate that evolution in the group involved radiation from a central ancestral type. There was no support for operation of a gradual stepwise process.

The third part of this series involved the three species that comprise *Helianthus* series *Microcephali* (Schilling et al., 1987). Extracts of ray florets of all three afforded coreopsin, sulfurein, and quercetin 7-*O*-glucoside. *Helianthus microcephalus* adds quercetin 3-*O*-glucoside to this list. Only the latter species possesses leaf glands, so it is understandable that flavone aglycones (luteolin, hispidulin, acerosin, hymenoxin, jaceosidin, nepetin, nevadensin, and an unidentified chalcone) were limited to this taxon. The presence of 6,8-dioxygenated flavones in this species (hymenoxin and nevadensin) was taken by these authors to indicate a closer relationship of this series to series *Angustifolii* than to series *Corona-solis*.

Hemizonia

Eighteen species (of about 33) of *Hemizonia* representing the four recognized sections (*Centromeris, Hemizonia, Madiomeris, Zonamra*) were subjected to an examination of their leaf exudate flavonoid profiles (Tanowitz et al., 1987). The compounds identified represented a diverse array of flavonoid types, flavanones, *O*-methylated kaempferol and quercetin derivatives, and 6-, 8-, and 6,8-oxygenated flavones and flavonols with various levels of *O*-methylation. With the exception noted below all taxa had unique profiles. Quercetin, quercetin 3-methyl ether, patuletin, patuletin 3-methyl ether, luteolin, and naringenin were present in nearly all species; all other compounds were of more or less scattered occurrence. The most highly substituted compounds, those with 6-, 8-, or 6,8-oxygenation, did not occur in any significant pattern. Numerical analysis of the results using ordination and dendrogram-generating procedures resulted in the delineation of three groups. Group I consisted of *H. australis* and *H. pungens*, which were the only two taxa in this study that gave identical profiles. The other member of section *Centromadia*, *H. fitchii*, had an altogether different flavonoid profile that allied it more closely with members of section *Madiomeris*, which comprised a rather spread-out group II in the ordination. The single species that comprises section *Hemizonia*, *H. luzulaefolia*, emerged as part of ordination group II, the other members of which are part of section *Madiomeris*. In general, while flavonoid patterns show a degree of concordance with the sectional taxonomy of *Hemizonia*, they are not in total accord with other sources of data. As the authors pointed out, further work is clearly indicated. Nonetheless, the flavonoid data do support the view that section *Madiomeris* is a heterogeneous assemblage of taxa as indicated by other data. The reader may also consult Baldwin (1998) and Carlquist et al. (2001) for additional comments on evolution within Madiinae.

Hulsea

All members of *Hulsea* were examined for their flavonoids as part of a systematic study of the genus by Wilken (1975). The taxa recognized in this study, 11 counting subspecies, were distinguishable on the basis of combinations of morphological differences, breeding barriers, and ecological features. The flavonoid profiles were highly uniform with regard to quercetin 3-*O*-glucoside, 3-*O*-diglucoside, 3-*O*-rutinoside, two unidentified quercetin 3-*O*-glycosides and luteolin 4'-methyl ether (diosmetin). Apigenin 7-*O*-glucoside and three additional unidentified quercetin 3-*O*-glycosides were of only scattered occurrence. Cyanidin 3-*O*-glucoside and 3-*O*-rutinoside occurred, together, in about half of the taxa. The flavonoid profile of this genus is highly conservative and offers no support for inferring relationships. The five subspecies of *H. vestita*, described in more detail in Chapter 23, comprise the most uniform group in the genus.

Hymenoxys scaposa-acaulis complex

During a study of flavonoids of *Hymenoxys*, Seeligmann and Alston (1967) observed that some taxa gave significantly different chromatographic patterns depending upon where in their range they were obtained. This geographic variation was most pronounced in the *Hymenoxys scaposa-acaulis* complex (includes *H. ivesiana*). This observation was all the more interesting in that the unusual flavonoid profiles were constant within a given area. The pigment profiles correlated more closely with an east-west geographical range than they did with identity of the plants as determined on morphological grounds. Owing to the lack of detailed chemical information at the time, the authors came to no specific conclusion but did suggest that regional selection for the different chemistries was possible, although some complex hybridization process may have been responsible.

Subsequent studies by Thomas and Mabry (1967, 1968a,b) established the structures of the compounds involved. The foliar flavonoids of specimens, regardless of geographical origin, were identified as patuletin 3-*O*-glucoside and 3-*O*-rutinoside, 5,7,4'-trihydroxy-6,8-dimethoxyflavone (demethoxysudachitin), 5,7,3'-trihydroxy-6,8,4',5'-tetramethoxyflavone (scaposin), and 5,7-dihydroxy-6,8,3',4'-tetramethoxyflavone (hymenoxin). The geographic differences showed up in the floral flavonoids. Plants from Texas had quercetagetin and patuletin 7-*O*-glucosides, whereas plants from Utah, Colorado, Wyoming, and northeastern Arizona had patuletin 3-*O*-glucoside and 3-*O*-rutinoside.

The original workers (Seeligmann and Alston, 1967) felt that the chemical differences were more likely to be a reflection of differentiation within a single, variable species than the result of parallel changes within two species. They thought it unlikely that two species would have undergone independent shifts from 3-*O*-glycosylation to 7-*O*-glycosylation in the same geographic area. The idea of hybridization was not pursued further, and the situation remains unresolved.

In order to arrive at a better understanding of the processes at work in situations such as the one just described, it is essential to examine the enzymes responsible. It would be necessary to determine the degree of relatedness of the flavonol *O*-glucosyltransferases responsible for making the isomeric glycosides. What is needed is information on protein structure and/or gene sequence data. An analogous situation exists in *Flaveria chloraefolia* where two sulphotransferases are known to exist, one that catalyzes formation of flavonol 3-sulphate and one that yields the 4'-isomer. Production of DNA clones coding for the two sulphotransferases allowed the authors (Varin et al., 1992) to determine their respective gene sequences and the deduced amino acid sequences. The two clones were shown to have an overall identity of 69% in 311 amino acid residues. Thus, the changes in protein structure were such that, while retaining the fundamental activity of transferring a sulphate group from cofactor to substrate, the site of attack on the substrate changed. (Incidentally, these data pointed to significant relationships between the *Flaveria* sulphotransferases and sulphotransferases from animal tissues.) It seems reasonable to suggest that the two glucosyltransferases in *H. scaposa* might be similarly related. Perhaps it might even be possible to determine the relationship of the enzymes in *Hymenoxys* to a larger family of glucosyltransferases.

Although the authors of the flavonoid studies referred to the species under study as *Hymenoxys scaposa*, and we have chosen to continue that usage for the purpose of this discussion, the taxon involved is, in fact, a member of the genus *Tetraneuris*. This situation was discussed briefly in Chapter 21.

Iva

The tortured taxonomic history of *Iva* was reviewed by Jackson (1960) in the introduction to his study of the morphology and cytology of the genus. Not only are the number of species rightfully included in the genus a matter of contention, but there has been disagreement upon how the species should be sorted into sections, although there seems to be some agreement that three sections probably are reasonable: *Iva*, *Linearbractea*, and *Cyclachaena*. Additional information has been brought to bear on these issues through a cladistic treatment (Bolick, 1985). That treatment resulted in the recognition of *Iva* and *Linearbractea* as sections within *Iva* s.s. with the five species comprising sect. *Cyclachaena* consigned to five monospecific genera.

It would be ideal if we were able to compare flavonoid profiles for representatives of the three putative sections, but, unfortunately this is not possible owing to the lack of flavonoid information for members of sect. *Linearbractea*. One interesting difference, however, did surface from the comparison of profiles of two members of sect. *Cyclachaena* (*I. acerosa* and *I. nevadensis*) with three members of sect. *Iva* (*I. axillaris*, *I. frutescens*, and *I. hayesiana*). *Iva acerosa* and *I. nevadensis* both accumulate *O*-methylated flavones that have extra oxygenation at both C-6 and C-8 (Farkas et al., 1966, 1967; Nogradi and Farkas, 1968). In contrast, *I. axillaris* (Herz et al., 1966a,c; Hörhammer et al., 1966), *I. frutescens* (Herz et al., 1972a), and *I. hayesiana* (Herz and Sudarsanam, 1970) have *O*-methylated flavones with only C-6 extra oxygenation. Some of the same 6-oxygenated flavones also occur in *I. acerosa* and *I. nevadensis*. Morphologically, *I. axillaris* has been placed in sect. *Iva*, but results of chloroplast DNA studies led Miao et al. (1995a,b) to segregate it in its own section, sect. *Rhizoma* B. Miao. Flavonoid data do not speak to this point. Further sampling, obviously including members of sect. *Linearbractea*, would be of interest in view of the other data now available for the entire "genus."

Lasthenia

In an epigram to his monograph of the genus, Ornduff (1966) cited the opinion, attributed to John Lindley and Thomas Moore, concerning *Lasthenia* Cass., that "They grow in wet places, and appear to be uninteresting weeds." Since many of the species of *Lasthenia* grow in great profusion in the wet meadows and vernal pools of California, the genus has gained the name "goldfields." We have no quarrel with their first comment concerning the plants' habitat choices, but we hope to show in this and following discussions of their flavonoids, and other features, that they are anything but uninteresting weeds!

The genus as presently understood includes species that at one time constituted three genera: *Baeria*, *Crockeria*, and *Lasthenia* (Ornduff, 1966). Ornduff concluded that the genus consists of 16 species with infraspecific taxa recognized in several. The species number has now been raised to 17 following a reappraisal of the status of *L. minor* subsp. *maritima* by Vasey (1985) who argued, as did Asa Gray, that *L. maritima* should be recognized at the specific level. The genus is mostly Californian with ranges of certain species reaching into Arizona, extreme southern Oregon, northern Baja California and, in the case of *L. maritima*, on offshore rocks and bird islands as far north as the northwestern corner of Vancouver Island, British Columbia. One species, *L. kunthii*, occurs in Chile.

The earliest study involving flavonoids in the genus was that of Gertz (1938) who included several species in his anthochlor (chalcone/aurone) survey. Detailed work by Shimokoriyama and Geissman (1960) on floral pigments of *L. californica* (reported as *L. chrysostoma*) confirmed the presence of aurones (maritimetin 6-*O*-glucoside and sulfuretin 6-*O*-glucoside) and chalcones (okanin 4'-*O*-glucoside and butein 6'-*O*-glucoside) in this species. The Gertz survey also showed that several species of *Lasthenia* did not give positive anthochlor tests. Since the negative species belonged to two of the sections of *Lasthenia*, as recognized by Ornduff, and that the species that gave negative tests all belonged to other sections, we undertook a detailed study of the entire genus. The distribution of anthochlor, flavone, and flavonol aglycones in the genus (using current nomenclature and Ornduff's sectional taxonomy) is summarized in Table 22–2. Although presentation of the data in this fashion discounts some of the variation, it does serve to show if clear-cut evolutionary trends can be detected among the sections. This goal is well met. As pointed out by Ornduff (1966), and reiterated in the flavonoid survey paper (Bohm et al., 1974), sections *Baeria* and *Burrielia* are considered to be the most primitive in the genus on morphological and cytological grounds. Members of section *Ptilomeris*, in contrast, are considered the most specialized. It is of interest then to see that

Table 22–2. Gametic chromosome numbers (n) and flavonoid aglycone distribution in species and sections of *Lasthenia*

Taxon	Number	Flavonoid[1]				
		Ant	Lut	Kae	Que	Pat
Sect. *Baeria*						
californica	8, 16, 24	+	+	–	+	–
macrantha						
ssp. *macrantha*	16, 24	+	+	–	–	–
ssp. *bakeri*	24	+	+	+	+	–
ssp. *prisca*	n.d.[2]	+	+	–	–	–
Sect. *Burrielia*						
debilis	4	+	–	–	–	
leptalea	8	+	+	–	+	–
microglossa	12	+	–	–	–	–
Sect. *Platycarpha*						
platycarpha	4	–	–	–	+	+
Sect. *Lasthenia*						
glaberrima	5	–	–	–	+	–
kunthii	5	–	–	–	+	–
Sect. *Hologymne*						
chrysantha	7	+	–	–	+	–
ferrisiae	7	+	–	–	+	–
glabrata						
ssp. *coulteri*	7	–	–	–	+	–
ssp. *glabrata*	7	–	–	–	+	–
Sect. *Ptilomeris*						
burkei	6	–	–	–	+	+
conjugens	6	–	–	–	+	+
fremontii	6	–	–	–	+	+
coronaria	5	–	–	–	+	–
maritima	4	–	–	–	+	–
minor	4	–	–	–	+	–

1 Ant = anthochlors; Lut = luteolin; Kae = kaempferol; Que = quercetin; Pat = patuletin.
2 Not determined.

production of anthochlors and luteolin is a property of these two sections (not all species, however). Section *Hologymne*, which is of intermediate position in the genus, consists of two species that produce anthochlors, but all four have lost the capacity to accumulate luteolin derivatives. Members of the remaining three sections, *Platycarpha*, *Lasthenia*, and *Ptilomeris*, have lost the capacity to accumulate either type. The overall evolutionary trends in this genus, then, involve initial loss of the capacity to make luteolin, followed by the loss of anthochlor accumulation. There appears to be a parallel increase in the diversity of quercetin and patuletin glycoside profiles. In some of the most specialized members of genus, *L. coronaria*, *L. minor*, and *L. maritima*, of section *Ptilomeris*, patuletin synthesis has been lost.

It is instructive to examine some of the interspecific relationships within *Lasthenia* in some detail. We can start by looking at sections *Baeria* and *Burrielia*, which are considered, as noted above, to be the most primitive groups in the genus. Members of these two sections exhibit close morphological similarities and have base chromosome numbers of $x = 4$ and 8, respectively. The only successful artificial intersectional hybrids (hybrids had 40% pollen viability) involved members of *Baeria* and *Burrielia* (Ornduff, 1966). The flavonoid chemistry of the two are also similar in that all members accumulate anthochlors, and most have both luteolin and quercetin glycosides as well.

Section *Hologymne*, consisting of *L. chrysantha*, *L. ferrisiae*, and two subspecies of *L. glabrata*, is considered by Ornduff (1966) to have its closest affinities with sections *Baeria* and *Burrielia*. The base number of section *Hologymne* is $x = 7$, a one chromosome reduction from the presumptive generic base number. The presence of anthochlors in both *L. chrysantha* and *L. ferrisiae* provide additional support for this connection. *Lasthenia ferrisiae* was newly described by Ornduff (1966) in his generic monograph. Traditionally, *L. ferrisiae* specimens had been referred to *L. glabrata* ssp. *coulteri*, but differences in achene pubescence were taken to be sufficient for species recognition. The different flavonoid profiles of the two species support this view.

Ornduff (1966) considers that *L. ferrisiae* could have resulted from a cross between *L. glabrata* ssp. *coulteri* and *L. chrysantha* followed by stabilization of the hybrid at the diploid level. Flavonoid profiles are in accord with this origin since the profile of *L. ferrisiae* to a substantial degree reflects the sum of the profiles of the putative parent taxa species. The aglycones sum perfectly, whereas there is a slight difference in the nature of the quercetin glycosides; *L. ferrisiae* has quercetin 3-*O*-glucoside, which was not seen in the putative parents.

The affinities of section *Platycarpha*, which consists of the single species *L. platycarpha*, are obscure. On the basis of leaf and pappus features, Chambers (1957) suggested that this species might be most closely related to section *Ptilomeris*. Ornduff (1966) countered this opinion with the view that an association with sections *Baeria* and *Burrielia* would seem more likely based on style and anther tip, phyllary, and pappus characters. *Lasthenia platycarpha* lacks the anthochlor pigments that characterize *Baeria* and *Burrielia*, but, interestingly, does possess patuletin which is otherwise seen only in some members of section *Ptilomeris*.

Lasthenia minor and *L. maritima* were dealt with in Ornduff's (1966) monograph as subspecies of *L. minor*. Although *L. maritima* was originally recognized as a distinct species by A. Gray (*Burrielia maritima*), Ornduff (1966) included the entity as a subspecies of *L. minor*. It occupies perhaps the most extreme habitats of any species in the genus, namely, seabird nesting sites in high salt-spray area. Its return to specific status was advocated by Vasey (1985). A study of allozymes of the two species showed that *L. maritima* is relatively recently derived from *L. minor*. Our original survey (Bohm et al., 1974) had shown that the two taxa shared quercetin 3-*O*-galactoside and 3-*O*-glucuronide while *L. minor* had a quercetin 3-*O*-diglycoside lacking in the profile of *L. maritima*. Although the difference is small, it is interesting to speculate that the further reduction of the already simple flavonoid profile of *L. minor* may have accompanied *L. maritima*'s specialization in such a severe habitat.

The original flavonoid survey (Bohm et al., 1974) used pooled samples in most cases, but a few individual plants were also analyzed. Some variation was seen in the pigment profiles of *L. macrantha* subsp. *macrantha* (anthochlors) and *L. macrantha* subsp. *bakeri* (all flavonoid classes). Further examples of interpopulation variation were subsequently reported for nine taxa of *Lasthenia* (Ornduff et al., 1974). This study showed that differences between populations may involve simple differences in glycosylation or larger differences such as presence or absence of entire classes of compounds.

Lasthenia section *Lasthenia* consists of two species, *L. glaberrima*, which occurs in California, Oregon, and extreme southern Washington, and *L. kunthii*, which occurs between ca. 32° and 38°S in Chile (Ornduff, 1966). The two species are morphologically very similar, they both have $n = 5$, and can be crossed to yield viable offspring. Although they occupy similar habitats, Ornduff (1966) points out that the South American taxon may be more tolerant of saline soils. *Lasthenia kunthii* is thought to have been derived by long distance dispersal from a North American source (Ornduff, 1963, 1966). The flavonoid profiles of both species consist solely of quercetin glycosides (Bohm et al., 1974). The two species have quercetin 3-*O*-galactoside in common, but are distinguished by the occurrence of a quercetin 3-*O*-galactosylglucoside in *L. kunthii*, and quercetin 3-*O*-glucuronide and a quercetin 3-*O*-rhamnosylgalactoside in *L. glaberrima*. These profiles are among the simplest seen in the genus and more closely resemble each other than most other profiles except those of *L. minor* and *L. maritima*. As we saw above, both of these taxa have flavonoid profiles that consist solely of quercetin 3-*O*-glycosides that also involve galactose and glucuronic acid as participating sugars. Any suggestion that these taxa might have been involved, however, would have to contend with problems of morphological and chromosome number difference: *L. minor* and *L. maritima* are both $n = 4$ species while *L. kunthii* has $n = 5$, as mentioned above.

Additional information on flavonoid profiles in *Lasthenia* can be found in Chapter 23 where subspecific differences in *L. macrantha* and in *L. glabrata* are discussed, Chapter 24 where extensive studies of variation in *L. californica* are reviewed, and in Chapter 25 where we examine the putative hybrid origin *of L. burkei.*

Parthenium

According to the description of *Parthenium* by Mears (1980b), the genus consists of 15 species in North America north of Guatemala, two species in South America, and one, *P. hysterophorus*, that has spread from tropical and subtropical America to many other parts of the world (where it has become a particularly aggressive weed). Mears (1980a,b) described the flavonoids of some two dozen or so taxa including varieties of four of the species (see Chapter 23). For our purposes in this chapter it is interesting, first, to look at the distribution of flavonoids as a function of the sectional taxonomy of the genus. One of the features of the flavonoid profile of the genus, as pointed out by Mears, is the presence in various combinations of two types of flavonoids, *O*-methylated aglycones based on kaempferol, quercetin, 6-hydroxykaempferol, and quercetagetin, and an array of glycosides based on kaempferol, quercetin, and isorhamnetin as well as on some of the simpler *O*-methylated flavonols. Four sections are recognized.

Section *Bolophyton* consists of three species of submontane perennials. This section stands out starkly from the other three in that its flavonoid profile consists of quercetagetin di-, tri- and tetramethyl ethers and 6-methoxykaempferol 3-methyl ether 7-*O*-glycoside. Simple flavonoids were not observed. In terms of number of compounds present, this is the simplest of the four sections. In terms of flavonoid structural complexity, however, members of this section make some of the more highly substituted compounds, e.g., quercetagetin 3,6,7,3'- and 3,6,7,4'-tetramethyl ethers.

The other three sections are much more similar to each other than any one of them is to section *Bolophyton*, although each has its peculiar structural feature. Section *Partheniastrum*, with species of grasslands and temperate deciduous forests, displays the most complex flavonoid array within the genus with methyl ethers of kaempferol, quercetin (including isorhamnetin), 6-hydroxykaempferol, and quercetagetin. Section *Parthenichaeta*, made up of arboreal species from tropical thorn forests, has an overall array similar to sect. *Partheniastrum*, but lacks isorhamnetin. Section *Argyrochaeta*, which comprises herbaceous annuals and perennials, has an overall flavonoid profile almost as complex as the former two, but can be distinguished from them in lacking kaempferol 7-methyl ether.

There is some difference of opinion surrounding the interpretation of the flavonoid differences among these sections. Using a combination of flavonoids, alkaloids (or tests therefore), and sesquiterpene lactones, Rodriguez (1977) concluded that the various combinations of secondary metabolites were a reflection of adaptive responses to physical and biotic factors, primarily herbivore pressure, in the environment. It is useful to review the points that Rodriguez (1977) made in his efforts to interpret the significance of the differences observed within the genus. In the first place, it is obvious that real differences in the chemical complexity of temperate vs. tropical taxa exist. Despite the variation seen in the secondary chemistry of these taxa, reliable sectional chemistries can be defined. Rodgriguez's second observation, which includes two parts, was that woody tropical members of the genus elaborate more complex sesquiterpene lactones than do herbaceous annuals and that herbaceous perennials contain as many lactones as tropical taxa. His third observation was that desert taxa contain insignificant concentrations of sesquiterpene lactones but do elaborate large amounts of methylated flavonoids as well as germacranes and triterpenes. And lastly, he pointed out that the montane perennials produce the smallest amounts of secondary compounds seen in *Parthenium*. Although he noted that there were problems with some of the interpretations and that cause and effect had not been established specifically, the overall relationships were consistent with co-evolutionary theory.

Based upon the studies of Levin (1976) and Levin and York (1978), Mears (1980b) pointed out that there should be a latitudinal correlation of flavonoid diversity for *Parthenium* in general. Mears failed to detect any such correlation. Support for the apparent absence of a latitudinal influence was taken from the fact that members of section *Bolophytum*, which are restricted to gypsum-based substrates in Colorado, Utah, and Wyoming, produce only a few flavonoids and at comparatively low concentration, while members of section *Partheniastrum* of similar latitude in eastern North America, produce the most complex array of flavonoids seen in the genus. Mears sees the correlation of flavonoid chemistry with substrate as significant and suggests that, if latitudinal effects are operating, they may be masked by the sensitivity of the taxa to edaphic factors. Mears' (1980b, p. 369) conclusion is worth quoting in its entirety: "If the species is a narrow endemic occurring on limestone or gypsum it will tend to have few flavonoids and these will be primarily methylated aglycones. If the species is widespread, occurring in a variety of primarily noncalcicolous habitats, then it will tend to express a high flavonoid diversity, through predominance and variation in flavonoid glycosides; and the aglycones will tend to be less methylated." Mears (1980b) ends his discussion with the suggestion that if such trends are related to habitat, or specifically, edaphic conditions, as they appear to be in *Parthenium*, the situation may exist for other systems. Implicit in this conclusion is the obvious need to understand how soil type influences flavonoid, or any secondary metabolite, expression in plants. We might add here that other situations involving different flavonoid profiles from plants growing on different substrates are known in the literature. One of these was observed by Reid and Bohm (1994) in the New Zealand endemic *Cassinia vauvilliersii* (see Chapter 12) where plants from two different soil types afforded different profiles which were maintained in plants grown in a common garden. Granted that this situation is different from the *Parthenium* system, in that the *Cassinia* specimens were clearly of the same species, but it does reinforce the idea that edaphic factors, and other physical factors involved therewith (e.g., water-holding properties, soil pH) may have profound influences on secondary chemistry.

Perityle

Relationships within subtribe Peritylinae have been the subject of several studies by Powell (1968, 1972, 1973a,b, 1974). Although *Perityle* itself is a fairly large genus, some 63 species according to Karis and Ryding (1994a), the study of flavonoids of nine of the species at least gives a hint that these compounds might be useful in examining sectional relationships within the genus (Crins et al., 1988b; Bohm et al., 1986a). All three sections, *Laphamia*, *Perityle*, and *Pappothrix*, were represented in that study. The compounds identified are methyl ethers of kaempferol, 6-hydroxykaempferol, and quercetagetin, and two flavanones, naringenin 7,4'-dimethyl ether and eriodictyol 7-methyl ether. Section *Perityle*, represented by *P. californica, P. emoryi*, and *P. vaseyi*, had the most complex array with up to 10 compounds. Section *Pappothrix*, represented by *P. cinerea, P. fosteri*, and *P. rupestris* var. *albiflora*, showed a much reduced profile with only three compounds seen in any of the species. Only two of these had any quercetagetin derivatives and neither flavanone was detected. Section *Laphamia*, represented by *P. bisetosa* var. *scalaris*, *P. lemmonii*, and *P. villosa*, had an even simpler array of compounds consisting of only one kaempferol and two 6-hydroxykaempferol methyl ethers. Powell (1973b) suggested that evolutionary advancement in *Perityle* has been accompanied by reduction in pappus, ray floret, phyllary number, pubescence, and disk floret number. He considered sections *Pappothrix* and *Laphamia* to be advanced relative to section *Perityle*. The flavonoid profiles determined for representatives of these three sections show a similar trend with reductions in both numbers of compounds per species/section and in the level of *O*-methylation seen in those that were present.

Robinsonia

The relationship between the Juan Fernandez Islands endemic genus *Robinsonia* and *Senecio* was commented upon briefly in Chapter 21. Here we will look at the flavonoid chemistry of all seven species, one of which previously constituted the monotypic genus *Rhetinodendron* (Pacheco et al., 1985; Crawford et al., 1998). The profiles were based upon glycosides of apigenin, luteolin, kaempferol, quercetin, isorhamnetin, naringenin, eriodictyol, and dihydroquercetin (taxifolin). Because of the larger array of flavonoid types and lower levels of interpopulational variation, flavonoid chemistry of *Robinsonia* offers more insights into relationships among the various component species than in either *Erigeron* or *Dendroseris*. In addition, ITS sequence data for several members of the genus (Sang et al., 1995) are in substantial agreement with the disposition of flavonoids. The flavonoid data are summarized in Table 22–3.

The uniform flavonoid profile of *R. gayana* and *R. thurifera* agrees perfectly with their appearance in a clade that corresponds to section *Robinsonia*.

Table 22–3. Flavonoids of *Robinsonia*

	Flavonoids[1]												
Taxon	A	B	C	D	E	F	G	H	I	J	K	L	M
Subgen. *Rhetinodendron*													
R. berteroi	+	+	–	–	–	–	–	–	–	–	–	–	–
Subgen. *Robinsonia*													
Sect. *Symphyochaeta*													
R. macrocephala	+	+	–	–	–	–	–	–	–	–	–	–	–
Sect. Robinsonia													
R. gayana	+	+	+	–	–	–	–	–	–	–	–	–	–
R. thurifera	+	+	+	–	–	–	–	–	–	–	–	–	–
Sect. *Eleutherolepis*													
R. evenia	+	–	+	–	–	–	–	–	+	+	–	–	–
R. masafuerae	–	+	+	+	–	–	–	+	–	–	–	–	–
R. gracilis	–	–	+	+	+	+	+	–	–	–	+	+	+

[1] A = Q-3-Gal; B = Q-3-DiGal; C = Q-7-Glc; D = Q-3,7-DiGlc; E = K-7-Glc; F = IR-3-Glc; G = IR-7-Glc; H = Lute-7-Glc; I = Lute-7-DiGly; J = Apig-7-DiGly; K = Erio-7-Glc; L = Nar-7-Glc; M = Tax-7-Glc, where Q = quercetin; K = kaempferol; IR = isorhamnetin; Apig = apigenin; Lute = luteolin; Nar = naringenin; Erio = eriodictyol; Tax = taxifolin; Gal = galactose; Glc = glucose; Gly = unspecified sugar.

Allozyme studies (Crawford et al., 1992a) suggested that this species pair is the most recently diverged set of taxa in the genus, at about 1.7 M.Y. It is interesting to note that the glycosidic profiles for these two species showed no variation among populations (seven of *R. thurifera* and 21 of *R. gayana*).

Section *Eleutherolepis*, which consists of *R. evenia, R. gracilis*, and *R. masafuerae*, emerged as a sister clade to section *Robinsonia*. Section *Eleutherolepis* was further divided into a clade consisting of *R. evenia* and *R. masafuerae* and a clade consisting solely of *R. gracilis*. The finding of quercetin and luteolin glycosides in both *R. evenia* and *R. masafuerae* is also in agreement with their linkage as suggested by the ITS data. *Robinsonia evenia* is further distinguished by the presence of apigenin derivatives. A more general statement concerning the cohesiveness of this clade is its unique accumulation of flavones. This certainly qualifies as a flavonoid synapomorphy in the genus. The absence of apigenin in the *R. masafuerae* profile should not be troubling since only a single collection of this species was available for flavonoid analysis. Thirteen specimens of *R. evenia* were studied from which apigenin was seen in only four for certain and possibly in three others. Three specimens of *R. evenia* lacked flavones altogether. If the absence of apigenin in *R. masafuerae* is real, and not the result of sampling error, then the flavonoid results could be taken to indicate its origin from an apigenin-lacking population of *R. evenia*.

Robinsonia gracilis stands apart in the genus in possessing kaempferol and isorhamnetin, the flavanones naringenin and eriodictyol, and the dihydroflavonol taxifolin. This unique flavonoid profile is consistent with the position occupied by *R. gracilis* in the cladogram, the sister taxon to the *evenia-masafuerae* pair. Pacheco et al. (1985) suggested that a trend toward more complex flavonoid profiles would account for their observations. They commented upon, but did not advocate, a gain mutation in *R. gracilis* to account for the accumulation of the flavanones and dihydroflavonol. The mechanism proposed involved a decrease in efficiency of the enzymes that convert flavanones to dihydroflavonols and then on to flavonols. Alternative explanations exist, e.g., an increase in the amount of carbon flowing through the system because of a change in control at the levels of phenylalanine ammonia lyase (PAL) or chalcone synthase (CHS), for example. Accumulation of intermediates would occur because the carrying capacity of the flavonoid biosynthetic enzymes would have been exceeded. These workers (Pacheco et al., 1985) did attribute the presence of flavones to a gain mutation. This seems an entirely reasonable suggestion. We would go one step further, however. *O*-Methylation is also considered to represent a gain mutation. In this system the presence of isorhamnetin (quercetin 3'-methyl ether) represents a unique condition, a flavonoid apomorphy, in *Robinsonia*.

At the highest level of branching, *R. berteroi*, the sole member of subgenus *Rhetinodendron*, is distinguished from the five species that constitute subgenus *Robinsonia*. The flavonoid profile of *R. berteroi* is the simplest seen in the genus with only two glycosides of quercetin. This profile is identical to that of *R. macrocephala*, which is the sole member of section *Symphyochaeta* of subgenus *Robinsonia*. We are not likely to resolve the issue of this taxon's position in the evolutionary scheme of things since *R. macrocephala* may be extinct (Sang et al., 1995).

Simsia

The floral flavonoid profiles of 23 species of *Simsia* were studied by Schilling and Spooner (1988). The compounds identified included apigenin and luteolin 7-*O*-glucosides, three quercetin glycosides, and the anthochlor pairs coreopsin/sulfuretin and marein/maritimetin. Quercetin 3-*O*-glucoside was present in all specimens; quercetin 3-*O*-glucuronide was present in almost all. Quercetin 7-*O*-glucoside was of scattered occurrence only. The rest of the compounds showed strong preference for one or the other of two groups of species, either the "*S. fruticulosa* group" or the "*S. amplexicaulis* group." Apigenin 7-*O*-glucoside was observed in two species from the *S. amplexicaulis* group but not in any of the 13 species tested of the *S. fruticulosa* group. Luteolin 7-*O*-glucoside occurred in all but one of the *S. amplexicaulis* group but was not seen in any species in the other group. The coreopsin/sulfuretin pair was identified in seven of the 10 species of the *S. amplexicaulis* group and in only one species from the *S. fruticulosa* group, *S. ghiesbreghtii*. *Simsia ghiesbreghtii* was further distinguished from all other species in possessing the marein/maritimetin anthochlor pair. In the only other work on flavonoids of *Simsia* of which we are aware, Maldonado et al. (1992a) reported nevadensin from *S. cronquistii*. The presence of flavones with the nevadensin-type substitution pattern in related genera, *Helianthus, Viguiera*, and *Tithonia*, was commented upon in Chapter 21.

Tithonia

La Duke (1982a,b) and Karis and Ryding (1994b) state that *Tithonia* consists of 11 species distributed primarily in Mexico and Central America. (Note that La Duke lists 12 species in both Tables 1 and 2 in the 1982a paper; in 1982b, however, he included *T. auriculata* as a variety of *T. calva*.) Evolutionary significance of the flavonoid profiles will be examined later; now let's look at the way in which the flavonoid classes assort themselves along infrageneric lines. La Duke (1982b) sees the genus as comprising two sections, sect. *Mirasolia* with three species (five taxa), and sect. *Tithonia* with three series, *Fruticosae* and *Grandifloreae* with two species each, and *Tithonia* with four.

Section *Mirasolia* exhibits the richest collection of normal flavonols and flavones (both have 5-OH), and is the only section to accumulate *C*-glycosylflavones. A further distinction can be made in so far as the *C*-glycosylflavones occur only in the two varieties of *T. calva*. Both species in sect. *Tithonia* series *Fruticosae* have a much reduced number of flavonols (two) and flavones (two) compared to sect. *Mirasolia*. The two species of series *Grandiflorae* have an even more reduced profile with only one flavone derivative (hispidulin) and no detectable flavonols. The four species that comprise series *Tithonia* show the greatest novelty in the genus and also the greatest apparent variation. The only uniformity in the series appears to be the absence of normal flavones. The novel compounds detected in this group are flavonols and flavones that lack the 5-OH group. Three species, *T. brachypappa, T. thurberi*, and *T. tubaeformis*, possess these interesting compounds, while *T. rotundifolia* lacks them. *Tithonia rotundifolia* is further distinguished within this group, and within the genus as a whole, by having the richest array of normal flavonols including the unique appearance of patuletin. One more difference should be noted: *T. rotundifolia* also lacks anthochlors, although this is not a particularly significant feature as pointed out below.

While *Tithonia* qualifies as an anthochlor-positive genus [Crawford and Stuessy (1981) cited La Duke's unpubl. data] the distribution of these compounds, as reported in the work under discussion here (La Duke, 1982a), is rather haphazard. Only in the case of series *Tithonia* do we see more than one species recorded as positive. Table 22–4 summarizes the variation in anthochlors and other flavonoids observed in the genus.

La Duke (1982a) superimposed the distribution pattern of the flavonoid types on a cladogram obtained from morphological studies in an effort to assess trends in flavonoid evolution. Certain trends were evident in the occurrence patterns, one of which involves the parallel loss of flavonols in three lines, once before the event that gave rise to series *Grandiflorae* (*T. diversifolia* and *T. koelzii*), and twice within series *Tithonia* (*T. thurberi* and *T. brachypappa*). As La Duke pointed out, it is easier to visualize three independent loses than it is to suggest that flavonol production developed independently nine times. There has also been a gradual loss in the number of positions bearing *O*-methyl groups. Dimethoxy and trimethoxy compounds predominate in the more basal groups with fewer such groups present in more advanced members of the genus. The most recently derived series, *Tithonia*, lacks *O*-methylated flavonoids. Another trend of interest is the

Table 22–4. Flavonoid variation in sections and series of *Tithonia*

Taxon	Flavonoid type[1]				
	Flols	Flones	*C*-Gly	5-Deoxy	Antho
Sect. *Mirasolia*	4/3[2]	4/3	1/3	0/3	1/3
Sect. *Tithonia*					
Ser. *Fruticosae*	2/2	2/2	0/2	0/2	1/2
Ser. *Grandiflorae*	0/2	1/2	0/2	0/2	1/2
Ser. *Tithonia*	2/4	0/4	0/4	2/4	2/4

1 Flols = normal flavonols; Flones = normal flavones; *C*-Gly = *C*-glycosylflavones; 5-Deoxy = 5-deoxyflavonoids; Antho = chalcones and/or aurones.

2 Denominator is the number of species in each section; numerator is the number of species in which each flavonoid type occurs.

gain of the 5-deoxy flavonoids, followed by their subsequent loss in *T. rotundifolia*. The appearance of this derived flavonoid structural type reinforces the view that the annual habit in these species represents the advanced condition in the genus. The loss of these compounds from the flavonoid profile of *T. rotundifolia* is in agreement with the view that this species possesses morphological traits considered advanced in the group, i.e., annual habit, 3-5-lobed leaves, long pales, and orange ray corollas.

This work serves as an excellent example of the value of combining data bases, in this case morphology and flavonoids, and subjecting the combined data to comparative analysis. In this case the comparative analysis simply involved superimposing one data set upon the other which resulted in a significantly improved view of the evolutionary changes that may have occurred in this group of species.

Tragopogon

In very few situations have all, or nearly all, of the techniques available to biosystematists been applied as successfully in understanding the evolutionary relationships between plants as we find in the study of *Tragopogon* in western North America. *Tragopogon* is an old world genus comprising some 50 species of annuals and perennials. Beltzer and Ownbey (1971) state that artificial crosses between *T. pratensis* and *T. porrifolius* had been made by Linnaeus, which, they pointed out, was one of the first hybridizations carried out for scientific purposes. [For more details on this topic, see Roberts (1929) and Zirkle (1935).]

Three species, *T. pratensis*, *T. porrifolius*, and *T. dubius*, have been introduced into North America. *Tragopogon pratensis* and *T. porrifolius* were first collected in eastern Washington in 1916; *T. dubius* was first found in the same area in 1928. Attention was focused on these species when Marion Ownbey, at Washington State University, noticed apparent intermediate forms. Detailed studies of these entities showed that hybridization had occurred (and likely continues to occur) in areas of sympatry and that two of the hybrid forms had become stabilized. The amphidiploids were named *T. mirus* and *T. miscellus* (Ownbey, 1950). Cytotaxonomic analysis suggested that cytoplasmic inheritance was involved and that the speciation event(s) had occurred repeatedly (Ownbey and McCollum, 1953, 1954).

The use of flavonoids in the study of natural plant hybrids, as dramatically demonstrated in the study of *Baptisia* (Alston and Turner, 1963b), suggested that similar analyses of *Tragopogon* might prove useful. That this was the case was demonstrated, initially, by Brehm and Ownbey in 1965a,b. They recognized early in their screening work that variation among individuals within populations would likely lead to confusing results. Profiles of populations, however, were very instructive. Species-specific flavonoid spot profiles were observed and additivity of these profiles in suspected hybrids allowed these workers to identify the parents. In some cases profiles were not clearly additive, with one or more spots present in higher or lower frequency than anticipated. These observations were taken as evidence for introgression. Evidence was also presented for separate origins of *T. mirus* in two locations.

The potential usefulness of characteristic flavonoid spot patterns notwithstanding, knowledge of the structures of the compounds represented by the spots is vitally important, because it allows one to assess the degree of biosynthetic differences (divergence) between two species or between two populations. Detailed structural studies on the *Tragopogon* flavonoids were first described in a paper by Kroschewsky et al. (1969). The compounds present in all five species were shown to be, predominantly, *C*-glycosylflavones with apigenin and luteolin present as well. With only one exception, the amphidiploid (derived) species exhibited flavonoid profiles that contained elements of both parental species. Thus, swertiajaponin and swertisin, both of which occur in *T. dubius* but not in *T. porrifolius*, and vitexin, which occurs in *T. porrifolius* but not in *T. dubius*, all occur in *T. mirus*. Similarly, vicenin-1 and lucenin-1 occur in *T. dubius* but not in *T. pratensis* while lucenin-2 occurs in the latter but not the former. Their hybrid offspring, *T. miscellus*, exhibits all three. The only apparent exception to additivity is seen in the case of *T. dubius* which accumulates *O*-xylosylisovitexin. In neither of the derived species was this compound encountered. In general, however, these results are in excellent agreement with the conclusions reached on the basis of morphological and cytotaxonomic evidence.

One of the fascinations of flavonoid genetic studies has been the possibility of finding hybrid molecules. These compounds would presumably be formed by enzymes unique to each parent, combined in the hybrid offspring, performing their special function on the same substrate. No hybrid compounds were observed in the *T. dubius* work, but such compounds were observed in a subsequent study of hybridization among several European species (Belzer and Ownbey, 1971). Two putative hybrid flavonoids appeared in the F_2 generation of a cross between *T. samaritanii* and *T. pterodes*. One of these compounds also appeared in the F_2 offspring of a cross between *T. balcaniscus* and *T. pterodes*. Hybrid compounds manifesting themselves in the F_1 offspring can be explained in terms of complementation. It is a much more difficult task to account for appearance of new compounds in the F_2 generation, however. Perhaps the compounds in question are in fact present in the parents but at such a low concentration that they are not detectable. Hybridization might involve alterations of control systems that would allow accumulation of larger (detectable) amounts of the compounds in question. The two compounds involved in these hybrids, swertiajaponin (orientin 7-methyl ether; their compound no. 10) and xylosylvitexin (their no. 13), are typical of the flavonoids identified in *Tragopogon* species and do not suggest the sudden appearance of unexpected processes. Unfortunately, this phenomenon has not been studied in sufficient detail to explain the formation of hybrid compounds at the macromolecular level. There is a case, however, where the two species involved make different *O*-methyl derivatives, 3'-hydroxy-4'-methoxy- in one species and 3'-methoxy-4'-hydroxy in the other. The "hybrid" 3',4'-dimethoxy compound presumably is not formed because the *O*-methyltransferases of the two species do not accept a monomethylated flavonoid as substrate.

Returning to the western American species for a moment, it is only necessary to point out that subsequent work at the macromolecular level has fully substantiated the early conclusions concerning multiple allopolyploid speciation. The more recent studies have involved allozymes (Roose and Gottlieb, 1976), chloroplast DNA (Soltis and Soltis, 1989), ribosomal DNA (Soltis and Soltis, 1991) and a meticulous geographic survey (Novak et al., 1991). A more recent study from the Soltis laboratory described additional surveys for genetic variation in the allotetraploids (Soltis et al., 1995). It is interesting to muse on what the next generation of botanists at Washington State University will find to do with "Ownbey's Tragopogons."

Viguiera

Viguiera, with something on the order of 180 species (Bremer, 1994), offers a wide variety of taxonomic problems. Schilling and his associates at the University of Tennessee have undertaken analyses of the flavonoid profiles of some infrageneric groups with very interesting results forthcoming in most cases. We can start by looking at their efforts with *Viguiera* series *Viguiera*, which consists of a dozen species distributed in parts of Mexico and southwestern North America. Nine species (12 taxa) belonging to this series were examined for their floral flavonoids by Rieseberg and Schilling (1985). The coreopsin/sulfurein anthochlor pair was universally present in this set of taxa, while all other compounds varied to a greater or lesser degree. The flavonoid profiles of several taxa stand out, especially *V. dentata* and *V. potosina*. In addition to the ubiquitous coreopsin/sulfurein pair, the floral flavonoid profile of *V. dentata* also had the marein/maritimein pair and quercetin 3-*O*-glucoside. The situation with regard to *V. potosina* is simpler in terms of flavonoid composition of the flowers compared to the other species tested in that they contain only the coreopsin/sulfurein pair. Moreover, this is the only species in which floral anthochlors appear to be restricted to disk florets. All others had anthochlors in at least the lower third of the ligule; several exhibited these pigments throughout the ligules.

Twenty specimens of *V. laciniata* were consistent in having luteolin 7-*O*-xyloside as a floral component. The 7-*O*-glucoside, in contrast, was present in at least some specimens of *V. deltoidea, V. lanata, V. subincisa*, and *V. tomentosa*. A considerable amount of variation in floral flavonoid pigments was seen in the four varieties of *V. deltoidea* studied. These data can be found in Chapter 23.

Somewhat better discrimination of groups within series *Viguiera* was realized with a later study of external flavonoid aglycones (Schilling, 1989). The first and most obvious distinction is the absence of exudate compounds on *V. dentata* and *V. potosina*. Each of the remaining species has a unique combination of exudate compounds, but a better appreciation of relationships within the group emerged from a cladistic analysis of the seven diploid species. Two groups emerged that are consistent with geographical distributions. One group, consisting of *V. lanata, V. microphylla, V. subincisa*, and *V. tomentosa*, is restricted to the Baja California peninsula. The second group consists of *V. laciniata, V. parishii*, and *V. reticulata* and occupies a range in the northern part of Baja California that extends into the United States. The key feature that distinguishes the more northerly taxa is the presence of isoliquiritigenin 4'-methyl ether. The feature by which the more southerly group is distinguished is 6-hydroxyluteolin 7-methyl ether, which, however, has evidently been lost from one of the taxa in the group, namely *V. subincisa*. Restriction fragment analysis of chloroplast DNA showed that the Baja California group of species comprise a well-defined clade (99% support) that is quite distinct from other groups of taxa within the genus (Schilling and Jansen, 1989).

According to Blake (1918), *Viguiera* series *Brevifoliae* consists of three species from central Mexico, *V. bicolor, V. greggii*, and *V. brevifolia*. Despite the small size of this set of species, the taxonomic problems that exist therewith are well addressed by the flavonoid chemistry (Schilling and Panero, 1988). First, the three species have very different flavonoid profiles with *V. brevifolia* totally lacking foliar aglycones, while *V. greggii* possesses a complex array, eight compounds in all, including a number characterized by extra oxygenation at C-6 and both C-6 and C-8. Several of these compounds have moderate levels of *O*-methylation as well. These differences are significant in that these two species are actually only pappose and epappose forms of the same taxon. The distinct flavonoid chemistry of each supports the view that they are species in their own right. Flavonoid chemistry actually supports the position that *V. greggii* is more closely related to *V. bicolor*, which produces only one aglycone, 5,7-dihydroxy-6,8,4'-trimethoxyflavone (nevadensin), but produces an unusually large amount of it (2 g from 50 g dried leaves). Nevadensin is one of the aglycones identified from extracts of *V. greggii*.

The final group for which we have flavonoid information in the genus is section *Maculatae*. Flavonoids for 10 of the 14 species (11 taxa) that constitute this section were described by Schilling et al. (1988). As in the case of most other species of *Viguiera*, each species in the current study displays a unique pigment profile. All taxa possess anthochlors in ray florets with some having quercetin glycosides as well. The situation with regard to foliar aglycones is not so straightforward. First, *V. neocronquistii* and *V. oaxacana* lack foliar glands and, thus, cannot have glandular flavonoids. *Viguiera splendens* also lacks foliar flavones but is set apart from the rest of the species in this section by its accumulation of the chalcones isoliquiritigenin and its 4-methyl ether (and possibly 4,4'-dimethyl ether) on its leaves. It is interesting to note that the simplest ray floret pigment profile, consisting of the coreopsin/sulfurein and butein/sulfuretin pairs, occurs in the three taxa just mentioned, those that lack glandular flavones. It might be suggested, then, that the only difference between *V. neocronquistii* and *V. oaxacana*, which lack the glandular apparatus all together, and *V. splendens*, which has only anthochlors in its exudate, is that the latter species has acquired the capacity for expression of its chalcone synthetic genes in leaves as well as in floral tissue. This would be an interesting system to examine at the macromolecular level.

Flavonoids are useful in assessing two other taxonomic problems within the group. *Viguiera adenophylla* and *V. quinquiradiata* are similar morphologically and can be confused. Their flavonoid floral profiles are distinctive, however. *Viguiera quinquiradiata* has only the two ubiquitous anthochlor pairs, while *V. adenophylla* also has isoliquiritigenin 4'-*O*-glucoside and quercetin 3-*O*- and 7-*O*-glucosides. In contrast, the two subspecies of *V. eriophora* have identical flavonoid profiles in both leaves and ray flowers, characters that could be used to maintain them as infraspecific taxa.

It remained only for Schilling et al. (1988) to compare the distribution of flavonoids in section *Maculatae* with the results of a cladistic analysis of the group done using parsimony methods (Panero and Schilling, 1988). This comparison is diagrammed in the 1988 flavonoids paper and shows, in general, that the pigment arrays agree with the relationships that emerged from the analysis. As those authors pointed out, the evolutionary tendency in this group of species has been towards loss of structural diversity. This, they went on to speculate, may be related to the narrow geographic distributions enjoyed by these species. These data, along with other information, lead to the suggestion that the species in this section may have been derived by fragmentation of some ancestral form followed by geographical isolation. It should be noted, as was done by Schilling et al. (1988), that the tendency of narrowly endemic taxa to have simpler pigment profiles is in agreement with observations of Mears (1980b) on *Parthenium*.

In concluding the discussion of work on *Viguiera* it should be pointed out that certain characteristic flavonoids emerged as more or less characteristic of the three groups as noted by the respective authors. Section *Maculatae* accumulated two distinctive compounds, 6-hydroxyluteolin and coreopsin 3-methyl ether, along with isoliquiritigenin 4'-*O*-glucoside, which suggests a possible link with a member of series *Brevifoliae* which also accumulates isoliquiritigenin derivatives. Series *Brevifoliae* is characterized by the presence of quercetagetin derivatives and flavonoids with 6,8,4'-trimethylation, a feature otherwise not known from the genus. Within series *Viguiera* 6-hydroxy-7-methoxyflavone may be characteristic of the "Baja California group." Since 6-hydroxyluteolin otherwise occurs in members of section *Maculatae*, this compound type may be a link between this section and the Baja group of species. For recent studies of cpDNA that includes *Viguiera* in the context of its subtribal position see Panero et al. (1999).

Wilkesia

Wilkesia consists of two species endemic to the well-aged volcanic soils of western Kauai. *Wilkesia gymnoxiphium* is the more widespread of the two taxa occurring in quite large populations, the most prominent of which is the main feature of the Iliau Loop Trail which borders the Waimea Canyon (Iliau is the Hawaiian name for the genus). The genus was considered monotypic until 1971 when Harold St. John described *W. hobdyi* from specimens collected on the westernmost point of Polihale Ridge (by R. Hobdy). It was subsequently found on adjacent Kaaweiki Ridge, but its existence is in peril through the destructive actions of feral goats. *Wilkesia gymnoxiphium* is an upright plant often reaching heights of five meters and is monocarpic, while *W. hobdyi* rarely reaches a meter in height and is polycarpic. Leaves of the former are longer and the plant is unbranched, while *W. hobdyi* is branched from the base (Carr, 1985).

Flavonoids of both species were studied (Bohm and Fong, 1990) as part of our examination of the tarweeds (Madiinae). Eleven *O*-methylated flavonoids, two flavonols, nine flavones, were identified. Five of these were held in common. The most striking feature of this array was the presence of 6-, 8-, and 6,8-dioxygenated flavonoids in both species. These features are shared with other genera in the subtribe, particularly the mainland genera *Calycadenia* and *Hemizonia* (Crins and Bohm, 1990). The main difference between the two Hawaiian species lies in the exact pattern of *O*-methylation since both have the capacity to achieve the same high level of oxygenation. The high degree of profile similarity clearly indicates a close relationship between these two endemic species. This is further borne out in the values for Nei's genetic-distance coefficient (0.033) and genetic-identity coefficient (0.967) calculated from allozyme studies of Witter and Carr (1988). Further documentation of close relationships between these species and among the three genera can be found in work on chloroplast DNA (Baldwin et al., 1991) and on the ITS region of 18-26S nuclear ribosomal DNA (Baldwin, 1992, 1998).

Wyethia

The flavonoid profiles of *Wyethia* and *Balsamorhiza* were compared in Chapter 21, where we saw that species of the former elaborate several unusual compound types, i.e., isoflavones, *C*-prenylated isoflavones, *C*-prenylated flavanones, and dihydroöxepines (where cyclization of the prenyl group results in a seven- rather than a six-membered ring). *Wyethia* also accumulates flavanones whereas *Balsamorhiza* does not. The patterns of occurrence of these compounds within *Wyethia*, however, is not uniform. The most extreme example comes from section *Agnorhiza* (McCormick et al., 1987). Isoflavones, in various combinations, occur in four of the six species investigated, *W. invenusta, W. ovata, W. reticulata*, and *W. scabra*.

They were not observed in either *W. bolanderi* or *W. elata*. The profile of *W. elata* consisted of dihydrokaempferol, 6-methoxyapigenin (hispidulin), and a selection of kaempferol and quercetagetin methyl ethers. *Wyethia bolanderi*, in contrast, exhibited a rich profile consisting of kaempferol, 6-hydroxykaempferol, quercetin, and quercetagetin methyl ethers. Flavonol methyl ethers were identified from the "isoflavone group," but the number of compounds found and their structural complexity were much reduced compared to the "non-isoflavone group." In the 10 species of *Wyethia* for which several compounds were reported (thus excluding *W. arizonica*), 10 different flavonoid profiles were observed. We can also note in passing that a different profile was observed for each of the five species of the related genus *Balsamorhiza* studied. See also Weber (1998) for additional hypotheses on limits in this generic complex.

A summing up

It seems reasonable to conclude that flavonoids have provided useful information in a variety of situations involving relationships at the specific level. For example, flavonoids served as important markers in following the evolution of species of *Arnica* in northwestern North America, allowing workers to speculate on the locations of glacial refugia and possible paths of postglacial migration. Flavonoids have also provided useful information on progenitor-daughter species pairs as in the cases of *Coreopsis nucensis* and *C. nucensoides*, species of *Stephanomeria*, *Lasthenia minor* and *L. maritima*, and *Lasthenia glaberrima* and *L. kunthii*. Studies of the last two species, which are native to California and Chile, respectively, are also representative of research efforts that address major disjunctions, i.e., between North America and South America in the example cited. Flavonoids have also been effectively used as markers in studies of the evolutionary relationships among insular species, as well as in efforts to identify their likely mainland ancestors; excellent examples come from work on *Erigeron*, *Dendroseris*, and *Robinsonia* from the Juan Fernandez Islands. Flavonoids have also proven an excellent source of information in studies of polyploid speciation, probably best exemplified by the work on *Tragopogon* in North America. And finally, simple structural differences between flavonoid pairs in closely related species, such as 3'-*O*- vs. 4'-*O*-methylation in *Achillea*, 3'-*O*- vs. 4'-*O*-sulphation in *Flaveria*, or 3-*O*- vs. 7-*O*-glycosylation in *Hymenoxys*, offer potentially powerful experimental systems for studying enzyme specificity or changes in enzyme and gene structure.

CHAPTER 23

Flavonoids at Infraspecific Levels

In this chapter we discuss patterns of flavonoid occurrence at the subspecific and varietal levels. There is no general agreement on how these two ranks should be used (e.g., Hamilton and Reichard, 1992), and we will not comment on this aspect of infraspecific taxonomy. For a detailed discussion see Stuessy (1990). Our sole interest here is to document the use of flavonoids at these levels. We also include a few examples of differences that have been reported among cultivars. These are treated together at the end of this chapter.

It will be noticed that comparatively few of the over 400 genera for which flavonoid data are available are represented in the discussions below. This arises from the regrettably small number of genera, or species complexes, that have been subjected to the sort of scrutiny that one finds has been done in such genera as *Achillea, Arnica, Bidens, Coreopsis, Dahlia, Helianthus*, *Lasthenia*, or *Parthenium*. Results of several flavonoid surveys have appeared in the literature that include occurrences at the varietal or subspecific level. However, often no comment is made about the significance of the data as they might apply to infraspecific relationships. These reports are included in the presentation below usually without comment as to their potential application to problems within their respective groups.

Several examples in this chapter involve geography as a significant factor. Geography will also bear on a number of systems in Chapter 24 where we will attend to populational variation.

The *Achillea ageratifolia* group

The *Achillea ageratifolia* group consists of three subspecies restricted to the Balkan Peninsula. Subspecies *ageratifolia* occurs at higher elevations (1200-2300 m) in central and northern Greece, Albania, Bulgaria, and in the southern part of what at that time was Yugoslavia. Subspecies *aizoon* also occurs at the higher elevations but can also be found at lower elevation

(200 m). Its range of occurrence is north-central Greece and Yugoslavia. Subspecies *serbica* occurs in southern Yugoslavia and Bulgaria and grows at lower elevation (200-600 m). Franzén (1988) investigated the flavonoid chemistry of all of the taxa, both aglycones and glycosides, and subjected the entire data set to multivariate analysis (18 populations by 36 compounds). The overall chemistry of the species consisted of 15 aglycones, *O*-methylated derivatives of 6-hydroxyapigenin (scutellarein), 6-hydroxyluteolin, 6-hydroxykaempferol, and 6-hydroxyquercetin (quercetagetin), and 19 glycosides involving flavonols, and two suspected as being *C*-glycosylflavones.

Principal components analysis resulted in an ordination that suggested that the distribution of flavonoids in *A. ageratifolia* follows neither subspecific delineation nor geography. Whereas four populations of subsp. *aizoon* clustered tightly together as did four populations of subsp. *serbica*, the populations of subsp. *ageratifolia* were highly dispersed. Four of the subsp. *ageratifolia* populations clustered with subsp. *aizoon*, two with subsp. *serbica*. The other four populations of subsp. *ageratifolia* appear as two clusters removed from other groupings. Factors other than geographic specialization must be at work in this system.

Plants of *A. ageratifolia* growing in the same general area but in different habitats have very similar flavonoid profiles despite being subject to quite different environmental factors. Populations of subsp. *serbica* and *ageratifolia* exhibit similar flavonoid profiles, again despite significant differences in local environments. Franzén (1988, p. 1649) concluded that "the different flavonoid profiles in *A. ageratifolia* are not of crucial importance in the adaptation of the plant to the local environment." He goes on to suggest that the differentiation of profiles may have occurred during the Pleistocene when alpine and montane habitats occurred at lower altitudes. Since the profiles do not appear to be environmentally sensitive, it was suggested that drift may have played an important part in their establishment.

Achillea sibirica

Valant-Vetschera included a number of subspecies of *Achillea* in her extensive studies of this genus. Interested readers should consult the cited papers for details of taxonomic history. Only sufficient information is given in each example to set the stage for appreciating the contribution of the flavonoid data.

The *Achillea sibirica* aggregation, which consists of a number of morphologically highly variable groups of plants, occurs in eastern Asia and in North America. Valant-Vetschera (1985a) studied representatives of five subspecies: *camtschatica, japonica, mongolica, ptarmicoides*, and *pulchra* as well as material from North America. *Achillea sibirica* subspp. *camtschatica, mongolica*, and *ptarmicoides* are similar in that they have flavonoid profiles dominated by orientin derivatives, although the arrays of individual compounds are different in each subspecies. The profiles of subspp. *mongolica* and *japonica* also have a quercetin 3-*O*-rhamnoglucoside, a compound not otherwise seen in the named subspecies she studied. Examination of the *A. sibirica* specimens from North America showed a flavonoid profile much akin to that of subsp. *mongolica* with the dominant presence of a particular orientin derivative, absence of kaempferol derivatives (seen in subsp. *ptarmicoides*), and presence of the quercetin 3-*O*-rhamnoglucoside. Valant-Vetschera (1985a) concluded that the North American material be assigned to subsp. *mongolica*. Another subspecific situation considered in that study (Valant-Vetschera, 1985a) dealt with the *A. ptarmica* group of taxa. *Achillea impatiens* subsp. *ledebouri* was represented in the study by a single collection, while *A. impatiens* without identification of subspecific status, was represented by five specimens. Although all members of this species have orientin derivatives as the dominant flavonoids with scattered vitexin, luteolin, and quercetin derivatives, *A. impatiens* subsp. *ledebouri* is apparently unique in having only orientin derivatives. This situation was not discussed in the text of that paper.

Allagopappus

Allagopappus consists of two species restricted to the Canary Islands. The flavonoids of two subspecies of *A. dichotomus* have been described by Gonzalez et al. (1995), subsp. *dichotomus*, which predominates on Tenerife and islands to the west, and subsp. *latifolius*, which is endemic on Gran Canaria. Subspecies *dichotomus* afforded chrysoeriol, eriodictyol, kaempferol 3-methyl ether, 5,7,4'-trihydroxy-3,6-dimethoxyflavone, and 5,7-dihydroxy-3,6,4'-trimethoxyflavone. The flavonoid profile of subsp. *latifolius* was simpler consisting of 5,7-dihydroxy-3,3',4'-trimethoxyflavone, quercetin 3,3'-dimethyl ether, and naringenin. Although comparable levels of *O*-methylation are achieved in both taxa, the presence of 6-oxygenated compounds in subsp. *dichotomus* represents a significant difference between the two. Any decision as to which state is the more derived one cannot be answered without further information.

Arnica

Arnica, both in Europe and North America, has been the subject of an impressive amount of study, by Merfort and her colleagues in Germany, and by Denford and his students in Canada. Work of the former group has, to a certain extent, been motivated by pharmaceutical issues, while the Canadian group has been concerned with understanding the evolutionary relationships among the many parts of this dominant group in North America, including those from circumboreal regions. Owing to the extensive sampling of North American *Arnica*, work from the Canadian group has yielded valuable information on subspecies relationships, population variation (Chapter 24), as well as general information about relationships within the genus itself (Chapter 22).

The first example of flavonoids and infraspecific taxonomy in *Arnica* involves subgenus *Chamissonia* (Merfort et al., 1986). Two subspecies of *A. chamissonis* were studied, subsp. *chamissonis* (indicated as var. *interior*), and subsp. *foliosa* (one sample of which was identified as var. *incana*). The flavonoid aglycone profiles of these taxa consisted of flavones, flavonols, and a flavanone. Several compounds exhibited both 6-oxygenation and a modest level of *O*-methylation. Variation was minimal. Although individual compounds may have been lacking, any given taxon exhibited all of the structural features, i.e., aglycone type, extra oxygenation, and *O*-methylation, seen for the group. The only qualitative difference in terms of aglycone type involves the presence of naringenin 4'-methyl ether in both specimens of subsp. *foliosa*.

A taxonomic re-evaluation of the *Arnica frigida* – *louiseana* complex, based on morphological, phytogeographical, and cytological data (Downie and Denford, 1986a) resulted in recognition of *A. frigida* with two subspecies, *frigida*, which occurs in Alaska, the Yukon, and British Columbia, and the newly proposed subsp. *griscomii*, which occurs in Quebec and Newfoundland. *Arnica louiseana*, recognized at specific rank, occurs in Alberta.

Flavonoid profiles of these taxa were described by the same workers (Downie and Denford, 1986b). The overall flavonoid profile consisted of apigenin 7-*O*-glucoside, kaempferol 3-*O*-glucoside and 3-*O*-galactoside, quercetin 3-*O*-galactoside, 3-*O*-diglucoside, and 3-*O*-gentiobioside (plus one unidentified compound seen in two populations of *A. louiseana*). Analysis of 38 populations of subsp. *frigida* showed the existence of seven different combinations of the six identified flavonoids. There was no correlation between flavonoid profile and ploidy level, which ranged from $2n = 38$ to $2n = 95$. Three populations of subsp. *griscomii* were studied. The profile of the one specimen from Quebec and the profile of the two from Newfoundland were different from each other, but each matched one of the profiles seen in the western subspecies. The close morphological similarity between the two subspecies, which is reflected in their flavonoid similarity, suggests that at one time a single taxon was widespread across North America and that nearly all of the plants in the center of the range were victims of Pleistocene glaciation. The western subspecies is well represented in central Alaska where the species likely survived glaciation. Subspecies *griscomii*, from the Gaspé Peninsula of Quebec and Newfoundland, then represents a palaeoendemic taxon found only in a few restricted sites. It is interesting to note, in this context, that the five populations of *A. louiseana* sampled in that work occur in an area of western Alberta where the two ice sheets, the Cordilleran and Laurentide (Pielou, 1991), met. Whether this was completely ice-free or whether it was simply a zone of early melting, it is known to be an area of high endemism.

Maguire (1943) recognized seven subspecies of *A. alpina* in his treatment of the genus. The considerable taxonomic confusion that has surrounded this complex was based, as Downie (1988) pointed out, on a combination of the polymorphic, apomictic, and polyploid nature of the group. His (Downie, 1988) reappraisal of the situation resulted in the recognition of *A. angustifolia* (there was a nomenclatural problem with *A. alpina*) with two subspecies: subsp. *angustifolia*, which combined Maguire's subspecies *alpina*, *angustifolia*, *attentuata*, *iljinii*, *intermedia*, *sornborgeri* (plus *A. plantaginea*), and subsp. *tomentosa*. Much like *A. frigida* discussed above, *A. angustifolia* is polymorphic, displays different chromosome number races, and is variable in its expression of flavonoid profiles. The flavonoid profiles are also very similar in that they consist of apigenin, luteolin, kaempferol, and quercetin glycosides. Eleven different flavonoid profiles were recorded for subsp. *angustifolia* representing populations in Canada and Alaska. There was no correlation with chromosome number. Subspecies *angustifolia* also occurs in Greenland, from which area eight populations were sampled. Two flavonoid profiles were observed, each of which matched a profile seen in the other subspecies. *Arnica angustifolia* subsp. *tomentosa*, from sites on the eastern face of the Rocky Mountains and in the Mackenzie River delta, exhibited two flavonoid profiles, each of which matched a profile seen in subsp. *angustifolia*. Again, the survival of populations of this species in unglaciated areas of Alaska and the Yukon suggests residence there in refugia during the Wisconsin glaciation, again, as suggested above for *A. frigida*.

Artemisia

Flavonoid information exists for subspecies of three species of *Artemisia*, *A. campestris*, *A. ludoviciana*, and *A. tridentata*, which we will look at in that order. The flavonoids of *A. campestris* subsp. *maritima* (syn. *A. crithmifolia*) have been described by three groups (Cavaleiro, 1986; Rauter et al., 1989; Sanz et al., 1991), all of whom reported flavanones and dihydroflavonols as the principal pigments, along with 5,4'-dihydroxy-6,7-dimethoxyflavone (cirsimaritin). The flavanone array consisted of naringenin, eriodictyol, their common methyl ethers, and mono- and dimethyl ethers of dihydrokaempferol and dihydroquercetin. Work by Hurabielle et al. (1982) and De Pascual Teresa et al. (1984, 1986) on *A. campestris* subsp. *glutinosa* afforded very similar profiles consisting of many of the same flavanone and dihydroflavonol derivatives. The only unusual compounds reported were the two B-ring deoxyflavanones, pinocembrin and pinostrobin (Hurabielle et al., 1982). The resident flavone in this taxon appears to be 6-methoxyapigenin (hispidulin). The similar flavonoid profiles of these two subspecies speak loudly of their close relationship. (Other descriptions of *A. campestris* flavonoids, but without subspecific designation, are known.)

Artemisia ludoviciana subsp. *ludoviciana* was shown by Liu and Mabry (1982) to elaborate a very complex array of flavone and flavonol derivatives including compounds with extra A-ring oxygenation, a moderately high level of *O*-methylation and, most remarkably, flavones that bore 2',4',5'- and 3',4',5'-oxygenation. The most complex flavonol encountered was quercetagetin 3,6,3',4'-tetramethyl ether (bonanzin). The only flavonoids reported from *A. ludoviciana* subsp. *mexicana* were 5,7,4'-trihydroxy-6,3'-dimethoxyflavone (jaceosidin) 5,7-dihydroxy-6,3',4'-trimethoxyflavone (eupatilin) (Ruiz-Cancino et al., 1993). Both compounds lie within the range of structural complexity that characterize subsp. *ludoviciana*. Direct comparison is not warranted, however, since work on the taxon was primarily directed toward sesquiterpene lactones.

Several *O*-methylated flavones and flavonols including axillarin, eupafolin, chrysosplenol-C, and penduletin, have been reported in various combinations from *A. tridentata* subspp. *tridentata*, *parishii*, *vaseyana*, and *wyomingensis* (Rodriguez et al., 1972b; Brown et al., 1975; Kelsey et al., 1973; Wilt et al., 1992). What differences may exist among these taxa could well be more apparent than real, since over half of the compounds detected by Wilt et al. (1992) in HPLC traces remained unidentified at the time. Apigenin 7-*O*-glucoside, kaempferol, quercetin, and quercetagetin were reported from subspp. *tridentata* and *vaseyana*; luteolin was not seen in subsp. *wyomingensis* (Wilt et al., 1992).

Bahia schaffneri

Studies of *Bahia schaffneri* var. *aristata* yielded 5,7,3'-trihydroxy-3,6,4'-trimethoxyflavone (Pérez and Romo De Vivar, 1994), while var. *schaffneri* afforded that compound along with 5,7-dihydroxy-3,6,3',4'-tetramethoxyflavone (Pérez and Romo De Vivar, 1989). Since these studies were aimed primarily at terpenoid constituents, the flavonoids shown do not represent the sole members of these varieties' profiles.

Bidens

Comparatively few comprehensive biosystematic studies of *Bidens* using secondary metabolites have been reported despite the size of the genus and its conspicuous place in most floras. However, biosystematic studies that have employed flavonoids, plus a number of purely natural products reports, combine to provide an excellent view of the overall biosynthetic capacities of the genus which were discussed in Chapter 21. One of the results of the biosystematic studies, of course, was a better understanding of the limits of infraspecific variation within certain target groups of *Bidens* species. Most of these studies resulted in taxonomic restructuring of these groups, which in some cases was quite drastic.

Bidens ferulaefolia complex

One of the earlier attempts to apply flavonoid occurrence patterns to problems within *Bidens* was Hart's (1979) study of the *B. ferulaefolia* complex. *Bidens ferulaefolia*, as viewed by Sherff (1937, 1955a), was a taxon that occurred from northern Guatemala to southwestern United States with the largest concentration of populations in Mexico. He recognized three varieties: var. *ferulaefolia*, var. *foeniculifolia*, and var. *ludens*. Hart's investigation consisted of morphological studies of fresh and dried specimens, cytological examinations, greenhouse and hybridization studies, and an examination of the flavonoids of all members of the assemblage. The differences observed were of such magnitude that three species were recognized: *B. ferulaefolia* (which includes the original varieties *foeniculifolia* and *ludens*), *B. carpodonta*, and *B. longistyla*. Distinguishing morphological features included red pigmentation of chaff, inner involucral bracts, and disc florets; style branch length, stamen structure and color; presence or absence of awns and their structure; and pubescence of the inner involucre. *Bidens ferulaefolia* was shown to have $n = 10$, while the other two species have n = 12. Attempts to make interspecific crosses invariably failed; all crosses between individuals within populations, between individuals from different populations of the same species, and between F_1 plants were successful.

The flavonoid chemistry of the three species is very rich, as is most often the case with *Bidens*. Chalcones and aurones are the prominent compounds with apigenin and luteolin glycosides representing the flavones, and kaempferol and quercetin the flavonols. Three flavanones were also reported. Although there were a few unidentified anthochlors on the list, most were derivatives of okanin, maritimein, butein, and sulfuretin. Each species presented a unique flavonoid profile. For example, *B. carpodonta* accumulates a set of "glucorhamnosyl" derivatives of the four anthochlors not seen in the other two species. Other interesting distinctions are the lack of flavonols in *B. ferulaefolia* and the absence of apigenin derivatives in both *B. carpodonta* and *B. longistyla*.

The *Bidens pilosa* complex

Bidens pilosa is a widely distributed weedy species with a center of diversity in Mexico recently reinvestigated by Ballard (1986). Included in the study was a detailed examination of flavonoid chemistry. Sherff (1937) based his recognition of six varieties on the highly variable nature of the taxon. Ballard's reinvestigation was predicated upon, among other factors, the lack of a comprehensive study of the group using fresh material.

Observations in the field, chromosome counts, a study of breeding systems and hybridization, a morphometric analysis, and identification of the flavonoid pigments of individuals of some 200 populations led Ballard (1986) to a reassessment of the taxonomy of the group. He recognized *B. odorata* with varieties *calcicola, chilpancingensis, oaxacensis*, and *odorata*, *B. alba* with varieties *alba* and *radiata*, and *B. pilosa*. A further distinction between northern and southern races of var. *oaxacensis* was made. The overall flavonoid chemistry of the group was based on the anthochlors maritimetin and okanin, which existed in a complex array of glycosides and methyl ethers, along with quercetin glycosides and a luteolin glycoside. Using Ballard's scheme of varieties, one sees some interesting differences. *Bidens alba* var. *alba* has a pair each of okanin and maritimetin monoglycosides and an acylated derivative of each anthochlor aglycone; var. *radiata* lacks all of these. Both varieties have okanin 3-methyl ether glycosides; var. *alba* has two while var. *radiata* has the same two plus a third. Variety *radiata* can be distinguished on the basis of its possession of two okanin 3,4-dimethyl ether 4'-*O*-glucosides; var. *alba* lacks these. The two varieties share one quercetin glycoside, differ in another two, and can be further distinguished by the unique presence of luteolin 7-*O*-glucoside in var. *radiata*.

Differences among the four varieties of *B. odorata* are equally striking. The southern race of var. *oaxacensis* is the only member of this group whose profile lacks maritimetin derivatives. Varieties *odorata* and *calcicola* share only four compounds, okanin 3'-*O*-glucoside, maritimetin 6-*O*-glucoside, and two quercetin 3-*O*-glycosides. Variety *calcicola* is distinguished by having a complex array of monoglycosides of both anthochlors, while var. *odorata* stands out in its accumulation of a series of okanin 3-methyl ether 3'- and 4'-*O*-glycosides. Variety *chilpancingensis* has the simplest profile of the four varieties with one monoglycoside of each of the anthochlors, a pair of okanin 3,4-dimethyl ether glycosides unique to the species, and a pair of quercetin 3-*O*-glycosides seen in all members of the species.

Based on overriding morphological considerations, Ballard (1986) maintained all four taxa within *B. odorata* despite geographical separation and the existence of reproductive barriers in Nature. In most instances specimens can be assigned to the appropriate variety on the grounds of morphology. Intermediate forms could only be determined with certainty by taking into consideration the source of the material and by determining the flavonoid profile.

Bidens in Hawai'i

The genus is represented in Hawai'i by 19 species and eight subspecies (Ganders et al., 1990). One of the features of the flavonoid occurrence pattern in the island species is that there may well be none. It was reported in the work just cited that the flavonoid profiles of two plants in the same population, in fact just a meter or so apart, were more different than the profiles of some species pairs. It is the subspecies picture that interests us here, and it should be noted at the outset that infrasubspecific differences are in some cases equally marked. Five species are involved which, in aggregate, show much the same sort of flavonoid chemistry as one would expect to see in this genus, flavonol glycosides as major compounds, flavones as lesser contributors, and anthochlors present in a variety of forms.

Several of the subspecific distinctions in Hawaiian *Bidens* involve populations limited to different islands, but populations on the same island can be almost totally isolated from one another owing to the rugged landscape and diverse habitats that characterize the islands. In the first example here we will examine *B. campylotheca* which consists of three subspecies, subsp.

campylotheca on Oahu and subspp. *pentamera* and *waihiensis* from Maui. The only compounds shared by all three subspecies are a suite of kaempferol and quercetin 3-*O*-glycosides. Subspecies *campylotheca* and *pentamera* have eriodictyol 7-*O*-glucoside in common, while subspp. *pentamera* and *waihoiensis* have three maritimetin derivatives in common as their sole anthochlor derivatives. Subspecies *campylotheca* afforded maritimetin as the aglycone, but none of the glycosides, and was further distinguished from the other two taxa by possessing a sulfuretin glycoside, and okanin and butein (both as aglycones).

The two subspecies of *B. forbesii*, subsp. *forbesii* and subsp. *kahiliensis*, both of which occur on Kauai, have fairly similar flavonoid profiles. Subspecies *forbesii* has the full complement of flavonol glycosides whereas subsp. *kahiliensis* lacks the kaempferol derivatives. *Bidens forbesii* subsp. *forbesii* also accumulates luteolin 7-*O*-glucuronide, eriodictyol 7-*O*-glucoside, and sulfuretin, none of which is seen in the other taxon.

Bidens menziesii subsp. *menziesii* is endemic to Maui, whereas subsp. *filiformis* appears to occur solely on the Island of Hawai'i. The taxon on Maui has the full complement of flavonol glycosides in contrast to the subspecies from Hawai'i where only the kaempferol diglycoside was seen. Subspecies *filiformis* is distinguished by virtue of its accumulation of eriodictyol 7-*O*-glucoside. The array of maritimetin glycosides in the two subspecies is very similar; subsp. *menziesii* has four while subsp. *filiformis* has three. Both accumulate okanin methyl ether(s).

Bidens micrantha is seen as comprising three subspecies, two of them, *micrantha* and *kalealaha*, on Maui and the other, *ctenophylla*, endemic to the Island of Hawai'i. *Micrantha* and *ctenophylla* have fairly similar flavonoid profiles differing mainly in the number of glycosides of any given anthochlor. Both also exhibited the eriodictyol derivative. Subspecies *kalealaha* differs significantly from the other two in that it lacks kaempferol glycosides, the eriodictyol compound, and the entire array of okanin and butein derivatives accumulated by the others.

The last set involves *B. sandvicensis* with subsp. *sandvicensis*, which is limited to Oahu and subsp. *confusa* known only from Kauai. Subspecies *confusa* is the only taxon in the group of island species to accumulate the flavone derivative diosmetin (luteolin 4'-methyl ether). It is further distinguished by lacking flavonols and any of the okanin derivatives exhibited by subsp. *sandvicensis*. It also has only a single maritimetin glucoside as opposed to four such glycosides in *sandvicensis*.

The evolutionary significance of these flavonoid differences, extreme ones in some cases, is unclear. Since most taxa of *Bidens* do not come into reproductive contact on the Hawaiian Islands, chance may play a major role in determining what any individual population, or individuals therein, will have in the way of a flavonoid profile. As concluded by the authors of the study (Ganders et al., 1990), it is clear that differentiation of flavonoids has not been involved in adaptive radiation and speciation in the islands. As in the case of a number of other systems in the Hawaiian flora, it appears that adaptive radiation has essentially involved only genetic differentiation for morphological and ecophysiological features. Perhaps, in systems such as we see in the Hawaiian Islands, the presence of some secondary metabolites is sufficient to perform whatever functions they serve, and that fine-tuning for optimum performance has not yet been accomplished. This may be the case with acetylenes (Marchant et al., 1984) and may now be argued for flavonoids as well. For further discussion on this point see Bohm (1998a).

Bidens section *Greenmanii*

According to Roseman (1986), *Bidens* section *Greenmanii*, which occurs throughout Mexico, Central America, and Jamaica, is a polyploid complex consisting of two diploids ($n = 6$), a tetraploid, two hexaploids, two octaploids and one species from Jamaica whose chromosome number was not known at the time of the preparation of the cited work. Relevant to the subject of this chapter are the several varieties into which Roseman (1986) attempted to partition the very considerable variation, cytological, morphological, and chemical, observed in the section. Varietal differences were not recognized for the two diploid species, *B. mexicana* and *B. reptans*. The flavonoid variation observed in *B. mexicana*, however, will be discussed in Chapter 24. It should be noted that it is not always easy, when reading Roseman's (1986) work, to keep straight differences between chemical races and the various clusters that resulted from multivariate analysis of the data. We will not comment on the clustering results here and only focus on comments in the text of that work that deal specifically with chemical races. In lieu of the publication of these studies, interested individuals should consult the dissertation itself.

The tetraploid *B. antiguensis* is seen as comprising three varieties, *antiguensis*, *savadorensis*, and *procumbens*. Variety *antiguensis* occurs at elevations in Guatemala where populations exhibited profiles comprising glycosides of naringenin, apigenin, luteolin, kaempferol, quercetin, quercetagetin, and common chalcones acylated or glycosylated at 4'-OH. *O*-Methylated chalcones were lacking. At higher elevations in the departments of Huehuetenango, Sololá, and Quiche, populations were found that exhibited a distinct profile with an okanin 4'-*O*-diglycoside but lacking kaempferol and quercetin glycosides. One of these populations also had a lanceoletin 4'-*O*-glycoside. Variety *salvadorensis* occurs on the Pacific coast in southern Guatemala and into El Salvador, Honduras, and Nicaragua. The flavonoid profile of this variety is much akin to that of var. *antiguensis.*

The third variety, var. *procumbens*, occurs on the Atlantic Ocean side of Guatemala and exhibits a unique profile that includes luteolin and diosmetin 7-*O*-diglycosides and a variety-specific luteolin 7-*O*-triglycoside. The profile lacks naringenin, kaempferol, quercetin, and quercetagetin glycosides.

Hexaploid members of section *Greenmanii* number only two, but they appear to be the most variable. The reader will find some information on *B. isabelensis* in Chapter 25 where a hybrid system involving this taxon is discussed. Here we will look at the other hexaploid, *B. squarrosa*, which is seen by Roseman as consisting of no less than seven varieties. Flavonoid data are available for all but one, var. *indivisa. Bidens squarrosa* var. *teraticaulis* occurs at low elevations along the eastern Gulf coast of Mexico and southeast to Yucatan and Belize. All populations of var. *teraticaulis* gave okanin 3-methyl ether glycosides, lanceolin glycosides, and quercetin 3-methyl glycosides. Three chemical races were then identified: the "Tamazunchale" race, which adds riparin 4-methyl ether and tomentin glycosides to the base profile; the "Chiltepec" race, which has quercetagetin and flavanone glycosides along with the base profile; and the "Yucatan" race, which consists essentially of the base array of compounds. The first two named chemical races correlate well with Roseman's "northern" morphological type.

Variety *atrostriata* can be found at lower elevations of the Sierra Madre in Michoacán and in Oaxaca. This variety is characterized by larger capitula than one finds in other varieties, except for var. *speciosa* whose ray florets are less numerous than in *atrostriata.* Chemically, varieties *atrostriata* and *chiapensis* are similar in flavonoid chemistry in having simple profiles based on lanceolin glycosides but lack luteolin and other *O*-methylated flavonoids. There are differences between the two, however. Variety *chiapensis* accumulates quercetagetin, riparin, and butein diglycosides, compounds not seen in var. *atrostriata.*

Variety *speciosa* occurs at higher elevations in Guatemala, on volcanic or other mountainous sites in the Sierra Madre de Chiapas, in the state of Vera Cruz, and disjunct in the state of Mexico. The base profile of this variety is dominated by riparin derivatives, both acylated derivatives and glycosides, and naringenin glycosides. Two morphologically identical chemical races were observed, one in Vera Cruz that accumulated *O*-methylated isoliquiritigenin and okanin derivatives, and the other in Guatemala that produced kaempferol and quercetin glycosides but lacked the *O*-methylated chalcones.

Bidens squarrosa var. *hondurensis* has its center of distribution in Honduras, but can also be found in El Salvador and in a few sites through Nicaragua and into northwestern Costa Rica. This variety has a flavonoid profile that consists of riparin derivatives, okanin 3-methyl ether glycosides, and isoliquiritigenin glycosides. Chemical races within this taxon were not observed.

The final entry in this group involves var. *squarrosa* which occurs in Costa Rica, Panama, and northern South America (Colombia and Venezuela). The flavonoid profile of this variety is unique among the hexaploids owing to the presence of butein 4'-*O*-mono- and diglycosides. Isoliquiritigenin 4-*O*-glycosides have also been detected as has lanceolin. No other *O*-methylated compounds were seen. No races *per se* were defined, but interpopulational differences involving luteolin and eriodictyol derivatives were noted.

Blennosperma nanum

Blennosperma nanum is a Californian species that consists of two varieties, var. *nanum*, the more widespread of the two, and var. *robustum*, known only from one area. Samples of var. *nanum* were obtained from Sutter, Sonoma, and Contra Costa Counties and a single specimen of var. *robustum* from Marin County were studied for their flavonoids by Ornduff et al. (1973b). These workers observed quercetin 3-*O*-galactoside and an isomeric quercetin 3-*O*-glucoside plus an unidentified quercetin 3-*O*-glycoside from all three samples of var. *nanum*. The specimen of var. *robustum* lacked the unidentified glycoside but exhibited a significant amount of luteolin 7-*O*-glucoside. The occurrence of luteolin 7-*O*-glucoside in some collections of the related genus *Crocidium* was commented upon in Chapter 21.

Carduus nutans

Carduus is an Old World genus where it occurs over a very wide range. It has been introduced to other parts of the world where it can become weedy. A biosystematic study of the *Carduus nutans* complex in Canada was undertaken in an effort to help sort out taxonomic difficulties that exist in the large-flowered thistle group (Bain and Desrochers, 1988; Desrochers et al., 1988). The *C. nutans* group in North America has been treated as either one species with four subspecies: subspp. *leiophyllus*, *macrocephalus*, *macrolepis*, and *nutans*, or as three species: *C. nutans*, with subsp. *nutans* and subsp. *macrolepis*, *C. thoermeri*, and *C. macrocephalus*. These workers' analysis of morphological, allozyme, and flavonoid characters led them to conclude that the patterns of variation observed should best be accounted for by recognizing two subspecies, subsp. *nutans* and *leiophyllus*.

With regard to flavonoids, these workers observed a uniform profile for subsp. *leiophyllus* that consisted of an apigenin 7-*O*-glycoside, luteolin, and three luteolin derivatives, 7-*O*-glucoside, 7-*O*-galactoside, and a 3-*O*-diglucoside. This profile homogeneity is especially interesting insofar as this taxon has a widely disjunct range. The western component includes populations in British Columbia, Alberta, Saskatchewan, and Manitoba, while the eastern component includes populations in southern Ontario and Quebec.

Subspecies *nutans* occurs in southern Ontario, Quebec, the Maritime Provinces, and as far east as the eastern coast of Newfoundland. Two populations each from Ontario and New Brunswick had invariant flavonoid profiles that consisted of luteolin and its 7-*O*-glucoside, 7-*O*-galactoside, and 7-*O*-rutinoside. Three populations from Ontario had luteolin, its 7-*O*-glucoside and 7-*O*-galactoside but were variable with regard to apigenin 7-*O*-glycoside, luteolin 7-*O*-rutinoside, and the luteolin 7-*O*-diglucoside. The ranges of the two subspecies overlap in Ontario and Quebec where hybridization is known to occur. Some of the individuals from these populations in the overlap zone exhibited a flavonoid profile that consisted of compounds normally found in one or the other subspecies, i.e., the apigenin 7-*O*-glycoside and luteolin 7-*O*-diglucoside from subsp. *leiophyllus* along with luteolin 7-*O*-rutinoside from subsp. *nutans*.

In addition to the Canadian specimens, these workers studied herbarium specimens from North Dakota and Montana, which represent an area adjacent to the Canadian Prairie Provinces. They found that plants from this area fit well within the circumscription of subsp. *leiophyllus*. As a final note, it should be pointed out that the genetic identities calculated from the allozyme data (I = 0.93) showed a level of similarity between subspecies that lies within the range expected for conspecific populations and/or subspecific taxa for outcrossing plants (Hamrick et al., 1979).

Cassinia fulvida

The New Zealand endemic *Cassinia fulvida* consists of two varieties, var. *fulvida* and var. *montana*. An examination of their exudate and vascular flavonoid profiles (Reid and Bohm, 1994) disclosed several differences. With regard to exudate compounds, they share 2',6'-dihydroxy-4'-methoxychalcone, the corresponding dihydrochalcone, and 5-hydroxy-7-methoxyflavanone. Variety *montana* is distinguished by having 2',4',6'-

trihydroxychalcone and eriodictyol. The two varieties share kaempferol and quercetin 3-*O*-glucosides and a quercetin 3-*O*-rhamnosylglucoside. Variety *fulvida* also has a kaempferol 3-*O*-rhamnosylglucoside while var. *montana* has eriodictyol 7-methyl ether.

Centaurea aspera

Centaurea aspera var. *stenophylla* accumulates apigenin, apigenin 7-*O*-glucuronate (as the ethyl ester), and 6-methoxyluteolin (Picher et al., 1984). *Centaurea aspera* var. *subinermis* also has 6-methoxyluteolin, but has 6-methoxyapigenin, 5,7-dihydroxy-6,4'-dimethoxyflavone (pectolinaringenin), and 5,7,4'-trihydroxy-6,3'-dimethoxyflavone as well (Cardona et al., 1991).

Cirsium japonicum

Markham (1983) published the corrected structures for two flavones from *C. japonicum* var. *takaoense* as 5,4'-dihydroxy-6,7-dimethoxyflavone (cirsimaritin) and its 4'-*O*-glucoside (cirsimarin). A more recent report described the identification of acacetin 7-*O*-rutinoside, 6-methoxyapigenin 7-*O*-neohesperidoside, and cirsimarin from *C. japonicum* var. *ussuriense* (Park et al., 1995).

Dahlia

The overall distribution of flavonoids in *Dahlia*, discussed in the preceding chapter, correlated well with sectional taxonomy of the genus (Giannasi, 1975a,b). An examination of *Dahlia australis*, thought to consist of four varieties according to Sorensen (1969), showed the existence of a very complicated array of flavonoid races (Giannasi, 1975b). An introduction to the varieties is in order. Variety *serratior*, $n = 16$, occurs in southwestern Guatemala, whereas var. *chiapensis* is found in central Chiapas, Mexico, some 130 miles away. Variety *australis* occurs further west in Oaxaca, Mexico and consists of a series of diploid ($n = 16$) and tetraploid ($n = 32$) populations. Variety *liebmannii* is known only from the type location and was not available for study.

Extensive collections of var. *australis* (Sorensen, 1969) had shown that despite differences in chromosome numbers no distinctive morphological features were evident. The immediate questions raised by Giannasi were whether flavonoid profiles would be equally homogeneous, and if flavonoid

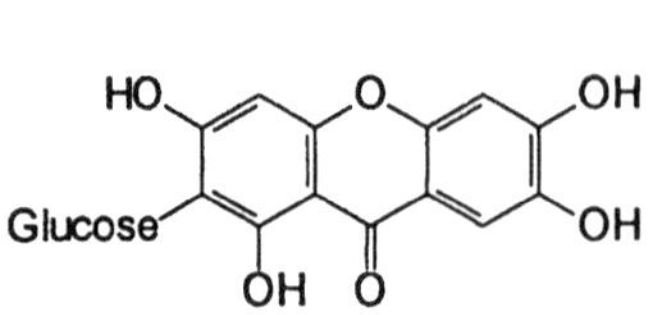

[23-1]

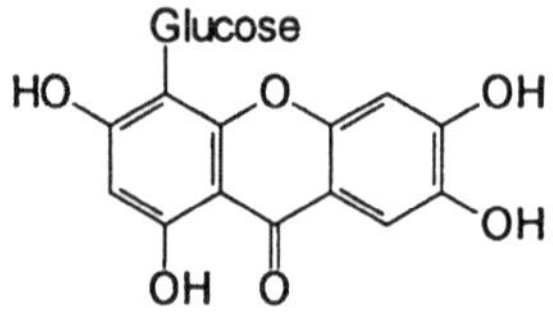

[23-2]

Fig. 23–1. *C*-Glucosylxanthones in *Dahlia australis* var. *australis*, mangiferin **[23–1]** and isomangiferin **[23–2]**

profiles would mirror the morphological, ecological, and geographic features by which the varieties were recognized. It was not to be.

Nine different flavonoid arrays were observed, two involving diploid var. *australis* and one diploid var. *serratior*, four involving tetraploid var. *australis*, and one each for tetraploid var. *serratior*, tetraploid var. *chiapensis*, and a population of *serratior* whose chromosome number was not determined. Some of the chemical differences reported in this group noted by Giannasi (1975b) include diploid populations of var. *australis* that occur north and south of Ciudad Oaxaca and populations of diploid var. *serratior* from Guatemala. Although superficially similar in that they share a suite of flavonols and both are anthochlor-positive, the two sets of populations are different. The single anthochlor in var. *australis* is a butein glycoside not seen in var. *serratior*. Variety *serratior* has two anthochlors and a flavone not seen in the diploid var. *australis* populations. Populations of the third var. *australis* have the most complex profiles seen in the diploids with quercetin 3,7-di-*O*-glycosides, several butein tri- and tetra-*O*-glycosides, and a *C*-glycosylflavone based on apigenin. Distinct diploid chemical races have thus evolved within each of the geographic areas.

One of the tetraploid races of var. *australis*, from the vicinity of Ciudad Oaxaca, has a flavonoid profile strongly reminiscent of that from the complex diploid var. *australis* just commented on. It is possible that these two cytotypes are directly related. The other tetraploids observed have profiles that do not point to any diploid population. Most of the rest of the tetraploids are quite similar including the presence of two *C*-glucosylxanthones (Fig. 23–1), mangiferin **[23–1]** and isomangiferin **[23–2]**, which at the time of that writing, at least, represented first reports from the family.

Encelia farinosa

Encelia farinosa is one of the dominant species of the Sonoran Desert, with a range that extends from near the cape of Baja California northward to the deserts of California, southern Nevada, southern Utah, Arizona, and in the states of Sonora and Sinaloa in Mexico. Three varieties are recognized (Kyhos, 1971): var. *radians* from the southernmost populations and vars. *phenicodonta* and *farinosa*, occurring further northward. It is the latter two that concern us here. These two varieties are distinguished by color of their disk florets; vars. *radians* and *phenicodonta* have brownish-purple disk florets whereas var. *farinosa* is characterized by bright yellow florets. Although we are unaware of chemical studies of the pigments involved, it seems reasonable to assume that the brownish-purple color results from anthocyanins laid down on a background of carotenoids. The possibility that anthochlors might be involved can be rejected based on the observation by Bohm and Ornduff (cited by Crawford and Stuessy, 1981) that *Encelia* does not accumulate anthochlors.

In the northern part of the species' range, var. *farinosa* becomes the predominant color form although mixed populations are known. It is the unusual distribution pattern that attracted attention (Kyhos, 1971). Populations consisting predominantly or exclusively of var. *phenicodonta* are restricted to major river valleys. Localized populations of var. *phenicodonta* also occur at higher elevations in the mountains of California and Arizona. The relationship between the occurrence of this variety and availability of water is seen in the limitation of this species to coastal habitats.

The question naturally arose as to the relationship between flower color and pollinator(s), on the one hand, and habitat selection on the other. It was first determined that pollen was shed by both varieties synchronously. Although visited by a number of insects, the most important pollinator of *E. farinosa* was the beetle *Tanaops abdominalis*. The key observation, however, was that the beetle visited flower heads indiscriminately. What, then, is the biological significance of the flower color polymorphism? By itself, flower color does not appear to have selective value. One alternative explanation put forth by Kyhos (1971) involves flower color expression as part of a pleiotropic expression of some other less obvious feature. A third suggestion was that the gene(s) controlling pigmentation (anthocyanin production?) in this plant is/are tightly linked with genes responsible for some other feature of adaptive importance, possibly temperature regulation.

Enceliopsis nudicaulis

Proksch et al. (1987) examined aerial and underground parts of two varieties of *E. nudicaulis*, var. *nudicaulis* and var. *corrugata*. Leaves of var. *nudicaulis* afforded gossypetin 3,8,3'-trimethyl ether, herbacetin 8-mono-, and 8,3'-dimethyl ethers, and three 8-methoxydihydroflavonols newly reported as natural products at that time. Roots of this variety apparently yielded only benzofuran derivatives. Leaves of var. *corrugata* gave the same flavonoids as var. *nudicaulis* and added gossypetin-3,8-dimethyl ether to the list. Roots of var. *corrugata* yielded the same array of benzofurans as seen in var. *nudicaulis* but had two of the dihydroflavonols as well.

Flaveria bidentis and *F. bidentis* var. *angustifolia*

A good deal of flavonoid information has been gathered for members of this genus no doubt owing to their unusual capacity to accumulate highly sulphated flavonols (Pereyra de Santiago and Juliani, 1972; Cabrera and Juliani, 1976, 1977, 1979; Suarez et al., 1979; Varin et al., 1986a,b; Barron et al., 1986, 1988; Hannoufa et al., 1994). From this accumulation of data it was evident that *F. bidentis* var. *bidentis* and var. *angustifolia* could be distinguished on the basis of the level of sulphation and the nature of the flavonols involved. Variety *bidentis* accumulates quercetin 3,7,3',4'-tetrasulphate, quercetin 3-O-acetyl-7,3',4'-trisulphate, both as major components, and quercetin 3,7,3'-trisulphate. The flavonod profile of var. *angustifolia* consists of quercetin 3- mono-, 3,4'-di-, and 3,7,4'-trisulphate as lesser components with isorhamnetin 3-mono- and 3,7-disulphate as the major components. The existence of morphological differences between the two taxa led to the suggestion by Dimitri and Orfila (1985) that var. *angustifolia* could be justifiably given specific status under the name *F. haumanii* Dim. & Orf. Flavonoid profile differences support this distinction. More recently, Agnese et al. (1999) have observed differences in the thiophene chemistry of the two taxa as well. These workers reported that var. *bidentis* contained α-terthienyl and 5-(3-buten-1-yl)-2,2'-bithienyl in both roots and aerial parts whereas var. *angustifolia* had only α-terthienyl in the roots. The combination of thiophene chemistry, flavonoid chemistry, and morphology was argued as providing further support for recognition of *F. haumanii*.

Hazardia squarrosa and *H. whitneyi*

Clark and Mabry (1978) undertook a study of flavonoids of all members of the genus *Hazardia* as part of their ongoing chemosystematic work with members of Astereae. Included in the study were two varieties of each species. The two varieties of *H. squarrosa* differ only in the nature of their *C*-glycosylflavone profile, that of var. *grindelioides* being the more complex. They are otherwise identical in their elaboration of a complex profile consisting of flavones, flavonols, and a number of *O*-methylated flavonols. *Hazardia whitneyi*, the sole member of section *Machaerantheroides*, differs from other members of the genus ecologically, geographically, and in chromosome number ($n = 4$ as opposed to $n = 5$). The two varieties, var. *whitneyi* and var. *discoidea*, have identical flavonoid profiles comprised solely of flavonol glycosides and flavonol methyl ethers.

Helianthus annuus

Helianthus annuus, the common sunflower, occurs widely in North America. Three subspecies are recognized, *annuus, lenticularis*, and *texanus*. A detailed study of the flavonoid aglycone chemistry of 265 individuals representing the three taxa, and including several individuals that were intermediate between subsp. *annuus* and subsp. *lenticularis*, was described by Rieseberg et al. (1987). Details of their study appear in Chapter 24, but it is relevant to mention here that the distribution patterns of the glandular and waxy cuticular flavonoids showed no correlation with subspecific taxonomy.

Hemizonia increscens

The two subspecies of *Hemizonia increscens* studied by Tanowitz et al. (1987), *increscens* and *villosa*, (apparently subsp. *foliosa* was not studied) showed quite dissimilar flavonoid profiles, differences that were of equal magnitude to those between other species pairs. The subspecies had eriodictyol, eriodictyol 7-methyl ether, quercetin, quercetin 3-methyl ether, kaempferol 4'-methyl ether, and luteolin in common. The major difference between the two taxa lies in the presence of 8-oxygenated flavonoids in subsp. *villosa*: 5,7,8,3',4'-pentahydroxy-6-methoxyflavone, 5,7,3',4'-tetrahydroxy-3,6,8-trimethoxyflavone, and 5,7,3',4'-tetrahydroxy-3,8-dimethoxyflavone. Subspecies *increscens* exhibited 5,3',4'-trihydroxy-3,6,7-trimethoxyflavone. The significance of the restriction of the 8-substituted compounds to subsp. *villosa* was not specifically addressed in the paper.

Hulsea vestita

Hulsea vestita consists of five subspecies, four restricted to California (*callicarpha, parryi, pygmaea,* and *vestita*), and one that occurs in both California and Nevada (*inyoensis*). All five were examined for flavonoids in Wilken's (1975) systematic study of the genus. The only qualitative difference observed was the absence of the cyanidin 3-*O*-glycosides from subsp. *inyoensis*. The cyanidin glycosides were major components of subspecies *parryi* and *pygmaea* but were present only as lesser components of subspecies *vestita* and *callicarpha*.

Lasthenia

Subspecies are recognized for three species of *Lasthenia* (Ornduff, 1966, 1971) all of which have been examined for flavonoids (Bohm et al., 1974). Variation in *L. macrantha* has been accommodated by recognizing three subspecies, *macrantha, bakeri,* and *prisca* (Ornduff, 1966, 1971). These taxa show a good deal of both chromosomal and morphological variation and in some parts of their ranges show a higher degree of similarity than in other parts. Ornduff (1966) suggested that this may be the result of intersubspecific hybridization. Flavonoid profiles of subspp. *macrantha* and *prisca* are very much alike with anthochlors and luteolin glycosides as the contributing types. They differ in the presence of the aurone sulfuretin 6-*O*-glucoside (sulfurein) in subsp. *macrantha*. The profile of subsp. *bakeri* consists of almost the same array of anthochlors, lacks luteolin 7-*O*-glucuronide, but adds three quercetin 3-*O*-glycosides and a kaempferol 3-*O*-glycoside. This intersubspecies difference is of the same magnitude as differences between species or even between sections within *Lasthenia*. The usual caveat is offered: a larger sampling of these taxa is in order.

The second example from this genus involves *L. glabrata* subspp. *glabrata* and *coulteri*. These two taxa, which differ in the nature of the achenes and in range, subsp. *glabrata* occurring in west-central California, subsp. *coulteri* in southern California, exhibited identical flavonoid profiles. Both appear to make only quercetin 3-*O*-glycosides.

The third example from *Lasthenia* involves taxa that were listed in Ornduff's (1966) monograph as *L. minor* subsp. *minor* and *L. minor* subsp. *maritima*. These taxa differ significantly in range. Subspecies *minor* occurs widely through central California, while subsp. *maritima* is restricted to offshore rocks and islands from central California to the northwestern tip of Vancouver Island that are frequented by seabirds. Subspecies *maritima* occurs in the most extreme habitat of any member of the genus. The flavonoid profiles of these two entities, although very simple, were shown to be different. Subspecies *maritima* afforded only quercetin 3-*O*-galactoside and 3-*O*-glucuronide while the profile of subsp. *minor* had these two compounds as well as a 3-*O*-galactosylglucoside. A detailed study of the maritime race was undertaken by Vasey (1985) who pointed out that much of the confusion surrounding the proper ranking of this taxon, which has involved consideration as a species, lay with its poor representation in herbaria. This situation has been rectified through an exhaustive re-examination of this taxon in the field resulting in the documentation of sufficient morphological differences to support recognition at specific rank (Vasey, 1985). The same year also saw publication of an allozyme study of the two taxa (Crawford et al., 1985). The level of genetic variation was similar to that expected for populations of the same species. Six unique alleles were detected in *L. minor*, only one in *L. maritima*, which suggests that the latter has diverged from the former only comparatively recently.

Leucocyclus

Leucocyclus, a monotypic Turkish genus, consists of *L. formosus* with two subspecies, *L. formosus* subsp. *formosus*, which occurs in the Taurus region, and subsp. *amanicus*, which is endemic in the Amanus region. Valant-Vetschera (1982) determined the flavonoid profiles of both taxa. The overall pigment profile of the genus consists of *C*-glycosylflavones, flavonol 3-*O*-glycosides, and trace amounts of a flavone 7-*O*-glycoside. A breakdown of the distribution of compounds by subspecies, however, shows one of the more clearly defined situations encountered in our survey of the family. To a first approximation subsp. *formosus* can be characterized by its accumulation of a rich array of *C*-glycosylflavones (seven or eight compounds), trace amounts of a luteolin 7-*O*-rhamnoglucoside, and virtual absence of flavonol 3-*O*-glycosides (one compound in one collection). Subspecies *amanicus*, on the other hand, can be characterized by a single *C*-glycosylflavone, two flavonol 3-O-glycosides, and no flavone *O*-glycosides. In more detail, subsp. *formosus* afforded the following *C*-glycosylflavones: isovitexin, isoschaftoside, shaftoside + vicenin-3 (nixture), isoörientin, isoörientin 2"-*O*-arabinoside, isoörientin caffeoyl ester, and lucenin. Subspecies *amanicus* had only isoschaftoside. The flav-nols involved are quercetin 3-*O*-rhamnosylglucoside, which was present in one collection of subsp. *formosus*, but all collections of subsp. *amanicus*, and kaempferol 3-*O*-rhamnosylglucoside, which was seen only in subsp. *amanicus*. Valant-Vetschera (1982) concluded that the distinctly different flavonoid profiles, coupled with achene differences, and different pigmentation patterns in the disc florets, argue for recognition of the two subspecies at the level of species. *Leucocyclus formosus* subsp. *amanicus* has been described as a species of *Achillea* (*A. amanica* Rech.) belonging to section *Arthrolepis*. By comparison, the flavonoid profile of subsp. *formosus* bears more resemblance to species of *Achillea* that lie in section *Santolinodea* (Valant-Vetschera, 1982).

Lipochaeta connata

Gardner's (1976) study of the Hawaiian endemic genus *Lipochaeta* included *L. connata* vars. *connata* and *acris*, both members of the tetraploid group of taxa in this genus. The two varieties share only three compounds, two luteolin glycosides and one quercetin glycoside. Variety *acris* accumulates two glycosides of luteolin (a third is variable) and one of apigenin that were not seen in var. *connata*. Similarly, var. *connata* has a glycoside each of apigenin and quercetin not seen in var. *acris*. Differences between the diploid and tetraploid lines within *Lipochaeta* were discussed in Chapter 11.

Parthenium

Mears' (1980a,b) extensive study of *Parthenium*, which was discussed in more general terms in Chapter 21, included analysis of the varieties that had been described for several species: *P. confertum, P. fruticosum, P. integrifolium*, and *P. tomentosum*. Mears described flavonoid profiles in terms of two structural classes, aglycones and glycosides. The aglycones were *O*-methyl derivatives of kaempferol, quercetin, 6-hydroxykaempferol, and quercetagetin. Most of the glycosides were derivatives of kaempferol, quercetin, and isorhamnetin, but a few members of the aglycone group also occurred in glycosylated forms. In some cases, the arrays of both flavonoid aglycones and glycosides were highly variable among varieties; in other cases somewhat less so. An extreme case involves the two varieties of *P. tomentosum* var. *tomentosum* and var. *stramonium*. According to Mears, these two entities have no flavonoids in common, although Rodriquez (1977) reported some minor compounds that might require this judgement to be altered. Mears (1980a,b) identified quercetagetin 3,6-dimethyl and 3,7-dimethyl ethers, and a quercetin 3-*O*-glycoside, plus an unidentified compound, from var. *tomentosum*, while var. *stramonium* afforded only kaempferol 7-methyl ether (variable or minor) and a kaempferol 3-*O*-glycoside.

Somewhat better profile agreement was reported between *P. fruticosum* var. *fruticosum* and var. *trilobatum*, although they share only two compounds, quercetagetin 3,7-dimethyl ether and a kaempferol 3-*O*-glycoside. Variety *trilobatum* accumulates kaempferol 7-methyl ether, quercetagetin 3,6-dimethyl ether, and a second kaempferol 3-*O*-glycoside as major compounds as well as four other *O*-methylated aglycones either as minor components or only in some individuals. It is interesting to note, as Mears indeed did, that the flavonoid profile of *P. fruticosum* var. *trilobatum* is much more similar to that of *P. lozanianum* than it is to the supposedly more closely related var. *fruticosum*. Mears did not expand on the possible significance of this similarity.

Five varieties of *P. integrifolium* appear in Mears' table, one of which is further represented by a form. The varieties are only weakly segregated according to Mears, but there are nonetheless some apparently significant differences between them, although, as mentioned, the differences are not as great as those seen in the two species discussed above. The five varieties share kaempferol 3-methyl ether and two kaempferol 3-*O*-glycosides as major components. Varieties *integrifolium* and *henryanum* are distinguished from the others by their possession of 6-hydroxykaempferol 3,7-dimethyl ether. Most of the other reports involve minor (or variable) compounds. Variety *hispidum* forma *repens* is unique in the species in its lack of kaempferol 3-methyl ether as well as some of the glycosides.

In contrast to the modest degree of flavonoid difference among the morphologically similar varieties of *P. integrifolium*, varieties of *P. confertum*, which appear to be edaphic specialists, exhibit the highest degree of flavonoid homogeneity within the genus. The four varieties share quercetagetin 3,7-dimethyl ether, a quercetin 3-*O*-glycoside, and two kaempferol 3-*O*-glycosides (variable in var. *microcephalum*). Varieties *confertum* and *lyratum* have quercetagetin-3,7,3'-trimethyl ether as major flavonoids; var. *microcephalum* has it as a variable or minor component and var. *divaricatum* lacks it all together. Variety *divaricatum*, on the other hand, has 6-hydroxykaempferol 3,7-dimethyl ether and kaempferol 3-methyl ether as major pigments.

Pyrrhopappus multicaulis

Harborne (1977b) reported apigenin and luteolin 7-*O*-glucosides in leaves and butein 4'-*O*-glucoside in ligules of all four tested species of *Pyrrhopappus*. One of these, *P. multicaulis*, was represented in the study by two varieties, *multicaulis* and *geiseri*. They gave identical results.

Senecio streptanthifolius

In a recent treatment of North American *Senecio*, Barkley (1978) pointed out that *S. streptanthifolius* consists of several "phases" and that it intergrades morphologically with nearly all of the other aureoid *Senecio* species with which it shares overlapping ranges. Earlier, he had described some of these phases as consisting of populations that had resulted from introgression involving *S. streptanthifolius* and related species (Barkley, 1962). Bain and Denford (1985) studied the flavonoids of this array of phases: *borealis* (2*x*), *borealis* (4*x*), *laetiflorus, leonardi, rubricaulis, streptanthifolius*, and *suksdorfii*. The flavonoid profile of the entire group is based on glycosides of apigenin, luteolin, kaempferol, quercetin, and isorhamnetin. The flavones occur as 7-*O*-glycosides while flavonol glycosides offer a note of variety, with 3-*O*-glycosides of kaempferol, quercetin, and isorhamnetin, 7-*O*-glycosides of isorhamnetin, and 3,7-di-*O*-glycosides of kaempferol and quercetin. The authors (Bain and Denford, 1985) chose to present their data, some 35 glycosides in all, in terms of classes rather than in terms of the array of individual compounds. This serves nicely to focus the reader's attention on unique biosynthetic properties of the taxa in question.

All seven phases exhibited quercetin 3-*O*-glycosides while only *borealis* (4*x*), *leonardi, rubricaulis*, and *streptanthifolius* had kaempferol 3-*O*-glycosides as well. Isorhamnetin 3-*O*-glycosides were restricted to phases *laetiflorus*, and *suksdorfii*. Quercetin 3,7-di-*O*-glycosides were of variable occurrence in *borealis* (2*x*) (3 of 14 populations), but were present in the three populations of *borealis* (4*x*) available for study. Isorhamnetin 7-*O*-glycosides were found to be present in all phases except *suksdorfii* and *laeiflorus*. Luteolin (tentative identification) 7-*O*-glycosides were identified from four phases, *borealis* (2*x*), *laetiflorus, streptanthifolius*, and *suksdorfii*, while apigenin 7-*O*-glycosides were seen only in *laetiflorus*. Considering the comparative simplicity of the flavonoid package in this group of taxa, it is rewarding to see that a characteristic combination of compound types exists for each taxon except for *rubricaulis* and *leonardii*. When the entire array of glycosides is considered, even this pair is presumably separable.

Although evidence is required to establish the position played by hybridization in arriving at the present assemblage of phases/taxa, the flavonoid occurrence patterns are at least consistent with the idea that gene flow between members of these phases is likely to have occurred. Adding to the complexity of the picture, as pointed out by those workers, is the very real possibility that hybridization may have occurred at different times during the evolution of the group and that the importance of these events is very difficult to assess, at least with currently available data. The present authors would like to point to this study as an excellent example of the usefulness of even a comparatively simple flavonoid array in helping to understand relationships among a complex group of organisms.

In the same paper, Bain and Denford (1985) reported the flavonoids of four varieties of *Senecio neomexicanus* which grow in Arizona, Colorado, New Mexico, and Utah. Several populations of each vars. *neomexicanus*, *metcalfei*, *mutabilis*, and *toumeyi* were analyzed. Quercetin 3-*O*-glycosides were present in members of all populations of all varieties. Isorhamnetin 7-*O*-glycoside was the only other compound reported, and that was seen only in var. *metcalfei*.

Sonchus asper

In a study of the flavonoids of *Sonchus* subgen. *Sonchus*, Giner et al. (1993b) looked at the pigments of two subspecies of *S. asper*, *asper* and *glaucescens*. Both had luteolin 7-*O*-glucoside, apigenin, apigenin-7-*O*-glucoside, and quercetin 3-*O*-glucoside. Subspecies *asper* lacked luteolin, which may not have any significance. A more substantive difference would appear to be the presence of quercetin 3-*O*-galactoside in subsp. *asper* which distinguished it from subsp. *glaucescens* and three other species in the subgenus.

Stephanomeria exigua

Stephanomeria consists of about a dozen species widely distributed in western North America (Gottlieb, 1971, 1972). Attention was drawn to the genus through the work of Gottlieb on the origin of *S. malheurensis* (Gottlieb, 1977 and citations therein). As part of the study of this system, we undertook an examination of the flavonoids of the annual species of *Stephanomeria* the prime goal of which was to see if any differences exist between *S. malheurensis* and its parent species, *S. exigua* subsp. *coronaria* (see Chapter 22) (Bohm and Gottlieb, 1989). In the course of the work four subspecies of *S. exigua* were examined: *coronaria, deanei, exigua*, and *macrocarpa*. The flavonoid profile of *S. exigua* consisted of apigenin, luteolin, kaempferol, quercetin, and eriodictyol glycosides along with the coumarins esculetin and cichoriin. The four subspecies of *S. exigua* were identical except for quercetin 3-*O*-rutinoside which was seen only in *S. exigua* subsp. *macrocarpa*.

Vernonia gigantea and *V. baldwinii*

The flavonoids of both of these species were included as part of a larger survey of the genus by Mabry et al. (1975). Two subspecies of *V. gigantea* are recognized, *gigantea* and *ovalifolia* (Urbatsch, 1972). These two taxa have 10 flavonoids in common including flavone monomethyl ethers, flavone glycosides, and flavonol glycosides, but differ in the accumulation of apigenin 7-methyl ether and luteolin 7,3'-dimethyl ether by subsp. *gigantea*. Since subsp. *ovalifolia* occurs on geologically younger substrate (James, 1961), it was suggested that the flavonoid differences may have resulted from loss mutations associated with range extension and intergradation as the taxa moved into more recently derived sites (Urbatsch, 1972; Mabry et al., 1975).

The two subspecies of *V. baldwinii*, *baldwinii* and *interior*, are morphologically similar, but differ significantly in their flavonoid profiles. Although the two taxa share 11 compounds, subsp. *baldwinii* is distinguished by having kaempferol and quercetin 3-*O*-rhamnoglucosides, which are lacking in the profile of subsp. *interior*. A further difference between the two taxa lies in the presence of three luteolin 7-*O*-glycosides in subsp. *interior* but not in subsp. *baldwinii*. Two points of view were expressed by different authors of the work (Mabry et al., 1975). One position taken suggested that the flavonoid differences between the two taxa could be used to argue for recognition of each at the specific level. It was the view of Prof. S. B. Jones, Jr. that secondary intergradation had occurred and that the two taxa were likely connected by a band of intermediate populations. Resolution of this problem appears not to have been achieved.

Viguiera deltoidea

Viguiera deltoidea, considered by Blake (1918) to be the most variable species in the genus, is currently seen as comprising four varieties, *chenopodina*, *deltoidea*, *parishii*, and *tastensis*. Rieseberg and Schilling (1985) showed that the varieties differ both in their floral flavonoids and in the nature of the ultraviolet pattern in their ligules. All specimens tested of all four varieties had quercetin 3-methyl ether 7-*O*-glucoside and the coreopsin/sulfurein pair. Variety *chenopodina* yielded only those two compounds with the anthochlors located in the lower 3/4 of the ligules. Variety *deltoidea* exhibited the same ligule pattern but had, in addition to the common pigments, variable occurrence of quercetin 7-*O*-glucoside and quercetin 3-methyl ether 7-*O*-glucuronide. Variety *parishii* exhibited anthochlor absorbance in the lower 1/3 of its ligules and had quercetin 3-methyl ether and quercetin 7-*O*-glucoside as variable floral components. The final variety, *tastensis*, exhibited anthochlor pigmentation in the lower half of its ligules and had the most complex overall pigment profile with luteolin and quercetin 7-*O*-glucosides in all five samples along with variable occurrences of quercetin 3-methyl ether 7-*O*-glucuronide and 7-*O*-galactoside.

By contrast, both subspecies of *V. eriophora*, *eriophora* and *poblana*, possess identical flavonoid profiles involving both aglycones from leaves and glycosidic compounds present in floral tissue (Schilling et al., 1988). This observation was suggested as support for continued recognition of these entities at the infraspecific level.

Xanthisma texanum

Xanthisma consists of one species, *X. texanum*, which is distributed in Texas, Oklahoma, and extreme eastern New Mexico (Bremer, 1994). Three infraspecific taxa have been defined: subsp. *drummondii*, which occurs throughout most of the range, and subsp. *texanum* var. *texanum* and subsp. *texanum* var. *orientalis*. Variety *orientalis* occurs in extreme southern Texas from the Corpus Christi area south, while var. *texanum* occurs in the San Antonio area to the west and slightly north of the range of *orientalis*. The species flavonoid profile is made up of luteolin and quercetin glycosides, a flavanone, and an anthocyanin plus some unidentified compounds. There is considerable variation, not only in the compounds present in the various taxa, but in the organs of the plant with leaves, receptacles, involucral bracts, ray floret bracts, and ray floret fruits and disc floret corollas all exhibiting different two dimensional chromatograms.

Subspecies *drummondii* was composed of four different flavonoid types, with nine, eight, seven, and six spots, respectively. The simplest one of these was seen in two populations, one in Oklahoma and one in central Texas northeast of Austin. Subspecies *texanum* var. *texanum* was represented in the study by six populations, five of which were identical consisting of an assortment of spots identical to the Oklahoma population of subsp. *drummondii*. Subspecies *texanum* var. *orientalis* had the most consistently simple profiles seen in the study, several of which consisted of only four spots.

Despite the level of variation observed in this complex, the authors (Semple and Averett, 1975) think the data are useful in selecting between two alternative origins of var. *orientalis*, which is intermediate in bract size and shape to var. *texanum* and subsp. *drummondii*. Perhaps var. *orientalis* arose from a hybridization between members of the other two taxa. Experimental and putative hybrids exhibit profiles that are more complex than non-hybrid plants suggesting that some other explanation, perhaps drift and isolation, is more likely.

Zaluzania augusta and *Z. megacephala*

This primarily Mexican genus consists of nine species two of which are further subdivided. *Zaluzania megacephala* vars. *megacephala* and *coahuilensis* exhibit similar flavonoid profiles (Olsen and Mabry, 1980) with quercetin 3-*O*-mono- and 3-*O*-diglycosides and a patuletin 3-O-glycoside in common. Variety *megacephala* also has a patuletin 3-*O*-diglycoside plus a trace of quercetagetin 3,6-dimethyl ether. The two varieties of *Z. augusta*, by contrast, have no compounds in common. The profile of var. *augusta* consists of a quercetin 3-*O*-glycoside, a quercetin 3,7-di-*O*-glycoside, quercetin 3,7-dimethylether, and quercetagetin 3,6-dimethyl ether. The profile of var. *rzedowskii* consists of the same array of glycosides as seen in *Z. megacephala* var. *megacephala* but lacks aglycones. The extreme differences in flavonoid chemistries of the two varieties of *Z. augusta* are interesting in view of the close morphological similarity between them.

Flavonoids of some cultivars

Artemisia dracunculus – tarragon

To non-botanists *Artemisia dracunculus* is known as the flavoring agent tarragon. This widely used herb is often dispensed, culinarily speaking, as a flavoring in vinegar. Alternatively, it is often used as crushed leaves in various fish dishes. We are informed by Vienne et al. (1989) that at least two varieties of tarragon are known in Europe, "French tarragon," generally considered to be the "true" tarragon, and "Russian tarragon." French tarragon is sterile while the so-called Russian form is fully fertile and is a much more robust plant. French tarragon is preferred for its aromatic character and is the more valuable of the two. The cited workers briefly review features that distinguish the two cultivars, including work on their essential oils, and then go on to describe their own comparative flavonoid study. The flavonoid profile of French tarragon is the simpler of the two consisting of quercetin 3-*O*-glucoside, quercetin 3-*O*-rhamnoglucoside, and quercetin 3-*O*-rhamnogalactoside. Russian tarragon afforded the same three quercetin glycosides, as well as quercetin itself, along with patuletin 3-*O*-glucoside, 3-*O*-rhamnoglucoside,and 3-*O*-rhamnogalactoside. It would seem unlikely that this sort of flavonoid difference would make any contribution to the culinary usefulness of the one variety over the other, but it would certainly offer a means by which adulteration of "true tarragon" with material of lesser value could be determined.

Dahlia

Giannasi (1975a) reported that an examination of 20 *Dahlia* cultivars (presumably *D. variabilis*) revealed, as one would expect, differences in floral anthocyanins, but also that a depauperate array of flavonoids exists in both floral and leaf tissue, and that the compounds present exist as simpler glycosides than seen in species in Nature. Giannasi speculated that selection for floral color, i.e., anthocyanins, may have affected the accumulation of other flavonoid types. This may be a more common effect as evidenced by the reduction of flavonoids and other bitter components in vegetable or citrus crops.

Giannasi (1975a) briefly discussed anthocyanins present in the species thought to be the progenitors of modern dahlias, *D. coccinea* and *D. pinnata*. The interested reader is referred to Sorensen's (1970) treatment of the history of cultivated dahlias.

Gerbera jamesonii – African Daisies

Gerbera jamesonii, known in the horticultural trade as the African, Barberton, or Transvaal Daisy, is much prized for the size of its flower head and for the variety of colors available. A quantitative study of 18 cultivars, ranging in color from orange through lavender, showed that pelargonidin and cyanidin 3-*O*-malonylglucosides are the major pigments with the nonacylated glucosides playing a lesser role (Asen, 1984). Co-pigments identified in the same study were apigenin and luteolin 7-*O*- and 4'-*O*-glucosides, apigenin 7-*O*-malonylglucoside, kaempferol and quercetin 3-*O*- and 4'-*O*-glucosides, and their respective malonylated derivatives. A combination of qualitative and quantitative information allows for ready identification of cultivars with similar colors.

Helianthus annuus

The sunflower, *Helianthus annuus*, is a very important crop plant in many parts of the world owing to desirable features of its seed oil. This has naturally led to efforts to select lines with improved oil production, better response to environmental factors, or different growth forms. In a study of the effects of selection on secondary metabolism, Sanlaville et al. (1988a,b) studied phenolic metabolites, flavonoids and phenolic acid derivatives, in a number of lines derived from such stock as the American line HA89 and a number of its daughter lines. Quantitative analyses using high performance liquid chromatography and multivariate analysis of the resulting data showed that the profiles of phenolic metabolites clearly responded to selection. The results indicate (Sanlaville et al., 1988a) that analysis of phenolic profiles "...could alone be practically equivalent to use of the large sum of morphological, physiological and agronomic characters on which recognition of lines is usually based." A recent study of genetic relationships among domesticated and wild sunflowers based on randomly amplified polymorphic DNA (RAPD) loci is available (Arias and Rieseberg, 1995).

Silybum marianum

Silybum marianum, the milk-thistle, has attracted a good deal of attention owing to the antihepatotoxic properties of its seeds. The active principals, called flavanolignans, are members of a group of compounds produced by phenolic coupling of a flavonoid and coniferyl alcohol. Several different products are possible (Chapter 4). A number of genetic lines are in cultivation which has led some workers to look at such features as total content of flavanolignans (Chiavari et al., 1991), contents and ratios of silybin, silydianin, and silychristin (Cappelletti and Caniato, 1983), and differences between white-flowered vs. purple-flowered types (Dranik et al., 1992).

A summing up

In most studies of infraspecific taxa, when sampling has been more or less representative, flavonoid profiles have provided generally reliable information on relationships. In some cases, however, such high levels of variation exist that the flavonoid profiles are not useful at all. Infraspecific taxa in Hawaiian *Bidens* fall into the latter category. Cases exist within that genus where differences between two members of the same population are greater than differences between two species. In the case of Hawaiian flora, and this is not restricted to *Bidens*, genetic differentiation for morphological and ecological features has occurred, but other features, such as secondary metabolites, occur in what seems to be a totally random fashion.

Perhaps any combination of these compounds is sufficient to achieve whatever function they fulfill for the plant. Insofar as we do not know what the function or functions of these compounds may be for any given system, we can do no more than speculate.

The question of flavonoid function was addressed briefly by Kyhos (1971) in his study of flavonoid variation in *Encelia farinosa*. In the absence of correlation of flower color with pollinator behavior, it was suggested that the elements controlling flavonoid formation might be tightly linked to other, unknown factors that have driven selection. Much more detailed work at the level of the genes, and factors involved in controlling their expression, is obviously required before situations of this sort can be understood. A further factor that requires study at the gene level is exemplified by the work of Proksch et al. (1987) on *Enceliopsis nudicaulis* where different pigment profiles exist in different organs, roots vs. leaves in this case.

Section V

Flavonoids as Indicators of the Evolutionary Process

In the late 1950s, when the efficacy of flavonoids in systematics was being ascertained, early studies in analysis of hybridization in *Baptisia* (Fabaceae) by Alston and Turner and colleagues at the University of Texas at Austin (e.g., Turner and Alston, 1959) gave confidence that there were species-specific profiles, and that these compounds could be used to help determine parentage of interspecific hybrids. These studies, in fact, helped usher in the great wave of interest in plant secondary product chemistry in systematic investigations.

From the very beginnings, therefore, flavonoids have been employed to help understand dynamics of the evolutionary process. Since that time, literally thousands of studies have documented the utility of flavonoids to reveal patterns of variation in different processes of speciation, populational differentiation, and hybridization. Because of the size, wide-spread occurrence, and the number of systematists working in the family, much attention has been given to flavonoids and evolutionary processes.

The following two chapters continue the case study approach used in other sections of this book by giving discussions on selected taxa in Asteraceae. These examples show how flavonoids have helped workers gain insights on complexities of populational differentiation (Chapter 24), including microgeographic patterns, and in revealing patterns of hybridization and introgression (Chapter 25). Although new DNA macromolecular techniques have added useful analytical tools for investigating dimensions of the evolutionary process, we believe that the included examples more than adequately argue for continued utility of flavonoids in attacking both these problems. We see a most interesting future synthesis involving use of DNA techniques to help provide more precision in analysis of specific processes, detailed ecological studies to understand the environmental context of these processes, and flavonoid differentiation in response to both genetic and ecological change.

CHAPTER 24

Flavonoids and Populational Variation

It is our purpose in this chapter to talk about variation in flavonoid profiles among populations within species. Most workers recognize that chemical variants occur naturally, and that such differences may be grounds for additional studies, which may in time lead to formal recognition of infraspecific taxa. Infraspecific taxa based solely on chemical variants, however, are not recommended. For an alternative view, the reader should consult the lengthy article by Tétényi (1970) who argued that chemical varieties (chemovars) ought to be recognized formally. Regardless of one's position on this point, it should be appreciated that any set of unusual or unexpected results, chemicals included, can lead to a better understanding of the system at hand. Re-evaluation of current taxonomy is not out of the question. An excellent example of this phenomenon, although it involves neither flavonoids nor interpopulational variation, is the discovery of the structure of betacyanins, which led to renewed investigation of Centrospermae (Mabry, 1966; Piattelli, 1981).

In addition, the existence of different chemotypes within a species may indicate gene flow not otherwise suspected. Examples of this phenomenon will appear in Chapter 25 where we review flavonoid data as they apply to studies of hybridization and introgression.

In order to arrive at a reasonable idea of the extent of variation in a given system, whether it is a species or any other taxon, it is necessary to become familiar with plants from the entire range of occurrence. Despite the apparent naivety of this statement, this has been a real problem throughout the history of chemosystematics. It is unfortunate that much (perhaps most) of the early work was done on bulked samples. The chemistry was often of the highest quality, but maximum biological significance was not realized. This situation was remedied in due course when researchers more versed in the intricacies of population biology began to use chemical characters as part of broad-spectrum biosystematic studies. This marked a fundamental change in philosophy.

Most of the studies discussed below reflect the broad-spectrum approach where chemical, cytological, and morphological data, often supplemented with studies of breeding biology, have been brought to bear on a problem. We start out with some of the best examples, the detailed studies of *Arnica* in North America by Keith Denford and his associates, then at the University of Alberta. The problem of overlap is immediately obvious since we have already talked about flavonoids and infraspecific taxonomy of *Arnica* in Chapter 23. However, the work is of such importance that it deserves restating from the present perspective.

Arnica

Arnica cordifolia, an apomictic, polyploid complex, occurs widely in western North America from the Yukon to northern New Mexico and central California with two outliers, one in Michigan and one in Ontario. Five chromosome races are known, with plants ranging from diploids with $2n = 38$ to hexaploids with $2n = 114$. Tetraploids are the most commonly found and occur throughout the entire range. Triploids occur in the front ranges of the Rocky Mountains with the other, rarer, cytotypes of more scattered occurrence (Wolf and Denford, 1983).

The flavonoids of 33 populations representing much of the range of the species were determined by Wolf and Denford (1983). The overall profile of the species consists of luteolin and quercetin *O*-glycosides, apigenin 7-methyl ether, quercetin 3-methyl ether, 6-methoxyapigenin, 6-methoxyluteolin 7-*O*-glucoside, and 6-methoxykaempferol 3-*O*-glucoside. Two quercetin 3-*O*-diglycosides were almost ubiquitous. Apigenin 7-methyl ether was seen in one population from California, while quercetin 3-methyl ether was seen in only two, one from British Columbia and one from Oregon. There was no correlation of flavonoid profile with chromosome number, which, the authors pointed out, was the same as had been reported by Glennie and coworkers (1971) in a study of *Senecio radicans* chromosome races. This appears to be the case with other species of *Arnica*, as we shall presently see.

There is, however, a degree of regionalism in the flavonoid distribution profiles. For example, the frequency of occurrence of 6-methoxyapigenin in Alberta (one of six populations) is lower than in other parts of the range. The Alberta populations also had disproportionately lower frequencies of the 6-methoxykaempferol glycoside, and kaempferol and quercetin 3-*O*-glucosides. In the case of luteolin 7-*O*-glucoside one sees an opposite pattern: the compound occurred in most populations from Alberta and British Columbia, all four samples from the Yukon, and one population from Washington. 6-Methoxyluteolin 7-*O*-glucoside also occurs in this overall pattern with one additional sighting from Montana. Otherwise, these compounds were lacking from populations sampled in California, Colorado, Idaho, Montana, Oregon, Utah, and from the specimen collected in Ontario. The Ontario population, separated from its nearest western neighbor by several hundred kilometres, had a flavonoid profile identical to one in Utah and one in Colorado and closely resembled others in Idaho and California.

The authors (Wolf and Denford, 1983) noted that luteolin 7-*O*-glucoside and 6-methoxyluteolin 7-*O*-glucoside are restricted to populations that occur near or north of the limits of advance of Pleistocene glaciation. To account for this distribution they suggested that there may have been two geographically isolated populations during the Pleistocene. It seems as likely that the populations present before glaciation may have had these two compounds, and that their respective ranges extended farther south than the glacial boundary, or at higher elevations, and that these populations served as sources for the northward migration once the ice front began to retreat. Perhaps the northern race has some physiological advantage over the southern race (possibly sensitivity to day length) that allows it to occupy the more northerly areas. The flavonoid differences between northern and southern populations may reflect differentiation at an earlier time in the evolution of the present situation as Wolf and Denford (1983) implied. Comparative DNA studies might offer some assistance in establishing directions of gene flow in this system. Such studies should also be able to tell if there were more than one center of genetic dispersion following glacial retreat.

Extensive variation has been reported for other North American *Arnica* species. A biosystematic study of the *A. frigida-louiseana* complex by Downie and Denford (1986a,b) involved collections of *A. frigida* subsp. *frigida* from 38 populations in western Canada and Alaska. Eleven different flavonoid profiles were observed based on combinations of apigenin 7-*O*-glucoside and glycosides of kaempferol and quercetin. *Arnica frigida* subsp. *griscomii*, from the Gaspé area of Quebec and Newfoundland, exhibited flavonoid profiles that were a subset of the subsp. *frigida* compounds. Two different flavonoid types were seen; both were also known from the western taxon. Eight populations in central Alaska afforded the simplest profile seen in the species with only quercetin 3-*O*-galactoside and 3-*O*-diglucoside in evidence. Profiles having those two compounds plus kaempferol 3-*O*-glucoside were seen in populations from Alaska and the Yukon and from the population of subsp. *griscomii* from Quebec. The related species, *A. louiseana*, occurs along the eastern edge of the Rocky Mountains in southwestern Alberta. Sampling of five populations of this taxon showed the existence of two chemotypes, which were among the more complex seen in the species complex. The overall flavonoid chemistry of the three taxa involved in this complex argues for a close relationship among them. There was no correlation with chromosome number in the group, but the variation in the occurrence of flavonoid types is generally concordant with the degree of morphological variation that characterizes this species complex.

There soon followed another study from the Alberta school, this time describing a study of the flavonoids of the *Arnica angustifolia* aggregate (Downie, 1988). Reappraisal of this group of taxa resulted in recognition of two subspecies, *angustifolia*, which occurs in Alaska, across much of northern Canada, and in coastal Greenland, and subsp. *tomentosa*, which occurs in southwestern Alberta and in the Mackenzie River Delta. Over 70 populations of subsp. *angustifolia* were analyzed with a result much akin to that which we just saw above: eleven different combinations of the two flavone and five flavonol glycosides. The simplest profiles consisted of two compounds, but there were three different combinations. Kaempferol 3-*O*-glucoside and quercetin 3-*O*-diglucoside were the sole compounds in the case of eight Alaskan and Canadian populations. Three populations from the same general area consisted of quercetin 3-*O*-galactoside and 3-*O*-diglucoside. A third simple profile, consisting of kaempferol 3-*O*-glucoside and 3-*O*-galactoside, was seen in a single population from the west, and in three populations from Greenland.

Examination of ten populations of *Arnica angustifolia* subsp. *tomentosa*, from the eastern foothills of the Rocky Mountains and Mackenzie River delta, showed the existence of two flavonoid types, one (four populations) with three compounds and the other with four compounds. These two profiles are the same as the two most common profiles (19 and 16 populations, respectively) seen in subsp. *angustifolia*. As in the case of *A. frigida*, there was no correlation between flavonoid composition and chromosome number, which in this group ranges from $2n = 38$ to $2n = 95$.

Eight species from *Arnica* subgenus *Austromontana* were subjected to flavonoid analysis by Wolf and Denford (1984b). A feature of this group that attracts particular attention is the existence of several rare, endemic species. Only four of the constituent species were sampled well enough to provide an idea of populational variation: *A. cordifolia*, *A. discoidea*, *A. latifolia*, and *A. spathulata*. The authors did not dwell on the significance of flavonoid variation in these taxa. Suffice it to say that the level of variation seen approximated what has been discussed for other taxa so far: a few compounds are essentially ubiquitous (e.g., quercetin 3-*O*-gentiobioside and 3-*O*-diglucoside), while most others (e.g., 6-substituted flavones and flavonols) are much more scattered in occurrence. Only two populations of each of the others, *A. cernua*, *A. nevadensis*, *A. venosa*, and *A. viscosa*, were available for study and they showed no interpopulational variation. The significance of flavonoid profiles of most of these species has been presented in Chapter 22.

Balsamorhiza sagittata

Balsamorhiza sagittata enjoys one of the largest ranges within the genus extending from south-central British Columbia and southwestern Alberta south through the Great Basin floristic province to the Rocky Mountains. Although flavonoid chemistry has been described for only two collections of this species, vastly different pigment profiles have been reported. A sample from a site in Modoc National Forest, Modoc County, California, afforded six compounds: 6-methoxykaempferol, 6-methoxykaempferol 4'-methyl ether (betuletol), quercetagetin 6-methyl ether (patuletin), quercetagetin 3,6-dimethyl ether (axillarin), quercetagetin 6,3'-dimethyl ether (spinacetin), and quercetagetin 3,6,3'-trimethyl ether (jaceidin) (Robson and McCormick, 1988). The other specimen, collected near Princeton, British Columbia, afforded only a single identifiable compound, 6-hydroxykaempferol 7-methyl ether (a trace of a possible *O*-methylated flavone was also detected) (Bohm et al., 1989a). It is interesting to note that the specimen employed by Robson and McCormick consisted of about 30g dry weight whereas the second set of workers employed several kilograms of dried plant material and commented that this was the lowest yielding member of *Balsamorhiza* (as well as in the related genus *Wyethia*) so far examined. Rarely do we find intraspecific differences, both in nature of compounds and in volume of material, as large as has been observed in this species. Further sampling is necessary to determine the extent of these chemotypes.

Recent studies of a related species, *B. macrophylla*, from northeastern Utah, have shown that flavonoid variation in the genus can occur over a much smaller distance. The original examination of this species (McCormick et al., 1985b) resulted in identification of quercetin 3-methyl ether, quercetin 4'-methyl ether (tamarexetin), quercetin 3,4'-dimethyl ether, and three quercetagetin derivatives, the 3,6-di- (axillarin), 6,3'-di (spinacetin), and 3,6,4'-trimethyl (centaureidin) ethers. Material of this species from a site some 30 km farther south was found to have quercetin 3-methyl ether, quercetin 3,4'-dimethyl ether, and centaureidin in common with the earlier study, as well as quercetin 3,3'-dimethyl ether and an unidentified flavanone (Onyilagha and Bohm, unpubl.). Axillarin, spinacetin, and tamarexetin were not detected.

All is not quiet within these two genera, however. See comments on taxonomic changes suggested by Weber (1998) that appear in Chapter 11 under *Wyethia*.

Bidens bouquetiensis

Bidens bouquetiensis is one of the octaploid members of *Bidens* section *Greenmanii* studied by Roseman (1986). It is a cloud forest species known from Costa Rica and Panama. This species and the hexaploid *B. squarrosa* var. *squarrosa* have similar flavonoid profiles, which led to speculation as to whether the latter was involved in the ancestry of *B. bouquetiensis*. Both have a base array consisting of isoliquiritigenin glycosides, butein glycosides, apigenin glycosides, and eriodictyol. Some populations of *B. bouquetiensis*, however, produce isoliquiritigenin 4,4'-dimethyl ether, okanin 3-methyl ether 4'-*O*-diglycoside, a 4'-acylated lanceolin derivative, and an apigenin 7-*O*-diglycoside. Roseman (1986) was cautious in her interpretation of these data. The populations with *O*-methylated flavonoids would have required an ancestor with the capacity for *O*-methylation, if we are to accept additivity as a reasonable process.

Flavonoid races and varieties in *B. squarrosa* were discussed in Chapter 23.

Calycadenia ciliosa

Calycadenia is a tarweed (Madiinae) genus of about a dozen species native to central California and extreme southern Oregon. The evolutionary relationships among members of the genus have been extensively studied (Carr, 1975, 1977, 1980; Carr and Carr, 1983; Baldwin, 1993). Of interest to us here is the study of flavonoid aglycone variation patterns in *C. ciliosa* (Emerson et al., 1986), which is a species that consists of several chromosome races and a high degree of structural heterozygosity (Carr and Carr, 1983). Five local chromosome races, "Corning," "Dry Creek," "Ciliosa," "Lewiston," and "Pillsbury" (names after Carr and Carr, 1983) were sampled in the field, and material from three structurally heterozygotic populations grown in cultivation were subjected to flavonoid analysis. Five 8-substituted, highly *O*-methylated flavonoids were identified ranging from 5,4'-dihydroxy-3,6,7,8-tetramethoxyflavone to the most highly substituted 5,4'-dihydroxy-3,6,7,8,3'-pentamethoxyflavone. One of these compounds was obtained only from a pooled sample and was not included in the description of variation patterns. Five different flavonoid patterns involving the four major compounds were observed varying from a simple profile of one compound (5,3',4'-trihydroxy-6,7,8-trimethoxyflavone) in the "Ciliosa" and "Pillsbury" races and cultivated material from the "Bowman Road" population. Two populations exhibited the complete array of compounds, "Dry Creek" from the field and "Redding" from cultivated material. Interpopulational and individual variation was minimal in all cases except the "Anderson" population, one of the populations in which structural heterozygosity had been observed. Cytological, morphological, and geographical data indicate that, among other relationships, the "Ciliosa" race to be basal with the "Pillsbury," "Corning," and "Dry Creek" races arising therefrom. Flavonoid data are concordant with the relationship between "Ciliosa" and "Pillsbury" but the relationships among the other three populations would require several flavonoid losses and gains. Although knowledge of the structures of minor components detected chromatographically might help to clarify the picture, it seems reasonable to conclude that selection for flavonoids has occurred (or is occurring) independently of changes in morphology or chromosome structure. This is yet another example of a system where detailed macromolecular studies of the enzymes directly involved in flavonoid biosynthesis would be useful.

The one highly variable system reported in this work involved the structurally heterozygotic race "Anderson." All of the 30 individuals examined had 5,3',4'-trihydroxy-3,6,7,8-tetramethoxyflavone, as noted above. However, 5,4'-dihydroxy-3,6,7,8-tetramethoxyflavone and 5,6,3',4'-tetrahydroxy-3,7,8-trimethoxyflavone varied among individuals. This population consists of several flower color morphs ranging from white through cream to yellow with some red pigmentation present proximally in ligules. There was no correlation between flower color and the pattern of occurrence of the *O*-methylated flavonoids.

Centaurea

One of the more impressive studies of flavonoid variation among populations comes from the work of Gonnet on *Centaurea* species in Europe (Gonnet, 1992, 1993). Gonnet was interested in determining the structural relationships within selected species growing under specifically defined conditions of community structure. In the initial work he (Gonnet, 1985, 1986) performed exhaustive analyses of flavonoid variation in *Chaerophyllum aureum* (Apiaceae) growing in two types of alpine plant community at the subalpine level (1500-2100 m) in the southern Alps. The communities studied were a high-altitude, manured meadow designated as a Triseto-Polygonion community, and a lower elevation tall grass prairie under *Larix* identified as an Adenostylion community. The original study had six targeted species among which were the umbel and *Centaurea montana*.

The overall flavonoid profile of *C. montana* consists of *C*-glycosyl flavones, *O*-glycosyl flavones, and a few flavonol derivatives. The majority of interesting variation involved the first two compound types, and it is the patterns of their occurrence that Gonnet discusses in detail. Two extreme chemotypes were defined, one that consisted of complex *C*-glycosides including *O*-glycosylated 6-*C*-glycosides based on apigenin, luteolin, and chrysoeriol; and one that consisted primarily of apigenin and luteolin 7-*O*- and 4'-*O*-monoglycosides. There are other chemotypes to be sure, but they represent various mixtures of the elements that make up the two extreme forms. Multivariate analysis showed that the polyphenolic polymorphism is not strictly structured, which was taken by Gonnet to indicate that the polymorphism may involve regulatory genes as much as it does structural ones.

Detailed studies using ramets of individuals from Nature maintained in growth facilities showed that there were no qualitative differences of flavonoid profiles between plants from Nature and those grown under cultivation. Some quantitative variation was noted in some instances. This indicated that the flavonoid profiles observed in individuals from the wild are under genetic control.

Analysis of individual plants from the two community types shows clear-cut differentiation, although it is in no way an either/or situation. About 75% of the individuals from the tall grass prairie (Adenostylion) community exhibited flavonoid profiles of the flavone *O*-glycoside type, while a comparable proportion (74%) of the individuals from the meadow (Triseto-Polygonion) community exhibited the *C*-glycosylflavone profile. The presence of about 25% of alternative chemotypes in each area suggests that differentiation as a function of site, essentially disturbed vs. non-disturbed habitats, is not complete. On the basis of current information, it is not possible to establish whether this mixture of chemotypes merely represents a stage in the eventual total domination of one by the other, or whether these systems are in some sort of steady-state. As Gonnet (1992) points out in his concluding remarks, long-term studies are necessary in order to understand the mechanism(s) involved in arriving at the current state of these community affairs. The apparent complexity of flavonoid biosynthesis and control in *C. montana* was documented further by a recent study of pigment profiles in wild progeny (Gonnet, 1996). Differences in glycoside composition, *C*-glycosylflavones (formed early in the pathway) in some plants vs. flavone *O*-glycosides (formed late in the pathway) in others, represent profound differences in the control of flavonoid biosynthesis.

In his second report dealing with *Centaurea* Gonnet (1993) investigated a species that has a close relationship with *C. montana*. *Centaurea triumfettii*, with $2n = 22$, has been suggested as the species from which *C. montana*, $2n = 44$, arose through polyploidization. Hybridization between the two is discussed in Chapter 25. Of relevance to us here is a comparison of the flavonoid profiles of *C. triumfettii* obtained from Switzerland and Yugoslavia. Further distinctions within these larger regions were also observed. Two chemotypes were identified in the species in Yugoslavia. Plants grown from seed collected in the Trebevic region were characterized by a strong accumulation, generally over 20% of the total flavonoids, of isoörientin 2"-*O*-glucoside and correspondingly low levels of apigenin 6,8-di-*C*-glucoside. The second group of plants was grown from seed that was collected in the area of Mt. Vlasic and Cincar/Golija. In these plants isoörientin 2"-*O*-glucoside never appeared at a level greater than 3%, or was not seen, while the apigenin diglucoside assumed major proportions, averaging about 22% of total flavonoids. In each group, however, differences were noted among populations, and then further variants were seen in the analysis of individual plants from these populations. Adding to the complexity was the observation that some of the samples from Trebevic gave different flavonoid profiles when grown under experimental garden conditions or when grown in an indoor growth facility. This clearly indicates a phenotypic response to growth conditions, although the collection of these seed lots in different years may have had an impact on the outcome of the growth studies.

Centaurea triumfettii collected in Switzerland, in the vicinity of Brig, Valais, showed an overall simpler flavonoid profile than seen in the Yugoslavian plants, with monoglycosides of apigenin present in abundance, up to 40% of total flavonoids in some instances. A problem with specimens from this area appears to lie with the likelihood of hybridization with *C. montana* with subsequent introgression, possibly in both directions. Nonetheless, Gonnet (1993) says that generalized variation in the flavonoids of *C. triumfettii* approaches that of *C. montana*, but manifests itself mainly in flavone *C*-glycosides.

The importance of understanding the flavonoid chemistry of *C. triumfettii* lies in the effect this taxon may have had on the flavonoid profile(s) of *C. montana* in the analysis of relationships between the tall-grass prairie and the meadow communities. *Centaurea montana* from those communities were typically *C. montana*, morphologically and cytologically ($2n = 44$). It was also thought that *C. triumfettii* did not occur in the general area containing the two study areas. However, Gonnet (1993) reported a collection of typical *C. triumfettii* approximately 5 km from stations previously studied. The potential for interaction between these can not be ruled out.

Cirsium palustre

Cirsium palustre is a predominantly purple-flowered thistle native to Europe and western Asia, but now widely distributed in North America and Australia. From time to time populations of *C. palustre* are encountered where plants with white or intermediate-colored flowers occur with high frequency. The anthocyanin(s) responsible for the purple color have not been established, as far as we are aware, but it would appear that this polymorphism is typical of flower color morphs in general. A report by Mogford (1974a) describes this system in detail including the results of crossing studies that showed the white form to be homozygous, which is normal for an early blockage of the flavonoid pathway (Chapter 5). That paper also discussed how selection forces were operating on sea cliff and mountainous populations in Great Britain. The accompanying paper (Mogford, 1974b) discusses the significance of the flower color polymorphism in terms of pollination. Selection for lighter-colored morphs appears to occur most strongly on sea cliffs and in mountain populations where pollinator activity is low. With lowered pollination, selfing occurs with greater frequency which results in an increase in the homozygous recessive combination. This results in the absence of anthocyanin production.

Clibadium

Flavonoid surveys of several species of *Clibadium*, a South and Central American genus of 29 species (Arriagada, 1994), revealed several interesting examples of variation among populations. An initial survey (Bohm and Stuessy, 1981a) revealed that the flavonol glycoside profiles of most of the species, for which multiple samples were available, were fairly constant. Two notable exceptions were evident, however. Nine populations of *C. leiocarpum* were available, eight from Costa Rica and one from Panama. Seven populations of this species from Costa Rica exhibited only 3-*O*-glycosides of kaempferol and quercetin with some variation seen in the presence of the rhamnosides. One specimen from Costa Rica also had a trace amount of quercetagetin 6,3'-dimethyl ether 7-*O*-glucoside. The single population from Panama exhibited the most complex profile consisting of the full array of kaempferol and quercetin glycosides, including the rhamnosides, and four quercetagetin derivatives that accounted for the bulk of the flavonoid fraction. The compounds were identified as quercetagetin, patuletin, quercetagetin 3'-methyl ether, and quercetagetin 6,3'-dimethyl ether, all of which were isolated as the 7-*O*-glucosides.

Fourteen populations of *C. surinamense*, from Costa Rica and Panama, were also investigated (Bohm and Stuessy, 1981a). No extra-oxygenated or *O*-methylated flavonoids were observed, but variation in occurrence of kaempferol and quercetin 3-*O*-rhamnosides, 3-*O*-rutinosides, and 3-*O*-diglucosides was noted. Material from an additional 41 populations from Colombia brought the number of populations studied to 55. Chromatographic analysis (Bohm et al., 1983) showed that all specimens exhibited kaempferol and quercetin 3-*O*-glucosides and quercetin 3-*O*-rutinoside. Kaempferol 3-*O*-rutinoside was seen in 34 specimens. Quercetin and kaempferol 3-*O*-rhamnosides were seen in 39 and 36 specimens, respectively. The 3-*O*-diglucosides were observed on chromatograms of only five and four specimens, respectively, and may well be more prevalent, but limitations in amount of plant material prevented a more certain analysis. The overall flavonoid profiles are similar to those reported for the Central American populations (Bohm and Stuessy, 1981a; Bohm et al., 1983). Two specimens from central Colombia exhibited a pair of spots on chromatograms lacking in the rest of the specimens. The compounds were identified as kaempferol and quercetin 3,7-di-*O*-glucosides. The sites from which the two samples were obtained occur in central Colombia and lie about 200 km apart. Several populations lying between these two sites gave the normal flavonoid profile for *C. surinamense*. These glycosides were also identified as constituents of *C. pilonicum*, a rare taxon known only from a very limited area in central Panama [recent studies by Arriagada (1994) suggested, however, that this aberrant form may be a local variant of *C. anceps*]. It was suggested, on the basis of no compelling evidence to the contrary, that the appearance of these compounds in two unrelated species likely represents a case of parallel evolution. It would be interesting to apply macromolecular methods to study the 7-*O*-rhamnosyltransferases responsible for the, presumably, second glycosylation reaction in these two systems. What is the macromolecular basis for such apparent parallel biochemical processes? It would seem to the present authors that a detailed study of systems such as this one might provide extremely useful insights into factors that control establishment of flavonoid profiles.

In the first example from *Clibadium* above we noted the occurrence of populations that differed significantly in their aglycone production. Another example of this was reported from *C. trianae*, sampled in Colombia (Bohm et al., 1983). Six of the eight collections afforded identical profiles consisting of kaempferol and quercetin 3-*O*-glucosides, 3-*O*-galactosides, 3-*O*-rutinosides, 3-*O*-rhamnosylgalactosides, and 3-*O*-diglucosides, essentially what appears to be the fundamental profile of the genus. A seventh population exhibited the base profile plus quercetagetin and patuletin 7-*O*-glucosides. The flavonoid profile of the eighth specimen also had the simple glycoside array, lacked the quercetagetin derivatives, however, but did present two compounds not otherwise found in the genus. These were identified as the anthochlor pair coreopsin (2',3',4',3,4-pentahydroxychalcone 4'-*O*-glucoside) and sulfurein (4,6,3',4'-tetrahydroxyaurone 6-*O*-glucoside). Trace amounts of two additional chalcones were also noted. Needless to say, no taxonomic significance was attributed to these observations, but the obvious question is why chalcone accumulation has become so prominent in this population? Chalcones are obviously formed in members of the other populations in the normal course of flavonoid biosynthesis, but in this particular one something has happened that allows, comparatively speaking, massive amounts of carbon to accumulate in compounds not otherwise known from the species, indeed from the genus. The operation of aurone synthetase in plants of this population raises other biosynthetic issues.

Dendroseris

Dendroseris, a member of Lactuceae, consists of 11 species restricted to the Juan Fernandez Islands, Chile. Pacheco and colleagues (1991) included flavonoids in their study of relationships among the species on the two islands, as well as relationships with possible mainland progenitors. The flavonoid profiles were based upon five apigenin and seven luteolin 7-*O*-glycosides and two quercetin 3-*O*-glycosides. Aglycones were also reported but are probably not significant owing to the possibility that they may be artifacts. Some variation was observed in the appearance of a few of the glycosides but the individual species profiles were sufficiently different from one another that use could be made of the flavonoid data regardless of the absence of any given compound in a population. See Chapter 21 for further discussion. For additional examples from the Juan Fernandez Islands see entries for *Erigeron* and *Robinsonia* below.

Erigeron in the Juan Fernandez Islands, Chile

Erigeron is represented on the Juan Fernandez Islands by six endemic species. Only one of these, *E. fernandezianus*, occurs on both islands; the other five occur only on the younger island, Masafuera. Compared to the other asteraceous species from the Juan Fernandez Islands that have been studied for flavonoids, *Dendroseris* and *Robinsonia*, species of *Erigeron* displayed a modest level of intrapopulational variation. The sampling of *E. fernandezianus* was very good with material available from 26 populations from Masatiera and 12 from Masafuera. Twelve populations of *E. rupicola* were sampled. Only two populations were sampled for *E. ingae, E. luteoviridis*, and *E. turricola* and only one for *E. stuessyi*. The flavonoid profile of the group, in sum, consists of quercetin and quercetagetin glycosides, a *C*-glycosylflavone, and glycosides of apigenin, luteolin, and 6-methoxyluteolin. Luteolin *per se* was detected in virtually all populations of all species.

The quercetagetin glycoside was one of the most variable compounds, present in 12 of 26 populations from Masatierra and only 3 of 12 from Masafuera. It was present in only half of the populations of *E. rupicola*, and one of two of *E. ingae*. The absence of quercetagetin from *E. luteoviridis*, and *E. turricola* (two populations each) and from *E. stuessyi* may reflect sampling error. The same problem exists with regard to the significance of the appearance of the luteolin glycoside and the *C*-glucosylflavone in only one specimen each of *E. ingae, E. luteoviridis*, and *E. turricola*, and their absence from *E. stuessyi*. This could have some impact on unravelling evolutionary relationships because these two compounds were present in most of the collections of *E. fernandeziana* from Masatierra but were not seen in any of the collections from Masafuera. Differences of this sort can provide useful clues as to the order of occupation of the islands.

Helianthus annuus

Helianthus annuus, the common sunflower, is probably the most readily recognized member of Asteraceae by non-botanists (with the possible exception of the dandelion). Sunflowers and their relatives have attracted a good deal of attention from the botanical community as well (Heiser, 1976). Of interest in this chapter is a study of flavonoid aglycones in 265 individuals from throughout the range of this species (Rieseberg et al., 1987). Representatives of subspecies *lenticularis* (169 individuals), *texanus* (46 individuals), and *annuus* (23 individuals), along with 27 individuals that were intermediate in form between *annuus* and *lenticularis* were examined for flavonoid aglycones. Two types of aglycones occur, flavones located in glandular trichomes and chalcones that are incorporated in the waxy cuticle.

Eight compounds were encountered, seven of which were identified, five flavones and two chalcones (the unidentified compound was likely another chalcone). The flavones were shown to be luteolin, 6-methoxyapigenin (hispidulin), 6-methoxyluteolin (nepetin), 5,7,4'-trihydroxy-6,3'-dimethoxyflavone (jaceosidin), and 5,7-dihydroxy-6,8,4'-trimethoxyflavone (nevadensin). The chalcones were identified as 2',4',4-trihydroxychalcone (isoliquiritigenin) and 2',4-dihydroxy-4'-methoxychalcone (isoliquiritigenin 4'-methyl ether).

Rieseberg et al. (1987) presented their data in terms of 10 geographic areas: Area 1, Washington, Oregon, and most of Idaho; Area 2, California and Nevada; Area 3, Alberta, Saskatchewan, Manitoba, Montana, and the Dakotas; Area 4, Wyoming, Utah, Colorado, and southeastern Idaho; Area 5, Arizona and New Mexico; Area 6, Texas, western Louisiana, and northeastern Mexico; Area 7, Nebraska, Iowa, Kansas, Missouri, Oklahoma, and northern Texas; Area 8, Illinois, Indiana, and Michigan; Area 9, southeastern United States; and Area 10, northeastern United States and Nova Scotia. Areas 1-5 correspond to subsp. *lenticularis*, Area 6 corresponds to subsp. *texanus*; Area 7 corresponds to a region of intergradation between subspecies *annuus* and *lenticularis*, and Areas 8-10 correspond to subsp. *annuus*. The very considerable variation seen in the data was presented in several ways including a tabulated array of maximum, minimum, and average number of compounds in any profile, but the most telling index of variation is probably their determination of the number of unique flavonoid profiles per area. Some of the values recorded were 18 for Area 1, 12 for Area 6, 9 for Area 2, and so forth, down to the lowest value of 3 for each of Areas 8 and 9. These values were discussed in terms of other studies where comparatively large numbers of flavonoid profiles have been recorded, e.g., *Phlox pilosa* ssp. *detonsa* (Levy, 1983) and *Menziesia ferruginea* (Bohm et al., 1984).

If further convincing is needed that the sunflower flavonoid aglycones occur in essentially a random fashion, it comes from multivariate analysis of the entire flavonoid data array. Principal components analysis showed no recognizable pattern in the data. As the authors pointed out, individuals representing all three subspecies clustered together at six of 24 points in the PCA display. It is also useful to note that in the UPGMA-derived cluster analysis the closest clustering involved Area 6, which corresponds to subsp. *texanus*, and Area 7, the zone of intergradation between subspecies *annuus* and *lenticularis*. Another pairing has Area 2, which consists of California and Nevada, and Area 8, Illinois, Indiana, and Michigan, together. This also involves subspecies *annuus* (Area 8) and *lenticularis* (Area 2).

Rieseberg et al. (1987) listed four reasons for the absence of flavonoid races in *H. annuus*: (1) dispersal by man; (2) hybridization; (3) random genetic variation; and (4) natural selection. Some comments on these are in order. Heiser (1954) suggested that the sunflower has been a travelling companion of people for hundreds or possibly thousands of years either as a crop plant or as a weed. This would have allowed contact between the three subspecies and subsequent opportunities for genetic exchange through extensive hybridization, and almost certainly introgression. It has been predicted (Loveless and Hamrick, 1984; Hamrick et al., 1979) that life history characteristics of a plant can have an important influence on its genetic structure. Most genetic variation in weedy outcrossers, to which group sunflower belongs, would be expected to occur within populations rather than among populations. Flavonoid variation in sunflower, at least with regard to the aglycones in the present study, is apportioned in this fashion, with 61% of profile variation arising from within and 39% from among populations. It is possible that different flavonoid aglycone profiles, or possibly the individual compounds themselves, have been selected for their function as feeding deterrents. Although Rieseberg et al. (1987) comment on this last aspect in terms of the toxicity of flavonoids, it might be prudent to point out that to be effective in preventing herbivory by vertebrates a compound does not have to be toxic, it need only taste bad (Bohm, 1998a).

It is only through the kind of thorough sampling and exhaustive analysis witnessed in this study of the sunflower that we gain insights into the processes at work in Nature. The use to which the flavonoid data have been put in this study exemplifies how research of this sort should be done.

Heterotheca

As part of a survey of genera in Astereae, Wollenweber et al. (1985b) undertook an examination of the resinous flavonoids of two species of *Heterotheca*, *H. grandiflora* and *H. psammophila*. One of the *H. grandiflora* specimens came, presumably, from cultivation (Pop. A, source cited as University of California, Irvine campus); the other from Nature (Pop. B, Cochise County, Arizona). Both specimens of *H. psammophila* were collected in Arizona, one in Cochise County (Pop. A), the other in Yavapai County (Pop. B). Although only two samples of each were available for investigation, the qualitative differences between them are marked. Both species accumulate methylated derivatives of kaempferol, 6-hydroxykaempferol, quercetin, quercetagetin, and 6-hydroxyapigenin (scutellarein). Population A of *H. grandiflora* afforded luteolin 7,3'-dimethyl ether, 6-hydroxyluteolin 6,7,3'-trimethylether, and ether. These compounds were not seen in the other sample of this species or in either sample of *H. psammophila*. Other distinctions between the species lie in the absence of quercetin and any of its methyl ethers from the profile of *H. psammophila* and in the array of methyl ethers elaborated by each. Differences between the populations of each species are also based upon level of *O*-methylation. The most extreme example involves kaempferol derivatives from *A. grandiflora*: population A plants had only one derivative, kaempferol 3,7-dimethyl ether while population B plants yielded kaempferol itself, the 3-, 7-, and 4'-monomethyl ethers, and the 3,7- and 3,4'-dimethyl ethers. A similar situation can be seen with quercetin derivatives where population B has quercetin and its 3-methyl, 7,3'-dimethyl, and 3,7,3'-trimethyl ethers in contrast to population A which has only the trimethyl ether. It is difficult to assess the significance of data such as these without a broader sampling, which the authors acknowledged, and without some idea of phenotypic plasticity. These differences suggest very different activities of the *O*-methyltransferases in these populations. Also, the significance of the differences between the two samples of *H. psammophila* has to be viewed cautiously because of the availability of only a "small test sample" as compared to the half kilogram sample of population A material available for the study.

Lasthenia californica

Lasthenia californica [*L. chrysostoma* in Ornduff (1966)] is the most wide ranging species in this mostly Californian genus. It is also one of the most variable with respect to morphology, cytology, and range of substrates upon which it grows (Ornduff, 1966). Preliminary studies of members of the genus had shown the existence of flavonoid races in several species (Saleh et al., 1971; Bohm et al., 1974; Bohm and Banek, 1986). Since most of these studies involved comparatively few sites, or limited numbers of individuals, it was considered timely to look at a very much larger sample and to examine a taxon that already was known to be variable. The first study involved a detailed examination of a large, undisturbed population of *L. californica* in the Jasper Ridge Biological Preserve of Stanford University, located in San Mateo County, California. Since the *L. californica* population in the Preserve grows on a serpentine-derived soil, choice of this area also afforded an opportunity to see if ultramafic soils exert any influence on the flavonoid chemistry of this species.

Preliminary surveys showed that three flavonoid profiles existed, a base type consisting of common flavonol glycosides, a second to which was added luteolin 7-*O*-glucoside, and a third that consisted of the base array plus eriodictyol 7-*O*-glucoside and flavonol 3,7-diglycoside sulphates. The quercetin component of this latter pair was the more evident of the two. Eventually, a fourth type was found, the 3,7-diglycosides that lacked sulphate groups, but this form was such a minor component (perhaps five sightings in 3000 individuals) that it did not figure in any of our analyses (Bohm et al., 1989b).

Owing to the size of the serpentine outcrop, it was decided to limit study to a more convenient area. Several transects were established in order to see if there was any pattern in the appearance of the three flavonoid types. It was demonstrated that a pattern did indeed exist. There was variation along the transect with one flavonoid type, the sulphated diglycosides, occupying one end of the transect and a mixture of the other two flavonoid types occupying the other end. This pattern repeated itself for a period of six years (Bohm et al., 1989b). Similar repeatable patterning was observed along the other transects. Subsequent studies (Desrochers and Bohm, 1993) showed that, after a gap of two years, and after further years of sampling, the pattern still existed in this population.

Other observations made during the course of these studies are relevant. First, seed from plants along the transects were germinated and grown under uniform conditions in ordinary potting soil (non-serpentine soil). The flavonoid chemistry of the offspring reflected the chemistry of the parent from which they had been obtained. It is obvious from these studies that the flavonoid profiles of these plants are under genetic control and that growth in non-serpentine soil did not alter the expression of the flavonoid phenotype. It was also observed that the two flavonoid types, those with sulphated diglycosides and those that lacked them, did not cross. This is in line with observations of Ornduff (1966) that crosses between individuals of different populations were difficult to obtain.

Part of the ongoing study of this genus by Desrochers and Bohm (1993) expanded the survey to include populations of *L. californica* from its entire range. To this end, 33 populations were compared. Although other mixed populations were encountered, most populations were either predominantly positive for the diglycoside sulphate or lacked it altogether. Only in the case of the Jasper Ridge population and a population from near Dunlap in Fresno County did mixed populations have a predominance of non-sulphated flavonoid-bearing plants. Four populations from the northern end of the species range, one from Tehama County and one from Shasta County, both in California, and two from Jackson County, Oregon, exhibited flavonoid profiles that lacked the sulphated diglycosides altogether. An additional population from San Luis Obispo County, California also lacked the sulphated compounds. All other populations from the vicinity of San Francisco south to northern Baja California and into Arizona consisted either predominantly or totally of the sulphated diglycoside flavonoid type. It would appear that finding both types of flavonoid in the Jasper Ridge population was fortuitous. Nonetheless, it was suggested in that paper that the Jasper Ridge population is in fact a mixed population with some individuals representing the northern flavonoid race and others representing the southern race. Further studies would show that additional distinctions between the two chemical races exist.

Plants collected from 36 populations, again representing the entire range of the species, were subjected to morphometric and electrophoretic analysis (Desrochers and Bohm, 1995). These studies demonstrated that two geographical races exist that can be distinguished on the basis of pappus shape and allozyme differences as well as on flavonoid profiles. All populations located north of San Francisco were dominated by the *b* allele at *Nadhdh* and by the *a*, *b*, and *c* alleles at *6Pgd-1*. Populations in the southern part of the range, except for one, were dominated by the *a* allele at *Nadhdh* and the slower alleles at *6Pgd-1*. Both types occur at the Jasper Ridge site and correlate well with the occurrence pattern of the sulphated diglycosides. Despite the existence of allozyme differences, pappus structure differences, and differences in flavonoid profiles, there are no reliable means by which the populations could be recognized as taxonomic groups. Adding to the overall variation is the existence of three cytotypes for *L. californica* (n = 8, 16, 24). Evidence does suggest, however, that the tetraploids have been derived through autopolyploidization. We can conclude this section by noting those workers' (Desrochers and Bohm, 1995) conclusion that the variation in the Jasper Ridge population, where the study began, is thought to have originated by two colonists, one representing the northern race, the other representing the southern race. It is interesting to note that, despite their proximity at that site, the two types appear to be living independently of one another in that their flowering times are not synchronous and that there is highly restricted pollen flow between them (Rajakaruna and Bohm, unpubl.). Studies have also revealed differences in soil pH, water holding capacity, soil texture, and metal ion content along the transects, and different capacities of the two flavonoid types to accumulate sodium ions (Rajakaruna and Bohm, 1999).

Perityle emoryi

Perityle, a genus in Helenieae, consists of over 60 species most of which occur in Mexico and in western United States. The range of one species extends into Peru and Chile. A study of flavonoid aglycones of Peritylinae (Crins et al., 1988b) included several species of *Perityle* including the South American *P. emoryi*. Four specimens from different parts of the range were available: (1) Peru, (2) Sonora, Mexico, (3) Baja California, and (4) California. The four specimens shared only one compound out of a total of 11 identified from the species as a whole. It is interesting to note that that compound, 5,7,4'-trihydroxy-3,6-dimethoxyflavone, was the only compound shared by all taxa included in the study: nine species of *Perityle*, and one species each of *Pericome* and *Amauria* (see Chapter 21). Kaempferol 3-methyl ether was found in specimens from sites in areas 1, 3, and 4, while kaempferol-3,4'-dimethyl and 6-hydroxykaempferol-3,6,4'-trimethyl ethers were reported from areas 1, 2, and 3. 6-Hydroxykaempferol-3,6,7-trimethyl ether was found only in plants from California, area 4. Quercetagetin-3,6-di-, 3,6,7-tri-, 3,6,3'-tri-, and 3,6,3',4'-tetramethyl ethers were obtained from material that originated in Baja California and California. Quercetagetin-3,6,4'-trimethyl ether was observed only in the Peruvian and Mexican material. Eriodictyol 7-methyl ether was seen only in the Californian specimen. Two patterns emerge from these data, the reality and significance of which are open to further investigation. The first is that a unique flavonoid aglycone profile exists for each area. This has to be tempered with the usual statement that a broader sampling in each area is needed. The second apparent pattern in the data is the possibility that the more complex profiles, consisting of quercetagetin derivatives and higher levels of *O*-methylation, characterize the species as one travels north through its range.

Robinsonia

Robinsonia, a member of Senecioneae, consists of seven dioecious species endemic to the Juan Fernandez Islands, Chile. An investigation of the flavonoid chemistry (Pacheco et al., 1985) involved sampling of 54 populations representing the entire genus. With the establishment of the limits of variation, it became possible to use the flavonoid profiles as reliable features in a biosystematic study of the group (see Chapter 21 for details). The pigment profiles consisted of, in various combinations, kaempferol, quercetin, and isorhamnetin glycosides, apigenin and luteolin glycosides, one glycoside each of naringenin and eriodictyol, and a glucoside of the dihydroflavonol taxifolin. Four species were sampled from more than one population: *R. evenia* (13), *R. gayana* (21), *R. gracilis* (15), and *R. thurifera* (7). The variance reported for quercetin *per se* in each species is probably not significant in so far as the appearance of quercetin glycosides is constant in each. The aglycone could very well be an artifact of the collection and/or analytical procedures. Variation in the occurrence of apigenin and luteolin 7-*O*-diglycosides in *R. evenia*, and naringenin 7-*O*-glucoside in *R. gracilis*, however, appear to be real since these are not trace components of the respective profiles. The authors pointed out these variable profiles but made no comment about their possible significance, other than to report that the profiles did not correlate with gender. The remainder of the profiles recorded within each species were remarkably constant with only a few individuals lacking any given compound. The overall invariance of the profiles established their potential value in discussing evolutionary relationships, which were discussed in Chapter 22.

Thelesperma simplicifolium

A study of 11 of 15 recognized taxa of *Thelesperma* revealed a high degree of chromatographic homogeneity with two exceptions (Melchert, 1966). One of these involved *T. simplicifolium*, which consists of three chromosomal races, $n = 10$, $n = 11$, and $n = 22$. The species occurs across central and western Texas and parts of adjacent Mexico. Tetraploid populations occur at the western end of the distribution (Trans-Pecos region) and are, with a single exception, geographically distinct from diploid populations, which occur on the Edwards Plateau. The diploids and tetraploids shared a basic set of flavonoids, but the tetraploids have an additional set. The flavonoid profiles of the two cytotypes are such that they can readily be distinguished on the basis of two-dimensional chromatograms.

These observations are of more than phenetic interest. The flavonoid profile of the tetraploids has much more in common with the profiles of the other species of *Thelesperma* than it does to diploid *T. simplicifolium.* This "generic" pigment profile suggests that the tetraploid cytotype was *not* derived by polyploidy (autopolyploidy) from a diploid progenitor from the Edwards Plateau. Melchert (1966) concluded by putting forth the suggestion that the tetraploid and diploid forms of *T. simplicifolium* may have had independent origins from some "biochemically typical" diploid race (or races).

CHAPTER 25

Hybridization and Introgression

The study of natural hybridization using secondary metabolites, particularly flavonoids, attracted a good deal of attention following the pioneering work of Ralph Alston and his colleagues at the University of Texas. Sorting out the complex patterns in populations of hybridizing *Baptisia* species by morphological means alone was extremely difficult bordering on impossible. The discovery that flavonoids could be used as markers led to a much deeper understanding of the system than was possible by other means, at least at that time (Alston and Turner, 1963b; McHale and Alston, 1964; Baetcke and Alston, 1968). It was not long before flavonoid analysis was applied to hybridization problems in Asteraceae. Early studies by Torres and Levin (1964) on *Zinnia* and on *Liatris* (Levin, 1967), were followed by work from Ownbey's group at Washington State University on *Tragopogon* (e.g., Beltzer and Ownbey, 1971).

Since hybrid individuals usually inherit both parental pigment patterns, it is often assumed that simple additive inheritance is the rule. Although additivity to a greater or lesser extent was seen, enough exceptions were noted to make workers aware that hybridization could result in modified pigment profiles, and in fact "hybrid compounds" were encountered on rare occasion. Another factor that enters into studies of hybridization is the potential for formation of higher level hybrids in Nature, and for backcrossing with one or both parental species. Artificially synthesized F_1 hybrids, as done in several of the studies below, serve as important controls, e.g., *Dubautia* (Crins et al., 1988a). The availability of F_1 plants also allows controlled backcross experiments to be performed as demonstrated with considerable success with *Liatris* (Levin, 1967).

Achillea clavennae group

In a recent flavonoid analysis of the *Achillea clavennae* group, Franzén (1988) reported the existence of a possible hybrid between *A. pindicola* and *A. umbellata* in central Greece. Although these species are allopatric, they make their closest approach to each other in this region. The two species share a close relationship as indicated by morphometric analysis and crossing experiments (Franzén, 1988). A population (identified as population M in the above papers) located on Mt. Vardousia exhibited a different flavonoid profile from those obtained with the two putative parental species. Although it is difficult to determine the exact nature of the flavonoids involved (the structures of some were apparently not established), the chromatographic pattern of the putative hybrid taxon, however, did show features of the parent species. The results are not as clear-cut as some of the examples below, but there were sufficient differences in the pigment patterns to suggest that something has happened in this area, possibly including backcrossing. A more detailed analysis of the plants in this population is necessary for resolution of the problem.

A possible instance of hybridization in *Achillea* comes from the work of Valant-Vetschera and Wollenweber (1988a) on the *A. nobilis* group. *Achillea nobilis* subsp. *sipylea* (one collection, presumably a cultivar) exhibited a different array of *O*-methylated 6-hydroxykaempferol and quercetagetin derivatives compared to the other subspecies, which were nearly invariant. There was no further discussion of this situation.

A further example from *Achillea*, based on somewhat better evidence, involves two members of the *A. ptarmica* group of species, *A. ptarmica* and *A. salicifolia*. *Achillea salicifolia* occurs mainly in Russia, but extends into Finland where some populations grow sympatrically with *A. ptarmica* where they are thought to hybridize. The two species differ in their respective flavonoid specializations: *A. ptarmica* is characterized by a predominance of *C*-glycosylflavone derivatives, whereas flavonol 3-*O*-glycosides are predominant in *A. salicifolia* (Valant-Vetschera, 1985a). The flavonoid profiles of hybrids are also dominated by flavonol 3-*O*-glycosides, but *C*-glycosylflavones are also present. Valant-Vetschera (1985a) concluded that the profiles of some of the possible parent plants did not differ significantly from those of the hybrids, implying that flavonoid data would not be useful in determining the hybrid nature of putative offspring. She did, however, suggest that the similarity might be taken as a sign of introgression. Examination of flavonoid data for that system suggests to the present authors, however, that hydridization may indeed have taken place. Leading to this conclusion is the observation that the hybrid flavonoid profiles approximate an additive array with only two apparent losses of compounds (but not compound types) in the hybrids. A larger sampling of pure and sympatric populations would help to clarify the situation, and attempts to obtain artificial hybrids for chemical study would be very useful.

Table 25–1. Flavonoids of *Arnica cordifolia*, *A. gracilis*, and *A. latifolia*

Compound[1]	Species of *Arnica* *cordifolia*	*gracilis*	*latifolia*
6-Methoxyapigenin	+	+	+
Apigenin 7-methyl ether[2]	+	–	–
Luteolin 4'-methyl ether	–	+	–
Quercetin 3-methyl ether	+	+	–
Patuletin	–	+	+
Apigenin 7-*O*-Glc	–	+	–
Luteolin 7-*O*-Glc	+	+	–
6-Methoxyluteolin 7-*O*-Glc	+	+	–
Kaempferol 3-*O*-Glc	+	+	+
Kaempferol 3-*O*-Gal	–	+	+
6-Methoxykaempferol 3-*O*-Glc	+	+	–
Quercetin 3-*O*-Glc	+	+	+
Quercetin 3-*O*-DiGlc	+	+	+
Quercetin 3-*O*-Gent	+	+	+
Patuletin 3-*O*-Glc	–	+	+

[1] Gal = galactose; Glc = glucose; Gent = gentiobiose.
[2] Seen in only one of 18 populations examined.

Arnica gracilis

Arnica gracilis occurs on exposed, rocky slopes in alpine areas of the Rocky Mountains from Alberta south to Wyoming and in the Cascades of Washington and British Columbia. Its taxonomic status has long been in question occasioned no doubt by its close morphological similarity to *A. latifolia* (Wolf and Denford, 1984b). With the flavonoid chemistry of *Arnica* subgen. *Austromontana* well in hand (Wolf, 1981), along with morphological, cytological, and ecological considerations, Wolf and Denford (1984a) set out to test the hypothesis that *A. gracilis* is of hybrid origin between *A. latifolia* and *A. cordifolia*. Infraspecific variation is sufficiently low and the individual flavonoid profiles of *A. latifolia* and *A. cordifolia* are sufficiently distinct that these compounds were considered ideal tools for studying putative hybrids. The flavonoids found by Wolf and Denford (1984a) are presented in Table 25–1. Flavonoids present in *A. gracilis* clearly show additivity with only three exceptions. Apigenin 7-methyl ether was not universally present in *A. cordifolia* and was not seen at all in *A. gracilis*. The other exceptions involved the observation of both luteolin 4'-methyl ether (diosmetin) and apigenin 7-*O*-glucoside in *A. gracilis* but not in either of the putative parental species.

Five compounds occur in all three taxa, four are shared by *A. cordifolia* and *A. gracilis*, and three are shared by *A. latifolia* and *A. gracilis*. The degree of additivity observed in the pigment profile of *A. gracilis*, along with other features, was taken as support for the formal recognition of *A. gracilis* (Wolf and Denford, 1984a). Numerical analysis of an expanded database involving the entire subgenus resulted in all 11 populations of *A. gracilis* sampled segregating as a cohesive group quite distinct from any other member of the subgenus (Wolf and Whitkus, 1987). This was taken as further support for recognition of *A. gracilis* as a species in its own right.

The degree of additivity seen in the hybrid member of this system is comparatively high, but there are two compounds present in the hybrid not seen in the parental species. What can be inferred about biosynthetic processes in the hybrid because of these anomalies? *O*-Methylation of apigenin at position-7 occurs in *A. cordifolia*, to yield genkwanin, which was not observed in either the other parent or in the hybrid. Since the hybrid accumulated apigenin 7-*O*-glucoside, in contrast to the other two taxa, it is tempting to speculate that, in the hybrid, glucosylation of apigenin occurs so efficiently that no apigenin remains to serve as substrate for the *O*-methyltransferase.

The hybrid species also uniquely accumulates luteolin 4'-methyl ether (diosmetin). Perhaps formation of the hybrid affected the *O*-methyltransferase such that site of attack on the flavonoid moiety was altered. Enzyme studies are required to resolve such problems.

Baccharis intermedia

The genus *Baccharis* is represented in Chile by 16 species which, according to Faini et al. (1991), have interacted in various combinations to produce 27 hybrids. Those workers undertook a detailed phytochemical study of the parental species, *B. linearis* and *B. macraei*. *Baccharis intermedia* is a widespread taxon in Chile whereas *B. macraei* is restricted to a narrow band along the coast. In the few places where the two species are sympatric, morphologically intermediate plants occur. Examination of the putative hybrids, recognized as *B. intermedia*, showed a high degree of intermediacy in their acetylenic, terpenoid, and phenolic chemistries. The two species and hybrid share only one flavonoid, kaempferol 3-methyl ether. The hybrid and *B. macraei* share 5,4'-dihydroxy-6,7-dimethoxyflavone (cirsimaritin), whereas the hybrid and *B. linearis* share two compounds, quercetin 3-methyl ether and quercetin 3,3'-dimethyl ether. *Baccharis macraei* has two compounds not seen in either of the other two taxa, 5-

hydroxy-6,7,4'-trimethoxyflavone (salvigenin) and 5,7-dihydroxy-6,4'-dimethoxyflavone (pectolinarigenin). The authors conclude that crossing between *B. linearis* and *B. macraei* has resulted in, and presumably continues to produce, a hybrid swarm of F_1 and backcross progeny, and that "*B. intermedia*" has to be cited as *Baccharis* x *intermedia* DC. (*pro spec.*).

Bidens isabelensis

Bidens isabelensis, a member of *Bidens* sect. *Greenmania*, has been examined in detail by Roseman (1986) in her biosystematic study of the section in Mexico, Central America, and Jamaica. This species, a hexaploid with $n = 36$, occurs in the Atlantic drainage of Guatemala in the departments of Alta Verapaz, Baja Verapaz, Zacapa, and Isabel, and in Belize. A population of *B. isabelensis* from Cobán differs from other populations both morphologically (smaller outer involucral bracts, increased pubescence, and mostly tripartite leaves) and in the nature of its flavonoids.

Several compounds in the flavonoid profile of *B. isabelensis* distinguish it either from other species or from other populations. Quercetagetin, for example, was not detected in the other hexaploid species, *B. squarrosa*. It was accompanied by the 3,7-dimethyl (tomentin) and 3,6,7,4'-tetramethyl (casticin) derivatives of quercetagetin. *Bidens isabelensis* lacks butein and kaempferol glycosides. Most interesting for our present purpose, however, is the observation that nine compounds in the Cobán population were not seen in other populations of this species. Four of these, isoliquiritigenin 4',4-dimethyl ether, 2',3',4',4-tetrahydroxychalcone (riparin) 3'-*O*-glucoside, an unidentified acylated derivative of riparin, and okanin 4'-*O*-diglycoside, were seen in two other hexaploid "groups." (The composition of these groups was not clear from the text.) The appearance of these compounds in the Cobán population led Roseman to ask if an individual from one of these other groups could have been involved in hybridization. In addition to that subset of flavonoids, four additional compounds were encountered that were unique to this population. Two of them, mangiferin and isomangiferin, are *C*-glucosylxanthones [**23–1** and **23–2**, respectively] and may simply represent novel chemistry of this population. [Giannasi (1975b) encountered this pair in tetraploid *Dahlia australis*.] The other two compounds, however, are methylated quercetagetins that may be intermediates in the biosynthesis of casticin (which has four *O*-methyl groups). Roseman (1986) suggested that their accumulation in detectable amounts might have resulted from a perturbation of flavonoid biosynthetic control.

Carduus hybrids in Canada

An overview of flavonoid studies of the *Carduus nutans* complex in Canada was discussed in Chapter 23. Briefly, Desrochers et al. (1988) showed that *C. nutans* is best considered to comprise two subspecies, subsp. *leiophyllus* and subsp. *nutans*. *Carduus nutans* subsp. *nutans* occurs from southern Ontario east into the Maritime Provinces; subsp. *leiophyllum* enjoys a disjunct distribution with representatives in the Prairie Provinces and western Canada, as well as in southern Ontario and Quebec. In this region of overlap, the two subspecies have been found to hybridize. Although the flavonoid profiles of the two subspecies have several elements in common, each presents a unique combination of compounds. Plants from the overlap area suspected of being of hybrid origin exhibited flavonoid profiles that showed pigment contribution from each of the subspecies, thus passing the "test of additivity."

Additional information on flavonoids of *Carduus* and a further example of natural hybridization come from a report by Bain and Desrochers (1988) on *C. nutans* and *C. acanthoides*. Like *C. nutans*, *C. acanthoides* is a Eurasian species that has become a widespread weed in North America. The two are known to hybridize and introgression has been postulated (Warwick and Bain, 1987 and citations therein). Although these authors did not examine hybrids from Nature, they did look at progeny of artificial crosses. First they established that the two species, although similar in their flavonoid synthetic capacities (only apigenin and luteolin 7-*O*-glycosides), can be distinguished because *C. acanthoides* has a luteolin 7-*O*-digalactoside that does not occur in *C. nutans*. They also showed that selfing of *C. acanthoides* gave progeny exhibiting only the profile seen in out-crossed individuals, and that selfing *C. nutans* gave offspring with a flavonoid profile that, although variable in some regards, did not contain the marker compound, i.e. the digalactoside. Using *C. nutans* as the pollen parent, as far as can be determined from the experimental description, F_1 hybrids were then produced that without exception had the digalactoside. This was taken by those authors that formation of this compound is controlled by a dominant allele in *C. acanthoides*. This is in agreement with the view that flavonoid glycosylation reactions are under simple genetic control (see Chapter 5 for details).

Centaurea triumfettii

Two species of *Centaurea* figured significantly in Gonnet's (1992, 1993) studies of community structure and dynamics in the mountains of southern

Europe, *C. montana* and *C. triumfettii* (see Chapter 24 for details). What concerns us in this chapter is the identity of some specimens of *C. triumfettii* collected near Brig (Valais) in southern Switzerland. First, a note about the relationship between these two species. The two species are thought to be closely related as indicated by a high degree of morphological similarity, similar geographical ranges, crossability, and chromosome similarities: *C. triumfettii* has $2n = 22$, *C. montana* has $2n = 44$. The combination of these features led to the suggestion that *C. montana* is a polyploid derivative of *C. triumfettii* [see Gonnet (1993) for a review of the taxonomic situation].

Two of the samples collected from the Brig area were originally described as being close to *C. montana*, based essentially on shape and ornamentation of capitular bracts, while others in the immediate vicinity were identified as *C. triumfettii*. The plants that were likened to *C. montana*, however, exhibited to a very substantial degree characteristics normally associated with *C. triumfettii*, including a chromosome number of $2n = 22$. The flavonoid profiles of these individuals were simpler than those of *C. montana*. The profiles were much akin to those obtained from pure *C. triumfettii*, dominated by apigenin 7-*O*- and 4'-*O*-glucosides with a much lesser contribution from *C*-glycosylflavones. A further individual that had been identified as *C. montana* cf. *burnati*, and considered to be *C. montana* introgressed by *C. triumfettii*, was in fact shown to be another specimen of *C. triumfettii*. In these cases, some introgression has likely occurred but after examination of morphological and cytological features, and now flavonoids, it is clear that the direction of introgression was the reverse (Gonnet, 1993).

Coreopsis mutica

As part of an extensive study of *Coreopsis*, Crawford (1972) examined synthetic hybrids between *C. mutica* var. *subvillosa* and *C. mutica* var. *carnosifolia*. Pure material was obtained from areas geographically remote ruling out natural hybridization. Both varieties had apigenin and luteolin 7-*O*-glycosides and 6-hydroxyluteolin 7-*O*-glucoside in common. Additionally, variety *carnosifolia* had two quercetin 3-*O*-glycosides and acylated derivatives of the anthochlor pair butein/sulfuretin. Twenty-eight F_1 hybrid individuals were analyzed. Overall, the hybrids exhibited flavonoid profiles containing elements of each parent, but there was a considerable degree of variation, as can be seen by inspection of Table 25–2. All F_1 hybrid individuals had luteolin 7-*O*-glucoside and two of the three anthochlor pairs. Most of the hybrids also had a third anthochlor pair. Two compounds, the luteolin 7-*O*-glycoside and a quercetin 3-*O*-glycoside, were seen in only

Table 25–2. Flavonoids of artificial hybrids between *Coreopsis mutica* var. *subvillosa* and *C. mutica* var. *carnosifolia*

Taxon or hybrid	Compound[1]									
	1	2	3	4	5	6	A	B	C	D
C. mutica var. *subvillosa*	+	+	+	+	−	−	−	−	−	−
C. mutica var. *carnosifolia*	+	+	+	−	+	+	−	+	+	+
F_1 Plants										
No. 1	−	+	−	+	−	−	−	+	+	+
No. 2	−	+	−	−	−	−	−	+	+	+
No. 3	−	+	−	+	+	−	−	+	+	+
No. 4	+	+	−	−	−	−	−	+	+	+
No. 5	+	+	+	+	−	−	−	+	+	+
No. 6	−	+	−	−	−	−	−	+	+	+
No. 7	+	+	−	−	−	−	−	+	+	+
No. 8	+	+	+	+	−	−	+	+	+	+
No. 9	−	+	−	−	−	−	+	−	+	+
No. 10	+	+	−	+	−	−	−	+	+	+
Backcross progeny										
No. 1	−	+	−	+	−	−	−	+	+	+
No. 2	−	+	−	−	−	−	−	+	+	+
No. 3	−	+	−	+	−	−	−	+	+	−
No. 4	−	+	−	−	−	−	+	−	−	+
No. 5	+	+	−	+	+	−	−	−	+	−
No. 6	+	+	−	+	−	−	−	−	+	−
F_2 Plants										
No. 1	+	+	−	−	−	−	+	+	+	+
No. 2	+	+	−	−	−	−	−	−	−	+
No. 3	−	−	+	−	−	−	−	+	+	+
No. 4	−	+	+	−	−	−	−	−	−	−
No. 5	−	+	−	−	−	−	−	−	−	+

[1] 1 = 6-Hydroxyluteolin 7-*O*-glucoside; 2 = Luteolin 7-*O*- glycoside; 3 = Luteolin 7-*O*-glycoside; 4 = Apigenin 7-*O*-glycoside; 5 = Quercetin 3-*O*-glycoside; 6 = Quercetin 3-*O*-glycoside; A = Coreopsin/Sulfurein; B = Acylated anthochlor pair; C = Acylated anthochlor pair; D = Butein/sulfuretin glycoside pair.

two of the 28 hybrid individuals. A second quercetin 3-*O*-glycoside, known from var. *carnosifolia*, was not detected in any of the hybrids.

The five F_2 hybrids obtained also exhibited a considerable degree of variation with some parental compounds missing altogether from the hybrids. Fourteen backcross hybrids were synthesized using var. *subvillosa* as the recurrent parent. Variation was again evident with significant flavonoid contribution from both parents. The majority of these backcross hybrids fell within the morphological limits of var. *subvillosa* and, as pointed out by Crawford, would have been identified as members of that variety had these plants been encountered in the field. It is clear that inheritance of morphological and chemical features (flavonoids at least) in this system are not closely associated.

Dubautia ciliolata × *D. scabra*

Dubautia, a member of Heliantheae subtribe Madiinae, consists of 21 species endemic to the Hawaiian Islands (Carr, 1985). The genus is further segregated into three sections, sect. *Venoso-reticulatae* (1 species), sect. *Dubautia* (10 species), and sect. *Railliardia* (10 species). Species in sections *Venoso-reticulatae* and *Dubautia*, characterized by chromosome number $n = 14$, tend to be found in more mesophytic sites and do not produce foliar secretions. Nine species of section *Railliardia* have $n = 13$; one, *D. scabra*, has $n = 14$. Members of this section are found in more xeric habitats and, with the exception of *D. scabra*, produce a foliar exudate that contains a complex mixture of compounds including flavonoid aglycones.

Hybridization between *Dubautia* species is common in zones of contact, but the future prospects of the offspring are not well understood. One such system involves *D. scabra* subsp. *scabra* ($n = 14$) and *D. ciliolata* subsp. *glutinosa* ($n = 13$). Two sites were selected for a study of the flavonoid chemistry of hybridizing populations (Crins et al., 1988a), one in the saddle region between Mauna Loa and Mauna Kea, and one near the Devastation Trail in the vicinity of Kilauea Iki. (Both of these sites are on the Island of Hawai'i.) Exudate flavonoids were identified from *D. ciliolata* and from suspected hybrid individuals. Vacuolar flavonoids were identified from both species and from suspected hybrids. Vacuolar flavonoids identified are derivatives of kaempferol, 6-hydroxykaempferol, quercetin and 6-hydroxyquercetin (quercetagetin), all of which occur as 3-*O*-glycosides, and eriodictyol 7-*O*-glucoside. The occurrence of these compounds in the two parent species and their putative hybrids is presented in Table 25–3.

Kaempferol, quercetin, and eriodictyol glucosides were the only compounds present in both species and hybrids. Since methoxyflavonoids occur in *D. scabra* but not in *D. ciliolata*, the presence of the entire array of flavonols in the hybrid plants means that they inherited, from *D. scabra*, the capacity to hydroxylate carbon-6 and methylate the resulting hydroxyl group. The hybrids produce 3-*O*-rhamnosylglucosides (probably rutinosides) of all four flavonols; these were not seen in either parent species. *Dubautia ciliolata* exhibited kaempferol and quercetin 3-*O*-rhamnosylgalactosides, but the capacity to use galactose in diglycoside formation appears to have been lost in the hybrids.

Table 25–3. Flavonoid glycosides of *Dubautia ciliolata* subsp. *glutinosa*, *D. scabra* subsp. *scabra*, and their putative hybrid

Compound	*Dubautia* taxon[1] *ciliolata*	hybrid	*scabra*
Kaempferol 3-*O*-glucoside	+	+	–
Kaempferol 3-*O*-rhamnoglucoside	–	+	–
Kaempferol 3-*O*-rhamnogalactoside	+	–	–
6-Methoxykaempferol 3-*O*-glucoside	–	+	+
6-Methoxykaempferol 3-*O*-diglucoside	–	–	+
6-Methoxykaempferol 3-*O*-rhamnoglucoside	–	+	–
Quercetin 3-*O*-glucoside	+	+	+
Quercetin 3-*O*-rhamnoglucoside	–	+	–
Quercetin 3-*O*-rhamnogalactoside	+	–	–
Patuletin 3-*O*-glucoside	–	+	+
Patuletin 3-*O*-diglucoside	–	–	+
Patuletin 3-*O*-rhamnoglucoside	–	+	–
Eriodictyol 7-*O*-glucoside	+	+	+

[1] *ciliolata* = *D. ciliolata* subsp. *glutinosa*; *scabra* = *D. scabra* subsp. *scabra*.

Examination of the leaf exudate of *Dubautia ciliolata* afforded apigenin, apigenin 7-metyl ether (genkwanin), scutellarein, luteolin 7-methyl ether and kaempferol 7-methyl ether. (*Dubautia scabra* lacks glandular structures and thus has no exudate flavonoids.) Three hybrid plants collected in 1986 and all 10 collected in 1987 from the Saddle site gave quercetin and luteolin 7-methyl ether but varied in their expression of the other aglycones. A single hybrid individual collected at the Devastation Trail site afforded only quercetin. Two synthetic F_1 hybrids (produced by G. Carr at the University of Hawaii) also gave only quercetin. Since the exudate chemistry of the Devastation Trail hybrid showed the same depauperate pattern as seen in synthetic F_1 hybrids, the authors concluded that that individual was likely an F_1 hybrid offspring. Exudate flavonoid profiles from the Saddle site were much more complex and resembled *D. ciliolata* to a greater extent. This was taken as evidence that first generation hybrids had backcrossed with the glandular parent. Difference in the age of the populations may also be of significance. The Devastation Trail site was created in 1959 by the eruption that created Kilauea Iki. The Saddle site is the older of the two existing on lava from 1935 and earlier eruptions. Additionally, the substrates in the Saddle area are more heterogeneous than those in the more recently derived site. In view of the heterogeneity of the site and the longer time available, it is not surprising to find evidence of introgression in these populations.

Lasthenia

Lasthenia, a small, mostly Californian genus, has been subjected to extensive flavonoid analyses including an examination of a number of artificial hybrids (Ornduff et al., 1973a). The flavonoid profiles of most species were comprised of glycosides of flavones and flavonols with aurone and chalcone derivatives playing a major role in a few. Each species exhibits a distinctive array of flavonoids. In general, anthochlor (aurones and chalcones), kaempferol, and most patuletin glycoside inheritance is additive. Luteolin and quercetin patterns are not always additive with some progenies producing glycosides not known from either parent. This sort of result has led to the suggestion by several workers that control mechanisms present in the parent organisms may have been upset by hybridization. Detailed molecular genetic studies would seem to offer solutions to some of these problems, but this would require a much better understanding of the enzymology of glycosylation.

In addition to the study of artificial hybrids, available flavonoid data allow us a look at some putative natural hybridizations. One such case, in which flavonoids behave almost perfectly, involves the possible origin of *L. ferrisiae* from a cross between *L. glabrata* subsp. *coulteri* and *L. chrysantha* (Ornduff, 1966). The profile of *L. glabrata* consists of quercetin 3-*O*-galactoside, 3-*O*-glucuronide, and a 3-*O*-rhamnosylgalactoside but lacks anthochlors. The *L. chrysantha* profile consists of the same three quercetin glycosides plus the anthochlors butein and sulfuretin. The *L. ferrisiae* profile consists of the same three glycosides plus quercetin 3-*O*-glucoside and the two anthochlors (Bohm et al., 1974). The appearance of quercetin 3-*O*-glucoside, not seen in either parent, could have resulted from some disruption in control of the glycosyltransferase system(s) with regard to substrate selectivity, glucose vs. galactose, or perhaps it is a reflection of an increase in concentration of the glucoside above the threshold of chromatographic visibility. Detailed biosynthetic and enzymological studies are needed to gain a better understanding of this phenomenon.

Lasthenia burkei, *L. conjugens*, and *L. fremontii* comprise a closely related group of species in section *Ptilomeris* (Ornduff, 1966). Several suggestions have been made to account for the close relationship between these species, one of which involved *L. burkei* being a stabilized hybrid derivative of the other two species (Ornduff, 1969, 1976). The availability of flavonoid data for these three species (Saleh et al., 1971; Bohm et al., 1974; Bohm and Banek, 1986) provided an opportunity to see if an additional data set might help clarify the picture. Adding to the complexity of the system, however, is the existence of chemical races of each of the three species. In many situations where flavonoids have been employed in analysis of hybridizations, deviation from additivity has involved changes in the nature of the sugars involved in glycoside formation. This appears to be the case again in the present situation where the flavonoid glycoside profile(s) of *L. burkei* are somewhat more complex than those seen in the putative parents.

Recent allozyme studies, however, do not support the hybrid origin of *L. burkei* (Crawford and Ornduff, 1989). Other hypotheses considered by these workers included the origin of *L. conjugens* from hybridization of *L. burkei* and *L. fremontii*, and the possibility that *L. fremontii* and *L. burkei* arose independently from *L. conjugens*. The bottom line in all of these analyses, morphological, cytological, and flavonoid, is that while one set of information may indicate one relationship, another set of data indicates another. It is obvious that the problem of relationships among members of this group is not solvable with the information available. Detailed gene sequence data might help in this case. If it were feasible to get some idea of the direction of evolution in this group, it might be possible to arrive at a clearer idea of what factors influence the formation of a given set of flavonoid glycosides, or other features for that matter. Until that time the question remains unresolved.

Liatris aspera and *L. spicata*

Hybridization between two species of *Liatris* in northeastern Illinois was studied in detail by Levin (1967). Hybridization between *L. aspera* and *L. spicata* is rare owing to different ecological requirements and different flowering times of the two species. However, an area of overlap was discovered in disturbed sand prairies at the southern end of Lake Michigan. Two dimensional paper chromatography showed a total of 14 compounds, the majority of which were taken to be flavonoids on the basis of UV behavior. Three compounds appeared solely in extracts of *L. aspera*, eight exclusively in *L. spicata*, and three occurred in both. Availability of synthetic F_1 hybrids provided an estimate of additivity of parental flavonoid profiles. Hybrids always exhibited some compounds typical of each parent, but variable patterns were noted. The occurrence of flavonoids in natural hybrids was measured in relation to a six-character morphological index. Pure *L. spicata* had an index value of 0; *L. aspera* 12, with hybrid values ranging from 1 to 11. Pure *L. spicata* lacked compounds 1, 2 and 3 and had variable frequencies of occurrence of the other 11. Hybrids with index values of 1, 2 and 3 also lacked the first three compounds and had variable occurrences of all the others. These hybrids and pure *L. spicata* are, in fact, very similar chemically. Pure *L. aspera* and hybrids with index of 11 contained only compounds 1–6. Individuals with indices of 8–10 had compounds 1–6 nearly uniformly and only a scattering of other compounds. The summed flavonoid chemistry of hybrid individuals with index values 4–7, which represents the range of expected additivity, have all compounds, although any given individual may not.

Parthenium incanum × *P. argentatum*

In parts of northern Mexico and western Texas *Parthenium incanum* and *P. argentatum* hybridize. The flavonoids of this system were studied by Mears (1980a,b) in his extensive examinations of the genus. Many of the desert populations in the areas noted exhibit combinations of features that distinguish the mesic form of *P. incanum* from the limestone form of *P. argentatum*. *Parthenium incanum* collected from an area where it was not influenced by the other species afforded only a single aglycone, kaempferol 3-methyl ether, and a set of eight glycosides based on kaempferol, quercetin, and quercetagetin. The "pure" form of *P. argentatum*, on the other hand, afforded a rich array of both aglycones and glycosides. The aglycone component comprised *O*-methyl derivatives, with up to three methoxy groups, based upon kaempferol, quercetin, 6-hydroxykaempferol, and quercetagetin. The glycosides were based upon kaempferol, quercetin, and quercetagetin. The intermediate forms, with a profile consisting of up to seven aglycones and as many as 13 glycosides, showed strong complementarity.

Simsia chaseae

Simsia is a genus of Heliantheae distributed from southwestern North America, through Central America to Argentina (Spooner, 1990). Of interest to us here is the origin of *S. chaseae*, which occurs in southern Veracruz and the Yucatan Peninsula (Spooner, 1990). It had been suggested, on the basis of morphological and crossing studies, that *S. chaseae* arose from diploid hybrid speciation involving *S. eurylepis* and *S. foetida* (Spooner, 1990).

The floral flavonoids of all members of the genus have been described by Schilling and Spooner (1988). The pigment profile of *S. chaseae* is very similar to those reported for other members of the "*S. amplexicaulis* group," which have, in various combinations, quercetin glycosides, flavone derivatives, and anthochlors. Other than stating that the floral pigments of *S. chaseae* are consistent with a hybrid origin from members of this group of species, the authors (Schilling and Spooner, 1988) do not comment directly

Table 25–4. Flavonoids of *Simsia chaseae* and its putative parents, *S. eurylepis* and *S. foetida*

Compound[1]	*Simsia* species *eurylepis*	*chaseae*	*foetida*
Ray floret anthochlors	No rays	+	+
Quercetin 3-*O*-glucoside	+	+	+
Quercetin 3-*O*-glucuronide	+	+	+
Quercetin 7-*O*-glucoside	+	+	+
Apigenin 7-*O*-glucoside	+	–	–
Luteolin 7-*O*-glucoside	+	+	+
Coreopsin/sulfuretin	–	+	+
Marein/maritimein	–	–	–

[1] Anthochlors = chalcones and/or aurones.

on the compounds themselves. We have summarized their results on the three taxa in question in Table 25–4. *Simsia chaseae* shares the quercetin 3-*O*-glycosides and luteolin 7-*O*-glucoside with both putative parents and has the coreopsin-sulfurein pair in common with *S. foetida* (*S. eurylepis* lacks ray florets). Two shared absences, apigenin 7-*O*-glucoside and quercetin 3-*O*-glucoside, are seen in the data as well: quercetin 7-*O*-glucoside, a component of *S. foetida*, was not detected in either *S. eurylepis* or *S. chaseae*, whereas the opposite situation is seen with regard to apigenin 7-*O*-glucoside, which was seen in *S. eurylepis* but not in the other two taxa. The biosynthetic ramifications of the flavonoid 7-*O*-glucoside situation are interesting to consider in view of the enzymes(s) required for 7-*O*-glycosylation. It is generally understood that flavonoid 7-*O*-glycosyltransferases are position-specific rather than substrate-specific, that is, they will accept flavonoids of various classes as long as the 7-hydroxyl group is available (see Chapter 5). The situation may be different in the species under consideration here where 7-*O*-glycosylation of flavones occurs; but the enzyme required for 7-*O*-glycosylation of quercetin is apparently missing from *S. eurylepis*. In the case of *S. chaseae* the situation is more complex since both apigenin and quercetin 7-*O*-glycosides are apparently missing whereas luteolin 7-*O*-glycoside is present. Could there be three enzymes responsible for glycosylation in this system? Or, could hybridization have altered the precursor balance such that the conversion of naringenin to eriodictyol occurs faster than the conversion of naringenin to apigenin? It should be borne in mind that apigenin may be formed (its glycoside too) but that its (their) concentration(s) may be below the limits of detection. Answers to questions such as these require detailed studies which are understandably beyond the interest of the respective workers. If a thorough understanding of the control of flavonoid biosynthesis is ever to be gained, however, systems such as this must be probed in detail.

Tragopogon in North America

Although *Tragopogon* consists of some 50 species world-wide, the focus of attention on this genus lies with the activities of a few species in eastern Washington and adjacent Idaho. Several generations of evolutionary botanists have turned their considerable talents and attention to this system. The story involves three introduced species: *T. dubius*, *T. porrifolius*, and *T. pratensis*. Hybrids involving certain combinations of these species were observed by Ownbey (1950) who described the stabilized products as two new species, *T. mirus* and *T. miscellus*, both of which were shown to be amphitetraploids. The tetraploid *T. mirus* has *T dubius* and *T. porrifolius* as parents; the diploid parents of *T. miscellus* are *T. dubius* and *T. pratensis*.

A paper chromatographic study of the three diploid, naturalized species, F_1 hybrids, and the two derived, tetraploid species showed that each of these units had its own characteristic flavonoid profile at the population level (Brehm and Ownbey, 1965a, b, 1968). A detailed re-examination of the system (Belzer and Ownbey, 1971) corroborated the earlier findings that the inheritance of flavonoids is usually additive in the F_1 generation. Analysis of F_2's showed that flavonoid biosynthetic capacities approximated 3:1 or 9:7 ratios. Parental compounds were missing occasionally in hybrids but "hybrid compounds," that is, compounds present in neither F_1 parent, appeared only in some F_2 plants.

It is comforting to note that allopolyploid speciation in *Tragopogon*, as indicated by morphological, cytological, crossing, and flavonoid studies, has been supported by recent macromolecular observations. Chloroplast DNA analyses (Soltis and Soltis, 1989) suggested that *T. mirus* has had at least three independent origins and that *T. porrifolius* was consistently the maternal parent. Plural origins of *T. miscellus* were also indicated, and it also was shown that *T. porrifolius* and *T. pratensis*, both comparatively rare, had served as the maternal parents in the majority of cases. *Tragopogon dubius*, the most common taxon of the group, was implicated as maternal parent in only two populations of *T. miscellus*. This study was followed by another (Soltis and Soltis, 1991) wherein reports of analysis of ribosomal DNA were described that wholly supported the idea that *T. mirus* had originated independently several, if not many, times. The final chapter of the story is yet to be written, however. Hybrids have now been observed between the tetraploid species.

Viguiera triangularis

The "Baja California group" of *Viguiera* series *Viguiera* contains three polyploid species that have a greater degree of similarity among their flavonoid profiles than was seen among the diploid species of this series (Schilling, 1989). Although the polyploids are not well differentiated on morphological grounds (they have often been lumped together with the diploid *V. parishii*), flavonoids from leaf resins provide useful indicators of possible ancestry, particularly in the case of *V. triangularis*. The differences between *V. triangularis* and *V. parishii* lie mainly in size differences of leaves, pales, and achenes. Their flavonoid profiles, however, are quite distinct. Major compounds from *V. parishii* are 6-hydroxyapigenin 7-methyl ether and the chalcone isoliquiritigenin 4'-methyl ether, with lesser amounts of apigenin, luteolin, and 6-methoxyapigenin. The flavonoid profile of *V. triangularis* is a good deal more complex with the major compounds identified as: luteolin, 6-methoxyapigenin, 6-hydroxyapigenin 7-methyl ether, the 6,4'- and 7,4'-dimethyl ethers of 6-hydroxyapigenin, the 6- and 7-methyl ethers of 6-hydroxyluteolin, and isoliquiritigenin 4'-methyl ether. Present in lesser amounts were apigenin 7-methyl ether and the 6,7-dimethyl ether of 6-hydroxyapigenin. Morphological similarities, as mentioned above, suggest that *V. parishii* may have supplied one of the genomes to the polyploid. Flavonoids support this suggestion since the pigment profile of *V. parishii* consists of a subset of the pigments identified from *V. triangularis*. Flavonoids can assist in identifying other possible contributors as well. *Viguiera microphylla* has morphological, geographic, and flavonoid profile similarities that suggest it as a possible parent. The major compounds identified from this species are luteolin and the 6- and 7-methyl ethers of both 6-hydroxyapigenin and 6-hydroxyluteolin. The 6,7-dimethyl ether of 6-hydroxyapigenin was present as a lesser component.

It was not possible to rule out *V. tomentosa* as a possible participant in the hybridization since it had appropriate flavonoids, i.e., luteolin and the 7-methyl ethers of 6-hydroxyapigenin and 6-hydroxyluteolin. It was possible to eliminate other species on the basis of their flavonoid profiles, however. *Viguiera laciniata* could be removed from contention on the basis of its specialization in accumulating flavonols, naringenin derivatives, and several anthochlors not seen in any other species in the series. *Viguiera subincisa* was discounted owing to its unique capacity to make the 6,3'-dimethyl ether of 6-hydroxyluteolin, and *V. reticulata* was eliminated on the basis of its capacity to accumulate apigenin 4'-methyl ether.

Zaluzania

Seven of the nine species that comprise *Zaluzania* have $n = 18$; the other two are polyploids, *Z. mollissima* with $n = 27$ and *Z. subcordata* with $n = 36$ (Olsen and Mabry, 1980). Those authors pointed out that the flavonoid chemistry was consistent with both polyploids having a close relationship with diploid *Z. augusta* or its ancestral stock. We are unaware of further studies of this system.

Zinnia subg. *Diplothrix*

The caespitose zinnias, subgenus *Diplothrix*, consist of seven taxa distributed in northern Mexico, Arizona, and Texas. Torres and Levin (1964) described a chromatographic study of all members of the group along with some artificial hybrids. They discussed the chromatographic results in relation to cytogenetic and morphological data already available for the group. *Zinnia juniperifolia* and *Z. oligantha* are $n = 10$ species, *Z. acerosa* exists in two forms, one $n = 10$ and one with $n = 20$. *Zinnia citrea* has $n = 20$, *Z. grandiflora* has $n = 21$ and *Z. anomala* has $n =$ ca. 42. *Zinnia anomala*, *Z. grandiflora*, *Z. juniperifolia*, and *Z. oligantha* gave a chromatographic pattern that consisted of four spots, designated $A_{1,3}$ (other spots were seen but were apparently invariant throughout the subgenus). The $n = 10$ form of *Z. acerosa* gave a pattern of three spots designated as A_2. *Zinnia citrea* and the $n = 20$ form of *Z. acerosa* gave a pattern, designated $A_{1,3}A_2$, that consisted of all seven spots. These data, along with the presumed genomes of the taxa, are presented in Table 25–5.

Table 25–5. Flavonoid patterns in caespitose species of *Zinnia*

Species	No.[1]	Spot number 1	2	3	4	5	6	7	Pattern
juniperifolia	10	–	–	–	+	+	+	+	$A_{1,3}$
acerosa	10	+	+	+	–	–	–	–	A_2
oligantha	10	–	–	–	+	+	+	+	$A_{1,3}$
acerosa	20	+	+	+	+	+	+	+	$A_{1,3}$ A_2
citrea	20	+	+	+	+	+	+	+	$A_{1,3}$ A_2
grandiflora	21	–	–	–	+	+	+	+	$A_{1,3}$
anomala	ca. 42	–	–	–	+	+	+	+	$A_{1,3}$

[1] Diploid chromosome number.

Combined data were used to address the origin of the four taxa that exhibited the higher chromosome numbers. The first of these that we will consider is the origin of the $4n$ form of *Z. acerosa*. Morphological and cytological analyses suggested that it is an allotetraploid with the genotype $A_2A_2A_3A_3$. The chromatographic pattern exhibited by this taxon was the sum of those seen for $2n$ *Z. acerosa* and either *Z. juniperifolia* or *Z. oligantha*. The presence of a dominant ray color marker in $4n$ *Z. acerosa* and in *Z. juniperifolia*, along with other structural features, clearly exclude *Z. oligantha* as a possible parent.

Tetraploid *Z. citrea* was thought to be of hybrid origin from a cross involving *Z. juniperifolia* and one of the other diploid species. The flavonoid pattern clearly indicates the presence of the compounds expected if diploid *Z. acerosa* were the other parent. The involvement of *Z. juniperifolia* was indicated on morphological grounds and, moreover, several artificial F_2 hybrids bore a strong phenotypic resemblance to *Z. acerosa*.

Zinnia grandiflora was thought to be the product of a cross between *Z. juniperifolia* and *Z. oligantha*. The chromatographic evidence supports this idea since the pigment profile of *Z. grandifolia* shows only spots typical of those two diploid species and none that are characteristic of *Z. acerosa*.

Zinnia anomala, with n = ca. 42, was thought to have originated more or less directly from *Z. grandiflora* ($n = 21$) by chromosome doubling. If this were the case, one would expect the chromatographic profile of *Z. anomala* to be the same as its precursor species. This is, in fact, the case, as both taxa exhibit the $A_{1,3}$ spot pattern. There was a noticeable increase in the concentration of one of the compounds in the $n = 42$ taxon, but there were no qualitative differences.

These examples clearly demonstrate the potential usefulness of flavonoid pattern analysis in the study of hybrid species. In most cases, species thought to be of hybrid origin exhibited additive pigment profiles in agreement with the generally held view that flavonoids are inherited in a codominant fashion. However, enough deviations from complete additivity have been observed, with patterns that can either be simpler or more complex than the parental patterns, to indicate that significant changes in biosynthetic capabilities can occur as a result of hybridization. Details of what causes these changes are lacking. A better understanding of the phenomenon could likely be achieved through the application of enzymological and molecular genetic techniques.

Section VI

Flavonoids and Phylogeny

Despite conclusions emphasized in Chapter 18–23 that flavonoids are, in general, more efficaceous at lower levels of the taxonomic hierarchy in Asteraceae, it is still compelling to cast an eye outward to familial relatives for possible hints on relationships. This is now feasible, because the general structure of phylogenetic relationships among families of Asteridae is reasonably well known from macromolecular data (Olmstead et al., 1992; Judd et al., 1999). We assess this situation in Chapter 26. New molecular phylogenies within the family at tribal levels (e.g., Jansen and Kim, 1996; Bayer and Starr, 1998) also offer opportunities to use this new framework of relationships for assessing evolution of the flavonoid system *within* the family. This is outlined in Chapter 27.

Because of variation in presence or absence of flavonoids within taxa (a point that has been emphasized repeatedly throughout this book), detailed quantitative analysis using either phenetic or cladistic algorithms is difficult, if not impossible, at this time. We have, in fact, attempted these analyses with little positive result. We have elected in these two chapters, therefore, to seek general trends and tendencies and paint relationships and correlations with a broad brush. Only by greater understanding of infrataxon variation of compounds, the genetic and enzymatic basis of biosynthetic pathways, and the potentials for interconversion of compounds under different ecological and/or physiological constraints, can we come to separate phyletic from ecological/genetic dimensions. We will then be in a much better position to use flavonoid data more effectively to assess phylogenetic relationships among families, and to interpret patterns and processes of flavonoid evolution. There is obviously still much to be done.

CHAPTER 26

Flavonoid Relationships with other Families

In the preceding chapters we attempted to bring together as much flavonoid occurrence data as possible in order to address the usefulness of the information in dealing with problems of relationships among various groups within Asteraceae. Flavonoid data have been very instructive in certain instances, particularly when dealing with situations at the lower levels in the taxonomic hierarchy, including hybrid systems. Absolute differences are rarely encountered at higher taxonomic levels, but significant tendencies in flavonoid occurrence frequencies are evident among subfamilies and among certain groups of tribes. Flavonoid patterns at the subtribal level leave much to be desired, a problem that wider sampling might help to alleviate.

There remains one additional task, however. In this chapter we will look at the use of flavonoids as indicators of relationships among Asteraceae and other families. This topic has been addressed within the context of larger flavonoid surveys of the angiosperms (Gornall et al., 1979) and the dicots (Giannasi, 1988, 2000). The treatment by Giannasi (1988) included a short, insightful, and information-rich summary of flavonoid profiles of Asteridae and their evolutionary implications. Data accumulated since that time do not materially alter the generalizations presented by him. A rehashing of those discussions is unnecessary considering the availability of the work in which his paper appeared.

Holub et al. (1987) presented a brief review of the use of various phytochemicals that have been used to address the question of relationship between Asteraceae and Apiaceae. They pointed out that although polyacetylenic compounds are common in both families, terminal groups typically observed in members of Asteraceae, i.e., cyclic, aromatic, or heteroaromatic, are rarely seen in species of Apiaceae. Sesquiterpene lactones also occur in both families but the stereochemical similarities observed are seen by these workers as representing parallel phylogeny in their respective subclasses.

Our purpose at this time is to examine flavonoid occurrence patterns with reference to newer views of relationships within the dicotyledons. A major revolution in our approach to understanding angiosperm phylogeny, based upon DNA sequence data, has provided the systematic community with a wealth of new phylogenetic hypotheses (e.g., Chase et al., 1993; Angiosperm Phylogeny Group, 1998; Judd et al., 1999). An entire issue of the Annals of the Missouri Botanical Garden [Vol.79(2), 1992] was dedicated to phylogeny of Asteridae. Of particular interest to us here are studies that addressed, either directly or indirectly, identification of the closest relatives of Asteraceae [e.g., Olmstead et al. (1992); Michaels et al. (1993)].

The Olmstead et al. (1992) work involved a parsimony analysis of *rbc*L data for a broadly based Asteridae. Linked together to form Asterales s.l. in one or more treatments were representatives of Asteraceae (*Vernonia, Lactuca, Senecio, Helianthus, Barnadesia,* and *Dasyphyllum*), Calyceraceae (*Boöpis*), Goodeniaceae (*Scaevola*), Campanulaceae (*Campanula* and *Lobelia*), Menyanthaceae (*Menyanthes* and *Villarsia*), Cornaceae (*Corokia*), and Escalloniaceae (*Escallonia*). The study by Michaels et al. (1993) was concerned specifically with interfamilial relationships and provided a phylogeny against which we can compare flavonoid biosynthetic capabilities. Representatives of Asteraceae for which *rbc*L data were obtained by this group are: *Senecio, Lactuca, Helianthus, Flaveria, Carthamnus, Barnadesia,* and *Dasyphyllum*. The first four genera constitute a sister group to *Carthamus*. Those five genera taken together emerge as the sister group to the *Barnadesia-Dasyphyllum* clade. The sister group to the Asteraceae clade consists of the Goodeniaceae representative (*Scaevola*) and the Calyceraceae representative (*Boöpis*). The next split involves the cornaceous genus *Corokia*, followed by *Villarsia* and *Menyanthes* (Menyanthaceae).

The studies cited above, coupled with restriction mapping studies (Downie and Palmer, 1992), and 18S rDNA sequences (Soltis et al., 1997) provide convincing evidence that Calyceraceae and Goodeniaceae are sister groups of Asteraceae (Gustafsson, 1996). The mapping studies uncovered extensive variation in sequence and length of cpDNA from Goodeniaceae, however. Significantly, Calyceraceae lacked the extensive rearrangements found in Goodeniaceae which led to the suggestion that Calyceraceae is the more likely sister group. Further information supporting this view was assembled by DeVore and Stuessy (1995) in their discussion of the place and time of origin of Asteraceae. It has also been independently determined cladistically by Gustafsson and Bremer (1995).

How do flavonoids fit into this picture? A recently published description of the flavonoid chemistry of Calyceraceae (Bohm et al., 1995) provided an opportunity to see if these compounds could add anything novel. Pigment profiles were determined for one species each of *Acicarpha, Boöpis, Gamocarpha,* and *Nastanthus,* and six of *Calycera*. The flavonoids identified included kaempferol and quercetin glycosides, 6-methoxykaempferol, and 6-methoxyquercetin (patuletin), although not all genera, nor species within *Calycera*, had all compounds. The compounds identified from Calyceraceae are known to occur widely in Asteraceae. Because of the common nature of the Calyceraceae flavonoids, it is not surprising that the pigment profile of the family falls easily within the bounds of structural variation seen in Asteraceae. Utilization of the flavonoid data alone provides no particular insight into assessing any relationship between the families, but combining the Calyceraceae profile with information from other sources proved to be very useful. One of those sources was a recent study of the flavonoids of Barnadesioideae (Bohm and Stuessy, 1995), considered to be the most primitive group in Asteraceae. The flavonoid profile of this subfamily consists almost solely of kaempferol and quercetin 3-*O*-glycosides and is the simplest seen in any other group of comparable size and taxonomic rank within the family. It is of interest, then, that the simplest flavonoid profile from the study of Calyceraceae belongs to *Acicarpha*, *Gamocarpha*, and one specimen of *Calycera leucanthema*, and consists of the same kaempferol and quercetin glycosides. The similarity of flavonoid profile of Calyceraceae and *Acicarpha* is of more than passing interest since *Acicarpha* has been suggested as representative of prototypical Calyceraceae (DeVore, 1994) based on its pattern of floral development (centripetal), diploid chromosome number (n = 8), which is the only diploid number so far recorded for the family, and its range of occurrence in the grasslands of Uruguay and nearby regions, which is similar to that of *Schlechtendalia* (Barnadesioideae). More recently, however, Carlquist and DeVore (1998) suggest cautiously, based on wood anatomy, that *Calycera* might be the basal genus of the family because it appears less specialized. In any event, possession of a very simple flavonoid profile, identical to that of Barnadesioideae, adds to the list of characters shared by primitive elements of Calyceraceae and ancestral members of Asteraceae.

Goodeniaceae are a principally Australian family comprising 16 genera and 430 species (Mabberley, 1987). So far as we are aware, the only flavonoid information available for the family comes from a study of *Goodenia* by Middleton and Jefferies (1968) who reported quercetin 3-methyl ether from *G. strophiolata*, and a study of *Scaevola* by Patterson (1984). In this latter study, seven endemic Hawaiian species were shown to possess an apigenin glycoside and several quercetin glycosides including the unusual quercetin-3-*O*-galactside-7-*O*-rhamnoside, which was found only in *S. coriacea*. The presence of an apigenin glycoside sets *Scaevola* apart from other taxa in this comparison as does the quercetin diglycoside. This represents a poor sampling of the family, however, as compared to Calyceraceae, and until additional information becomes available flavonoids certainly cannot be mustered as convincing evidence for a particularly close relationship between Asteraceae and Goodeniaceae. The same proviso should include Brunoniaceae, a monogeneric Australian family, that at times has been considered as part of Goodeniaceae. Some information on chemistry of *Brunonia* might be informative insofar as the family emerged as part of a well-formed clade including Asteraceae, Calyceraceae, Goodeniaceae, and Brunoniaceae (Gustafsson and Bremer, 1995).

An unexpected result of the *rbc*L sequence studies was the emergence of Menyanthaceae, a family that had never been associated with Asteraceae on the basis of traditional data. A study of flavonoids of Menyanthaceae (Bohm et al., 1986b) provides us with an opportunity to compare that family's biosynthetic capacities to those of Asteraceae. Members of Menyanthaceae accumulate a variety of flavonols: kaempferol, kaempferol 7-methyl ether, quercetin, quercetin 7-methyl ether (rhamnetin), quercetin 3'-methyl ether (isorhamnetin), and quercetin 3,7-di- and 7,3'-dimethyl ethers. Several of these occur as a complex mixture of glycosides. As was noted in that paper, the absence of *C*-glycosylflavones and xanthones sets Menyanthaceae apart from Gentianaceae, the family with which it is most often allied. Menyanthaceae is a flavonol specialist with a modest level of *O*-methylation. Singly *O*-methylated compounds occurred in all genera tested except *Liparophyllum* and the dimethyl ethers, along with rhamnetin, were found only in a few species of *Nymphoides*. Menyanthaceae lack all of the other flavonoid types seen in Asteraceae and appear not to have the capacity for extra oxygenation of the A-ring. Flavonoids do not provide support for the association of Menyanthaceae with Calyceraceae or Goodeniaceae as suggested by the *rbc*L data. Other features do, however, as summarized by Lammers (1992): presence of *seco*-loganin, inulin as a storage carbohydrate, multinuclear tapetal cells, absence of endosperm haustoria, and chromosome numbers based on $n = 8$ or 9. Menyanthaceae emerged as a significant factor in a recent phylogenetic study of 23 families that comprise or have been associated with the "Asterales-Campanulales complex" (Gustafsson and Bremer, 1995). The analysis utilized 46 characters and included the presence of iridoids and inulin as a storage carbohydrate. Asteraceae, Calyceraceae, Brunoniaceae, and Goodeniaceae constituted a well-formed clade with Calyceraceae as the sister group to Asteraceae. Menyanthaceae emerged as a sister-group to the set of four families. The attention given to Menyanthaceae came about because of the similarity of *rbc*L sequences. Other data suggest that this relationship may, indeed, be real. It would seem entirely appropriate to examine this possibility with additional studies.

Cornaceae, as represented by *Corokia*, also emerged unexpectedly from the *rbc*L studies. As reviewed by Hegnauer (1964, 1989), the flavonoid chemistry of Cornaceae consists of glycosides of kaempferol and quercetin plus several anthocyanidins. The only data specifically involving *Corokia*, according to Hegnauer (1964), is the presence of leucoanthocyanidins in several species. The presence of such simple flavonol glycosides in this family can hardly be taken as indicating any relationship with Asteraceae.

Some comment on the flavonoids of Campanulaceae and Lobeliaceae, which constituted the sister group to the assemblage of families discussed above, would seem in order. Referring again to Hegnauer's summary (1989), we find that two lines exist in Campanulaceae, one characterized by glycosides of the common flavonols kaempferol, quercetin, rhamnetin, and isorhamnetin, and one by glycosides of the simple flavones apigenin and luteolin. Species of *Lobelia* accumulate the luteolin derivative diosmin. There is no mention of other types of flavonoids nor of any flavonoids having extra oxygenation or other structural modifications. The level of *O*-methylation seen in these families is modest at best. Thus, flavonoid profiles of Campanulaceae, Cornaceae, Lobeliaceae, and Menyanthaceae are of no value in isolation for assessing degrees of relationship with Asteraceae. Rather, it is the combination of chemical, cytological, ecological, geographic, and morphological features that allows for hypotheses of relatedness to be tested.

CHAPTER 27

Evolution of the Flavonoid System in Asteraceae

In addition to examination of presence and absence and frequencies of occurrence of flavonoid data in Asteraceae for purposes of testing subfamilial, tribal, and subtribal delimitations (Chapters 18-20), which comprised essentially a phenetic perspective, it is also appropriate to consider the data in a phylogenetic context. Use of flavonoids to interpret phylogeny has received numerous comments over many decades (e.g., Harborne, 1977a; Gornall and Bohm, 1978). Several workers have also considered use of secondary products in a cladistic context (Estabrook, 1980; Humphries and Richadson, 1980; Richardson, 1983a, 1983b; Seaman and Funk, 1983; Richardson and Young, 1982; Stuessy and Crawford, 1983). Our approach is to view the patterns in a phyletic context, attempting to gain maximum evolutionary insights from the data (Stuessy and Crawford, 1983; Stuessy, 1990). For these considerations, we will use data from Table 18–1, concepts of biosynthesis of major structural types (Chapter 5), and discussion of particular structural adornments. The existence of this sizeable data base at the tribal level for the entire family, recent identification of the basal group of Asteraceae (subfamily Barnadesioideae; Jansen and Palmer, 1987b), acceptance of Calyceraceae as sister family to Asteraceae (DeVore and Stuessy, 1995; Gustaffson and Bremer, 1995), recent studies of flavonoids of both these taxa (Barnadesioideae, Bohm and Stuessy, 1995; Calyceraceae, Bohm et al., 1995), and recent morphological (Bremer, 1994, 1996), and macromolecular (Jansen et al., 1990; Jansen and Kim, 1996; Bayer and Starr, 1998) cladograms among tribes of Asteraceae (see Figs. 2–3 to 2–8), all argue for attempts at phyletic interpretations of the evolution of flavonoids within the family.

One important aspect of flavonoid data in phyletic reconstructions is the large amount of information on biosynthesis of major structural types available in the literature. Although these pathways have not been confirmed in many genera of the family, the data so far are sufficiently consistent to assume that this is a general biosynthesis for the entire family (Chapter 5), and that the possession of the same classes of compounds reflects homologous levels of biochemical specialization. Previous workers have used different levels of biosynthetic specialization as a basis of determining evolutionary affinities (e.g., Levy, 1977; Gottlieb, 1982; Emerenciano et al., 1987).

The basis of all previous attempts to designate evolutionary advancement in secondary products of angiosperms is that simpler compounds tend to occur earlier in the biosynthetic pathways and more complex molecules later on. Taxa with more complex chemistries, therefore, are considered more derived. Because of the step-wise patterns of compound biosynthetic modification, however, loss of an enzyme early in the pathway can result in accumulation of compounds in early stages of biosynthesis. Advanced taxa, therefore, may also exhibit simple chemical profiles (Gornall and Bohm, 1978). Further, because of the limited *basic* structural types of flavonoids, the likelihood of such evolutionary modification is even greater. Within this cautious context, therefore, it is obvious that possession of some of the flavonoids early in the biosynthetic pathways will not necessarily occur in the most primitive tribes of the family. The most conspicuous example involves the anthochlors, one member of which, the chalcones, are precursors to all other flavonoid types. Anthochlors have been reported from some members of all tribes except Anthemideae, Calenduleae, Tageteae, Liabeae and Vernonieae. At the other extreme are the biosynthetically more derived flavones, which also occur in many tribes. Even more telling is that Barnadesioideae, judged evolutionarily basal by morphological (Bremer, 1994) and macromolecular data (Jansen and Palmer, 1987b), contain only flavonols and relatively unspecialized flavanones. These two classes of compounds are also found in all tribes (only exceptions are lack of flavanones in Calenduleae and Tageteae). Essentially, the only meaningful evolutionary perspective that can be gleaned from the distribution of basic structural types among tribes is the restriction of the highly derived isoflavones to three (or four) tribes in Asteroideae, namely Heliantheae (and Helenieae), Plucheeae and Mutisieae. Heliantheae have been judged to be one of the most derived tribes of the family based on DNA data. Even more derived, however, are Eupatorieae, which lack isoflavones but which do enjoy a comparatively high level of flavonoid "adornment," which we will address next.

More meaningful insights regarding evolutionary positions of tribes and subfamilies within the family come from the "adornment" configurations, i.e., represented as columns 9–20 in Tables 18–1 and 18–2. Some of these are structurally unusual and restricted in distribution, which enhance their evolutionary potential. In a general way, Barnadesioideae contain *none* of these features (except for *O*-methylation in a single compound, which is also found in all other tribes). This supports its presumptive primitive status. Table 18–2 also reveals that 6,8-dioxygenation, other B-ring substitution patterns, sulphates, and phenolic acylation are *largely* confined to subfamily Asteroideae, suggesting more advanced status for this group. These data are compatible with those given by Jansen et al. (1991a) for occurrence of more advanced structural types of other secondary products of the family.

One particular structural feature, *O*-methylation, offers additional evolutionary insights (Table 18–3). Flavonoids with at least one *O*-methyl group occur in some members of all tribes. Flavonoids with two *O*-methyl groups occur with comparatively high frequency in members of Asteroideae, with the exception of Calenduleae, where sampling is poor, and in Tageteae. Di-*O*-methylated flavonoids also occur in some members of Cichorioideae, particularly in Cynareae; tri- and tetra-*O*-methylated flavonoids each occur only in two genera. Higher levels of *O*-methylation have not been reported from members of this subfamily. Vernonieae have two genera that make di-*O*-methylated flavonoids and only one that makes flavonoids with four *O*-methyl groups. Tri-, tetra-, and penta-*O*-methylated flavonoids, however, occur widely in tribes of Asteroideae. Hexa- and hepta-*O*-methylated compounds occur in 13 and three genera, respectively. Flavonoids with eight *O*-methyl groups have been reported from one genus of Eupatorieae.

Examination of the values for *O*-methylation shows that 68% of the genera of Asteroideae accumulate *O*-methylated compounds, while only 42% of genera of Cichorioideae have *O*-methylated flavonoids. An even lower level of *O*-methylation is seen in Barnadesioideae (14%). Quite clearly, therefore, Asteroideae contain much higher levels of *O*-methylation, reinforcing the idea that this subfamily is more derived than the other two. This same result has been documented by Emerenciano et al. (1987). They also show a general increase in levels of oxygenation in both the A and B rings from Cichorioideae to Asteroideae.

A recent survey of flavonoids in genera of the related family Calyceraceae (Bohm et al., 1995) reveals a good deal of structural diversity (Chapter 26). Within Calyceraceae, however, following one hypothesis *Acicarpha* and *Boöpis* may be regarded morphologically as the more primitive genera (DeVore, 1994). These two taxa have very simple flavonoid profiles similar to those seen in Barnadesioideae, which reinforces the concept of relatively simple flavonoid composition in the common ancestor to both families (DeVore and Stuessy, 1995). Much structural and substitutional flavonoid evolution has occurred in the evolution of Asteraceae, therefore, but the basic changes have been confounded by numerous modifications to basic biosynthetic pathways during 30 million or more years of existence of the family (DeVore and Stuessy, 1995; Kim and Jansen, 1995a).

Addendum

This section consists of flavonoid reports in the family that were published or came to our attention after the manuscript had been sent to press. The information is arranged alphabetically by tribe and within tribes alphabetically by genus in the same format as used in the text.

Flavonoid biosynthesis (Chapter 5)

Ogata et al. (1998) isolated a UDP-glucose: cyanidin 3-*O*-glucosyltransferase from flower buds of *Senecio* ×*hybridus*, purified it 260-fold, and determined that it has a pH optimum of 7.5. Km values for cyanidin and UDPG of 0.33 and 0.20 mM, respectively, required no cofactors, and had an approximate molecular weight of 52 kDa.

Anthemideae (Chapter 7)

Artemisia: An examination of *A. annua* for sesquiterpenes by Sy and Brown (1998) also resulted in identification of known *O*-methylated flavonols based upon 6-hydroxykaempferol, quercetin, and quercetagetin.

Tanacetum: A study of *Tanacetum parthenium* (feverfew) (Williams et al., 1999b) resulted in the isolation of a number of 6-hydroxyflavonol methyl ethers, one of whose anti-inflamatory activity was found to be significant. More recently, Williams et al. (1999a) identified a number of lipophilic and polar compounds from nine species of *Tanacetum*. Lipophilic compounds identified were: 6-hydroxykaempferol 3,6-dimethyl ether, 6-hydroxykaempferol 3,6,4'-trimethyl ether, quercetagetin 3,6-dimethyl, 3,6,3', and 3,6,4'-trimethyl ethers, apigenin, luteolin, chrysoeriol, scutellarein 6-methyl-, and 6,4'-dimethyl ethers, 6-hydroxyluteolin 6-methyl, 6,3'-di-, and 6,7,4'-trimethyl ethers. Not all compounds were observed in all nine taxa. Polar flavonoid derivatives identified were: apigenin 7-*O*-glucoside, 7-*O*-glucuronide, 7-*O*-glucosylglucuronide, and 7-*O*-diglucuronide, the *C*-glucosylflavone schaftoside, luteolin 7-*O*-glucoside, 7-*O*-glucuronide,

7-*O*-gentiobioside, 6-hydroxyluteolin 7-*O*-glucoside, chrysoeriol 7-*O*-glucuronide, diosmetin 7-*O*-glucuronide, and quercetin 7-*O*-glucoside and 7-*O*-glucuronide. Again, not all taxa had all compounds.

Astereae (Chapter 8)

Baccharis: Faini et al. (1999) examined the components of epicuticular leaf waxes of *B. linearis* for seasonal variation. Involved in this study were the flavonoids kaempferol 3-methyl ether, quercetin 3-methyl ether, and isorhamnetin 3-methyl ether. The latter two compounds were present in leaves collected during each of the four seasons, whereas kaempferol 3-methyl ether was found only in leaves collected in autumn and winter.

Chrysothamnus: Stevens et al. (1999) reported a rich array of aglycones from the leaf exudates of three species of *Chrysothamnus* collected in eastern Oregon, *C. humilis, C. nauseosus,* and *C. viscidiflorus.* The results from that study coupled with literature information indicate that each of these taxa accumulates a characteristic array of compounds by which they could be easily distinguished from one another. *Chrysothamnus nauseosus* is characterized by an array of flavones, mostly *O*-methylated, along with quercetin and isorhamnetin, whereas *C. viscidiflorus* would be easily recognised by its rich array of quercetin methyl ethers, flavanones, and derivatives of dihydroquercetin (taxifolin) 3-acetate-7-methyl ether and another, unidentified, taxifolin derivative. *Chrysothamnus humilis* exhibited a comparatively simpler profile consisting of kaempferol and quercetin methyl ethers. Although the flavonoid data are helpful in distinguishing between these three species, they do not provide support for inclusion of *Chrysothamnus* in *Ericameria* as has been proposed on the basis of earlier morphological data (see Stevens et al. paper for leading references).

Heliantheae s.l. (Chapter 11)

Arnica (Hel): Schmidt and Willuhn (2000) examined the flavonoid (and other) constituents of flower heads of *Arnica angustifolia* Vahl [formerly referred to *A. alpina* (L.) Olin.)]. A rich array of compounds was reported including flavanones, flavones, and flavonols with significant differences observed depending upon which of the subspecies, subsp. *angustifolia,* subsp. *attenuata*, or subsp. *tomentosa,* was the source. Subspecies *attenuata* provided the most diverse profile, which consisted of the flavanones naringenin-7,4'-dimethyl ether, sakuranetin, and eriodictyol-7,4'-

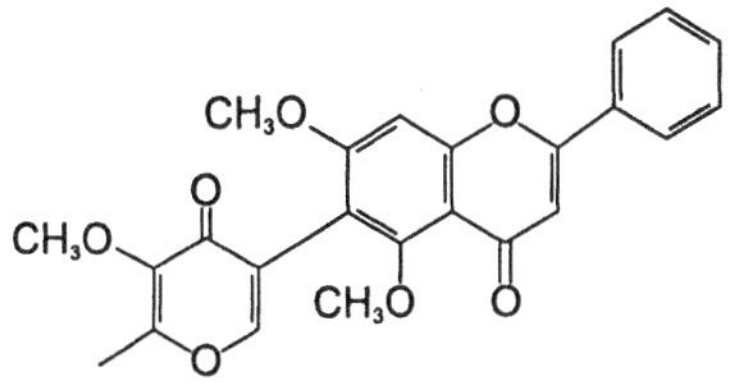

Fig. 1. 5-*O*-Methylhoslandin, an unusual compound from *Bidens pilosa*

dimethyl ether (persicogenin), the flavones apigenin-7,4'-dimethyl ether, luteolin-7,4'-dimethyl ether (pilloin), 5,7-dihydroxy-6,4'-dimethoxyflavone (pectolinarigenin), 5-hydroxy-6,7,4'-trimethoxyflavone (salvigenin), and the flavonols kaempferol, kaempferol-7-methyl ether (rhamnocitrin), kaempferol-7,4'-dimethyl ether, isorhamnetin, rhamnazin, 6-methoxykaempferol, 3,5,7-trihydroxy-6,4'-dimethoxyflavone (betuletol), 3,5,4'-trihydroxy-6,7-dimethoxyflavone (eupalitin), and quercetagetin-6,3',4'-trimethyl ether. Subspecies *angustifolia* yielded naringenin-7,4'-dimethyl ether, sakuranetin, persicogenin, and eriodictyol-7,3'-dimethyl ether, and the flavones apigenin-7,4'-dimethyl ether and luteolin-7,3',4'-trimethyl ether. Subspecies *tomentosa* had a slightly simpler profile differing from subsp. *angustifolia* only in lacking eriodictyol-7,3'-dimethyl ether. Both of the latter two subspecies differed from subsp. *attentuata* in possessing the luteolin trimethyl ether. They also differed form the former by the total absence of flavonols.

Bidens: Sarkar et al. (2000) described the rare flavonol 5-*O*-methylhoslundin (Fig. 1) from *B. pilosa*. This compound has not previously been reported from Asteraceae. Until this report, it appears to have been known only from *Hoslundia,* a genus in Lamiaceae.

Dubautia (Hel): *Dubautia arborea* (Madiinae), an alpine and desert shrub endemic to the Island of Hawaii, has been shown to accumulate apigenin, luteolin 7-methyl ether, luteolin 4'-methyl ether (diosmetin), 6-methoxyluteolin (hispidulin), naringenin 7-methyl ether (sakuranetin), eriodictyol 7-methyl ether, and 5,4'-dihydroxy-6,7-dimethoxyflavone (cirsimaritin) in its leaf exudate (Bohm, 1999).

Helianthus: Macias et al. (1999), in a continuation of their search for allelochemicals from sunflower cultivars, isolated 13 compounds from *H. annuus* cv. Peredovic. Four of these were identified as flavonoids: 2',4-dihydroxy-3',4'-dimethoxychalcone (heliannone A), 2',4',4-trihydroxy-3'-methoxychalcone (kulkulanine B), 7,8-dimethoxy-4'-hydroxyflavanone (heliannone B), and 3,5-dihydroxy-7,8,4'-trimethoxyflavone (tambulin). The biologically active compounds isolated by these workers were shown to be hitherto unknown sesquiterpene lactones.

Iva: Eriodictyol 5-methyl ether was reported as a constituent of *I. frutescens* subsp. *frutescens* by Ahmed et al. (1998).

Inuleae (Chapter 12)

Buphthalamum: *Buphthalamum salicifolium* L. (one of two species in this southern and central European genus) has been shown by Heilmann et al. (1999) to elaborate a rich array of flavonoids. Of primary interest to us here are a set of unusual acylated flavonol glucosides: three quercetagetin 7-*O*-ß-D-glucosides acylated at position-6" with isobutyric acid, isovaleric acid, and 2-methylbutyric acid, respectively, and patuletin 7-*O*-(6"-isobutyryl)-ß-D-glucoside. Also identified were the unacylated 7-*O*-glucosides of luteolin, 6-hydroxykaempferol, patuletin, and quercetagetin, the 3-*O*-glucosides of kaempferol, quercetin, isorhamnetin, and patuletin, and eriodictyol 7-*O*- and 3'-*O*-monoglucosides.

Inula: Park et al. (2000) have reported three new acylated flavonol glucoside derivatives from *I. britannica* using bioassay-guided fractionation (DPPH assay and cytochrome-*c* reductase assay using HL-60 cell culture). The new compounds are patuletin 7-*O*-(6"-isobutyryl)glucoside, patuletin 7-*O*-[6"-(2-methylbutyryl)]glucoside, and patuletin 7-*O*-(6"-isovaleryl)glucoside. Known compounds also identified were patuletin (6-methoxyquercetin), patulitrin (patuletin 7-*O*-glucoside), nepetrin (6-methoxyluteolin 7-*O*-glucoside), and axillarin (quercetagetin 3,6-dimethyl ether).

Mutisieae and Barnadesioideae (Chapter 14)

Gochnatia: Ybarra et al. (1994) reported eriodictyol 7-methyl ether and genkwanin from *G. glutinosa,* along with a number of terpene derivatives.

Chuquiraga: Mendiondo et al. (2000) reported the presence of kaempferol and quercetin 3-*O*-monoglucosides and 3-*O*-rutinosides from five species (six taxa) of *Chuquiraga*: *C. avellanedae, C. erinacea, C. erinacea* subsp. *hystrix, C. incana, C. oppositifolia,* and *C. rosulata.* Although not all taxa had all four compounds, the results were in accord with other published data for the genus.

Senecioneae (Chapter 15)

Senecio: In a study of leaf surface flavonoid chemistry, Urzua et al. (2000) reported 5,4'-dihydroxy-3,7-dimethoxyflavone (kumatakenin), 5,7,4'-trihydroxy-3-methoxyflavone (isokaempferide), 5,7,4'-trihydroxy-3,3'-dimethoxyflavone (isorhamnetin 3-methyl ether), and 5,4'-dihydroxy-3,7,3'-trimethoxyflavone (pachypodol) from *S. viscossimus* Colla. An extraction of leaf exudate of *S. cerberoanus* Remy, however, revealed no flavonoids.

Tageteae (Chapter 16)

Tagetes: Aerial parts of *Tagetes subulata* afforded quercetin 3-methyl ether, quercetagetin 3,6,3',4'-tetramethyl ether (the authors said "quercetin" which cannot have a 6-methyl ether), and quercetagetin 3,5,6,7,3'-pentamethyl ether (Abdala and Seeligmann, 1997). In a later paper *Tagetes lucida* was described as accumulating isorhamnetin and its 7-*O*-glucoside, a quercetagetin 3-*O*-arabinosylgalactoside, and patuletin (Abdala, 1999).

References

Abdala L. R. (1999) Flavonoids of the aerial parts from *Tagetes lucida* (Asteraceae). Biochem. Syst. Ecol. 27:753–754.

Abdala L. R., Seeligmann P. (1997) Chemosystematic interpretation of flavonoids in *Tagetes subulata* Cerv. (Asteraceae). Biochem. Syst. Ecol. 25:567–568.

Ahmed A. A., Balboul B. A. A., Scott A. I., Williams H. H., Miao B., Mabry T. J. (1998) Eudesmane derivatives from *Iva frutescens.* Phytochemistry 47:411–413.

Bohm B. A. (1999) Major exudate flavonoids of *Dubautia arborea* (Asteraceae). Biochem. Syst. Ecol. 27:755–757.

Faini F., Labbe C., Coll J. (1999) Seasonal changes in chemical composition of epicuticular waxes from the leaves of *Baccharis linearis.* Biochem. Syst. Ecol. 27:673–679.

Heilmann J., Müller E., Merfort I. (1999) Flavonoid glucosides and dicaffeoylquinic acids from flower heads of *Buphthalamum salicifolium.* Phytochemistry 51:713–718.

Macias F. A., Oliva R. M., Varela R. M., Torres A., Molinillo J. M. G. (1999) Allelochemicals from sunflower leaves cv. Peredovick. Phytochemistry 52:613–621.

Mendiondo M. E., Juarez B. E., Seeligmann P. (2000) Flavonoid profiles of some Argentine species of *Chuquiraga* (Asteraceae). Biochem. Syst. Ecol. 28:283–285.

Ogata J., Teramoto S., Yoshitama K. (1998) Isolation and characterization of UDP-glucose: cyanidin 3-*O*-glucosyltransferase from the flower heads of *Senecio hybridus.* J. Plant Res. 111:213–216.

Park E. J., Kim Y., Kim J. (2000) Acylated flavonol glycosides form the flower of *Inula britannica.* J. Nat. Prod. 63:34–36.
Sarkar S. D., Bartholomew B., Nash R. J., Robinson N. (2000) 5-*O*-Methylhoslundin: an unusual flavonoid from *Bidens pilosa* (Asteraceae). Biochem. Syst. Ecol. 28:591–593.
Schmidt T. J., Willuhn G. (2000) Sesquiterpene lactone and flavonoid variability of the *Arnica angustifolia* aggregate (Asteraceae). Biochem. Syst. Ecol. 28:133–142.
Stevens J. F., Wollenweber E., Ivancic M., Hsu V. L., Sandberg S., Deinzer M. L. (1999) Leaf surface flavonoids of *Chrysothamnus.* Phytochemistry 51:771–780.
Sy L.-K., Brown G. D. (1998) Three sesquiterpenes from *Artemisia annua.* Phytochemistry 48:1207–1211.
Urzua A., Andrade L., Mendoza L., Bustos F. (2000) Comparative leaf surface chemistry from *Senecio cerberoanus* and *Senesio viscosissimus.* Biochem. Syst. Ecol. 28:399–401.
Williams C. A., Harborne J. B., Eagles J. (1999a) Variations in lipophilic and polar flavonoids in the genus *Tanacetum.* Phytochemistry 52:1301–1306.
Williams C. A., Harborne J. B., Geiger H., Hoult J. R. S. (1999b) The flavonoids of *Tanacetum parthenium* and *T. vulgare* and their anti-inflammatory properties. Phytochemistry 51:417–423.
Ybarra M. I., Catalan C. A. N., Diaz J. G., Herz W. (1994) Two homofarnesane derivatives and other constituents from *Gochnatia glutinosa.* Phytochemistry 35:1585–1587.

References

Abad M. J., Villar B. A., Valverde S. (1993) Anti-inflammatory activity of two flavonoids from *Tanacetum microphyllum*. J. Nat. Prod. 56:1164–1167.

Abdala L. R. (1997) Chemotaxonomic differentiation between *Tagetes mandonii* Sch. Bip. and *T. dianthiflora* H.B.K. (Asteraceae). Biochem. Syst. Ecol. 25:679–680.

Abdala L. R. (1999) Flavonoids of the aerial parts from *Tagetes lucida* (Asteraceae). Biochem. Syst. Ecol. 27:753–754.

Abdala L. R., Seeligmann R. (1995a) Natural distribution of flavonoids in *Tagetes minuta*. Biochem. Syst. Ecol. 23:567–568.

Abdala L. R., Seeligmann R. (1995b) Flavonoids in *Tagetes zipaquirensis* and their chemosystematic significance. Biochem. Syst. Ecol. 23:871–872.

Abdala L. R., Seeligmann R. (1995c) The determination of the flavonoids in *Tagetes dianthiflora* H.B.K. (Compositae). Lilloa 38: 169–171.

Abdala L. R., Seeligmann R. (1997) Chemosystematic interpretation of flavonoids in *Tagetes subulata* Cerv. (Asteraceae). Biochem. Syst. Ecol. 25:567–568.

Abdallah O. M., Ramadan M. A., El-Shanwany M. A. (1989) Phytochemical study of *Carduus nutans* L. (Asteraceae). Bull. Fac. Sci., Assuit Univ. 18:69.

Abdel Salam N. A., Mahmoud Z. F., Kassem F. K. (1986) Sesquiterpene lactones, coumarins and flavonoids of *Launaea tenuiloba* Boiss. grown in Egypt. Egypt. J. Pharm. Sci. 27:275–277.

Abdel-Mogib M., Dawidar A. M., Metwally M. A., Abou-Elzahab M. (1989) Flavonoids of *Pulicaria undulata*. Die Pharmazie 44:801.

Abegaz B. M., Herz W. (1991) A nor-monoterpene from *Artemisia schimperi*. Phytochemistry 30:1011–1012.

Abegaz B. M., Keige A. W., Diaz J. D., Herz W. (1994) Sesquiterpene lactones and other constituents of *Vernonia* species from Ethiopia. Phytochemistry 37:191–196.

Abou-Zaid M. M., El-Karemy Z., El-Negoumy S. I., Altosaar I., Saleh N. A. M. (1991) The flavonoids of *Psiadia punctata*. Bull. Chem. Soc. Ethiop. 5:37–40.

Abou-Zied E. N., Rizk A. M. (1973) Phytochemical investigation of *Anthemis nobilis* growing in Egypt. Qual. Plant. Mater. Veg. 22:171–173.

Abraham R. J., Takagi S., Rosenstein R. D., Shiono R., Wagner H., Hörhammer L., Seligmann O., Farnsworth N. R. (1970) The structure of silydianin, an isomer of silymarin, by X-ray analysis. Tetrahedron Letters 2675–2678.

Adanson M. (1763) *Familles des Plantes*. Paris.

Adekenov S. M., Aitugamov K. A., Golovtsov N. I. (1987a) Artausin, a new sesquiterpene lactone from *Artemisia austriaca*. Chem. Nat. Cmpds. 23:127 [148]. Note: The bracketed number in these citations refers to the page in the original Russian article in Khim. Prir. Soed.

Adekenov S. M., Gafurov N. M., Turmukhambetov A. Zh., Ivlev V. I. (1987b) Terpenes of *Achillea micrantha*. Chem. Nat. Cmpds. 23:258 [305].

Adekenov S. M., Kadirberlina G. M., Sadykova V. I., Kupriyanova T. I., Kagarlitskii A. D. (1986) Biologically active compounds of *Centaurea pseudomaculosa*. Izv. Akad. Nauk. Kaz. SSR, Ser. Khim. 65–69.

Adekenov S. M., Mukhametzhanov M. N., Dagarlitskii A. D., Turmukhambetov A. Zh. (1984) A chemical investigation of *Achillea nobilis*. Chem. Nat. Cmpds. 20:568 [603].

Adesogan E. K., Okunade A. L. (1979) A new flavone from *Ageratum conyzoides*. Phytochemistry 18:1863–1864.

Adler W., Oswald K., Fischer R. (1994) *Exkursionsflora von Österreich*. Stuttgart: Eugen Ulmer.

Afifi F. U. (1991) Flavonoids from *Varthemia iphionoides*. Alexandria J. Pharm. Sci. 5:1–3.

Agnese A. M., Montoya S. N., Espinar L. A., Cabrera J. L. (1999) Chemotaxonomic features in Argentinia species of *Flaveria* (Compositae). Biochem. Syst. Ecol. 27: 739–742.

Aguinagalde I. (1982) Comparative study of the flavonoids of some *Artemisia* species. Parodiana 1:311–322.

Ahluwalia V. K., Rani N. (1976) Constitution and synthesis of gnaphalin – a new chalkone glycoside from *Gnaphalium multiceps* Wall. Indian J. Chem. 14B:594–595.

Ahmad V. U., Ismail N. (1991a) 5-Hydroxy-3,6,7,2',5'-pentamethoxyflavone from *Inula grantioides*. Phytochemistry 30:1040–1041.

Ahmad V. U., Ismail N. (1991b) Two new flavones from *Inula grantioides*. Z. Naturforsch. 46b:950–952.

Ahmad V. U., Ismail N. (1993) Chemical constituents of *Inula grantioides*. Fitoterapia 64:381.

Ahmad V. U., Ismail N., Khan M. A., Baqai F. T., Tareen R. B. (1995) Santoflavone, a 5-deoxyflavonoid from *Achillea santolina*. Phytochemistry 38:1305–1307.

Ahmed A. A., Ali A. A., Mabry T. J. (1989a) Flavonoid aglycones from *Jasonia montana*. Phytochemistry 28: 665–667.

Ahmed A. A., El-Sayed N. H., El-Negoumy S. I., Mabry T. J. (1987a) Flavonoids of *Cotula cinerea*. J. Nat. Prod. 50:519–520.

Ahmed A. A., El-Sayed N. H., Mabry T. J. (1989b) Flavonoids of *Achillea santolina* and *A. fragrantissima* (Compositae, Anthemideae). Rev. Latinoamer. Quim. 20:5.

Ahmed A. A., Ishak M. S., Michael H. N., El-Ansari M. A., El-Sissi H. I. (1991) Flavonoids of *Astericus graveolens*. J. Nat. Prod. 54:1092–1093.

Ahmed A. A., Jakupovic J., Eid F., Ali A. A. (1988a) 11-Hydroxyjasionone, a new sesquiterpene type from *Jasonia montana*. Phytochemistry 27:3875–3877.

Ahmed A. A., Mabry T. J. (1987) Flavonoids of *Iphiona scabra*. Phytochemistry 26:1517–1518.

Ahmed A. A., Mabry T. J., Matlin S. A. (1989c) Flavonoids of the flowers of *Silybum marianum*. Phytochemistry 28:1751–1753.

Ahmed A. A., Mahmoud A. A., Tanaka T., Iinuma M. (1994) Two methylated flavonols from *Jasonia candicans*. Phytochemistry 35:241–243.

Ahmed A. A., Melek F. R., El-Din A. A. S., Mabry T. J. (1988b) Polysulphated flavonoids from *Iphiona mucronata*. Rev. Latinoamer. Quim. 19:107–109.

Ahmed A. A., Melek F. R., Mabry T. J. (1987b) Sulfated and non-sulfated flavonoids from *Pluchea dioscoridis*. J. Nat. Prod. 50:311.

Ahmed A. A., Norris J. A., Mabry T. J. (1986) Flavonoids of *Brickellia vernicosa*. Phytochemistry 25:1501–1502.

Ahmed A. A., Norris J. A., Mabry T. J. (1988c) Flavonoids of *Brickellia diffusa*. Biochem. Syst. Ecol. 16:15–16.

Ahmed A. A., Shalaby A. M., Melek F. R., Mabry T. J. (1988d) Swertisin 2''-arabinoside, a new *C*-glycosylflavone from *Achillea fragrantissima*. J. Nat. Prod. 51:971–972.

Ahmed A. A., Spaller M., Mabry T. J. (1992) Flavonoids of *Pallenis spinosa*. Biochem. Syst. Ecol. 20:785–786.

Ahmed N., Bibi R. (1979) Chemical investigation of *Centaurea iberica*. Fitoterapia 50:199–200.

Ahmed N., Bibi R. (1981) Chemical investigation of *Centaurea phyllocephala*. Fitoterapia 52:187–188.

Ahmed Z. F., Rimpler H., Rizk A. M., Hammouda F. M., Ismail S. I. (1970a). The flavonoid constituents of certain *Centaurea* species grown in Egypt. Phytochemistry 9:1595–1601.

Ahmed Z. F., Rizk A. M., Hammouda F. M. (1970b) Postep Dziedzinie Leku Rosl., Pr. Ref. Dosw. Wygloszone Symp. 20–23. Cited by Al-Easa and Rizk (1992).

Airy Shaw H. K. (1964) Diagnoses of new families, new names, etc. for the seventh edition of Willis's "Dictionary". Kew Bulletin 18:249–273.

Airy Shaw H. K. (1983) *Willis' A Dictionary of the Flowering Plants & Ferns*. ed. 8. Cambridge: Cambridge Univ. Press.

Akashi T., Sato N., Hirota H., Ayabe S. (1997) Anthocyanin-producing dandelion callus as a chalcone synthase source in recombinant polyketide reductase assay. Phytochemistry 46:283–287.

Akhmedov I. S., Kasymov Sh. Z., Sidyakin G. P. (1969) A study of the chemical composition of *Artemisia absinthium*. Chem. Nat. Cmpds. 5:50 [57–58].

Akkal S., Benayache F., Benayache S., Jay M. (1997) Flavonoids from *Centaurea incana* (Asteraceae). Biochem. Syst. Ecol. 25:361–362.

Al-Easa H. S. (1990) Ph.D. thesis. Univ. Reading, England.

Al-Easa H. S., Kamel A., Rizk A. F. M. (1992) Flavonoids from *Centaurea sinaica*. Fitoterapia 63:468–469.

Al-Easa H. S., Rizk A. M. (1992) Constituents of *Centaurea* species. Qatar Univ. Sci. J. 12:27–57.

Alfatafta A. A., Mullin C. A. (1992) Epicuticular terpenoids and an aurone from flowers of *Helianthus annuus*. Phytochemistry 31:4109–4113.

Al-Hazimi H. M. G., Alkhathlan H. Z. (1993) A new diterpene and flavonoids from *Pulicaria somalensis*. J. King Saud Univ., Sci. 5:141–146.

Ali E., Bagchi D., Pakrashi S. C. (1979) New flavonoids from *Anaphalis araneosa*: Phytochemistry 18:356–357.

Ali Y. E., Omar A. A., Sarg T. M., Slatkin D. J. (1987) Chemical constituents of *Centaurea pallescens*. Planta Medica 53:503–504.

Alieva Sh. A., Omurkamginova V. B., Glyzin V. I. (1979) Flavonoids of *Doronicum macrophyllum* and *Doronicum oblongifolium*. Chem. Nat. Cmpds. 15:82 [96].

Aljancic I., Vajs V., Menkovic N., Karadzic I., Juranic N., Milosavljevic S., Macura S. (1999) Flavones and sesquiterpene lactones from *Achillea atrata* L. subsp. *multifida*: antimicrobial activity. J. Nat. Prod. 62:909–911.

Al-Khubaizi M. S. (1977) Sulfated and non-sulfated flavonoids from *Flaveria*, *Sartwellia* and *Haploëstes*. Ph.D. Dissertation, Univ. of Texas, Austin.

Al-Khubaizi M. S., Mabry T. J., Bacon J. (1978) 6-Methoxykaempferol 3-*O*-glucoside from *Flaveria brownii*. Phytochemistry 17:163.

Allan H. H. (1961) Flora of New Zealand. Wellington: Government Printer.

Allen G. A. (1988) Flavonoids of *Aster bracteolus*. Biochem. Syst. Ecol. 16:139–141.

Alroy J. (1995) Continuous track analysis: a new phylogenetic and biogeographic method. Syst. Biol. 44:152–179.

Alston R. E, Turner B. L. (1963a) *Biochemical Systematics*. Englewood Cliffs, New Jersey: Prentice Hall.

Alston R. E., Turner B. L. (1963b) Natural hybridization among four species of *Baptisia* (Leguminosae). Amer. J. Bot. 50:159–173.

Alvarez L., Velasco A., Villarreal M. L., Degaldo G., Chavez M. I., Del Rio F. (1992) Constituents of a population of *Ambrosia hispida* and cytotoxic evaluation of its major constituent, hispidulin. Fitoterapia 63:366–367.

Al-Yahya M. A., Hifnawy M. S., Mossa J. S., El-Feraly F. S., McPhail D. R., McPhail A. T. (1987) X-Ray structure of psiadiarabin, a flavone from *Psiadia arabica*. Phytochemistry 26:2648–2649.

Amaro J. M., Delgado-Mendez P. (1993) Flavonoids from the leaves of *Chromolaena subscandens*. J. Nat. Prod. 56:610–612.

Amaro J. M., Hung M. (1985) Phytochemical study of the Venezuelan Andean flora. III. Flavonoids and steroids of *Stevia lucida* Lagasca. Rev. Fac. Farm., Univ. Los Andes 25:125–128.

Amaro J. M., Morales-Mendez A. (1983) Phytochemical study of Venezuelen Andean flora. II. Flavonoids of *Chromolaena meridensis*. Rev. Latinoamer. Quim. 14: 86.

Amer M. M. A., Salama O. M., Omar A. A. (1984) Methylated flavonoidal aglycones from *Centaurea alexandrina*. Acta Pharm. Jugosl. 34:257–259.

Amer M. M. A., Salama O. M., Omar A. A. (1985) Flavonoids of *Carduus pycnocephalus*. Fitoterapia 56:61.

Anderberg A. A. (1989) Phylogeny and reclassification of the tribe Inuleae (Asteraceae). Can. J. Bot. 67:2277–2296.

Anderberg A. A. (1991a) Taxonomy and phylogeny of the tribe Inuleae (Asteraceae). Plant Syst. Evol. 176:75–123.

Anderberg A. A. (1991b) Taxonomy and phylogeny of the tribe Gnaphalieae Rydb. Opera Bot. 104: 1–195.

Anderberg A. A. (1991c) Taxonomy and phylogeny of the tribe Plucheeae (Asteraceae). Plant Syst. Evol. 176:145–177.

Anderberg A. A. (1994) Tribe Inuleae. In: Bremer K. (ed.) *Asteraceae: Cladistics and Classification*. Portland, Oregon: Timber Press. 273–291.

Anderberg A. A. (1996) "Inuleae"-a paraphyletic assemblage. In: Hind D. J. N., Beentje H. J. (eds.) *Compositae: Systematics. Proceedings of the International Compositae Conference, Kew (1994) Vol. 1*. Kew: Royal Botanic Gardens. 569–573.

Anderson L. C., Reveal J. L. (1966) *Chrysothamnus bolanderi*, an intergeneric hybrid. Madroño 18:225–233.

Anderson M. C. (1992) An analysis of variability in seed settling velocities of several wind-dispersed Asteraceae. Amer. J. Bot. 79:1087–1091.

Anderson M. C. (1993) Diaspore morphology and seed dispersal in several wind-dispersed Asteraceae. Amer. J. Bot. 80:487–492.

Angiosperm Phylogeny Group (1998) An ordinal classification for the families of flowering plants. Ann. Missouri Bot. Gard. 85:531–553.

Anyos T., Steelink C. (1960) Fluorescent petal constituents of *Chrysanthemum coronarium*. Arch. Biochem. Biophys. 90:63–67.

Appendino G. P., Gariboldi P. (1982) The stereochemistry of matricin and 4-epimatricin, proazulene sesquiterpene lactones from *Artemisia arborescens*. Phytochemistry 21:2555–2557.

Appendino G. P., Gariboldi P., Nano G. M. (1983) Isomeric hydroperoxyeudesmanolides from *Artemisia umbelliformis*. Phytochemistry 22:2767–2772.

Appendino G. P., Menichini F., Puntillo D. (1988) Aromatic acetylenes and flavanones from *Artemisia variabilis*. Fitoterapia 59:425.

Aqil M. (1995) Flavonoid glycosides from *Eupatorium africanum*. Ultra Sci. Phys. Sci. 7:1–6.

Aquino R., D'Agostino M., Domestico M., Senatore F., Schettino O. (1988) Flavonoid glycosides from *Eupatorium cannabinum* subsp. *cannabinum*. Fitoterapia 59:236.

Arene E. O., Pettit G. R., Ode R. H. (1978) The isolation of isosakuranetin methyl ether from *Eupatorium odoratum*. Lloydia 41:186–189.

Arias D. M., Rieseberg L. H. (1995) Genetic relationships among domesticated and wild sunflowers (*Helianthus annuus*, Asteracceae). Economic Botany 49:239–248.

Arisawa M., Ishiwara Y., Nakaoki T., Sekino S., Takakuwa T. (1969) Unutilized resources. III. Components of *Juncus* genus plants, the leaves of *Aesculus turbinata* and the petals of *Chrysanthemum morifolium*. Shoyakugaku Zasshi 23:49–52.

Aritomi M., Kawasaki T. (1974) Dehydro-*para*-asebotin, a new chalcone glucoside in the flowers of *Helichrysum affine* D. Don. Chem. Pharm. Bull. 22:1800–1805.

Aritomi M., Shimojo M., Mazaki T. (1964) Constituents in flowers of *Gnaphalium affine*. J. Pharm. Soc. Japan 84:895–896.

Arriagada J. E. (1994) Systematics of *Clibadium* L. (Asteraceae, Heliantheae). Ph.D. Dissertation. Ohio State University.

Arriagada J. E. (1995) Ethnobotany of *Clibadium* L. (Compositae, Heliantheae) in Latin America. Economic Botany 49:328–332.

Arriagada J. E. (1998) The genera of Inuleae (Compositae; Asteraceae) in the Southeastern United States. Harvard Papers in Botany 3:1–48.

Arriagada J. E., Miller N. G. (1997) The genera of Anthemideae (Compositae; Asteraceae) in the Southeaastern United States. Harvard Papers in Botany No. 11:1–46

Arriaga-Giner F. J., Borges-del-Castillo J., Manresa-Ferrero M. T., Pena de Recinos S., Rodriguez Luis F. (1982) Compounds from Salvadoran plants. III. Components of *Baccharis trinervi* Pers. Rev. Latinoam. Quim. 13:47–48.

Arriaga-Giner F. J., Borges-del-Castillo J., Manresa-Ferrero M. T., Vazquez-Bueno P., Rodriguez-Luis F., Vales-Iraheta S. (1983) Eudesmane derivatives from *Pluchea odorata*. Phytochemistry 22:1767–1769.

Arriaga-Giner F. J., Wollenweber E., Hradetzky D. (1986) New flavonoids from the exudate of *Baccharis bigelovii* (Asteraceae). Z. Naturforsch. 41c:946–948.

Asen S. (1967) Anthocyanins and other flavonoids in red cornflowers. Proc. Am. Soc. Hort. Sci. 91:653–659.

Asen S. (1984) HPLC analysis of flavonoid chemical markers in petals from *Gerbera* flowers as an adjunct for cultivar and germplasm identification. Phytochemistry 23:2523–2526.

Asen S., Horowitz R. M. (1974) Apigenin 4'-*O*-ß-D-glucoside 7-*O*-ß-D-glucuronide: The copigment in the blue pigment of *Centaurea cyanus*. Phytochemistry 13:1219–1223.

Asen S., Jurd L. (1967) The constitution of a crystalline blue cornflower pigment. Phytochemistry 6: 577–584.

Ashlock P. D. (1971) Monophyly and associated terms. Syst. Zool. 20:63–69.

Ashlock P. D. (1984) Monophyly: its meaning and importance. In: Duncan T., Stuessy T. F. (eds.) *Cladistics: Perspectives on the Reconstruction of Evolutionary History.* New York: Columbia Univ. Press. 39–46.

Ashlock P. D. (1991) Weighting and anagenetic analysis. In: Mayr E., Ashlock P. D. (eds.) *Principles of Systematic Zoology*, 2nd ed. New York: McGraw-Hill. 236–241.

Ashtakala S. S., Schwartz A. M. (1971) Identification of phenolic compounds in flowers of *Cineraria* and their relation to flower color. J. Amer. Hort. Sci. 96:805.

Ates N., Ulubelen A., Clark W. D., Brown G., Mabry T. J., Dellamonica G., Chopin J. (1982) Flavonoids of *Haplopappus scrobiculatus* and *Haplopappus sericeus*. J. Nat. Prod. 45:189–190.

Ateya A.-M. A., Okarter T. U., Knapp J. E., Schiff P. L. Jr., Slatkin D. J. (1982) Flavonoids of *Thelesperma megapotanicum* Spreng. Planta Medica 45:247–248.

Atindehou K. K., Reynaud J., Koukoua G., Nguessan Y. T., Anoma G., Fournier J. (1993) [1994] Phytochemical study of *Vernonia nigritiana* Oliv. & Hiern. Plantes Medicin. Phytother. 26:356–361.

Augier J., DuMerac M.-L. R. (1951) La phylogénie des composées. La Revue Scient. 89:167–182.

Averett J. E., Beaman W. C. (1975) Chemotaxonomy and flavonoids of *Gaillardia* section *Hollandia* (Asteraceae-Helenieae). Plant Syst. Evol. 123:237–240.

Ayanoglu E., Ulubelen A., Clark W. D., Brown G. K., Kerr R. R., Mabry T. J. (1981) Myricetin and quercetin methyl ethers from *Haplopappus integerrimus* var. *punctatus*. Phytochemistry 20:1715–1717.

Ayyad S.-E. D. N., Hanna M. A., Abdel-Mogib M. (1994) Constituents of *Onopordum alexandrinum*. Egypt. J. Pharm. Sci. 35:553–559.

Babcock E. B. (1947) The genus *Crepis*. Part one. The taxonomy, phylogeny, distribution, evolution of *Crepis*. Univ. Calif. Pub. Bot. 21:1–197.

Bacon J. D., Urbatsch L. E., Bragg L. H., Mabry T. J., Neuman P., Jackson D. W. (1978) The flavonoids of *Tetragonotheca* (Compositae). Phytochemistry 17:1939–1943.

Bader G., Tuja D., Wray V., Hiller K. (1993) Flavonol glycosides from *Heteropappus altaicus*. Planta Medica 59:284–285.

Baetcke K. P., Alston R. E. (1968) The composition of a hybridizing population of *Baptisia sphaerocarpa* and *Baptisia leucophaea*. Evolution 22:157–165.

Baeva R. T., Nabizade L. I., Zapesochnaya G. G., Karryev M. O. (1988) Flavonoids of *Artemisia annua*. Chem. Nat. Cmpds. 24:256 [298–299].
Baez D. H., De Los Rios C., Crescente O., Caserta A. (1998) Antibacterial and chemical evaluation of *Chromolaena moritziana*. J. Ethnopharmacol. 59: 203–206.
Bain J. F. (1988) Taxonomy of *Senecio streptanthifolius* Greene. Rhodora 90:277–312.
Bain J. F., Denford K. E. (1985) Flavonoid variation in the *Senecio streptanthifolius* complex. Can. J. Bot. 63:1685–1690.
Bain J. F., Desrochers A. M. (1988) Flavonoids of *Carduus nutans* and *C. acanthoides*. Biochem. Syst. Ecol. 16:265–268.
Baker M. M., Turner B. L. (1986) Taxonomy of *Flyriella* (Asteraceae-Eupatorieae). Sida 11:300–317.
Balboul B. A. A. A., Ahmed A. A., Otsuka H., Bando M., Kido M., Takeda Y. (1997) A guianolide and a germacranolide from *Achillea santolina*. Phytochemistry 46:1045–1049.
Baldwin B. G. (1992) Phylogenetic utility of the internal transcribed spacers of nuclear ribosomal DNA in plants: An example from the Compositae. Mol. Phylogeny and Evol. 1:3–16.
Baldwin B. G. (1993) Molecular phylogenetics of *Calycadenia* (Compositae) based on its sequences of nuclear rDNA: Chromosomal and morphological evolution reexamined. Am. J. Bot. 80:222–238.
Baldwin B. G. (1996) Phylogenetics of the California tarweeds and the Hawaiian silversword alliance (Madiinae; Heliantheae *sensu lato*). In: Hind D. J. N., Beentje H. J. (eds.) *Compositae: Systematics. Proceedings of the International Compositae Conference, Kew, (1994) Vol. 1*. Kew: Royal Botanic Gardens. 377–391.
Baldwin B. G. (1998) Evolution in the endemic Hawaiian Compositae. In: Stuessy T. F., Ono M. (eds.) *Evolution and Speciation of Island Plants*. Cambridge: Cambridge Univ. Press. 49–73.
Baldwin B. G., Bainbridge S. J. (1993) *Layia*. In: Hickman J. C. (ed.) *The Jepson Manual: Higher Plants of California*. Berkeley: Univ. California Press. 300–303.
Baldwin B. G., Kyhos D. W., Dvorak J. (1990) Chloroplast DNA evolution and adaptive radiation in the Hawaiian silversword alliance (Asteraceae-Madiinae). Ann. Missouri Bot. Gard. 77:96–109.
Baldwin B. G., Kyhos D. W., Dvorak J., Carr G. D. (1991) cpDNA evidence for a North American origin of the Hawaiian silversword alliance. Proc. Natl. Acad. Sci. U.S.A. 88:1840–1843.
Ball E. A., Harborne J. B., Arditti J. (1972) Anthocyanins of *Dimorphotheca*: pigments in flowers, stems and callus tissues. Amer. J. Bot. 59:924–930.
Ballard R. (1986) *Bidens pilosa* complex (Asteraceae) in North and Central America. Amer. J. Bot. 73:1452–1465.
Ballard R. (1988) A new species of *Bidens* (Asteraceae) from San Luis Potosi, Mexico. Syst. Bot. 13:184–186.
Balza F., Jamieson L., Towers G. H. N. (1985) Chemical constituents of the aerial parts of *Artemisia dracunculus*. J. Nat. Prod. 48:339–340.
Balza F., Towers G. H. N. (1984) Dihydroflavonols of *Artemisia dracunculus*. Phytochemistry 23:2333–2337.

Bandoni A. L., Medina J. E., Rondina R. V. D., Coussio J. D. (1978) Genus *Baccharis* L. 1. Phytochemical analysis of a non-polar fraction from *B. crispa*. Planta Medica 34:328–331.

Bandopadhyay M., Pardeshi N. P., Seshadri T. R. (1972) Components of *Silybum marianum*. Indian J. Chem. 10:808–809.

Bandyukova V. A. (1967) The quercimeritrin content of some species of *Centaurea*. Chem. Nat. Cmpds. 3:49 [58].

Bandyukova V. A., Khalmatov Kh. Kh. (1967) Isolation of scutellarein from *Centaurea depressa* M. B. Chem. Nat. Cmpds. 3:48 [57].

Bandyukova V. A., Khalmatov Kh. Kh., Alimov Kh. I. (1969) Flavonoids of *Centaurea depressa*. Chem. Nat. Cmpds. 5:274 [324].

Bannerjee N. R., Seshadri T. R. (1956) Isolation and constitution of patulitrin. Proc. Indian Acad. Sci. 44A:284–286.

Baranska K. (1963) Certain flavonoids occurring in *Bidens tripartitus*. Acta Polon. Pharm. 20:357–364.

Barbera O., Marco J. A., Sanz J. F., Sanchez-Parareda J. (1986) 3-Methoxyflavones and coumarins from *Artemisia incanescens*. Phytochemistry 25:2357–2360.

Bardón A., Catalán C. A. N., Gutiérrez A. B., Herz W. (1988) Guaianolides and other constituents from *Vernonia nitidula*. Phytochemistry 27:2691–2694.

Bardón A., Kamiya N. I., de Ponce de Leon C. A., Catalán C. A. N., Diaz J. G., Herz W. (1992) Glaucolides and related sesquiterpene lactones from *Vernonia nudiflora* and *Chrysolaena propinqua*. Phytochemistry 31:609–613.

Barkley T. M. (1962) A revision of *Senecio aureus* and allied species. Trans. Kansas Acad. Sci. 65:318–408.

Barkley T. M. (1978) *Senecio*. In: Rogerson C. T. (ed.) *North American Flora*, Series II, Part 10. New York: New York Botanical Garden.

Barkley T. M. (1985) Generic boundaries in the Senecioneae. Taxon 34:17–21.

Barkley T. M., Clark B. L., Funston A. M. (1996) The segregate genera of *Senecio sensu lato* and *Cacalia sensu lato* (Asteraceae: Senecioneae) in Mexico and Central America. In: Hind D. J. N., Beentje H. J. (eds.) *Compositae: Systematics. Proceedings of the International Compositae Conference, Kew, (1994) Vol. 1*. Kew: Royal Botanic Gardens. 613–620.

Barrero A. F., Alvarez-Manzaneda E. J. R., Alvarez-Manzaneda R. (1990a) Bisabolene derivatives and other constituents from *Achillea odorata*. Phytochemistry 29:3213–3216.

Barrero A. F., Alvarez-Manzaneda E. J. R., Alvarez-Manzaneda R. (1990b) Flavonoids from the genus *Achillea*. Ars Pharm. 31:1–9.

Barrero A. F., Sanchez J. F., Alvarez-Manzaneda E. J., Alvarez-Manzaneda R. (1988) Di-*O*-acyl derivatives of shikimic acid from *Senecio nebrodensis*. Phytochemistry 27:1191–1193.

Barron D., Colebrook L. D., Ibrahim R. K. (1986) An equimolar mixture of quercetin 3-sulphate and patuletin 3-sulphate from *Flaveria chloraefolia*. Phytochemistry 25:1719–1721.

Barron D., Ibrahim R. K. (1986) Novel flavonol sulfates from *Flaveria chloraefolia*. Bull. Liaison – Groupe Polyphenols 13:540–541.

Barron D., Ibrahim R. K. (1987a) Quercetin and patuletin 3,3'-disulphates from *Flaveria chloraefolia*. Phytochemistry 26:1181–1184.

Barron D., Ibrahim R. K. (1987b) 6-Methoxyflavonol 3-monosulphates from *Flaveria chloraefolia*. Phytochemistry 26:2085–2088.

Barron D., Ibrahim R. K. (1988) Ombuin 3-sulphate from *Flaveria chloraefolia*. Phytochemistry 27:2362–2363.

Barron D., Varin L., Ibrahim R. K., Harborne J. B., Williams C. A. (1988) Sulphated flavonoids—an update. Phytochemistry 27:2375–2395.

Barua N. C., Sharma R. P. (1992) (2*R*,3*R*)-7,5'-Dimethoxy-3,5,2'-trihydroxyflavanone from *Blumea balsamifera*. Phytochemistry 31:4040.

Barua N. C., Sharma R. P., Thyagarajan G., Herz W. (1978) Flavonoids of *Chromolaena odorata*. Phytochemistry 17:1807–1808.

Barua N. C., Sharma R. P., Thyagarajan G., Herz W., Govindan S. V. (1979) New flavonoids from *Inula cappa*. Phytochemistry 18:2003–2006. [Actually plant is *Blumea balsamifera*; see Goswami et al. (1984) Phytochemistry 23: 367]

Basargin D. D., Tsiklauri G. Ch. (1990) Phenolic compounds of *Saussurea pulchella* (Fisch.) Fisch. Rastit. Resur. 26:68–71.

Bastos M. M. S. M., Kijjoa A., Pinto M. M. M. (1994) Constituents of *Centaurea ornata* ssp. *ornata*. Fitoterapia 65:191.

Bate-Smith E. C. (1962) The phenolic constituents of plants and their taxonomic significance. I. Dicotyledons. Bot. J. Linn. Soc. 58:95–173.

Bate-Smith E. C. (1968) The phenolic constituents of plants and their taxonomic significance. II. Monocotyledons. Bot. J. Linn. Soc. 60:325–356.

Bate-Smith E. C., Geissman T. A. (1951) Benzalcoumaranones. Nature 167:688.

Bate-Smith E. C., Swain T. (1953) The isolation of 2',4,4'-trihydroxychalcone from yellow varieties of *Dahlia variabilis*. J. Chem. Soc. 2185–2187.

Bate-Smith E. C., Swain T., Nordström C. G. (1955) Chemistry and inheritance of flower colour in the *Dahlia*. Nature 176:1016–1018.

Batyuk V. S., Kol'tsova L. F. (1968) Isoquercetin – a new flavone glycoside of *Solidago canadensis* L. Chem. Nat. Cmpds. 4:323 [381].

Batyuk V. S., Kol'tsova L. F. (1969) Flavonols of *Solidago canadensis* L. Chem. Nat. Cmpds. 5:104 [121].

Batyuk V. S., Kovaleva S. N. (1985) Flavonoids of *Solidago canadensis* and *S. virgaurea*. Chem. Nat. Cmpds. 21:533 [566].

Bauer L., Bacha C. T. M., Gilberto A., Silva A. B., De Siqueira N. C. S., Sant'Ana B. M. S. (1985) Isolation of gnaphalin from *Achyrocline alata* (H.B.K.) DC. Cienc. Nat. (St. Maria, Braz.). 7:45–47.

Bauer L., Simoes C. M. O., De Siqueira N. C. S., Schenkel E. P., Silva G. A. A. B., Alice C. B. (1987) Flavonoids of the inflorescences of *Achyrocline alata*. Rev. Latinoamer. Quim. 18:91.

Bauer L., Simoes C. M. O., De Siqueira N. C. S., Schenkel E. P., Silva G. A. A. B., Alice C. B. (1989) Flavonoids of the inflorescences of *Achyrocline alata*. Rev. Latinoamer. Quim. 20:68–69.

Bayer E. (1958) Über die blauen Farbstoffe der Kornblume. I. Natürliche und synthetische Anthocyan-Metallkomplexe. Chem. Ber. 91:1115–1122.

Bayer E., Nether K., Egeter H. (1960) Natürliche und synthetische Anthocyankomplexe. III. Synthese der blauen, im Kornblumenfarbstoff enthaltenen Chelate. Chem. Ber. 93:2871–2879.
Bayer R. J. (1988) Patterns of isozyme variation in western North American *Antennaria* (Asteraceae: Inuleae). I. Sexual species of sect. *Dioicae*. Syst. Bot. 13:525–537.
Bayer R. J., Starr J. R. (1998) Tribal phylogeny of the Asteraceae based on two non-coding chloroplast sequences, the *trn*L intron and *trn*L/*trn*F intergenic spacer. Ann. Missouri Bot. Gard. 85:242–256.
Beal J. (1949) Determination of quercetin-like substances in several mid-western plants. Trans. Kansas Acad. Sci. 52:209–218.
Beauverd G. (1910) Contribution à l'étude des composées. 4. Recherches sur la tribu des Gnaphaliées. Bull. Soc. Bot. Genève, ser. 2, 2:207–253.
Becchi M., Adzet T., Iglesias J., Combier H. (1976) Heterosides flavoniques de feuilles de *Santolina chamaecyparissus* L. Pl. Medica Phytother. 10:139–143.
Becchi M., Carrier M. (1980) 6-Methoxyflavones of *Santolina chamaecyparissus*. Planta Medica 38:267–268.
Becchi M., Combier H. (1976) Sur deux glucosides flavoniques des feuilles de *Santolina chamaecyparissus* L. Planta Medica 30:388–394.
Belenovskaya L. M. (1996) *Artemisia*: The flavonoids and their systematic value. In: Hind D. J. N., Beentje H. J. (eds.) *Compositae: Systematics. Proceedings of the International Compositae Conference, Kew, (1994) Vol. 1.* Kew: Royal Botanic Gardens. 253–259.
Belenovskaya L. M., Markova L. P. (1979) Phenolic compounds of *Ajania achilleoides*. Chem. Nat. Cmpds. 15:201 [232].
Belenovskaya L. M., Markova L. P. (1992) Flavonoids of some *Artemisia* L. species of the flora of Mongolia. Rastit. Resur. 28:122–125.
Belenovskaya L. M., Markova L. P., Kapranova G. I. (1982) Phenolic compounds of *Artemisia xerophytica*. Chem. Nat. Cmpds. 18:115 [121–122].
Belenovskaya L. M., Markova L. P., Nadezhina T. P., Ligea U. (1977) Flavonoids of *Ajania fruticulosa*. Chem. Nat. Cmpds. 13:480 [575].
Belenovskaya L. M., Musaev K. L., Kukenov M. K. (1992) Flavonoids of some species of subgenus *Seriphidium*, genus *Artemisia*. Rastit. Resur. 28:91–94.
Bellini A. A., de Oliveira D. R., Vichnewski W. (1999) Steroidal saponin, 7-oxostigmasterol and diterpenes from *Aspilia montevidensis*. Biochem. Syst. Ecol. 27:317–319.
Belzer N. F., Ownbey M. (1971) Chromatographic comparison of *Tragopogon* species and hybrids. Amer. J. Bot. 58:791–802.
Benayache S., Benayache F., Dendougui H., Jay M. (1991) Flavonoids of *Inula viscosa* L. Pl. Medica Phytother. 25:170–176.
Bennett R. N., Wallsgrove R. W. (1994) Secondary metabolites in plant defence mechanisms. New Phytol. 127:617–633.
Bentham G. (1873) Notes on the classification, history, and geographical distribution of the Compositae. Bot. J. Linn. Soc. 13:335–577.
Bentham G., Hooker J. D. (1862–1883) *Genera Plantarum*. 3 vols. London: Reeve & Co.

Berezovskaya T. P., Velikhanova V. I. (1968) Phytochemical analysis of *Artemisia siversiana* grass gathered in the vicinity of Tomsk. Tr. Nauch. Konf. Tomsk. Otd. Vses. Khim. Obshchest, 272–273.

Berkhey J. F. van (1760) Expositio characteristica structurae florum qui dicuntur Compositi. Leiden (Lugdunum Batavorum).

Bessey C. E. (1897) The phylogeny and taxonomy of angiosperms. Botanical Gaz. 24:1–34.

Bessey C. E. (1914) Revisions of some plant phyla. Univ. Nebraska Studies 14(1):37–109.

Bessey C. E. (1915) The phylogenetic taxonomy of flowering plants. Ann. Missouri Bot. Gard. 2:109–164.

Besson E., Dombris A., Raynaud J., Chopin J. (1979) Corymbosiode, nouvelle di-*C*-glycosylflavone des racines de *Carlina corymbosa*. Phytochemistry 18:1899–1900.

Betancourt S., Proliac A. (1988) Quatre mono-*O*-glycosyl flavonoides des tiges feuillees de *Senecio sanguisorba* DC. Die Pharmazie 43:142–143.

Betto P., Casinovi C. G., Gabriele R., Grandolini G., Menichini F. (1988) Systematic study of the minor components of *Artemisia arborescens* (Compositae) of Calabria. Rev. Latinoamer. Quim. 19:40–42.

Beutler J. A., Cardellina J. H. II, Lin C. M., Hamel E., Cragg G. M., Boyd M. R. (1993) Centaureidin, a cytotoxic flavone from *Polymnia fruticosa*, inhibits tubulin polymerization. Bioorg. Medica Chem. Lett. 3:581–584.

Bhardwaj D. K., Bisht M. S., Uain S. C., Mehta C. K., Sharma G. C. (1980) Quercetagetin 5-methyl ether from the petals of *Tagetes patula*. Phytochemistry 19:713–714.

Bhardwaj D. K., Jain S. C., Sharma G. C. (1977) Synthesis of 3,5-dihydroxy-6,7,3',4'-tetramethoxyflavone. Indian J. Chem. 15B:860–861.

Bhargava K. K., Krishnaswamy N. R., Seshadri T. R. (1970) Isolation of demethylwedelolactone and its glucoside from *Eclipta alba*. Indian J. Chem. 8:664–665.

Bhatia V. K., Kagan J. (1971) Quercetin 3-ß-glucoside from *Ambrosia trifida*. Phytochemistry 10:1401.

Bierner M. W. (1973) Chemosystematic aspects of flavonoid distribution in twenty-two taxa of *Helenium*. Biochem. Syst. Ecol. 1:55–57.

Bierner M. W. (1974) A systematic study of *Dugaldia* (Compositae). Brittonia 26:385–392.

Bierner M. W. (1978) Flavonoids of *Plummera*: Taxonomic implications at the generic level within the Gaillardiinae. Biochem. Syst. Ecol. 6:293–295.

Bierner M. W. (1979) Flavonol 3-glycosides in *Amblyolepis setigera*. Phytochemistry 18:358.

Bierner M. W. (1994a) Pectolinarigenin from *Dugaldia pinetorum* (Standl.) Bierner. Biochem. Syst. Ecol. 22:109–110.

Bierner M. W. (1994b) Submersion of *Dugaldia* and *Plummera* in *Hymenoxys* (Asteraceae: Heliantheae: Gaillardiinae). Sida 16:1–8.

Bierner M. W., Dennis W. M., Wofford B. E. (1977) Flavonoid chemistry, chromosome number and phylogenetic relationships of *Helenium chihahuensis*. Biochem. Syst. Ecol. 5:23–28.

Bierner M. W., Jansen R. K. (1998) Systematic implications of DNA restriction site variation in *Hymenoxys* and *Tetraneuris* (Asteraceae, Helenieae, Gaillardiinae). Lundellia 1:17–26.

Bilyk A., Sapers G. M. (1985) Distribution of quercetin and kaempferol in lettuce, kale, chive, garlic, leek, horseradish, red raddish and red cabbage tissues. J. Agric. Food Chem. 33:226–228.

Bingöl S., Çubukçu B. (1984) Flavonoids of *Helichrysum noeanum* Boiss. (Asteraceae). Sci. Pharm. 52:127–130.

Biolley J. P., Jay M., Barbe J. P. (1992) Chemometric approach (flavonoids) in an automatic recognition of modern rose cultivars. Biochem. Syst. Ecol. 20:697–705.

Biryuk V. A., Chernobai V. T. (1972) Flavonoids of *Calendula officinalis*. Farm. Zh. Kiev. 27:44–49.

Bittmann M. (1990a) Die Gattung *Adenocaulon* (Compositae): I. Morphologie. Candollea 45:389–420.

Bittmann M. (1990b) Die Gattung *Adenocaulon* (Compositae): II. Ökologie, Verbreitung und Systematik. Candollea 45:493–518.

Bittner M., Jakupovic J., Bohlmann F., Silva M. (1989) Coumarins and guaianolides from further Chilean representatives of the subtribe Nassauviinae. Phytochemistry 28:2867–2868.

Bittner M., Silva M., Rozas Z., Papastergiou F., Jakupovic J. (1994) Sesquiterpenes and other constituents from Chilean Mutisieae. Phytochemistry 36:695–698.

Bittner M., Silva M., Vargas J., Bohlmann F. (1983) Biologically active flavones from *Gutierrezia resinosa*. Phytochemistry 22:1523–1524.

Bittner M., Watson W. H. (1982) Phytochemical study of *Haplopappus ciliatus*. Rev. Latinoamer. Quim. 13:27–28.

Bjorkman O., Holmgren P. (1958) Anthocyanins and other flavonoids and respiration rates in different ecotypes of *Solidago virgaurea*. Physiol. Plant. 11:154–157.

Blake S. F. (1918) A revision of the genus *Viguiera*. Contr. Gray Herb. 54:1–205.

Blake S. F. (1924) New American Asteraceae. Contrib. U. S. Natl. Herb. 22:587–661.

Blake S. F. (1937) *Tracyina*, a new genus of Asteraceae from northern California. Madroño 4:73–77.

Bloom M., Vickery R. K. (1973) Pattern partitioning of floral flavonoids in the *Mimulus luteus* complex. Phytochemistry 12:165–167.

Bloor S. J. (1999) Novel pigments and copigmentation in the blue marguerite daisy. Phytochemistry 50: 1395–1399.

Boehrer I. (1968) Flavonoids from fructus cardui. Tet. Lett. 6219–6222.

Boeker R., Bohlmann F. (1986) Flavanones from *Chromolaena heteroclinium*. Rev. Latinoamer. Quim. 17:45–46.

Bogacheva N. G., Glyzin G. S., Libizov N. I. (1971) Flavonoids of the inflorescences of *Callistephus chinensis*. Chem. Nat. Cmpds. 353 [375].

Bogs H. U., Bogs U. (1967) Über die Inhaltsstoffe von *Onopordum acanthium*. Die Pharmazie 22:54–58.

Bohlmann F., Abraham W.-R. (1979a) Neue, chlorsubstituierte Thiophenacetylenverbindungen mit ungewöhnlicher Struktur aus *Helichrysum*-Arten. Phytochemistry 18:839–842.

Bohlmann F., Abraham W.-R. (1979b) Neue Diterpene und weitere Inhaltsstoffe aus *Helichrysum calliconum* und *Helichrysum heterolasium*. Phytochemistry 18:889–891.

Bohlmann F., Abraham W.-R. (1979c) Neue Diterpene aus *Helichrysum acutatum*. Phytochemistry 18:1754–1756.

Bohlmann F., Abraham W.-R. (1979d) Neue Prenylflavanone aus *Helichrysum hypocephalum*. Phytochemistry 18:1851–1853.

Bohlmann F., Abraham W.-R. (1981) Aplotaxene epoxide from *Cirsium hypoleucum*. Phytochemistry 20:855–856.

Bohlmann F., Abraham W.-R., King R. M., Robinson H. (1980b) A new labdane derivative and geranylophloroglucinols from *Achyrocline alata*. Phytochemistry 19:2475–2477.

Bohlmann F., Abraham W.-R., King R. M., Robinson H. (1981a) Thiophene acetylenes and flavanols from *Pterocaulon virgatum*. Phytochemistry 20:825–827.

Bohlmann F., Abraham W.-R., Sheldrick W. S. (1980a) Weitere Diterpene mit Helifulvangerust und andere Inhaltsstoffe aus *Helichrysum chionosphaerum*. Phytochemistry 19:869–871.

Bohlmann F., Adler A., King R. M., Robinson H. (1982a) Germacranolides from *Mikania grazielae*. Phytochemistry 21:1169–1170.

Bohlmann F., Ahmed M., Jakupovic J., King R. M., Robinson H. (1983a) Dimeric sesquiterpene lactones and kolavane derivatives from *Gochnatia paniculata*. Phytochemistry 22:191–195.

Bohlmann F., Ates (Gören) N. (1984) Three prenylated flavanoids from *Helichrysum athrixiifolium*. Phytochemistry 23:1338–1341.

Bohlmann F., Banerjee S., Jakupovic J., Grenz M., Misra L. N., Schmeda-Hirschmann G., King R. M., Robinson H. (1985a) Clerodane and labdane diterpenoids from *Baccharis* species. Phytochemistry 24:511–515.

Bohlmann F., Banerjee S., Jakupovic J., King R. M., Robinson H. (1984a) New cadinene derivatives from *Baccharis viminea*. Rev. Latinoamer. Quim. 15:71–73.

Bohlmann F., Banerjee S., Wolfrum C., Jakupovic J., King R. M., Robinson R. (1985b) Sesquiterpene lactones, geranylnerol and tremetone derivatives from *Ageratina* species. Phytochemistry 24:1319–1321.

Bohlmann F., Bapuji M., Jakupovic J., King R. M., Robinson H. (1982c) Unusual diterpenes from *Brickellia eupatoriodes*. Phytochemistry 21:181–186.

Bohlmann F., Bapuji M., King R. M., Robinson H. (1982b) Kolavenol derivatives from *Goyazianthus etrastichus*. Phytochemistry 21:939–941.

Bohlmann F., Burkhardt T., Zdero C. (1973) *Naturally Occurring Acetylenes*. New York: Academic Press.

Bohlmann F., Czerson H., Schoneweiss S. (1977) Neue Inhaltsstoffe aus *Inula viscosa* Chem. Ber. 110:1330–1334.

Bohlmann F., Dutta L., Robinson H., King R. M. (1979a) Neue Labdan-Derivate aus *Chrysothamnus nauseosus*. Phytochemistry 18:1889–1892.

Bohlmann F., Fritz U. (1979) Neue Flavone aus *Heteromma simplicifolium*. Phytochemistry 18:1080–1081.

Bohlmann F., Grenz M. (1972) Notiz über die Isolierung von Hautriwa-Säure aus *Conyza ivaefolia* Less. Chem. Ber. 105:3123–3125.

Bohlmann F., Grenz M. (1977) Über neue Inhaltsstoffe aus Vertretern der *Eupatorium*-Gruppe. Chem. Ber. 110:1321–1329.

Bohlmann F., Grenz M., Dahr A. K., Goodman M. (1981b) Labdane derivatives and flavones from *Gutierrezia dracunculoides*. Phytochemistry 20:105–107.

Bohlmann F., Grenz M., Jakupovic J., King R. M., Robinson H. (1984b) New labdane derivatives and other constituents from *Fleischmania viscidipes*. Rev. Latinoamer. Quim. 15:1–3.

Bohlmann F., Grenz M., Niemeyer M., Dillon M. O. (1990) A lignane derivative from *Lucilia nivea*. Rev. Latinoamer. Quim. 21:75–76.

Bohlmann F., Gupta R. K. (1982) Guaianolides and furanosesquiterpenes from *Ursinia nana*. Phytochemistry 21:1309–1312.

Bohlmann F., Gupta R. K., Jakupovic J., Robinson H., King R. M. (1981c) Three germacranolides and other constituents from *Eremanthus* species. Phytochemistry 20:1609–1612.

Bohlmann F., Gupta R. K., King R. M., Robinson H. (1981d) Prostaglandin-like fatty acid derivative from *Chromolaena morii*. Phytochemistry 20:1417–1418.

Bohlmann F., Hoffmann E. (1979) Cannabigerol-ähnliche Verbindungen aus *Helichrysum umbraculigerum*. Phytochemistry 18:1371–1374.

Bohlmann F., Jakupovic J. (1978) Über neue Chromene und andere Inhaltsstoffe von *Lagascea rigida*. Phytochemistry 17:1677–1678.

Bohlmann F., Jakupovic J. (1979) Neue Sesquiterpene, Triterpene, Flavanone und andere aromatische Verbindungen aus *Flourensia heterolepis*. Phytochemistry 18:1189–1194.

Bohlmann F., Jakupovic J. (1990) Progress in the chemistry of the Vernonieae (Compositae). Plant Syst. Evol., Suppl. 4:3–43.

Bohlmann F., Jakupovic J., Ahmed M., Schuster A. (1983b) Sesquiterpene lactones and other constituents from *Schistostephium* species. Phytochemistry 22:1623–1636.

Bohlmann F., Jakupovic J., King R. M., Robinson H. (1980c) Chromones and flavans from *Marshallia obovata*. Phytochemistry 19:1815–1820.

Bohlmann F., Jakupovic J., King R. M., Robinson H. (1981e) New germacranolides, guaianolides and rearranged guaianolides from *Lasiolaena santosii*. Phytochemistry 20:1613–1622.

Bohlmann F., Jakupovic J., Schuster A. (1985c) Germacranolides from *Perymenium klattianum* and *Perymeniopsis ovalifolia*. Phytochemistry 24:495–499.

Bohlmann F., Knauf W., King R. M., Robinson H. (1979b) Ein neues Diterpen und weitere Inhaltsstoffe aus *Baccharis*-Arten. Phytochemistry 18:1011–1014.

Bohlmann F., Knoll K.-H., Robinson H., King R. M. (1980d) Neue Eudesmanolide aus *Steiractina mollis*. Phytochemistry 19:971–972.

Bohlmann F., Kramp W., Grenz M., Robinson H., King R. M. (1981f) Diterpenes from *Baccharis* species. Phytochemistry 20:1907–1913.

Bohlmann F., Kramp W., Jakupovic J., Robinson H., King R. M. (1982d) Diterpenes from *Baccharis* species. Phytochemistry 21:399–403.

Bohlmann F., Ludwig G.-W., Jakupovic J., King R. M., Robinson H. (1984c) Neue Spirosesquiterpenlactone, Germacranolide und Eudesmanolide aus *Wunderlichia mirabilis*. Liebig's Ann. 228–239.

Bohlmann F., Mahanta P. K. (1979) Eine neue Diterpenesäure aus *Centipeda orbicularis*. Phytochemistry 18:1067–1068.
Bohlmann F., Mahanta P. K., Zdero C. (1978a) Neue Chalkon-Derivate aus Südafrikanischen *Helichrysum*-Arten. Phytochemistry 17:1935–1937.
Bohlmann F., Misra L. N. (1984) New prenylflavanones and chalcones from *Helichrysum rugulosum*. Planta Medica 50:271–272.
Bohlmann F., Misra L. N., Jakopovic J. (1984d) Weitere Phloroglucin- und a-Pyrone-Derivate aus *Helichrysum*-Arten. Planta Medica 50:174–176.
Bohlmann F., Misra L. N., Jakopovic J. (1985d) *seco*-Helenanolides from *Dugaldia integrifolia*. Phytochemistry 24:1378–1380.
Bohlmann F., Rosenberg E., King R. M., Robinson H. (1980e) Neue Labdan-Derivate aus *Aristeguietia buddleaefolia*. Phytochemistry 19:977–979.
Bohlmann F., Scheidges C., Zdero C., King R. M., Robinson H. (1984e) *ent*-Labdanes from *Baccharis sternbergiana*. Phytochemistry 23:1109–1111.
Bohlmann F., Schmeda-Hirschmann G., Jakupovic J., King R. M., Robinson H. (1984f) Germacranolides from *Gochnatia vernonioides*. Phytochemistry 23:1989–1993.
Bohlmann F., Singh P., Jakupovic J., King R. M., Robinson H. (1982e) Three cadinene derivatives and a prostaglandin-like acid from *Chromolaena* species. Phytochemistry 21:371–374.
Bohlmann F., Singh P., Jakupovic J., Robinson H., King R. M. (1982f) An epoxygermacranolide and further constituents from *Mikania* species. Phytochemistry 21:705–707.
Bohlmann F., Suwita A. (1979a) Ein neues Bisabolen-Derivat und ein neues Dihydrokaffeesaure-Derivat aus *Tarchonanthus trilobus*. Phytochemistry 18:677–678.
Bohlmann F., Suwita A. (1979b) Ein neues Guajanolid und ein Secoguajanolid aus *Helichrysum spendidum*. Phytochemistry 18:885–886.
Bohlmann F., Suwita A. (1979c) Weitere Phloroglucin-Derivate aus *Helichrysum*-Arten. Phytochemistry 18:2046–2049.
Bohlmann F., Suwita A., Jakupovic J., King R. M., Robinson H. (1981g) Trixikingolides and germacrene derivatives from *Trixis* species. Phytochemistry 20:1649–1655.
Bohlmann F., Suwita A., Zdero C. (1978b) Neuartige Ester von Eremophilen-Derivaten aus Südafrikanischen *Senecio* Arten. Phytochemistry 17:1763–1767.
Bohlmann F., Wegner P., Jakupovic J. (1982g) Unusual diterpenes and sesquiterpene xylosides from *Nidorella hottentotica*. Phytochemistry 21:1109–1114.
Bohlmann F., Zdero C. (1967) Über Flavone aus *Centaurea*-Arten. Tet. Lett. 3239–3242.
Bohlmann F., Zdero C. (1969) Neue Sesquiterpene aus *Brickellia guatemaliensis*. Tet. Lett. 5109–5110.
Bohlmann F., Zdero C. (1977) Inhaltsstoffe der Gattung *Polymnia*. Phytochemistry 16:492–493.
Bohlmann F., Zdero C. (1978) New furanosesquiterpenes from *Eumorphia* species. Phytochemistry 17:1155–1159.
Bohlmann F., Zdero C. (1979a) Neue C-10-Saureamide, furanoeremophilane und andere Inhaltsstoffe aus Bolivianischen *Senecio*-Arten. Phytochemistry 18:125–128.
Bohlmann F., Zdero C. (1979b) Neue Anol-Derivate aus *Ophryosporus angustifolius*. Phytochemistry 18:145–147.

Bohlmann F., Zdero C. (1980) Neue Geranylphloroglucin-Derivate aus *Helichrysum monticola*. Phytochemistry 19:683–684.
Bohlmann F., Zdero C. (1982) Chemotaxonomy of the genus *Pleiotaxis*. Phytochemistry 21:1434–1435.
Bohlmann F., Zdero C. (1983) Flavanones from *Helichrysum thapsus*. Phytochemistry 22:2877–2878.
Bohlmann F., Zdero C., Abraham W.-R., Suwita A., Grenz M. (1980f) Neue Diterpene und neue Dihydrochalkon-Derivative sowie weitere Inhaltstoffe aus *Helichrysum* Arten. Phytochemistry 19:873–879.
Bohlmann F., Zdero C., Grenz M., Dhar A. K., Robinson H., King R. M. (1981h) Five diterpenes and other constituents from nine *Baccharis* species. Phytochemistry 20:281–286.
Bohlmann F., Zdero C., Gupta R. K., King R. M., Robinson H. (1980g) Diterpenes and tetranorditerpenes from *Acritopappus* species. Phytochemistry 19:2695–2705.
Bohlmann F., Zdero C., Hoffmann E., Mahanta P. K., Dorner W. (1978c) Neue Diterpene und Sesquiterpene aus südafrikanischen *Helichrysum*-Arten. Phytochemistry 17:1917–1922.
Bohlmann F., Zdero C., Huneck S. (1985e) Diterpenes from *Heteropappus altaicus*. Phytochemistry 24:1027–1030.
Bohlmann F., Zdero C., King R. M., Robinson H. (1979d) Neue Cadinen-Derivate und andere Inhaltsstoffe aus *Chromolaena*-Arten. Phytochemistry 18:1177–1179.
Bohlmann F., Zdero C., King R. M., Robinson H. (1979e) New prenylflavonoids from *Marshallia grandiflora*. Phytochemistry 18:1246–1247.
Bohlmann F., Zdero C., King R. M., Robinson H. (1980h) Sesquiterpene lactones from *Eremanthus* species. Phytochemistry 19:2663–2668.
Bohlmann F., Zdero C., King R. M., Robinson H. (1980i) Seven guaianolides from the tribe Vernonieae. Phytochemistry 19:2669–2673.
Bohlmann F., Zdero C., King R. M., Robinson H. (1981i) Thirteen kolavane derivatives from *Symphyopappus* species. Phytochemistry 20:1657–1663.
Bohlmann F., Zdero C., King R. M., Robinson H. (1984g) Kingidiol, a kolavane derivative from *Baccharis kingii*. Phytochemistry 23:1511–1512.
Bohlmann F., Zdero C., King R. M., Robinson H. (1984h) Neue Aromadendron-Derivate aus *Wyethia arizonica*. Planta Medica 50:195–196.
Bohlmann F., Zdero C., Robinson H., King R. M. (1980j) Caryophyllene derivatives and a heliangolide from *Lychnophora* species. Phytochemistry 19:2381–2385.
Bohlmann F., Zdero C., Robinson H., King R. M. (1981j) A diterpene, a sesquiterpene quinone and flavanones from *Wyethia helenioides*. Phytochemistry 20:2245–2248.
Bohlmann F., Zdero C., Schoneweiss S. (1976) Über die Inhaltsstoffe aus *Stevia*-Arten. Chem. Ber. 109:3366–3370.
Bohlmann F., Zdero C., Ziesche J. (1979c) Neue Flavone und Phloroglucin-Derivate aus *Helichrysum herbaceum* und *Helichrysum chrysargyrum*. Phytochemistry 18:1375–1378.
Bohlmann F., Ziesche J. (1980) Neue Diterpene aus *Gnaphalium* Arten. Phytochemistry 19:71–74.

Bohlmann F., Ziesche J., Mahanta P. K. (1979f) Neue Chalkon-Derivate und Humulon-ähnliche Verbindungen aus *Helichrysum*-Arten. Phytochemistry 18:1033–1036.

Bohm B. A. (1987) Intraspecific flavonoid variation. Botanical Review 53:197–279.

Bohm B. A. (1988) Flavonoid systematics of the Datiscaceae. Biochem. Syst. Ecol. 16:151–155.

Bohm B. A. (1994) The Minor Flavonoids. In: J. Harborne B. (ed.) *Flavonoids: Advances in Research since (1986).* London: Chapman and Hall.

Bohm B. A. (1998a) Secondary compounds and evolutionary relationships of island plants. In: Stuessy T. F., Ono M. (eds.) *Evolution and Speciation of Island Plants.* Cambridge: Cambridge Univ. Press. 233–295.

Bohm B. A. (1998b) *Introduction to Flavonoids.* Amsterdam: Harwood Academic Publishers.

Bohm B. A. (1999) Major exudate flavonoids of *Dubautia arborea* (Asteraceae). Biochem. Syst. Ecol. 27:755–757.

Bohm B. A., Averett J. E., Powell A. M. (1986a) Non-polar flavonoids of *Perityle vaseyi* (Asteraceae). Phytochemistry 25:2551–2553.

Bohm B. A., Banek H. M. (1986) Flavonoid variation in *Lasthenia burkei.* Biochem. Syst. Ecol. 15:57–59.

Bohm B. A., Banek H. M., Maze J. R. (1984) Flavonoid variation in North American *Menziesia* (Ericaceae). Syst. Bot. 9:324–345.

Bohm B. A., Berlow S., Stuessy T. F. (1983) Flavonoid variation in *Clibadium trianae* and *C. surinamense*. Phytochemistry 22: 2743–2744.

Bohm B. A., Chalmers G., Bhat U. G. (1988) Flavonoids and the relationship of *Itea* to the Saxifragaceae. Phytochemistry 27:2651–2653.

Bohm B. A., Chan J. (1992) Flavonoids and affinities of Greyiaceae with a discussion of the occurrence of B-ring deoxyflavonoids in dicotyledonous families. Syst. Bot. 17:272–281.

Bohm B. A., Choy J. B. (1987) Flavonoids of *Balsamorhiza deltoidea* and *Wyethia mollis*. Biochem. Syst. Ecol. 15:541–542.

Bohm B. A., Choy J. B., Lee A. Y.-M. (1989a) Flavonoids of *Balsamorhiza* and *Wyethia*. Phytochemistry 28:501–503.

Bohm B. A., Crins W. J., Wells T. C. (1994) External flavonoids of *Holocarpha* (Asteraceae: Madiinae). Biochem. Syst. Ecol. 22:859.

Bohm B. A., Fong C. E. (1990) Nonpolar flavonoids of *Wilkesia* and *Argyroxiphium*. Phytochemistry 29:1175–1176.

Bohm B. A., Fong C., Hiebert M., Jamal A., Crins W. J. (1992) Nonpolar flavonoids of *Calycadenia*, *Lagophylla* and *Madia*. Phytochemistry 31:1261–1263.

Bohm B. A., Gottlieb L. D. (1989) Flavonoids of annual *Stephanomeria* (Asteraceae). Biochem. Syst. Ecol. 17:451–453.

Bohm B. A., Herring A., Nicholls K. W., Bohm L. R., Ornduff R. (1989b) A six year study of flavonoid variation in a population of *Lasthenia californica*. Amer. J. Bot. 76:157–163.

Bohm B. A., Nicholls K. W., Ornduff R. (1986b) Flavonoids of the Menyanthaceae: Intra- and interfamilial relationships. Amer. J. Bot. 73:204–213.

Bohm B. A., Reid A., DeVore M. L., Stuessy T. F. (1995) Flavonoid chemistry of Calyceraceae. Canad. J. Bot. 73:1962–1965.

Bohm B. A., Saleh N. A. M., Ornduff R. (1974) The flavonoids of *Lasthenia* (Compositae). Amer. J. Bot. 61:551–561.

Bohm B. A., Stuessy T. F. (1981a) Flavonol derivatives of the genus *Clibadium* (Compositae). Phytochemistry 20:1053–1055.

Bohm B. A., Stuessy T. F. (1981b) Flavonol derivatives of *Desmanthodium* (Compositae). Phytochemistry 20:1573–1575.

Bohm B. A., Stuessy T. F. (1982) Flavonol derivatives of *Ichthyothere terminalis*. Phytochemistry 21:2761–2763.

Bohm B. A., Stuessy T. F. (1985) Further studies of flavonols of *Clibadium* (Compositae). Phytochemistry 24:2134–2136.

Bohm B. A., Stuessy T. F. (1991) Flavonoid variation in *Melampodium*. Biochem. Syst. Ecol. 19:677–679.

Bohm B. A., Stuessy T. F. (1995) Flavonoid chemistry of Barnadesioideae (Asteraceae). Syst. Bot. 20:22–27

Bohm B. A., Yang J. Y., Page J. E., Soltis D. E. (1998) A comparison of flavonoid and DNA sequence data as indicators of relationships of *Pterostemon* and *Itea*. Biochem. Syst. Ecol. 27:79–83.

Bolick M. R. (1983) A cladistic analysis of the Ambrosiinae Less. and Engelmanniinae Stuessy. In: Platnick N. I., Funk V. A. (eds.) *Advances in Cladistics, Vol. 2*. New York: Columbia University Press. 125–141.

Bolick M. R. (1985) Cladistic analysis of *Iva*: A case in point. Taxon 34:81–84.

Bombardelli E., Gabetta B., Martinelli E. M. (1977) Gas liquid chromatographic and mass spectral investigation on *Cynara scolymus* extracts. Fitoterapia 48:143–152.

Bondarenko V. G., Glyzin V. I., Ban'kovskii A. I., Shelyuto V. L. (1974) Isocynaroside – A new flavone glycoside. Chem. Nat. Cmpds. 10:680 [665].

Bondarenko V. G., Glyzin V. I., Shelyuto V. L. (1973) Flavonoids of the flowers of *Sonchus arvense* L. Chem. Nat. Cmpds. 9:522 [554].

Bondarenko V. G., Glyzin V. I., Shelyuto V. L. (1978) Sonchoside – A new flavonoid glycoside from *Sonchus arvensis*. Chem. Nat. Cmpds. 14:340 [403].

Bondarenko V. G., Glyzin V. I., Shelyuto V. L. (1983) Flavonoids of the flowers of *Sonchus oleraceus*. Chem. Nat. Cmpds. 19:228 [234].

Bondarenko V. G., Glyzin V. I., Shelyuto V. L., Smirnova L. P. (1976) Flavonoids of *Sonchus arvensis*. Chem. Nat. Cmpds. 12:484 [542].

Borella J. C., Lopes J. L. C., Vichnewski W., Cunha W. R., Herz W. (1998) Sesquiterpene lactones, triterpenes and flavones form *Lychnophora ericoides* and *L. pseudovillosissima*. Biochem. Syst. Ecol. 26:671–676.

Borisov M. I., Isakova T. I., Serbin A. G. (1980) Flavonoids of *Bidens cernua*. Chem. Nat. Cmpds. 15:197 [229].

Borkosky S., Kowalewski Z., Skrzypczakowa L. (1966) Chemical composition of inflorescences of some *Arnica* species. I. Analysis of flavonoid fractions. Diss. Pharm. Pharmacol. 18:367–374.

Borkosky S., Valdés D. A., Bardon A., Diaz J. G., Herz W. (1996) Sesquiterpene lactones and other constituents of *Eirmocephala megaphylla* and *Cyrtocymura cincta*. Phytochemistry 42:1637–1639.

Borkowski, B., Pasich B. (1961) Isolation of naringenin 5-ß-D-glucoside from *Helichrysum arenarium*. Acta Polon. Pharm. 18:91–96.

Bose P. K., Barna A. K., Chakraborti P. A. (1968) A revised structure for erianthin, a flavonol from *Blumea eriantha*. J. Indian Chem. Soc. 45:851–852.

Bose P. K., Chakrabarti P., Chakravarti S., Dutta S. P., Barua A. K. (1973) Flavonoid constituents of *Eupatorium odoratum*. Phytochemistry 12:667–668.

Boulos L. (1974) Révision systématique du genre *Sonchus* L. s.l. VI. Sous-genera 3. *Origosonchus*. Genres *Embergeria, Babcockia* et *Taeckholmia*. Bot. Notiser 127:402–441.

Boulos L. (1976) Launea. In: Tutin T. G., Heywood V. H., Burges N. A., Moore D. M., Valentine D. H., Walters S. M., Webb D. A. (eds.) *Flora Europaea, Vol. 4. Plantaginaceae to Compositae (and Rubiaceae)*. Cambridge: Cambridge University Press. 326–327.

Boulter D., Gleaves J. T., Haslett B. G., Peacock D., Jensen U. (1978) The relationships of eight tribes of the Compositae as suggested by plastocyanin amino acid sequence data. Phytochemistry 17:1585–1589.

Bouzid N., Fouraste I., Voirin B., Favre-Bonvin J., Lebreton P. (1982a) A methylated flavone from *Artemisia mesatlantica*. Phytochemistry 21:803–804.

Bouzid N., Moulis C., Fouraste I. (1982b) Flavones libre de *Artemisia mesatlantica*. Planta Medica 44:157–158.

Bramwell D., Dakshini K. M. M. (1971) Luteolin 7-glucosides and hydroxycoumarins in Canary Islands *Sonchus* species. Phytochemistry 10:2245–2246.

Brandao M. G. L., Nery C. G. C., Mamao M. A. S., Krettli A. U. (1998) Two methoxylated flavone glycosides from *Bidens pilosa*. Phytochemistry 48:397–399.

Brasseur T., Angenot L., Pincemail J., Deby C. (1987) Antiradical effect of flavonoids and extracts of *Chrysanthellum indicum*. Pl. Medica Phytother. 21:131–135.

Brehm B. G., Krell D. (1975) Flavonoid localization in epidermal papillae of flower petals; a specific adaptation for ultraviolet absorption. Science 190:1221–1223.

Brehm B. G., Ownbey M. (1965a) Variation in chromatographic patterns in the *Tragopogon dubius-pratensis-porrifolius* complex. Amer. J. Bot. 52:811–818.

Brehm B. G., Ownbey M. (1965b) Genetic analysis of 2-dimensional chromatographic pattern components in *Tragogpogon*. Amer. J. Bot. 52:647. Abstr.

Brehm B. G., Ownbey M. (1968) Inheritance of *C*-glycosylflavonoids in interspecific hybridizations of *Tragogpogon*. Amer. J. Bot. 55:738. Abstr.

Breitwieser I., Ward J. W. (1993) Systematics of New Zealand Inuleae (Compositae-Asteraceae) 3. Numerical phenetic analysis of leaf anatomy and flavonoids. New Zealand J. Bot. 31:43–58.

Bremer K. (1987) Tribal interrelationships of the Asteraceae. Cladistics 3:210–253.

Bremer K. (1992) Ancestral areas: a cladistic reinterpretation of the center of origin concept. Syst. Biol. 41:436–445.

Bremer K. (1994) *Asteraceae: Cladistics and Classification*. Portland, Oregon: Timber Press.

Bremer K. (1996) Major clades and grades of the Asteraceae. In: Hind D. J. N., Beentje H. J. (eds.) *Compositae: Systematics. Proceedings of the International Compositae Conference, Kew, (1994) Vol. 1*. Kew: Royal Botanic Gardens. 1–7.

Bremer K., Anderberg A. A., Karis P. O., Lundberg J. (1994) Tribe Eupatorieae. In: Bremer K. (ed.) *Asteraceae: Cladistics and Classification*. Portland, Oregon: Timber Press. 625–680.

Bremer K., Gustafsson M. H. G. (1997) East Gondwana ancestry of the sunflower alliance of families. Proc. Natl. Acad. Sci. U.S.A. 94:9188–9190.

Bremer K., Humphries C. (1993) Generic monograph of the Asteraceae-Anthemideae. Bull. Nat. Hist. Mus. Lond. (Bot.) 23:71–177.

Bremer K., Jansen R. K. (1992) A new subfamily of the Asteraceae. Ann. Missouri Bot. Gard. 79:414–416.

Bremer K., Jansen R. K., Karis P. O., Källersjö M., Keeley S. C., Kim K.-J., Michaels H. J., Palmer J. D., Wallace R. S. (1992) A review of the phylogeny and classification of the Asteraceae. Nord. J. Bot. 12:141–148.

Breton Fumes J. L., Garcia Marrero B., Gonzalez Gonzalez A. (1968) Flavonoids from *Centaurea arguta*. An. Quim. 64:187–191

Breton Fumes J. L., Gonzalez Gonzalez A., Rodriquez Rincones M. (1969) Flavonoids of *Centaurea arbutifolia*. An. Quim. 65:297–300.

Bridle P., Loeffler R. S. T., Timberlake C. F., Self R. (1984) Cyanidin 3-malonylglucoside in *Cichorium intybus*. Phytochemistry 23:2968–2969.

Briquet J. (1917) Quelques points de l'organisation des élichryses stoechadinés. Biblioth. Universelle Rev. Suisse, ser. 4, 43:253–259.

Briquet J. (1918) L'appareil; staminal des composées; structure et fonctions de ses diverses parties. Bull. Soc. Vaud. Sci. Nat. 51:208–210.

Britton N. L., Brown A. (1896–1898) *An Illustrated Flora of the Northern United States*, 3 vols. New York: Charles Scribner's Sons.

Broussalis A. M., Ferraro G. E., Coussio J. D. (1993) Phenolic constituents of *Achyrocline alata*. Biochem. Syst. Ecol. 21:306.

Brown D., Asplund R. O., McMahon V. A. (1975) Phenolic constituents of *Artemisia tridentata* ssp. *vaseyana*. Phytochemistry 14:1083–1084.

Brown J. H., Gibson A. C. (1983) *Biogeography*. St. Louis: Mosby.

Brown J. H., Lomolino M. V. (1998) *Biogeography*, 2nd Edition. Sunderland, Massachusetts: Sinauer Associates.

Brown R. K., Schaak C. S. (1972) Two new species of *Tragopogon* for Arizona. Madroño 21:304.

Bruno M., Herz W. (1988a) Guaianolides and other constituents of *Achillea ligustica*. Phytochemistry 27:1871–1872.

Bruno M., Herz W. (1988b) Sesquiterpene lactones and flavones from *Centaurea cineraria* ssp. *umbrosa*. Phytochemistry 27:1873–1875.

Budesinsky M., Perez Souto N., Holub M. (1994) Sesquiterpene lactones of some species of *Vernonia* Schreb. Collect. Czech. Chem. Communs. 59:913–928.

Budzianowski J. (1990) Two di-*C*-glycosylflavones from *Solidago graminifolia*. Sci. Pharm. 58:413–416.

Bukharov V. G., Shcherbak S. P. (1967) Study of the chemical composition of *Rhaponticum cartamoides*. Chem. Nat. Cmpds. 3:235 [280].

Bulinska-Radomska Z., Norris J. A., Hosage D. A., Mabry T. J. (1985a) Flavonoids from *Berlandiera texana* var. *texana*. J. Nat. Prod. 48:667.

Bulinska-Radomska Z., Norris J. A., Mabry T. J. (1985b) Flavonoids from *Brickelliastrum fendleri*. J. Nat. Prod. 48:144.
Burnett W. C., Jr., Jones S. B., Jr., Mabry T. J., Padolina W. G. (1974) Sesquiterpene lactones—insect feeding deterrents in *Vernonia*. Biochem. Syst. Ecol. 2:25–29.
Burrows F. M. (1973) Calculation of the primary trajectories of plumed seeds in steady winds with variable convection. New Phytol. 72:647–664.
Cabrera A. L. (1949) El género *Senecio* en Chile. Lilloa 15:27–501.
Cabrera A. L. (1954) Las especies del género *Nardophyllum*. Notas Museo de La Plata, Botánica 17:55–66.
Cabrera A. L. (1965) Revisión del género *Mutisia* (Compositae). Opera Lilloana 13:1–227.
Cabrera A. L. (1971) Compositae. In: Correa, M. N. (ed.) *Flora Patagonica, parte VII*. Buenos Aires: Instituto Nacional de Tecnología Agropecuaria.
Cabrera A. L. (1977) Mutisieae—systematic review. In: Heywood V. H., Harborne J. B., Turner B. L. (eds.) *The Biology and Chemistry of the Compositae*, Vol. 2. London: Academic Press. 1039–1066.
Cabrera J. L., Juliani H. R. (1976) Quercetin-3-acetyl-7,3',4'-trisulphate from *Flaveria bidentis*. J. Nat. Prod. 39:253–254.
Cabrera J. L., Juliani H. R. (1977) Isorhamnetin 3,7-disulphate from *Flaveria bidentis*. Phytochemistry 16:400.
Cabrera J. L., Juliani H. R. (1979) Two new quercetin sulphates from leaves of *Flaveria bidentis*. Phytochemistry 18:510–511.
Cabrera J. L., Juliani H. R., Gros E. G. (1985) Quercetin-3,7,3'-trisulphate from *Flaveria bidentis*. Phytochemistry 24:1394–1395.
Caesalpino A. (1583) *De Plantis*. Florence.
Cain A. J. (1995) Linnaeus's natural and artificial arrangements of plants. Bot. J. Linn. Soc. 117:73–133.
Calanasan C. A., MacLeod J. K. (1998) A diterpenoid sulphate and flavonoids from *Wedelia asperrima*. Phytochemistry 47:1093–1099.
Calatayud P. A., Rahhe Y., Delobel B., Khuong-Huu F., Tertuliano M., Ru B. L. (1994) Influence of secondary compounds in the phloem sap of cassava on expression of antibiosis towards mealy bug *Phenacoccus manihoti*. Entomol. Exp. Appl. 72:47–57.
Caldwell M. M., Robberecht R., Flint S. D. (1983) Internal filters: prospects for UV-acclimation in higher plants. Physiol. Plant. 58:445–450.
Camin J. H., Sokal R. P. (1965) A method for deducing branching sequences in phylogeny. Evolution 19:311–326.
Candolle A. L. P. P. de (1844–1873) *Prodromus Systematis Naturalis Regni Vegetabilis*. Vols. 8–17. Paris.
Candolle A. P. de (1824–1838) *Prodromis Systematis Naturalis Regni Vegetabilis*. Vols. 1–7 (Asteraceae vols. 5–7). Paris.
Candolle A. P. de (1838) Observations sur la structure et la classification de la famille des Composées. 9th Memoire. Paris.
Candy H. A., Laing M., Weeks C. M., Kruger G. J. (1975) The crystal and molecular structure of helichrysoside, a new acylated flavonoid glycoside from *Helichrysum kraussii*. Tet. Lett. 1211–1214.

Cannon J. F. M., Marshall J. B. (1976) *Serratula*. In: Tutin T. G., Heywood V. H., Gurges N. A., Moore D. M., Alentine D. H., Walters S. M., Webb D. A. (eds.) *Flora Europaea, Vol. 4. Plantaginaceae to Compositae (and Rubiaceae)*. Cambridge: Cambridge Univ. Press. 250–252.

Canonica L., Rindone B., Scolastico C., Ferrari G., Casagrande C. (1969) Constituenti estrattivi della *Psiadia altissima* Benth. & Hook. Gazz. Chim. Ital. 99:260–275.

Cappelletti E. M., Caniato R. (1983) Chemical characterization of wild Italian populations of *Silybum marianum* (L.) Gaertn. Pl. Medica Phytother. 17:209–214.

Cappelletti E. M., Caniato R., Caniato R., Appendino G. (1986) Localization of the cytotoxic hydroperoxyeudesmanolides in *Artemisia umbelliformis*. Biochem. Syst. Ecol. 14:183–190.

Cardona L., Aleman R. A., Garcia B., Pedro J. R. (1992) Sesquiterpenes, flavonoids and lignans from *Onopordum acaulon*. Phytochemistry 31:3630–3632.

Cardona L., Fernandez I., Pedro J. R., Valle A. A. (1987) Flavonoids and other constituents from *Onopordum macracanthum*. Planta Medica 53:506.

Cardona M. L., Fernandez I., Pedro J. R., Perez B. (1991) Sesquiterpene lactones and flavonoids from *Centaurea aspera*. Phytochemistry 30:2331–2333.

Cardona M. L., Garcia B., Pedro J. R., Sinisterra J. F. (1990) Flavonoids, flavonolignans and a phenylpropanoid from *Onopordum corymbosum*. Phytochemistry 29: 629–631.

Carlquist S. (1957) Systematic anatomy of *Hesperomannia*. Pacific Sci. 11:207–215.

Carlquist S. (1961) *Comparative Plant Anatomy*. New York: Holt, Rinehart and Winston.

Carlquist S. (1966) Wood anatomy of Compositae: a summary wilth comments on factors controlling wood evolution. Aliso 6:25–44.

Carlquist S. (1974) *Island Biology*. New York: Columbia Univ. Press.

Carlquist S. (1976) Tribal interrelationships and phylogeny of the Asteraceae. Aliso 8:465–492.

Carlquist S. (1980) *Hawaii. A Natural History*. Lawai, Kauai, Hawaii: Pacific Tropical Botanical Garden.

Carlquist S. (1981) Chance dispersal. Amer. Sci. 69:509–516.

Carlquist S., Carr G. D., Baldwin B. (eds.) (2001) *Madiinae*. Berkeley: Univ. California Press.

Carlquist S., DeVore M. L. (1998) Wood anatomy of Calyceraceae with reference to ecology, habit, and systematic relationships. Aliso 17:63–76.

Carman N. J., Watson T., Bierner M. W., Averett J., Sanderson S., Seaman F. C., Mabry T. J. (1972) 6-Methoxyapigenin from thirty-four species of Compositae. Phytochemistry 11:3271–3272.

Carpenter K. E. (1993) Optimal cladistic and quantitative evolutionary classifications as illustrated by fusilier fishes (Teleostei: Caesionidae). Syst. Biol. 42:142–154.

Carr G. D. (1975) Chromosome evolution and aneuploid reduction in *Calycadenia pauciflora* (Asteraceae). Evolution 29:681–699.

Carr G. D. (1977) A cytological conspectus of the genus *Calycadenia* (Asteraceae): an example of contrasting modes of evolution. Amer. J. Bot. 64:694–703.

Carr G. D. (1980) Experimental evidence for saltational chromosome evolution in *Calycadenia pauciflora* Gray (Asteraceae). Heredity 45:107–112.

Carr G. D. (1985) Monograph of the Hawaiian Madiinae (Asteraceae): *Argyroxiphium*, *Dubautia*, and *Wilkesia*. Allertonia 4:1–123.

Carr G. D. (1995) A fully fertile intergeneric hybrid derivative from *Argyroxiphium sandwicense* ssp. *macrophyllum* x *Dubautia menziesii* (Asteraceae) and its relevance to plant evolution in the Hawaiian Islands. Amer. J. Bot. 82:1574–1581.

Carr G. D., King R. M., Powell A. M., Robinson H. (1999) Chromosome numbers in Compositae. XVIII. Amer. J. Bot. 86:1003–1013.

Carr R. L., Carr G. D. (1983) Chromosome races and structural heterozygosity in *Calycadenia ciliosa* Greene (Asteraceae). Amer. J. Bot. 70:744–755.

Cassini H. (1814) Precis d'un second memoire sur les synanthereres contenant l'analyse des etamines. J. de Physique, de Chim., d'Hist. Nat. et des Arts, 78:274–291.

Cassini H. (1816) Sur une nouvelle famille de plantes (les Boopidées). Bull. Sci. Soc. Philom. Paris 1816:160–161.

Cassini H. (1819) Sur la famille de Synathérées, contenant les charactères des tribus. J. Phys., de Chim., d'Hist. Nat. et des Arts 88:189–204.

Cassini H. (1828) Vernoniées. In: Cuvier F. (ed.) *Dictionaire des science naturelles*, Vol. 57, Paris. 338–340. [from *Cassini on Compositae*, 1975, compiled by King R. M., Davies H. W., Oriole Editions, New York].

Cassini H. (1834) *Opuscules Phytologiques*, Vol. 3. Paris.

Catalan C. A. N., De Iglesias D. I. A., Kavka J., Sosa V. E., Herz W. (1988) Glaucolides and related sesquiterpene lactones from *Vernonia chamaedrys*. Phytochemistry 27:197–202.

Caula S. A., Villar S. I., Martino V. S., Coussio J. D., Ferraro G. E. (1991) Polyphenols isolated from *Eupatorium buniifolium*. Rev. Latinoamer. Quim. 22:1–3.

Cavaleiro J. A. S. (1986) Isosakuranetin, a flavanone from *Artemisia campestris* subsp. *maritima*. Fitoterapia 57:278

Cave M. S. (ed.) (1958–1965) *Index to Plant Chromosome Numbers* (1956–1964 & Supplement). Chapel Hill: Univ. North Carolina Press.

Cekan Z., Herout V. (1956) Über Pflanzenstoffe. IV. Die Isolierung von 5-Hydroxy-3,6,7,3',4'-pentamethoxyflavone aus Wermut *Artemisia absinthium* L. Coll. Czech. Chem. Communs. 21:79–85.

Cezard C., Torck M., Pinkas M. (1984) *Senecio cineraria* DC.: etude preliminaire. Pl. Medica Phytother. 18:160–164.

Chaboud A., Dellamonica G., Raynaud J. (1988) Neocorymboside, a di-*C*-glycosylflavone from *Atractylis gummifera*. Phytochemistry 27:2360–2361.

Chaboud A., Proliac A., Raynaud J., Dellamonica G. (1990) New *O*-glycosyl *C*-glycosyl flavones of *Atractylis gummifera* L. (Compositae). Pharm. Acta Helv. 65:286.

Chakraborti S. K., Banerjee A. K., Bose P. K. (1959–60) Natural flavones. VI. Yellow pigment of the flower of *Helichrysum annuum*. Trans. Bose Res. Inst. (Calcutta) 23:59.

Chambers K. L. (1957) Taxonomic notes on some Compositae of the western United States. Contrib. Dudley Herb. 5:57–61.

Chandrasekharan I., Khan H. A., Ghanin A. (1981) Flavonoids from *Artemisia scoparia*. Planta Medica 43:310–311.

Chandrashekar V., Krishnamutri M., Seshadri T. R. (1965) Chemical investigation of *Artemisia sacrorum*. Curr. Sci. 34:609.

Charaux C., Rabaté J. (1940) *Centaurea scabiosa* L. J. Pharm. Chim. 1:155–162.

Chase M. W., Soltis D. E., Olmstead R. E., Morgan D., Les D. H., Mishler B. D., Duvall M. R., Price R. A., Wells H. G., Qiu Y.-L., Kron K. A., Rettig J. H., Conti E., Palmer J. D., Manhart J. R., Sytsma K. J., Michaels H. J., Kress W. J., Karol K. G., Clark W. D., Hedrén M., Gaut B. S., Jansen R. K., Kim K.-J., Wimpee C. F., Smith J. F., Furnier G. R., Strauss S. H., Xiang Q.-Y., Plunkett G. M., Soltis P. S., Swensen S. M., Williams S. E., Gadek P. A., Quinn C. J., Eguiarte L. E., Goldenberg E., Learn G. H., Jr., Graham S. W., Barrett S. C. H., Dayanandan S., Albert V. A., Cobley T. (1993) Phylogenetics of seed plants: an analysis of nucleotide sequences from the plastid gene *rbc*L. Ann. Mo. Bot. Gard. 80:528–580.

Chater A. O. (1976) *Kleinia* Miller. In: Tutin T. G., Heywood V. H., Burges N. A., Moore D. M., Valentine D. H., Walters S. M., Webb D. A. (eds.) *Flora Europaea, Vol. 4, Plantaginaceae to Compositae (and Rubiaceae)*. Cambridge: Cambridge Univ. Press. 205–206.

Chater A. O., Walters S. M. (1976) *Senecio* L. In: Tutin T. G., Heywood V. H., Burges N. A., Moore D. M., Valentine D. H., Walters S. M., Webb D. A. (eds.) *Flora Europaea, Vol. 4, Plantaginaceae to Compositae (and Rubiaceae)*. Cambridge: Cambridge Univ. Press. 191–205.

Chatterjee A., Sarkar S., Saha S. K. (1981) Acacetin 7-O-ß-D-galactopyranoside from *Chrysanthemum indicum*. Phytochemistry 20:1760–1761.

Chaudhuri P. K. (1988) Constituents of *Echinatus echinacea*. Fitoterapia 59:150–151.

Chaudhuri P. K., Thakur R. S. (1986) An acylated flavone, apigenin 7-O-ß-D-4''-*cis*-*p*-coumaroyl glucoside from *Echinops echinatus*. Phytochemistry 25:1770–1771.

Chaumont J. P. (1969) A propos des polyphénols flavonoides, acides-phénols, coumarines. des capitules de Camomille Romaine (*Anthemis nobilis* L.). Pl. Medica Phytother. 3:167–174.

Chemesova I. I. (1987) Flavonol composition of *Artemisia xanthochroa*. Chem. Nat. Cmpds. 23:120 [141].

Chemesova I. I., Belenovskaya L. M., Markova L. P. (1978) Phenolic compounds of *Artemisia palustris*. Chem. Nat. Cmpds. 14:338 [401].

Chemesova I. I., Belenovskaya L. M., Markova L. P. (1982) Phenolic compounds of *Artemisia sieversiana*. Chem. Nat. Cmpds. 18:492 [521].

Chemesova I. I., Belenovskaya L. M., Markova L. P. (1983a) Phenolic compounds of *Artemisia gmelinii*. Chem. Nat. Cmpds. 19:364 [384].

Chemesova I. I., Belenovskaya L. M., Markova L. P. (1984a) Flavonoids of *Artemisia rutifolia*. Chem. Nat. Cmpds. 20:232 [249].

Chemesova I. I., Belenovskaya L. M., Markova L. P. (1984b) Flavonoids of *Artemisia xanthochroa*. Chem. Nat. Cmpds. 20:748 [789].

Chemesova I. I., Belenovskaya L. M., Nadezhina T. P. (1983b) Phenolic compounds of *Artemisia adamsii*. Chem. Nat. Cmpds. 19:365 [385].

Chemesova I. I., Belenovskaya L. M., Stukov A. N. (1987) Antitumor activity of flavonoids from some species of *Artemisia* L. Rastit. Resur. 23:100–103.

Chemesova I. I., Burkeeva T. V., Boiko E. V. (1990) Phenols of *Artemisia laciniata*. Khim. Prir. Soedin. 26:115.

Chemesova I. I., Markova L. P. (1986) Flavonoid compounds of *Artemisia halodendron*. Chem. Nat. Cmpds. 22:725 [780].
Cheminat A., Brouillard R., Guerne P., Bergmann P., Rether B. (1989) Cyanidin 3-malonylglucoside in two *Echinacea* species. Phytochemistry 28:3246–3247.
Chen J., Huang S., Li F., Fang S., Chen Y. (1992) Chemical constituents in *Inula salsoloides*. Zhiwu Xuebao 34:62–65.
Chevrier M., Raynaud J. (1973) Phytochemical study of *Echinops ritro*. Bull. Trav. Soc. Pharm. Lyon 17:72–75.
Chiang M. T., Silva M. (1978) Anticancer agents from *Pluchea chingoyo* DC. Rev. Latinoamer. Quim. 9:102–103.
Chiappini I., Fardella G., Menghini A., Rossi C. (1982) Flavonoids from *Dittrichia viscosa* L. Planta Medica 44:159–161.
Chiavari G., Galletti G. C., Marotti M., Piccaglia R. (1991) Silymarin content of different *Silybum marianum* cultivars. Herba Hungarica 30:23–34.
Chivers H., Corbett R. E., Mitchell R. E. M. (1966) Extractives from the leaves of *Olearia paniculata*. J. Chem. Soc. C. 1814–1816.
Chmiel E., Sütfeld R., Wiermann R. (1983) Conversion of phloroglucinyl-type chalcones by purified chalcone isomerase from tulip anthers and from *Cosmos* petals. Biochem. Physiol. Pflanzen 178:139–146.
Christensen L. P. (1992) Acetylenes and related compounds in Anthemideae. Phytochemistry 31:7-49.
Christensen L. P., Lam J. (1990) Acetylenes and related compounds in Cynareae. Phytochemistry 29:2753–2785.
Christensen L. P., Lam J. (1991a) Acetylenes and related compounds in Heliantheae. Phytochemistry 30:11–49.
Christensen L. P., Lam J. (1991b) Acetylenes and related compounds in Astereae. Phytochemistry 30:2453–2476.
Christensen L. P., Lam J. (1991c) Flavones and other constituents from *Centaurea* species. Phytochemistry 30:2663–2665.
Christensen L. P., Lam J. (1991d) Acetylenes and other constituents from *Centaurea* species. Phytochemistry 30:3289–3292.
Christy C. (1985) Analysis of artificial hybrids in *Hieracium* (Compositae, Lactuceae). Amer. J. Bot. 72:946.
Chumbalov T. K., Fadeeva O. V. (1967) Flavonoids of *Artemisia transiliensis*. Chem. Nat. Cmpds. 3: 235–236 [281].
Chumbalov T. K., Fadeeva O. V. (1969) Flavonoids of *Artemisia transiliensis*. Chem. Nat. Cmpds. 5:364 [439].
Chumbalov T. K., Fadeeva O. V. (1970) Glycoflavonoids of *Artemisia transiliensis*. IV. Chem. Nat. Cmpds. 6:364–365 [364–365].
Chumbalov T. K., Fadeeva O. V. (1974) Flavonoids of some Kazakhstan wormwoods. Prikl. Teor. Khim. 5:71–76.
Chumbalov T. K., Fadeeva O. V., Chanysheva I. S. (1969a) Flavonoids of *Artemisia transiliensis*. II. A new flavonoid glycoside transilin. Chem. Nat. Cmpds. 5:201–202 [236].

Chumbalov T. K., Fadeeva O. V., Nikischenko T. K. (1972a) Flavonoids of *Senecio subdentatus*. I. Chem. Nat. Cmpds. 8:384 [390].

Chumbalov T. K., Fadeeva O. V., Nikischenko T. K. (1974a) Flavonoids of *Senecio subdentatus*. II. Chem. Nat. Cmpds. 10:89 [89].

Chumbalov T. K., Fadeeva O. V., Nikischenko T. K. (1975) Flavonoids of *Senecio subdentatus*. III. Chem. Nat. Cmpds. 11:816 [802].

Chumbalov T. K., Glyzin V. I., Zhubaeva R. A. (1974b) Flavonoids of *Ajania fastigiata*. Chem. Nat. Cmpds. 10:814 [791].

Chumbalov T. K., Mukhamed'yarova M. (1970) Flavonoids of *Artemisia dracunculus*. II. Chem. Nat. Cmpds. 6:645 [629].

Chumbalov T. K., Mukhamed'yarova M., Fadeeva O. V. (1969b) The flavonoids of *Artemisia dracunculus*. Chem. Nat. Cmpds. 5:273 [323].

Chumbalov T. K., Zhubaeva R. A. (1976) *C*-Glycosides of *Ajania fastigiata*. Chem. Nat. Cmpds. 12:594 [661].

Chumbalov T. K., Zhubaeva R. A., Nurgalieva G. M. (1972b) Flavonoids of *Ajania fastigiata*. Chem. Nat. Cmpds. 9:111 [117].

Cisowski W., Dembinska-Migas W., Luczkiewicz M. (1993) Flavonoids from *Rudbeckia hirta* L. herb. Pol. J. Chem. 67:829–836.

Clark L. E., Clark W. D. (1988) Seasonal variation in flavonoid aglycones of *Isocoma acradenia*. Amer. J. Bot. 75 (6 part 2):128.

Clark L. E., Clark W. D. (1990) Seasonal variation in leaf exudate flavonoids of *Isocoma acradenia*. Biochem. Syst. Ecol. 18:145–148.

Clark L. E., Mabry T. J. (1978) Systematic implications of flavonoids in *Hazardia*. Biochem. Syst. Ecol. 6:19–20.

Clark L. E., Urbatsch L. E., Hartman R. L., Mayes R., Mabry T. J. (1980) Systematic implications of flavonoid patterns in *Haplopappus* segregates. Biochem. Syst. Ecol. 8:257–259.

Clark L. E., Wollenweber E. (1984) Flavonoids of *Ericameria laricifolia*. Z. Naturforsch. 39c:1184–1185.

Clark L. E., Wollenweber E. (1985) Flavonoid aglycones in the resin of *Hazardia squarrosa* var. *grindelioides*. Phytochemistry 24:1122–1123.

Clark L. E., Wollenweber E., Clark W. D. (1987) Flavonoid similarities among five species of *Isocoma*. Biochem. Syst. Ecol. 15:643–646.

Clark W. D. (1979) The taxonomy of *Hazardia* (Compositae: Astereae). Madroño 26:105–127.

Coe E. H., McCormick S. M., Modena S. A. (1981) White pollen in maize. J. Heredity 72:318–320.

Coile N. C., Jones S. B., Jr. (1981) *Lychnophora* (Compositae:Vernonieae), a genus endemic to the Brazilian Planalto. Brittonia 33:528–542.

Coleman P. C., Potgieter D. J. J., Van Aswegen C. H., Vermeulen N. M. J. (1984) Flavonoids of *Geigeria*. Phytochemistry 23:1202–1203.

Collado I. G., Macias F. A., Massanet G. M., Luis F. R. (1985) Flavonoids from *Centaurea clementei*. J. Nat. Prod. 48:819–822.

Constabel F., Shyluk J. P., Gomborg O. L. (1971) The effects of hormones on anthocyanin accumulation in cell cultures of *Haplopappus gracilis*. Planta 96:306–316.

Constance L. (1964) Systematic botany—an unending synthesis. Taxon 13:257–273.

Constantinescu D. G., Tarpo C., Dinh Duc Tien (1962) The localization and dynamic study of the flavonoids from *Tagetes patula* and *T. erecta*. Acad. Rep. Populaire Romane, Studii Cercetari Biol. Ser. "Biol. Veg." 14:299–310.

Cordano G. F., de la Pena A. M., Medina J. J. (1982) Scopoletin and hispidulin in *Baccharis magellanica*. J. Nat. Prod. 45:653.

Correa J., Cervera M. L. (1971) Isolation and structure of tithonine, 4',7-dimethoxy-3'-hydroxyflavone. Bull. Soc. Chim. France 475–477.

Correll D. S., Johnston M. C. (1970) *Manual of the Vascular Plants of Texas*. Renner, Texas: Texas Research Foundation.

Cox P. B., Urbatsch L. E. (1987) A preliminary flavonoid survey of the genus *Rudbeckia* L. (Heliantheae, Asteraceae). Amer. J. Bot. 74:768–769.

Craw R. C. (1989) Quantitative panbiogeography: introduction to methods. New Zealand J. of Zoology 16:484–494.

Craw R. C., Grehan J. R., Heads M. J. (1999) *Panbiogeography. Tracking the History of Life*. Oxford: Oxford Univ. Press.

Crawford D. J. (1970) *Coreopsis mutica*: flavonoid chemistry, chromosome numbers, morphology and hybridization. Brittonia 22:93–111.

Crawford D. J. (1971) Morphology, chromosome number, and flavonoid chemistry of *Bidens cordylocarpa*. Madroño 21:41–48.

Crawford D. J. (1972) The morphology and flavonoid chemistry of synthetic infraspecific hybrids in *Coreopsis mutica* (Compositae). Taxon 21:27–35.

Crawford D. J. (1978) Okanin 4'-*O*-diglucoside from *Coreopsis petrophiloides* and comments on anthochlors and evolution in *Coreopsis*. Phytochemistry 17:1680–1681.

Crawford D. J. (1981) A new variety of *Coreopsis mutica* (Compositae) from Mexico. Brittonia 33:547–551.

Crawford D. J., Ornduff R. (1989) Enzyme electrophoresis and evolutionary relationships among three species of *Lasthenia* (Asteraceae: Heliantheae). Amer. J. Bot. 76:289–296.

Crawford D. J., Ornduff R., Vasey M. C. (1985) Allozymic variation within and between *Lasthenia minor* and its derivative species *L. maritima* (Asteraceae). Amer. J. Bot. 72:1177–1184.

Crawford D. J., Palmer J. D., Kobayashi M. (1991) Chloroplast DNA restriction site variation, phylogenetic relationships, and character evolution among sections of North American *Coreopsis* (Asteraceae). Syst. Bot. 16:211–224.

Crawford D. J., Sang T., Stuessy T. F., Kim S.-C., Silva O. M. (1998) *Dendroseris* (Asteraceae: Lactuceae) and *Robinsonia* (Asteraceae: Senecioneae) on the Juan Fernandez Islands: similarities and differences in biology and phylogeny. In: Stuessy T. F., Ono M. (eds.) *Evolution and Speciation of Island Plants*. Cambridge: Cambridge Univ. Press. 97–119.

Crawford D. J., Smith E. B. (1980) Flavonoid chemistry of *Coreopsis grandiflora* (Compositae). Brittonia 32:154–159.

Crawford D. J., Smith E. B. (1983a) The distribution of anthochlor floral pigments in North American *Coreopsis* (Compositae): taxonomic and phyletic interpretations. Amer. J. Bot. 70:355–362.

Crawford D. J., Smith E. B. (1983b) Leaf flavonoid chemistry of North American *Coreopsis* (Compositae): intra- and intersectional variation. Bot. Gaz. 144:577–583.

Crawford D. J., Smith E. B. (1985) Leaf flavonoid chemistry and taxonomy of *Coreopsis* sect. *Coreopsis*. Biochem. Syst. Ecol. 13:115–118.

Crawford D. J., Smith E. B., Mueller A. M. (1980) Leaf flavonoid chemistry of *Coreopsis* (Compositae) sect. *Palmatae*. Brittonia 32:452–463.

Crawford D. J., Smith E. B., Roberts M. L., Benkowski M., Hoffman M. (1990) Phylogenetic implications of differences in number of plastid phosphoglucose isomerase isozymes in North American *Coreopsis* (Asteraceae: Heliantheae: Coreopsidinae). Amer. J. Bot. 77:54–63.

Crawford D. J., Stuessy T. F. (1981) The taxonomic significance of anthochlors in the subtribe Coreopsidinae (Compositae: Heliantheae). Amer. J. Bot. 68:107–117.

Crawford D. J., Stuessy T. F., Cosner M. B., Haines D. W., Silva O. M., Baeza M. (1992b) Evolution of the genus *Dendroseris* (Asteraceae: Lactuceae) on the Juan Fernandez Islands: evidence from chloroplast and ribosomal DNA. Syst. Bot. 17:676–682.

Crawford D. J., Stuessy T. F., Haines D. W., Cosner M. B., Silva O. M., Lopez P. (1992a) Allozyme diversity within and divergence among four species of *Robinsonia* (Asteraceae: Senecionee), a genus endemic to the Juan Fernandez Islands, Chile. Amer. J. Bot. 79:962–966.

Crawford D. J., Whitkus R. (1988) Allozyme divergence and the mode of speciation for *Coreopsis gigantea* and *C. maritima* (Compositae). Syst. Bot. 13:256–264.

Crepet W. L., Stuessy T. F. (1978) A reinvestigation of the fossil *Viguiera cronquistii* (Compositae). Brittonia 30:483–491.

Crins W. J., Bohm B. A. (1987) Flavonoid aglycones of *Holocarpha obconica*. Phytochemistry 26:2128–2129.

Crins W. J., Bohm B. A. (1988) Flavonoids of *Adenothamnus validus*. Phytochemistry 27:2647–2649.

Crins W. J., Bohm B. A. (1990) Flavonoid diversity in relation to the systematics and evolution of the tarweeds. Ann. Missouri Bot. Gard. 77:73–83.

Crins W. J., Bohm B. A., Carr G. (1988a) Flavonoids as indicators of hybridization in a mixed population of lava-colonizing Hawaiian tarweeds (Asteraceae: Heliantheae: Madiinae). Syst. Bot. 13:567–571.

Crins W. J., Bohm B. A., Powell A. M., Guppy C. S. (1988b) Evolutionary considerations based on flavonoid aglycones in Peritylinae. Biochem. Syst. Ecol. 16:273–278.

Crins W. J., Thomas S., Bohm B. A. (1988c) Nonpolar phenolics of seven species of *Layia*. Biochem. Syst. Ecol. 16:467–469.

Crisci J. V. (1976) Revisión del género *Leucheria* (Compositae: Mutisieae). Darwiniana 20:9–126.

Crisci J. V., Stuessy T. F. (1980) Determining primitive character states for phylogenetic reconstruction. Syst. Bot. 5:112–135.

Croizat L. (1958) *Panbiogeography, or An Introductory Synthesis of Zoogeography, Phytogeography, and Geology; with Notes on Evolution, Systematics, Ecology, Anthropology, etc.* Caracas: Publ. by Author.

Cronquist A. (1943) The separation of *Erigeron* from *Conyza*. Bull. Torrey Bot. Club. 70:629–632.

Cronquist A. (1947) Revision of the North American species of *Erigeron*, north of Mexico. Brittonia 6:121–302.
Cronquist A. (1955) Phylogeny and taxonomy of the Compositae. Amer. Midl. Natur. 53:478–511.
Cronquist A. (1968) *The Evolution and Classification of Flowering Plants*. Boston: Houghton Mifflin.
Cronquist A. (1977) The Compositae revisited. Brittonia 29:137–153.
Cronquist A. (1980) Asteraceae. In: Radford A. E. (ed.) *Vascular FLora of the Southeastern United States, Vol. 1*. Chapel Hill: Univ. of North Carolina Press.
Cronquist A. (1981) *An Integrated System of Classification of Flowering Plants*. New York: Columbia Univ. Press.
Cronquist A. (1985) History of generic concepts in the Compositae. Taxon 34:6–10.
Cronquist A. (1988) *The Evolution and Classification of Flowering Plants*, 2nd Ed. New York: The New York Botanical Garden.
Cronquist A. (1994) Asterales. In: Cronquist A., Holmgren A. H., Holmgren N. H., Reveal J. L., Holmgren P. K. (eds.) *Intermountain Flora: Vascular Plants of the Intermountain West, U.S.A., Vol. 5*. Bronx, New York: New York Botanical Garden.
Cronquist A., Keck D. D. (1957) A reconstitution of the genus *Machaeranthera*. Brittonia 9:231–239.
Crovello T. J. (1981) Quantitative biogeography: an overview. Taxon 30:563–575.
Cruickshank I. A. M., Perrin D. R. (1960) Isolation of a phytoalexin from *Pisum sativum* L. Nature (London) 187:799–800.
Cuadra P. J., Harborne B., Waterman P. G. (1997) Increases in surface flavonoids and photosynthetic pigments in *Gnaphalium luteo-album* in response to UV-B radiation. Phytochemistry 45:1377–1383.
Cuatrecasas J. (1969) Prima flora Colombiana, 3: Compositae-Astereae. Webbia 24:1–335.
Cuatrecasas J. (1986) Un género nuevo de Astereae, Compositae, de Colombia. An. J. Bot. Madrid 42:415–426.
Çubukçu B. (1976) Sur les derives flavoniques et coumariniques d'*Helichrysum orientale* (L.) Gaertn. Pl. Medica Phytother. 10:44–53.
Çubukçu B. (1982) Studies on lipophilic flavonoids of *Helichrysum* species growing in Anatolia. Doga, Seri A, 6:83–90.
Çubukçu B., Bingöl S. (1981) A further investigation on the flavonoids of *Helichrysum orientale*. Istanbul Univ. Eczacilik Fak. Mecm. 17:86–90.
Çubukçu B., Bingöl S. (1984) Pharacognostical investigations on *Helichrysum pallasii* (Sprengel) Ledeb. Pl. Medica Phytother. 18:28–35.
Çubukçu B., Damatyan B. (1980) An Anatolian folk medicine, *Helichrysum graveolens*. Planta Medica 39:258.
Çubukçu B., Damatyan B. (1986) Flavonoides d'*Helichrysum graveolens*. Fitoterapia 57:124.
Çubukçu B., Melikoglu G. (1995) Flavonoids of *Artemisia austriaca*. Planta Medica 61:488.
Çubukçu B., Mericli A. H. (1977) Flavonoids d'*Helichrysum plicatum* DC. Pl. Medica Phytother. 11:294–302.

Çubukçu B., Mericli A. H. (1979) Flavonoids d'*Helichrysum plicatum* subspecies *polyphyllum*. Pl. Medica Phytother. 13:107–115.

Çubukçu B., Yuksel V. (1982) Constituents of Anatolian medicinal plants: flavonoids of *Helichrysum armenium*. J. Nat. Prod. 45:137–139.

Cunha W. R., Lopes J. L. C., Vichnewski W., Diaz J. G., Herz W. (1995) Eremantholides and a guaianolide from *Lychnophora rupestris*. Phytochemistry 39:387–389.

da Costa F. B., Dias D. A., Lopes J. L. C., Vichnewski W. (1993) Flavonoids and heliangolides from *Lychnophora diamantinana*. Phytochemistry 34:261–263.

D'Agostino M., De Feo V., De Simone V., Pizza C. (1990a) Chemical constituents of *Eupatorium grayanum* L. Fitoterapia 61:375.

D'Agostino M., De Feo V., De Simone V., Pizza C. (1991b) Three chalcones from *Senecio pseudotites*. Phytochemistry 30:2440–2441.

D'Agostino M., De Feo V., Pizza C., De Simone F. (1994) Isolation of 3,5,7,2',3',4'-hexahydroxy 3-*O*-D-glucoside from *Eupatorium sternbergianum*. Fitoterapia 65:472–473.

D'Agostino M., De Simone F., Dini A., Ramundo E., Zollo F. (1990b) Flavonol glycosides from *Eupatorium tinifolium*. Phytochemistry 29:353–354.

D'Agostino M., De Simone F., Zhou Z. L., Pizza C. (1992) Flavonol glycosides from *Tagetes elliptica*. Phytochemistry 31:4387–4388.

D'Agostino M., De Simone F., Zollo F., Pizza C. (1991a) Isolation of 5,7-dimethoxy-6-methoxykaempferol 3-glucoside from *Eupatorium tinifolium*. Fitoterapia 62:445–446.

D'Agostino M., De Simone F., Zollo F., Pizza C. (1991c) Constituents of *Mikania cordifolia*. Fitoterapia 62:461.

D'Agostino M., Senatore F., De Feo V., De Simone F. (1995) Flavonol glycosides from *Jungia paniculata*. Fitoterapia 66:283.

Dahlgren G. (1989) An updated angiosperm classification. Bot. J. Linn. Soc. 100:197–203.

Dahlgren R. (1975) A system of classification of the angiosperms to be used to demonstrate the distribution of characters. Bot. Notiser 128:119–147.

Dahlgren R. (1980) A revised system of classification of the angiosperms. Bot. J. Linn. Soc. 80:91–124.

Dahlgren R. (1983) General aspects of angiosperm evolution and macrosystematics. Nord. J. Bot. 3:119–149.

Dahlgren R., Clifford H. T., Yeo P. F. (1985) *The Families of the Monocotyledons: Structure, Evolution, and Taxonomy*. Berlin: Springer.

Dahlgren R., Rosendahl-Jensen S., Nielson B. J. (1981) A revised classification of the angiosperms with comments on correlation between chemical and other characters. In: Young D. A., Seigler D. S. (eds.) *Phytochemistry and Angiosperm Phylogeny*. New York: Praeger. 149–204.

Daily A., Seligmann O., Nonnenmacher G., Fessler B., Wong S.-M., Wagner H. (1988) New chromone, coumarin and coumestan derivatives from *Mutisia acuminata* var. *hirsuta*. Planta Medica 54:50–52; erratum 270.

Dakora F. D., Joseph C. M., Phillips D. A. (1993) Alfalfa (*Medicago sativa* L.) root exudates contain isoflavonoids in the presence of *Rhizobium meliloti*. Plant Physiol. 101:819–824.

Daniewski W. M., Skibicki P., Bloszyk E., Budesinsky M., Holub M. (1993) Constituents of *Helianthus mollis* and their antifeedant activity. Pol. J. Chem. 67:1255–1259.

Darlington C. D., Wylie A. P. (1955) *Chromosome Atlas of Flowering Plants*, 2nd ed. London: George Allen & Unwin.

Darwin C. (1859) *On the Origin of Species by Means of Natural Selection*. London: Murray.

Das C., Tripathi A. K., Jogi S. R. (1996) Chrysoeriol 7-*O*-(6-*O*-a-L-arabinofuranosyl)-ß-D-glucopyranoside from *Tagetes patula*. Orient. J. Chem. 12:327–328.

Dashbalyn T. S., Glyzin V. I. (1978) Flavonoid glycosides of edelweiss, *Leontopodium ochroleucum* Beauv. Chem. Nat. Cmpds. 14:690 [807].

Dawidar A. M., Metwally M. A., Abou-Elzahab M., Abdel-Mogib M. (1989) Chemical constituents of two *Centaurea* species. Die Pharmazie 44:735–736.

De Feo V., D'Agostino M., De Simone F., Pizza C. (1990) Constituents of *Tessaria integrifolia*. Fitoterapia 61:474.

de Gutierrez A. N., Catalan C. A. N., Diaz J. G., Herz W. (1992) Sesquiterpene lactones and other constituents of *Stevia jujuyensis*. Phytochemistry 31:1818–1820.

de Gutierrez A. N., Catalan C. A. N., Diaz J. G., Herz W. (1995) Sesquiterpene lactones, a labdane and other constituents of *Urolepis hecatantha* and *Chromolaena arnottiana*. Phytochemistry 39:795–800.

De Israilev L. R. A. (1992) [1993] Flavonoids as taxonomic interspecific markers in *Tagetes*. Lilloa 38:101.

De Israilev L. R. A., Del Pero de Martinez M. A., Seeligmann P. (1991) Myricetin in *Tagetes*: chemotaxonomic significance. Phytochemistry 30:4037–4038.

De Israilev L. R. A., Gonzalez J. A. (1994) The flavonoid content of *Perezia ocorzonera*, a medical high subtropical Andean plant. Biochem. Syst. Ecol. 22:761.

De Israilev L. R. A., Meniondo M. E., Seeligmann P. (1992) [1993] Flavonoids of *Schkuhria multiflora* Hooker and Arnott. Lilloa 38:93–94.

De Israilev L. R. A., Seeligmann P. (1983) Distribution of leaf flavonoids in three species of *Tagetes* and their chemosystematic significance. Lilloa 36:5.

De Israilev L. R. A., Seeligmann P. (1990a) Flavonoids in leaves of *Pectis sessiliflora* (Less.) Sch. Bip. (Compositae). Lilloa 37:63–66.

De Israilev L. R. A., Seeligmann P. (1990b) Flavonoids from *Tagetes campanulata*. Lilloa 37:71–73.

De Israilev L. R. A., Seeligmann P. (1992) [1993] Flavonoids of *Tagetes mendocina* Phil. (Compositae). Lilloa 38:95–97.

De Israilev L. R. A., Seeligmann P. (1994) Flavonoids from *Tagetes rupestris* (Asteraceae): some chemosystematic implications. Biochem. Syst. Ecol. 22:431.

de La Fuente J. R., Del Valle Morales I., Sosa V. E. (1994) Furanoheliangolides and flavonoids from *Viguiera mollis*. An. Asoc. Quim. Argent. 82:61–64.

de La Fuente J. R., Diaz O. I., Morales I., Oberti J. C., Sosa V. E., Herz W. (1986) Sesquiterpene lactones and flavones from *Artemisia copa*. Planta Medica 52:243.

De La Puerta R., Garcia M. D., Saenz M. T. (1994) Phenolics from *Helichrysum picardii*. Fitoterapia 65:375.

De La Puerta R., Garcia M. D., Saenz M. T., Gil A. M. (1990) Phytochemistry of *Helichrysum picardii* Boiss. & Reuter. Pl. Medica Phytother. 24:258–263.

de Laubenfels D. J. (1969) A revision of the Malesian and Pacific rainforest conifers. I. Podocarpaceae, in part. J. Arnold Arboretum 50:274–369.

de Laubenfels D. J. (1978) The genus *Prumnopitys* (Podocarpaceae) in Malesia. Blumea 24:189–190.

De Luca V., Ibrahim R. K. (1982) Characterization of three distinct flavonol *O*-methyltransferases from *Chrysosplenium americanum*. Phytochemistry 21:1537–1540.

De Luca V., Ibrahim R. K. (1985a) Enzymatic synthesis of polymethylated flavonols in *Chrysosplenium americanum*. I. Partial purification and some properties of *S*-adenosyl-L-methionine:flavonol 3-, 6-, 7- and 4'-*O*-methyltransferases. Arch. Biochem. Biophys. 238:596–605.

De Luca V., Ibrahim R. K. (1985b) Enzymatic synthesis of polymethylated flavonols in *Chrysosplenium americanum*. II. Substrate interaction and product inhibition studies of flavonol 3-, 6-, and 4'-*O*-methyltransferases. Arch. Biochem. Biophys. 238:606–618.

De Luengo D. H., Mabry T. J. (1986) 6-Methoxyflavonoids from *Decachaeta ovatifolia*. J. Nat. Prod. 49:183–184.

de Pascual Teresa J., Gonzalez M. S., Muriel M. R., Arcocha A. D., Bellido I. S. (1986) Flavonoids from *Artemisia campestris* ssp. *glutinosa*. J. Nat. Prod. 49:177.

de Pascual Teresa J., Gonzalez M. S., Muriel M. R., Bellido I. S. (1984) Phenolic derivatives from *Artemisia campestris* subsp. *glutinosa*. Phytochemistry 23:1819–1821.

De Tommasi N., De Feo V., De Simone F., Pizza C., Zhou Z. L. (1992) Constituents of *Bidens parviflora*. Fitoterapia 63:470.

de Villarraga A. A., de Perez C. E. (1994) 6-Methoxyflavonoids from flowers of *Ageratina angustifolia* R. M. King & H. Robinson (Compositae). Rev. Colomb. Quim. 23:45–51.

Debenedetti S. L., Ferraro G. E., Coussio J. D. (1983) Flavonoids isolated from *Pterocaulon virgatum* L. Acta Farm. Bonaerense 2:1–3.

Debenedetti S. L., Martino V., Palacios P., Coussio J. D. (1987a) 6-Methoxy flavonoids from *Acanthospermum australe*. J. Nat. Prod. 50:325.

Debenedetti S. L., Nadinic E. L., Gomez M. A., Coussio J. D. (1987b) Polyphenols isolated from *Pterocaulon purpurescens*. I. 6-Hydroxyflavonoids. J. Nat. Prod. 50:512–513.

Debenedetti S. L., Palacios P. S., Wilson E. G., Coussio J. D. (1994) Polyphenols of *Pterocaulon polystachium*. Fitoterapia 65:188–189.

Delgado G., Alvarez L., Romo de Vivar A. (1984) Terpenoids and a flavan-3-ol from *Viguiera quinqueradiata*. Phytochemistry 23:675–678.

Delpino F. (1871) *Studi sopra un Lignaggio Anemofilo delle Composte ossia sopra il Gruppo delle Artemisiacee*. Florence.

Dem'yanenko V. G., Dranik L. I. (1973) Flavonoids of *Cichorium intybus*. Chem. Nat. Cmpds. 9:115 [119].

Dembinska-Migas W., Gill S., Handke E., Worobiec S. (1983) Flavonoid compounds in *Zinnia elegans* Jacq. Herba Pol. 29:197–202.

Dement W. A., Raven P.H. (1974) Pigments responsible for UV patterns in flowers of *Oenothera* (Onagraceae). Nature (London) 252:705–706.

Desrochers A. M., Bain J. F., Warwick S. I. (1988) A biosystematic study of the *Carduus nutans* complex in Canada. Can. J. Bot. 66:1621–1631.

Desrochers A. M., Bohm B. A. (1993) Flavonoid variation in *Lasthenia californica* (Asteraceae). Biochem. Syst. Ecol. 21:449–453.

Desrochers A. M., Bohm B. A. (1995) Biosystematic study of *Lasthenia californica* (Asteraceae). Syst. Bot. 20:65–84.

DeVore M. L. (1994) Systematic studies of Calyceraceae. Ph.D. Dissertation, The Ohio State Univ., Columbus.

DeVore M. L., Stuessy T. F. (1995) The place and time of origin of the Asteraceae, with additional comments on the Calyceraceae and Goodeniaceae. In: Hind D. J. N., Pope G. V., Jeffrey C. (eds.) *Advances in Compositae Systematics*. Kew: Royal Botanic Gardens. 23–40.

DeWolf G. P., Jr. (1995) Review of *Asteraceae: Cladistics and Classification*, by K. Bremer. Rhodora 97:176–178.

Diaz J. G., Barba B., Herz W., Bierner M. W. (1992) Sesquiterpene lactones and monoterpene glucosides of *Tetraneuris linearifolia* var. *arenicola*. Phytochemistry 31:3471–3477.

Didry N., Pinkas M., Torck M. (1980) Studies on the polyphenols from coltsfoot. Ann. Pharm. Fr. 38:237–241.

Dillon M. O. (1984) A systematic study of *Flourensia* (Asteraceae, Heliantheae). Fieldiana Bot., ser. 2, 16:1–66.

Dillon M. O., Mabry T. J. (1977) Flavonoid aglycones from *Flourensia*. Phytochemistry 16:1318–1319.

Dillon M. O., Mabry T. J., Besson E., Bouillant M.-L., Chopin J. (1976) New flavonoids from *Flourensia cernua*. Phytochemistry 15:1085–1086.

Dillon M. O., Sagástegui A. (1991a) Flora of Peru. Family Asteraceae. Fieldiana Botany, ser. 2, 26:1–70.

Dillon M. O., Sagástegui A. (1991b) Sinopsis de los géneros de Gnaphaliinae (Asteraceae – Inuleae) de Sudamérica. Arnaldoa 1:5–91.

Dimitri M. J., Orfila E. N. (1985) Un nuevo taxon de *Flaveria* (Compositae) de la Flora Argentina, *Flaveria haumanii* Dim. et Orf.: Tratado de morfología y sistemática vegetal. Buenos Aires: ACME. 466–467.

Dimitrova K., Nakov N., Akhtardzhiev Kh. (1981) Flavonoid content of *Senecio pancicii* Deg. Farmatsiya Sofia 31:25.

Dittrich M. (1977) Cynareae—systematic review. In: Heywood V. H., Harborne J. B., Turner B. L. (eds.) *The Biology and Chemistry of the Compositae*. London: Academic Press. 999–1015.

Dittrich M. (1984) Neukombinationen in der Gattung *Stemmacantha* Cass. (Compositae) mit Bemerkungen zur Typisierung einiger ihrer Arten. Candollea 39:45–49.

Djermanovic M., Jokic A., Mladenovic S., Stefanovic M. (1975) Quercetagetin 6,7,3',4'-tetramethyl ether: a new flavonol from *Artemisia annua*. Phytochemistry 14:1873.

Do J. C., Jung K. Y., Son K. H. (1994) Isolation of pectolinarin from the aerial parts of *Cirsium nipponicum*. Saengyak Hakhoechi 25:73–75.

Dobberstein R. H., Tin-Wa N., Fong H. H. S., Crane F. A. (1977) Flavonoid constituents of *Eupatorium altissimum* L. J. Pharm. Sci. 66:600–602.

Dobberstein R. H., Tin-Wa N., Fong H. H. S., Crane F. A., Farnsworth N. R. (1974) Cytotoxic flavones from *Eupatorium altissimum* L. Lloydia 37:640.

Dombris A., Raynaud J. (1977) Recherches chimiotaxinomique sur les plantes vasculaires. Sur la presence de la di-*C*-glucosyl-6,8-luteoline chez *Carlina corymbosa* L. (Composées). Pl. Medica Phytother. 11:109–111.

Dombris A., Raynaud J. (1980) Les glycosides flavoniques des feuilles de *Carlina macrocephala* M. (Composées). Pl. Medica Phytother. 14:69–72.

Dombris A., Raynaud J. (1981a) Les heterosides flavoniques de fleurs de *Carlina macrocephala* M. (Composées). Die Pharmazie 36:384.

Dombris A., Raynaud J. (1981b) Les heterosides flavoniques des tiges feuilles de *Carlina corymbosa* M. (Composées). Pl. Medica Phytother. 15:21–24.

Dombrowicz E., Greiner M. (1968) Chromatographic comparison of the extracts from flowers of *Arnica montana* and *Inula brittanica*. Farm. Pol. 24:471–474.

Dominguez X. A., Cardenas G. E. (1975) Achillin and deacetylmatricarin from two *Artemisia* species. Phytochemistry 14:2511–2512.

Dominguez X. A., de la Fuente E. R. (1973) Sakuranetin and pulcherryl acetate from *Eupatorium havanense*. Phytochemistry 12:2060.

Dominguez X. A., el Dahmi S., Rombold C. (1986c) Hydroxy-acetophenone derivatives from *Baccharis glutinosa*. J. Nat. Prod. 49:143–144.

Dominguez X. A., Franco R., Bohlmann F., Bohlmann R. (1980) A pseudoguaianolide from *Hymenoxys linearifolia*. Phytochemistry 19:2204–2205.

Dominguez X. A., Gonzalez A., Zamudio M. A., Garza A. (1974) Taraxasterol from *Stevia berlandieri* and *Cirsium texanum*. Phytochemistry 13:2001.

Dominguez X. A., Gutierrez M., Aragon R. (1976) Isolation of baileyolin, a tumor inhibitor and antibiotic sesquiterpene lactone from *Baileya multiradiata*. Planta Medica 30:356–359.

Dominguez X. A., Jakupovic J., Pathak V. P., Sanchez H., King R. M., Robinson H. (1986a) A chemical study of *Aster tanacetifolius*. Rev. Latinoam. Quim. 17:207–210.

Dominguez X. A., Torre B. (1974) Two pentamethoxylated flavonoids from *Gymnosperma glutinosum* Phytochemistry 13:1624–1625.

Dominguez X. A., Verde J. S., N. E. Guerra R., Ellenmauer E., Jakupovic J. (1986b) A *nor*-diterpene and other constituents from *Isocoma coronopifolia*. Phytochemistry 25:2893–2895.

Dong X.-P., Che C.-T., Farnsworth N. R. (1987) Cytotoxic flavonols from *Gutierrezia microcephala*. J. Nat. Prod. 50:337.

Dora G., Edwards J. M. (1991) Taxonomic status of *Lanaria lanata* and isolation of a novel biflavone. J. Nat. Prod. 54:798–801.

Dormer K. J. (1962) The fibrous layer in the anthers of Compositae. New Phytol. 61:150–153.

Dostál J. (1976) *Centaurea*. In: Tutin T. G., Heywood V. H., Gurges N. A., Moore D. M., Valentine D. H., Walters S. M., Webb D. A. (eds.) *Flora Europaea, Vol. 4. Plantaginaceae to Compositae (and Rubiaceae)*. Cambridge: Cambridge Univ. Press. 254–301.

Downie S. R. (1988) Morphological, cytological, and flavonoid variability of the *Arnica angustifolia* aggregate (Asteraceae). Can. J. Bot. 66:24–39.

Downie S. R., Denford K. E. (1986a) The taxonomy of *Arnica frigida* and *A. louiseana* (Asteraceae). Can. J. Bot. 64:1355–1372.

Downie S. R., Denford K. E. (1986b) The flavonoids of *Arnica frigida* and *A. louiseana* (Asteraceae). Can. J. Bot. 64:2748–2752.

Downie S. R., Denford K. E. (1987) The biosystematics of *Arnica fulgens* and *A. sororia* (Asteraceae). Can. J. Bot. 65:559–570.

Downie S. R., Denford K. E. (1988) Flavonoid variation in *Arnica* subgenus *Arctica*. Biochem. Syst. Ecol. 16:133–137.

Downie S. R., Palmer J. D. (1992) Restriction site mapping of the chloroplast DNA inverted repeat: a molecular phylogeny of the Asteridae. Ann. Missouri Bot. Gard. 79:266–283.

Dranik L. I. (1966) Phenol compounds from some plants of the Compositae. Artichoke (*Cynara scolymus*). Fenol'nye Soedin. Ikh. Biol. Funkts., Mater. Vses. Simp., 1st, 53.

Dranik L. I., Chernobai V. T. (1966) A new flavonoid isolated from the leaves of *Cynara scolymus* L. Chem. Nat. Cmpds. 2:12 [16].

Dranik L. I., Chernobai V. T., Kolesnikov D. G. (1964) Polyphenolic compounds of *Cynara scolymus*. Medica Prom. SSSR 18:23.

Dranik L. I., Dolganenko L., Grysodub A., Levin M., Eshami I. M. (1992) A preliminary investigation of flavanolignans of *Silybum marianum* (L.) Gaertn. Bull. Pharm. Sci., Assiut Univ. 15:22–26.

Dranik L. I., Dolganenko L., Grysodub A., Levin M., Eshami I. M. (1993) Flavanolignans from the fruits of red- and white-flowering *Silybum marianum*. Farm. Zh. (Kiev) 4:83–85.

Drinnen A. N., Crane P. R. (1990) Cretaceous paleobotany and its bearing on the biogeography of austral angiosperms. In: Taylor T. N., Taylor E. L. (eds.) *Antarctic Paleobotany*. New York: Springer. 192–219.

Drury D. G. (1970) A fresh approach to the classification of the genus *Gnaphalium* with particular reference to the species present in New Zealand (Inuleae-Compositae). N. Z. J. Bot. 8:222–248.

Duistermaat H. (1996) Monograph of *Arctium* L. (Asteraceae): generic delimitation (including *Cousinia* Cass. p.p.), revision of the species, pollen morphology and hybrids. Gorteria, Suppl. 3:1–143.

Dupre S., Grenz M., Jakupovic J., Bohlmann F., Niemeyer H. M. (1991) Eremophilane, germacrane and shikimic acid derivatives from Chilean *Senecio* species. Phytochemistry 30:1211–1220.

Dutta P., Chaudhuri R. P., Sharma R. P. (1993) Insect feeding deterrents from *Tithonia diversifolia* (Hemsl.) Gray. J. Environ. Biol. 14:27–33.

Ebel J., Hahlbrock K. (1982) Biosynthesis. In: Harborne J. B., Mabry T. J. (eds.) *The Flavonoids: Advances in Research*. London: Chapman and Hall. 641–679.

Ebert M., Merfort I., Willuhn G. (1988) Flavonoid distribution in *Arnica* subgenera *montana* and *austromontana*. Phytochemistry 27:3849–3851.

Echeverri F., Cardona G., Torres F., Pelaez C., Quiñones W., Renteria E. (1991) Ermanin: an insect deterrent flavonoid from *Passiflora foetida* resin. Phytochemistry 30:153–155.

El-Ansari M. A. I., El-Negoumy S. I., Saleh N. A. M. (1988) The flavonoids of *Cynara sibthorpiana*. Biochem. Syst. Ecol. 16:285.

Eldenäs P., Anderberg A. A., Källersjö A. (1998) Molecular phylogenetics of the tribe Inuleae s. str. (Asteraceae), based on ITS sequences of nuclear ribosomal DNA. Plant Syst. Evol. 210:159–173.

El-Din A. A. S. (1987) Flavone glycosides from *Carthamus glaucus* Bieb. emend. Hanelt ssp. *alexandrinus* (Boiss. & Heldr.) Hanelt. Acta Pharm. Jugosl. 37:97.

El-Din A. A. S. (1989) Phytochemical study of the root of *Otanthus maritimus* growing in Egypt. Alexandria J. Pharm. Sci. 3:14–16.

El-Din A. A. S., El-Ghazouly M. G., Abou-Donia A. (1987) Phytochemical study of *Lactuca saligna* L. growing in Egypt. Alexandria J. Pharm. Sci. 1:5–7.

El-Din A. A. S., El-Sebakhy N., El-Ghazouly M. (1985) Methoxylated flavonoids of *Tanacetum santalinoides* and *Jasonia montana*. Acta Pharm. Jugosl. 35:283–287.

El-Emary N. A., Ali A. A. (1983) Revised phytochemical study of *Tagetes erecta*. Fitoterapia 54:9.

El-Emary N. A., Bohlmann F. (1980) Zwei neue Germacranolide aus *Artemisia argentea*. Phytochemistry 19:845–847.

El-Emary N. A., Kobayashi Y., Ogihara Y. (1983) A simple flavonoid from *Centaurea cineraria* L. Fitoterapia 54:133–134.

El-Feraly F. S., Mossa J. S., Al-Yahya M. A., Hifnawy M. S., Hafez M. M., Hufford C. D. (1990) Two flavones from *Psiadia arabica*. Phytochemistry 29:3372–3373.

Elgamel M. H. A., Hanna A. G., Duddeck H. (1991a) Constituents of *Achillea santolina*. Fitoterapia 62:359.

Elgamel M. H. A., Yassin F. Y., Duddeck H. (1991b) Constituents of *Artemisia monosperma*. Fitoterapia 62:360.

El-Ghazouly M. G., Omar A. A. (1984) Flavonoid constituents of *Artemisia campestris*. Fitoterapia 55:115.

El-Karemy Z. A. R., Mansouri R. M. A., Fayed A. A., Saleh N. A. M. (1987) The flavonoids of local members of Astereae (Compositae). Biochem. Syst. Ecol. 15:53–59.

El-Keiy M. A., Marwish M., Moustafa M. A. (1963) Flavonoid hunting in plants growing in Egypt. I. Buckwheat *Fagopyrum esculentum*, German camomile *Matricaria chamomilla*, and guava *Psidium guajava*. Bull. Fac. Pharm. Cairo Univ. 2:107–111.

El-Lakany A. M., Abdel-Kader M. S., Hammoda H. M., Ghazy N. M., Mahmoud Z. F. (1995) Flavonoids from *Carduus pycnocephalus* L. Alexandria J. Pharm. Sci. 9:41–43.

El-Lakany A. M., Aboul Ela M. A., Hammoda H. M., Ghazy N. M., Mahmoud Z. F. (1996) New methoxylated flavonols from *Inula crithmoides*. Die Pharmazie 51:435–436.

Ellison W. L. (1964) A systematic study of the genus *Bahia* (Compositae). Rhodora 66:67–86, 177–215, 281–311.

Ellison W. L. (1966) A comparative study of *Picradeniopsis oppositifolia* and *P. woodhousei* (Compositae) in northeastern New Mexico and the Oklahoma panhandle. Southwest. Nat. 11:456–466.

Ellmauerer E., Pathak V. P., Jakupovic J., Bohlmann F., Dominguez X. A., King R. M., Robinson H. (1987) 6ß-lactonized xanthanolides from *Ratibida* species. Phytochemistry 26:159–163.

El-Masry S., Omar A. A., Abou-Shoer M. I. A., Saleh M. R. I. (1980) Flavonoid constituents of *Aegialophila pumila*. J. Drug. Res. 12:173–175.

El-Masry S., Omar A. A., Abou-Shoer M. I. A., Saleh M. R. I. (1981) Flavonoid constituents of *Aegialophila pumila* syn. *Centaurea*. Planta Medica 42:199–201.

El-Naggar S. F., Doskotch R. W. (1979) Patuletrin and acetyl patuletrin, flavonol glycosides from *Lagascea mollis*. J. Nat. Prod. 42:126–128.

El-Negoumy S. I., El-Sayed N. H., Saleh N. A. M. (1987) Flavonoid glycosides of *Cynara scolymus*. Fitoterapia 58:178.

El-Negoumy S. I., Mansour R. M. A., Saleh N. A. M. (1982) Flavonols of *Pulicaria arabica*. Phytochemistry 21:953–954.

El-Sayed N. H., El-Khrisy E. A. M., Khadiga M. A., Mabry T. J. (1991) Flavonoids of *Conyza linifolia*. Rev. Latinoamer. Quim. 22:89–90.

El-Sayed N. H., Ishak M. S., Kandil F. E., Mabry T. J. (1992) Flavonoids of *Otanthus maritimus*. Rev. Latinoamer. Quim. 23:1.

El-Sayed N. H., Lenherr A., Sundberg S., Mabry T. J. (1987) Flavonoids of *Aster subulatus*. Biochem. Syst. Ecol. 15:549–550.

El-Sayed N. H., Norris J., Ahmed A. A., Mabry T. J. (1990) Flavonoids of *Brickellia longifolia*. Phytochemistry 29:2364–2365.

El-Sayed N. H., Norris J., Mabry T. J. (1988) Non 6-methoxyflavonoids from *Brickellia coulteri* and 6-methoxylated flavonoids from *B. microphylla*. Biochem. Syst. Ecol. 16:189–190.

Emerenciano V. de P., Ferreira Z. S., Kaplan M. A. C., Gottlieb O. R. (1987) A chemosystematic analysis of tribes of Asteraceae involving sesquiterpene lactones and flavonoids. Phytochemistry 26:3103–3115.

Emerenciano V. de P., Kaplan M. A. C., Gottlieb O. R. (1985) Evolution of sesquiterpene lactones in Angiosperms. Biochem. Syst. Ecol. 13:145–166.

Emerenciano V. de P., Kaplan M. A. C., Gottlieb O. R., de M. Bonfanti M. R., Ferreira Z. S., Comegno L. M. A. (1986) Evolution of sesquiterpene lactones in Asteraceae. Biochem. Syst. Ecol. 14:585–589.

Emerson J., Carr R., McCormick S., Bohm B. A. (1986) 8-*O*-Methylated flavonoids from *Calycadenia ciliosa*: inter- and intrapopulational variation. Biochem. Syst. Ecol. 14:29–32.

Engler A. (1891) In: Engler A., Prantl K. (eds.) *Die natürlichen Pflanzenfamilien*, III. 2a. Leipzig: Engelmann. 41.

Engler A. (1928) In: Engler A., Prantl K. (eds.) *Die natürlichen Pflanzenfamilien*, ed. 2, 18a. Leipzig: Engelmann. 74.

Erbar C. (1993) Studies on the floral development and pollen presentation in *Acicarpha tribuloides* with a discussion of the systematic position of the family Calyceraceae. Bot. Jahrb. Syst. 115:325–350.

Erbar C., Leins P. (1995) Portioned pollen release and the syndromes of secondary pollen presentation in the Campanulales-Asterales-complex. Flora 190:323–338.

Erdtman H. (1963) Some aspects of chemotaxonomy. In: Swain T. (ed.) *Chemical Plant Taxonomy*. New York: Academic Press. 89–125.

Escarria R. S., Torrenegra R. D., Angarita B. (1977) Colombian plants of the genus *Gnaphalium*. Phytochemistry 16:1618.

Espitia de Perez C., Salgado de Pinilla G. (1990) Methoxyflavones of *Ageratina glyptophlebia*. Rev. Colomb. Quim. 19:101–103.

Espitia de Perez C., Villegas Molina E. (1995) Glycosidic flavonoids of *Ageratina arbutifolia*. Rev. Colomb. Quim. 24:39–45.

Estabrook G. F. (1980) The compatibility of occurrence patterns of chemicals in plants. In: Bisby F. A., Vaughn J. G., Wright C. A. (ed.) *Chemosystematics: Principles and Practice*. London: Academic Press. 379–397.

Estabrook G. F. (1986) Evolutionary classification using convex phenetics. Syst. Zool. 35:560–570.

Esteban M. D., Collado I. G., Macias F. A., Massanet G. M., Luis F. R. (1986) Flavonoids of *Artemisia lanata*. Phytochemistry 25:1502–1504.

Estilai A. (1984) Inheritance of flower color in guayule (*Parthenium argentatum*). Crop Sci. 24:760–762.

Exner J., Reichling J., Becker H. (1980) Flavonoids in *Matricaria chamomilla*. Planta Medica 39:219.

Exner J., Reichling J., Cole T. C. H., Becker H. (1981) Methylated flavonoid aglycones from "Matricariae Flos". Planta Medica 41:198–200.

Fadeeva O. V., Nikischenko T. K. (1979) Flavonoids of *Senecio subdentatus*. Chem. Nat. Cmpds. 15:503 [576].

Faini C. L. F., Castillo M. (1990) Diterpenoids from Chilean *Baccharis* species. Phytochemistry 29:324–325.

Faini F., Castillo M., Torres M. R. (1982) Flavonoids of *Baccharis incarum*. J. Nat. Prod. 45:501–502.

Faini F., Hellwig F., Labbe C., Castillo M. (1991) Hybridization in the genus *Baccharis*: *Baccharis linearis* × *B. macraei*. Biochem. Syst. Ecol. 19:53–57.

Faini F., Labbe C., Coll J. (1999) Seasonal changes in chemical composition of epicuticular waxes from the leaves of *Baccharis linearis*. Biochem. Syst. Ecol. 27:673–679.

Faini F., Labbe C., Polanco M. I. (1992) Diterpenoids from *Baccharis neaei*. Phytochemistry 31:3274–3276.

Faini F., Labbe C., Salgado I., Coll J. (1997) Chemistry, toxicity and antifeedant activity of the resin of *Flourensia thurifera*. Biochem. Syst. Ecol. 25:189–193.

Faini F., Rivera P., Mahu M., Castillo M. (1987) *neo*-Clerodane diterpenoids and other constituents from *Baccharis* species. Phytochemistry 26:3281–3283.

Faini F., Torres R., Castillo M. (1984) Costunolide and flavonols from *Gochnatia foliosa* var. *fascicularis*. J. Nat. Prod. 47:552–562.

Falk A. J., Smolenski S. J., Bauer L., Bell C. L. (1975) Isolation and identification of three new flavones from *Achillea millefolium* L. J. Pharm. Sci. 64:1838–1842.

Fang N., Leidig M., Mabry T. J. (1985a) Highly oxygenated flavonoid aglycones from *Gutierrezia grandis*. Phytochemistry 24:2693–2698.

Fang N., Leidig M., Mabry T. J. (1985b) Six 2'-hydroxy-flavonols from *Gutierrezia microcephala*. Phytochemistry 24:3029–3034.

Fang N., Leidig M., Mabry T. J. (1986a) Flavonoids from *Stephanodoria tomentella*. J. Nat. Prod. 49:737–738.

Fang N., Leidig M., Mabry T. J. (1986b) Fifty-one flavonoids from *Gutierrezia microcephala*. Phytochemistry 25:927–934.

Fang N., Leidig M., Ngo L.-V. (1986c) Flavonoids from *Gutierrezia grandis*. Phytochemistry 25:235–238.

Fang N., Mabry T. J. (1986) Flavonoid aglycones from *Ageratina tomentella*. J. Nat. Prod. 49:738–739.

Fang N., Yu S., Mabry T. J. (1986d) Flavonoids from *Gutierrezia wrightii*. J. Nat. Prod. 49:739–741.

Fang N., Yu S., Mabry T. J. (1986e) Flavonoids from *Ageratina calophylla*. Phytochemistry 25:2684–2686.

Farag S. F., El-Emary N. A., Makboul M. A., Hasanean H. A., Niwa M. (1997) Phytochemical investigation of *Wedelia prostrata* (Hook. et Arn.) Hemsl. II. The ethyl acetate soluble fraction of the methanol extract. Bull. Pharm. Sci., Assiut Univ. 20: 37–46.

Farkas L., Hörhammer L., Wagner H., Rosler H. (1964c) Untersuchungen über die Glykoside von *Centaurea jacea* L. II. Die Struktur des Centaurein und dessen Synthese aus dem Aglucon und Acetobromglucose. Chem. Ber. 97:1666–1668.

Farkas L., Hörhammer L., Wagner H., Rosler H., Gurniak R. (1964a) Magy. Kem. Folyoirat 70:310–312. Cited by Al-Easa and Rizk (1992).

Farkas L., Hörhammer L., Wagner H., Rosler H., Gurniak R. (1964b) Untersuchungen über die Glykoside von *Centaurea jacea* L. I. Die Struktur des Jacein und dessen Synthese aus dem Aglucon und Acetobromglucose. Chem. Ber. 97:610–615.

Farkas L., Nogradi M., Sudarsanam V., Herz W. (1966a) Constituents of *Iva* species-V. Isolation, structure and synthesis of nevadensin, a new flavone from *Iva nevadensis* and *Iva acerosa*. J. Org. Chem. 31:3228–3232.

Farkas L., Nogradi M., Sudarsanam V., Herz W. (1967) Constituents of *Iva* species-IX. Isolation, structure proof and synthesis of acerosin from *Iva acerosa* (Nutt.) Jackson. Tetrahedron 23:3557–3561.

Farkas L., Pallos L. (1965) Synthese und endgültiger Strukturbeweis des Bracteins, eines Glucosids aus *Helichrysum bracteatum* (Vent.) Willd. Chem. Ber. 98:2930–2932.

Farnsworth N. R., Wagner H., Hörhammer L., Hörhammer H.-P. (1968) *Aster pilosus* (Compositae). I. Isolation of hyperoside quercetin-3-ß-D-galactoside from the leaves. J. Pharm. Sci. 57:1059–1061.

Farrag N. M., Abd El Aziz E. M., El Domiaty M. M., El Shafea A. M. (1993) Phytochemical investigation of *Centaurea araneosa* growing in Egypt. Zagazig J. Pharm. Sci. 2:29–45.

Fassina G., Veronese G. (1961) Analysis of extracts of *Atractylis gummifera*. IV. Chromatographic studies of the pigments of the rhizome. Boll. Soc. Ital. Biol. Sper. 37:518–520.

Favier L. S., Nieto M., Giordana O. S., Tonn C. E. (1997) Diterpenoids and flavonoids from *Ophyrosporus charrua*. Phytochemistry 45:1469–1474.

Fedorov A. A. (ed.) (1969) *Khromosomnye Chisla Tsvetkovykh Rastenii (Chromosome Numbers of Flowering Plants)*. Leningrad: Acad. Sci. U.S.S.R.

Feeny P., Sachdev K., Rosenberry L., Carter M. (1988) Luteolin 7-*O*-(6"-*O*-malonyl)-8-glucoside and *trans*-chlorogenic acid: oviposition stimulants for the black swallowtail butterfly. Phytochemistry 27:3439–3448.

Feráková V. (1977) *The genus* Lactuca *L. in Europe*. Bratislava: Univerzita Komenskéo.

Fernandez I., Garcia B., Grancha F. J., Pedro J. R. (1989) Sesquiterpene lactones, flavonoids and coumarins from *Centaurea collina*. Phytochemistry 28:2405–2407.

Fernandez I., Garcia B., Pedro J. R., Varea A. (1991) Lignans and flavonoids from *Carduus assoi*. Phytochemistry 30:1030–1032.

Ferracini V. L., Gao F., Bierner M. W., Mabry T. J. (1994) Sesquiterpene lactones and one flavonoid from *Hymenoxys scaposa* (DC.) K. Parker var. *linearis* (Nutt.) K. Parker. Biochem. Syst. Ecol. 22:111–112.

Ferracini V. L., Roewer I., Gao F., Mabry T. J. (1989) *ent*-Labdane diterpenoids, tremetone and chromene derivatives and flavonoids from *Ophryosporus heptanthus*. Phytochemistry 28:1463–1465.

Ferraro G. E., Boussalis A. M., Van Baren C. M., Muschietti L. V., Coussio J. D. (1988a) Polyphenolic compounds isolated from *Conyza bonariensis*. Rev. Latinoamer. Quim. 19:141–143.

Ferraro G. E., Coussio J. D. (1973) Flavonoids from *Eupatorium subhastatum*. Phytochemistry 12:1825.

Ferraro G. E., Martino V. S., Borrajo G., Coussio J. D. (1987) 5,7,3',4'-Tetrahydroxy-6-methoxyflavanone from *Eupatorium subhastatum*. Phytochemistry 26:3092–3093.

Ferraro G. E., Martino V. S., Coussio J. D. (1977) New flavonoids from *Eupatorium inulaefolium*. Phytochemistry 16:1618–1619.

Ferraro G. E., Martino V. S., Coussio J. D. (1986) 4',4''-Dimethoxycupressuflavanone, a new biflavanone from *Eupatorium subhastatum*. Planta Medica 52:526.

Ferraro G. E., Martino V. S., Coussio J. D. (1988b) 4',4''-Dimethylcuppressuflavanone from *Eupatorium subhastatum*. J. Nat. Prod. 51:586–587.

Ferraro G. E., Martino V. S., Villar S. I., Coussio J. D. (1985) Flavonoids from *Achyrocline tomentosa*. J. Nat. Prod. 48:817–818.

Ferraro G. E., Norbedo C., Coussio J. D. (1981) Polyphenols from *Achyrocline satureioides*. Phytochemistry 20:2053–2054.

Ferrer M. A., Pedreño M. A., Muñoz R., Ros Barcelo A. (1992) Constitutive isoflavones as modulators of IAA oxidase activity of acidic cell wall isoperoxidases from lupin hypocotyls. Phytochemistry 31:3681–3684.

Ferreres F., Tomas F. (1980) Bull. Liason-Groupe polyphenols (9): 294–301. Cited by Al-Easa and Rizk (1992).

Ferreres F., Tomas F., Guirado A., Tomas F. A. (1980) Afinidad 37(368):337–338. Cited by Al-Easa and Rizk (1992).

Ferreres F., Tomas-Lorente F., Tomas-Barberan F. A., Rivera D., Obon C. (1989) Biochemical identification of *Sideritis serrata* × *S. bourgaeana* hybrids by HPLC analyses of flavonoids. Z. Naturforsch. 44c:568–572.

Ferry S. (1977) Sur les polyphenols, acides-phenols, flavonoides de quelques senecons indigenes. Pl. Medica Phytother. 11:25–39.

Fiasson J. L., Gluchoff-Fiasson K., Mugnier C., Barghi N., Siljak-Yakovlev S. (1991) Flavonoid analysis of European species of the genus *Hypochoeris* (Asteraceae). Biochem. Syst. Ecol. 19:157–162.

Fiebig M., Wagner H. (1984) Neue antihepatotoxische wirksame Flavonolignane aus einer weissbluhender *Silybum*-Varietät. Planta Medica 50:310–313.

Finnegan R. A., Bachman P. L., Knutson D. (1972) Occurrence of 5-hydroxy-4',6,7-trimethoxyflavone in *Psathyrotes annua*. Lloydia 35:457–460.

Fischer D., Stich K., Britsch L., Grisebach H. (1988) Purification and characterization of (+)–dihydroflavonol (3-hydroxyflavanone) 4-reductase from flowers of *Dahlia variabilis*. Arch. Biochem. Biophys. 264:40–47.

Flamini G., Ghelli G. C., Pistelli L., Morelli I. (1994) Phenolic compounds from *Santolina pinnata*. Planta Medica 60:97.

Fleak L. S., Dunn D. B. (1971) Nomenclature of the *Lupinus sericeus* L. complex (Papilionaceae). Trans. Missouri Acad. Sci. 5:85–88.

Flores S. E., Herran J. (1958) Structure of pendulin and penduletin. New flavonol glucoside from *Brickellia pendula*. Tetrahedron 2:308–315.

Flores S. E., Herran J. (1960) Atanasin. A new flavone isolated from *Brickellia squarrosa*. Chem. & Ind. 291.

Flyr D. (1968) New names and records in *Brickellia* (Compositae). Sida 3:252–256.

Forey P. L., Humphries C. J., Kitching I. J., Scotland R. W., Siebert D. J., Williams D. M. (1992) *Cladistics: a practical course in systematics*. Oxford: Clarendon Press.

Forgacs P., Desconclois J. F., Dubec J. (1981) Flavones et sesquiterpene lactones de *Volutarella divaricata*. Planta Medica 42:284–287.

Forkmann G. (1983) 5,7,3',4',5'-Pentahydroxyflavanone in the bracts of *Helichrysum bracteatum*. Z. Naturforsch. 38c:891–893.

Forkmann G. (1989) Gene-enzyme relations and genetic manipulation of anthocyanin biosynthesis in flowering plants. In: Styles D. E., Gavazzi G. A., Racchi M. L. (ed.) *The Genetics of Flavonoids*. Milan: Edizioni Unicopli. 49–60.

Forkmann G. (1994) Genetics of flavonoids. In: Harborne J. B. (ed.) *The Flavonoids: Advances in Research*. London: Chapman and Hall. 536–564.

Forkmann G., Stotz G. (1984) Selection and characterization of flavanone 3-hydroxylase mutants of *Dahlia, Streptocarpus, Verbena* and *Zinnia*. Planta 161:261–265.

Forsyth W. G. C., Simmonds N. W. (1954) A survey of the anthocyanins of some tropical plants. Proc. Royal. Soc. B142:549–564.

Fosberg F. R., Sachet M.-H. (1980) Systematic studies of Micronesian plants. Smithsonian Contrib. Bot. 45:1–40.

Francisco-Ortega J., Crawford D. J., Santos-Guerra A., Jansen R. K. (1997) Origin and evolution of *Argyranthemum* (Asteraceae) in Macaronesia. In: Givnish T. J., Sytsma K. J. (eds.) *Molecular Evolution and Adaptive Radiation*. Cambridge: Cambridge Univ. Press. 407–431.

Franz G., Grun M. (1983) Chemistry, occurrence and biosynthesis of *C*-glycosyl compounds in plants. Planta Medica 47:131–140.

Franzén R. (1988) Flavonoid diversification in the *Achillea ageratifolia* and *Achillea clavennae* groups (Asteraceae). Amer. J. Bot. 75:1640–1654.

Freitag H. J. (1979) Zwei neue Inhaltstoffe aus *Helichrysum italicum* G. Don. Ph.D. Dissertation, Berlin.

Friedrich H. (1962a) Über das Vorkommen von Isorhamnetin-Glycosiden in den Blüten von *Calendula officinalis* L. Arch. Pharm. 295:59–67.

Friedrich H. (1962b) Isorhamnetin-Glykosiden in den Blüten von *Calendula officinalis*. Arch. Pharm. 295:464–471.

Friedrich H. (1962c) Isorhamnetin und Astilbin in den Blüten von *Arnica montana*. Naturwissen. 49:541–542.
Frontera M. A., Tomas M. A., Brunengo M. C., Badajos M. A., Tombesi O. L., Ponce M. A., Gros E. G. (1991) Chemical study of *Centaurea calitrapa*. Rev. Latinoamer. Quim. 22:29–31.
Fuchs L., Iliev V. (1949) Isolation of quercetin from *Solidago virga aurea*, *S. serotina* and *S. canadensis*. An old medicinal plant in a new light. Sci. Pharm. 17:128–131.
Funk V. A. (1985) Cladistics and generic concepts in the Compositae. Taxon 34:72–80.
Funk V. A., Raven P. H. (1980) Polyploidy in *Montanoa* Cerv. (Compositae, Heliantheae). Taxon 29:417–419.
Furuya M. A., Galston A. W., Stowe B. B. (1962) Isolation from peas of co-factor and inhibitors of indole-3-acetic acid oxidase. Nature (London) 193:456–457.
Fusiak F., Schilling E. E. (1984) Systematics of the *Prenanthes roanensis* complex (Asteraceae: Lactuceae). Bull. Torrey Bot. Club 111:338–348.
Gadeschi E., Jorge Z. D., Massanet G. M., Luis F. R. (1989) Two derivatives of costic acid from *Centaurea arguta*. Phytochemistry 28:2204–2206.
Gage D. (1985) Chemical data and their bearing upon generic delineations in the Eupatorieae. Taxon 34:61–71.
Gajhede M., Encarnacion R., Leal G. C., Patino J. C., Christophersen C., Nielsen P. H. (1989) X-Ray structure of 5-hydroxy-3,7,4'-trimethoxyflavone from *Haplopappus sonorensis*. Acta Crystallogr., Sect. C. 45:2012–2014.
Gallego M. J., Talavera S., Silvestre S. (1980) Revisión del género *Reichardia* Roth (Compositae). Lagascalia 9:159–217.
Ganders F. R., Bohm B. A., McCormick S. P. (1990) Flavonoid variation in Hawaiian *Bidens*. Syst. Bot. 15:231–239.
Ganzer U., Jakupovic J. (1990) Schkuripinnatolides, unusual sesquiterpene lactones from *Schkuhria pinnata*. Phytochemistry 29:535–539.
Gao F., Wang H., Mabry T. J. (1987) Sesquiterpene lactones and flavonoids from *Helianthus* species. J. Nat. Prod. 50:23–29.
Gao F., Wang H., Mabry T. J., Abboud K. A., Simonsen S. H. (1989) Sesquiterpene lactones, flavanones and a diterpene acid from *Viguiera laciniata*. Phytochemistry 28:2409–2414.
Gao F., Wang H., Mabry T. J., Kinghorn A. D. (1990) Dihydroflavonol sweeteners and other consituents from *Hymenoxys turneri*. Phytochemistry 29:2865–2869.
Garcia B., Skaltsa H., Navarro F. I., Pedro J. R., Lazari D. (1996) Sesquiterpene lactones and elemane derivatives from *Onopordum myriacanthum*. Phytochemistry 41:1113–1117.
Garcia E. E., Guerreiro E. (1988) Sesquiterpene lactones from *Gochnatia palosanto* and coumarins from *G. argentina*. Phytochemistry 27:288–290.
Gardner R. C. (1973) Acacetin-7-*O*-rutinoside and pectolinarin from *Cirsium coloradense*. Phytochemistry 12:223.
Gardner R. C. (1976) Evolution and adaptive radiation in *Lipochaeta* (Compositae) of the Hawaiian Islands. Syst. Bot. 1:383–391.
Gardner R. C. (1977) Chromosome numbers and their systematic significance in *Lipochaeta* (Compositae: Heliantheae). Amer. J. Bot. 64:810–813.

Gardner R. C. (1979) Revision of *Lipochaeta* (Compositae: Heliantheae) of the Hawaiian Islands. Rhodora 81:291–343.

Gastony G. J. (1986) Electrophoretic evidence for the origin of fern species by unreduced spores. Amer. J. Bot. 73:1563–1569.

Gavina F., Delgado J., Gonzalbes M., Villar E. (1974) An. Quim. 70:1035–1036. Cited by Al-Easa and Rizk (1992).

Geissman T. A. (1941a) Anthochlor pigments. The pigments of *Coreopsis douglasii*. J. Am. Chem. Soc. 63:656–658.

Geissman T. A. (1941b) Anthochlor pigments. II. The pigments of *Coreopsis gigantea*. J. Am. Chem. Soc. 63:2689–2690.

Geissman T. A. (1942) Anthochlor pigments. III. The pigments of *Cosmos sulphureus*. J. Am. Chem. Soc. 64:1704–1717.

Geissman T. A. (ed.) (1962) *The Chemistry of Flavonoid Compounds*. New York: The Macmillan Co.

Geissman T. A., Harborne J. B., Seikel M. K., (1956) Anthochlor pigments. XI. The constituents of *Coreopsis maritima*. Reinvestigation of *Coreopsis gigantea*. J. Am. Chem. Soc. 78:825–829.

Geissman T. A., Heaton C. D. (1944a) Anthochlor pigments. IV. The pigments of *Coreopsis grandiflora* Nutt. I. J. Am. Chem. Soc. 65:677–683.

Geissman T. A., Heaton C. D. (1944b) Anthochlor pigments. V. The pigments of *Coreopsis grandiflora* Nutt. II. J. Am. Chem. Soc. 66:486–487.

Geissman T. A., Mojé W. (1951) Anthochlor pigments. VIII. The pigments of *Coreopsis grandiflora* Nutt. III. J. Amer. Chem. Soc. 73:5765–5768.

Geissman T. A., Mukherjee R., Sim K. Y. (1967) Constituents of *Helichrysum viscosum* var. *bracteatum* DC. Phytochemistry 6:1575–1581.

Geissman T. A., Steelink C. (1957) Flavonoid petal constituents of *Chrysanthemum segetum* L. J. Org. Chem. 22: 946–948.

Gentzbittel L., Perrault A., Nicolas P. (1992) Molecular phylogeny of the *Helianthus* genus based on nuclear restriction-fragment-length polymorphism (RFLP). Mol. Biol. Evol. 9:872–892.

Gerats A. G. M., Mol J. N. M., Stuitje A. R. (1989) Flavonoid synthesis in *Petunia hybrida*, In: Styles D. E., Gavazzi G. A., Racchi M. L. (eds.) *The Genetics of Flavonoids*. Milan: Edizioni Unicopli.

Gershenzon J., Rossiter M., Mabry T. J. (1985) Insect antifeedant tepenoids in wild sunflower (*Helianthus*): a possible source of resistance to the sunflower moth (*Homesoma electellum*). In: Nedin P. A. (ed.) *Bioregulators for pest control*. A.C.S. Symposium Series, Washington, D.C.: American Chemical Society.

Gertz O. (1938) Über die Verbreitung des Anthochlors bei den Compositen. Kungl. Fysiogr. Sallsk. Lund Forh. 8:62–70.

Ghafoor, A., Al-Turki T.A. (1997) A synopsis of the genus *Anthemis* L. (Compositae-Anthemideae) in Saudi Arabia. Candollea 52:457–474.

Ghanin A., Zaman A., Kidwai A. R. (1963) Chemical examination of *Elephantopus scaber*. Indian J. Chem. 1:320–321.

Ghazy N. M., Mahmoud Z. F. (1987) Phytochemical study of the roots of *Hyoseris lucida* L. Egypt. J. Pharm. Sci. 28:321–326.

Gianello J. C., Giordano O. S. (1984) Chemical examination of six species of *Baccharis*. Rev. Latinoamer. Quim. 15:84–86.
Gianello J. C., Giordano O. S. (1987) Constituents of *Baccharis grisebachii*. An. Asoc. Quim. Argent. 75:1–5.
Gianello J. C., Giordano O. S. (1989a) Chemical components from *Baccharis triangularis*. An. Asoc. Quim. Argent. 77:353–355.
Gianello J. C., Giordano O. S. (1989b) Chemical components of *Baccharis triangularis*. An. Asoc. Quim. Argent. 78:87–89.
Gianello J. C., Pestchanker M. J., Tonn C. E., Guo M., Giordano O. S. (1990) 2a,3a-Dihydroxycativic acid, a diterpenoid from *Baccharis petiolata*. Phytochemistry 29:656–659.
Giannasi D. (1975a) Flavonoid chemistry and evolution in *Dahlia* (Compositae). Bull. Torrey Bot. Club 102:404–412.
Giannasi D. (1975b) The flavonoid systematics of the genus *Dahlia*. Mem. New York Bot. Gard. 26:1–125.
Giannasi D. (1978) Generic relationships in the Ulmaceae based on flavonoid chemistry. Taxon 27:331–344.
Giannasi D. (1988) Flavonoids and evolution in the dicotyledons. In: Harborne J. B. (ed.) *The Flavonoids. Advances in Research since 1980*. London: Chapman and Hall. 479–504.
Giannasi D. (2000) *Molecular Systematics and Secondary Metabolites: Perspectives in Plant Systematics and Evolution*. Boca Raton, Florida: CRC Press.
Giannasi D., Niklas K. J. (1977) Flavonoid and other chemical constituents of fossil miocene *Celtis* and *Ulmus* (Succor Creek Flora). Science 197:765–767.
Gil R. R., Del A., Pacciaroni V., Oberti J. C., Diaz J. G., Herz W. (1992) A rearranged germacranolide and other sesquiterpene lactones from *Stevia jujuyensis*. Phytochemistry 31:593–596.
Gil R. R., Pastoriza J. A., Oberti J. C., Gutierrez A. B., Herz W. (1989) Guaianolides from *Stevia sanguinea*. Phytochemistry 28:2841–2843.
Gilbert M. G., Jeffrey C. (1988) A revision of the genus *Ethulia* (Compositae). Kew Bull. 43:165–193.
Gill S., Zielinska-Stasiek M., Miecznikowska D. (1987) Flavonoid compounds in *Baeria californica*. Herba Pol. 33:75–82.
Giner R. M., Cuéllar M. J., Recio M. C., Mañez S., Rios J. L. (1992a) Chemical constituents of *Urospermum picroides*. Z. Naturforsch. 47c:531–534.
Giner R. M., Diaz J., Mañez S., Recio M. C., Soriano C., Rios J. L. (1992b) Phenolics of Spanish *Launaea* species. Biochem. Syst. Ecol. 20:187–188.
Giner R. M., Recio M., Cuellar M. J., Mañez S., Peris J. B., Stübing G., Mateu I., Rios J.-L. (1993a) A taxonomical study of the subtribe Leontodontinae based on the distribution of phenolic compounds. Biochem. Syst. Ecol. 21:613–616.
Giner R. M., Recio M., Villalba M., Mañez S., Rios J. L. (1992c) Phenolic compounds of *Hieracium compositum* and *Hieracium amplexicaule*. Die Pharmazie 47:308.
Giner R. M., Ubeda A., Just M. J., Serrano A., Mañez S., Rios J. L. (1993b) A chemotaxonomic survey of *Sonchus* subgenus *Sonchus*. Biochem. Syst. Ecol. 21:617–620.

Given D. (1996) Flora. In: Preece A. (ed.) *The Chatham Islands: Heritage and Conservation*. Christchurch: Canterbury Univ. Press. 80–92.

Gleason H. A., Cronquist A. (1991) *Manual of Vascular Plants of Northeastern United States and Adjacent Canada,* 2. ed. Bronx: New York Botanical Garden.

Glennie C. W., Harborne J. B. (1971) Flavone and flavonol 5-glucosides. Phytochemistry 10:1325–1329.

Glennie C. W., Harborne J. B. (1972) Flavonoids of *Pyrethrum*. Pyrethrum Post 11:82.

Glennie C. W., Harborne J. B., Rowley G. D., Marchant C. J. (1971) Correlations between flavonoid chemistry and plant geography in the *Senecio radicans* complex. Phytochemistry 10:2413–2417.

Glennie C. W., Jain S. C. (1980) Flavonol 3,7-diglycosides of *Verbesinia encelioides*. Phytochemistry 19:157–158.

Glinkowska G., Strzelecka H. (1987) Polyphenolic compounds of *Erigeron canadensis*. Acta Pol. Pharm. 44:476–477.

Gluchoff-Fiasson K., Favre-Bonvin J., Fiasson J. L. (1991) Glycosides and acylated glycosides of isoëtin from European species of *Hypochoeris*. Phytochemistry 30:1673–1675.

Glyzin V. I., Ban'kovskii A. I., Mel'nikova T. M. (1982) 3-*O*-Methylquercetin from *Serratula inermis*. Chem. Nat. Cmpds. 8:383 [389].

Glyzin V. I., Senov P. L. (1965) Flavonoids of *Nardosmia laevigata*. Aptechn. Delo 14:34.

Goldblatt P. (ed.) (1981, 1984, 1985, 1988) *Index to Plant Chromosome Numbers 1975–1978, 1979–1981, 1982–1983, 1984–1985*. St.Louis: Missouri Botanical Garden.

Goldblatt P., Johnson D. E. (eds.) (1991, 1994 & 1996) *Index to Plant Chromosome Numbers 1988–1989, 1990–1991, 1992–1993*. St.Louis: Missouri Botanical Garden.

Gonnet J. F. (1985) Individual variation of flavonoid glycosides in *Chaerophyllum aureum*. Biochem. Syst. Ecol. 13:313–317.

Gonnet J. F. (1986) Individual variation of flavonoid glycosides in *Chaerophyllum aureum*. Biochem. Syst. Ecol. 14:409–415.

Gonnet J. F. (1992) Flavonoid glycoside variation in wild specimens of *Centaurea montana* (Compositae). Biochem. Syst. Ecol. 20:149–161.

Gonnet J. F. (1993) Flavonoid glycoside variation in wild specimens of *Centaurea triumfettii* (Compositae) and comments on its relationships with *Centaurea montana* based on flavonoid fingerprints. Biochem. Syst. Ecol. 21:389–396.

Gonnet J. F. (1996) Flavonoid glycoside variations in the progeny of wild specimens of *Centaurea montana* and comments on the origin of their natural diversity. Biochem. Syst. Ecol. 24:447–460.

Gonzalez A. G., Aguiar Z. E., Grillo T. A., Luis J. G., Rivera A., Calle J. (1991) Methoxyflavones from *Ageratum conyzoides*. Phytochemistry 30:1269–1271.

Gonzalez A. G., Barrera J. B., Barrera J. L. B., Massanet G. M. (1973) Chemistry of plant constituents. XIV. Flavonoids of *Eupatorium macrocephalum*. An. Quim. 69:229–234.

Gonzalez A. G., Barrera J. B., Diaz J. G., Perez E. M., Yanes A. C., Rauter P., Pozo J. (1990) Diterpenes and other constituents of *Eupatorium salvia*. Phytochemistry 29:321–323.

Gonzalez A. G., Barrera J. B., Garcia T. Z., Rosas F. E. (1984a) Sesquiterpene lactones from *Centaurea* species. Phytochemistry 23:2071–2072.

Gonzalez A. G., Barrera J. B., Mendez J. T., Martinez J. E., Sanchez M. L. (1992a) Germacranolides from *Allagopappus viscosissimus*. Phytochemistry 31:330–331.

Gonzalez A. G., Barrera J. B., Mendez J. T., Sanchez M. L., Martinez J. L. E. (1992b) Sesquiterpene lactones and other constituents of *Tanacetum* species. Phytochemistry 31:1821–1822.

Gonzalez A. G., Bermejo J., Breton J. L., Triana J. (1971) Flavonoids of *Centaurea hyssopifolia*. An. Quim. 67:795–799.

Gonzalez A. G., Bermejo J., Estevez F., Velazquez R. (1983) Phenolic derivatives from *Artemisia glutinosa*. Phytochemistry 22:1515–1516.

Gonzalez A. G., Bermejo J., Triana J., Eiroa J. L., Lopez M. (1995) Sesquiterpene lactones and other constituents of *Allagopappus dichotomus*. J. Nat. Prod. 58:432–437.

Gonzalez A. G., Fraga B. M., Garcia V. P., Hernandez M. G. (1984b) Sterbina, a new flavanone from *Eupatorium sternbergianum*. Rev. Latinoamer. Quim. 14:115–117.

Gonzalez A. G., Fraga B. M., Hernandez M. G., Garcia V. P. (1982) Eupatarone, a 2-acetylbenzofuran from *Eupatorium sternbergianum*. Phytochemistry 21:1826–1827.

Goodwin R. S., Rosler K.-H. A., Mabry T. J., Varma S. (1984) Flavonoids from *Brickellia glutinosa*. J. Nat. Prod. 47:711–714.

Gora J., Swiatek L., Kalemba D., Kurowska A. (1979) Chemical substances from inflorescences of *Arnica montana* L. and *Calendula officinalis* L. Planta Medica 36:286–287.

Gören N., Bozok-Johansson C., Jakupovic J., Lin L.-J., Shieh H.-L., Cordell G. A., Celik N. (1992) Sesquiterpene lactones with antibacterial activity from *Tanacetum densum* ssp. *sivasicum*. Phytochemistry 31:101–104.

Gören N., Jakupovic J. (1990) Glaucolide-like sesquiterpene lactones from *Tanacetum albipannosum*. Phytochemistry 29:3031–3032.

Gören N., Öksüz S., Ulubelen A. (1988) A sesquiterpene lactone, sintenin, from *Achillea sintenisii*. Phytochemistry 27:2346–2347.

Gören N., Tahtasakal E. (1994) Constituents of *Tanacetum densum* subsp. *eginense*. Phytochemistry 36:1281–1282.

Gornall R. J., Bohm B. A. (1978) Angiosperm flavonoid evolution: a reappraisal. Syst. Bot. 3:353–363.

Gornall R. J., Bohm B. A., Dahlgren R. (1979) The distribution of flavonoids in the angiosperms. Bot. Notiser 132:1–30.

Goto T., Kondo T., Kawai T., Tamura H. (1984) Structure of cinerarin, a tetra-acylated anthocyanin isolated from the blue garden cineraria, *Senecio cruentus*. Tet. Lett. 6021–6024.

Gottlieb L. D. (1971) Evolutionary relationships in the outcrossing diploid annual species of *Stephanomeria* (Compositae). Evolution 25:312–329.

Gottlieb L. D. (1972) A proposal for classification of the annual species of *Stephanomeria* (Compositae). Madroño 21:463–481.

Gottlieb L. D. (1977) Phenotypic variation in *Stephanomeria exigua* ssp. *coronaria* (Compositae) and its recent derivative species "malheurensis." Amer. J. Bot. 64:873–880.

Gottlieb L. D., Bennett J. P. (1983) Interference between individuals in pure and mixed cultures of *Stephanomeria malheurensis* and its progenitor. Amer. J. Bot. 70:276–284.

Gottlieb O. R. (1982) *Micromolecular Evolution, Systematics and Ecology, an Essay into a Novel Botanical Discipline*. Berlin: Springer.

Gottlieb O. R. (1988) From molecules to morphology. Biochem. Syst. Ecol. 16:249–250.

Govindachari T. R., Nagarajan K., Pai B. P. (1956a) Chemical examination of *Wedelia calendulacea*. I. Study of wedelolactone. J. Chem. Soc. 629–632.

Govindachari T. R., Nagarajan K., Pai B. P. (1956b) Wedelolactone from *Eclipta alba*. J. Sci. Ind. Research (India) 15B:664–665.

Govindachari T. R., Premila M. S. (1985) The benzofuran norwedelic acid from *Wedelia calendulacea*. Phytochemistry 24:3068–3069.

Graham A. (1996) A contribution to the geologic history of the Compositae. In: Hind D. J. N., Beentje H. J. (eds.) *Compositae: Systematics. Proceedings of the International Compositae Conference, Kew, (1994), Vol. 1*. Kew: Royal Botanic Gardens. 123–140.

Grancai D., Nagy M., Suchy V., Novomesky P. (1994) Cynarin from fresh flower buds of *Cynara cardunculus*. Fitoterapia 65:282.

Granda P. A. (1997) Una nueva especie de *Chuquiraga* (Asteraceae-Mutisieae) del Peru. Kurtziana 25:151–156.

Grande M., Piera F., Cuenca A., Torres P., Bellido I. S. (1985) Flavonoids from *Inula viscosa*, Planta Medica 51:414–419.

Grant V. (1981) *Plant Speciation*, 2. ed. New York: Columbia Univ. Press.

Grau J. (1977) Astereae—systematic review. In: Heywood V. H., Harborne J. B., Turner B. L. (eds.) *The Biology and Chemistry of the Compositae*, Vol. 1. New York: Academic Press. 539–566.

Greger H. (1969) Flavonoide und Systematik der Anthemideae (Asteraceae). Naturwissen. 56:467–468.

Greger H. (1975) Laubblatt-Flavonoide und Systematik bei *Matricaria* und *Tripleurospermum* (Asteraceae-Anthemideae). Plant Syst. Evol. 124:35–55.

Greger H. (1977) Anthemideae—chemical review. In: Heywood V. H., Harborne J. B., Turner B. L. (eds.) *The Biology and Chemistry of the Compositae*, Vol. 2. New York: Academic Press.

Greger H. (1978) Comparative phytochemistry and systematics of *Anacyclus*. Biochem. Syst. Ecol. 6:11–17.

Greger H. (1982) New chemical markers within *Artemisia* (Compositae-Anthemideae). In: Margaris N., Koedam A., Vokou D. (eds.) *Aromatic Plants: Basic and Applied Aspects*. The Hague: Martinus Nijhoff.

Greuter W., McNeill J., Barrie F. R., Burdet H. M., Demoulin V., Filgueiras T. S., Nicolson D. H., Silva P. C., Skog J. E., Trehane P., Hawksworth D. L. (2000) *International Code of Botanical Nomenclature (St. Louis Code)*. Königstein: Koeltz Scientific Books.

Greuter W., Brummitt R. K., Farr E., Kilian N., Kirk P. M., Silva P. C. (1993) *NCU-3. Names in Current Use for Extant Plant Genera*. Königstein: Koeltz Scientific Books.

Grisebach H., Brander G. (1961) Biogenesis of isoflavones. IV.The biogenesis of biochanin-A and formononetin in the chick pea. Z. Naturforsch. 16b:2–5.

Grisebach H., Grambow H. J. (1968) Biosynthesis of flavonoids – XV. Occurrence and biosynthesis of flavonoids in *Datisca cannabina*. Phytochemistry 7:51–56.

Grudzinskaya I. A. (1965) The Ulmaceae and reasons for distinguishing the Celtidoideae as a separate family Celtidaceae. Bot. Zhurn. 52:1723–1749.

Grzybowski J., Jerzmanowska Z. (1954) Studies on *Eupatorium cannabinum*. III. Chemical analysis of roots and stems with leaves. Rocz. Chem. 28:213–231.

Gu H., Hoch P. C. (1997) Systematics of *Kalimeris* (Asteraceae: Astereae). Ann. Missouri Bot. Gard. 84:762–814.

Gu L., Wang S., Li X., Zhu T. (1987) Antibacterial constituents from *Gerbera anandria* (L.) Sch. Bip. Yaoxue Xuebao 22:272.

Gu Y., Tu Y. (1993) Chemical constituents of Japanese wormwood (*Artemisia japonica*). Zhongcaoyao 24:122–124.

Gudej J., Nazaruk J. (1997) Apigenin glycoside esters from flowers of *Bellis perennis* L. Acta Pol. Pharm. 54:233–235.

Guédon D., Abbe P., Lamaison J. L. (1993) Leaf and flower head flavonoids of *Achillea millifolium* L. subspecies. Biochem. Syst. Ecol. 21:607–611.

Guerreiro E. (1989) Coumarins and flavonoids from *Dolichlasium glanduliferum*. Rev. Latinoamer. Quim. 20:47–49.

Guerreiro E., Kavka J., Giordano O. S. (1973) Flavonoids from *Tessaria dodoneifolia* Cabr. An. Asoc. Quim. Argent. 61:161–164.

Guerreiro E., Kavka J., Giordano O. S. (1982) 5,8-Dihydroxy-3,6,7-trimethoxyflavone from *Gnaphalium gaudichaudianum*. Phytochemistry 21:2601–2602.

Guerreiro E., Kavka J., Giordano O. S., Gros E. G. (1979) Sesquiterpenoids and flavonoids from *Flourensia oolepis*. Phytochemistry 18:1235–1237.

Guerreiro E., Pestchanker M. J., Del Vitto L., Giordano O. S. (1990) Sesquiterpenes and flavonoids from *Tessaria* species. Phytochemistry 29:877–879.

Guerrero R. C., de Lourdes Flores S. (1990) The flavonoids of *Flourensia laurifolia*. Rev. Latinoamer. Quim. 21:95–96.

Gunasegaran R., Ubeda A., Alcarez M. J., Jayaprakasam R., Nair A. G. R. (1993) Acylated flavonol glycosides and other phenolics of *Helichrysum bracteatum*. Die Pharmazie 48:230–231.

Gunasingh C. B. C., Nagarajan S. (1981) Flavonoids of the flowers of *Vernonia cinerea*. Indian J. Pharm. Sci. 43:114.

Gunatilaka A. A. L., Dhanabalasingham B., Paredes L., Jakupovic J., Bohlmann F., Adikaram N. K. B. (1987) Microglossic acid, an alicyclic diterpene and other constituents of *Microglossa zeylandica*. Phytochemistry 26:2408–2409.

Guppy G. A., Bohm B. A. (1976) Flavonoids of five *Hieracium* species of British Columbia. Biochem. Syst. Ecol. 4:231–234.

Gupta D. R., Ahmed B. (1985) Asplenetin, a flavone and its glycoside from *Launaea asplenifolia*. Phytochemistry 24:873–875.

Gupta D. R., Bkhustan R., Ahmed B. (1985a) Components of *Launea asplenifolia*. Chem. Nat. Cmpds. 21:382 [408].

Gupta D. R., Dhiman R. P., Ahmed B. (1985b) Constituents of *Launaea asplenifolia* Hook. Die Pharmazie 40:273–274.

Gustafsson M. H. G. (1996) Phylogenetic hypotheses for Asteraceae relationships. In: Hind D. J. N., Beentje H. J. (eds.) *Compositae: Systematics. Proceedings of the International Compositae Conference. Kew (1994), Vol. 1*. Kew: Royal Botanic Gardens. 9–19.

Gustafsson M. H. G., Backlund A., Bremer B. (1996) Phylogeny of the Asterales *sensu lato* based on *rbc*L sequences with particular reference to the Goodeniaceae. Plant Syst. Evol. 199:217–242.

Gustafsson M. H. G., Bremer K. (1995) Morphology and phylogenetic interrelationships of the Asteraceae, Calyceraceae, Campanulaceae, Goodeniaceae, and related families (Asterales). Amer. J. Bot. 82:250–265.

Gutierrez A. B., Herz W. (1988) Guaianolides and other constituents of *Helianthus microcephalus*. Phytochemistry 27:2225–2228.

Haag-Berrurier M., Duquenois P. (1962) Extraction and physicochemical characterization of luteolin in flowers of *Hieracium pilosella*. Compt. Rend. Seances Acad. Sci. 254:3419–3421.

Haag-Berrurier M., Duquenois P. (1963) Luteolin 7-ß-glucoside in the leaves of *Hieracium pilosella*. Compt. Rend. Seances Acad. Sci. 257:3239–3241.

Haag-Berrurier M., Duquenois P. (1969) Les flavonoids de l'*Hieracium murorum* ssp. *grandidens* var. *minoriceps*. Phytochemistry 8:681–683.

Hafez S., Jakupovic J., Bohlmann F., Sarg T. M., Omar A. A. (1989), Transformation products of furoeremophilanes and other constituents from *Euryops arabicus*. Phytochemistry 28:843–847.

Hahlbrock K., Grisebach H. (1975) Biosynthesis of flavonoids. In: Harborne J. B., Mabry T. J., Mabry H. (eds.) *The Flavonoids*. London: Chapman and Hall. 866–915.

Hahn G., H., Lehmann D., Khrten M., Uebel K., Vogel G. (1968) Zur Pharmakologie und Toxikologie von Silymarin, des antihepatotoxischen Wirkprinzipes aus *Silybum marianum* (L.) Gaertn. Arzneim. Forsch. 18:698–700.

Halbach G., Winkler W. (1971) Flavonoide Inhaltsstoffe in den Früchten von *Silybum eburneum*. Z. Naturforsch. 26b:971–972.

Halfon B., Öksüz S., Cirpici A. (1989) Flavonoids of *Centaurea cariensis* Boiss. Doga Turk Saglik Bilimleri Derg 13:138–140.

Hall A. V. (1988) A joint phenetic and cladistic approach for systematics. Biol. J. Linn. Soc. 33:367–382.

Hall A. V. (1991) A unifying theory for methods of systematic analysis. Biol. J. Linn. Soc. 42:425–456.

Hall A. V. (1995) Classification of evolutionary groups with uneven samplings and patchy extinctions. Taxon 44:319–332.

Hall H. M. (1928) *The genus* Haplopappus. *A phylogenetic study in the Compositae*. Carnegie Institution of Washington Publ. No. 389:1–391.

Hamilton C. W., Reichard S. H. (1992) Current practice in the use of subspecies, variety, and forma in the classification of wild plants. Taxon 41:485–498.

Hamrick J. L., Linhart Y. B., Mitton J. B. (1979) Relationships between life history characteristics and electrophoretically detectable genetic variation in plants. Ann. Rev. Ecol. Syst. 10:173–200.

Hanks D. L., McArthur E. D., Stevens R., Plummer A. P. (1973) Chromatographic characteristics and phylogenetic relationships of *Artemisia* sect. *Tridentatae*. USDA Forest Service Research Paper INT-141.

Hannoufa A., Brown R. H., Ibrahim R. K. (1994) Variation in flavonoid sulphate patterns in relation to photosynthetic types of five *Flaveria* species. Phytochemistry 36:353–356.

Hannoufa A., Varin L., Ibrahim R. K. (1991) Spatial distribution of flavonoid conjugates in relation to glucosyltransferase and sulfotransferase activities in *Flaveria bidentis*. Plant Physiol. 97:259–263.

Hänsel R., Çubukçu B. (1972) 3,5-Dihydroxy-6,7,8-trimethoxy-flavone aus *Helichrysum graveolens*. Phytochemistry 11:2632.

Hänsel R., Cybulski E.-M., Çubukçu B., Mericli A. H., Bohlmann F., Zdero C. (1980) Neue pyron-Derivate aus *Helichrysum*-Arten. Phytochemistry 19:639–644.

Hänsel R., Heise D. (1959) Zwei diastereomere Naringen-ß-D-glukoside aus Flores Stoechados. Arch. Pharm. 292:398–411.

Hänsel R., Khaliefi F., Pelter A. (1981) 3,5-Dihydroxy-6,7,8-trimethoxyflavone from *Helichrysum graveolens*: confirmation of the constitution. Z. Naturforsch. 36b:1171–1172.

Hänsel R., Langhammer L. (1963) *Helichrysum bracteatum*: Über die Identität von natürlichen Bracteatin mit synthetischen 4,6,3',4',5'-Pentahydroxy-auron. Arch. Pharm. 296:619–622.

Hänsel R., Langhammer L., Albrecht A. G. (1962) A new aurone glucoside from *Helichrysum bracteatum*. Tet. Lett. 599–603.

Hänsel R., Langhammer L., Frenzel J., Rauft G. (1963) Thin layer chromatographic investigation with anthochlor pigments. Comparison of natural bractein aglucone with synthetic 4,6,3',4',5'-pentahydroxyaurone. J. Chromatog. 11:369–375.

Hänsel R., Ohlendorf D. (1969) Im B-Ring unsubstituierte Flavone aus *Gnaphalium obtusifolium*. Tetr. Lett. 431–432.

Hänsel R., Ohlendorf D. (1971) Ein neues im Ring B unsubstituiertes Flavon aus *Achyrocline satureoides*. Arch. Pharm. 304:893–896.

Hänsel R., Ohlendorf D., Pelter A. (1970) Obtusifoline, a flavanone with a biogenetically unusual C_9 unit. Z. Naturforsch. 25b:989–994.

Hänsel R., Pinkewitz G., Langhammer L., Heise D. (1960) Das gelbe Pigment der Flores Stoechados. Arch. Pharm. 293: 485–489.

Hänsel R., Rimpler H. (1968) Lignans and flavonoids, Deut. Apoth.-Ztg. 108:1985–1987.

Hänsel R., Rimpler H., Schwarz R. (1967) Zur Frage des "*Helichrysum*-Auronols": Eine Berichtigung. Tetr. Lett. 735–738.

Hänsel R., Rimpler H., Walther K. (1966) Ein lipophiles Flavon aus der Kamille (*Matricaria chamomilla* L.) Naturwissen. 53:19–20.

Hänsel R., Schöpflin G. (1967) Bemerkungen zu einem Formelvorschlag für "Silybin". Tetr. Lett. 3645–3648.

Hänsel R., Schulz J., Pelter A. (1972) Structure of silybin. J. Chem. Soc. Chem. Commun. 195–196.
Hansen A., Sunding P. (1979) *Flora Macaronesia, Checklist of Vascular Plants. 2nd revised ed. Part II: Synonym Index*. Oslo: Botanical Garden and Museum.
Hansen H. V. (1991) Phylogenetic studies in Compositae tribe Mutisieae. Opera Bot. 109:1–50.
Hansen H. V. (1992) Studies in the Calyceraceae with a discussion of its relationship to Compositae. Nord. J. Bot. 12:63–75.
Harborne J. B. (1960) Two new naturally occurring anthocyanidins. Chem. Ind. 229–230.
Harborne J. B. (1967a) Comparative biochemistry of the flavonoids. VI. Flavonoid patterns in the Bignoniaceae and the Gesneriaceae. Phytochemistry 6:1643–1651.
Harborne J. B. (1967b) *Comparative Biochemistry of the Flavonoids*. New York: Academic Press.
Harborne J. B. (1977a) Flavonoids and the evolution of the angiosperms. Biochem. Syst. Ecol. 5:7–22.
Harborne J. B. (1977b) Variations in pigment patterns in *Pyrrhopappus* and related taxa of the Cichorieae. Phytochemistry 16:927–928.
Harborne J. B. (1977c) Inuleae – chemical review. In: Heywood V. H., Harborne J. B., Turner B. L. (eds.) *The Biology and Chemistry of the Compositae*, Vol. 1. London: Academic Press. 603–619.
Harborne J. B. (1978) The rare flavone isoëtin as a yellow pigment in *Heywoodiella oligocephala* and in other Cichorieae. Phytochemistry 17:915–917.
Harborne J. B. (ed.) (1987) *The Flavonoids. Advances in Research since 1980*. London: Chapman and Hall.
Harborne J. B. (1988) *Introduction to Ecological Biochemistry*, 3. ed. New York: Academic Press.
Harborne J. B. (1991) Revised structures for three isoëtin glycosides, yellow flower pigments in *Heywoodiella oligocephala*. Phytochemistry 30:1677–1678.
Harborne J. B. (ed.) (1993) *The Flavonoids. Advances in Research since (1986)*. London: Chapman and Hall.
Harborne J. B. (1996) Chemotaxonomy of anthocyanins and phytoalexins in the Compositae. In: Hind D. J. N., Beentje H. J. (eds.) *Compositae: Systematics. Proceedings of the International Compositae Conference, Kew (1994), Vol. 1*. Kew: Royal Botanic Gardens. 207–218.
Harborne J. B., Boardley M. (1983) Identification of two gossypetin monomethyl ethers as yellow flower pigments in the Rutaceae. Z. Naturforsch. 38c:148–150.
Harborne J. B., Boardley M. (1985) The widespread occurrence in nature of anthocyanins as zwitterions. Z. Naturforsch. 40c:305–308.
Harborne J. B., Geissman T. A. (1956) Anthochlor pigments. XII. Maritimein and marein. J. Am. Chem. Soc. 78:829–832.
Harborne J. B., Girija A. R., Devi H. M., Lakshmi N. (1983) Anthochlor pigments from the petals of *Mussaenda hirsutissima* and *Zinnia linearis*. Phytochemistry 22:2741–2742.
Harborne J. B., Greenham J., Eagles J. (1990) Malonylated chalcone glycosides in *Dahlia*. Phytochemistry 29:2899–2900.

Harborne J. B., Greenham J., Eagles J., Wollenweber E. (1991) 6-Hydroxyflavonol glycosides from *Chrysactinia mexicana*. Phytochemistry 30:1044–1045.

Harborne J. B., Heywood V. H., King L. (1976) Evolution of yellow flavonols in flowers of Anthemideae. Biochem. Syst. Ecol. 4:1–4.

Harborne J. B., Heywood V. H., Saleh N. A. M. (1970) Chemosystematics of the Compositae: flavonoid patterns in the *Chrysanthemum* complex of the tribe Anthemideae. Phytochemistry 9:2011–2017.

Harborne J. B., Mabry T. J. (eds.) (1982) *The Flavonoids. Advances in Research.* London: Chapman and Hall.

Harborne J. B., Mabry T. J., Mabry H. (eds.) (1975) *The Flavonoids.* Chapman and Hall: London.

Harborne J. B., Nash R. J. (1983) Flavonoid pigments responsible for ultraviolet patterning in petals of the genus *Potentilla.* Biochem. Syst. Ecol. 12:315–319.

Harborne J. B., Saleh N. A. M., Smith D. M. (1978) On the natural occurrence of gossypetin 7- and 8-monomethyl ethers. Phytochemistry 17:589–591.

Harborne J. B., Sherratt A. S. A. (1957) Variations in the glycosidic pattern of anthocyanins. Part II. Experientia 13:486–487.

Harborne J. B., Smith D. M. (1978) Anthochlors and other flavonoids as honey guides in the Compositae. Biochem. Syst. Ecol. 6:287–291.

Harborne J. B., Turner B. L. (1984) *Plant Chemosystematics.* New York: Academic Press.

Harborne J. B., Williams C. A. (1977) Vernonieae – chemical review. In: Heywood V. H., Harborne J. B., Turner B. L. (eds.) *The Biology and Chemistry of the Compositae*, Vol. 1. London: Academic Press. 523–537.

Harling G. (1991) Compositae: Mutisieae. In: Harling G., Andersson L. (eds.) *Flora of Ecuador, No. 42.* Goteborg: Department of Syst. Bot., Univ. of Goteborg.

Harling G. (1995) The genus *Jungia* L. fil. (Compositae-Mutisieae). Acta R. Soc. Scient. Litt. Gothoburg., Bot. 4:1–133.

Harraz F. M., Amer M. E., Yang S. S., Mabry T. J. (1989) Flavonoids of *Centaurea alexandrina* Del. Rev. Latinoamer. Quim. 20:119–120.

Harraz F. M., Kassem F. F., El-Shaer N. S. (1994) Sesquiterpene lactones and flavonoids from *Centaurea bruguierana.* Alexandria J. Pharm. Sci. 8:219–222.

Harris E. M. (1994) Developmental evidence for the derivation of syncephalia in *Lagascea* (Heliantheae; Asteraceae). Amer. J. Bot. 81:1139–1148.

Hart C. R. (1978) A new species of *Bidens* (Compositae) from Mexico. Brittonia 30:289–292.

Hart C. R. (1979) The systematics of the *Bidens ferulaefolia* complex (Compositae). Syst. Bot. 4:130–147.

Hartman R. L. (1990) A conspectus of *Machaeranthera* (Asteraceae: Astereae). Phytologia 68:439–465.

Hartman R. L., Lane M. A. (1991) A natural intergeneric hybrid in the $x = 6$ group of the Astereae (Asteraceae). Sida 14:321–329.

Hartwig U. A., Maxwell C. A., Joseph C. M., Phillips D. A. (1989) Interactions among flavonoid *nod* gene inducers released from alfalfa seeds and roots. Plant Physiol. 91:1138–1142.

Harvey P. H., Pagel M. D. (1991) *The Comparative Method in Evolutionary Biology*. Oxford: Oxford Univ. Press.

Hasenstein K. H., Pesacreta T. C., Sullivan V. I. (1993) Thigmonasticity of thistle staminal filaments. II. Mechanoelastic properties. Planta 190:58–64.

Haslett B. G., Gleaves T., Boulter D. (1977) *N*-Terminal amino acid sequences of plastocyanins from various members of the Compositae. Phytochemistry 16:363–365.

Hatam N. A. R., Seifert K. (1994) Flavonoids from *Achillea micrantha*. Planta Medica 60:600.

Hattori K. (1992) Inheritance of anthocyanin pigmentation in the flower color of chrysanthemum. Japanese. J. Genet. 67:253–258.

Hattori M., Huang X.-L., Che Q.-M., Kawata Y., Tezuka Y., Kikuchi T., Namba T. (1992) 6-Hydroxykaempferol and its glycosides from *Carthamus tinctorius* petals. Phytochemistry 31:4001–4004.

Hawksworth F. L. (1994) It is no longer necessary to change botanical names in most ranks for non-taxonomic reasons. The Linnean 10:12–15.

Hawksworth F. L. (ed.) (1991) *Improving the Stability of Names: Needs and Options*. Königstein: Koeltz Scientific Books.

Hayashi K. (1933) Vereinfachte Darstellungsmethode der Anthocyanpräparate aus Dahlienblüten. Bot. Mag. Tokyo 47:394–399.

Hayashi K. (1941) Anthocyanin of *Cosmos sulphureus* Cav. Acta Phytochim. Japan. 12:65–67.

Hayashi K., Saito N., Mitsui K. (1961) Anthocyanins. XXXIV. Metallic components in newly crystallized specimens of Bayer's protocyanin, a blue metalloanthocyanin from the cornflower. Proc. Japan Acad. 37:393–397.

Hegnauer R. (1964) *Chemotaxonomie der Pflanzen*, Vol. 3. Basel: Birkhäuser Verlag.

Hegnauer R. (1989) *Chemotaxonomie der Pflanzen*, Vol. 8. Basel: Birkhäuser Verlag.

Hehl F. W. (1919) The yellow coloring substances of ragweed pollen. J. Amer. Chem. Soc. 41:1285.

Heiser C. B. (1954) Variation and subspeciation in the common sunflower, *H. annuus*. Amer. Midl. Naturalist 51:287–305.

Heiser C. B. (1976) *The Sunflower*. Norman: Univ. of Oklahoma Press.

Heiser C. B., Smith D. M., Clevenger S. B., Martin W. C., Jr. (1969) The North American sunflowers (*Helianthus*). Mem. Torr. Bot. Club 22:1–218.

Heller W., Forkmann G. (1988) Biosynthesis. In: Harborne J. B. (ed.) *The Flavonoids: Advances in Research since 1980*. London: Chapman & Hall. 399–425.

Heller W., Forkmann G. (1994) Biosynthesis. In: Harborne J. B. (ed.) *The Flavonoids: Advances in Research since 1986*. London: Chapman & Hall. 499–535.

Hellwig F. H. (1990) Die Gattung *Baccharis* L. (Compositae-Astereae) in Chile. Mitt. Bot. Staats. München 29:1–156.

Hellwig F. H. (1993) The genera *Pingraea* Cassini and *Neomolina* Hellwig (Compositae-Astereae). Candollea 48:203–219.

Hellwig F. H. (1996a) Taxonomy and evolution of Baccharidinae (Compositae). In: Hind D. J. N., Beentje H. J. (eds.) *Compositae: Systematics. Proceedings of the International Compositae Conference, Kew (1994), Vol. 1*. Kew: Royal Botanic Gardens. 575–590.

Hellwig F. H. (1996b) Untersuchungen zur Phylogenie der Cardueae-Centaureinae (Compositae) unter Verwendung molekulärer und morphologisch-anatomischer Merkmale. Göttingen: Habilitationsschrift.

Hennig W. (transl. Davis D. D., Zangerl R.). (1966) *Phylogenetic Systematics*. Urbana: Univ. of Illinois Press.

Hérisset A., Paris R. R., Chaumont J.-P. (1971) Les flavonoides de la Camomille Romaine (*Anthemis nobilis* L.). Pl. Medica Phytother. 5:234–239.

Hernández L. R., Catalan C. A. N., Cerda-Garcia-Rojas C. M., Joseph-Nathan P. (1996) Sesquiterpene lactones from *Stevia vaga*. Phytochemistry 42:1369–1373.

Herz W. (1961) Isolation of 5-hydroxy-3,6,7,3',4'-pentamethoxyflavone from *Kuhnia eupatorioides* L. var. *pyramidalis*. J. Org. Chem. 26:3014–3015.

Herz W. (1996) Terpenoid chemistry of the Astereae. In: Hind D. J. N., Beentje H. J. (eds.) *Compositae: Systematics. Proceedings of the International Compositae Conference, Kew (1994), Vol. 1*. Kew: Royal Botanic Gardens. 261–293.

Herz W., Anderson G. D., Wagner H., Mauer G., Flores G., Farkas L. (1973a) Isolierung, Synthese und Strukturbeweis von 5,7-dihydroxy-3,4',6,8-tetramethoxyflavone, einem neuen Flavon aus *Ambrosia grayi* Nelson. Tetr. Lett. 2571–2572.

Herz W., Aota K., Hall A. L. (1970a) New pseudoguaianolides from *Hymenoxys* species. A new type of lactone closure. J. Org. Chem. 35:4117–4121.

Herz W., Aota K., Holub M., Samek Z. (1970b) Sesquiterpene lactones and lactone glycosides from *Hymenoxys* species. J. Org. Chem. 35:2611–2624.

Herz W., Bhat S. V. (1970) Isolation and structure of two new germacranolides from *Polymnia uvedalia* (L.) L. J. Org. Chem. 35:2605–2611.

Herz W., Bhat S. V. (1972) Woodhousin, a new germacranolide from *Bahia woodhousei* (Gray) Gray. J. Org. Chem. 37:906–912.

Herz W., Bhat S. V., Crawford H., Wagner H., Maurer G., Farkas L. (1972b) Bahifolin, a new sesquiterpene lactone, and 5,7-dihydroxy-3,3',4',6-tetramethoxyflavone, a new flavone from *Bahia oppositifolia*. Phytochemistry 11:371–375.

Herz W., Bhat S. V., Santhanam P. S. (1970c) Coumarins of *Artemisia dracunculoides* and 3',6-dimethoxy-4',5,7-tri-hydroxyflavone in *A. arctica*. Phytochemistry 9:891–894.

Herz W., Bhat S. V., Sudarsanam V. (1972a) Sesquiterpene lactones and flavones of *Iva frutescens*. Phytochemistry 11:1829–1831.

Herz W., Bruno M. (1987a) Helenanolides from *Gaillardia powellii*. Phytochemistry 26:201–204.

Herz W., Bruno M. (1987b) Flavonoids, inositol esters and pantolactone homologues from *Marshallia tenuifolia*. Phytochemistry 26:1175–1180.

Herz W., de Groot R., Murari R., Kuman N., Blunt J. F. (1979) Sesquiterpene lactones of *Eupatorium serotinum*. J. Org. Chem. 44:2784–2788.

Herz W., De Groote R. (1977) Desacetyleupaserrin and nevadensin from *Helianthus pumilus*. Phytochemistry 16:1307–1308.

Herz W., Farkas L., Sudarsanam V., Wagner H., Hörhammer L., Rüger R. (1966a) Isolierung und Struktur von Axillarin, einem neuen Flavon aus *Iva axillaris* Pursh ssp. *robustior*. Chem. Ber. 99:3539–3543.

Herz W., Fitzhenry B., Anderson G. D. (1973b) 5,4'-Dihydroxy-3,7-dimethoxyflavone from *Ambrosia eriocentra*. Phytochemistry 12:1181–1182.

Herz W., Gibaja S. (1972) 5-Hydroxy-6,7,4'-trimethoxyflavone from *Ageratina gilbertii*. Phytochemistry 11:2625.

Herz W., Gibaja S., Bhat S. V., Srinivasan A. (1972c) Dihydroflavonols and other flavonoids of *Eupatorium* species. Phytochemistry 11:2859-2863.

Herz W., Govindan S. V., Kumar N. (1981) Sesquiterpene lactones and other constituents of *Eupatorium lancifolium* and *E. semiserratum*. Phytochemistry 20:1343–1347.

Herz W., Govindan S. V., Riess-Maurer I., Kreil B., Wagner H., Farkas L., Strelisky J. (1980) Isolation and synthesis of two new flavones from *Conoclinium coelestinum*. Phytochemistry 19:669–672.

Herz W., Kulanthaivel P. (1982) Flavones from *Eupatorium leucolepis*. Phytochemistry 21:2363–2366.

Herz W., Kulanthaivel P. (1983) Sesquiterpene lactones from *Liatris acidota, L. aspera* and *L. mucronata*. Phytochemistry 22:513–521.

Herz W., Kulanthaivel P. (1984) *ent*-Pimaranes, *ent*-kaurenes, heliangolides and other constituents of three *Helianthus* species. Phytochemistry 23:1453–1459.

Herz W., Kulanthaivel P. (1985) Trihydroxy-C-18-acids and a labdane from *Rudbeckia fulgida*. Phytochemistry 24:89–91.

Herz W., Kulanthaivel P., Watanabe K. (1983) *ent*-Kauranes and other constituents of three *Helianthus* species. Phytochemistry 22:2021–2025.

Herz W., Kumar N. (1980) Sesquiterpene lactones of *Calea zacatechichi* and *C. urticifolia*. Phytochemistry 19:593–597.

Herz W., Kumar N. (1981a) Sesquiterpene lactones from *Helianthus grosseserratus*. Phytochemistry 20:99–104.

Herz W., Kumar N. (1981b) Minor sesquiterpene lactones of *Helianthus pumilus*. Phytochemistry 20:1339–1341.

Herz W., Pethtel K. D., Raulais D. (1991) Isoflavones, a sesquiterpene lactone-monoterpene adduct and other constituents of *Gaillardia* species. Phytochemistry 30:1273–1279.

Herz W., Pilotti A.-M., Soderholm A.-C., Shuhama I. K., Vichnewski W. (1977) New *ent*-clerodane-type diterpenoids from *Baccharis trinerva*. J. Org. Chem. 42:3913–3917.

Herz W., Rajappa S., Roy S. K., Schmid J. J., Mirrington R. N. (1966b) Constituents of *Gaillardia* species-IV. The sesquiterpene lactones of *Gaillardia fastigiata* Greene. Tetrahedron 22:1907–1915.

Herz W., Ramakrishnan G., Murari R. (1978) A glycosidic germacradienolide from *Eupatorium altissimum*. Phytochemistry 17:1953–1955.

Herz W., Santhanam P. S., Wagner H., Hoer R., Hörhammer L., Farkas L. (1969) Isolation, structure and synthesis of 4',5,6,7-tetrahydroxy-3'-methoxyflavone batatifolin., a new flavone from *Mikania batatifolia* DC. Tetr. Lett. 3419–3421.

Herz W., Santhanam P. S., Wagner H., Hoer R., Hörhammer L., Farkas L. (1970d) Isolierung, Struktur und Synthese von 5,6,7,4'-Tetrahydroxy-3'-methoxy-flavon Batatifolin, einem neuen Flavon aus *Mikania batatifolia*. Chem. Ber. 103:1822–1827.

Herz W., Sosa V. E. (1988) Sesquiterpene lactones and other constituents of *Arnica acaulis*. Phytochemistry 27:155–159.

Herz W., Sudarsanam V. (1970) Flavones of *Iva hayesiana*. Phytochemistry 9:895–896.

Herz W., Sudarsanam V., Schmid J. J. (1966c) Constituents of *Iva* species. VII. New guaianolides from *Iva axillaris* Pursh. ssp. *robustior*. J. Org. Chem. 31:3232–3236.

Herz W., Sumi Y. (1964) Constituents of *Ambrosia hispida* Pursh. J. Org. Chem. 29:3438–3439.

Heywood V. H. (1972) *Sideritis*. In: Tutin T. G., Heywood V. H., Burges N. A., Moore D. M., Valentine D. H., Walters S. M., Webb D. A. (eds.) *Flora Europaea*, Diapensiaceae to Myoporaceae, Vol. 3. Cambridge: Cambridge Univ. Press. 138–143.

Heywood V. H., Harborne J. B., Turner B. L. (eds.) (1977) *The Biology and Chemistry of the Compositae*, 2 vols. London: Academic Press.

Heywood V. H., Humphries C. J. (1977) Anthemideae—systematic review. In: Heywood V. H., Harborne J. B., Turner B. L. (eds.) *The Biology and Chemistry of the Compositae*, Vol. 2. London: Academic Press. 851–898.

Hikino H., Kiso Y., Wagner H., Fiebig M. (1984) Antihepatotoxic actions of flavonolignans from *Silybum marianum* fruits. Planta Medica 50:248–250.

Hiller K., Gil-Rjong R., Otto A. (1979) Über die Flavonoide von *Solidago virgaurea* L. var. *leiocarpa* (Benth.) A. Gray. Die Pharmazie 34:571.

Hilliard O. M., Burtt B. L. (1981) Some generic concepts in Compositae-Gnaphaliinae. Bot. J. Linn. Soc. 82:181–232.

Hind D. J. N. (ed.) (1996) *Compositae: Systematics (Vol. 1), Biology and Utilization (Vol. 2), Proceedings of the International Compositae Conference, Kew (1994)*. Kew: Royal Botanic Gardens.

Hind D. J. N., Jeffrey C., Pope G. V. (eds.) (1995) *Advances in Compositae Systematics*. Kew: Royal Botanic Gardens.

Hino Y., Sasaki K., Oshima Y., Hikino H. (1982) Liver protective drugs. 5. Validity of the oriental medicines. 42. Structure of arcapillan, an antihepatotoxic principle of *Artemisia capillaris*. Heterocycles 19:1615–1617.

Hinou J., Harvala C., Philianos S. (1989) Polyphenolic substances of *Cynara scolymus* L. leaves. Ann. Pharm. Fr. 47:95–98.

Hirose Y., Hayashi S., Hayashida T., Tanaka Y., Kawagishi E., Hori T., Nomura Y., Matsuoka C. (1962) The utilization of plant products IV. Components occurring in the roots of *Solidago altissima*. Kumamoto Pharm. Bull. 5:44–47.

Hirschmann G. S. (1984) The constituents of *Achyrocline satureioides* DC. Rev. Latinoamer. Quim. 15:134–135.

Hitchcock C. L., Cronquist A., Ownbey M., Thompson J. W. (1955) *Vascular Plants of the Pacific Northwest*, Vol. 5. Seattle: Univ. of Washington Press.

Ho L. K., Lu G. L., Huang S. C. (1989) Chemical studies of *Ixeris laevigata*. Hua Hsueh 47:134–138.

Hoeneisen M., Silva M., Jakupovic J., Papastergiou F., Peter M. G. (1993) Flavanones of *Lophopappus tarapacanus* and triterpenoids of *Pachylaena atriplicifolia*. Phytochemistry 34:1653.

Hoffmann B., Herrmann K. (1982) Flavonoid glycosides of worm-wood (*Artemisia vulglaris* L.), tarragon (*A. dracunculus* L.), and absinthe (*A. absinthium* L.). Z. Lebensm.-Unters. Forsch. 174:211–215.

Hoffmann B., Hölzl J. (1988a) New chalcones from *Bidens pilosa*. Planta Medica 54:52–54.

Hoffmann B., Hölzl J. (1988b) A methylated chalcone glucoside from *Bidens pilosa*. Phytochemistry 27:3700–3701.

Hoffmann B., Hölzl J. (1988c) Weitere acylierte Chalkone aus *Bidens pilosa*. Planta Medica 54:450–452.

Hoffmann B., Hölzl J. (1989a) Chalcone glucosides from *Bidens pilosa*. Phytochemistry 28:247–249.

Hoffmann B., Hölzl J. (1989b) Acylated compounds from *Bidens pilosa*. Planta Medica 55:108–110.

Hoffmann O. (1890) Compositae. In: Engler A., Prantl K. (eds.) *Die natürlichen Pflanzenfamilien*, Vol. 4(5). Leipzig.

Hogale M. B., Uthale A. C. (1988) Chemical constituents of *Echinops echinatus*. Orient. J. Chem. 4:334–335.

Holbo H. R., Mozingo H. N. (1965) The chromatographic characterization of *Artemisia* sect. *Tridentatae*. Amer. J. Bot. 52:970–978.

Holeman M., Ilidrissi A., Berrada M. (1991) Flavonoids and sesquiterpene lactones of *Artemisia mesatlantica* L. Planta Medica 57:198–199.

Holmes W. C. (1995) A review preparatory to an infrageneric classification of *Mikania* (tribe: Eupatorieae). In: Hind D. J. N., Jeffrey C., Pope G. V. (eds.) *Advances in Compositae Systematics*. Kew: Royal Botanic Gardens. 239–254.

Holmgren N. H. (1994) Addenda to Asterales by Cronquist. In: Cronquist A., Holmgren A. H., Holmgren N. H., Reveal J. L., Reveal P. K. (eds.) *Intermountain FLora: Vascular Plants of the Intermountain West, U.S.A.*, Vol. 5. Bronx: New York Botanical Garden. 472–480.

Holub M., Toman J., Herout V. (1987) The phylogenetic relationships of the Asteraceae and Apiaceae based on phytochemical characters. Biochem. Syst. Ecol. 15:321–326.

Homma T., Naruse A., Matsuhisa Y. (1990) Studies of the flavonoids from fresh seedlings of safflower (*Carthamus tinctoris* L.). J. Adv. Sci. 2:77–80.

Hörhammer L., Farkas L., Wagner H., Ostermayer J. (1964) Isolierung des Luteolin-7-glucosids aus *Achillea millefolium* L. und eine neue Synthese des Glucosids. Acta Chim. Hung. 40:463–469.

Hörhammer L., Wagner H. (1962) Recent work on the isolation, analysis and phytochemistry of flavones and isoflavones. In: Gore T. S., Joshi B. S., Sunthankar S. V., Tilak B. D. (eds.) *Natural and Synthetic Colouring Matters*. New York: Academic Press. 315–330.

Hörhammer L., Wagner H., Farkas L., Rhger R., Herz W., Sudarsanam V. (1966) Abstr. 4th Int. Natural Products Symposium Stockholm. 43.

Hörhammer L., Wagner H., Salfner B. (1963) Neue Flavon-glykoside aus der Kamille *Matricaria chamomilla* L. Arzneim. Forsch. 13:33–36.

Hörhammer L., Wagner H., Wilkomirsky M. T., Iyengar M. A. (1973) Flavonoide in einigen chilenischen Heilpflanzen. Phytochemistry 12:2068–2069.

Horie T., Tominaga H., Kawamura Y. (1993) Revised structure of a natural flavone from *Ageratum conyzoides*. Phytochemistry 32:1076–1077.

Horie T., Tsukayama M., Kawamura Y., Yamamoto S. (1988) 3,5-Dihydroxy-7,8-dimethoxyflavones and revised structures for some natural flavones. Phytochemistry 27:1491–1495.

Horovitz A., Cohen Y. (1972) Ultraviolet reflectance characteristics in flowers of crucifers. Amer. J. Bot. 59:706–713.

Hort A. (1916) *Theophrastus' Enquiry into Plants* (English translation), 2 vols. London: William Heinemann.

Howe T. D. (1975) The female gametophyte of three species of *Grindelia* and of *Prionopsis ciliata* (Compositae). Amer. J. Bot. 62:273–279.

Hradetzky D., Wollenweber E., Roitman J. N. (1987) Flavonoids from the leaf resin of snakeweed, *Gutierrezia sarothrae*. Z. Naturforsch. 42c:73–76.

Hsu K. K., Hong W. H. (1964) The constituents of the flower of *Chrysanthemum morifolium*. Taiwan Ke Hsueh 18:102–104.

Hu C.-Q., Chen K., Shi Q., Kilkuskie R. E., Cheng Y.-C., Lee K.-H. (1994) Anti-AIDS agents. 10. Acacetin 7-*O*-ß-D-galactopyranoside, an anti-HIV principle from *Chrysanthemum morifolium* and a structure-activity correlation with some related flavonoids. J. Nat. Prod. 57:42–51.

Huber W., Leuchtmann A. (1992) Genetic differentiation of the *Erigeron* species (Asteraceae) in the Alps: a case of unusual allozymic uniformity. Plant Syst. Evol. 183:1–16.

Humphries C. J. (1977) A new genus of the Compositae from North Africa. Bot. Notiser 130:155–161.

Humphries C. J. (1979) A revision of the genus *Anacyclus* L. (Compositae: Anthemideae). Bull. Br. Mus. Nat. Hist. (Bot.) 7:83–142.

Humphries C. J., Parenti L. R. (1986) *Cladistic Biogeography*. Oxford: Clarendon Press.

Humphries C. J., Richardson P. M. (1980) Hennig's methods and phytochemistry. In: Bisby F. S., Vaughn J. G., Wright C. A. (eds.) *Chemosystematics: Principles and Practice*. London: Academic Press. 353–378.

Huneck S., Bohlmann F., Banerjee S., Hartono L., Ang W., Khaidav T. (1985) Inhaltstoffe aus vier Vertretern der Tribus Anthemideae aus der Mongolie. Die Pharmazie 40:365–366.

Hurabielle M., Eberle J., Paris M. (1982) Étude des flavonoides d'*Artemisia campestris* sous-espèces *glutinosa*. Planta Medica 46:124–125.

Hutchinson J. (1916) Aquatic Compositae. Gardener's Chronicle No. 1537:305 (plus 306, figs. only).

Hutchinson J. (1926) *The Families of Flowering Plants I. Dicotyledons. Arranged According to a New System Based on their Probable Phylogeny*. London: Macmillan and Co.

Ibrahim A.-R. S., Galal A. M., Mossa J. S., El-Feraly F. S. (1997) Glucose-conjugation of the flavones of *Psiadia arabica* by *Cunninghamella elegans*. Phytochemistry 46:1193–1195.

Ibrahim R. K., De Luca V., Khouri H., Latchinian L., Brisson L., Charest P. M. (1987) Enzymology and compartmentation of polymethylated flavonol glucosides in *Chrysosplenium americanum*. Phytochemistry 26:1237–1245.

Ickes G. R., Fong H. H. S., Schiff P. L., Jr., Perdue R. E., Jr., Farnsworth N. R. (1973) Antitumor activity and preliminary photochemical examination of *Tagetes minuta* (Compositae). J. Pharm. Sci. 62:1009–1011.

Igile G. O., Oleszek W., Jurzysta M., Burda S., Fafunso M., Fasanmade A. A., (1994) Flavonoids from *Vernonia amygdalina* and their antioxidative properties. J. Agric. Food Chem. 42:2445–2448.

Iinuma M., Roberts M. F., Matlin S. A., Stacey V. E., Timmermann B. N., Mabry T. J., Brown R. (1985) Synthesis and revised structure of the flavone brickellin. Phytochemistry 24:1367–1368.

Ikramov M. T., Khalmatov Kh. Kh., Batirov E. Kh., Malikov V. M. (1986) Flavonoids of *Anaphalis velutina*. Chem. Nat. Cmpds. 22:230 [245].

Imai K., Mayama S. (1953) Chemical constituents of *Erigeron* species (Compositae). II. *Erigeron annuus*. J. Pharm. Soc. Japan 73:128–131.

Imai K., Mayama S. (1954) Chemical constituents of *Erigeron* species (Compositae). IV. *Erigeron bonariensis*. J. Pharm. Soc. Japan 74:109–110.

Inderjit K. M., Dakshini M. (1991) Hesperetin 7-rutinoside hesperidin. and taxifolin 3-arabinoside as germination and growth inhibitors in soils associated with the weed *Pluchea lanceolata* (DC.) C.B. Clarke (Asteraceae). J. Chem. Ecol. 17:1585–1591.

Inderjit K. M., Dakshini M. (1994) Allelopathic potential of the phenolics from the roots of *Pluchea lanceolata*. Physiol. Plant. 92:571–576.

Inderjit K. M., Dakshini M. (1995) Quercetin and quercetrin from *Pluchea lanceolata* and their effect on growth of asparagus bean. ACS Symp. Ser. 582(Allelopathy): 86–93.

Ino I., Nishiyama H., Yamaguchi M.-a. (1993) Malonylation of anthocyanins by extracts of flower buds in *Dendranthema morifolium* cultivars. Phytochemistry 32:1425–1426.

Ino I., Yamaguchi M.-a. (1993) Acetyl-coenzyme A: Anthocyanidin 3-glucoside acetyltransferase from flowers of *Zinnia elegans*. Phytochemistry 33:1415–1417.

Isakova T. I., Serbin A. G., Belikov V. V., Chushenko V. N. (1986) Flavonoids and polysacharides of *Bidens* L. Species. Rastit. Resur. 22:517–522.

Ismail S. I., Hammouda F. M., Rimpler H., Abdallah S. M. (1986) Flavonoids and sesquiterpene lactones, constituents of *Centaurea alexandrina*. Planta Medica 52:527.

Ismail S. I., Hammouda F. M., Rizk A. M., El-Kasy S. E. (1979) Flavonoid constituents of *Conyza dioscoridis* L. and *Conyza aegyptica* Ait. Die Pharmazie 34:112.

Ismail S. I., Rizk A. M., Hammouda F. M., Hassan N. M. (1989) Methylated flavonoids from *Artemisia monosperma*. Qatar Univ. Sci. Bull. 79–80.

Isman M. B., Rodriguez E. (1983) Larval growth inhibitors from species of *Parthenium* (Asteraceae). Phytochemistry 22:2709–2713.

Itakura Y., Imoto T., Kato A., Yagashita K. (1975) Flavonoids in the flowers of *Gnaphalium affine*. Agric. Biol. Chem. 39:2237–2238.

Ito M., Soejima A., Watanabe K. (1998) Phylogenetic relationships of Japanese *Aster* (Asteraceae, Astereae) sunsu lato based on chloroplast-DNA restriction site mutations. J. Plant Research 111:217–223.

Ivancheva S., Behar M. (1995) Flavonoids in *Tanacetum vulgare* L. Fitoterapia 66:373.

Ivancheva S., Kuzmanov B. (1986) Flavonoids in the genus *Achillea*, *Achillea nobilis* group. Bull. Liason-Groupe Polyphenols 13:577–579.

Ivancheva S., Kuzmanov B. A., Gerasimov L. P. (1990) Epicuticular flavonoid aglycones in two species of *Achillea*. Farmatsiya Sofia 40:20–22.

Ivancheva S., Nedelcheva A. (1995) Flavonoids in the genus *Achillea* sect. *Filipendula*. Colloq.-Inst. Natl. Rech. Agron. (Polyphenols 94), 303–304.

Ivancheva S., Stancheva B. (1996) Flavonoids in genus *Achillea* section *Ptarmica* (Asteraceae). Polyphenolics Communications 96, Bordeaux (France), July 1996, 97–98.

Ivancheva S., Van Sumere C., Kuzmanov B. (1987) Flavonoids of *Achillea urumoffii*. F.E.C.S Int. Conf. Chem. Biotechnol. Biol Act. Nat. Prod. [Proc.], 3rd, 1985, #4, 427.

Ivancheva S., Zdravkova M. (1993) Flavonoids in *Tagetes patula*. Fitoterapia 64:555.

Iwashina T., Ito T., Ootani S. (1988) Identification and peculiarity of flavonoid compounds in leaves of Japanese *Cirsium* species. Ann. Tsukuba Bot. Gard. 7:149–158.

Iwashina T., Ito T., Ootani S. (1989) Characterization and identification of 5-glucosides of flavones obtained from Japanese *Cirsium* species. Ann. Tsukuba Bot. Gard. 8:15–19.

Iwashina T., Kadota Y. (1999) Flavonoids from *Schmalhausenia nidulans* (Compositae): a taxon endemic to the Tien Sian Mountains. Biochem. Syst. Ecol. 27:97–98.

Iwashina T., Kadota Y., Ueno T., Ootani S. (1995) Foliar flavonoid composition in Japanese *Cirsium* species (Compositae), and their chemotaxonomic significance. J. Japanese Bot. 70:280–290.

Iwashina T., Kamenosono K., Ueno T. (1999) Hispidulin and nepetin 4'-glucosides from *Cirsium oligophyllum*. Phytochemistry 51:1109–1111.

Iwashina T., Ootani S. (1998) Foliar flavonoids from *Cirsium hachijoense* (Compositae). Res. Inst. Evolut. Biol. Sci. Rep. 9:11–19.

Iwu M. M., Chiori C. O. (1984) Antibacterial components of *Eupatorium odoratum*. Fitoterapia 55:354.

Iyengar M. A. (1975) Flavonoids of *Liatris pauciflora*. J. Health Sci. (India) I, 74. Cited by Iyengar et al. 1976a.

Iyengar M. A., Bhat U. G., Katti S. B., Wagner H., Seligmann O., Herz W. (1976a) Mearnsetin, a rare flavonol from *Liatris pauciflora*. Indian J. Chem. 14B:714.

Iyengar M. A., Singri B. P., Wagner H., Seligmann O., Herz W. (1976b) Eupatolitin-3-*O*-rhamnoside in *Rudbeckia mollis* (Compositae). Indian J. Chem. 14B:906.

Jackson B. D. (1895) *Index Kewensis Plantarum Phanerogamarum*. Oxford: Clarendon Press.

Jackson R. C. (1960) A revision of the genus *Iva* L. Univ. Kansas Sci. Bull. 41:793–876.

Jacobs H., Bunbury M., Hylands P. J., McLean S. (1986b) Flavonoids and a heliangolide, niveusin-C, from *Melampodium camphoratum*. J. Nat. Prod. 49:1163.

Jacobs H., Bunbury M., McLean S. (1986a) Tricin from *Vernonia remotiflora*. J. Nat. Prod. 49:1164.

Jacobs M., Rubery P. H. (1988) Naturally occurring auxin transport regulators. Science 241:346–349.

Jain S. C. (1977) Chemical investigation of *Tagetes* tissue cultures. Planta Medica 31:68–70.

Jakupovic J., Chen Z.-L., Bohlmann F. (1987a) Artamolide, a dimeric guaianolide and phenylalanine derivatives from *Artemisia anomala*. Phytochemistry 26:2777–2779.

Jakupovic J., Ganzer U., Bohlmann F., King R. M. (1991a) Kaurane and beyerane derivatives from *Calocephalus knappii*. Phytochemistry 30:2651–2652.

Jakupovic J., Grenz M., Bohlmann F., Mungai G. M. (1990a) Carvotacetone derivatives and eudesman-12,6ß-olides from *Sphaeranthus* species. Phytochemistry 29:1213–1217.

Jakupovic J., Grenz M., Bohlmann F., Mungai G. M. (1990b) 12ß-Hydroxyabieta-7,13-diene and other constituents from East African *Helichrysum* species. Phytochemistry 29:1589–1590.

Jakupovic J., Grenz M., Miski M., Bohlmann F., Boldt P. E. (1990c) Geranyl derivatives of *p*-coumaric acid from *Chrysothamnus pulchellus*. Phytochemistry 29:617–620.

Jakupovic J., Jaensch M., Bohlmann F., Dillon M. O. (1988a) Eudesmanolides, 5,10-*bis-epi*-eudesmanes and oplopanone derivatives from *Ambrosia artemisioides*. Phytochemistry 27:3551–3556.

Jakupovic J., Klemeyer H., Bohlmann F., Graven E. H. (1988b) Glaucolides and guaianolides from *Artemisia afra*. Phytochemistry 27:1129–1133.

Jakupovic J., Kuhnke J., Schuster A., Metwally M. A., Bohlmann F. (1986) Phloroglucinol derivatives and other constituents from South African *Helichrysum* species. Phytochemistry 25:1133–1142.

Jakupovic J., Lehmann L., Bohlmann F., King R. M., Robinson H. (1988c) Sesquiterpene lactones and other constituents from *Cassinia, Actinobole* and *Anaxeton* species. Phytochemistry 27:3831–3839.

Jakupovic J., Paredes L., Bohlmann F., Watson L. (1988e) Prenyl flavanes from *Marshallia* species. Phytochemistry 27:3273–3275.

Jakupovic J., Pathak V. P., Bohlmann F., King R. M., Robinson H. (1987b) Obliquin derivatives and other constituents from Australian *Helichrysum* species. Phytochemistry 26:803–807.

Jakupovic J., Schuster A., Bohlmann F., Ganzer U., King R. M., Robinson H. (1989a) Diterpenes and other constituents from Australian *Helichrysum* and related species. Phytochemistry 28:543–551.

Jakupovic J., Schuster A., Chau-Thi T. V., Bohlmann F., Dominguez X. A. (1988d) Guaianolides and homoditerpenes from *Heliopsis helianthoides*. Phytochemistry 27:2235–2240.

Jakupovic J., Schuster A., Wasshausen D. C. (1991b) Acetylenes and labdanes from *Baccharis pedunculata*. Phytochemistry 30:2785–2787.

Jakupovic J., Tan R. X., Bohlmann F., Boldt P. E., Jia Z. J. (1991c) Sesquiterpene lactones from *Artemisia ludoviciana*. Phytochemistry 30:1573–1577.

Jakupovic J., Tsichritzis F., Tamayo-Castillo G., Castro V., Bohlmann F., Boldt P. E. (1989b) Diterpenes from *Fleischmannia hymenophylla* and *Brickellia laciniata*. Phytochemistry 28:2741–2744.

Jakupovic J., Zdero C., Grenz M., Tsichritzis F., Lehmann L., Hashemi-Nejad S. M., Bohlmann F. (1989c) Twenty-one acylphloroglucinol derivatives and further constituents from South African *Helichrysum* species. Phytochemistry 28:1119–1131.

James C. W. (1961) Endemism in Florida. Brittonia 13:225–244.

Janiak B., Hänsel R. (1960) Phytochemisch-Pharmakognostische Untersuchungen über *Fructus Cardui mariae*. 1. Planta Medica 8:71–84.

Jansen R. K. (1985) The systematics of *Acmella* (Asteraceae-Heliantheae). Syst. Bot. Monographs 8:1–115.

Jansen R. K., Holsinger K. E., Michaels H. J., Palmer J. D. (1990) Phylogenetic analysis of chloroplast DNA restriction site data at higher taxonomic levels: an example from the Asteraceae. Evolution 44:2089–2115.

Jansen R. K., Kim K.-J. (1996) Implications of chloroplast DNA data for the classification and phylogeny of the Asteraceae. In: Hind D. J. N., Beentje H. J. (eds.) *Compositae: Systematics. Proceedings of the International Compositae Conference, Kew (1994), Vol. 1.* Kew: Royal Botanic Gardens. 317–339.

Jansen R. K., Michaels H. J., Palmer J. D. (1991a) Phylogeny and character evolution in the Asteraceae based on cpDNA restriction site mapping. Syst. Bot. 16:98–115.

Jansen R. K., Michaels H. J., Wallace R. S., Kim K.-J., Keeley S. C., Watson L. E., Palmer J. D. (1992) Chloroplast DNA variation in the Asteraceae: phylogenetic and evolutionary implications. In: Soltis P. S., Soltis D. E., Doyle J. J. R. (eds.) *Molecular Systematics of Plants*. New York: Chapman and Hall. 252–279.

Jansen R. K., Palmer J. D. (1987a) Chloroplast DNA from lettuce and *Barnadesia* (Asteraceae): structure, gene localization, and characterization of a large inversion. Curr. Genet. 11:553–564.

Jansen R. K., Palmer J. D. (1987b) A chloroplast DNA inversion marks an ancient evolutionary split in the sunflower family, Asteraceae. Proc. Natl. Acad. Sci., U.S.A. 84:5818–5822.

Jansen R. K., Palmer J. D. (1988) Phylogenetic implications of chloroplast DNA restriction site variation in the Mutisieae (Asteraceae). Amer. J. Bot. 75:753–766.

Jansen R. K., Smith E. B., Crawford D. J. (1987c) A cladistic study of North American *Coreopsis* (Asteraceae: Heliantheae). Plant Syst. Evol. 157:73–84.

Jansen R. K., Stuessy T. F. (1980) Chromosome counts of Compositae from Latin America. Amer. J. Bot. 67:585–594.

Jansen R. K., Wallace R. S., Kim K. J., Chambers K. L. (1991b) Systematic implications of chloroplast DNA variation in the subtribe Microseridinae (Asteraceae: Lactuceae). Amer. J. Bot. 78:1015–1027.

Jares E., Pomilio A. B. (1987) Pyrrolizidine alkaloids and other components of *Senecio vira-vira.* J. Nat. Prod. 50:514.

Jauhari P. K., Sharma S. C., Tandon J. S., Dhar M. M. (1979) A new flavonol glycoside from *Rudbeckia bicolor.* Phytochemistry 18:359–360.

Jeffrey C. (1966) Notes on Compositae: I. The Cichorieae in East Tropical Africa. Kew Bull. 18:427–486.

Jeffrey C. (1977) Corolla forms in Compositae—some evolutionary and taxonomic speculations. In: Heywood V. H., Harborne J. B., Turner B. L. (eds.) *The Biology and Chemistry of the Compositae*, Vol. 1. London: Academic Press. 111–118.

Jeffrey C. (1978) Compositae. In: V. H. Heywood (ed.) *Flowering Plants of the World.* New York: Mayflower Books. 263–268.

Jeffrey C. (1979) Note on the lectotypification of the names *Cacalia* L., *Matricaria* L. and *Gnaphalium* L. Taxon 28:349–351.

Jeffrey C. (1986) The Senecioneae in East Tropical Africa. Kew Bull. 41:873–943.

Jeffrey C. (1988) The Vernonieae in East Tropical Africa. Notes on Compositae: V. Kew Bull. 43:195–277.

Jeffrey C. (1992) The tribe Senecioneae (Compositae) in the Mascarene Islands with an annotated world check-list of the genera of the tribe. Notes on Compositae. VI. Kew Bull. 47:49–109.

Jeffrey C., Chen Y.-L. (1984) Taxonomic studies on the tribe Senecioneae (Compositae) of Eastern Asia. Kew Bull. 39:205–446.

Jeffrey C., Halliday P., Wilmot-Dear M., Jones S. W. (1977) Generic and sectional limits in *Senecio* (Compositae): I. Progress report. Kew Bull. 32:47–67.

Jefferies P. R., Knox J. R., Price K. R., Scaf B. (1974) Constituents of the tumor-inhibitory extract of *Olearia muelleri*. Aust. J. Chem. 27:221–225.

Jepson W. L. (1925) *A Manual of the Flowering Plants of California*. Berkeley: Associated Students Store.

Jeremic D., Stefanovic M., Dokovic D., Milosavljevic S. (1979) Flavonols from *Artemisia annua*. Glas. Hem. Drus. Beograd 44:615–618.

Jerga C., Merfort I., Willuhn G. (1990) Flavonoidaglyka aus den Blüten von *Heterotheca inuloides*. Planta Medica 56:122–123.

Jerzmanowska Z. I., Grzybowska J. (1958) Flavonoids in the flowers of *Helichrysum arenarium*. Acta Polon. Pharm. 15:13.

Jerzmanowska Z. I., Grzybowska J. (1960) Flavonoid compounds in the inflorescence of *Helichrysum arenarium*. Nature 186:807–808.

Jeske F., Huneck S., Jakupovic J. (1996) Further sesquiterpene lactones from *Inula salsoloides*. Phytochemistry 41:1539–1542.

Jia Z., Fei H., Li Y., Zhu Z. (1986) Studies on the constituents of *Saussurea medusa* Maxim. I. Gaodeng Xueziao Huaxue Xuebao 7:789.

Jia Z., Gong N., Du M. (1990) Chemical constituents of *Saussurea medusa* Maxim. Gaodeng Xueziao Huaxue Xuebao 11:202–204.

Jia Z., He K., Du M., Li Y., Chu T. (1988) Constituents of *Saussurea involucrata* Kar. & Kin. IV. Gaodeng Xueziao Huaxue Xuebao 9:198–200.

Johnson R. R. (1969) Monograph of the plant genus *Porophyllum* (Compositae: Helenieae). Univ. Kansas Sci. Bull. 48:225–267.

Jolad S. D., Timmermann B. N., Hoffmann J. J., Bates R. B., Camou F. A., Siahaan T. J. (1988) Kolavane diterpenoids of *Vanclevea stylosa*. Phytochemistry 27:505–515.

Jones A. G. (1980) A classification of the New World species of *Aster* (Asteraceae). Brittonia 32:230–239.

Jones A. G. (1985) Chromosomal features as generic criteria in the Astereae. Taxon 34:44–54.

Jones A. G., Young D. A. (1983) Genetic concept of *Aster* (Asteraceae): a comparison of cladistic, phenetic, and cytological approaches. Syst. Bot. 8:71–84.

Jones S. B. (1977) Vernonieae—systematic review. In: Heywood V. H., Harborne J. B., Turner B. L. (eds.) *The Biology and Chemistry of the Compositae, Vol. 1*. New York: Academic Press. 503–522.

Jordan C., Reimann-Philipp R. (1983) Z. Pflanzenzücht. 91:111.

Joysey K. A., Friday A. E. (eds.) (1982) *Problems of Phylogenetic Reconstruction*. London: Academic Press.

Juarez B. E., Mendiondo M. E., Seeligmann P. (1990) The floral flavonoids of *Porophyllum lanceolatum* DC. (Compositae). Lilloa 37:75–78.

Juarez B. E., Mendiondo M. E., Seeligmann P. (1995) Flavonoids from leaves and flowers *of Liabum polymnioides* and *L. candidum* (Asteraceae). Chemotaxonomical significance. Biochem. Syst. Ecol. 23:335–336.
Judd W. S., Campbell C. S., Kellog E. A., Stevens P. F. (1999) *Plant Systematics: A Phylogenetic Approach.* Sunderland: Sinauer.
Julian E. A., Crawford D. J. (1972) Sulphuretin glycosides of *Coreopsis mutica.* Phytochemistry 11:1841–1843.
Jussieu A.-L. de (1789) *Genera Plantarum.* Paris.
Kagan J. (1968) The flavonoid pigments of *Liatris spicata.* Phytochemistry 7:1205–1207.
Kahler A. L., Lay C. L. (1985) Genetics of electrophoretic variants in the annual sunflower. J. Hered. 76:335–340.
Kakegawa K., Hattori E., Koike K., Takeda K. (1991) Induction of anthocyanin synthesis and related enzyme activities in cell cultures of *Centaurea cyanus* by UV-light irradiation. Phytochemistry 30:2271–2273.
Kakegawa K., Kaneko Y., Hattori E., Koike K., Takeda K. (1987) Cell cultures of *Centaurea cyanus* produce malonated anthocyanin in UV light. Phytochemistry 26:2261–2263.
Källersjö M. (1986) Fruit structure and generic delimitation of *Athanasia* and related South African genera. Nord. J. Bot. 5:527–542.
Källersjö M. (1991) The genus *Athanasia* (Compositae-Anthemideae). Opera Bot. 106:1–75.
Kaloga M. (1981a) Isosilychristin, a new flavonolignan from *Silybum marianum* (L.) Gaertn. Z. Naturforsch. 36b:262–265.
Kaloga M. (1981b) Isolation of dihydrokaempferol from *Silybum marianum* (L.) Gaertn. Z. Naturforsch. 36b:524–525.
Kaloshina N. A., Glyzin V. I., Ban'kovskii A. I. (1975a) Flavonoids from *Anthemis tinctoria.* Chem. Nat. Cmpds. 11:685 [654].
Kaloshina N. A., Glyzin V. I., Ban'kovskii A. I. (1975b) The flavonoids of *Carduus nutans.* Chem. Nat. Cmpds. 11:686 [654].
Kaloshina N. A., Mazulin A. V. (1983a) Flavonoids of seeds of plants of the genus *Tagetes.* Chem. Nat. Cmpds. 19:101 [104].
Kaloshina N. A., Mazulin A. V. (1983b) Flavonoids of *Achillea pannonica.* Chem. Nat. Compds. 19:102 [104].
Kaloshina N. A., Mazulin A. V. (1988) Flavonoids of *Carduus nutans.* Chem. Nat. Cmpds. 24: 389 [453].
Kamanzi K., Raynaud J. (1977) Les pigments anthocyaniques des fleurs de *Centaurea montana* et de *Centaurea lugdunensis* (Composées). Pl. Medica Phytother. 11:289–293.
Kamanzi K., Raynaud J., Voirin B. (1982a) Les aglycones flavoniques de fleurs de *Centaurea collina* L. (Composées). Die Pharmazie 37:454.
Kamanzi K., Raynaud J., Voirin B. (1982b) Les glycosides flavonique des fleurs et des feuilles de *Centaurea collina* L. Die Pharmazie 37:523–524.
Kamanzi K., Raynaud J., Voirin B. (1982c) Les aglycones flavoniques des feuilles de *Centaurea collina* L. (Composées). Pl. Medica Phytother. 16:30–33.

Kamanzi K., Raynaud J., Voirin B. (1982d) Les glycosides flavonique de *Centaurea melitensis* L. (Composées). Die Pharmazie 37:454.

Kamanzi K., Raynaud J., Voirin B. (1983a) Les *C*-glycosyl flavonoides des fleurs de *Centaurea solstitialis* L. (Composées). Die Pharmazie 38:494.

Kamanzi K., Raynaud J., Voirin B. (1983b) Les *C*-glycosyl flavonoides des fleurs de *Centaurea melitensis* L. (Composées). Pl. Medica Phytother. 17:47–51.

Kamanzi K., Raynaud J., Voirin B. (1983c) Les *O*-hétérosides flavoniques de fleurs de *Centaurea solstitialis* L. (Composées). Pl. Medica Phytother. 17:57–60.

Kamanzi K., Voirin B., Raynaud J. (1983d) Les flavonoides méthoxylés des fleurs de *Centaurea solstitialis* L. (Composées). Pl. Medica Phytother. 17:52–56.

Kametaka T., Perkin A. G. (1910) Carthamine. I. J. Chem. Soc. 97:1415–1427.

Kamsteeg J., van Brederode J., Hommels C. H., van Nigtevect G. (1980) Identification, properties and genetic control of hydroxycinnamoyl-coenzyme A:anthocyanidin 3-rhamnosyl(1→6)glucoside 4'''-hydroxycinnamoyl transferase isolated from petals of *Silene dioica*. Biochem. Physiol. Pflanzen 175:403–411.

Kaneta M., Hikichi H., Endo S., Sugiyama N. (1978) Identification of flavones in sixteen Compositae species. Agric. Biol. Chem. 42:475–477.

Kapoor S. K., Khan H., Zaman A. (1970) Chemical investigation of *Carthamus oxyacantha*. Indian J. Appl. Chem. 33:262.

Karikome H., Ogawa K., Sashida Y. (1992) New acylated glucosides of chalcones from the leaves of *Bidens frondosa*. Chem. Pharm. Bull. 40:689–691.

Karis P. O. (1989) Systematics of the genus *Metalasia* (Asteraceae-Gnaphalieae). Opera Bot. 99:1–150.

Karis P. O. (1993a) Morphological phylogenetics of the Asteraceae-Asteroideae, with notes on character evolution. Plant Syst. Evol. 186:69–93.

Karis P. O. (1993b) Heliantheae *sensu lato* (Asteraceae), clades and classification. Plant Syst. Evol. 188:139–195.

Karis P. O. (1995) Cladistics of the subtribe Ambrosiinae (Asteraceae: Heliantheae). Syst. Bot. 20:40–54.

Karis P. O. (1996) Phylogeny of the Asteraceae-Asteroideae revisited. In: Hind D. J. N., Beentje H. J. (eds.) *Compositae: Systematics. Proceedings of the International Compositae Conference, Kew (1994), Vol. 1*. Kew: Royal Botanic Gardens. 41–47.

Karis P. O., Källersjö M., Bremer K. (1992) Phylogenetic analysis of the Cichorioideae (Asteraceae), with emphasis on the Mutisieae. Ann. Missouri Bot. Gard. 79:416–427.

Karis P. O., Ryding O. (1994a) Tribe Helenieae. In: Bremer K. (ed.) *Asteraceae: Cladistics and Classification*. Portland: Timber Press. 521–558.

Karis P. O., Ryding O. (1994b) Tribe Heliantheae. In: Bremer K. (ed.) *Asteraceae: Cladistics and Classification*. Portland: Timber Press. 559–624.

Karrfalt E. E. (1975) The nature of the cortical bundles of *Adenocaulon* (Compositae). Bot. Gaz. 136:236–245.

Katinas L., Stuessy T. F. (1997) Revision of *Doniophyton* (Compositae, Barnadesioideae). Plant Syst. Evol. 206:33–45.

Kaufmann F., Lam J. (1967) Chemical constituents of the genus *Dahlia*. II. The isolation of two aromatic compounds: naringenin trimethyl ether and fraxidin. Acta Chem. Scand. 21:311–313.

Kaufmann H. P., El Baya A. E. W. (1969) Pro- and antioxidants in the field of fats. XXIV. Their effect on the biosynthesis of aurones in the blossoms. Fette, Seifen, Anstrichm. 71:25–28.

Kaul V. K., Shawl A. S., Bindra R. L. (1989) Flavonoids and coumarins of *Artemisia salsoloides* and *A. laciniata*. Ind. J. Pharm. Sci. 51:111–112.

Kavka J., Guerreiro E., Giordano O. S. (1977) Flavonoids of *Tessaria dodoneifolia* (Hook. & Arn.) Cabr. An. Quim. 73:305–306.

Kavka J., Guerreiro E., Giordano O. S., Romo J. (1973) Flavonoides del *Baccharis salicifolia* (R. et P.) Persoon. Rev. Latinoamer. Quim. 4:101–104.

Kawase K., Tsukamoto Y., Saito N., Osawa Y. (1970) Flower colour in *Chrysanthemum morifolium* – anthocyanins. Plant Cell. Physiol. 11:349–353.

Kazuma K., Shirai E., Wada M., Umeo K., Sato A., Atsushi M. K., Matsumoto T., Okuno T. (1995) Structure of precarthamnin, a biosynthetic precursor of carthamin. Biosci. Biotechnol. Biochem. 59:1588–1590.

Keeley S. C., Jansen R. K. (1991) Evidence from chloroplast DNA for the recognition of a new tribe, the Tarchonantheae, and the tribal placement of *Pluchea* (Asteraceae). Syst. Bot. 16:173–181.

Keeley S. C., Turner B. L. (1990) A preliminary cladistic analysis of the genus *Vernonia* (Vernonieae:Asteraceae). Plant Syst. Evol. (Suppl. No. 4) 45–66.

Kelsey R. G., Shafizadah F. (1979) Sesquiterpene lactones and systematics of the genus *Artemisia*. Phytochemistry 18:1591–1611.

Kelsey R. G., Stevenson T. T., Scholl J. P., Watson T. J., Shafizadeh F. (1973) The chemical composition of the litter and soil in a community of *Artemisia tridentata* ssp. *vaseyana*. Biochem. Syst. Ecol. 6:193–200.

Kerber V. A., Miguel O. G., Moreira E. A. (1993) Flavonoids from *Heterothalamus psiadioides*. Fitoterapia 64:185.

Kerr K. M., Mabry T. J., Yoser S. (1981) 6-Hydroxy-and 6-methoxyflavonoids from *Neurolaena lobata* and *N. macrocephala*. Phytochemistry 20:791–794.

Kery A., Tawaij H. A. A., Al-Kazraji N. K. (1985) Herba Hung. 24:183–194. Cited by Al-easa and Rizk (1992).

Kevan P. G. (1972) Floral colors in the high arctic with reference to insect-flower relations and pollination. Can. J. Bot. 50:2289–2316.

Khafagy S. M., Al-Yahya M. A., Ziesche J., Bohlmann F. (1983) An unusual nerolidol derivative from *Artemisia inculta*. Phytochemistry 22:1821–1822.

Khafagy S. M., El-Ghazooly M. G., Metwally A. M. (1979a) Isolation and characterization of two methoxylated flavones from *Artemisia monosperma*. Die Pharmazie 34:748–749.

Khafagy S. M., El-Salam N. A. A., Hamid R. A. (1981a) Constituents of *Silybum marianum* (L.) Gaertn. Sci. Pharm. 49:157.

Khafagy S. M., Halim A. R., Gohar A. A. (1981b) Chemical study of *Cirsium syriacum*. Fitoterapia 52:21.

Khafagy S. M., Nazmisabri N., El-Salem N. A. A., Seifieldin A. A. (1981c) Constituents of *Otanthus maritimus* Hoffmgg. & Link. Die Pharmazie 36:507–508.

Khafagy S. M., Sabri N. N., El-Salam N. A. A., El-Din A. A. S. (1979b) Isolation of sesamin-like compound and acacetin 7-*O*-neohesperidoside from *Otanthus maritimus*. Planta Medica 35:186–187.

Khafagy S. M., Sabri N. N., El-Salam N. A. A., El-Din A. A. S. (1980) Flavonoidal contents of *Carthamus glaucus* Bieb. emend Hanelt ssp. *alexandrinum* Hanelt. Egypt J. Pharm. Sci. 19:307–312.

Khafagy S. M., Sabri N. N., El-Salam N. A. A., El-Din A. A. S., Blessington B. (1979c) Flavonoids of *Carthamus glaucus* subsp. *alexandrinus* Hanelt. Acta Pharm. Jugosl. 29:161–165.

Khafagy S. M., Sabri N. N., Soliman F. S. G., Abou-Donia A. H., Mosandl A. (1976) Isolation of two flavonoids from *Achillea santolina* L. growing in Egypt. Die Pharmazie 31:894–895.

Khalifa A. A., Aballah O. M., Mesbah M. K. (1991) Sesquiterpene lactones and flavonoids from *Chrysanthemum vulgare*. Bull. Fac. Pharm. Cairo Univ. 29:63–66.

Khan I. A., Tali T., Sticher O. (1993) Flavonoids and caffeic acid esters from *Dichrocephala bicolor*. Planta Medica 59:288.

Kharlamov I. A., Khazanovich R. L., Khalmatov K. K. (1968) Flavonoids of *Arctium*. Farmatsiya Moscow 17:45.

Khouri H. E., Ibrahim R. K. (1987) Resolution of five position-specific flavonoid *O*-methyltransferases by fast protein liquid chromatofocusing. J. Chromatog. 407:291–297.

Khvorost P. P. (1966) Polyphenolic compounds from some plants of the Compositae family. Common tansy (*Tanacetum vulgare*). Fenol'nye Soedin. Ikh. Biol. Funkts, Mater. Vses. Simp., 1st, 85–89.

Khvorost P. P., Chernobai V. T., Kolesnikov D. G. (1964) Flavonoidal compounds of *Tanacetum vulgare*. Zh. Obshch. Khim. 34:4108–4111.

Khvorost P. P., Chernobai V. T., Kolesnikov D. G. (1966) Flavonoid compounds of *Tanacetum vulgare*. Med. Prom. SSSR 20:19–21.

Kiang A. K., Sim K. Y., Goh J. (1965) Constituents of *Mikania cordata* (Burm. f.) Robins. B. L. (Compositae). Isolation of mikanin, epifriedelinol and fumaric acid; the structure of mikanin. J. Chem. Soc. 6371–6374.

Kikuchi M., Mori M. (1992) Components from flower buds of *Tussilago farfara* L. III. Phenolic compounds. Annu. Rep. Tohoku Coll. Pharm. 39:69–73.

Kim H.-G., Keeley S. C., Jansen R. K. (1996) Phylogenetic position of the Hawaiian endemic *Hesperomannia* (Mutisieae) based on *ndh*F sequence data. Amer. J. Bot. 83 (6, suppl.):167. Abstr.

Kim K.-J., Jansen R. K. (1995a) Molecular clocks and the origin of the sunflower family (Asteraceae). Amer. J. Bot. 82 (6, suppl.):405. Abstr.

Kim K.-J., Jansen R. K. (1995b) *ndh*F sequence evolution and the major clades in the sunflower family. Proc. Natl. Acad. Sci. U.S.A. 92:10379–10383.

Kim K.-J., Jansen R. K., Wallace R. S., Michaels H. J., Palmer J. D. (1992) Phylogenetic implications of *rbc*L sequence variation in the Asteraceae. Ann. Missouri Bot. Gard. 79:428–445.

Kim S.-C., Crawford D. J., Francisco-Ortega J., Santos-Guerra A. (1995) Phylogenetic relationships in subtribe Sonchinae (Asteraceae) based on ribosomal DNA sequences from the ITS region. Amer. J. Bot. 82 (6, suppl.):141. Abstr.

Kim S.-C., Crawford D. J., Jansen R. K. (1996) Phylogenetic relationships among the genera of the subtribe Sonchinae (Asteraceae): evidence from ITS sequences. Syst. Bot. 21:417–432.

Kim S.-C., Crawford D. J., Jansen R. K., Santos-Guerra A. (1999a) The use of a non-coding region of chloroplast DNA in phylogenetic studies of the subtribe Sonchinae (Asteraceae: Lactuceae). Plant Syst. Evol. 215:85–99.

Kim S.-C., Crawford D. J., Tadesse M., Berbee M., Ganders F. R., Pireyedi M., Esselman E. J. (1999b) ITS sequences and phylogenetic relationships in *Bidens* and *Coreopsis* (Asteraceae). Syst. Bot. 24:480–493.

King B. L., Jones S. B. (1982) Chemosystematics of *Vernonia* series *Flexuosae* (Vernonieae: Compositae). Bull. Torrey Bot. Club 109:279–286.

King R. M., Dawson H. W. (1975) Introduction. In: King R. M., Dawson H. W. (eds.) *Cassini on Compositae.* New York: Oriole Editions.

King R. M., Janaske P. C., Lellinger D. B. (1995a) *Cassini on Compositae II. Collected from the Bulletin des Sciences par la Société Philomatique de Paris.* St. Louis: Missouri Botanical Garden.

King R. M., Janaske P. C., Lellinger D. B. (1995b) *Cassini on Compositae III. Collected from the Journal de Physique, de Chimie, d'Histoire Naturelle et des Arts and from the Annales des Sciences Naturelles.* St. Louis: Missouri Botanical Garden.

King R. M., Krantz E. (1975) Ultraviolet reflectance patterns in the Asteraceae. I. Local and cultivated species. Phytologia 31:66–85.

King R. M., Robinson H. (1972a) Studies in the Eupatorieae (Asteraceae) LXXVIII. A new genus, *Brickelliastrum.* Phytologia 24:63–64.

King R. M., Robinson H. (1972b) Studies in the Eupatorieae (Asteraceae) LXXX. A new genus, *Flyriella.* Phytologia 24:67–69.

King R. M., Robinson H. (1977) The new synantherology. Taxon 19:6–11.

King R. M., Robinson H. (1982) Studies in the Eupatorieae (Asteraceae) CCXII. Additions to *Austroeupatorium, Flyriella,* and *Teixeiranthus.* Phytologia 50:379–381.

King R. M., Robinson H. (1987) *The genera of the Eupatorieae (Asteraceae).* Monographs Syst. Bot. (Missouri Bot. Gard.) 22:1–581.

Kisiel W. (1975) Phytochemical examination of *Vernonia flexuosa.* 1. Isolation of 5,4'-dihydroxy-7,3'-flavone. Polish J. Pharmacol. Pharm. 27:339–343.

Kisiel W. (1976) Flavonoids of *Vernonia flexuosa.* Herba Polonica 22:3–10.

Kiso Y., Ogasawara S., Hirota K., Watanabe N., Oshima Y., Konno C., Hikino H. (1984) Antihepatotoxic principles of *Artemisia capillaris* buds. Planta Medica 50:81–85.

Kitagawa I., Fukuda Y., Yoshihara M., Yamahara J., Yoshikawa M. (1983) Capillartemisin A and B, two choleretic principles from *Artemisia capillaris* Thunb. Chem. Pharm. Bull. 31:352–355.

Kitamura S. (1937) Compositae Japonicae, Pars Prima. Mem. Coll. Sci. Kyoto Imperial Univ., Ser. B, 13:1–421.

Kitching I. J., Forey P. L., Humphries C. J., Williams D. M. (1998) *Cladistics: the Theory and Practice of Parsimony Analysis*, ed. 2. Oxford: Oxford Univ. Press.

Klyshev L. K., Baltabaeva G. R. (1978) Resources and biochemical characteristics of *Artemisia cina* Berg. Izv. Akad. Nauk. Kaz. SSR, Ser. Biol. 16:7–9.

Knobloch I. W. (1972) Intergeneric hybridization in flowering plants. Taxon 21:97–103.

Knuth P. (1891) Die Einwirkung der Blütenfarben auf die photographische Platte. Bot. Zentralb. 48:160, 314.

Koch M. F. (1930a) Studies in the anatomy and morphology of the composite flower. I: the corolla. Amer. J. Bot. 17:938–952.

Koch M. F. (1930b) Studies in the anatomy and morphology of the composite flower. II: the corollas of the Heliantheae and Mutisieae. Amer. J. Bot. 17:995–1010.

Kodama O., Miyakawa J., Akatsuka T., Kiyosawa S. (1992) Sakuranetin, a flavanone phytoalexin from UV-irradiated rice leaves. Phytochemistry 31:3807–3809.

Komissarenko N. F., Chernobai V. T., Derkach A. I. (1988) Flavonoids of inflorescences of *Calendula officinalis*. Chem. Nat. Cmpds. 24:675 [795].

Komissarenko N. F., Derkach A. I., Kovalyov I. P., Bublik N. P. (1994) Diterpene glycosides and phenyl propanoids of leaves of *Stevia rebaudiana* Bertoni (Asteraceae). Rastit. Resur. 30:53–64.

Komiya T., Naruse Y., Oshio H. (1976) Studies on "Inchinko." II. Studies on the compounds related to capillarisin and flavonoids. Yakugaku Zasshi 96:855–862.

Kondo T., Ueda M., Tamura H., Yoshida K., Isobe M., Goto T. (1994) Structure of protocyanin, a self-organized supramolecular pigment from the cornflower *Centaurea cyanus*. Angew. Chem. 106:1039–1040. [see also: English ed. 33(9):978–979 (1994)].

Konopleva M. M., Glyzin V. I., Shelyuto V. L. (1978) Flavonoids of *Gnaphalium sylvaticum*. Chem. Nat. Cmpds. 14:339 [402].

Konopleva M. M., Smirnova L. P., Glyzin V. T. (1979b) Flavonoids of *Gnaphalium uliginosum*. Chem. Nat. Cmpds. 15:420 [483].

Konopleva M. M., Smirnova L. P., Glyzin V. T., Shelyuto V. L. (1979a) New acylated flavone glycoside from *Gnaphalium uliginosum*. Chem. Nat. Cmpds. 15:269 [311].

Kornkven A. B., Watson L. E., Estes J. R. (1998) Phylogenetic analysis of *Artemisia* section *Tridentatae* (Asteraceae) based on sequence from the internal transcribed spacers (ITS) of nuclear ribosomal DNA. Amer. J. Bot. 85:1787–1795.

Kotaiah Y., Lakshmi N. K., Rao E. V., Rao D. V. (1976) Chemical constitution of the flowers and leaves of *Notonia grandiflora*. Indian J. Pharm. 38:130–131.

Kowalewska K., Lutomski J. (1978) Flavonoids in the inflorescences of *Inula helenium* L. Herba Pol. 24:107–113.

Kowalska M. (1968) Isolation of quercetin and rutin from flowers of *Senecio jacobea*. Rocz. Wyzsz. Szk. Roln. Poznanium 39:49.

Kranen-Fiedler U. (1956) Some ingredients of *Artemisia abrotanum*. Arznm.-Forsch. 6:475–479.

Krishnamoorthy V., Seshadri T. R. (1966) Occurrence of 3,4,2',4',6'-pentahydroxychalcone in the petals of *Helichrysum bracteatum*. Curr. Sci. 35:609.

Krolikowska M., Wolbis M. (1979) Polyphenolic compounds in *Inula britannica*. Acta Pol. Pharm. 36:395.

Krolikowska M., Wolbis M. (1981) Polyphenolic compounds in *Inula britannica*. Acta Pol. Pharm. 38:107–114.

Kroschewsky J. R., Mabry T. J., Markham K. R., Alston R. E. (1969) Flavonoids from the genus *Tragopogon* (Compositae). Phytochemistry 8:1495–1498.

Kuhn B., Forkmann G., Seyffert W. (1978) Genetic control of chalcone-flavanone isomerase activity in *Callistephus chinensis* (L.) Nees. Planta 138:199–203.

Kulanthaivel P., Benn M. H. (1986) A new truxillate and some flavonoid esters from the leaf gum of *Traversia baccharoides* Hook. f. Can. J. Chem. 64:514–519.
Kulkarni M. M., Rojatkar S. R., Nagasampagi B. A. (1987) Four 6-hydroxyflavonols from *Blumea malcomii*. Phytochemistry 26:2079–208.
Kumazawa T., Kanenari D., Kunimatsu A., Amano Y., Sato S., Matsube S., Obara H., Onodera J., Suzuki M. (1994b) Structure of the precursor of carthamin. Tennen Yuki Kagobutsu Toronkai Koen Yoshiku 36:665–672.
Kumazawa T., Sato S., Kaneuari D., Kunimatsu A., Hirose R., Matsuba S., Obara H., Suzuki M., Sato M., Onodera J.-i. (1994a) Precursor of carthamin, a constituent of safflower. Chem. Lett. (12) 2343–2344.
Kunde R., Isaac O. (1979) Über die Flavone der Kamille (*Matricaria chamomilla* L.) und ein neues acetyliertes Apigenin-7-glucosid. Planta Medica 37:124–130.
Kupchan S. M., Bauerschmidt E. (1971) Cytotoxic flavonols from *Baccharis sarothroides*. Phytochemistry 10: 664–666.
Kupchan S. M., Knox J. R., Udayamurthy M. S. (1965) Tumor inhibitors VIII. Eupatorin, a new cytotoxic flavone from *Eupatorium semiserratum*. J. Pharm. Sci. 54:929–930.
Kupchan S. M., Sigel C. W., Hemingway R. J., Knox J. R., Udayamurthy M. S. (1969a) Tumor inhibitors-XXXIII. Cytotoxic flavones from *Eupatorium* species. Tetrahedron 25:1603–1615.
Kupchan S. M., Sigel C. W., Knox J. R., Udayamurthy M. S. (1969b) Tumor inhibitors. XXXVI. Eupatin and eupatoretin, two cytotoxic flavonols from *Eupatorium semiserratum*. J. Org. Chem. 34:1460–1463.
Kurihara T., Kikuchi M. (1972) Constituents of the flower stalks of *Petasites japonicus*. VIII. Water-soluble components. Tohoku Yakka Daigaku Kenkyu Nempo 55:63.
Kurkin V. A., Zapesochnaya G. G. (1987) Flavolignans and other natural lignoids: problems of structural analysis. Chem. Nat. Cmpds. 23:7–23.
Kuroda C. (1930a) The constitution of carthamnin. Part I. J. Chem. Soc. 752–765.
Kuroda C. (1930b) The constitution of carthamnin. Part II. J. Chem. Soc. 765–767.
Kuroyanagi M., Fujita K., Kazaoka M., Matsumoto S., Ueno A., Fukushima S., Katsuoka M. (1985) Studies on the constituents of *Baccharis genistelloides*. Chem. Pharm. Bull. 33:5075–5078.
Kustrak D., Males Z. (1993) Isolation of flavonoids from leaves of *Solidago gigantea* Ait. Farm. Glas. 49:129–133.
Kyhos D. W. (1967) Natural hybridization between *Encelia* and *Geraea* (Compositae) and some related experimental investigations. Madroño 19:33–64.
Kyhos D. W. (1971) Evidence of different adaptations of flower color variants of *Encelia farinosa* (Compositae). Madroño 21:49–61.
La Duke J. C. (1982a) Flavonoid chemistry and systematics of *Tithonia* (Compositae). Amer. J. Bot. 69:784–792.
La Duke J. C. (1982b) Revision of *Tithonia*. Rhodora 84:453–522.
Labbe C., Rovirosa J., Faini F., Mahu M., San-Martin A., Castillo M. (1986) Secondary metabolites from Chilean *Baccharis* species. J. Nat. Prod. 49:517–518.
Lack H. W. (1978) Die Gattung *Heywoodiella* Svent. & Bramw. (Asteraceae, Lactuceae). Willdenowia 8:329–339.

Lam J., Wrang P. (1975) Flavonoids and polyacetylenes in *Dahlia tenuicaulis*. Phytochemistry 14:1621–1623.

Lamboy W. F. (1994) The accuracy of the maximum parsimony method for phylogenetic reconstruction with morphological characters. Syst. Bot. 19:489–505.

Lammers T. G. (1992) Circumscription and phylogeny of the Campanulales. Ann. Missouri Bot. Gard. 79:388–413.

Lane M. A. (1979) Taxonomy of the genus *Amphiachyris* (Asteraceae; Astereae). Syst. Bot. 4:178–189.

Lane M. A. (1985) Features observed by electron microscopy as generic criteria. Taxon 34:38–43.

Lane M. A. (1996) Pollination biology of Compositae. In: Calgari P. D. S., Hind D. J. (eds.) *Compositae: Biology and Utilization. Proceedings of the International Compositae Conference, Kew 1994, Vol. 2.* Kew: Royal Botanic Gardens. 61–80.

Lane M. A., Hartman R. L. (1996) Reclassification of North American *Haplopappus* (Compositae: Astereae) completed: *Rayjacksonia* gen. nov. Amer. J. Bot. 83:356–370.

Lane M. A., Morgan D. R., Suh Y., Simpson B. B., Jansen R. K. (1996) Relationships of North American genera of Astereae, based on chloroplast DNA restriction site data. In: Hind D. J. N., Beentje H. J. (eds.) *Compositae: Systematics. Proceedings of the International Compositae Conference, Kew (1994), Vol. 1.* Kew: Royal Botanic Gardens. 49–77.

Lao A., Fujimoto Y., Tatsuno T. (1983) Studies on the constituents of *Artemisia rubripes* Nakai. Yakugaku Zasshi 103:696–699.

Larson R. A. (1988) The antioxidants of higher plants. Phytochemistry 27:969–978.

Laskaris G. G., Gournelis D. C., Kokkalon E. (1995) Phenolics from *Picnomon acarna*. J. Nat. Prod. 58:1248–1250.

Lawrence G. H. M. (1951) *Taxonomy of Vascular Plants*. New York: Macmillan.

Lazari D., Skaltsa H., Harvala C. (1998) Flavonoids of *Onopordum sibthorpianum* and *L. laconicum*. Biochem. Syst. Ecol. 26:105–107.

Le Quesne P. W., Menachery M. D., Raffauf R. F. (1979) Antitumor plants. Part IX. Structural reassignments for three flavonoids from *Lychnophora affinis* Gardn. J. Nat. Prod. 42:320–321.

Le Quesne P. W., Pastore M. P., Raffauf R. F. (1976) The cytotoxic flavonoids of *Lychnophora affinis*. J. Nat. Prod. 39:391–394.

Lee H. B., Kwak J. H., Zee O. P., Yoo S. J. (1994) Flavonoids from *Cirsium rhinoceros*. Arch. Pharmacal. Res. 17:273–277.

Lee K. H., Simpson R. F., Geissman T. A. (1969) Sesquiterpenoid lactones of *Artemisia*. Constituents of *A. cana* ssp. *cana*. The structure of canin. Phytochemistry 8:1515–1521.

Lee K.-H., Anuforo D. C., Huang E.-S., Piantadosi C. (1972) Antitumor agents. I: Angustibalin, a new cytotoxic sesquiterpene lactone from *Balduina angustifolia* (Pursh.) Robins. J. Pharm. Sci. 61:626–628.

Lee Y. C. (1967) Constituents of *Chrysanthemum sibiricum*. Yakhak Hoeji 11:7–10.

Lenherr A., Fang N., Mabry T. J. (1986) Flavonols from *Gutierrezia alamanii* var. *megalocephala*, J. Nat. Prod. 49:185–186.

Leonhardt R. (1949) Phylogenetisch-systematische Betrachtungen. I. Betrachtungen zur Systematik der Compositen. Österr. Bot. Zeitschrift 96:293–324.
Lessing C. F. (1832) *Synopsis Generum Compositarum*. Berlin.
Le-Van N., Pham T. V. C. (1979) Two new flavones from *Eupatorium coelestinum*. Phytochemistry 18:1859–1861.
Levin D. A. (1967) An analysis of hybridization in *Liatris*. Brittonia 19:248–260.
Levin D. A. (1976) The chemical defenses of plants to pathogens and herbivores. Annu. Rev. Ecol. Syst. 7:121–159.
Levin D. A. (1993) Local speciation in plants: the rule not the exception. Syst. Bot. 18:197–208.
Levin D. A., York B. M. (1978) The toxicity of plant alkaloids: an ecogeographic perspective. Biochem. Syst. Ecol. 6:61–76.
Levy M. (1977) Minimum biosynthetic-step indices as measures of comparative flavonoid affinity. Syst. Bot. 2:89–98.
Levy M. (1983) Flavone variation and subspecific divergence in *Phlox pilosa* (Polemoniaceae). Syst. Bot. 8:118–126.
Lewis H. (1962) Catastrophic selection as a factor in speciation. Evolution 16:257–271.
Li H., Meng J. C., Cheng C. H. K., Higa T., Tanaka J., Tan R. X. (1999) New guaianolides and xanthin oxidase inhibitory flavonoids from *Ajania fruticulosa*. J. Nat. Prod. 62:1053–1055.
Li J., Ou-Lee T.-M., Raba R., Amundson R. G., Last R. L. (1993) *Arabidopsis* flavonoid mutants are hypersensitive to UV-B irradiation. The Plant Cell 5:171–179.
Li R., Ding Z., Dong J. (1997) Chemical constituents from *Eupatorium adenophorum*. Yunnan Zhiwu Yanjiu 19:196–200.
Li R., Fang N., Mabry T. J. (1986) Flavonoids from *Brickellia scoparia*. J. Nat. Prod. 49:732–733.
Li R., Fang N., Mabry T. J. (1987) Flavonoids from *Gutierrezia texana* var. *texana*. Phytochemistry 26:2831–2833.
Li R., Fang N., Mabry T. J. (1988) Flavonoids and a coumarin from *Gutierrezia sphaerocephala*. Phytochemistry 27: 1556–1559.
Li S., Ding J. (1994) The flavonoids of *Laggera pterodonta*. Yunnan Zhiwu Yanjiu 16:434–436.
Li Y., Hu Y., Shi Y. (1994a) Alkaloids and flavonoids from Roxburgh wormwood (*Artemisia roxburghiana*). Zhongcaoyao 25:622–623.
Li Y., Shi Y., Hu Y. (1994b) Chemical constituents of *Artemisia roxburghiana*. Indian J. Chem. 33B:302–304.
Lima M. A. S., Silveira E. R., Marques M. S. L., Santos R. H. A., Gambardela M.T. P. (1996) Biologically active flavonoids and terpenoids from *Egletes viscosa*. Phytochemistry 41:217–223.
Lin C.-H., Arisawa M., Shimizu M., Morita N. (1978) The constituents of *Cirsium japonicum* DC. var. *takaoense* Kitamura. Chem. Pharm. Bull. 26:2036–2039.
Lin J. H. (1993) Anaphaloside, a new acyl flavonol glycoside from the flowers of *Anaphalis contorta*. J. Chin. Chem. Soc. (Taipei) 40:93–95.
Lin J.-H., Chou C.-J., Liu K.-C. (1975) Studies on the constituents of *Glossogyne tenuifolia* (Labill.) Cass. Taiwan Yao Hsueh Tsa Chih 27:98–100.

Ling Y.-r. (1994) The genera *Artemisia* L. and *Seriphidium* (Bess.) Poljak. in the world. Comp. Newsl. 25:39–45.
Ling Y.-r. (1995) Survey of *Artemisia* and *Seriphidium* (Asteraceae-Anthemideae) in the Himalayan Mountains and the South Asian subcontinent. Comp. Newsl. 26:49–60.
Linnaeus C. (1751) *Philosophia Botanica.* Stockholm.
Linnaeus C. (1753) *Species Plantarum.* Stockholm
Lipschitz S. (1979) Genus *Saussurea* DC. (Asteraceae). Leningrad: Academia Scientiarum URSS.
List P. H., Fribel B. (1974) Neue Inhaltsstoffe der Blüten von *Arnica montana* L. Arzneim. Forsch. 24:148–151.
Litvinenko V. I., Bubenchikova V. N. (1988) Phytochemical study of *Centaurea cyanus.* Chem. Nat. Cmpds. 24:672 [792].
Liu H. M., Li G.-L., Wu H. W. (1981) Studies on the chemical constituents of *Artemisia annua* L. Yao Hsueh T'ung Pao 16:65–67.
Liu H. M., Li K.-L., Wo W.-C. (1980) Studies on the constituents of *Artemisia annua.* Yao Hsueh T'ung Pao 15:39.
Liu Y., Yu D. (1985) Studies on the chemical constituents of *Artemisia rupestris* L. Yaoxue Xuebao 20:514–518.
Liu Y.-L., Mabry T. J. (1981a) Two methylated flavones from *Artemisia frigida.* Phytochemistry 20:309–311.
Liu Y.-L., Mabry T. J. (1981b) Flavonoids from *Artemisia frigida.* Phytochemistry 20:1389–1395.
Liu Y.-L., Mabry T. J. (1982) Flavonoids from *Artemisia ludoviciana* var. *ludoviciana.* Phytochemistry 21:209–214.
Liu Y.-L., Mabry T. J., Lardy C., Chopin J. (1982) *C*-Glycosylflavonoids from *Artemisia frigida,* Rev. Latinoamer. Quim. 13:56–57.
Liu Y.-L., Timmermann B. N., Hoffmann J. J., McLaughlin S. P. (1993) A flavonol triglycoside from *Baccharis thesioides.* Phytochemistry 33:1549–1551.
Liu Z., Jia Z., Chu T. (1989) Chemical constituents of *Saussurea hieracioides* Hook. f. Gaodeng Xueziao Huaxue Xuebao 10:1090–1094.
Lloyd F. E. (1911) Guayule (*Parthenium argentatum* Gray), a rubber-plant of the Chihuahuan Desert. Carnegie Instit. Wash. Publ. No. 139:1–213.
Lois R. (1994) Accumulation of UV-absorbing flavonoids induced by UV-B radiation. I. Mechanisms of UV-resistance in *Arabidopsis.* Planta 194:498–503.
Lois R., Buchanan B. B. (1994) Severe sensitivity to ultraviolet radiation in an *Arabidopsis* mutant deficient in flavonoid accumulation. II. I. Mechanisms of UV-resistance in *Arabidopsis.* Planta 194:504–509.
Loveless M. D., Hamrick J. L. (1984) Ecological determinants of genetic structure in plant populations. Annu. Rev. Ecol. Syst. 15:65–95.
Loyola L. A., Morales G., Mancilla A., Borquez J., Pedreros S., Gallardo O., Trujillo R. (1995) Secondary metabolites of three species of genus *Parastrephia.* Bol. Soc. Chil. Quim. 40:293–296.
Loyola L. A., Naranjo S. J., Morales B. G. (1985) 5,7-Dihydroxy-3,8,3',4'-tetramethoxyflavone from *Parastrephia quadrangularis.* Phytochemistry 24:1871–1872.

Lukovnikova G. A. (1965) Some biologically active compounds of vegetables. Prikl. Biokhim. i Mikrobiol. 1:594–597.

Lunardello M. A., Tomaz J. C., Vichnewski W., Lopes J. L. C., Gutierrez A. B., Diaz J. G., Herz W. (1995) Sesquiterpene lactones and flavonoids from *Eremanthus matto-grossensis* and *E. eriopus*. J. Braz. Chem. Soc. 6:307–311.

Lutz F. E. (1924) Apparently non-selective characters and combinations of characters including a study of ultraviolet in relation to the flower-visiting habits of insects. Ann. New York Acad. Sci. 29:181–283.

Lynn D. G., Chang M. (1990) Phenolic signals in cohabitation: implications for plant development. Annu. Rev. Plant Physiol. Plant Mol. Biol. 41:497–526.

Lynn D. G., Steffens J. C., Kamut V. S., Graden D. W., Shabanowitz J., Riopel J. L. (1981) Isolation and characterization of the first host recognition substances for parasitic angiosperms. J. Amer. Chem. Soc. 103:1868–1870.

Ma J.-Y., Wang Z.-T., Xu L.-S., Xu G.-J. (1999) A sesquiterpene lactone glucoside from *Ixeris denticulata* f. *pinnatipartita*. Phytochemistry 50: 113–115.

Ma T., Li J., Yuan J. (1991) Isolation and identification of flavonoid glycosides from the leaf of small flower beggar ticks. Zhongcaoyao 22:531–533.

Mabberley D. J. (1987) *The Plant-book*. Cambridge: Cambridge Univ. Press. [ed. 2, 1997]

Mabry T. J. (1964) The betacyanins, a new class of red-violet pigments, and their phylogenetic significance. In: Leone C. A. (ed.) *Taxonomic Biochemistry and Serology*. New York: Ronald Press. 239–254.

Mabry T. J. (1966) The betacyanins and betaxanthins. In: Swain T. (ed.) *Comparative Phytochemistry*. New York: Academic Press. 231–244.

Mabry T. J., Abdel-Baset Z., Padolina W. G., Jones S. B., Jr. (1975) Systematic implications of flavonoids and sesquiterpene lactones in species of *Vernonia*. Biochem. Syst. Ecol. 2:185–192.

Mabry T. J., Bohlmann F. (1977) Summary of the chemistry of the Compositae. In: Heywood V. H., Harborne J. B., Turner B. L. (eds.) *The Biology and Chemistry of the Compositae*, Vol. 2. London: Academic Press. 1097–1118.

Mabry T. J., Markham K. R., Thomas M. B. (1970) *The Systematic Identification of Flavonoids*. New York: Springer.

Mabry T. J., Timmermann B. N., Heil N., Powell A. M. (1981) Systematic implications of the flavonoids and chromosomes of *Flyriella* (Compositae-Eupatorieae). Plant Syst. Evol. 137:275–280.

Mabry T. J., Timmermann B. N., Roberts M. F., Ulubelen A., Mues R. (1980) Cytotoxic 6-methoxyflavonoids from *Brickellia*. Planta Medica 39:220.

Macias F. A., Fischer N. H. (1992) Melampolides from *Lecocarpus pinnatifidus* Decne. Phytochemistry 31:2747–2754.

Macias F. A., Molinillo J. M. G., Fischer N. H. (1993) Melampolides and *cis, cis*-germacranolides from *Lecocarpus lecocarpoides*. Phytochemistry 32:127–131.

Macias F. A., Molinillo J. M. G., Torres A., Varela R. M., Castellano D. (1997) Bioactive flavonoids from *Helianthus annuus* cultivars. Phytochemistry 45:683–687.

MacLeish N. F. F. (1984) Eight new combiantions in *Vernonia* (Compositae: Vernonieae). Syst. Bot. 9:133–136.

MacLeish N. F. F. (1987) Revision of *Eremanthus* (Compositae: Venonieae). Ann. Missouri Bot. Gard. 74:265–290.

Macleod J. K., Rasmussen H. B. (1999) A hydroxy-ß-caryophyllene from *Pterocaulon serratulum*. Phytochemistry 50:105–108.

Maffei M. (1996) Chemotaxonomic significance of leaf wax alkanes in the Compositae. In: Hind D. J. N., Beentje H. J. (eds.) *Compositae: Systematics. Proceedings of the International Compositae Conference, Kew (1994), Vol. 1.* Kew: Royal Botanic Gardens. 141–158.

Maguire B. (1943) A monograph of the genus *Arnica*. Brittonia 4:386–510.

Mahmoud Z. F., Masem F. F., Abdel Salam N. A. (1986) Sesquiterpene lactones and flavonoids of *Centaurea ragusina* L. subsp. *ragusina* growing in Egypt. Egypt. J. Pharm. Sci. 27:283–289.

Mahran G. H., Ahmed M. S., Ansary S. M. (1975) A study of the flavonoid content of *Cotula cinerea* Del. Bull. Fac. Pharm. Cairo Univ. 14:237–239.

Makino T. (1989) *Revised Makino's New FLora of Japan* [revised by M. Ono, H. Oba, M. Nishida]. Tokyo: The Hokuryukan Co.

Maldonado E., Flores A. M., Ortega A. (1986) Constituents of *Helenium scorzoneraefolium*. J. Nat. Prod. 49:1152.

Maldonado E., Flores A. M., Ortega A. (1988a) A germacranolide from *Gochnatia hypoleuca* and its transformation into a melampolide. Phytochemistry 27:861–863.

Maldonado E., Hernandez E., Ortega A. (1992a) Amides, coumarins and other constituents from *Simsia cronquistii*. Phytochemistry 31:1413–1414.

Maldonado E., Toscano R. A., Mancera C., Teresa Tripp M., Ortega A. (1992b) Acylated flavonols and other constituents from *Galeana pratensis*. Phytochemistry 31:1003–1007.

Maldonado Z., Hoeneisen M., Silva M. (1988b) A dithiophene from *Aphyllocladus denticulatus*. Phytochemistry 27:2993–2994.

Maldonado Z., Hoeneisen M., Silva M. (1993) Constituents of *Haplopappus bezanillanus* and *H. hirtellus*. Bol. Soc. Chil. Quim. 38:43–48.

Manadilova A. M., Adiyatova Zh. F., Kunaeva R. M. (1987) Antimicrobial activity of phenolic compounds from *Artemisia sublessingiana* Kell, Krasch et Poljaj. Izv. Akad. Nauk. Kaz. SSR, Ser. Biol. (2) 26–29.

Manerikar S. V., Kulkarni A. B. (1978) Chemical investigation of *Tricholepis glaberrima* DC. Indian J. Chem. 16B:439–440.

Mañez S., Giner R. M., Recio M. C., Terencio M. C., Rios J. L. (1992) Phenolics of *Crepis* and their taxonomic implications. Planta Medica 58 (Suppl. 1): A698–699.

Mañez S., Recio M. C., Giner R. M., Sanz M. J., Terencio M. C., Peris J. B., Stübing G., Rios J.-L. (1994) A chemotaxonomic review of the subtribe Crepidinae based on its phenolic constituents. Biochem. Syst. Ecol. 22:297–305.

Mani M. S., Saravanan J. M. (1999) *Pollination Ecology and Evolution in Compositae* (Asteraceae). Enfield, New Hampshire: Science Publ.

Mansour R. M. A., Abdalla M. F., Saleh N. A. M. (1988) Quercetin-3,7-disulfate from *Centaurea alexandrina*. Bull. Chem. Soc. Ethiopia 2:45–46.

Mansour R. M. A., Ahmed A. A., Melek F. R., Saleh N. A. M. (1990) The flavonoids of *Pulicaria incisa*. Fitoterapia 61:186.

Mansour R. M. A., Ahmed A. A., Saleh N. A. M. (1983a) Flavone glycosides of some *Launaea* species. Phytochemistry 22:2630–2631.
Mansour R. M. A., Saleh N. A. M. (1981) Flavonoids of three local *Senecio* species. Phytochemistry 20:1180–1181
Mansour R. M. A., Saleh N. A. M., Boulos L. (1983b) A chemosystematic study of the phenolics of *Sonchus*. Phytochemistry 22:489–492.
Marambio O., Silva M. (1989) New compounds isolated from *Haplopappus taeda* Reiche. Bol. Soc. Chil. Quim. 34:105–106.
Marchant Y. Y., Ganders F. R., Wat C.-K., Towers G. H. N. (1984) Polyacetylenes in Hawaiian *Bidens*. Biochem. Syst. Ecol. 12:167–178.
Marchishyn S. M., Komissarenko N. F. (1981) Components of *Arnica montana* L. and *A. foliosa*. Khim. Prir. Soed. 662.
Marco J. A., Barbera O. (1990) Natural products from the genus *Artemisia* L. In: Atta-ur-Rahman (ed.) *Studies in Natural Products Chemistry*, Vol. 7. Amsterdam: Elsevier Science Publishers. 201–264.
Marco J. A., Barbera O., Rodriguez S., Domingo C., Adell J. (1988) Flavonoids and other phenolics from *Artemisia hispanica*. Phytochemistry 27:3155–3159.
Marco J. A., Barbera O., Sanchez-Parareda J. (1987) Flavonol and coumarin glycosides from *Artemisia incanescens*. J. Nat. Prod. 50:774–775.
Marco J. A., Sanz J. F., Bea J. F., Barbera O. (1990) Phenolic compounds from *Artemisia annua*. Die Pharmazie 45:382–383.
Marco J. A., Sanz-Cervera J. F., Yuste A., Oriola M. C. (1994) Sesquiterpene lactones and dihydroflavonols from *Andryala* and *Urospermum* species. Phytochemistry 36:725–729.
Marini-Bettolo G. B., Chiavarelli S., Casinovi C. G. (1957) Constitution of polycladin, the yellow pigment of *Lepidophyllum quadrangularis*. Gazz. Chim. Ital. 87:1185–1190.
Markham K. R. (1982) *Techniques of Flavonoid Identification*. New York: Academic Press.
Markham K. R. (1983) Revised structures for the flavones cirsitakaoside and cirsitakaogenin. Phytochemistry 22:316–317.
Markham K. R. (1989) A reassessment of the data supporting the structures of *Blumea malcolmii* flavonols. Phytochemistry 28:243–244.
Markham K. R., Webby R. F., Whitehouse L. A., Molloy B. P. J., Villain C., Mues R. (1985) Support from flavonoid glycoside distribution for the division of *Podocarpus* in New Zealand. New Zealand J. Bot. 23:1–13.
Marsh C. A. (1957) Glucuronide metabolism in plants. 2. The isolation of flavone glucosiduronic acids from plants. Biochem. J. 59:58–64.
Martens S., Forkmann G. (1996) Enzymology and genetics of flavonoid biosynthesis in *Gerbera* hybrids. Polyphenols Communications 96, Bordeaux (France), March 1996, 545–546.
Martens S., Forkmann G. (1998) Genetic control of flavone synthase II activity in flowers of *Gerbera* hybrids. Phytochemistry 49: 1953–1958
Marticorena C., Quezada M. (1985) Catálogo de la flora vascular de Chile. Gayana Bot. 42:1–157.

Martin C., Gerats T. (1993) The control of flower coloration. In: Jordan B. R. (ed.) *The Molecular Biology of Flowering*. Wallingford, U.K.: C.A.B. International. 219–255.
Martin S. S., Townsend C. E., Lenssen A. W. (1994) Induced isoflavonoids in diverse populations of *Astragalus cicer*. Biochem. Syst. Ecol. 22:657–661.
Martinez J., Silván A. M., Abad M. J., Bermejo P., Villar A. (1997) Isolation of two flavonoids from *Tanacetum microphyllum* as PMA-induced ear edema inhibitors. J. Nat. Prod. 60:142–144.
Martinez M., Barbera O., Sanchez-Parareda J., Marco J. A. (1987b) Phenolic and acetylenic metabolites from *Artemisia assoana*. Phytochemistry 26:2619–2624.
Martinez M., Naveda-Diaz E., Joseph-Nathan P. (1988) Derivados del timol de la *Calea zacatechichi* Schdl. Rev. Latinoamer. Quim. 19:56.
Martinez M., Sanchez A., Joseph-Nathan P. (1987a) Thymol derivatives from *Calea nelsonii*. Phytochemistry 26:2577–2579.
Martinez-Vasquez M., Vasquez Garcia H. M., Toscano R. A., Perez G. E. (1993) Methylated flavones from *Conoclinium greggii*. J. Nat. Prod. 56:1410–1413.
Martino V. S., Ferraro G. E., Coussio J. D. (1976) A new flavonoid from *Pluchea sagittalis*. Phytochemistry 15: 1086–1087.
Martino V., Caula S., Villar S., Ferrar G., Coussio J. (1988) Polyphenolic compounds isolated from *Achyrocline tomentosa* Rusby (Compositae). Rev. Latinoamer. Quim. 19:111–113.
Maruyama M., Hayasaka K., Sasaki S.-i., Hosokawa S., Uchiyama H. (1974) A new chalcone glucoside from *Gnaphalium multiceps*. Phytochemistry 13:286–288.
Masquelier J., Michaud J. (1956) Note on the polyphenolic constituents of *Cynara scolymus*. Bull. Soc. Pharm. Bordeaux 95:65–67.
Masterova I., Gramcaiova Z., Suchy V., Grancai D., Ubik K. (1993) Phenolic substances in flowers of *Anthemis tinctoria*. Fitoterapia 64:277.
Masterova I., Gramcaiova Z., Uhrinova S., Suchy V., Ubik K., Nagy M. (1991) Flavonoids in flowers of *Calendula officinalis* L. Chem. Pap. 45:105–108.
Mata R., Rojas A., Acevedo L., Estrada S., Calzada F., Rojas I., Bye R., Linares E. (1997) Smooth muscle relaxing flavonoids and terpenoids from *Conyza filaginoides*. Planta Medica 63:31–35.
Mathe I., Tyihak E. (1962) The spread of *Matricaria matricarioides* in Hungary and study of its active ingredients. Acta Pharm. Hung. 32:43–48.
Matsueda S., Nagaki M., Tanaka K., Shimoyama M. (1980) Chemotaxonomical study of *Artemisia* in the Far East area: habitat variation of natural coumarins. Sci.-Rep. Hirosaki Univ. 27:17–23.
Matsuo T., Kanamori H., Sakamoto I. (1986) Nonsweet glucosides in the leaves of *Stevia rebaudiana*. Hiroshima-ken Eisei Kenkyusho Kenkyu Hokoku No. 33:25–29.
Matthei J. O. (1995) *Manual de la Malezas que Crecen en Chile.* Santiago: Publ. by author.
Maxwell C. A., Hartwig U. A., Joseph C. M., Phillips D. A. (1989) A chalcone and two related flavonoids released from alfalfa roots induce *nod* genes of *Rhizobium meliloti*. Plant Physiol. 91:842–847.
Mayr E. (1995) Systems of ordering data. Biol. and Philosophy 10:419–434.
Mazur Y., Meisels A. (1955) The isolation of 5-hydroxy-3,6,7,3',4'-pentamethoxyflavone from *Artemisia arborescens*. Bull. Res. Council Israel 5A:67–69.

Mazza G., Gao L. (1994) Malonylated anthocyanins in purple sunflower seeds. Phytochemistry 35:237–239.

McArthur E. D., Plummer A. P. (1978) Biogeography and management of native western shrubs: a case study, section *Tridentatae* of *Artemisia*. Great Basin Nat. Mem. 2:229–243.

McArthur E. D., Sanderson S. C. (1999) Cytogeography and chromosome evolution of subgenus *Tridentatae* of *Artemisia* (Asteraceae). Amer. J. Bot. 86:1754–1775.

McClure J. W. (1975) Physiology and functions of flavonoids. In: Harborne J. B., Mabry T. J., Mabry H. (eds.) *The Flavonoids*. London: Chapman and Hall. 970–1055.

McClure J. W. (1986) Physiology of flavonoids in plants. In: Cody V., Middleton E., Harborne J. B. (eds.) *Plant Flavonoids in Biology and Medicine, Progress in Clinical and Biological Research*. New York: Alan R. Liss.

McCormick S. P., Bohm B. A. (1986) Flavonoids of *Artemisia lindleyana*. J. Nat. Prod. 49:167.

McCormick S. P., Bohm B. A., Ganders F. R. (1984) Methylated chalcones from *Bidens torta*. Phytochemistry 23:2400–2401.

McCormick S. P., Robson K., Bohm B. A. (1985a) Flavonoids from *Wyethia glabra*. Phytochemistry 24:1614–1616.

McCormick S. P., Robson K., Bohm B. A. (1985b) Methylated flavonols from *Wyethia bolanderi* and *Balsamorhiza macrophylla*. Phytochemistry 24:2133.

McCormick S. P., Robson K., Bohm B. A. (1986) Flavonoids of *Wyethia angustifolia* and *Wyethia helenioides*. Phytochemistry 25:1723–1726.

McCormick S. P., Robson K., Maze J. R., Bohm B. A. (1987) Flavonoids of *Wyethia* sect. *Agnorhiza*. Phytochemistry 26:2421–2423.

McGowan S. G., Wallace J. W. (1972) Flavonoids and phenolic acids from *Cirsium lanceolatum*. Phytochemistry 11:1503.

McHale J., Alston R. E. (1964) Utilization of chemical patterns in the analysis of hybridization between *Baptisia leucantha* and *Baptisia sphaerocarpa*. Evolution 18:304–311.

McVaugh R. (1984) Compositae, *Flora Nueva Galiciana: A Descriptivve Account of the Vascular Plants of Western Mexico*, Vol. 12. Ann Arbor: Univ. Michigan Press.

Mears J. A. (1973) Flavonoids of *Parthenium* sect. *Bolophytum*. Phytochemistry 12:2265–2268.

Mears J. A. (1979) Chemistry of polyploids: a summary with comments on *Parthenium* (Asteraceae-Ambrosiinae). In: Lewis W. H. (ed.) *Polyploidy, Biological Relevance*. New York: Plenum Press. 77–101.

Mears J. A. (1980a) The flavonoids of *Parthenium* L. J. Nat. Prod. 43:708–716.

Mears J. A. (1980b) Flavonoid diversity and geographic endemism in *Parthenium*. Biochem. Syst. Ecol. 8:361–370.

Medjroubi K., Benayache F., Benayache S., Akkal S., Kaabeche M., Tillequin F., Seguin E. (1998) Eudesmanolide from *Centaurea granata*. Phytochemistry 49: 2425–2427.

Meikle R. D. (1979) *Rhagadiolus*: a misunderstood genus. Taxon 28:133–141.

Melchert T. E. (1966) Chemo-demes of diploid and tetraploid *Thelesperma simplicifolium* (Heliantheae, Coreopsidinae). Amer. J. Bot. 53:1015–1020.

Melek F. R., Ahmed A. A., Mabry T. J. (1985) 6-Methoxylated flavones from *Helianthus hirsutus* and *H. gracilentus*. Rev. Latinoamer. Quim. 16:27.

Melek F. R., El-Ansari M., Hassan A., Regaila A., Ahmed A. A., Mabry T. J. (1988) Methoxylated flavonoid aglycones from *Pulicaria arabica*. Rev. Latinoamer. Quim. 19:119–120.

Melo C. L., Maia S. B., Souza M. F., Rao V. N., Lima M. A. S., Silveira E. R. (1993) CNS actions of ternatin, a flavonoid from *Egletes viscosa*. Fitoterapia 64:306–309.

Mendiondo M. E., De Israilev L. R. A., Seeligmann P. (1990a) *Schkuhria pinnata* (Lam.) O.K. (Compositae). A study of the floral flavonoids. Lilloa 37: 81–82.

Mendiondo M. E., Juárez B. E., Seeligmann P. (1990b) The floral flavonoids of *Porophyllum ruderale* (Jacq.) Cassini and *P. obscurum* (Spreng.) DC. (Compositae). Lilloa 37:79–80.

Mendiondo M. E., Juárez B. E., Seeligmann P. (1997) Flavonoid patterns of some Barnadesioideae (Asteraceae). Eventual chemosystematic significance. Biochem. Syst. Ecol. 25:673–674.

Mendiondo M. E., Seeligmann P. (1985) Two flavonoids in the leaves of *Stevia mercedensis*. Lilloa 36:177.

Menichini F., Appendino G. (1987) Constituents of *Artemisia variabilis* L. Essenze Deriv. Agrum. 57:44–50.

Menichini F., Statti G., Delle Monache F. (1994) Flavonoid glycosides from *Scorzonera columnae*. Fitoterapia 65:555–556.

Merfort I. (1984) Methylierte Flavonoide aus *Arnica montana* und *Arnica chamissonis*. Planta Medica 50:107–108.

Merfort I. (1985) Flavonoide aus *Arnica montana* und *A. chamissonis*. Planta Medica 51:136–138.

Merfort I. (1986) Flavonol glycosides of Arnicae Flos DAB9. Planta Medica 52:427.

Merfort I. (1988a) Acylated flavonoid glycosides and dicaffeoylquinic acid from Arnicae Flos DAB 9. Planta Medica 54:571.

Merfort I. (1988b) Acetylated and other flavonoid glycosides from *Arnica chamissonis*. Phytochemistry 27:3281–3284.

Merfort I. (1994) Flavonoid glycosides in flowers of *Arnica acaulis*, subgenus *montana*. Biochem. Syst. Ecol. 22:229.

Merfort I., Marcinek C., Eggert A. (1986) Flavonoid distribution in *Arnica* subgenus *Chamissonis*. Phytochemistry 25:2901–2903.

Merfort I., Passreiter C. M., Willuhn G. (1995) Flavonoid glycosides from *Arnica amplexicaulis* and *A. mollis*. Biochem. Syst. Ecol. 23:681–682.

Merfort I., Wendisch D. (1987) Flavonoid glycoside aus *Arnica montana* und *Arnica chamissonis*. Planta Medica 53:434–437.

Merfort I., Wendisch D. (1988) Flavonolglucuronide aus die Blüten von *Arnica montana*. Planta Medica 54:247–250.

Merfort I., Wendisch D. (1992) New flavonoid glycosides from Arnicae Flos DAB9. Planta Medica 58:355–357.

Mericli A. H. (1980) Flavonoids from *Gnaphalium luteo-album* L. Istanbul Univ. Eczacilik Fak. Mecm. 16:84–87.

Mericli A. H. (1983) Constituents of *Antennaria dioica*. J. Nat. Prod. 46:941.

Mericli A. H. (1988) Flavonolignans, kaempferol 3-sulfate and other flavonoids from *Silybum marianum* subsp. *anatolicum*. Planta Medica 54:44–45.

Mericli A. H., Çubukçu B., Dortunc T. (1984) Flavonoids and anthocyanins of *Helichrysum sanguineum*. Fitoterapia 55:112.

Mericli A. H., Dellamonica G. (1983) Flavonoids of *Notobasis syriaca* (L.) Cass. Fitoterapia 54:35.

Mericli A. H., Ergezen K., Çubukçu B. (1992) Constituents of *Helichrysum stoechas* ssp. *barrelieri*. Fitoterapia 63:475.

Mericli A. H., Korkmaz Y. (1992) Flavonoids of *Matricaria chamomilla* var. *recutita*. Acta Pharm. Turc. 34:71–73.

Mericli A. H., Miski M., Korkmaz Y., Tuzlaci E. (1991) Phenolic compounds of *Onopordum tauricum*. Acta Pharm. Turc. 33:83–84.

Mericli F., Mericli A. H., Jakupovic J., Bohlmann F., Sanchez H., Dominguez X. A. (1989) Labdane derivatives from *Brickellia schaffneri*. Rev. Latinoamer. Quim. 20:15–16.

Merxmüller H., Leins P., Roessler H. (1977) Inuleae – systematic review. In: Heywood V. H., Harborne J. B., Turner B. L. (eds.) *The Biology and Chemistry of the Compositae*, Vol. 1. London: Academic Press. 577–602.

Meselhy M. R., Kadota S., Hattori M., Namba T. (1993a) Metabolism of safflor yellow by human intestinal bacteria. J. Nat. Prod. 56:39–45.

Meselhy M. R., Kadota S., Momose Y., Hatekeyama N., Kusai A., Hattori M., Namba T. (1993b) Two new quinochalcone yellow pigments from *Carthamus tinctorius* and Ca^{++} antagonistic activity of tinctormine. Chem. Pharm. Bull. 41:1796–1802.

Mesfin T. (1984a) *Microlecane* (Sch. Bip.) Bentham, a congener of *Bidens* (Compositae—Heliantheae) Nord. J. Bot. 4:737–746.

Mesfin T. (1984b) The genus *Bidens* (Compositae) in NE tropical Africa. Acta Univ. Ups. Symb. Bot. 24:1–138.

Mesfin T. (1986) The morphological basis for inclusion of African species of *Coreopsis* L. in *Bidens* L. (Compositae-Heliantheae). Acta Univ. Ups. Symb. Bot. 26:189–203.

Mesfin T., Crawford D. J., Smith E. B. (1995) Comparative capitular morphology and anatomy of *Coreopsis* L. and *Bidens* L. (Compositae), including a review of generic boundaries. Brittonia 47:61–91.

Mesquita A. A. L., de B. Correa D., de Padua A. P., Guedes M. L. O., Gottlieb O. R. (1986) Flavonoids from four Compositae species. Phytochemistry 25:1255–1256.

Metwally A. M., Ekejiuba E. C. (1981) Methoxylated flavonols and flavanones from *Eupatorium odoratum*. Planta Medica 42:403–405.

Metwally A. M., Saleh M. R. I., Amer M. M. A. (1973) Isolation of two flavonoidal principles from *Urospermum picroides*. Planta Medica 23:94–98.

Metwally M. A. (1995) Chemical constituents of *Baccharis calvescens* DC. Boll. Chim. Farm. 134:281–282.

Metwally M., Dawidar M. A.-A., Metwally S. (1986) A new thymol derivative from *Pulicaria undulata*. Chem. Pharm. Bull. 34:378–379.

Meuer B., Mabry T. J. (1987) 6-Methoxylated flavones from *Carphochaete bigelovii*. J. Nat. Prod. 50:775–776.

Meusel H., Kästner A. (1990) Lebensgeschichte der Gold- und Silberdisteln. Monographie der mediterran-mitteleuropäischen Compositen-Gattung *Carlina*. Band 1. Merkmalsspektrum und Lebensräume der Gattung. Wien: Springer.

Miao B., Turner B., Mabry T. (1995a) Chloroplast DNA variations in sect. *Cyclachaena* of *Iva* (Asteraceae). Amer. J. Bot. 82:919–923.

Miao B., Turner B., Mabry T. (1995b) Molecular phylogeny of *Iva* (Asteraceae, Heliantheae) based on chloroplast DNA restriction site variation. Plant Syst. Evol. 195:1–12.

Miao B., Turner B., Simpson B., Mabry T. (1995c) Chloroplast DNA study of the genera *Ambrosia* s.l. and *Hymenoclea* (Asteraceae): systematic implications. Plant Syst. Evol. 194:241–255.

Michaels H. J., Scott K. M., Olmstead R. G., Szaro T., Jansen R. K., Palmer J. D. (1993) Interfamilial relationships of the Asteraceae: insights from *rbc*L sequence variation. Ann. Missouri Bot. Gard. 80:742–751.

Michaluk A. (1962) Identification of flavones in the baskets of *Achillea millefolium*. Diss. Pharm. Pharmakol. 14:347–353.

Michaluk A., Sendra J. S. (1964) Identification of flavones of *Tanacetum vulgare*. Diss. Pharm. Pharmakol. 16:183–187.

Middleton J., Jefferies P. R. (1968) Substances derived from *Goodenia strophiolata* F. Muell. Aust. J. Chem. 21:2349–2351.

Miles D. H. (1993) Potential agrochemicals from the leaves of *Wedelia biflora*. Phytochemistry 32:1427–1429.

Miller J. S. (1996) Collecting methodologies for plant samples for pharmaceutical research. In: Stuessy T. F., Sohmer S. H. (eds.) *Sampling the Green World: Innovative Concepts of Collection, Preservation, and Storage of Plant Diversity.* New York: Columbia Univ. Press. 74–87.

Milosavljevic S., Aljancic I., Macura S., Milinkovic D., Stefanovic M. (1991) Sesquiterpene lactones from *Achillea crithmifolia*. Phytochemistry 30:3464–3466.

Milosavljevic S., Macura S., Stefanovic M., Aljancic I., Milinkovic D. (1994) Sesquiterpene lactones from *Achillea crithmifolia*. J. Nat. Prod. 57:64–67.

Milovanovic M., Djermanovic M. (1994) Constituents of *Helminthia echioides*. Fitoterapia 65:377–378.

Mionskowski H., Gill S. (1973) Flavonoids from *Ageratum mexicanum*. Acta Pol. Pharm. 30:105–107.

Miserez F., Potterat O., Marston A., Mungai G. M., Hostettmann K. (1996) Flavonol glycosides from *Vernonia galamensis* ssp. *nairobiensis*. Phytochemistry 43:283–286.

Mishima H., Kurabayashi M., Hirai K. (1971) Total synthesis of silymarin and silybin. Ann. Sankyo Res. Lab. 23:70–75.

Mishler B. D., Albert V. A., Chase M. W., Karis P. O., Bremer K. (1996) Character-state weighting for DNA restriction site data: asymmetry, ancestors and the Asteraceae. Cladistics 12:11–19.

Miski M., de Luengo D. H., Mabry T. J. (1987) Guaiane sesquiterpenes from *Decachaeta scabrella*. Phytochemistry 26:199–200.

Miski M., Gage D. A., Mabry T. J. (1985) Quercetagetin 6,7,4'-trimethyl ether and 3-sulphate from *Decachaeta haenkeana*. Phytochemistry 24:3078–3080.

Misra L. N. (1986) Arteannuin-C, a sesquiterpene lactone from *Artemisia annua*. Phytochemistry 25:2892–2893.

Miyaichi Y., Matsuura M., Tomimori T. (1995) Phenolic compounds from the roots of *Cirsium japonicum* DC. Nat. Med. 49:92–94.

Miyakado M., Kato T., Ohno N., Mabry T. J. (1976) Pinocembrin and (+)–ß-eudesmol from *Hymenoclea monogyra* and *Baccharis glutinosa*. Phytochemistry 15:846.

Mladenova K., Tsankova E., Van Hung D., Budesinsky M. (1988) Constituents of *Chrysanthemum indicum* var. *tuneful*. Dokl. Bolg. Akad. Nauk. 41:81–90.

Mo Y., Nagel C., Taylor L. P. (1992) Biochemical complementation of chalcone synthase mutants defines a role for flavonols in functional pollen. Proc. Natl. Acad. Sci. U.S.A. 89:7213–7217.

Mogford D. J. (1974a) Flower colour polymorphism in *Cirsium palustre*. 1. Heredity 33:241–256.

Mogford D. J. (1974b) Flower colour polymorphism in *Cirsium palustre*. 2. Pollination. Heredity 33:257–263.

Molloy B. P. J. (1995) Two new species of *Leucogenes* (Inuleae: Asteraceae) from New Zealand and typification of *L. grandiceps*. New Zealand J. Bot. 33:53–63.

Monea M. (1986) Phytochemical study of *Grossheimia macrocephala*. Farmacia (Bucharest) 34:187–190.

Moore D. M. (1983) *Flora of Tierra del Fuego*. London: Anthony Nelson.

Moore M. O. (1990) A systematic study of eastern North American *Vitis* L. (Vitaceae) north of Mexico. Ph. D. Dissertation, Univ. Georgia, Athens.

Moore M. O., Giannasi D. E. (1987) Foliar flavonoids of selected *Vitis* taxa in the Southeastern United States. Biochem. Syst. Ecol. 15:79–83.

Moore R. J. (ed.) (1970–1977) Index to plant chromosome numbers *(1968–1974)*. Regnum Veg. 68:1–119; 77:1–116; 84:1–138; 91:1–108; 96:1–157.

Morales G., Mancilla A., Gallardo O., Trujillo R., Loyola L., Breton J. L. (1990) Diterpenoids and flavonoids from *Baccharis boliviensis*. Bol. Soc. Chil. Quim. 35:257–263.

Morales-Mendez A., Nunez O., Rosquete C., Montilla A. (1984) Flavones from *Baccharis decussata* (Klatt) Hieron. An. Quim., Ser. C. 80:98–99.

Morales-Mendez A., Rosquete C. (1988) Flavonoids from *Ageratina stevioides* Steyermark. Rev. Latinoamer. Quim. 19:47–49.

Morazzoni P., Bombardelli E. (1995) *Silybum marianum* (*Carduus marianus*). Fitoterapia 66:3-42.

Moreno B., Delle Monache G., Della Monache F., Marini-Bettolo G. B. (1979) Flavones and kauranoid diterpenes from *Eupatorium tinifolium* H.B.K. Il Farmaco, Ed. Sci. 35:457–464.

Morgan D. R. (1997) Reticulate evolution in *Machaeranthera* (Asteraceae). Syst. Bot. 22:599–615.

Morgan D. R., Simpson B. B. (1992) A systematic study of *Machaeranthera* (Asteraceae) and related groups using restriction site analysis of chloroplast DNA. Syst. Bot. 17:511–531.

Morgan D. R., Soltis D. E. (1993) Phylogenetic relationships among members of Saxifragaceae *sensu lato* based on *rbc*L sequence data. Ann. Missouri Bot. Gard. 80:631–660.

Morgenstern T., King R. M., Jakupovic J. (1996) Sesquiterpene lactones from *Bejaranoa balansae* species. Phytochemistry 41:1543–1546.

Morita N., Shimizu M., Arisawa M. (1973) Two new flavone glycosides from *Cirsium lineare*. Phytochemistry 12:421–423.

Morrone J. J., Crisci J. V. (1995) Historical biogeography: introduction to methods. Annu. Rev. Ecol. Syst. 26:373–401.

Mossa J. S., El-Domiaty M. M., Al-Meshal I. A., El-Feraly F. S., Hufford C. D., McPhail D. R., McPhail A. T. (1992) A flavone and diterpene from *Psiadia arabica*. Phytochemistry 31:2863–2868.

Mossa J. S., Hifnawy M. S., Al-Yahya M. A., Hafez M. M., Shehata A. A., El-Feraly F. S. (1988) Flavonoids and coumarins from three Saudi Arabian Compositae species. Int. J. Crude Drug Res. 26:181–184.

Mues R., Timmermann B. N., Ohno N., Mabry T. J. (1979) 6-Methoxyflavonoids from *Brickellia californica*. Phytochemistry 18:1379–1383.

Mungarulire J. (1990) A phytochemical investigation of medicinal plants of the Compositae from Rwanda. Herba Hung. 29:73–79.

Murav'eva D. A., Batuk V. S., Chernova G. P. (1974) Flavonoids of the herb *Senecio platyphylloides*. Chem. Nat. Cmpds. 10:808 [786].

Murti V. V. S., Raman P. V., Seshadri T. R., Thakur R. S. (1962) Components of the ivory-white flowers of *Carthamus tinctorius*. J. Sci. Indust. Res. 21B:80–83.

Musaev K. L., Belenovskaya L. M. (1989) Phenols of *Artemisia gorjaevii* and *A. saissanica*. Khim. Prir. Soedin. (3): 430.

Muschietti L., Martino V., Ferraro G., Coussio J. (1990) Polyphenolic compounds isolated from *Eupatorium buniifolium*. An. Asoc. Quim. Argent. 78:329–331.

Muschietti L., Martino V., Ferraro G., Coussio J. (1993) 2'-Oxygenated flavonoids from *Eupatorium buniifolium*. Planta Medica 59 (Suppl. 1): A606.

Muschietti L., Martino V., Ferraro G., Coussio J. (1994) 5,7,5'-Trihydroxy-3,6,2',4'-tetramethoxyflavone from *Eupatorium buniifolium*. Phytochemistry 36:1085–1086.

Nair A. G. R., Gunasegaran R., Joshi B. S. (1982) Chemical investigation of certain South Indian plants. Indian J. Chem. 21B:979–980.

Nair A. G. R., Gunasegaran R., Krishnan S., Bayet C., Voirin B. (1995) Flavonol glycosides from leaves of *Eupatorium glandulosum*. Phytochemistry 40:283–285.

Nair A. G. R., Jayaprakasam R., Gunasekaran R., Bayet C., Voirin B. (1993) 6-Hydroxykaempferol 7-(6''-caffeoylglucoside) from *Eupatorium glandulosum*. Phytochemistry 33:1275–1276.

Nair A. G. R., Jayaprakasam R., Madhusudanan K. P. (1989) Flavonoids from *Helichrysum buddleioides* DC. J. Indian Chem. Soc. 66:493.

Nair A. G. R., Rao S. A., Voirin B., Favre-Bonvin J. (1985) Polyphenolic compounds from the leaves of *Acanthospermum hispidum*. Fitoterapia 56:249.

Nair A. G. R., Sivakumar R. (1990) 6-Hydroxykaempferol 6,4'-dimethyl ether 3-galactoside from *Eupatorium glandulosum*. Phytochemistry 29:1011–1012.

Nakaoki T. (1936) Constituents of *Cosmos bipinnatus* Cav. IV. Glucoside of the flavone of the white flower. J. Pharm. Soc. Japan 55:967–978.

Nakaoki T. (1938) Constituents of *Dahlia variabilis*. J. Pharm. Soc. Japan 58:639–647.

Nakaoki T. (1941) Constituents of *Zinnia elegans*. Chem. Zentral. II, 892.

Nakaoki T., Morita N. (1956) Kaempferitin from *Tagetes erecta.* J. Pharm. Soc. Japan 76:347.

Nakaoki T., Morita N. (1958) Constituents of *Eupatorium stoechadosmum.* J. Pharm. Soc. Japan 78:557.

Nakaoki T., Morita N. (1959) Medicinal resources. XIII. Flavonoids of *Cirsium microspicatum, C. otayae, C. yoshizawae, C. japonicum* and *C. purpuratum.* J. Pharm. Soc. Japan 79:1338–1340.

Nakaoki T., Morita N. (1960a) Medicinal resources. XIV. Flavonoids of *Cirsium.* 2. Components of the leaves of *C. kagamontanum, C. indundatum* and *C. matsumurae.* J. Pharm. Soc. Japan 80:1296–1299.

Nakaoki T., Morita N. (1960b) Medicinal resources. XVI. Flavonoids of the leaves of *Castanea pubinervis, Hydrocotyle wilfordi, Sanguisorba hakusanensis, Euptelaea polyandra, Carthamus tinctorius, Lactuca repens, Daucus carota, Ilex integra, Smilax china* and *Smilax medica* J. Pharm. Soc. Japan 80:1473–1478.

Nakaoki T., Morita N., Nagata Y., Oguri H. (1961) Medicinal resources. XIX. Flavonoids of the leaves of *Nelumbo nucifera, Cosmos bipinnatus* and *Foeniculum vulgare.* J. Pharm. Soc. Japan 81:1158–1159.

Nakashima C., Vichnewski W., Diaz J. G., Herz W. (1994) Two flavones from *Graziela mollissima.* Phytochemistry 37:285–286.

Nakayama M., Horie T., Lin C. N., Arisawa M., Shimizu M., Morita N. (1983) The synthesis of 5,7-dihydroxy-4',8- dimethoxyflavone: revised structures for cirsitaogenin and cirsitakaoside. J. Pharm. Soc. Japan 103:675–678.

Nakayama M., Koshioka M., Shibata M., Hiradate S., Sugre H., Yamaguchi M.-a. (1997) Identification of cyanidin 3-*O*-(3",6"-*O*-dimalonyl-ß-glucopyranoside) as a flower pigment of chrysanthemum (*Dendranthema grandiflorum*). Biosci., Biotechnol., Biochem. 61:1607–1608.

Namba T., Hattori M., Takehana Y., Tsunezuka M., Tomimori T., Kizu H., Miyaichi Y. (1983) A flavone from *Artemisia capillaris.* Phytochemistry 22:1057–1058.

Nanayakkara N. P. D., Hussain R. A., Pezzuto J. M., Soejarto D. D., Kinghorn A. D. (1988) Potential sweetening agents of plant origin. 13. An intensely sweet dihydroflavonol derivative based on a natural product lead compound. J. Med. Chem. 31:1250–1253.

Narain N. K. (1977) Flavonoids from *Vernonia fasciculata.* Isolation of genkwanin and a new flavone disaccharide fascicularin. J. Chem. Soc., Perkin Trans. I, 1018–1020.

Narayanan P. N., Zechmeister K., Röhrl M., Hoppe W. (1970) The crystal structural analysis of obtusifolin, a flavanone. Tetr. Lett. 3643–3644.

Nedel'ko E. S., Nikonov G. K. (1987) Methyl ethers of quercetin from *Artemisia heptapotamica.* Chem. Nat. Cmpds. 23:254 [300–301].

Negrete R. E., Backhouse N., Bravo B., Erago S., Garcia R., Avendao S. (1987) Quelques flavonoids de *Centaurea floccosa* Hook. et Arn. Pl. Medica Phytother. 21:168–172.

Negrete R. E., Backhouse N., Prieto P., Mejius H., Camargo R. C., Cassels B. K., Breitmaier E., Hartmann R. (1989) Constituents of *Centaurea melitensis.* Pl. Medica Phytother. 23:293–304.

Negrete R. E., Latorre I., Backhouse N., Peña R., Delporte C. (1988) Anatomical and phytochemical studies: flavonoids and a lactone from *Centaurea chilensis*. Pl. Medica Phytother. 22:1–5.

Nelson A. (1899) Some species of *Tetraneuris* and allies. Bot. Gaz. 28:126–130.

Nelson G., Platnick N. (1981) *Systematics and Biogeography: Cladistics and Vicariance*. New York: Columbia Univ. Press.

Nelson G., Rosen D. (eds.) (1981) *Vicariance Biogeography*. New York: Columbia Univ. Press.

Neshta I. D., Kaloshina N. A. (1971) Flavonoids of *Achillea cartilaginea*. Chem. Nat. Cmpds. 7:643 [664].

Neshta I. D., Kaloshina N. A., Zapesochnaya G. G., Ban'kovskii A. I. (1972) Rutin from *Achillea millefolium*. Chem. Nat. Cmpds. 8:664 [676]

Neshta I. D., Zapesochnaya G. G., Kaloshina N. A., Ban'kovskii A. I. (1973) Flavonoids of *Achillea cartilaginea*. II. Chem. Nat. Cmpds. 9:760 [796].

Nesom G. L. (1988) *Baccharis* sect. *Baccharidastrum* (Compositae: Astereae), including two monoecious and one dioecious species. Phytologia 65:169–173.

Nesom G. L. (1989a) The separation of *Trimorpha* (Compositae: Astereae) from *Erigeron*. Phytologia 67:61–66.

Nesom G. L. (1989b) Infrageneric taxonomy of New World *Erigeron* (Compositae: Astereae). Phytologia 67:67–93.

Nesom G. L. (1989c) Review of "*The Genera of the Eupatorieae (Asteraceae)*" by Robert Merrill King and Harold Robinson. Syst. Bot. 14:263–266.

Nesom G. L. (1990a) Further definition of *Conyza* (Asteraceae: Astereae). Phytologia 68:229–233.

Nesom G. L. (1990b) Studies in the systematics of Mexican and Texas *Grindelia* (Asteraceae: Astereae). Phytologia 68:303–332.

Nesom G. L. (1990c) Taxonomy of *Achyrocline* (Asteraceae: Inuleae) in México and Central America. Phytologia 68:181–185.

Nesom G. L. (1991) Taxonomy of *Isocoma* (Compositae: Astereae). Phytologia 70:69–114.

Nesom G. L. (1993) Synopsis of *Parastrephia* (Asteraceae: Astereae). Phytologia 75:347–357.

Nesom G. L. (1994a) Subtribal classification of the Astereae (Asteraceae). Phytologia 76:193–274.

Nesom G. L. (1994b) Review of the taxonomy of *Aster sensu lato* (Asteraceae: Astereae), emphasizing the New World species. Phytologia 77:141–297.

Nesom G. L. (1997) Synopsis of *Stephanodoria* (Asteraceae: Astereae). Phytologia 82:107–113.

Nesom G. L., Baird G. I. (1993) Completion of *Ericameria* (Asteraceae: Astereae), diminution of *Chrysothamnus*. Phytologia 75:74–93.

Nesom G. L., Morgan D. R. (1990) Reinstatement of *Tonestus* (Asteraceae: Astereae). Phytologia 68:174–180.

Nesom G. L., Suh Y., Morgan D. R., Simpson B. B. (1990) *Xylothamia* (Asteraceae: Astereae), a new genus related to *Euthamia*. Sida 14:101–116.

Nguyen T. D. T., Le V. N. P., Nguyen X. D. (1993) Chalcones of *Eupatorium odoratum* L. from Vietnam. Tap Chi Hoa Hoc 31:79–80.

Nhia N. T. (1975) Pharmacognostic study of Bulgarian *Senecio nemorensis* varieties. III. Farmatsiya (Sofia). 25:39–40.

Nicholls K. W., Bohm B. A. (1979) Flavonoids and affinities of *Coreopsis bigelovii*. Phytochemistry 18:1076.

Nicholls K. W., Bohm B. A. (1982) Quantitative flavonoid variation in *Lupinus sericeus*. Biochem. Syst. Ecol. 10:225–231.

Nicholls K. W., Bohm B. A. (1987) Flavonoids of *Lupinus* sects. *Platycarpos* and *Lupinellus* (Fabaceae). Syst. Bot. 12:320–323.

Nicholls K. W., Bohm B. A., Wells E. F. (1986) The flavonoids of *Mitella*. Canad. J. Bot. 64:525–530.

Nieto M., Gianello J. C., Tonn C. E. (1994) Chemical constituents of three *Parastrephia* species. An. Asoc. Quim. Argent. 82:105–108.

Niklas J. J., Giannasi D. E. (1977a) Flavonoids and other chemical constituents of fossil Miocene *Zelkova* (Ulmaceae). Science 196:877–878.

Niklas J. J., Giannasi D. E. (1977b) Geochemistry and thermolysis of flavonoids. Science 197:767–769

Nikonova L. P., Nikonov G. K. (1975) 4',5,6-Trihydroxy-3,7-dimethoxyflavone from *Inula grandis*. Chem. Nat. Cmpds. 11:104 [96].

Nkunya M. H. H., Weene H., Kinabo L. S. (1992) Constituents of *Artemisia afra* Willd. Fitoterapia 63:279–280.

Nogradi M., Farkas L. (1968) Structure determination, by means of nuclear magnetic resonance and by synthesis, of some flavones isolated from various species of *Iva*. Kem. Kozlem. 29:25–34.

Norbaek R., Holm M. L., Brandt K. (1996) Effects of anthocyanin composition on flower colors in *Alstroemeria* and *Aster*. Polyphenols Communications 96, Bordeaux (France), July, 1996, 45–46.

Nordenstam B. (1977) Senecioneae and Liabeae—systematic review. In: Heywood V. H., Harborne J. B., Turner B. L. (eds.) *The Biology and Chemistry of the Compositae*, Vol. 2. London: Academic Press. 799–830.

Nordenstam B. (1978) Taxonomic studies in the tribe Senecioneae (Compositae). Opera Bot. 44:1–84.

Nordenstam B. (1985) *Acrisione* (Compositae-Senecioneae), a new cacalioid genus from Chile. Bot. Jahrb. Syst. 107:581–589.

Nordenstam B. (1994) Calenduleae. In: Bremer K. (ed.) *Asteraceae: Cladistics and Classification*. Portland: Timber Press. 365–376.

Nordenstam B. (1996) Recent revision of Senecioneae and Calenduleae systematics. In: Hind D. J. N., Beentje H. J. (eds.) *Compositae: Systematics. Proceedings of the International Compositae Conference, Kew (1994), Vol. 1*. Kew: Royal Botanic Gardens. 591–596.

Norbedo C., Ferraro G., Coussio J. D. (1982) A new flavanone from *Achyrocline flaccida*. J. Nat. Prod. 45:635–636.

Norbedo C., Ferraro G., Coussio J. D. (1984) Flavonoids from *Achyrocline flaccida*. Phytochemistry 23:2698–2700.

Nordström C. G., Swain T. (1953) The flavonoid glycosides of *Dahlia variabilis*. Part I. General introduction. Cyanidin, apigenin and luteolin glycosides from the variety "Dandy". J. Chem. Soc. 2764–2773.

Nordström C. G., Swain T. (1956) The flavonoid glycosides of *Dahlia variabilis*. Part II. Glycoside of yellow varieties "Pius IX" and "Coton". Arch. Biochem. Biophys. 60:329–344.

Norlindh T. (1943) Studies in the Calenduleae. I. Monograph of the genera *Dimorphotheca*, *Castalis*, *Osteospermum*, *Gibbaria* and *Chrysanthemoides*. Lund: C. W. K. Gleerup.

Norlindh T. (1977) Calenduleae—systematic review. In: Heywood V. H., Harborne J. B., Turner B. L. (eds.) *The Biology and Chemistry of the Compositae*, Vol. 2. London: Academic Press. 961–988.

Norris J. A., Mabry T. J. (1985) 6-Methoxyflavonoids of *Brickellia monocephala*. J. Nat. Prod. 48:668.

Northington D. K. (1974) Chemosystematic studies of the genus *Pyrrhopappus* (Cichorieae). Spec. Publ. Mus. Texas Tech. Univ. 6:1–38.

Novak S. J., Soltis D. E., Soltis P. S. (1991) Ownbey's Tragopogons: 40 years later. Amer. J. Bot. 78:1586–1600.

Noyes R. D., Rieseberg L. H. (1999) ITS sequence data support a single origin for North American Astereae (Asteraceae) and reflect deep geographic divisions in *Aster* s.l. Amer. J. Bot. 86:398–412.

Nuñez-Alarcón J., Dolz H., Quiñones M. H., Carmona M. T. (1993) Epicuticular flavonoids from *Haplopappus baylahuen* and the hepatoprotective effect of isolated 7-methyl aromadendrin. Bol. Soc. Chil. Quim. 38:15–22.

Nuñez-Alarcón J., Quiñones M. (1995) Flavonoids and coumarins of *Haplopappus multifolius*. Biochem. Syst. Ecol. 23:453–454.

Nuttall T. (1841) Descriptions of new species and genera of plants in the natural order of the Compositae, collected in a tour across the continent to the Pacific, a residence in Oregon, and a visit to the Sandwich Islands and Upper California, during the years 1834 and 1835. Trans. Amer. Philosoph. Soc., ser. 2, 7:283–453.

Obara H., Onodera J.-i. (1979) Structure of carthamin. Chem. Lett. 201–204.

Obara H., Onodera J.-i., Abe S. (1974) The synthesis and properties of two analogs of carthamin. Chem. Lett. 335–338.

Ober A. G., Fischer N. H., Parodi F. (1986) Jamaicolides A-D, four sesquiterpene lactones from *Calea jamaicensis*. Phytochemistry 25:877–881.

Ober A. G., Fronczek F. R., Fischer N. H. (1985a) Sesquiterpene lactones of *Calea divaricata* and the molecular structure of leptocarpin acetate. J. Nat. Prod. 48:302–306.

Ober A. G., Urbatsch L. E., Fischer N. H. (1985b) Guaianolides and chromenes from *Calea* species. Phytochemistry 24:795–799.

Oberprieler C. (1998) The systematics of *Anthemis* L. (Compositae, Anthemideae) in western and central North Africa. Bocconea 9:1–328.

Oganesyan E. T., Smirnova L. P., Dzhumyrko S. F., Kechatova N. A. (1976) Axillaroside – a new flavonol glycoside from *Artemisia taurica*. Chem. Nat. Cmpds. 12:537 [599].

Ognyanov I., Todorova M. (1983) Sesquiterpene lactones and flavonoids in flowers of *Tanacetum vulgare*. Planta Medica 48:181–183.

Ogurtsova L. N., Syuzeva Z. F. (1974) Flavonoids of *Cirsium heterophyllum*. Chem. Nat. Cmpds. 10:90 [90].

Ohlendorf D., Schwarz R., Hänsel R. (1971) 3,5,7-Trihydroxy-6,8-dimethoxyflavon aus *Gnaphalium obtusifolium*. Arch. Pharm. 304:213–215.

Ohno N., Gershenzon J., Neuman P., Mabry T. J. (1981) Diterpene carboxylic acids and a heliangolide from *Helianthus angustifolius*. Phytochemistry 20:2393–2396.

Ohwi J. (1965) *Flora of Japan* (in English; edited by Meyer F. G., Walker E. W.). Washington, D.C.: Smithsonian Institution Press.

Öksüz S. (1987) Flavonoidal compounds of *Inula viscosa*. Part II. Planta Medica 31:270–273.

Öksüz S. (1990) Sesquiterpenoids and other constituents from *Tanacetum cilicium*. Phytochemistry 29:887–890.

Öksüz S., Ayylidiz H. (1985) Chemical investigations of sesquiterpene lactones and flavonoids of *Centaurea* species. Doga Bilim Dergisi, A1,9, 58–66.

Öksüz S., Ayylidiz H. (1986) Sesquiterpene lactones from *Centaurea coronopifolia*. Phytochemistry 25:535–537.

Öksüz S., Ayylidiz H., Johansson C. (1984) 6-Methoxylated and *C*-glycosyl flavonoids from *Centaurea* species. J. Nat. Prod. 47:902–903.

Öksüz S., Gümüs S., Alpinar K. (1991) Sesquiterpenoids and flavonoids of *Achillea* species. Biochem. Syst. Ecol. 19:439.

Öksüz S., Halfon B., Terem B. (1988) Flavonoids of *Centaurea cuneifolia*. Planta Medica 54:89.

Öksüz S., Putun E. (1987) Flavonoids of *Centaurea kotschyi* var. *kotschyi*. Doga: Kim. Serisi 11:66–71.

Öksüz S., Topcu G. (1987) Triterpene fatty acid esters and flavonoids from *Inula britannica*. Phytochemistry 26:3082–3084.

Öksüz S., Topcu G. (1992) A eudesmanolide and other constituents from *Inula graveolens*. Phytochemistry 31:195–197.

Öksüz S., Ulubelen A., Aynechi Y., Wagner H. (1982) A guaianolide from *Centaurea behen*. Phytochemistry 21:2747–2749.

Öksüz S., Ulubelen A., Clark W. D., Brown G. K., Mabry T. J. (1981) Flavonoids of *Haplopappus canescens*. Rev. Latinoamer. Quim. 12:12.

Öksüz S., Ulubelen A., Tuzlaci E. (1990) Constituents of *Achillea teretifolia*. Fitoterapia 61:283.

Okuno I., Uchida K., Nakamura M., Sakurawi K. (1988) Studies on choleretic constituents in *Artemisia capillaris* Thunb. Chem. Pharm. Bull. 36:769–775.

Olechowska-Baranska K., Lamer E. (1962) Identification and determination of certain flavonoids in coltsfoot flowers. Acta Polon. Pharm. 19:199–207.

Olmstead R. G., Jansen R. K., Michaels H. J., Downie S. R., Palmer J. D. (1990) Chloroplast DNA and phylogenetic studies in the Asteridae. In: Kawano S. (ed.) *Biological Approaches and Evolutionary Trends in Plants*. London: Academic Press. 119–134.

Olmstead R. G., Michaels H. J., Scott K. M., Palmer J. D. (1992) Monophyly of the Asteridae and identification of its major lineages inferred from *rbc*L sequences. Ann. Missouri Bot. Gard. 79:249–265.

Olsen J. (1979) Systematics of *Zaluzania* (Asteraceae: Heliantheae). Rhodora 81:449–501.

Olsen J. (1985) Synopsis of *Verbesina* sect. *Ochractinia* (Asteraceae). Plant Syst. Evol. 149:47–63.

Olsen J., Mabry T. J. (1980) Systematic implications of flavonoids in *Zaluzania*. Biochem. Syst. Ecol. 8:261–262.

Omar A. A. (1980) Apigenin-7-*O*-glucoside and echinopsine alkaloid from *Echinops spinossimus*. Sci. Pharm. 48:43–45.

Omar A. A., El-Ghazouly M. G. (1984) Flavonoids of *Artemisia campestris*. Poster, Biochem. of Plant Phenolics meeting, Ghent. [Cited by Wollenweber and Jay (1988)].

Onodera J.-i., Kawamoto K.-i., Matsuba S., Sato S., Kojima H., Kaneya Y., Obara H. (1995) Biotransformation of safflower yellow B to carthamin, a coloring matter of safflower. Chem. Lett. 901–902.

Onodera J.-i., Obara H., Osone M., Maruyama Y., Sato S. (1981a) The structure of safflomin-A, a component of safflower yellow. Chem. Lett. 433–436.

Onodera J.-i., Saito T., Obara H. (1979) Hydrolysis of carthamin. Chem. Lett. 1327–1330.

Onodera J.-i., Sato S., Obara H. (1981b) Methylated derivatives from the hydrolysis products of safflomin-A. Chem. Lett. 887–890.

Opitz L., Ohlendorf D., Hänsel R. (1971) 5,7-Dihydroxy-3,8-dimethoxyflavon aus *Helichrysum italicum*. Phytochemistry 10:1948.

Ornduff R. (1963) Experimental studies in two genera of Helenieae (Compositae): *Blennosperma* and *Lasthenia*. Quart. Rev. Biol. 38:141–150.

Ornduff R. (1966) A biosystematic survey of the gold-field genus *Lasthenia*. Univ. Calif. Publ. Bot. 40:1–92.

Ornduff R. (1969) The origin and relationships of *Lasthenia burkei* (Compositae). Amer. J. Bot. 56:1042–1047.

Ornduff R. (1971) A new tetraploid subspecies of *Lasthenia* (Compositae) from Oregon. Madroño 21:96–98.

Ornduff R. (1976) Speciation and oligogenic differentiation in *Lasthenia* (Compositae). Syst. Bot. 1:91–96.

Ornduff R. (ed.) (1967–1969) Index to plant chromosome numbers (1965–1967). Regnum Veg. 50:1–128; 55:1–126; 59:1–129.

Ornduff R., Bohm B. A. (1975) Flavonoids and affinities of *Tracyina* and *Rigiopappus* (Compositae). Madroño 23:53–55.

Ornduff R., Bohm B. A., Saleh N. A. M. (1973a) Flavonoids of artificial interspecific hybrids in *Lasthenia*. Biochem. Syst. Ecol. 1:147–151.

Ornduff R., Mosquin T. (1970) Variation in the spectral qualities of flowers in the *Nymphoides indica* complex (Menyanthaceae) and its possible adaptive significance. Can. J. Bot. 48:603–605.

Ornduff R., Saleh N. A. M., Bohm B. A. (1973b) The flavonoids and affinities of *Blennosperma* and *Crocidium* (Compositae). Taxon 22:407–412.

Ornduff R., Saleh N. A. M., Bohm B. A. (1974) Intraspecific variation of flavonoids in *Lasthenia*. Brittonia 26:411–420.

Ortega A., Del J., Lopez C., Maldonado E. (1989) A tris-norsesquiterpene lactone and other sesquiterpenes from *Calea crocinervosa*. Phytochemistry 28:2735–2736.

Oshima Y., Cordell G. A., Fong H. H. S. (1986) Studies on zoapatle, III. Flavonoid glycosides from *Montanoa tomentosa* ssp. *tomentosa*. J. Nat. Prod. 49:552–553.

Oskay E., Yesilada A. (1984) Four flavonoids and three other constituents from *Achillea biebersteinii*. J. Nat. Prod. 47:742.

Oswiecimska M., Miedzobrodzka J. (1966) Flavonoid compounds in the polyploidal complex of *Achillea millefolium*. Diss. Pharm. Pharmacol. 18:601–610.

Oswiecimska M., Polak A., Seidel O., Sendra J. (1965) Comparative study of chromatograms of the flavonoid fractions from herbs of some species of the genus *Artemisia*. Diss. Pharm. Pharmacol. 17:503–511.

Ovdienko O. A., Salo V. P., Pakaln D. A., Spiridonov V. N., Litvinenko V. I., Prokopenko A. P., Shreter A. I. (1977) Comparative phytochemical study of flowers of various everlasting species. Khim. Farm. Zu. 11:102–105.

Ownbey M. (1950) Natural hybridization and amphiploidy in the genus *Tragopogon*. Amer. J. Bot. 37:487–499.

Ownbey M., McCollum G. D. (1953) Cytoplasmic inheritance and reciprocal amphiploidy in *Tragopogon*. Amer. J. Bot. 40:788–796.

Ownbey M., McCollum G. D. (1954) The chromosomes of *Tragopogon*. Rhodora 50:9–21.

Ownbey M., Weber W. A. (1943) Natural hybridization in the genus *Balsamorhiza*. Amer. J. Bot. 30:179–187.

Ozawa A. T., Rivera P. A., Romo de Vivar A. (1983) Active principles of the toxic plant *Helenium integrifolium*. Rev. Latinoamer. Quim. 14:40–43.

Pacciaroni A. del V., Sosa V. E., Espinar L. A., Oberti J. C. (1995) Sesquiterpene lactones from *Schkuria pinnata*. Phytochemistry 39:127–131.

Pacheco P., Crawford D. J., Stuessy T. F., Silva O. M. (1985) Flavonoid evolution in *Robinsonia* (Compositae) of the Juan Fernandez Islands. Amer. J. Bot. 72:989–998.

Pacheco P., Crawford D. J., Stuessy T. F., Silva O. M. (1991) Flavonoid evolution in *Dendroseris* (Compositae, Lactuceae) from the Juan Fernandez Islands, Chile. Amer. J. Bot. 78:534–543.

Palacios P., Gutkind G., Rondina R. V. D., de Torres R., Coussio J. D. (1983) Genus *Baccharis* II. Antimicrobial activity of *B. crispa* and *B. notosergila*. Planta Medica 49:128.

Pandey U. C., Singhal A. K., Barua N. C., Sharma R. P., Baruah J. N., Watanabe K., Kulanthaivel P., Herz W. (1984) Stereochemistry of strictic acid and related furanoditerpenes from *Conyza japonica* and *Grangea maderaspatana*. Phytochemistry 23:391–397.

Panero J. L., Jansen R. K., Clevinger J. A. (1999) Phylogenetic relationships of subtribe Ecliptinae (Asteraceae: Heliantheae) based on chloroplast DNA restriction site data. Amer. J. Bot. 86:413–427.

Panero J. L., Schilling E. E. (1988) Revision of *Viguiera* sect. *Maculatae* (Asteraceae: Heliantheae). Syst. Bot. 13:371–406.

Pankhurst C. E., Biggs D. R. (1980) Sensitivity of *Rhizobium* to selected isoflavonoids. Can. J. Microbiol. 26:542–545.

Pares J. O., Öksüz S., Ulubelen A., Mabry T. J. (1981) 6-Hydroxyflavonoids from *Pulicaria dysenterica* (Compositae). Phytochemistry 20:2057.

Park J. C., Lee J. H., Choi J. S. (1995) A flavone diglycoside from *Cirsium japonicum* var. *ussuriense*. Phytochemistry 39:261–262.

Parker E. (1972) Biosystematic study of the genus *Balduina*. Ph.D. Thesis, Univ. Georgia.

Parodi F. J., Fischer N. H. (1988) Guaianolides from *Baccharis salicina*. Phytochemistry 27:2987–2988.

Paschenko M. M., Pivnenko G. P., Litvinenko V. T. (1966) Phytochemical study of *Xanthium spinosum*. Farm. Zh. Kiev 21:44–49.

Passreiter C. M., Merfort I., Willuhn G. (1990) Flavonoids of *Arnica sachalinensis* subgen. *Andropurpurea*. Biochem. Syst. Ecol. 18:17–18.

Patterson R. (1984) Flavonoid uniformity in diploid species of Hawaiian *Scaevola* (Goodeniaceae). Syst. Bot. 9:263–265.

Payne W. W. (1964) A re-evaluation of the genus *Ambrosia* (Compositae). J. Arn. Arbor. 45:401–430.

Payne W. W. (1976) Biochemistry and species problems in *Ambrosia* (Asteraceae-Ambrosieae). Plant Syst. Evol. 125:169–178.

Pelter A., Hänsel R. (1968) The structure of silybum, *Silybum* substance E-6, the first flavonolignan. Tetr. Lett. 2911–2916.

Peneva P., Ivancheva S., Terzieva L. (1989) Essential oil and flavonoids in the racemes of camomile (*Matricaria recutita*). Rast. Nauki 26:25–33.

Peneva P., Vitkova A., Ivancheva S. (1988) Phenolic compounds from *Artemisia santonicum* L. Rastenievud. Nauki 25:35–42.

Pereira P. S., Dias D. A., Vichnewski W., Nasi A. M. T. T., Herz W. (1997) Sesquiterpene lactones form Brazilian *Tithonia diversifolia*. Phytochemistry 45:1445–1448.

Pereyra de Santiago O. J., Juliani H. R. (1972) Isolation of quercetin 3,7,3',4'-tetrasulphate from *Flaveria bidentis*. Experientia 28:380.

Perez A. L., Lara P., Romo de Vivar A. (1992) Sesquiterpenoids and diterpenoids from *Tithonia longiradiata*. Phytochemistry 31:4227–4231.

Perez A. L., Nava L., Romo de Vivar A. (1987) Absinthifolide, a sesquiterpene glycoside from *Bahia absinthifolia* var. *absinthifolia*. Phytochemistry 26:765–767.

Perez A. L., Nieto D. A., Romo de Vivar A. (1990) Sesquiterpenoid and other metabolites from two *Bahia* species. Phytochemistry 29:901–903.

Perez A. L., Ortega A., Romo de Vivar A. (1988) An acyclic diterpene and sesquiterpene lactones from *Tithonia pedunculata*. Phytochemistry 27:3897–3901.

Perez A. L., Romo de Vivar A. (1989) Triacyl glucopyranoses from *Bahia schaffneri*. Phytochemistry 28:1639–1643.

Perez A. L., Romo de Vivar A. (1994) The monoterpene 9-hydroxythymol from *Bahia schaffneri* var. *aristata*. Phytochemistry 36:1081–1082.

Perez-Castorena A.-L., Castro A., Romo de Vivar A. (1997a) A dirhamnoside from *Psacalium megaphyllum*. Phytochemistry 46:1297–1299.

Perez-Castorena A.-L., Martinez-Vazquez M., Romo de Vivar A. (1997b) Diterpenes of *Bahia glandulosa*. Phytochemistry 46:729–734.

Perkin A. G. (1913) Quercetagetin. J. Chem. Soc. 103:209–210.

Pesacreta T. C., Stuessy T. F. (1996) Autofluorescent walls of connective bases in anthers of Barnadesioideae (Asteraceae), and systematic implications. Taxon 45:473–485.

Pesacreta T. C., Sullivan V. I., Hasenstein K. H. (1993) The connective base of *Cirsium horridum* (Asteraceae): description and comparison with the viscoelastic filament. Amer. J. Bot. 80:411–418.

Pesacreta T. C., Sullivan V. I., Hasenstein K. H., Durand J. M. (1991) Thigmonasticity of thistle staminal filaments I. Involvement of a contractile cuticle. Protoplasma 163:174–180.

Petenatti E. M., Pestchanker M. J., Del Vitto L. A., Guerreiro E. (1996) Chemotaxonomy of the Argentinian species of *Gaillardia*. Phytochemistry 42:1367–1368.

Peterson K. M., Payne W. W. (1973) The genus *Hymenoclea* (Compositae: Ambrosieae). Brittonia 25:243–256.

Petit D. P. (1999) Generic interrelationships of the Cardueae (Compositae): a cladistic analysis of morphological data. Plant Syst. Evol. 207:173–203.

Petit D. P., Mathez J., Qaid A. (1996) Early differentiation of the Cardueae *sensu lato*: morphology and pollen. In: Hind D. J. N., Beentje H. J. (eds.) *Compositae: Systematics. Proceedings of the International Compositae Conference, Kew (1994), Vol. 1.* Kew: Royal Botanic Gardens. 79–93.

Petrak F. (1979) *Cirsium.* In: Rechinger K. H. (ed.) *Flora Iranica, Compositae, III-Cynareae.* Graz: Akademische Druck- und Verlagsanstalt. 231–280.

Petrovic S. D., Gorunovic M. S., Wray V., Merfort I. (1999a) A taraxasterol derivative and phenolic compounds from *Hieracium gymnocephalum*. Phytochemistry 50:293–296.

Petrovic S. D., Löscher R., Gorunovic M. S., Merfort I. (1999b) Flavonoid and phenolic acid patterns in seven *Hieracium* species. Biochem. Syst. Ecol. 27: 651–656.

Pfeil R. M., Gage D. A., Lee E. F., Miski M., Mabry T. J., Powell A. M. (1987) Sesquiterpene lactones from *Perityle vaseyi*. Phytochemistry 26:195–198.

Phillips D. A., Tsai S. M. (1992) Flavonoids as plant signals to rhizosphere microbes. Mycorrhiza 1:55–58.

Piattelli M. (1981) The betalains: structure, biosynthesis, and chemical taxonomy. In: Conn E. E. (ed.) *Secondary Plant Products* [Vol. 7 of *The Biochemistry of Plants. A Comprehensive Treatise*, Stumpf P. K., Conn E. E. (eds.)]. New York: Academic Press. 557–575.

Picher M. T., Seone E., Tortajada A. (1984) Flavones, sesquiterpene lactones and glycosides isolated from *Centaurea aspera* var. *stenophylla*. Phytochemistry 23:1995–1997

Piegay I., Reynaud J. (1986) Les flavonoides de *Senecio aureus* L. Die Pharmazie 41:524–525.

Pielou E. C. (1991) *After the Ice Age*. Chicago: Univ. Chicago Press.

Pietta P., Manera E., Ceva P. (1987) Simultaneous isocratic HPLC determination of flavones and coumarins in *Matricaria chamomilla* extracts. J. Chromatography 404:179–281.

Pinhey J. T., Simpson R. F., Batey I. L. (1972) The constituents of *Olearia heterocarpa*. The structure of olearin, a diterpene lactone of the cascarillin group. Austral. J. Chem. 24:2621–2637.

Pinkas M., Torck M. (1989) Classification des *Arnica*: Heliantheae ou Senecioneae? Role chimiotaxonomique des flavonoides. Pl. Medica Phytother. 23:136–158.

Pinkas M., Torck M., Bezanger-Beauquesne L. (1973) Recherches sur les flavonoids des fleuers d'*Helichrysum stoechas* DC. Pl. Medica Phytother. 7:332–335.

Pinkas M., Torck M., Bezanger-Beauquesne L. (1977) Recherches sur les flavonoids des l'*Grindelia squarrosa* et l'*Grindelia aphanactis*. Compt. rend. Hebd. Seances Acad. Sci., 284D:1593–1596.

Pinkas M., Torck M., Bezanger-Beauquesne L., Cazin J. C. (1978) Study on polyphenols from some *Grindelia* species. Ann. Pharm. Fr. 36:97–100.

Pinkas M., Torck M., Perrin E. (1975) Chemistry of field marigold *Calendula arvensis*. Compt. rend. Hebd. Seances Acad. Sci., 281D:447–450.

Plekhanova T. I., Bandyukova V. A., Mikhailova G. A. (1977) Flavonoids of *Galinsoga parviflora*. Chem. Nat. Cmpds. 13:728 [862].

Plouvier V. (1970) Compt. rend. Acad. Sci., Ser. D. 270:2710–2713. Cited by Al-Easa and Rizk (1992).

Poethke W., Bulin P. (1969) Phytochemical study of a newly cultured *Matricaria chamomilla*. Pharm. Zentralh. 108:733–747.

Poljakov P. P. (1967) *Systematics and the Origin of the Compositae*. Alma-Ata: Nauka. (In Russian.)

Pontedera J. (1720) *Methodus Compositarum. Dissertationes XI, Botanicae*. Patavini.

Poplawski J., Holub M., Samek Z., Herout V. (1971) On terpenes. CCIX. Arnicolides-sesquiterpenic lactones from the leaves of *Arnica montana* L. Coll. Czech. Chem. Communs. 36:2189–2199.

Popova T. P., Litvinenko V. I. (1993) Isohelichrysin of *Helichrysum raceme*. Farm. Zh. (Kiev) (1) 60–65.

Powell A. M. (1968) Additional discussions pertaining to the congeneric status of *Perityle* and *Laphamia* (Compositae). Sida 3:270–278.

Powell A. M. (1972) Taxonomy of *Amauria* (Compositae-Peritylinae). Madroño 21:516–525.

Powell A. M. (1973a) Taxonomy of *Pericome* (Compositae-Peritylinae). Southwestern Naturalist 18:335–339.

Powell A. M. (1973b) Taxonomy of *Perityle*, section *Laphamia* (Compositae-Helenieae). Sida 5:61–128.

Powell A. M. (1974) Taxonomy of *Perityle* section *Perityle* (Compositae-Peritylinae). Rhodora 76:229–306.

Powell A. M. (1979) Systematics of *Flaveria* (Flaveriinae-Asteraceae). Ann. Missouri Bot. Gard. 65:590–636.

Powell A. M. (1985) Crossing data as generic criteria in the Asteraceae. Taxon 34:55–60.

Prassler M. (1967) Revision der Gattung *Ursinia*. Mitt. Bot. München 6:363–478.

Price J. R. (1939) The yellow colouring matter of *Dahlia variabilis*. J. Chem. Soc. 1017–1018.

Prokopenko A. P., Spiridonov V. N., Litvinenko V. I., Chernobai V. T. (1972) Flavonoid aglycones of the inflorescences of *Helichrysum arenarium*. Chem. Nat. Cmpds. 8:620 [649].

Proksch P. (1985) Vorkommen und biologische Bedeutung von Benzopyranen (Chromenen) und Benzofuranen in den Asteraceae. Plant Syst. Evol. 150:89–100.

Proksch P., Breuer M., Mitsakos A., Budzikiewicz H. (1987) New dihydroflavonols from *Enceliopsis* and *Geraea*. Planta Medica 53:334–337.

Proksch P., Budzikiewicz H., Tanowitz B. D., Smith D. (1984) Flavonoids from the external leaf resins of four *Hemizonia* species (Asteraceae). Phytochemistry 23:679–680.

Proksch P., Kunze A. (1996) Chemosystematic evidence from prenylated acetophenones—conclusions at the tribal, inter- and intrageneric level. In: Hind D. J. N., Beentje H. J.

(eds.) *Compositae: Systematics. Proceedings of the International Compositae Conference, Kew (1994), Vol. 1.* Kew: Royal Botanic Gardens. 295–305.
Proksch P., Mitsakos A., Bodden J., Wollenweber E. (1986) Benzofurans and methylated flavonoids of *Geraea* (Asteraceae). Phytochemistry 25:2367–2369.
Proksch P., Politt U., Wollenweber E., Wray V., Clark C. (1988) Epicuticular flavonoids from *Encelia.* Planta Medica 54:542–546.
Proksch P., Proksch M., Rundel P. W., Rodriguez E. (1982) Ecological significance of the chemistry of the leaf resin of *Elytropappus rhinocerotis.* Biochem. Syst. Ecol. 10:49–53.
Proksch P., Rodriguez E. (1983) Chromenes and benzofurans of the Asteraceae, their chemistry and biological significance. Phytochemistry 22:2335–2348.
Proliac A. (1972) Sur la presence de glucosyl-7 luteoline dans *Catananche caerulea* L. Pl. Medica Phytother. 6:275.
Proliac A., Chaboud A., Raynaud J. (1991) Isolation and identification of two *C*-glycoflavones of the roots of *Atractylis gummifera* L. Pharm. Acta Helv. 66:316–317.
Proliac A., Combier H. (1974) Presence of apigenin 7-glucuronide in the flowers of *Catananche caerulea.* Fitoterapia 45:193.
Proliac A., Raynaud J., Combier H., Bouillant M.-L., Chopin J. (1973) Sur les di-*C*-glycosylflavones de *Catananche caerulea* L. (Composées). Compt. rend. Hebd. Seances Acad. Sci. Paris., Ser. D, 277:2813–2815.
Proliac A., Raynaud, J. (1977) Sur le presence du carlinoside ou 6-*C*-ß-D-glucopyranosyl-8-*C*-a-L-arabinopyranosyl-luteoline chez *Catananche caerulea.* Planta Medica 32:68–69.
Proliac A., Raynaud J. (1983) Isolement et identification de la *C*-galactosyl-8 apigenine dans les tiges feuilles de *Carlina acanthifolia.* Helv. Chim. Acta 66:2412–2416.
Prosovskii M. A., Oleshko G. I. (1985) Phenolic compounds of *Matricaria discoidea.* Chem. Nat. Cmpds. 21:671 [712].
Puppo A. (1992) Effect of flavonoids on hydroxyl radical formation by Fenton-type reactions: influence on the iron chelator. Phytochemistry 31:85–88.
Puri B., Seshadri T. R. (1954) Survey of anthoxanthins. IV. Chemical study of yellow garden flowers and constitution of coreopsin. J. Sci. Ind. Res. India. 13B:321–325.
Puttock C. F. (1994) Re-analysis of Anderberg's Gnaphalieae data matrix. Compositae Newsletter. No. 25:1–14.
Putun A. E., Ogretir C., Putun E. (1989) *C*- and *O*-Glycosyl flavonoids of *Centaurea thracia* (Janka) Hayek. Doga Turk Kim. Derg. 13:293–298.
Putun A. E., Putun E. (1990) Flavonoid compounds of *Centaurea pichleri* Boiss. subsp. *pichleri.* Chim. Acta Turc. 18:225–231.
Qiu L., Lian M., Ma Z., He G. (1989) Components of *Saussurea gossypiphora* D. Don. Zhiwu Xuebao 31:398–400.
Qu G., Liu J., Li X., Wang S., Wu L. (1995) Flavonoids of lieyejumaicai. Zhongcaoyao 26:233–235.
Qu G., Wang S., Wu L., Li X. (1993) Chemical constituents of *Sonchus arvensis.* Zhongguo Zhongyao Zazhi 18:101–102.
Quicke D. L. J. (1993) *Principles and Techniques of Contemporary Taxonomy.* London: Blackie Academic & Professional.

Quijano L., Calderon J. S., Gomez F., Escobar E., Rios T. (1985) Octa-substituted flavones from *Ageratum houstonianum*. Phytochemistry 24:1085–1088.
Quijano L., Calderon J. S., Gomez F., Rios T. (1982a) Four flavonoids from *Ageratum strictum*. Phytochemistry 21:2575–2579.
Quijano L., Calderon J. S., Gomez F., Rios T. (1982b) Two polymethoxyflavones from *Ageratum houstonianum*. Phytochemistry 21:2965–2967.
Quijano L., Calderon J. S., Gomez F., Soria I. E., Rios T. (1980) Highly oxygenated flavonoids from *Ageratum corymbosum*. Phytochemistry 19:2439–2442.
Quijano L., Calderon J. S., Gomez-Garibay F., Escobar E., Rios T. (1987) Further polysubstituted flavones from *Ageratum houstonianum*. Phytochemistry 26:2075–2078.
Quijano L., Gomez -Garibay F., Garcia-Perez G., Rios T. (1992) Flavonoids from *Ageratum corymbosum*. Phytochemistry 31:2859–2862.
Quijano L., Malanco F., Rios T. (1970) The structure of eupalin and eupatolin, two new flavonol rhamnosides isolated from *Eupatorium ligustrinum* DC. Tetrahedron 26:2849–2859.
Quijano L., Rios T., Fronczek F. R., Fischer N. H. (1998) The molecular structure of maniladiol from *Baccharis salicina*. Phytochemistry 49: 2065–2068.
Quijano L., Trejo R. I., Collera O., Rios T. (1994) Sesquiterpene lactones from *Montanoa leucantha* ssp. *leucantha*. Phytochemistry 36:1443–1448.
Rabakonandrianina E. (1980) Infrageneric relationships and the origin of the Hawaiian endemic genus *Lipochaeta* (Compositae). Pacific Sci. 34:29–39.
Rabakonandrianina E., Carr G. D. (1981) Intergeneric hybridization, induced polyploidy, and the origin of the Hawaiian endemic *Lipochaeta* from *Wedelia* (Compositae). Amer. J. Bot. 68:206–215.
Rajakaruna N., Bohm B. A (1999) The edaphic factor and patterns of variation in *Lasthenia californica* (Asteraceae). Amer. J. Bot. 86:1576–1596.
Rajbhandari A., Roberts M. F. (1983) The flavonoids of *Stevia rebaudiana*. J. Nat. Prod. 46:194–195.
Rajbhandari A., Roberts M. F. (1984) Flavonoids of *Stevia nepetifolia*. J. Nat. Prod. 47:559–560.
Rajbhandari A., Roberts M. F. (1985a) The flavonoids of *Stevia microchaeta*, *Stevia monardifolia* and *Stevia origanoides*. J. Nat. Prod. 48:502–503.
Rajbhandari A., Roberts M. F. (1985b) The flavonoids of *Stevia cuzcoensis*, *S. galeopsidifolia*, *S. serrata*, and *S. soratensis*. J. Nat. Prod. 48:858–859.
Rakotoarimanga J. V., Krebs H. C., Habermehl G. (1992) Constituents of *Vernonia glutinosa*. Fitoterapia 63:271–272.
Ram S. N., Roy R., Singh B., Singh R. P., Pandey V. B. (1996) An acylflavone glucoside of *Echinops echinatus*. Planta Medica 62:187.
Randeria A. J. (1960) The composite genus *Blumea*, a taxonomic revision. Blumea 10:176–317.
Randriaminahy M., Proksch P., Witte L., Wray V. (1992) Lipophilic phenolic constituents from *Helichrysum* species endemic to Madagascar. Z. Naturforsch. 47c:10–16.
Rao A. S. (1990) Root flavonoids. Botanical Review 56:1–84.
Rao C. B., Rao T. N., Muralikrishna B. (1977) Flavonoids from *Blumea lacera*. Planta Medica 31:235–237.

Rao D. V., Rao E. V. (1972) Phytochemical investigation on the leaves of *Notonia grandiflora* DC. Planta Medica 22:205–208.

Rao M. M., Kingston D. G. I., Spittler T. D. (1970) Flavonoids from *Flourensia cernua*. Phytochemistry 9:227–228.

Rao P. R., Seshadri T. R., Sharma P. (1973) Polyphenolic constituents of Pyrethrum flowers (*Chrysanthemum cinerariifolium*). Curr. Sci. 42:811–812.

Rasolojaona L., Raynaud J. (1979) Les heterosides flavoniques de *Carlina vulgaris* L. (Composées). Ann. Pharm. France 37:469–473.

Rauschert S. (1974) Nomenklatorische Probleme in der Gattung *Matricaria* L. Folia Geobot. Phytotax. Praha 9:249–260.

Rauter A. P., Branco I., Tostao Z., Pais M. S., Gonzalez A. G., Bermejo J. B. (1989) Flavonoids from *Artemisia campestris* subsp. *maritima*. Phytochemistry 28:2173–2175.

Raven P. H., Axelrod D. I. (1974) Angiosperm biogeography and past continental movements. Ann. Missouri Bot. Gard. 61:593–673.

Raven P. H., Kyhos D. W. (1961) Chromosome numbers in Compositae. II. Helenieae. Amer. J. Bot. 48:842–850.

Ray J. (1682) *Methodus Plantarum*. London.

Raynaud J., Dombris A. (1978) Contribution a l'etude des *C*-glycosyl flavonoides des feuilles d'*Atractylis gummifera* L. Composées. Pl. Medica Phytother. 12:220–224.

Raynaud J., Dombris A., Prum N. (1981) Les heterosides flavoniques des fleurs de *Carlina corymbosa* L. Die Pharmazie 36:444–445.

Raynaud J., Gorunovic M., Lebreton P. (1971) Contribution a l'etude biochimique de *Staehelina dubia*. Planta Medica 20:199–204.

Raynaud J., Rasolojaona L. (1976) Sur la presence d'un nouveau di-*C*-glycosyl flavonoide: Le carlinoside chez *Carlina vulgaris* L. Composées. Compt. rend. Hebd. Seances Acad. Sci. Paris., Ser. D 282:1059–1061.

Raynaud J., Rasolojaona L. (1978) Les *C*-glycosyl flavonoides des feuilles de *Carlina acanthifolia* L. Composées. Pl. Medica Phytother. 12:281–286.

Raynaud J., Rasolojaona L. (1979) Flavonoides des feuilles de *Carlina acaulis*. Planta Medica 37:168–171.

Recio M. C., Giner R. M., Hermenegildo M., Peris J. B., Manez S., Rios J. L. (1992) Phenolics of *Reichardia* and their taxonomic significance. Biochem. Syst. Ecol. 20:449–452.

Recio M. C., Terencio M. C., Arenas J. A., Giner R., Sanz M. J., Manez S., Rios Canavate J. L. (1993) Luteolin 7-*O*-galactosylglucuronide, a new flavonoid from *Andryala ragusina*. Die Pharmazie 48:228–229.

Redaelli C., Formentini L., Santaniello E. (1980) Apigenin 7-glucoside and its 2"- and 6"-acetates from ligulate flowers of *Matricaria chamomilla*. Phytochemistry 19:985–986.

Redaelli C., Formentini L., Santaniello E. (1982) Apigenin 7-glucoside diacetates in ligulate flowers of *Matricaria chamomilla*. Phytochemistry 21:1828–1830.

Redl K., Davis B., Bauer R. (1993) Chalcone glycosides from *Bidens campylotheca*. Phytochemistry 32:218–220.

Rees S., Harborne J. B. (1984) Flavonoids and other phenolics of *Cichorium* and related members of the Lactuceae (Compositae). Bot. J. Linn. Soc. 89:313–319.

Reichling J., Beiderbeck R., Becker H. (1979) Vergleichende Untersuchungen über sekundäre Inhaltsstoffe bei Pflanzentumoren, Blüte, Kraut und Wurzel von *Matricaria chamomilla* L. Planta Medica 36:322–332.

Reid A. R., Bohm B. A. (1993) External and vacuolar flavonoids of *Brachyglottis cassinioides*. Biochem. Syst. Ecol. 21:746.

Reid A. R., Bohm B. A. (1994) Vacuolar and exudate flavonoids of New Zealand *Cassinia*. Biochem. Syst. Ecol. 22:501.

Reid A. R., Bohm B. A. (1995) Flavonoids of *Raoulia* Hook. f. ex Raoul (Asteraceae:Gnaphalieae). Biochem. Syst. Ecol. 23:209–210.

Rendle A. B. (1938) *The Classification of Flowering Plants*, Vol. 2. Cambridge: Cambridge Univ. Press.

Rendon W., Vila J. L. (1995) Phytochemical study of *Baccharis trimera*. Rev. Boliv. Quim. 12:13–15.

Rendyuk T. D., Shreter A. I., Shelyuto V. L., Glyzin V. I. (1977) A flavonoid glycoside from *Cirsium arvense*. Chem. Nat. Cmpds. 13:244 [282].

Repcak M., Martonfi P. (1995) The variability pattern of apigenin glucosides in *Chamomilla recutita* diploid and tetraploid cultivars. Pharmazie 50:696–699.

Repcak M., Svehlikova V., Imrich J., Pihlaja K. (1999) Jaceidin and chrysosplenetin chemotypes of *Chamomilla recutita* (L.) Rauschert. Biochem. Syst. Ecol. 27:727–732.

Reveal J. K., Keener C. S. (1981) *Virgulus* Raf. (1837), an earlier name for *Lasallea* Greene (1903) (Asteraceae). Taxon 30:648–651.

Reveal J. L. (1995) Vascular Plant Suprageneric Names in Asteraceae. Mimeo. 16 pp. [See also the Indices Nominum Supragenericorum Plantarum Vascularium Project on the Internet (WWW) at:
http://matrix.nal.usda.gov.8080/star/suprагenericname.html]

Reveal J. L. (1997) Early suprageneric names in Asteraceae. Compositae Newsletter 30:29–45.

Reyes A., Navarrete A., Martinez R., Miranda N. (1981) Some compounds from *Senecio yegua* Cabr. var. *depilis* (Phil.) Cabr. Bol. Soc. Chil. Quim. 26:8.

Reyes A., Peruzzo G. L., Oyazun G. D., Romero M. A. (1986) A quercetin triglycoside from *Senecio subdiscoides*. Rev. Latinoamer. Quim. 17:10–12.

Reyes A., Reyes M., Lemarie M., Romero M. (1990) Phenolic components of two Chilean species of *Baccharis*. Rev. Latinoamer. Quim. 21:1–2.

Reyes A., Vicuna P., Garcia H., Silva M. (1977) Flavonoides en algunas especies de *Senecio* en Chile. Rev. Latinoamer. Quim. 8:134–136.

Reynaud J., Becchi M., Raynaud J. (1985) Presence de 6''-*O*-acetyl-glucosyl-3-quercetine chez *Doronicum grandiflorum* Lam. (Compositae). Die Pharmazie 40:132–133.

Reynaud J., Couble A., Raynaud J. (1992) Flavonoids from *Centaurea macrophylla* Muss. Puschk. ex Willd. Die Pharmazie 47:51–52.

Reynaud J., Raynaud J. (1984) Sur la presence d'Onopordum chez *Doronicum grandiflorum* Lam. (Compositae). Die Pharmazie 39:126.

Reynaud J., Raynaud J. (1986) Les flavonoides de *Doronicum grandiflorum*. Biochem. Syst. Ecol. 14:191–193.

Reynaud J., Raynaud J., Voirin B. (1983) Sur le presence de deux flavone methoxyles rares chez *Doronicum grandiflorum* Lam. (Compositae). Die Pharmazie 38:628–629.

Rice E. L. (1974) *Allelopathy.* New York: Academic Press.
Rice E. L. (1984) *Allelopathy*, 2nd ed. Orlando, Florida: Academic Press.
Rice E. L. (1995) *Biological Control of Weeds: Advances in Applied Allelopathy.* Norman, Oklahoma: Univ. Oklahoma Press.
Richard L. C. (1820) Mémoire sur une famille de plantes, dites les Calycérées. Mém. Mus. Hist. Nat. 6:28–82.
Richardson P. M. (1983a) Flavonoids and phylogenetic systematics. In: Platnick N. I., Funk V. A. (eds.) *Advances in Cladistics*, Vol. 2. New York: Columbia Univ. Press. 495–499.
Richardson P. M. (1983b) Methods of flavonoid data analysis. In: Felsenstein J. (ed.) *Numerical Taxonomy.* Berlin: Springer. 495–499.
Richardson P. M., Young D. A. (1982) The phylogenetic content of flavonoid point scores. Biochem. Syst. Ecol. 10:251–255.
Rieseberg L. H., Beckstrom-Sternberg S. M., Liston A., Arias D. M. (1991) Phylogenetic and systematic inferences from chloroplast DNA and isozyme variation in *Helianthus* sect. *Helianthus* (Asteraceae). Syst. Bot. 16:50–76.
Rieseberg L. H., Hanson M., Philbrick C. T. (1992) Androdioecy is derived from dioecy in Datiscaceae: evidence from restriction site mapping of PCR-amplified chloroplast DNA fragments. Syst. Bot. 17:324–336.
Rieseberg L. H., Schilling E. E. (1985) Floral flavonoids and UV patterns in *Viguiera* (Compositae). Amer. J. Bot. 72:999–1004.
Rieseberg L. H., Soltis D. E., Arnold D. (1987) Variation and localization of flavonoid aglycones in *Helianthus annuus* (Compositae). Amer. J. Bot. 74:224–233.
Rimpler H., Hänsel R. (1965) Zwei neue Chalkonepigmente aus *Helichrysum bracteatum.* Arch. Pharm. 298:838–847.
Rimpler H., Langhammer L., Frenzl H.-J. (1963) Farbsorten von *Helichrysum bracteatum*: Verteilung des C-15-Körper innerhalb der Pflanzen. Planta Medica 11:325–331.
Rios J.L., Giner R. M., Cuellar M. J., Recio M. C., Serrano A. (1992) Phenolics from some species of the subtribe Leontodontinae. Planta Medica 58 (Suppl. 1):A700.
Rivera A. P., Faini F., Castillo M. (1988) 15a-Hydroxy-ß-amyrin and patagonic acid from *Baccharis magellanica* and *Baccharis patagonica.* J. Nat. Prod. 51:155–157.
Rizk A. F. M., Hammouda F. M., Ismail S. I., Hussiney H. A. (1993) Constituents of plants growing in Qatar. XXIII. Flavonoids of *Francoeuria crispa.* Qatar Univ. Sci. J. 13:51–52.
Rizk A. F. M., Ismail S. I. (1982) Flavonoids of *Francoeuria crispa.* Planta Medica 45:146.
Robberecht R., Caldwell M. M. (1983) Protective mechanisms and acclimation to solar ultraviolet-B radiation in *Oenothera stricta.* Plant, Cell and Environment 6:477–485.
Roberts H. F. (1929) *Plant Hybridization before Mendel*, facsimile ed. (1965). New York: Hafner Publ. Co.
Roberts M. L. (1980) The flavonoids of *Megalodonta beckii* and isolation of 2',3-dihydroxy-4-methoxy-4'-glucosyl chalcone. Biochem. Syst. Ecol. 8:115–118.
Roberts M. L., Timmermann B. N., Mabry T. J. (1980) 6-Methoxyflavonols from *Brickellia veronicaefolia* (Compositae). Phytochemistry 19:127–129.
Roberts M. L., Timmermann B. N., Mabry T. J., Brown R., Matlin S. A. (1984) Brickellin, a novel flavone from *Brickellia veronicaefolia* and *B. chlorolepis.* Phytochemistry 23:163–165.

Robertson C. (1928) *Flowers and Insects. Lists of visitors of four hundred and fifty-three flowers.* Carlinville, Illinois: Publ. by author.

Robin V., Boustie J., Amoros M., Girre L. (1998) In-vitro antiviral activity of seven *Psiadia* species (Asteraceae): isolation of two antipoliovirus flavonoids from *Psiadia dentata*. Pharm. & Pharmacol. Comm. 4:61–64.

Robinson G. M., Robinson R., Todd A. R. (1934) Experiments on the synthesis of anthocyanins. Part XIX. 5-Glucosidylapigeninidin, believed to be identical with gesnerin, an anthocyanin of *Gesneria fulgens*. J. Chem. Soc. 809–813.

Robinson H. (1978a) Studies in the Heliantheae (Asteraceae). XI. A new genus *Kingianthus*. Phytologia 38:415–416.

Robinson H. (1978b) Studies in the Heliantheae (Asteraceae). XII. Re-establishment of the genus *Smallanthus*. Phytologia 39:47–53.

Robinson H. (1979) Studies in the Heliantheae (Asteraceae). XVIII. A new genus *Helianthopsis*. Phytologia 44:257–269.

Robinson H. (1980a) Studies in the Heliantheae (Asteraceae). XXV. A new species of *Unxia*. Phytologia 46:289–292.

Robinson H. (1980b) Re-establishment of the genus *Critoniopsis* (Vernonieae: Asteraceae). Phytologia 46:437–442.

Robinson H. (1981) A revision of the tribal and subtribal limits of the Heliantheae (Asteraceae). Smithsonian Contr. Bot. 51:1–102.

Robinson H. (1983a) Five new species of *Lychnophora* from Bahia, Brasil (Vernonieae: Asteraceae). Phytologia 53:369–384.

Robinson H. (1983b) A generic review of the tribe Liabeae (Asteraceae). Smithsonian Contr. Bot. 54:1–69.

Robinson H. (1984) Style rotation in the Asteraceae. Taxon 33:400–404.

Robinson H. (1987) Some suggestions regarding the significance of chloroplast DNA variation in the Asteraceae. Phytologia 65:316–324.

Robinson H. (1988a) Studies in the *Lepidaploa* complex (Vernonieae: Asteraceae) IV. The new genus *Lessingianthus*. Proc. Biol. Soc. Wash. 101:929–951.

Robinson H. (1988b) Studies in the *Lepidaploa* complex (Vernonieae: Asteraceae) V. The new genus *Chrysolaena*. Proc. Biol. Soc. Wash. 101:952–958.

Robinson H. (1988c) Studies in the *Lepidaploa* complex (Vernonieae: Asteraceae) VI. The new genus *Aynia*. Proc. Biol. Soc. Wash. 101:959–965.

Robinson H. (1990a) Six new combinations in *Baccharoides* Moench and *Cyanthillium* Blume (Vernonieae: Asteraceae). Proc. Biol. Soc. Wash. 103:248–253.

Robinson H. (1990b) Studies in the *Lepidaploa* complex (Vernonieae: Asteraceae) VII. The genus *Lepidaploa*. Proc. Biol. Soc. Wash. 103:464–498.

Robinson H. (1992a) The Asteraceae of the Guianas, III: Vernonieae and restoration of the genus *Xiphochaeta*. Rhodora 94:348–361.

Robinson H. (1992b) A new genus *Vernonanthura* (Vernonieae, Asteraceae). Phytologia 73:65–76.

Robinson H. (1993a) A review of the genus *Critoniopsis* in Central and South America (Vernonieae: Asteraceae). Proc. Biol. Soc. Wash. 106:606–627.

Robinson H. (1993b) New species of *Ericentrodea* from Bolivia and Colombia (Asteraceae, Coreopsidinae, Heliantheae). Novon 3:75–78.

Robinson H. (1994) *Colobus*, *Pseudopiptocarpha*, and *Trepadonia*, three new genera from South America (Vernonieae: Asteraceae). Proc. Biol. Soc. Wash. 107:557–568.

Robinson H. (1995) Two new species of *Ichthyothere* (Heliantheae: Asteraceae) from Ecuador and Peru. Sida 16:731–736.

Robinson H. (1996a) The status of generic and subtribal revisions in the Vernonieae. In: Hind D. J. N., Beentje H. J. (eds.) *Compositae: Systematics. Proceedings of the International Compositae Conference, Kew (1994) Vol. 1*. Kew: Royal Botanic Gardens. 511–529.

Robinson H. (1996b) Recent studies in the Heliantheae and Eupatorieae. In: Hind D. J. N., Beentje H. J. (eds.) *Compositae: Systematics. Proceedings of the International Compositae Conference, Kew (1994) Vol. 1*. Kew: Royal Botanic Gardens. 627–653.

Robinson H. (1999) Generic and subtribal classification of American Vernonieae. Smithson. Contr. Bot. 89:1–116.

Robinson H., Brettell R. D. (1973a) Tribal revisions in the Asteraceae. III. A new tribe, Liabeae. Phytologia 25:404–407.

Robinson H., Brettell R. D. (1973b) Tribal revisions of Asteraceae. V. The relationships of *Rigiopappus*. Phytologia 26:69–70.

Robinson H., Brettell R. D. (1973c) Tribal revisions in the Asteraceae. VIII. A new tribe, Ursinieae. Phytologia 26:76–85.

Robinson H., Brettell R. D. (1973d) Studies in the Senecioneae (Asteraceae), III: the genus *Psacalium*. Phytologia 27:254–264.

Robinson H., Cuatrecasas J. (1973) The generic limits of *Pluchea* and *Tessaria* (Inuleae, Asteraceae). Phytologia 27:277–285.

Robinson H., Cuatrecasas J. (1992) *Thelechitonia* Cuatrecasas, an older name for *Complaya* Strother (Ecliptinae-Heliantheae-Asteraceae). Phytologia 72:141–143.

Robinson H., Kahn B. (1986) Trinervate leaves, yellow flowers, tailed anthers, and pollen variation in *Distephanus* Cassini (Vernonieae: Asteraceae). Proc. Biol. Soc. Wash. 99:493–501.

Robinson H., King R. M. (1977) Eupatorieae—systematic review. In: Heywood V. H., Harborne J. B., Turner B. L. (eds.) *The Biology and Chemistry of the Compositae*, Vol. 1. London: Academic Press. 437–486.

Robinson H., King R. M. (1985) Comments on generic concepts in the Eupatorieae. Taxon 34:11–16.

Robson K. A., McCormick S. P. (1988) 6-Methoxyflavonoids from *Balsamorhiza* section *Artorhiza*. Biochem. Syst. Ecol. 16:411–412.

Rodriguez E. (1977) Ecogeographic distribution of secondary constituents in *Parthenium* (Compositae). Biochem. Syst. Ecol. 5:207–218.

Rodriguez E., Carman N. J., Chavez P., Mabry T. J. (1972a) Quercetagetin 3,3'-dimethyl ether, a new flavonoid from *Parthenium tomentosum*. Phytochemistry 11:1507–1508.

Rodriguez E., Carman N. J., Vander Velde G., McReynolds J. H., Mabry T. J., Irwin M. A., Geissman T. A. (1972b) Methoxylated flavonoids from *Artemisia*. Phytochemistry 11:3509–3514.

Roitman J. N., James L. F. (1985) Chemistry of toxic range plants. Highly oxygenated flavonol methyl ethers from *Gutierrezia microcephala*. Phytochemistry 24:835–848.

Romo de Vivar A., Reyes B., Delgado G., Schlemper E. O. (1982) Constituents of *Pluchea sericea*. Structure and stereochemistry of (11*S*)-11,13-dihydrotessaric acid. Chem. Lett. 957–960.

Romussi G., Ciarallo G. (1978) Flavonoidverbindungen aus *Scolymus hispanicus* L. Die Pharmazie 33:685–686.

Romussi G., Pagani F. (1970) Constituents of *Bidens frondosa*. Boll. Chim. Farm. 109:467–475.

Roose M. L., Gottlieb L. D. (1976) Genetic and biochemical consequences of polyploidy in *Tragopogon*. Evolution 30:818–830.

Ros Barcelo A., Muñoz R. (1989) Epigenetic control of a cell wall scopoletin peroxidase by lupisoflavone in *Lupinus*. Phytochemistry 28:1331–1333.

Roseman R. R. (1986) A Systematic Study of *Bidens* sect. *Greenmania* in Mexico, Central America and Jamaica: Chemotaxonomy, Cytotaxonomy and Phenetics. Ph.D. Dissertation, University of Iowa.

Rosler H., Goodwin R. S., Mabry T. J., Varma S. D., Norris J. (1984) Flavonoids with anticataract activity from *Brickellia arguta*. J. Nat. Prod. 47:316–319.

Rosler H., Star A. E., Mabry T. J. (1971) New 6-methoxyflavonols from *Centaurea jacea*. Phytochemistry 10:450–451.

Ross S. A., Sayed H. M., El-Sayyad S. M. (1984) Phytochemical studies on *Aster squamatus* L. Part. I. Flavonoid constituents of the flowers. Egypt. J. Pharm. Sci. 25:195–202.

Rossomando P. C., Guerreiro E., Giordano O. S. (1990) Phytochemical study of four *Baccharis* species that were included in the genus *Psila*. Rev. Latinoamer. Quim. 21:18–20.

Ruangrungsi N., Tappayuthpijarn P., Tantivatana P., Borris R. P., Cordell G. A. (1981) Traditional medicinal plants of Thailand. I. Isolation and structure elucidation of two new flavonoids, (2*R*,3*R*)-dihydroquercetin-4'-methyl ether and (2*R*,3*R*)-dihydroquercetin-4',7-dimethyl ether from *Blumea balsamifera*. J. Nat. Prod. 44:541–545.

Rubery P. H., Jacobs M. (1988) Auxin transport and its regulation by flavonoids. In: Pharis R. P., Rood S. B. (eds.) *Plant Growth Substances (International Conference on Plant Growth Substances)*. Berlin: Springer. 428–440.

Rubio B., Diaz A. M. (1995) Flavonol glycosides in leaves and flowers of *Scolymus hispanicus*. Die Pharmazie 50:629–631.

Rubio B., Villaescusa L., Diaz A. M., Fernandez L., Martin T. (1995) Flavonol glycosides from *Scolymus hispanicus* and *Jasonia glutinosa*. Planta Medica 61:583.

Rubio-Garcia B., Elias R., Faure R., Diaz-Lanza A. M., Debrauwer L., Balansard G. (1994) Flavonol glucuronides from *Scolymus hispanicus* leaves. Pharm. Pharmacol. Lett. 3:207–208.

Rücker G., Kehrbaum S., Sakulas H., Lawong B., Goeltenboth F. (1994) Acetylated aurone glucosides from *Microglossa pyrifolia*. Planta Med. 60:288–289.

Ruhnau B., Forkmann G. (1988) Flavan-3,4-diols in anthocyanin biosynthesis, enzymatic formation with flower extracts from *Callistephus chinensis*. Phytochemistry 27:1035–1039.

Ruiz S. O., Guerreiro E., Giordano O. S. (1981) Flavonoids in three species of *Grindelia*. An. Asoc. Quim. Argent. 69:293–295.

Ruiz S. O., Guerreiro E., Giordano O. S. (1992) Flavonoids of *Thymophylla belenidium* DC. Fitoterapia 63:276–277.

Ruiz-Cancino A., Cano A. E., Delgado G. (1993) Sesquiterpene lactones and flavonoids from *Artemisia ludoviciana* ssp. *mexicana*. Phytochemistry 33:1113–1115.

Rustaiyan A., Niknejad A., Aynehchi Y. (1982) Chemical constituents of *Centaurea brugueriana*. Planta Medica 44:185–186.

Rwangabo P. C., Laekeman G., Claeys M., Tote J., Pieters L., Vanden Berghe D., Herman A., Vlietinck A. J. (1986) Phytochemical and pharmacological investigation of the biologically active fraction from the flowers of *Vernonia amygdalina*. Planta Medica 52:P34.

Ryakhovskaya T. V., Alyukina L. S. (1973) Flavonoid components of *Artemisia scoparia*. Fenol'nya Soed. Ikh. Fiziol. Svoistva, 76–79.

Ryakhovskaya T. V., Alyukina L. S. (1980), Flavonoid characteristics of some wormwoods of the subgen. *Seriphidium* Bess. Rouy, Izv. Akad. Nauk. Kaz. SSR, Ser. Biol, 4:7–11.

Ryakhovskaya T. V., Manadilova A. M., Sapko O. A. (1985) Flavonoids of *Artemisia sublessingiana*. Chem. Nat. Cmpds. 21:381 [407].

Ryakhovskaya T. V., Sapko O. A. (1985) Flavonoids of *Artemisia cina*. Chem. Nat. Cmpds. 21:380 [406].

Ryakhovskaya T. V., Sapko O. A., Manadilova A. M., Ushbaeva G. G. (1987) Polyphenols of sagebrush and biological activity of its phenolic complex. Proc. 3rd F.E.C.S. Int. Conf. Chem. Biotechnol. Biol. Act. Nat. Prod. 4:380.

Rydberg P. A. (1914) Helenieae. Bahianae. North American Flora 34:31–43.

Rydberg P. A. (1922) Carduales. Family 1. Ambrosiaceae. North American Flora 33:3–44.

Ryding O., Bremer K. (1992) Phylogeny, distribution and classification of the Coreopsideae (Asteraceae). Syst. Bot. 17:649–659.

Saad J. R., Davicino J. G., Giordana O. S. (1988) A diterpene and flavonoids of *Baccharis flabellata*. Phytochemistry 27:1884–1887.

Saad J. R., Pestchanker M. J., Giordano O. S. (1987) Furanoditerpenes from *Baccharis thymifolia*. Phytochemistry 26:3033–3035.

Sagareishvili T. G. (1985) Flavonoids of *Eupatorium micranthum*. Chem. Nat. Cmpds. 21:668 [710].

Sagareishvili T. G., Alaniya M. D., Kemertilidze E. P. (1983) An apigenin glucuronide from *Leucanthemum vulgare*. Chem. Nat. Cmpds. 19:614 [647].

Sagareishvili T. G., Alaniya M. D., Kikoladze V. S., Kemertilidze E. P. (1982) Nivyaside–a new glycoside from *Leucanthemum vulgare*. Chem. Nat. Cmpds. 18:408 [442].

Sagareishvili T. G., Alaniya M. D., Tsitsishvili V. G., Kemertilidze E. P. (1981) Micranthoside – a new glycoside from *Eupatorium micranthum*. Chem. Nat. Cmpds. 17:225 [298].

Saitbaeva I. M., Mallabaev A., Sidyakin G. P. (1983) Components of *Artemisia leucodes*. Chem. Nat. Cmpds. 19:741 [780].

Saito K. (1974) Distribution of flavonoids and related compounds in various parts of *Cosmos bipinnatus*. Z. Pflanzenphysiol. 71:80–82.

Saito K. (1976) Flavone glycosides in the ray florets of *Cosmos pinnatus*. Planta Medica 30:349–355.

Saito K., Kanehira T., Horimoto M., Moritome N., Komamine A. (1993) Biosynthesis of carthamin in florets and cultured cells of *Carthamus tinctorius*. Biol. Plant. 35:537–546.

Saito N., Harborne J. B. (1992) Correlations between anthocyanin type, pollinator and flower colour in Labiatae. Phytochemistry 31:3009–3015.

Saito N., Ishizuka K., Osawa Y. (1970) Paper chromatographic identification of flavonoids from a scarlet-flowering dahlia and crystallization of pelargonidin and butein. Bot. Mag. Tokyo. 83:229–232.

Saleh A. A., Cordell G. A., Farnsworth N. R. (1976) Isolation of 3,6-dimethoxy-4',5,7-trihydroxyflavone from *Acanthospermum glabratum*. J. Nat. Prod. 39:456–458.

Saleh A. A., Cordell G. A., Farnsworth N. R. (1980) Potential anticancer agents. Part 13. Cytotoxic constituents of *Acanthospermum glabratum* (Asteraceae). J. Chem. Soc. Perkin Trans. I 1090–1097.

Saleh M. M., Rizk A. M. (1974) Flavonoids and coumarins of *Anthemis pseudocotula*. Planta Medica 25:60–62.

Saleh M. R. I., Habib A. A. M., El-Ghazouly M. G., Gabr O. M. K., El-Fiky F. K. (1988) Chemical constituents from *Launaea resedifolia*. Egypt. J. Pharm. 29:507–510.

Saleh M. R. I., Metwally A. M., Amer M. M. A. (1975) Isolation of a flavonoidal substance from *Cichorium pumilum* Jacq. Die Pharmazie 30:404.

Saleh N. A. M., Bohm B. A. (1971) Flavonoids of *Arctium minus* (Compositae). Experientia 27:1494.

Saleh N. A. M., Bohm B. A., Ornduff R. (1971) Flavonoid constituents of *Lasthenia fremontii* and *L.conjugens*. Phytochemistry 10:611–614.

Saleh N. A. M., El-Negoumy I., Abd-Alla M. F., Abou-Zaid M. M., Dellamonica G., Chopin J. (1985) Flavonoid glycosides of *Artemisia monosperma* and *A. herba-alba*. Phytochemistry 24:201–203.

Saleh N. A. M., El-Negoumy S. I., Abou-Zaid M. M. (1987) Flavonoids of *Artemisia judaica, Artemisia monosperma* and *Artemisia herba-alba*. Phytochemistry 26:3059–3064.

Saleh N. A. M., Mansour R. M. A., El-Kareemy Z. A. R., Fayed A. A. (1988) The chemosystematics of local members of the subtribe Gnaphaliinae (Compositae). Biochem. Syst. Ecol. 16:615–617.

San Feliciano A., Medarde M., Gordaliza M., des Oro E., del Corral J. M. M. (1989) Sesquiterpenoids and phenolics of *Pulicaria paludosa*. Phytochemistry 28:2717–2721.

San Martin A., Rovirosa J., Castillo M. (1983) Diterpenoids from *Baccharis tola*. Phytochemistry 22:1461–1463.

Sanders R. W., Stuessy T. F., Marticorena C., Silva O. M. (1987) Phytogeography and evolution of *Dendroseris* and *Robinsonia*, tree-Compositae of the Juan Fernandez Islands. Opera Bot. 92:195–215.

Sando C. E. (1926a) Anthocyanin formation in *Helianthus annuus*. J. Biol. Chem. 64:71–74.

Sando C. E. (1926b) The coloring matter, quercemeritrin, from the double chrysanthemum-flowered *Helianthus annuus*. J. Biol. Chem. 68:407–414.

Saner A., Leupin K. (1966) The composition of *Artemisia montana*. Pharm. Acta Helv. 41:431–435.

Sang T., Crawford D. J., Kim S.-C., Stuessy T. F. (1994) Radiation of the endemic genus *Dendroseris* (Asteraceae) on the Juan Fernandez Islands: evidence from sequences of the ITS regions of nuclear ribosomal DNA. Amer. J. Bot. 81:1494–1501.

Sang T., Crawford D. J., Stuessy T. F., Silva O. M. (1995) ITS sequences and the phylogeny of the genus *Robinsonia* (Asteraceae). Syst. Bot. 20: 55–64.

Sanlaville C., Jay M., Guiard J. (1988a) Phenolic variation in the progeny of the cultivated sunflower. Biochem. Syst. Ecol. 16:269–272.

Sanlaville C., Jay M., Guiard J. (1988b) Utilisation des composes phenoliques pour le marquage de pools geniques de Tournesol cultive. Agronomie 8:341–345.

Santavy F., Potesilova H., Kubicek R. (1957) Isolation of rutin from *Senecio erraticus* ssp. *barbaraeifolius*. Chem. Listy 51:1767–1769.

Santavy F., Potesilova H., Kubicek R. (1959) Isolation of rutin from *Senecio erraticus* ssp. *barbaraeifolius*. Coll. Czech. Chem. Communs. 24:646–648.

Santos S. M. B. P., Palma F. B., Grande M. B. (1992) Flavonoids from *Centaurea rothmalerana*, an endemic species of Portugal. Bull. Liaison—Groupe Polyphenol 16(Pt. 1):153–156.

Sanz J. F., Barbera O., Alberto Marco J. (1989) Sesquiterpene lactones from *Artemisia hispanica*. Phytochemistry 28:2163–2167.

Sanz J. F., Garcia-Lliso V., Marco J. A., Valles Xirau J. (1991) A cadinene derivative from *Artemisia crithmifolia*. Phytochemistry 30:4167–4168.

Sanz J., Martinez-Castro I., Pinar M. (1985) Chemical constituents of *Achillea santolinoides*. J. Nat. Prod. 48:993.

Sanz M. J., Terencio M. C., Manez S., Rios J. L. (1993) A new quercetin acylglucuronide from *Scolymus hispanicus*. J. Nat. Prod. 56:1995–1998

Sarg T. M. (1977) Phytochemical investigation of *Pulicaria crispa* (Forsk). B. & H. growing in Saudi Arabia. Egypt. J. Pharm. Sci. 16:421–427.

Sarg T. M., Ateya A. M., Dora G. A. (1987) 3,4-Dihydroscopoletin, a new compound from *Launaea spinosa* growing in Egypt. Fitoterapia 58:133.

Sarg T. M., El-Domiaty M. M., Ateya A.-M. M., El-Dahmy S. I., El-Shazly A. M. (1993) Phytochemical investigation of *Centaurea eryngioides* Lam. growing in Egypt. Alexandria J. Pharm. Sci. 7:50–54.

Sarg T. M., Omar A. A., Ateya A. M., Hafiz S. S. (1984) Phenolic constituents of *Launaea nudicalis* (L.). Hook. Egypt. J. Pharm. Sci. 25:35–38.

Sashida Y., Nakata H., Shimomura H., Kagaya M. (1983) Sesquiterpene lactones from Pyrethrum flowers. Phytochemistry 22:1219–1222.

Sato M., Moreira E. O. (1985) Flavonoids in *Eupatorium littorale* Cabrera. Trib. Farm. 53:13.

Sayed H. M., Abdel-Hafez M. A., Anton R. (1991) Phytochemical study of *Conyza dioscoridis*. I. Steroids, terpenes and flavonoids from the leaves. Bull. Fac. Sci., Assiut Univ. 20:23–30.

Sayed H. M., Ross S. A. (1986) Phytochemical studies on *Aster squamatus* L. Part III: Constituents of the leaves. Bull. Pharm. Sci., Assiut Univ. 9:149–163.

Schenkel E. P., Sobottka A. M., Lankhorst P. P., Schripsema J. (1997) An unusual diterpene and other constituents from *Baccharis ochracea*. Phytochemistry 44:755–757.

Schilling E. E. (1983) Flavonoids of *Helianthus* series *Angustifolii*. Biochem. Syst. Ecol. 11:341–344.

Schilling E. E. (1989) External flavonoid aglycones of *Viguiera* series *Viguiera*. Biochem. Syst. Ecol. 17:535–538.

Schilling E. E., Jansen R. K. (1989) Restriction fragment analysis of chloroplast DNA and the systematics of *Viguiera* and related genera (Asteraceae: Heliantheae). Amer. J. Bot. 76:1769–1778.

Schilling E. E., Mabry T. J. (1981) Flavonoids of *Helianthus* series *Corona-solis*. Biochem. Syst. Ecol. 9:161–163.

Schilling E. E., Panero J. L. (1988) Flavonoids of *Viguiera* series *Brevifoliae*. Biochem. Syst. Ecol. 16:417–418.

Schilling E. E., Panero J. L. (1991) Evidence for a close relationship between *Iostephane* and *Viguiera* (Asteraceae: Heliantheae). Amer. J. Bot. 78:1054–1062.

Schilling E. E., Panero J. L. (1996) Relationships in Heliantheae subtribe Helianthinae based on chloroplast DNA restriction site analysis. In: Hind D. J. N., Beentje H. J. (eds.) *Compositae: Systematics. Proceedings of the International Compositae Conference, Kew (1994) Vol. 1*. Kew: Royal Botanic Gardens. 361–376.

Schilling E. E., Panero J. L., Bohm B. A. (1988) Flavonoids of *Viguiera* section *Maculatae*. Biochem. Syst. Ecol. 16:413–416.

Schilling E. E., Panero J. L., Storbeck T. A. (1987) Flavonoids of *Helianthus* series *Microcephali*. Biochem. Syst. Ecol. 15:671–672.

Schilling E. E., Rieseberg L. H. (1985) Foliar flavonoids of *Phoebanthus*. Biochem. Syst. Ecol. 13:403–404.

Schilling E. E., Spooner D. M. (1987) Flavonoids of *Simsia*. Amer. J. Bot. 74:704. Abstr.

Schilling E. E., Spooner D. M. (1988) Floral flavonoids and the systematics of *Simsia* (Asteraceae, Heliantheae). Syst. Bot. 13:572–575.

Schmeda G., Pacheco P., Silva M. (1983) Flavonoids of *Eupatorium laeve* DC. Rev. Latinoamer. Quim. 13:36.

Schoch R. M. (1986) *Phylogeny Reconstruction in Paleontology*. New York: Van Nostrand-Reinhold.

Scholz E., Heinrich M., Hunkler D. (1994) Caffeoylquinic acids and some biological activities of *Pluchea symphytifolia*. Planta Medica 60:360–364.

Schrall R., Becker H. (1977) Callus- und Suspensionskulturen von *Silybum marianum*. II. Umsetzung von Flavonoiden mit Coniferylalkohol zu Flavonolignanen. Planta Medica 32:27–32.

Schröder G., Zähringer U., Heller W., Ebel J., Grisebach H. (1979) Biosynthesis of antifungal isoflavones in *Lupinus albus*. Enzymatic prenylation of genistein and 2'-hydroxygenistein. Arch. Biochem. Biophys. 194:635–636.

Schulte K. E., Rucker G., Muller F. (1968) Einige Inhaltsstoffe der Blütenköpfchen von *Pulicaria dysenterica*. Arch. Pharm. 301:115–119.

Schulte K. E., Rucker G., Reithmayr K. (1969) Einige Inhaltsstoffe von *Arnica chamissonis* und anderer *Arnica*-Arten. Lloydia 32:360–368.

Schuster N., Christiansen C., Jakupovic J., Mungai M. (1993) An unusual [2+2] cycloadduct of terpenoid coumarin from *Ethulia vernonioides* (Schweinf.) M. Gilbert. Phytochemistry 34:1179–1181.

Schwendimann J.-M., Tabacchi R., Jacot-Guillarmod A. (1974) Identification des composes phenoliques dans les feuilles de *Homogyne alpina*, *Petasites albus*, *Petasites hybridus* et *Adenostyles alliariae*. Helv. Chim. Acta 57:552–557.

Scogin R. (1976) Floral UV patterns and anthochlor pigments in the genus *Coreopsis* (Asteraceae). Aliso 8:425–431.

Scogin R. (1978) Floral UV-absorption patterns and anthochlor pigments in the Asteraceae. Southwest. Nat. 23:371–374.

Scogin R. (1983) Natural variation and regulation of floral ultraviolet absorption in *Coreopsis* and *Bidens*. Aliso 10:443–448.

Scogin R., Young D. A., Jones C. E., Jr. (1977) Anthochlor pigments and pollination biology. II. The ultraviolet floral pattern of *Coreopsis gigantea* (Asteraceae). Bull. Torrey Bot. Club 104:155–159.

Scogin R., Zakar K. (1976) Anthochlor pigments and floral UV patterns in the genus *Bidens*. Biochem. Syst. Ecol. 4:165–167.

Scotland R. W., Siebert D. J., Williams D. M. (eds.) (1994) *Models in Phylogeny Reconstruction*. Oxford: Clarendon Press.

Scott R. W. (1985) Microcharacters as generic markers in the Eupatorieae. Taxon 34:26–30.

Seaman F. C. (1982) Sesquiterpene lactones as taxonomic characters in the Asteraceae. Botanical Review 48:121–595.

Seaman F. C., Funk V. A. (1983) Cladistic analysis of complex natural products: developing transformation series from sesquiterpene data. Taxon 32:1–27.

Seaman F. C., Rodriguez E., Carman N. J., Mabry T. J. (1972) A new flavonoid from *Ambrosia dumosa*. Phytochemistry 11:2626–2627.

Seeligmann P. (1996) Flavonoids of the Compositae as evolutionary parameters in the tribes which synthesize them: a critical approach. In: Hind D. J. N., Beentje H. J. (eds.) *Compositae: Systematics. Proceedings of the International Compositae Conference, Kew (1994) Vol. 1*. Kew: Royal Botanic Gardens. 159–167.

Seeligmann P., Alston R. E. (1967) Complex chemical variation and the taxonomy of *Hymenoxys scaposa* (Compositae). Brittonia 19:205–211.

Segal R., Cohen D., Sokoloff S., Zaitschek D. V. (1973) A new flavone from *Artemisia herba-alba*. Lloydia 36:103–105.

Seida A. A., El-Gohary H. M., Azzam S. M., El-Deeb K. S. (1990) Flavonoids of *Achillea fragrantissima*. Bull. Fac. Pharm. Cairo Univ. 28:59–61.

Seidl P. R., Gottlieb O. R., Kaplan M. A. C. (eds.) (1995) *Chemistry of the Amazon: Biodiversity, Natural Products, and Environmental Issues*. Washington, D.C.: American Chemical Society.

Seikel M. K., Geissman T. A. (1950) Anthochlor pigments. VI. The pigments of *Coreopsis stillmanii*. J. Am. Chem. Soc. 72:5720–5724.

Sell P. D. (1976) *Cichorium*. In: Tutin T. G., Heywood V. H., Burges N. A., Moore D. M., Valentine D. H., Walters S. M., Webb D. A. (eds.) *Flora Europaea, Vol. 4, Plantaginaceae to Compositae (and Rubiaceae)*. Cambridge: Cambridge Univ. Press. 304–305.

Semple J. C. (1996) A revision of *Heterotheca* sect. *Phyllotheca* (Nutt.) Harms (Compositae: Astereae): the prairie and montane goldenasters of North America. Univ. Waterloo Biol. Ser. No. 37:1–164.

Semple J. C., Averett J. E. (1975) Flavonoid variation in *Xanthisma texanum*: infrapopulation and interpopulation variation. Biochem. Syst. Ecol. 3:11–14.

Semple J. C., Brouillet L. (1980) A synopsis of North American asters: the subgenera, sections and subsections of *Aster* and *Lasallea*. Amer. J. Bot. 67:1010–1026.

Semple J. C., Chinnappa C. C. (1984) Observations on the cytology, morphology, and ecology of *Bradburia hirtella* (Compositae—Astereae). Syst. Bot. 9:95–101.

Sen A. K., Mahato S. B., Dutta N. L. (1976) Flavonoids of *Conyza stricta* (Compositae): isolation of 5,6-dihydroxy-3,8,4'-trimethoxy- and 5,7-dihydroxy-3,8,3',4',5'-pentamethoxyflavones. Indian J. Chem. 14B:849.

Senatore F., D'Agostino M., Dini I. (1999) Two new quercetagetin *O*-glucosides from *Tagetes mandonii*. Biochem. Syst. Ecol. 27:309–311.

Sendra J., Walczak Z. (1967) Flavonoid compounds in Herba Cnici. Diss. Pharm. Pharmacol. 19:661–668.

Sepulveda S., Delhvi S., Koch B., Zilliken F., Cassels B. K. (1994a) Constituents of *Senecio hollermayerii*. Fitoterapia 65:88.

Sepulveda S., Delhvi S., Zilliken F., Cassels B. K. (1994b) Constituents of *Centaurea chilensis*. Fitoterapia 65:88–89.

Serbin A. G., Borisov M. I., Chernobai V. I. (1972a) Flavonoids of *Bidens tripartita*. I. Chem. Nat. Cmpds. 8:119 [121].

Serbin A. G., Borisov M. I., Chernobai V. I. (1972b) Flavonoids of *Bidens tripartita*. II. Chem. Nat. Cmpds. 8:439 [440].

Serbin A. G., Borisov M. I., Chernobai V. I., Kovalev I. P., Gordienko V. (1973) Flavonoids of *Bidens tripartita*. III. Chem. Nat. Cmpds. 9:160 [144].

Serbin A. G., Borisov M. I., Chernobai V. I., Kovalev I. P., Gordienko V. (1975) Flavonoids of *Bidens tripartita*. Absolute configuration of isocoreopsin. Farm. Zh. Kiev. 30:88.

Seshadri T. R. (1956) Occurrence of chalcones and flavanones in plant products. Sci. Proc. Royal Dublin Soc. 27:77–86.

Seshadri T. R., Thakur R. S. (1960) The colouring matter of the flowers of *Carthamus tinctorius*. Curr. Sci. 29:54–55.

Sezik E., Akdemir Z. (1986) Flavonoids of *Helichrysum pamphylicum* Davis-Kupicha. Acta Pharm. Turc. 28:141–144.

Sham'yanov I. D., Batirov E. Kh., Yuldashev M. P., Mallabaev A. (1983) Components of *Saussurea elegans*. Chem. Nat. Cmpds. 19:763 [796].

Shannon R. K., Wagner W. L. (1997) *Oparanthus* (Asteraceae, subtribe Coreopsidinae) revisited. Allertonia 7:273–295.

Shaw S. C., Sharam V., Agrhari S. L. (1990) Synthesis of 5,6,7,8,3',4',5'-heptamethoxyflavone isolated from *Eupatorium coelestinum* and *Ageratum conyzoides*. Acta Cienc. Indica, Chem. 16C:233–236.

Shawl A. S. (1993) Constituents of *Tanacetum dolichophyllum* and *Tanacetum gracile*. Fitoterapia 64:284.

Sheldon J. C., Burrows F. M. (1973) The dispersal effectiveness of the achene-pappus units of selected Compositae in steady winds with convection. New Phytol. 72:665–675.

Shelyuto V. L., Ban'kovskii A. I. (1970) Flavonoids of *Cirsium arvense*. Chem. Nat. Cmpds. 6:366 [365–366].

Shelyuto V. L., Bubon N. T., Al'khimovich V. N., Smirnova L. P. (1976a) Flavonoids from *Hieracium umbellatum*. Chem. Nat. Cmpds. 12:660–661 [592].

Shelyuto V. L., Bubon N. T., Piotukh V. I., Smirnova L. P. (1976b) Flavonoids from *Arnica foliosa*, Chem. Nat. Cmpds. 12:483 [541].

Shelyuto V. L., Glyzin V. I., Ban'kovskii A. I. (1970) Flavonoids of *Cirsium arvense*. Chem. Nat. Cmpds. 6:366 [365].

Shelyuto V. L., Glyzin V. I., Ban'kovskii A. I., Bubon N. T. (1971a) The flavonoids of the flowers of *Cirsium oleraceum*, Chem. Nat. Cmpds. 7:347 [371].

Shelyuto V. L., Glyzin V. I., Ban'kovskii A. I., Bubon N. T. (1971b) Flavonol glycosides of *Cirsium oleraceum*. Chem. Nat. Cmpds. 7:349 [372].

Shelyuto V. L., Glyzin V. I., Bubon N. T. (1972a) 3-*O*-Methyl-kaempferol from the flowers of *Cirsium arvense*. Chem. Nat. Cmpds. 8:114 [118].

Shelyuto V. L., Glyzin V. I., Bubon N. T. (1972b) Flavonoid glucuronides from *Cirsium arvense*. Chem. Nat. Cmpds. 8:236 [240].

Shelyuto V. L., Glyzin V. I., Bubon N. T. (1972c) The flavonoids of *Cirsium palustre*. Chem. Nat. Cmpds. 8:238 [241].

Shelyuto V. L., Glyzin V. I., Kruglova E. P., Smirnova L. P. (1977) Flavonoids from *Hieracium pilosella*. Chem. Nat. Cmpds. 13:727 [860].

Shen M. C., Rodriguez E., Kerr K., Mabry T. J. (1976) Flavonoids of four species of *Parthenium* (Compositae). Phytochemistry 15:1045–1047.

Sherff E. E. (1936) Revision of the genus *Coreopsis*. Field Mus. Nat. Hist., Bot. 11:279–475.

Sherff E. E. (1937) The genus *Bidens*, parts I and II. Field Mus. Nat. Hist., Bot. 16:1–709.

Sherff E. E. (1955a) *Bidens*. In: Sherff E. E., Alexander E. J. (eds.) *North American Flora*, Series II, Part 2. New York: New York Botanical Garden.

Sherff E. E. (1955b) *Coreopsis*. In: Sherff E. E., Alexander E. J. (eds.) *North American Flora*, Series II, Part 2. New York: New York Botanical Garden.

Shi J., Jia Z., Li Y. (1991) Chemical constituents of *Saussurea japonica* (Thumb.) DC. Gaodeng Xueziao Huaxue Xuebao 12:906–909.

Shi Y.-p., Li Y., Zhang H.-c. (1992) Chemical constituents of *Artemisia subdigitata* Mattf. Gaodeng Xueziao Huaxue Xuebao 13:1258–1261.

Shilin Y., Roberts M. F., Phillipson J. D. (1989) Methoxylated flavones and coumarins from *Artemisia annua*. Phytochemistry 28:1509–1511.

Shimokoriyama M. (1957) Anthochlor pigments of *Coreopsis tinctoria*. J. Am. Chem. Soc. 79:214–220.

Shimokoriyama M. (1962) Flavanones, chalcones and aurones. In: T. Geissman (ed.) *The Chemistry of Flavonoid Compounds*. New York: The Macmillan Co.

Shimokoriyama M., Geissman T. A. (1960) Anthochlor pigments. XIV. The pigments of *Viguiera multiflora* Nutt. and *Baeria chrysostoma* F. and M. J. Org. Chem. 25:1956–1959.

Shimokoriyama M., Hattori S. (1953) Anthochlor pigments of *Cosmos sulphureus*, *Coreopsis lanceolata* and *Coreopsis saxicola*. J. Am. Chem. Soc. 75:1900–1904.

Shishkin B. K., Bobrov E. G. (1961) *Compositae. Flora URSS (Flora Unionis Rerumpublicarum Socialisticarum Sovieticarum)*, Vol. 26. Moscow: Academiae Scientiarum URSS. (In Russian.)

Shmatova V. V., Komissarenko N. F. (1987) Flavonoids of Turkmenian species of the genus *Achillea*. Chem. Nat. Cmpds. 23:255 [301–302].

Si J., Gau G., Chen D. (1994) Phytochemical study of *Vernonia nigritiana* Oliv. et Hiern (Compositae). Tianran Chanwu Yanjiu Yu Kaifa 6:49–51.

Sigstad E., Catalan C. A. N., Diaz J. G., Herz W. (1993) Diprenylated derivatives of *p*-hydroxyacetophenone from *Ophryosporus macrodon*. Phytochemistry 33:165–169.

Silva G. A. B., Henriques A., Alice C. B., Gleye J., Moulis C. (1985) Methoxyflavones from *Baccharis patens*. J. Nat. Prod. 48:861.

Silva G. L., del V. Pacciaroni A., Oberti J. C., Espinar L. A., Diaz J. G., Herz W. (1992a) Helenanolides, guaianolide glucosides and other constituents of two *Helenium donianum* varieties. Phytochemistry 31:1621–1630.

Silva G. L., Oberti J. C., Herz W. (1992b) Sesquiterpene lactones and other constituents of Argentine *Ambrosia* species. Phytochemistry 31:859–861.

Silva M., Mundaco J. M., Sammes P. G. (1971) Flavonoid and triterpene constituents of *Baccharis rhomboidalis*. Phytochemistry 10:1942–1943.

Silva M., Wiesenfeld A., Sammes P. G., Tyler T. W. (1977) New sesquiterpenes from *Pleocarphus revolutus*. Phytochemistry 16:379–385.

Simöes F., Nacimento J. (1990) Constituents of *Dittrichia viscosa* subsp. *viscosa*. Fitoterapia 61:553.

Simpson B. B. (1983) An historical phytogeography of the high Andean flora. Revista Chilena Hist. Nat. 56:109–122.

Singh K. N., Pandey V. B. (1986) Isorhamnetin 7-glucoside from *Cnicus wallichi*. Phytochemistry 25:2683.

Singh K. N., Pandey V. B. (1989) Constituents of *Echinops echinatus*. Orient. J. Chem. 5:342.

Singh K. N., Pandey V. B. (1993) Antifungal flavonoids of *Echinops echinatus*. Orient. J. Chem. 9:149–151.

Singh K. N., Pandey V. B., Banerjee S., Bohlmann F., Keinan E. (1986) Novel acylated flavone glycosides from *Echinops echinatus*. Chem. Ind. (London) 713.

Singh K. N., Pandey V. B., Shah A. H. (1988) Constituents of *Cnicus wallichi*. Fitoterapia 59:78.

Singh R. P., Pandey V. B. (1994) Further flavonoids of *Echinops niveus*. Fitoterapia 65:374.

Singh R. P., Singh K. N., Pandy V. B. (1990) Constituents of *Echinops niveus*. Fitoterapia 61:279.

Singh U. P., Pandey V. B., Singh K. N., Singh R. D. (1988) Antifungal activity of some new flavones and flavone glycosides of *Echinops echinatus*. Can. J. Bot. 66:1901–1903.

Sinha S., Khanna R. K., Srivastava S. N., Singh A. (1986) Occurrence of apigenin and its glucoside in the flowers of *Chrysanthemum morifolium*. Himalayan Chem. Pharm. Bull. 3:8–9.

Sivakumar R., Nair A. G. R., Madhusudhanan K. P. (1995) Polyphenolic compounds of flowers of *Helichrysum bracteatum*. Curr. Sci. 69:23–24.

Skibinski A., Merfort I., Willuhn G. (1994) Thirty-seven flavonoids from flowers of *Arnica longifolia*. Phytochemistry 37: 1635–1636.

Skrzypczakowa L. (1961) Flavonoids in *Solidago virgaurea*, I. Isolation and identification of rutin. Acta Polon. Pharm. 18:39–40.

Skvarla J. J., Turner B. L. (1966) Pollen wall ultrastructure and its bearing on the systematic position of *Blennosperma* and *Crocidium* (Compositae). Amer. J. Bot. 53:555–563.

Skvarla J. J., Turner B. L., Patel V. C., Tomb A. S. (1977) Pollen morphology in the Compositae and in morphologically related families. In: Heywood V. H., Harborne J. B., Turner B. L. (eds.) *The Biology and Chemistry of the Compositae, Vol. 1*. London: Academic Press. 141–265.

Small J. (1919) *The Origin and Development of the Compositae*. New Phytologist Reprint No. 11. London: William Wesley & Son.

Small J. K. (1903) *Flora of the Southeastern United States*. New York: Publ. by author.

Smith A. B. (1994) *Systematics and the Fossil Record*. London: Blackwell Sci. Publ.

Smith E. B. (1974) *Coreopsis nuecensis* (Compositae) and a related new species from southern Texas. Brittonia 26:161–171.

Smith E. B. (1975) The chromosome numbers of North American *Coreopsis* with phyletic interpretations. Bot. Gaz. 136:78–86.

Smith E. B. (1989) A biosystematic study and revision of the genus *Coreocarpus* (Compositae). Syst. Bot. 14:448–472.

Smith E. B., Crawford D. J. (1981) Comparative leaf flavonoid chemistry of *Coreopsis nuecensoides* and *C. nuecensis* (Compositae), a progenitor-derivative species pair. Bull. Torrey Bot. Club 108:7–12.

Smith L. C. (1944) *Lupinus*. Stanford: Stanford Univ. Press.

Smith-White S. (1968) *Brachycome lineariloba*: a species for experimental cytogenetics. Chromosoma (Berlin) 23:359–364.

Smolarz H., Krzaczek T. (1988) Phytochemical studies of the herb *Tragopogon orientalis* L. (Asteraceae) II. Components of the methanol extract. Acta Soc. Bot. Pol. 57:93–105.

Sneath P. H. A., Sokal R. R. (1973) *Numerical Taxonomy: The Principles and Practice of Numerical Classification*. San Francisco: Freeman.

Sober E. (1983) Parsimony in systematics: philosophical issues. Annu. Rev. Ecol. Syst. 14:335–358.

Sober E. (1988) *Reconstructing the Past: Parsimony, Evolution, and Inference*. Cambridge, Massachusetts: MIT Press.

Soicke H., Leng-Peschlow E. (1987) Characterization of flavonoids from *Baccharis trimera* and their antihepatotoxic properties. Planta Medica 53:37–39.

Sokal R. R., Sneath P. H. A. (1963) *Principles of Numerical Taxonomy*. San Francisco: Freeman.

Solbrig O. T. (1963) Subfamilial nomenclature of Compositae. Taxon 12:229–235.

Solbrig O. T. (1977) Chromosomal cytology and evolution in the family Compositae. In: Heywood V. H., Harborne J. B., Turner B. L. (eds.) *The Biology and Chemistry of the Compositae, Vol. 1*. London: Academic Press. 267–281.

Solomko O. V., Dzhumyrko S. F., Kompantsev V. A. (1978) Flavonoids of *Achillea nobilis*. Chem. Nat. Cmpds. 14:224 [266].

Soltis D. E., Soltis P. S. (1989) Allopolyploid speciation in *Tragopogon*: insights from chloroplast DNA. Amer. J. Bot. 76:1119–1124.

Soltis D. E., Soltis P. S. (1993) Molecular data and the dynamic nature of polyploidy. Crit. Rev. Plant Sci. 12:243–273.

Soltis D. E., Soltis P. S., Nickrent D. L., Johnson L. A., Hahn W. J., Hoot S. B., Sweere J. A., Kuzoff R. K., Kron K. A., Chase M. W., Swensen S. M., Zimmer E. A., Chaw S.-M., Gillespie L. J., Kress W. J., Sytsma K. J. (1997) Angiosperm phylogeny inferred from 18S ribosomal DNA sequences. Ann. Missouri Bot. Gard. 84:1–49.

Soltis P. S., Plunkett G. M., Novak S. J., Soltis D. E. (1995) Genetic variation in *Tragopogon* species: additional origins of the allotetraploids *T. mirus* and *T. miscellus* (Compositae). Amer. J. Bot. 82:1329–1341.

Soltis P. S., Soltis D. E. (1991) Multiple origins of the allotetraploid *Tragopogon mirus* (Compositae): rDNA evidence. Syst. Bot. 16:407–413.

Sørensen N. A. (1977) Polyacetylenes and conservation of chemical characters in the Compositae. In: Heywood V. H., Harborne J. B., Turner B. L. (eds.) *The Biology and Chemistry of the Compositae, Vol. 1.* London: Academic Press. 385–409.

Sorensen P. D. (1969) Revision of the genus *Dahlia* (Compositae, Heliantheae, Coreopsidinae). Rhodora 71:309–365, 367–416.

Sorensen P. D. (1970) The *Dahlia*: an early history. Arnoldia 30:121–138.

Sorensen P. D. (1980) New taxa in the genus *Dahlia* (Asteraceae, Heliantheae, Coreopsidinae). Rhodora 82:353–360.

Sorensen P. D. (1987) *Dahlia congestifolia*, section *Entemophyllon* (Asteraceae: Heliantheae, Coreopsidinae), new from Hidalgo, Mexico. Rhodora 89:197–203.

Sorm F., Cekan Z., Herout V., Raskova H. (1952) Isolation of spasmolytic substances from *Matricaria chamomilla.* Chem. Listy 46:308–309.

Sorm F., Cekan Z., Herout V., Raskova H. (1953) Isolation of spasmolytic substances from *Matricaria chamomilla.* Coll. Czech. Chem. Commun. 18:127–130.

Sosa V. E., Gil R., Oberti J. C., Kulanthaivel P., Herz W. (1985) Sesquiterpene lactones and flavones from *Stevia procumbens.* J. Nat. Prod. 48:340–341.

Sosa V. E., Oberti J. C., Prasad J. S., Herz W. (1984) Flavonoids and eupahakonenin B from *Stevia satureiaefolia.* Phytochemistry 23:1515–1516.

Soule J. A. (1996) Infrageneric systematics of *Tagetes.* In: Hind D. J. N., Beentje H. J. (eds.) *Compositae: Systematics. Proceedings of the International Compositae Conference, Kew (1994), Vol. 1.* Kew: Royal Botanic Gardens. 435–443.

Souleles C., Laskaris G. (1988) Flavonoids from *Picnomon acarna* (L.) Cass. Planta Medica 54:47–48.

Southwick L., Mabry T. J., Averett J., Powell A. M. (1972) Penduletin 4'-*O*-methyl ether from *Perityle vaseyi.* Phytochemistry 11:2351.

Spjut R. W. (1994) A systematic treatment of fruit types. Mem. New York Bot. Gard. 70:1–182.

Spooner D. M. (1990) Systematics of *Simsia* (Compositae-Heliantheae). Syst. Bot. Monographs 30:1–90.

Spring O. (1991) Trichome microsampling of sesquiterpene lactones for the use of systematic studies. In: Fischer N. H. (ed.) *Recent Advances in Phytochemistry, Vol. 25, Modern Phytochemical Methods.* New York: Plenum Press. 319–345.

Spring O., Heil N., Eliasson U. (1999) Chemosystematic studies on the genus *Scalesia* (Asteraceae). Biochem. Syst. Ecol. 27: 277–278.

Spring O., Heil N., Vogler B. (1997) Sesquiterpene lactones and flavanones in *Scalesia* species. Phytochemistry 46:1369–1373.

St. John H. (1971) The status of the genus *Wilkesia* (Compositae), and a discovery of a second Hawaiian species. Occas. Papers Bernice P. Bishop Mus. 24:128–137.

Stafford H. A. (1990) *Flavonoid metabolism.* Boca Raton, Florida: CRC Press.

Stafleu F. A. (1963) Adanson and his "Families des plantes." In: Lawrence G. H. M. (ed.) *Adanson, the Bicentennial of Michel Adanson's "Familles des plantes".* Pittsburgh: Hunt Botanical Library and Carnegie Inst. of Technol. 123–264.

Stafleu F. A. (1966) Adanson's *Familles des plantes* (Introduction to reprint edition, v–xv). Lehre: J. Cramer.

Stahl B., Lewis G. P., Kitgaard B. B., Lozano P. C. (1999) New records of *Fulcaldea* (Compositae–Barnadesieae) and the importance of local herbaria for floristic inventory in the tropics. Comp. Newsl. 33:39–45.

Stangl R., Greger H. (1980) Monoterpene und Systematik der Gattung *Artemisia.* Plant Syst. Evol. 136:125–136.

Stebbins G. L. (1953) A new classification of the tribe Cichorieae, family Compositae. Madroño 12:65–81.

Stebbins G. L. (1971) *Chromosomal Evolution in Higher Plants.* London: Edward Arnold.

Stebbins G. L. (1993) *Stephanomeria.* In: Hickman J. C. (ed.) *The Jepson Manual: Higher Plants of California.* Berkeley: Univ. California Press. 345, 346, 348.

Stefanovic M., Djermanovic V., Gorunvic M., Djermanovic M., Macura S., Milosavljevic S. (1989) Sesquiterpene lactones from *Achillea abrotanoides.* Phytochemistry 28:1765–1767.

Stefanovic M., Djermanovic V., Verencevic M. (1982) Chemical investigation of the plant species *Artemisia vulgaris.* Glas. Khem. Drush. Beogr. 47:7–12.

Stefanovic M., Krstic L., Jokic A., Rihter B., Mladenovic S. (1979) Synthesis of 3,5-dihydroxy-3',4',6,7-tetramethoxyflavone. Glas. Khem. Drush. Beogr. 44:497–502.

Stefanovic M., Solaja B., Dermanovic M., Milovanovic M. (1986) Chemical investigations of the plant species *Eupatorium cannabinum* L. (Compositae). J. Serb. Chem. Soc. 51:575–581.

Stenlid G., Saddik K. (1962) The effect of some growth regulators and uncoupling agents upon oxidative phosphorylation in mitochondria of cucumber hypocotyls. Physiologia Plant. 15:369–379.

Stenlid G., Saddik K. (1963) The effects of flavonoid compounds on oxidative phosphorylation and on the enzymic destruction of indoleacetic acid. Physiologia Plant. 16:110–120.

Stepanova T. A., Sheichenko V. I., Smirnova L. P., Glyzin V. I. (1981) Flavanones of *Tanacetum sibiricum.* Chem. Nat. Cmpds. 17:520 [721].

Stevens J. F., Elema E. T., Wollenweber E. (1995) Exudate flavonoids of *Eupatorium cannabinum.* Biochem. Syst. Ecol. 23:451–452.

Stevens J. F., Wollenweber E., Ivancic M., Hsu V. L., Sandberg S., Deinzer M. L. (1999) Leaf surface flavonoids of *Chrysothamnus.* Phytochemistry 51:771–780.

Stevens P. F. (1994) *The Development of Biological Systematics: Antoine-Laurent de Jussieu, Nature, and the Natural System.* New York: Columbia Univ. Press.

Stewart G. R., Press M. C. (1990) The physiology and biochemistry of parasitic angiosperms. Annu. Rev. Plant Physiol. Plant Mol. Biol. 41:127–151.

Stich K., Ebermann R., Forkmann G. (1988) Effect of cytochrome P-450 specific inhibitors on the activity of flavonoid 3'-hydroxylase and flavone synthase II in certain plants. Phyton (Austria) 28:237–247.

Stich K., Halbwirth H., Wurst F., Forkmann G. (1994) Formation of 6'-deoxychalcone 4'-glucosides by enzyme extracts from petals of *Dahlia variabilis*. Z. Naturforsch. 49c:737–741.

Stich K., Halbwirth H., Wurst F., Forkmann G. (1996) UDP-Glucose: flavonol 7-*O*-glucosyltransferase activity in flower extracts of *Chrysanthemum segetum*. Z. Naturforsch. 52c:153–158.

Stickland R. G., Sunderland N. (1972) Production of anthocyanins, flavonols and chlorogenic acids by cultured callus tissue of *Haplopappus gracilis* Nutt. Ann. Bot. 36:443–457.

Stirton J. Z., Harborne J. B. (1980) Two distinctive anthocyanin patterns in the Commelinaceae. Biochem. Syst. Ecol. 8:285–287.

Stosic D., Gorunovic M., Djermanovic M. (1988) Flavonoids of *Achillea coarctata* Poir., Asteraceae. Pl. Medica Phytother. 22:165–170.

Stotz G., Forkmann G. (1981) Oxidation of flavanones to flavones with flower extracts of *Antirrhinum majus* (snapdragon). Z. Naturforsch. 36c:737–741.

Stoutamire W. P. (1960) Relation of anthocyanins to *Gaillardia* taxonomy. Pap. Mich. Acad. Sci. Arts Lett. 45:35–39.

Straley G. B. (1980) Systematics of *Arnica* subgenus *Austromontana* and a new subgenus *Callarnica* (Asteraceae:Senecioneae). Ph.D. Dissertation, University of British Columbia, Vancouver.

Strother J. L. (1969) Systematics of *Dyssodia* Cavanilles (Compositae: Tageteae). Univ. Calif. Publ. Bot. 48:1–88.

Strother J. L. (1977) Tageteae—systematic review. In: Heywood V. H., Harborne J. B., Turner B. L. (eds.) *The Biology and Chemistry of the Compositae,* Vol. 2. London: Academic Press. 769–784.

Strother J. L. (1983) *Pionocarpus* becomes *Iostephane* (Compositae: Heliantheae): a synopsis. Madroño 30:34–38.

Strother J. L. (1986) Renovation of *Dyssodia* (Compositae: Tageteae). Sida 11:371–378.

Strother J. L. (1991) Taxonomy of *Complaya, Elaphandra, Iogeton, Jefea, Wamalchitamia, Wedelia, Zexmenia,* and *Zyzyxia* (Compositae-Heliantheae-Ecliptinae). Syst. Bot. Monographs 33:1–111.

Strother J. L. (1999) Compositae—Heliantheae s.l. In: Breedlove D. (ed.) *Flora of Chiapas.* San Francisco: California Academy of Sciences.

Strother J. L., Pilz G. (1975) Taxonomy of *Psathyrotes* (Compositae: Senecioneae). Madroño 23:24–40.

Stubbe H. (1996) *Genetik und Cytologie von Antirrhinum L. Sect. Antirrhinum.* Jena: G. Fischer.

Stuckey R. L. (1972) Taxonomy and distribution of the genus *Rorippa* (Cruciferae). Sida 4:279–430.

Stuessy T. F. (1969) Re-establishment of the genus *Unxia* (Compositae—Heliantheae). Brittonia 21:314–321.

Stuessy T. F. (1971) Chromosome numbers and phylogeny in *Melampodium* (Compositae, Heliantheae). Amer. J. Bot. 58:732–736.

Stuessy T. F. (1972) Revision of the genus *Melampodium* (Compositae: Heliantheae). Rhodora 74:161–219.

Stuessy T. F. (1975) A revision of *Moonia* (Compositae, Heliantheae, Coreopsidinae). Brittonia 27:97–102.

Stuessy T. F. (1977) Heliantheae-systematic review. In: Heywood V. H., Harborne J. B., Turner B. L. (eds.) *The Biology and Chemistry of the Compositae,* Vol. 2. London: Academic Press. 621–671.

Stuessy T. F. (1978) Revision of *Lagascea* (Compositae, Heliantheae). Fieldiana Bot. 38:75–133.

Stuessy T. F. (1979) Cladistics of *Melampodium* (Compositae). Taxon 28:179–195.

Stuessy T. F. (1987) Explicit approaches for evolutionary classification. Syst. Bot. 12:251–262.

Stuessy T. F. (1988) Generic relationships of *Oparanthus* and *Petrobium*, especially with reference to *Bidens* (Compositae, Heliantheae, Coreopsidinae). Brittonia 40:195–199.

Stuessy T. F. (1990) *Plant Taxonomy: The Systematic Evaluation of Comparative Data.* New York: Columbia Univ. Press.

Stuessy T. F. (1997) Classification: more than just branching patterns of evolution. Aliso 15:113–129.

Stuessy T. F., Crawford D. J. (1983) Flavonoids and phylogenetic reconstruction. Plant Syst. Evol. 143:83–107.

Stuessy T. F., Foland K. A., Sutter J. F., Sanders R. W., Silva O. M. (1984a) Botanical and geological significance of potassium-argon dates from the Juan Fernandez Islands. Science 225:49–51.

Stuessy T. F., Garver D. (1996) The defensive role of pappus in heads of Compositae. In: Caligari P. D. S., Hind D. J. N. (eds.) *Compositae: Biology and Utilization, Proceedings of the International Compositae Conference, Kew (1994) Vol. 2.* Kew: Royal Botanic Gardens. 81–91.

Stuessy T. F., Irving R. S., Ellison W. L. (1973) Hybridization and evolution in *Picradeniopsis* (Compositae). Brittonia 25:40–56.

Stuessy T. F., Ono M. (eds.) (1998) *Evolution and Speciation of Island Plants.* Cambridge: Cambridge Univ. Press.

Stuessy T. F., Sagástegui A. A. (1993) Revisión de *Arnaldoa* (Compositae, Barnadesioideae), género endémico del norte del Peru. Arnaldoa 1:9–21.

Stuessy T. F., Sanders R. W., Silva M. (1984b) Phytogeography and evolution of the flora of the Juan Fernandez Islands: a progress report. In: Radovsky F. J., Raven P. H., Sohmer S. H. (eds.) *Biogeography of the Tropical Pacific.* Lawrence, Kansas: Assoc. Syst. Collections and B. P. Bishop Museum. 55–69.

Stuessy T. F., Sang T., DeVore M. (1996) Phylogeny and biogeography of the subfamily Barnadesioideae with implications for early evolution of the Compositae. In: Hind D. J., Beentje H. J. (eds.) *Compositae: Systematics. Proceedings of the International*

Compositae Conference, Kew (1994) Vol. 1. Kew: Royal Botanic Gardens. 463–490.
Stuessy T. F., Spooner D. M. (1988) The adaptive and phylogenetic significance of receptacular bracts in the Compositae. Taxon 37:114–126.
Stuessy T. F., Spooner D. M., Evans K. A. (1986) Adaptive significance of ray corollas in *Helianthus grosseserratus* (Compositae). Amer. Midl. Natur. 115:191–197.
Stuppner H., Müller E. P. (1994) Rare flavonoid aglycones from *Flourensia retinophylla*. Phytochemistry 37:1185–1187.
Suarez S. S., Cabrera J. L., Juliani H. R. (1979) Flavonoids from *Flaveria bidentis* (L.) O. K. and *F. bidentis* var. *angustifolia* O. K. (Compositae). An. Asoc. Quim. Argent. 67:229–231.
Sulyok G., Laszlo-Bencsik A. (1985) Cyanidin 3-(6-succinyl glucoside)-5-glucoside from flowers of seven *Centaurea* species. Phytochemistry 24:1121–1122.
Sundberg S. (1985) Micromorphological characters as generic markers in the Astereae. Taxon 34:31–37.
Sundberg S. D., Stuessy T. F. (1990) Isolating mechanisms and implications for modes of speciation in Heliantheae (Compositae). Plant Syst. Evol. (Suppl. 4):77–97.
Susanna A., Garnatje T., Garcia-Jacas N. (1999) Molelcular phylogeny of *Cheirolophus* (Asteraceae: Cardueae—Centaureinae) based on ITS sequences of nuclear ribosomal DNA. Plant Syst. Evol. 214:147–160.
Susanna A., Jacas N. G., Soltis D. E., Soltis P. S. (1995) Phylogenetic relationships in tribe Cardueae (Asteraceae) based on ITS sequences. Amer. J. Bot. 82:1056–1068.
Suttisri R., Kinghorn A. D., Wright A. D., Sticher O. (1994) *neo*-Clerodane diterpenoids and other constituents from *Baccharis genistelloides* Pers. Phytochemistry 35:443–446.
Sventenius E. R., Bramwell D. (1971) *Heywoodiella* genus novum. Acta Phytotax. Barcinonensia 7:5–8.
Sweitzer E. M. (1971) Comparative anatomy of Ulmaceae. J. Arnold Arb. 52:523–585.
Swensen S. M., Mullin B. C., Chase M. W. (1994) Phylogenetic affinities of Datiscaceae based on an analysis of nucleotide sequences from the plastid *rbc*L gene. Syst. Bot. 19:157–168.
Swenson U. (1995) Systematics of the Blennospermatinae (Asteraceae, Senecioneae). Ph.D. Dissertation, Univ. Uppsala.
Swenson U., Bremer K. (1997) Pacific biogeography of the Asteraceae genus *Abrotanella* (Senecioneae, Blennospermatinae). Syst. Bot. 22:493–508.
Swenson U., Yong J. Y., Bohm B. A. (1999) Flavonoid chemistry of Blennospermatinae, a trans-Pacific disjunct subtribe of Senecioneae (Asteraceae). Plant Syst. Evol. 216:231–241.
Sy L.-K., Brown G. D. (1998) Three sesquiterpenes from *Artemisia annua*. Phytochemistry 48:1207–1211.
Syrchina A. I., Chernousova A. V., Vereshchagin A. L., Semenov A. A. (1993) Constituents of *Saussurea controversa*. Khim. Prir. Soedin. 5:765–766.
Syrchina A. I., Semenov A. A., Zinchenko S. V. (1997) Chemical investigation of *Cirsium setosum*. Chem. Nat. Cmpds. 33:212.
Tahara S., Ibrahim R. K. (1995) Prenylated isoflavonoids-an update. Phytochemistry 38:1073–1094.

Taillade C., Susplugas P., Balansard G. (1980) Sur les flavonoides d'*Inula viscosa* Ait. (Composées). Pl. Medica Phytother. 14:26–28.

Takahashi Y., Miyasaka N., Tasaka S., Miura I., Urano S., Ikura M., Hikichi K., Matsumoto T., Wada M. (1982) Constitution of two coloring matters in the flower petals of *Carthamnus tinctorius* L. Tet. Lett. 5163–5166.

Takahashi Y., Saito K., Yamagiya M., Ikura M., Hikichi K., Matsumoto T., Wada M. (1984) Chemical constituents of safflower yellow B, a quinochalcone *C*-glycoside from the flower petals of *Carthamnus tinctorius* L. Tet. Lett. 25:2471–2474.

Takeda K., Harborne J. B., Self R. (1986) Identification and distribution of malonated anthocyanins in plants of the Compositae. Phytochemistry 25:1337–1342.

Takeda K., Kumegawa C., Harborne J. B., Self R. (1988) Pelargonidin 3-(6"-succinyl glucoside)-5-glucoside from pink *Centaurea cyanus* flowers. Phytochemistry 27:1228–1229.

Takeda K., Tominaga S. (1983) The anthocyanins in blue flowers of *Centaurea cyanus*. Bot. Mag., Tokyo 96:359–363.

Takeda K., Yoshitomo N., Haruji O. (1976) Flavonoids of *Artemisia capillaris*. J. Pharm. Soc. Japan 96:855.

Takemoto T., Ikegawa S., Nomoto K. (1975) Constituents of *Silybum marianum*. I. New flavonolignans named 2,3-dehydrosilymarin and 2,3-dehydrosilychristin. J. Pharm. Soc. Japan 95:1017–1021.

Takhtajan A. (1969) *Flowering Plants: Origin and Dispersal*. Trans. C. Jeffrey. Edinburgh: Oliver & Boyd.

Talapatra S. K., Bhar D. S., Talapatra B. (1974) Flavonoid and terpenoid constituents of *Eupatorium odoratum*. Phytochemistry 13:284–285.

Tamainschyan S. G. (1959) *Conyzanthus*. In: Komorov L. (ed.) *Flora S.S.S.R.*, Vol. 25. Moscow: Akademija Nauk. 185–188, 583.

Tamayo-Castillo G., Jakupovic J., Bohlmann F, King R. M., Boldt P. E. (1990) Germacranolides and diterpenes from *Viguiera* species. Rev. Latinoamer. Quim. 21:67–69.

Tamayo-Castillo G., Jakupovic J., Bohlmann F., King R. M., Robinsom H. (1989) *ent*-Clerodane derivatives from *Chromolaena connivens*. Phytochemistry 28:641–642.

Tamayo-Castillo G., Jakupovic J., Bohlmann F., Rojas A., Castro V., King R. M. (1988) Germacranolides and other constituents from *Ageratina* species. Phytochemistry 27:2893–2897.

Tamura H., Kondo T., Kato Y., Goto T. (1983) Structures of a succinyl anthocyanin and a malonyl flavone, two constituents of the complex blue pigment of cornflower, *Centaurea cyanus*. Tet. Lett. 5749–5752.

Tan R., Jia Z. (1992) Eudesmanolides and other constituents from *Artemisia argyi*. Planta Medica 58:370–372.

Tanbug R., Halfon B., Ululeben A. (1997) Flavonoids from *Senecio racemosus*. Sci. Pharm. 65:175–179.

Tandon S., Rastogi R. P. (1977) Conyzatin – a new flavone from *Conyza stricta*. Phytochemistry 16:1455–1456.

Tanowitz B. D., Leeder G., Ross P. N., Proksch P. (1987) Foliar flavonoid exudates and sectional taxonomy of *Hemizonia*. Biochem. Syst. Ecol. 15:535–540.

Tarpo E. (1967) Patuletin isolation from *Tagetes patula* flowers. Farmacia (Bucharest) 15:685–688.

Tarpo E. (1968) Isolation of patuletrin from flowers of *Tagetes patula*. Farmacia (Bucharest) 16:93–98.

Tarpo E. (1969a) Flavones of *Tagetes patula*. Farmacia (Bucharest) 17:607–616.

Tarpo E. (1969b) Isolation of quercetagetin from *Tagetes patula* flowers. Farmacia (Bucharest) 17:739–744.

Tarpo E. (1971) Isolation of quercetagetin from *Tagetes patula* flowers. Farmacia (Bucharest) 19:25–30.

Taylor A. E., Wong E. (1965) C-10 Substituted 19-norsteroids-VIII. 10-Cyanosteroids with an oxygen at C-3. Tet. Lett. 3675–3680.

Taylor L. P., Jorgensen R. (1992) Conditional male fertility in chalcone synthase-deficient petunia. J. Heredity 83:11–17.

Terahara N., Toki K., Honda T. (1993) Acylated anthocyanins from flowers of cineraria, *Senecio cruentus*, Red cultivar. Z. Naturforsch. 48c:430–435.

Terencio M. C., Giner R. M., Sanz M. J., Manez S., Rios J. L. (1993) On the occurrence of caffeoyltartronic acid and other phenolics in *Chondrilla juncea*. Z. Naturforsch. 48c:417–419.

Terencio M. C., Sanz M. J., Fonseca M. L., Manez S., Rios J. L. (1992) Phenolic compounds from *Lactuca viminea* L. Z. Naturforsch. 47c:17–20.

Tereschuk M. L., Riera M. V. Q., Castro C. R., Abdala L. R. (1997) Antimicrobial activity of flavonoids from leaves of *Tagetes minuta*. J. Ethnopharmacol. 56:227–232.

Tesevic V., Djokovic D., Vajs V., Marin P., Milosavljevic S. (1994) Constituents of the roots of *Centaurea scabiosa*. J. Serb. Chem. Soc. 59:979–981.

Tétényi P. (1970) *Infraspecific Chemical Taxa of Medicinal Plants*. Budapest: Akadémiai Kiadó.

Tettamanzi M. C., Jares E. A., Iannone L. M., Pomilio A. B. (1994) Constituents of *Senecio crassiflorus*. Fitoterapia 65:468.

Teusch M., Forkmann G. (1987) Malonyl-coenzyme A: anthocyanidin 3-glucoside malonyltransferase from flowers of *Callistephus chinensis*. Phytochemistry 26:2181–2183.

Teynor T. M., Ascher P. D., Widmer R. E., Luby J. J. (1989a) Inheritance of flower color in *Dendranthema grandiflora* Tzvelev. (*Chrysanthemum morifolium* Ramat.) using cultivars and inbreds. I. Plastid pigmentation. Euphytica 42:199–208.

Teynor T. M., Ascher P. D., Widmer R. E., Luby J. J. (1989b) Inheritance of flower color in *Dendranthema grandiflora* Tzvelev. (*Chrysanthemum morifolium* Ramat.) using cultivars and inbreds. II. Vacuole pigmentation. Euphytica 42:297–305.

Thiele E.-M. (1988) Bau und Funktion des Antheren-Griffel-Komplexes der Compositen. Dissertat. Botan. 117:1–169.

Thomas M. B., Mabry T. J. (1967) Isolation, structure and synthesis of hymenoxin, a new flavone from *Hymenoxys scaposa* (Compositae). J. Org. Chem. 32:3254–3256.

Thomas M. B., Mabry T. J. (1968a) Flavonoid constituents of *Hymenoxys scaposa* DC. (Compositae). Phytochemistry 7:787–790.

Thomas M. B., Mabry T. J. (1968b) Isolation, strtucture and synthesis of scaposin, a new flavone from *Hymenoxys scaposa* DC. (Compositae). Tetrahedron 24:3675–3679.

Thompson W. R., Meinwald J., Aneshansley D., Eisner T. (1972) Flavonols: Pigments responsible for UV absorption in nectar guides of flowers. Science 177:528–530.

Thorne R. F. (1968) Synopsis of a putatively phylogenetic classification of the flowering plants. Aliso 6:57–66.

Thorne R. F. (1976) A phylogenetic classification of the Angiospermae. Evol. Biol. 9:35–106.

Thorne R. F. (1983) Proposed new realignments in the angiosperms. Nord. J. Bot. 3:85–117.

Thorne R. F. (1989) Phylogeny and biogeography. Rhodora 91:10–24.

Thorne R. F. (1992) Classification and geography of the flowering plants. Bot. Rev. 58:225–348.

Thorne R. F. (1996) Some guiding principles of biogeography. Telopea 6:845–850.

Timmermann B. N., Graham S. A., Mabry T. J. (1981) Sulfated and 6-methoxyflavonoids from *Brickellia baccharidea* (Compositae). Phytochemistry 20:1762.

Timmermann B. N., Hoffmann J. J., Jolad S. D., Bates R. B., Siahaan T. J. (1986) Diterpenoids and flavonoids from *Grindellia discoidea.* Phytochemistry 25:723–727.

Timmermann B. N., Hoffmann J. J., Jolad S. D., Schram K. H. (1985) Grindelane diterpenoids from *Grindelia squarrosa* and *G. camporum*. Phytochemistry 24:1031–1034.

Timmermann B. N., Mabry T. J. (1983) 6-Methoxy-flavonoids from disjunct populations of *Brickellia cylindracea* (Compositae). Biochem. Syst. Ecol. 11:37–39.

Timmermann B. N., Mues R., Mabry T. J., Powell A. M. (1979) 6-Methoxyflavonoids from *Brickellia laciniata* (Compositae). Phytochemistry 18:1855–1858.

Timmermann B. N., Wollenweber E., Dörr M., Armbruster S., Valant-Vetschera K., Fuentres E. R. (1994) External flavonoids in two *Grindelia* species. Z. Naturforsch. 49c:395–397.

Tira S., Galeffi C., di Modica G. (1970) Flavonoids of *Leontopodium alpinum*. Experientia 26:1192.

Tira S., Galeffi C., Monache E. M. (1969) NMR of flavonoids, flavones from *Antennaria dioica.* Ann. Chim. Roma 59:284–290.

Toki K., Saito N., Honda T. (1991) Three cyanidin 3-glucuronylglucosides from red flowers of *Bellis perennis.* Phytochemistry 30:3769–3771.

Toki K., Saito N., Kawano H., Shigihara A., Honda T. (1995) Acylated pelargonidin 3,7-glycosides from pink flowers of *Senecio cruentus*. Phytochemistry 38:1509–1512.

Toman J. (1972) A taxonomic survey of the genera *Petasites* and *Endocellion.* Folia Geobot. Phytotax. 7:381–406.

Tomas-Barberan F. A., Gil M. I., Ferreres F., Tomas-Lorente F. (1991) Correlations between flavonoid composition and infrageneric taxonomy of some European *Galeopsis* species. Phytochemistry 30:3311–3314.

Tomas-Barberan F. A., Maillard M., Hostettmann K. (1988) Antifungal flavonoids from the leaf surfaces of *Helichrysum nitens* and from the stem bark of *Erythrina berteroana*, In: Cody V. et al. (eds.) *Progress in Clinical and Biological Research, Vol. 280, Plant Flavonoids in Biology and Medicine, II, Biochemical, Cellular and Medicinal Properties.*

Tomas-Barberan F. A., Msonthi J. D., Hostettmann K. (1988) Antifungal epicuticular fully methylated flavonoids from *Helichrysum nitens*. Phytochemistry 27:753–755.

Tomas-Lorente F., Iniesta-Sanmartin E., Tomas-Barberan F. A., Guirado A. (1991) Antimicrobial phenolics from *Helichrysum picardii*. Fitoterapia 62:521.

Tomas-Lorente F., Iniesta-Sanmartin E., Tomas-Barberan F. A., Trowitzsch-Kienast W., Wray V. (1989) Antifungal phloroglucinol derivatives and lipophilic flavonoids from *Helichrysum decumbens*. Phytochemistry 28:1613–1615.

Tomb A. S. (1977) Lactuceae—systematic review. In: Heywood V. H., Harborne J. B., Turner B. L. (eds.) *The Biology and Chemistry of the Compositae, Vol. 2*. New York: Academic Press. 1067–1080.

Tonn C. E., Rossomando P. C., Giordano O. S. (1982) Batudic acid and flavonoids from *Baccharis tucumanensis*. Phytochemistry 21:2599–2600.

Torck M., Pinkas M., Didry N., Bezanger-Beauquesne L. (1976) Polypenols du *Grindelia aphanactis*. Pl. Medica Phytother. 10:188–191.

Toropkina R. Y., Minina S. A. (1976) Flavonoids of *Petasites georgicus*. Chem. Nat. Cmpds. 12:487 [544–545].

Torrance S. J., Steelink C. (1974) New monoterpenes from *Artemisia filifolia* Torr. Structure, synthesis, rearrangements, and biosynthesis. J. Org. Chem. 39:1068–1074.

Torrenegra R. D., Bautista A. R., Pendrozo J. A. (1985) Chemical studies on *Eupatorium micropillum* L. Rev. Latinoamer. Quim. 16:64–66.

Torrenegra R. D., Escarria S., Bogota C., Achenbach H. (1979) Colombian plants of the genus *Gnaphalium*. Part II. Rev. Latinoamer. Quim. 10:83–84.

Torrenegra R. D., Escarria S., Dominguez X. A. (1978) Flavonoids of *Gnaphalium pellitum*. Rev. Latinoamer. Quim. 9:101.

Torrenegra R. D., Escarria S., Raffelsberger B., Achenbach H. (1980) 5,7-Dihydroxy-3,6,8-trimethoxyflavone from the flowers of *Gnaphalium elegans*. Phytochemistry 19:2795–2796.

Torrenegra R. D., Escarria S., Tenorio E., Achenbach H. (1982) Phytochemical study of *Achyrocline bogotensis*. Rev. Latinoamer. Quim. 13:75–76.

Torrenegra R. D., Pedorozo J. A., Escarria S. (1984) Chemical study of some plants of the Bogota Savanna. Flavonoids from *Eupatorium gracile*. Rev. Latinoamer. Quim. 15:129–130.

Torrenegra R. D., Pedorozo J. A., Robles J., Fuentes O. (1990) Palmitolactone from *Ageratina viscosa*. Phytochemistry 29:305–306.

Torrenegra R. D., Pedorozo J. A., Rojas C. P., Carrisoza S. (1987) Colombian plants of the genus *Gnaphalium* IV. Rev. Latinoamer. Quim. 18:116–118.

Torrenegra R. D., Robles J., Pedrozo J. (1995) Terpenes and flavonoids from *Ageratina fastigiata* (H.B.K.) King & Robinson. Rev. Colomb. Quim. 24:9–16.

Torrenegra R. D., Tellez A. N. A. (1995) Chemotaxonomic value of melampolides in *Espeletia* species (Asteraceae). Biochem. Syst. Ecol. 25:449–450.

Torrenegra R. D., Tellez A. N. A. (1996) Phytochemistry of *Espeletia killipii* Cuatr. and gibberellic activity of some of the isolated compounds. Rev. Latinoamer. Quim. 24:2–6.

Torrenegra R. D., Tellez A. N. A., Garcia G. (1994) Chemical investigation of the species of the genus *Espeletia*. Part I. Chemistry of *Espeletia kilipii* and *E. tunjana*. Rev. Colomb. Quim. 23:29–35.

Torres A. M. (1963) Taxonomy of *Zinnia*. Brittonia 15:1–25.
Torres A. M., Levin D. A. (1964) A chromatographic study of cespitose zinnias. Amer. J. Bot. 51:639–643.
Torres G. T. (1985) Plantas fósiles en la Antártica. Bol. Antart. Chil. 5:17–31.
Torres L. M. B., Akisue M. K., Roque N. F. (1987) Quercitrin from *Solidago microglossa* DC., a Brazilian substitute for *Arnica*, Rev. Pharm. Bioquim. Univ. Sao Paulo 23:33–40.
Triana J. (1984) Furanoheliangolides and flavonoids from *Lourteigia ballotaefolia*. Phytochemistry 23:2072–2074.
Tripathy A. K., Gupta K. R. (1991) Plant phenolics of *Tagetes erecta*. Fitoterapia 62:91.
Tripathy A. K., Paliwal M. K., Singh J. (1991) Isorhamnetin 7-*O*-ß-D-galactoside from *Tagetes patula*. J. Indian Chem. Soc. 68:674.
Troshchenko A. T., Limasova T. I. (1966) 5-*O*-Desmethylnobiletin from *Aster altaicus*. Chem. Nat. Cmpds. 2:357 [436].
Tsankova E., Ognyanov I. (1985), Chemical constituents of *Achillea depressa*. Planta Medica 51:180.
Tschan G. M., König G. M., Wright A. D., Sticher O. (1996) Chamaemeloside, a new flavonoid glycoside from *Chamaemelum nobile*. Phytochemistry 41:643–646.
Tschan G. M., Sticher O. (1993) Quantitative analysis of flavonoids in Roman chamomile by RP-HPLC. Planta Medica 59 (Suppl. 1): A628.
Tsichritzis F., Jakupovic J. (1990) Diterpenes and other constituents from *Relhania* species. Phytochemistry 29:3173–3187.
Tu Y-y., Ni M.-y., Zhong Y.-r., Li L.-n., Cui S.-i., Zhang M.-q., Wang X.-z., Ji X., Liang X.-t. (1982) Studies on the constituents of *Artemisia annua*. Part II. Planta Medica 44:143–145.
Tu Y-y., Yin J., Ji L., Huang M., Liang X. (1985a) Chemical constituents of sweet wormwood *Artemisia annua*. III. Zhongcaoyao 16:200–204.
Tu Y-y., Zhu Q., Shen X. (1985b) Constituents of young *Artemisia annua*. Zhongyao Tongbao 10:419–420.
Tunmann P., Isaac O. (1957) Die Konstitution des Artemisetin, eines neuen Flavonols. Arch. Pharm. 290:37–43.
Turner B. L. (1972) *Strotheria* (Compositae-Tageteae) a new monotypic genus from north-central Mexico. Amer. J. Bot. 59:180–182.
Turner B. L. (1977a) Fossil history and geography. In: Heywood V. H., Harborne J. B., Turner B. L. (eds.) *The Biology and Chemistry of the Compositae*, Vol. 1. London: Academic Press. 21–39.
Turner B. L. (1977b) Summary of the biology of the Compositae. In: Heywood V. H., Harborne J. B., Turner B. L. (eds.) *The Biology and Chemistry of the Compositae*, Vol. 2. London: Academic Press. 1105–1118.
Turner B. L. (1977c) *Henricksonia* (Asteraceae-Coreopsidinae), a newly discovered genus with a paleaceous pappus from north-central Mexico. Amer. J. Bot. 64:78–80.
Turner B. L. (1985a) A summing up. Taxon 34:85–88.
Turner B. L. (1985b) Two new species of *Eupatorium* (Asteraceae) from northeastern Mexico. Brittonia 37:373–377.

Turner B. L. (1987) Taxonomy of *Carphochaete* (Asteraceae–Eupatorieae). Phytologia 64:145–162.

Turner B. L. (1988a) Submergence of the genera *Asanthus* and *Dyscritogyne* within *Steviopsis* (Asteraceae, Eupatorieae) including new combinations. Phytologia 64:259–262.

Turner B. L. (1988b) A new species of *Melampodium* (Asteraceae-Heliantheae) from Oaxaca, Mexico. Phytologia 64:445–447.

Turner B. L. (1993a) A new species of *Melampodium* (Asteraceae, Heliantheae) from Jalisco, Mexico. Phytologia 75:136–139.

Turner B. L. (1993b) Revision of South American species of *Perymenium* (Asteraceae, Heliantheae). Phytologia 75:204–217.

Turner B. L. (1993c) A new species of *Psathyrotopsis* (Asteraceae, Helenieae) from Coahuila, Mexico. Phytologia 75:143–146.

Turner B. L. (1995) A new species of *Flaveria* (Asteraceae, Helenieae) from Oaxaca, Mexico. Phytologia 78:400–401.

Turner B. L. (1996) Revision of *Desmanthodium* (Asteraceae). Phytologia 80:257–272.

Turner B. L. (1997) *The Comps of Mexico. A Systematic Account of the Family Asteraceae*, Vol. 1. Huntsville, Texas: Phytologia Memoirs, Vol. II.

Turner B. L., Alston R. E. (1959) Segregation and recombination of chemical constituents in a hybrid swarm of *Baptisia laevicaulis* × *B. viridis* and their taxonomic implications. Amer. J. Bot. 46:678–686.

Turner B. L., Kim K.-J. (1990) An overview of the genus *Pyrropappus* (Asteraceae: Lactuceae) with emphasis on chloroplast DNA restriction site data. Amer. J. Bot. 77:845–850.

Turner B. L., Powell M. A. (1977) Helenieae—systematic review. In: Heywood V. H., Harborne J. B., Turner B. L. (eds.) *The Biology and Chemistry of the Compositae,* Vol. 2. London: Academic Press. 699–737.

Turner B. L., Powell M. A., Watson T. J., Jr. (1973) Chromosome numbers in Mexican Asteraceae. Amer. J. Bot. 60:592–596.

Turner B. L., Zippin D. (1992) Taxonomic study of *Venegasia* (Asteraceae: Helenieae). Sida 15:223–229.

Tutin T. G., Heywood V. H., Burges N. A., Moore D. M., Valentine D. H., Walters S. M., Webb D. Q. (eds.) (1976) *Flora Europaea, Plantaginaceae to Compositae (and Rubiaceae)*, Vol. 4. Cambridge: Cambridge Univ. Press.

Twaij H. A. A., Kery A., Al-Khazraji N. K. (1983) Some pharmacological, toxicological and phytochemical investigations on *Centaurea phyllocephala*. J. Ethnopharm. 9:299–314.

Tyihak E., Sarkany-Kiss I., Vergar-Petri G. (1962) Phytochemical investigation of apigenin glycosides of *Matricaria chamomilla*. Die Pharmazie 17:301–304.

Tzakou O., Couladis M., Verykokidou E., Loukis A. (1995) Leaf flavonoids of *Achillea ligustica* and *A. holosericea*. Biochem. Syst. Ecol. 23:569–570.

Ulubelen A., Ayanoglu E., Clark W. D., Brown G. K., Mabry T. J. (1982) Flavonoids from *Haplopappus foliosus*. J. Nat. Prod. 45:363–364.

Ulubelen A., Goren N., Öksüz S. (1989) Constituents of *Achillea spirokensis*. Fitoterapia 60:93.

Ulubelen A., Kerr K. M., Mabry T. J. (1980a) New 6-hydroxyflavonoids and their methyl ethers and glycosides from *Neurolaena oaxacana*. Phytochemistry 19:1761–1766.

Ulubelen A., Kerr K. M., Mabry T. J. (1981) Flavonoids of *Neurolaena venturana*. Planta Medica 43:95–96.

Ulubelen A., Mabry T. J. (1981a) Flavonoids of *Haplopappus rengifoanus*. J. Nat. Prod. 44:294–295.

Ulubelen A., Mabry T. J. (1981b) 6-Oxygenated flavonoids from *Neurolaena macrophylla*. J. Nat. Prod. 44:457–458.

Ulubelen A., Mabry T. J., Aynehchi Y. (1979) Flavonoids of *Anvilla garcini*. J. Nat. Prod. 42:624–626.

Ulubelen A., Öksüz S. (1982) Cytotoxic flavones from *Centaurea urvillei*. J. Nat. Prod. 45:373.

Ulubelen A., Öksüz S., Gören N. (1987a) Sesquiterpene acids from *Inula viscosa*. Phytochemistry 26:1223–1224.

Ulubelen A., Öksüz S., Mericli A. (1988) Palmitic acid ester of sitosteryl 3-ß-glucoside from *Centaurea regia*. Phytochemistry 27:3964–3965.

Ulubelen A., Öksüz S., Tuzlaci E. (1987b) Flavonoids and coumarins from *Achillea schischkinii*. Planta Medica 53:507.

Ulubelen A., Timmermann B. N., Mabry T. J. (1980b) Flavonoids from *Brickellia chlorolepis* and *B. dentata*. Phytochemistry 19:905–908.

Ulubelen A., Tuzlaci E. (1988) Flavonoids and terpenoids from *Cousinia eriocephala*. Fitoterapia 59:350.

Urbatsch L. E. (1972) Systematic study of the Altissimae and Giganteae species groups of the genus *Vernonia* (Compositae). Brittonia 24:229–238.

Urbatsch L. E. (1978) The Chihuahuan desert species of *Ericameria* (Compositae: Astereae). Sida 7:298–303.

Urbatsch L. E., Bacon J. D., Mabry T. J. (1975) Flavonol methyl ethers from *Chrysothamnus viscidiflorus*. Phytochemistry 14:2279–2282.

Urbatsch L. E., Jansen R. K. (1995) Phylogenetic affinities among and within the coneflower genera (Asteraceae, Heliantheae), a chloroplast DNA analysis. Syst. Bot. 20:28–39.

Urbatsch L. E., Mabry T. J., Miyakado M., Ohno N., Yoshioka H. (1976) Flavonol methyl ethers from *Ericameria diffusa*. Phytochemistry 15:440–441.

Urtubey E. (1999) Revisión del género *Barnadesia* (Asteraceae, Barnadesioideae). Ann. Missouri Bot. Gard. 86:57–117.

Urzua A., Andrade L., Muñoz E., Rodriguez M. E., Belmonte E. (1997) Flavonoids in the trichome resinous exudate of *Diplostephium cinereum*. Biochem. Syst. Ecol. 25:681–682.

Urzua A., Cuadra P. (1989) Flavonoids from the resinous exudate of *Gnaphalium robustum*. Bol. Soc. Chil. Quim. 34:247–251.

Urzua A., Cuadra P. (1990) Acylated flavonoid aglycones from *Gnaphalium robustum*. Phytochemistry 29:1342–1343.

Urzua A., Cuadra P., Rodriguez R., Mendoza L. (1991) Flavonoid aglycones in the resinous exudate of some Chilean plants. Fitoterapia 62:358–359.

Urzua A., Mendoza E. (1993) Chemical composition of the resinous exudate from *Haplopappus velutinus*. Bol. Soc. Chil. Quim. 38:89–93.

Urzua A., Torres R., Bueno C., Mendoza L. (1995) Flavonoids and diterpenoids in the trichome resinous exudates from *Pseudognaphalium cheiranthifolium, P. heterotrichium* and *P. viravira*. Biochem. Syst. Ecol. 23:459.

Vaccari A., Pifferi P. G., Zaccherini G. (1982) Anthocyanins of sunflower (*Helianthus annuus*). J. Food Sci. 47:40–42.

Vaillant S. (1718–1743) De l'établissement de nouveaux caracteres de classes de plantes à fleurs cômposées. Memoires de Hist. L'Acad. Roy. des Sci., Paris 1718:143–191; 1719:277–318; 1720:277–339; 1721:174–224; 1743:143–191.

Valadon L. R. G. (1977) Arctoteae and Calenduleae—chemical review. In: Heywood V. H., Harborne J. B., Turner B. L. (eds.) *The Biology and Chemistry of the Compositae*. London: Academic Press. 989–998.

Valant K. (1978) Charakteristische Flavonoidglykoside und verwandtschaftliche Gliederung der Gattung *Achillea*. Naturwissens. 65:437–438.

Valant K., Besson E., Chopin J. (1978) *C*-Glycosylflavones from the genus *Achillea*. Phytochemistry 17:2136–2137.

Valant K., Besson E., Chopin J. (1980) Isoorientin 7,3'-dimethyl ether, a new *C*-glycosylflavone from *Achillea cretica*. Phytochemistry 19:156.

Valant-Vetschera K. M. (1981) Vergleichende Flavonoidchemie und Systematik der Gattungen *Achillea* und *Leucocyclus* (Asteraceae-Anthemideae). Ph.D. Dissertation, Univ. of Vienna.

Valant-Vetschera K. M. (1982) Flavonoid pattern and systematics of the genus *Leucocyclus*. Phytochemistry 21:1067–1069.

Valant-Vetschera K. M. (1984) Laubblattflavonoide der *Achillea millefolium*-Gruppe 1: Infraspezifische Variabilität bei *A. setacea* W.& K. und verwandten Arten. Sci. Pharm. 52:307–311.

Valant-Vetschera K. M. (1985a) Flavonoid diversification in *Achillea ptarmica* and allied taxa. Biochem. Syst. Ecol. 13:15–21.

Valant-Vetschera K. M. (1985b) *C*-Glycosylflavones in roots and aerial parts of some *Achillea* species. Biochem. Syst. Ecol. 13:119–121.

Valant-Vetschera K. M. (1985c) *C*-Glycosylflavones as an accumulation tendency: a critical review. Botanical Review 51:1–52.

Valant-Vetschera K. M. (1987) Flavonoid glycoside accumulation trends of *Achillea nobilis* L. Biochem. Syst. Ecol. 15:45–52.

Valant-Vetschera K. M., Wollenweber E. (1986) External accumulation of flavonoid aglycones in *Achillea*. In: Farkas L., Gabor M., Kallay F. (eds.) *Studies in Organic Chemistry 23, Flavonoids and Bioflavonoids 1985*. Amsterdam: Elsevier. 213–221.

Valant-Vetschera K. M., Wollenweber E. (1988a) Distribution of externally accumulated flavonoid aglycones within the *Achillea nobilis* group. Biochem. Syst. Ecol. 16:403–409.

Valant-Vetschera K. M., Wollenweber E. (1988b) Leaf flavonoids of the *Achillea millefolium* group. Part II: Distribution patterns of free aglycones in leaf exudates. Biochem. Syst. Ecol. 16:605–614.

Valant-Vetschera K. M., Wollenweber E. (1989) Flavonoid aglycones in the leaf exudate of the genus *Leucocyclus* (Compositae). Z. Naturforsch. 44c:323–324.

Valant-Vetschera K. M., Wollenweber E. (1994) Exudate flavonoid aglycones of *Achillea* sect. *Achillea* and sect. *Babounya*: a comparative study. Biochem. Syst. Ecol. 22:609–620.

Valant-Vetschera K. M., Wollenweber E. (1995) Flavonoid aglycones from the leaf surfaces of some *Artemisia* spp. (Compositae–Anthemideae). Z. Naturforsch. 50c:353–357.

Valant-Vetschera K. M., Wollenweber E. (1996a) Comparative analysis of leaf exudate flavonoids in *Achillea* subsect. *Filipendulinae*. Biochem. Syst. Ecol. 24:435–446.

Valant-Vetschera K. M., Wollenweber E. (1996b) Leaf exudate flavonoids of *Achillea clusiana* Tausch and related species. Biochem. Syst. Ecol. 24:477–478.

Valant-Vetschera K. M., Wollenweber E. (1999) Leaf exudate flavonoids of *Achillea sibirica* subsp. *mongolica* and *A. latiloba*. Biochem. Syst. Ecol. 27: 523–525.

Valdebenito H., Stuessy T. F., Crawford D. J. (1990a) Synonymy in *Peperomia berteroana* (Piperaceae) results in biological disjunction between Pacific and Atlantic Oceans. Brittonia 42:121–124.

Valdebenito H., Stuessy T. F., Crawford D. J. (1990b) A new biogeographic connection between islands in the Atlantic and Pacific Oceans. Nature (London) 347:549–550.

Valdebenito H., Stuessy T. F., Crawford D. J. (1992a) Evolution of *Erigeron* (Compositae) in the Juan Fernandez Islands, Chile. Syst. Bot. 17:470–480.

Valdebenito H., Stuessy T. F., Crawford D. J., Silva O. M. (1992b) Evolution of *Peperomia* (Piperaceae) in the Juan Fernandez Islands. Plant Syst. Evol. 182:107–119.

van Baren C. M., Juarez B., Ferraro G. E., Coussio J. D. (1994) Phenolic constituents of *Porophyllum obscurum*. Biochem. Syst. Ecol. 22:543.

Van Horn G. S. (1973) The taxonomic status of *Pentachaeta* and *Chaetopappa* with a revision of *Pentachaeta*. Univ. Calif. Publ. Bot. 65:1–41.

Van Puyvelde L., De Kimpe N., Costa J., Munyjabo V., Nyirankuliza S., Hakizamungu E., Schamp N. (1989) Isolation of flavonoids and a chalcone from *Helichrysum odoratissimum* and synthesis of helichrysetin. J. Nat. Prod. 52:629–633.

Varga E., Sarik G., Hajdu Z., Szendrei K., Pelczer I., Jerkovich G. (1990) Flavonoids from *Leuzea carthamoides* DC. Herba Hung. 29:51.

Varga E., Szendrei K., Reisch J. (1982) Flavonol 5-glycosides from *Rhaponticum carthamoides*. Fitoterapia 53:9.

Varin L., Barron D., Ibrahim R. (1986a) Identification, biosynthesis and enzymology of sulfated flavonoids of *Flaveria bidentis*. Bull. Liaison-Groupe Polyphenols 13:71–72.

Varin L., Barron D., Ibrahim R. (1986b) Identification and biosynthesis of glucosylated and sulfated flavonoids in *Flaveria bidentis*. Zeit. Naturforsch. 41c:813–819.

Varin L., DeLuca V., Ibrahim R. K., Brisson N. (1992) Molecular characterization of two plant flavonol sulfotransferases. Proc. Natl. Acad. Sci. U.S.A. 89:1286–1290.

Varin L., Ibrahim R. K. (1989) Partial purification and characterization of three flavonol-specific sulfotransferases from *Flaveria chloraefolia*. Plant Physiol. 90:977–981.

Varin L., Ibrahim R. K. (1991) Partial purification and some properties of flavonol 7-sulfotransferase from *Flaveria bidentis*. Plant Physiol. 95:1254–1258.

Varin L., Ibrahim R. K. (1992) Novel flavonol 3-sulfotransferase, purification, kinetic properties, and partial amino acid sequence. J. Biol. Chem. 267:1858–1863.

Vasconcelos J. M. J., Silva A. M. S., Cavaleiro J. A. S. (1998) Chromones and flavanones from *Artemisia campestris* subsp. *maritima*. Phytochemistry 49: 1421–1424.

Vasey M. C. (1985) The specific status of *Lasthenia maritima* (Asteraceae), an endemic of seabird-breeding habitats. Madroño 32:131–142.

Vasquez M., Quijano M., Fronczek F. R., Macias F., Urbatsch L. E., Cox P. B., Fischer N. H. (1990) Sesquiterpene lactones and lignanes from *Rudbeckia* species. Phytochemistry 29:561–565.

Veldkamp J. F., Kreffer L. A. (1991) Notes on southeast Asian and Australian Coreopsidinae (Asteraceae). Blumea 35:459–482.

Verykokidou-Vitsaropoulos E., Becker H. (1983) Flavonoide aus *Spilanthes oleracea*. Arch. Pharm. 316:815–816.

Vichnewski W., Skrochy C. A., Nasi A. M. T. T., Lopes J. L. C., Herz W. (1999) 15-Hydroxyeremantholide B and derivatives from *Eremanthus arboreus*. Phytochemistry 50:317–320.

Vichnewski W., Takahashi A. M., Nasi A. M. T., Goncalves D. C. R. G., Dias D. A., Lopes J. N. C., Goedken V. L., Gutierrez A. B., Herz W. (1989) Sesquiterpene lactones and other constituents from *Eremanthus seidilii*, *E. goyazensis* and *Vanillosmopsis erythropappa*. Phytochemistry 28:1441–1451.

Vidal-Ollivier E., Elias R., Crespin F., Diaz Lanza A. M., Maillard C., Balansard G., (1991) Identification of the major flavonoids of *Calendula officinalis* L. by HPLC as a function of cultivated variety and date of harvest. Pharm. Acta Helv. 66:318–320.

Vidal-Ollivier E., Elias R., Faure F., Babadjamian A., Crespin F., Balansard G., Boudon G. (1989) Flavonol glycosides from *Calendula officinalis* flowers. Planta Medica 55:73–74.

Vieira L. M., Nascimento M. S. J., Pereira J., Kijjoa A. (1997) Anticomplementary activity and constituents of *Achillea ageratum*. Pharm. Pharmacol. Lett. 7:131–134.

Vienne M., Braemer R., Paris M., Couderc H. (1989) Chemotaxonomic study of two cultivars of *Artemisia dracunculus* L.: "French" and "Russian" tarragon. Biochem. Syst. Ecol. 17:373–374.

Vilegas W., Boralle N., Cabrera A., Bernardi A. C., Pozetti G. L., Arantes S. F. (1995) Coumarins, a flavonoid from *Pterocaulon alopecuroides*. Phytochemistry 38:1017–1019.

Villaescusa L., Diaz A. M., Barolome C. (1995) *Jasonia glutinosa* DC. and its relation with other species of *Jasonia*. Die Pharmazie 50:639–640.

Villaescusa L., Diaz A. M., Lapica B. C., Fernandez L. (1992) Patuletin and quercetin derivatives in *Jasonia glutinosa*. Ars Pharm. 33:469–472.

Vogt R., Oberprieler C. (1996) The genus *Nivellea* B. H. Wilcox, K. Bremer & C. Humphries (Compositae, Anthemideae). Bot. J. Linn. Soc. 122:123–135.

Vries I.M. de, van Raamsdonk L. W. D. (1994) Numerical morphological analysis of lettuce cultivars and species (*Lactuca* sect. *Lactuca*, Asteraceae). Plant Syst. Evol. 193:125–141.

Vrkoc J., Dolejs L., Budesinsky M. (1975) Methylene-*bis*-2*H*-pyran-2-ones and phenolic constituents from the root of *Helichrysum arenarium*. Phytochemistry 14:1383–1384.

Vrkoc J., Herout V., Sorm F. (1959) Über Pflanzenstoffe X. Isolierung der Krystallinen Bestandteile der Sandstrohblume *Helichrysum arenarium* Mch. Coll. Czech. Chem. Commun. 24:3938–3954.

Vrkoc J., Ubik K., Sedmera P. (1973) Phenolic extractives from the achenes of *Helichrysum arenarium*. Phytochemistry 12:2062.

Vyas A. V., Mulchandani N. B. (1984) Structure reinvestigation of conyzorigun, a new chromene from *Ageratum conyzoides*. J. Chem. Soc. Perkin Trans. 1:2945–2947.

Vyas A. V., Mulchandani N. B. (1986) Polyoxygenated flavones from *Ageratum conyzoides*. Phytochemistry 25:2625–2627.

Waddell T. G. (1973) Hymenoxin: a flavone from *Helianthus angustifolius*. Phytochemistry 12:2061.

Waddell T. G., Elkins S. K., Mabry T. J., Norris J. A., Ulubelen A. (1985) Constituents of *Rudbeckia serotina*. J. Nat. Prod. 48:163–164.

Waddell T. G., Thomasson M. H., Moore M. W., White H. W., Swanson-Bean D., Green M. E., Van Horn G. S., Fales H. M. (1982) Isoflavone, wax and triterpene constituents of *Wyethia mollis*. Phytochemistry 21:1631–1633.

Wagenitz G. (1972) Beiträge zur Kenntnis der Gattung *Centaurea* L. 1. Zur Taxonomie türkischer Arten der Sektionen *Acrolophus* und *Acrocentron*. Willdenowia 6:479–508.

Wagenitz G. (1976) Systematics and phylogeny of the Compositae (Asteraceae). Plant Syst. Evol. 125:29–46.

Wagenitz G. (1986) *Centaurea* in South-West Asia: patterns of distribution and diversity. Proc. Roy. Soc. Edinburgh 89B:11–21.

Wagenitz G., Bedarff U. (1989) Taxonomic notes on some species of the genus *Cichorium* (Compositae-Lactuceae). In: Tan K. (ed.) *The Davis & Hedge Festschrift*. Edinburgh: Edinburgh Univ. Press. 11–21.

Wagenitz G., Gamal-Eldin E. (1985) Zur Kenntnis der griechischen *Centaurea*-Arten der Sektion *Acrocentron*. Bot. Jahrb. Syst. 107:95–127.

Wagenitz G., Hellwig F. H. (1996) Evolution of characters and phylogeny of the Centaureinae. In: Hind D. J. N., Beentje H. J. (eds.) *Compositae: Systematics. Proceedings of the International Compositae Conference, Kew (1994) Vol. 1*. Kew: Royal Botanic Gardens. 491–510.

Wagner H., Flores G., Geissman T. A. (1972a) Eupatilin aus *Tanacetum vulgare*. Phytochemistry 11:451.

Wagner H., Flores G., Hörhammer L., Farkas L., Strelisky J. (1971a) Über die Flavonoide von *Eupatorium*-Arten II. Synthesis von Eupatilin, Eupatoretin, Eupatin und Eupalitin. Tet. Lett. 1233–1236.

Wagner H., Geyer B., Kiso Y., Hikino H., Rao G. S. (1986) Coumestans as the main active principles of liver drugs *Eclipta alba* and *Wedelia calendulacea*. Planta Medica 52:370–378.

Wagner H., Hoer R., Murakami T., Farkas L. (1973a) Isolierung, Strukturaufklärung und Synthese von 4',5,7-Trihydroxy-3',6-dimethoxy-7-mono-ß-D-glucopyranoside (Jaceoside), einem neuen Flavonglycosid aus den Wurzeln von *Centaurea jacea*. Chem. Ber. 106:20–27.

Wagner H., Hörhammer L., Herz W. (1972b) Flavonol-3-glucosides in *Baccharis angustifolia* and *Berlandiera pumila*. Phytochemistry 11:444–445.

Wagner H., Hörhammer L., Herz W. (1972c) Flavone-*C*-glycosides in *Helenium* species. Phytochemistry 11:446.

Wagner H., Hörhammer L., Herz W. (1972d) Flavone *C*-glycosides in a coastal race of *Gaillardia pulchella*. Phytochemistry 11:851.

Wagner H., Hörhammer L., Herz W. (1973b) Flavonoids in ten *Liatris* species. Phytochemistry 12:2063–2064.

Wagner H., Hörhammer L., Hoer L., Farkas L. (1967) Notiz über die Synthese des Mikanins, eines Flavonols aus *Mikania cordata* (Burm. f.) B. L. Robins. Chem. Ber. 100:1768–1770.

Wagner H., Hörhammer L., Hoer L., Murakami T., Farkas L. (1969) Untersuchungen über Glykoside von *Centaurea jacea* L. III. Isolierung, Struktur, und Synthesis von 4',5,7-Trihydroxy-3',6-dimethoxy-flavon-7-mono-ß-D-Glucopyranosid, einem neuen Flavonglykosid aus den Wurzeln von *Centaurea jacea* L. Tet. Lett. 3411–3414.

Wagner H., Hörhammer L., Kirchner W. (1960) Über weitere Vorkommen von Pectolinarin und Linarin im Pflanzenreich. Arch. Pharm. 293:1053–1062.

Wagner H., Hörhammer L., Munster R. (1965) Über die Struktur eines neuen Flavonoids aus den Früchten von *Carduus marianus* L. Naturwissen. 52:305.

Wagner H., Hörhammer L., Munster R. (1968) Zur Chemie des Silymarins (Silybin), ein Wirkprinzip der Früchte von *Silybum marianum* (L.) Gaertn. (*Carduus marianus* L.). Arzneim. Forsch. 18:688–696.

Wagner H., Iyengar M. A. (1972) Flavonol 3-glycosides in eight *Hymenoxys* species. Phytochemistry 11: 3087–3088.

Wagner H., Iyengar M. A., Dull P., Herz W. (1972e) Flavon *O*-und *C*-Glykoside in *Eupatorium serotinum*. Phytochemistry 11:1506.

Wagner H., Iyengar M. A., Hörhammer L., Herz W. (1971b) Patuletin-3-*O*-ß-glucosid in *Flaveria* Arten. Phytochemistry 10:2824–2825.

Wagner H., Iyengar M. A., Hörhammer L., Herz W. (1972f) Flavonol-3-glycosides in eight *Eupatorium* species. Phytochemistry 11:1504–1505.

Wagner H., Iyengar M. A., Hörhammer L., Herz W. (1972g) Flavonoids of *Helenium brevifolium*. Phytochemistry 11:1857.

Wagner H., Iyengar M. A., Hörhammer L., Herz W. (1972h) Flavonol-3-glycosides in eight *Hymenoxys* species. Phytochemistry 11:3087–3088.

Wagner H., Iyengar M. A., Hörhammer L., Herz W. (1972i) Flavonol-3-glycosides in *Plummera ambigens*. Phytochemistry 11:2633.

Wagner H., Iyengar M. A., Michahelles E., Herz W. (1971c) Quercetin-3-*O*-acetyl-ß-D-glucopyranosid in *Plummera floribunda* und *Helenium hoopesii*. Phytochemistry 10: 2547–2548.

Wagner H., Iyengar M. A., Seligmann O., Herz W. (1974) Rhamnocitrin-3-glucuronid in *Verbesina myriecephala*. Phytochemistry 13:493–494.

Wagner H., Iyengar M. A., Seligmann O., Hörhammer L., Herz W. (1972j) Chrysoeriol-7-glucuronide in *Grindelia squarrosa*. Phytochemistry 11:2350.

Wagner H., Iyengar M. A., Seligmann O., Hörhammer L., Herz W. (1972k) Flavonoids in *Vernonia* Arten. Phytochemistry 11:3086–3087.

Wagner H., Iyengar M. A., Seligmann O., Hörhammer L., Herz W. (1972l) Isorhamnetin-3-glykoside in *Eupatorium leptophyllum*. Phytochemistry 11:2630–2631.

Wagner H., Kirmayer W. (1957) Über das Vorkommen von Apigenin Glykosiden in einigen Compositenbluten. Naturwissen. 44:307.
Wagner H., Seitz R., Lotter H., Herz W. (1978) New furanoid *ent*-clerodanes from *Baccharis tricuneata*. J. Org. Chem. 43:3339–3345.
Wagner H., Seligmann O., Hörhammer L., Seitz M., Sonnenbichler J. (1971e) Zur Struktur von Silychristin, einem zweiten Silymarin-Isomeren aus *Silybum marianum*. Tet. Lett. 22:1895–1899.
Wagner W. L., Fletcher R. A., Shannon R. K. (1999) Another look at *Hymenoxys* subgenus *Plummera* (Asteraceae: Heliantheae: Gaillardiinae) from Arizona and New Mexico. Brittonia 51:79–86.
Wagner W. L., Herbst D. R., Sohmer S. H. (1990) *Manual of the Flowering Plants of Hawai'i*, Vol. 1. Honolulu: Univ. Hawaii Press and Bishop Museum Press.
Wallace J. W. (1974) Tricin-5-*O*-glucoside and other flavonoids of *Cirsium arvense*. Phytochemistry 13:2320–2321.
Wallace J. W., Bohm B. A. (1971) A new flavone glycoside from *Cirsium brevistylum*. Phytochemistry 10:452–453.
Wallace J. W., Grisebach H. (1973) The *in vivo* incorporation of a flavanone into *C*-glycosylflavones. Biochim. Biophys. Acta 304:837–841.
Wallace J. W., Mabry T. J., Alston R. E. (1969) On the biogenesis of flavone *O*-glycosides and *C*-glycosides in the Lemnaceae. Phytochemistry 8:93–99.
Wallace R. S., Jansen R. K. (1990) Systematic implications of chloroplast DNA variation in the genus *Microseris* (Asteraceae: Lactuceae). Syst. Bot. 15:606–616.
Wang J., Hui Q., Qin H., Zhu J., Ishii H., Harayama T. (1992) Chemical constituents of Spanish needles (*Bidens bipinnata*). Zhongcaoyao 23:229–231.
Wang J., Yang H., Lin Z.-W., Sun H.-D. (1997) Flavonoids from *Bidens pilosa* var. *radiata*. Phytochemistry 46:1275–1278.
Wang W. (1997) Determination of the content of flavonoids in tissue cultural media of *Saussurea medusa* cells by RP-HPLC. Zhongcaoyao 28:654–656.
Wang Y., Hamburger M., Gueho J., Hostettmann K. (1989b) Antimicrobial flavonoids from *Psiadia trinervia* and their methylated and acetylated derivatives. Phytochemistry 28:2323–2327.
Wang Y., Hamburger M., Hostettmann K. (1989a) Antifungal flavonoids from *Psiadia trinervia*. Planta Medica 55:109.
Wang Y., Hostettmann K. (1990) Polyphenols from *Psiadia trinerva*. Bull. de Liason-Groupe Polyphenols 15:167–170.
Warning U., Bohlmann F., King R. M., Haegi L. (1988) Diterpenes from *Olearia* species. J. Nat. Prod. 51:513–516.
Warning U., Bohlmann F., Sanchez V. H., Del Rio S. E., Dominguez X. A. (1986) New constituents of *Baccharis salicifolia*. Rev. Latinoamer. Quim. 17:199–200.
Warwick S. I., Bain J. F. (1987) Hybridization and introgression in *Carduus nutans* L. and *Carduus acanthoides* L. Amer. J. Bot. 74:764. [Abstr.]
Wassel G. M. (1975) Constituents of the wild Egyptian plant *Onopordum alexandrinum*. Z. Naturforsch. 30c:349.
Watanabe K., Ohno N., Yoshioka H., Gershenzon J., Mabry T. J. (1982) Sesquiterpene lactones and diterpenoids from *Helianthus argophyllus*. Phytochemistry 21:709–713.

Watanabe K., Smith-White S. (1987) Phyletic and evolutionary relationships of *Brachyscome lineariloba* (Compositae). Plant Syst. Evol. 157:121–141.

Watson L. E., Estes J. R. (1990) Biosystematic and phenetic analysis of *Marshallia* (Asteraceae). Syst. Bot. 15:403–414.

Watson L. E., Jansen R. K., Estes J. R. (1991) Tribal placement of *Marshallia* (Asteraceae) using chloroplast DNA restriction site mapping. Amer. J. Bot. 78:1028–1035.

Watson S. (1873) Revision of the extra-tropical North American species of the genera *Lupinus*, *Potentilla* and *Oenothera*. Proc. Amer. Acad. Arts 8:517–618.

Weber W. A. (1946) A taxonomic and cytological study of the genus *Wyethia*, family Compositae, with notes on the related genus *Balsamorhiza*. Amer. Midl. Natur. 35:400–452.

Weber W. A. (1998) New names and combinations in Asteraceae: Heliantheae—Ecliptinae. Phytologia 85:19–21.

Weberling F. (1989) *Morphology of Flowers and Inflorescences*. Cambridge: Cambridge Univ. Press.

Weiss M. R. (1995) Floral color change: a widespread functional convergence. Amer. J. Bot. 82:167–185.

Welle R., Schröder J. (1992) Expression cloning in *E. coli* and preparative isolation of the reductase coacting with chalcone synthase during the key step in the biosynthesis of soybean phytoalexins. Arch. Biochem. Biophys. 293:377–381.

Welle R., Schröder J., Schlitz E., Grisebach H., Schröder J. (1991) Induced plant responses to pathogen attack. Analysis and heterologous expression of the key enzyme in the biosynthesis of phytoalexin in soybean (*Glycine max* L. Merr. cv. Harosoy 63). Eur. J. Biochem. 196:423–430.

Wells J. R. (1965) A taxonomic study of *Polymnia* (Compositae). Brittonia 17:144–159.

Welsh S. L. (1993) New taxa and new nomenclatural combinations in the Utah flora. Rhodora 95:392–421.

Werner K. (1976) *Cirsium* and *Notobasis*. In: Tutin T. G., Heywood V. H., Burges N. A., Moore D. M., Valentine D. H., Walters S. M., Webb D. A. (eds.) *Flora Europaea, Plantaginaceae to Compositae (and Rubiaceae)*, Vol. 4. Cambridge: Cambridge Univ. Press. 232–242.

Whang W. K., Oh I. S., Lee M. T., Yang D. S., Kim I. H. (1994) Pharmaco-constituents of *Taraxacum hallaisanensis*. (I) Phenolic compounds from aerial part of *Taraxacum hallaisanensis*. Saengyak Hakhoechi 25:209–213.

Whitton J., Wallace R. S., Jansen R. K. (1995) Phylogenetic relationships and patterns of character change in the tribe Lactuceae (Asteraceae) based on chloroplast DNA restriction site variation. Canad. J. Bot. 73:1058–1073.

Wichlacz M., Daniewski W. M., Danikiewicz W., Gumulka M., Drozdz B., Grabarczyk H. (1994) Constituents of *Cladanthus arabicus*. Pol. J. Chem. 68:2147–2152.

Wilcox B. H. R. (1984) Flavonoid distribution patterns in *Leucanthemum* and related species from North Africa. Biochem. Syst. Ecol. 12:357–361.

Wiley E. O. (1981) *Phylogenetics: The Theory and Practice of Phylogenetic Systematics*. New York: John Wiley and Sons.

Wilken D. H. (1975) A systematic study of the genus *Hulsea* (Asteraceae). Brittonia 27:228–244.

Wilkomirski B., Batko A. (1993) Presence of flavonoids and intrafamilial classification of the Compositae. Acta Soc. Bot. Polon. 62:47–51.

Williams C. A., Goldstone F., Greenham J. (1996) Flavonoids, cinnamic acids and coumarins from the different tissues and medicinal preparations of *Taraxacum officinale*. Phytochemistry 42:121–127.

Williams C. A., Harborne J. B., Geiger H., Hoult J. R. S. (1999) The flavonoids of *Tanacetum parthenium* and *T. vulgare* and their anti-inflamatory properties. Phytochemistry 51:417–423.

Williams C. A., Hoult J. R. S., Harborne J. B., Greenham J., Eagles J. (1995) A biologically active lipophilic flavonol from *Tanacetum parthenium*. Phytochemistry 38:267–270.

Willis J. C. (1918) The age and area hypothesis. Science 47:626.

Willis J. C. (1922) *Age and Area*. Cambridge: Cambridge Univ. Press.

Willstätter R., Bolton E. K. (1916) Über ein Anthocyan der Winteraster *Chrysanthemum*. Liebig's Ann. Chem. 412:136–148.

Willstätter R., Everest A. E. (1913) The colouring matter of cornflower. Liebig's Ann. Chem. 401:189–231.

Willuhn G., Merfort I., Passreiter C. M., Schmidt T. J. (1995) Chemistry and systematics of the genus *Arnica*. In: Hind D. J. N., Jeffrey C., Pope G. V. (eds.) *Advances in Compositae Systematics*. Kew: Royal Botanic Garden. 167–195.

Wilson P. G. (1992) The classification of Australian species currently included in *Helipterum* and related genera (Asteraceae: Gnaphalieae), part 1. Nuytsia 8:379–438.

Wilt F. M., Geddes J. D., Tamma R. V., Miller G. C., Everett R. L. (1992) Interspecific variation of phenolic concentrations in persistent leaves among six taxa from subgenus *Tridentata* of *Artemisia* (Asteraceae). Biochem. Syst. Ecol. 20:41–52.

Wilt F. M., Miller G. C. (1992) Seasonal variation of coumarin and flavonoid concentrations in persistent leaves of Wyoming big sagebrush (*Artemisia tridentata* ssp. *wyomingensis*: Asteraceae). Biochem. Syst. Ecol. 20:53–67.

Wimmer G., Halbwirth H., Wurst F., Forkmann G., Stich K. (1998) Enzymic hydroxylaton of 6'-deoxychalcones with protein preparations from petals of *Dahlia variabilis*. Phytochemistry 47:1013–1016.

Witter M. S., Carr G. D. (1988) Adaptive radiation and genetic differentiation in the Hawaiian silversword alliance (Compositae: Madiinae). Evolution 42:1278–1287.

Woeldecke M., Herrmann K. (1974) Isolation and identification of the hydroxyflavone glycosides of endive (*Cichorium endivia*) and lettuce (*Lactuca sativa*). Z. Naturforsch. 29c:355–359.

Wolbis M., Krolikowska M., Bednarek P. (1993) Polyphenolic compounds in *Taraxacum officinale*. Acta Pol. Pharm. 50:153–158.

Wolf S. J. (1981) A biosystematic revision of *Arnica* L. (Compositae) subgenus *Austromontana* Maguire. Ph.D. Dissertation, Univ. Alberta.

Wolf S. J., Denford K. E. (1983) Flavonoid variation in *Arnica cordifolia*: an apomictic polyploid complex. Biochem. Syst. Ecol. 11:111–114.

Wolf S. J., Denford K. E. (1984a) *Arnica gracilis* (Compositae), a natural hybrid between *A. latifolia* and *A. cordifolia*. Syst. Bot. 9:12–16.

Wolf S. J., Denford K. E. (1984b) Flavonoid diversity and endemism in *Arnica* subgen. *Austromontana*. Biochem. Syst. Ecol. 12:183–188.

Wolf S. J., Whitkus R. A. (1987) A numerical analysis of flavonoid variation in *Arnica* subgenus *austromontana* (Asteraceae). Amer. J. Bot. 74:1577–1584.

Wollenweber E. (1984) Exkret-Flavonoide bei *Hieracium intybaceum* (Asteraceae), einen alpinen Habichtskraut. Z. Naturforsch. 39c:833–834.

Wollenweber E. (1985) Exkret-Flavonoide bei höheren Pflanzen arider Gebiete. Plant Syst. Evol. 150:83–88.

Wollenweber E. (1988) Flavones and flavonols. In: Harborne J. B. (ed.) *The Flavonoids-Advances in Research since 1986*. London: Chapman and Hall. 259–335.

Wollenweber E. (1991) External leaf flavonoids of *Centaurea macrocephala*. Fitoterapia 62:364–365.

Wollenweber E., Arriaga-Giner F. J., Rumbero A., Greenaway W. (1989a) New phenolics from *Baccharis* leaf exudate. Z. Naturforsch. 44c:727–730.

Wollenweber E., Dörr M., Fritz H., Valant-Vetschera K. M. (1997) Exudate flavonoids in several Asteroideae and Cichorioideae (Asteraceae). Z. Naturforsch. 52c:137–143.

Wollenweber E., Dörr M., Muniappan R. (1995a) Exudate flavonoids in a tropical weed, *Chromolaena odorata*. Biochem. Syst. Ecol. 23:873–874.

Wollenweber E., Dörr M., Muniappan R., Roitman J. N., Schilling E. (1995b) External flavonoids of three species of *Viguiera*, section *Hypargyrea* (Asteraceae). Z. Naturforsch. 50c:588–590.

Wollenweber E., Dörr M., Timmermann B. N., Strand J., Fuentes E. R. (1993a) Exudate flavonoids from *Grindelia tarapacana*. Z. Naturforsch. 48c:533–534.

Wollenweber E., Favre-Bonvin J., Waton H., Hauteville M., Rustaiyan A. (1992) A phloracetophenone derivative and flavonoid aglycones from the lipophilic exudate of *Artemisia aucheri*. Phytochem. Life Sci. Adv. 11:105–107.

Wollenweber E., Fritz H., Henrich B., Jakupovic J., Schilling G., Roitman J. N. (1993b) Rare flavonoid aglycones from *Anaphalis margaritacea* and two *Gnaphalium* species. Z. Naturforsch. 48c:420–424.

Wollenweber E., Hradetzky D., Mann K., Roitman J. N., Yatskievych G., Proksch M., Proksch P. (1987a) Exudate flavonoids from aerial parts of five *Ambrosia* species. J. Plant Physiol. 131:37–43.

Wollenweber E., Jay M. (1988) Flavones and flavonols. In: Harborne J. B. (ed.) *The Flavonoids – Advances in Research since 1980.* London: Chapman and Hall. 233–302.

Wollenweber E., Mann K. (1987) Internal and external flavonoids of *Pericome caudata*. Phytochemistry 26:3249–3250.

Wollenweber E., Mann K. (1989) Exudate flavonoids in three essential oil plants from the Ciskei (South Africa). Fitoterapia 60:249–251.

Wollenweber E., Mann K., Arriaga F. J., Yatskievych G. (1985a) Flavonoids and terpenoids from the leaf resin of *Pluchea odorata*. Z. Naturforsch. 40c:321–324.

Wollenweber E., Mann K., Arriaga-Giner F. J., Roitman J. N., West J. G. (1993c) Exudate flavonoids from two Australian Asteraceae, *Bracteantha viscosa* and *Cassinia quinquefaria*. Phytochemistry 33:871–873.

Wollenweber E., Mann K., Valant-Vetschera K. M. (1989b) External flavonoid aglycones in *Artemisia* and further Anthemideae (Asteraceae). Fitoterapia 60:460–462.

Wollenweber E., Mayer K. (1991) Exudate flavonoids of *Anthemis nobilis* and *A. tinctoria*. Fitoterapia 62:365–366.

Wollenweber E., Mayer K., Roitman J. N. (1991) Exudate flavonoids of *Inula viscosa*. Phytochemistry 30:2445–2446.

Wollenweber E., Roitman J. N. (1996) Marionol, a novel methyl ether of quercetagetin from *Chromolaena odorata* leaf exudate. Biochem. Syst. Ecol. 24:479.

Wollenweber E., Rustaiyan A. (1991) Exudate flavonoids in three Persian Asteraceae species. Biochem. Syst. Ecol. 19: 673–675.

Wollenweber E., Schober I., Clark W. D., Yatskievych G. (1985b) Flavonoid aglycones from leaf resins of two species of *Heterotheca* (Compositae). Phytochemistry 24:2129–2131.

Wollenweber E., Schober I., Dostal P., Hradetzky D., Arriaga-Giner F. J., Yatskievych G. (1986) Flavonoids and terpenoids from the exudates of some *Baccharis* species. Z. Naturforsch. 41c:87–93.

Wollenweber E., Stevens J. F., Ivancic M. (1998) Flavonoid aglycones and a thiophene derivative from *Helichrysum cassianum*. Phytochemistry 47:1441–1443.

Wollenweber E., Valant-Vetschera K. M. (1996) New results with exudate flavonoids in Compositae. In: Hind D. J. N., Beentje H. J. (eds.) *Compositae: Systematics. Proceedings of the International Compositae Conference, Kew (1994), Vol. 1.* Kew: Royal Botanic Gardens. 169–185.

Wollenweber E., Valant-Vetschera K. M., Ivancheva S., Kuzmanov B. (1987b) Flavonoid aglycones from the leaf surfaces of some *Achillea* species. Phytochemistry 26:181–182

Wollenweber E., Yatskievych G. (1985) Flavonoides en la resina foliar de *Flourensia resinosa*, compuesta endémica de Hidalgo, Mexico. Rev. Latinoamer. Quim. 16:45–46.

Wright W. G. (1976) South African plant extractives. Part III. Helichrysin, a new chalcone glucoside from a *Helichrysum* species. J. Chem. Soc. Perkin Trans 1:1819–1820.

Wu C., Tu Y. (1985) Studies on chemical constituents of *Artemisia* species. III. Isolation and identification of the lipophyllic constituents from *Artemisia argyi*. Zhongyao Tongbao 10:31–32.

Wu J., Musadillin S., Fen S., Kong F., Xu R. (1991) Chemical constituents of *Vernonia anthelmintica* Willd. Huaxue Xuebao 49:1018–1022.

Wu J.-B., Chun Y.-T., Ebizuka Y., Sankawa U. (1991) Biologically active constituents of *Centipeda minima*: sesquiterpenes of potential anti-allergy activity. Chem. Pharm. Bull. 39:3272–3275.

Wussow J. R., Urbatsch L. E., Sullivan G. A. (1985) *Calea* (Asteraceae) in Mexico, Central America and Jamaica. Syst. Bot. 4:241–267.

Xiang C., Semple J. C. (1996) Molecular systematic study of *Aster sensu lato* and related genera (Asteraceae: Astereae) based on chloroplast DNA restriction site analyses and mainly North American taxa. In: Hind D. J. N., Beentje H. J. (eds.) *Compositae: Systematics. Proceedings of the International Compositae Conference, Kew (1994), Vol. 1.* Kew: Royal Botanic Gardens. 393–423.

Xiao Y., Tu Y. (1984) Isolation and identification of the lipophilic constituents from *Artemisia anomela* S. Moore. Yaoxue Xuebao 19:909–913.

Xiao Y., Tu Y. (1986) Chemical constituents of *Artemisia anomala* S. Moore. Zhiwu Xuebao 28:307–310.

Xu L., Liu A. (1987) Coulometric titration of acaciin in *Chrysanthemum indicum*. Yaoxue Xuebao 22:318–319.

Yadev R. N., Barsainya D. (1997) A novel 8,5'-dimethoxy-3',4'-methylenedioxy-3,7-dihydroxyflavone from seeds of *Centratherum anthelminticum* Kuntze. J. Inst. Chem. (India) 69:60–62.

Yahara S., Morita Y., Nohara T. (1990) Studies on the constituents of *Chrysanthemi Flos*. Shoyakugaku Zasshi 44:335–338.

Yamaguchi M.-A., Kawanobu S., Maki T., Ino I. (1996) Cyanidin 3-malonylglucoside and malonyl-coenzyme A: anthocyanidin malonyltransferase in *Lactuca sativa* leaves. Phytochemistry 42:661–663.

Yamaguchi M.-A., Maki T., Ohishi T., Ino I. (1995) Succinyl-coenzyme A: anthocyanidin 3-glucoside succinyltransferase in flowers of *Centaurea cyanus*. Phytochemistry 39:311–313.

Yamaguchi M.-A., Terahara N., Shizukuishi K.-i. (1990) Acetylated anthocyanins in *Zinnia elegans* flowers. Phytochemistry 29:1269–1270.

Yang S. S., El-Sayed N. H., Whittemore A. T., Mabry T. J. (1989) Flavonoids of *Critonia daleoides*. Rev. Latinoamer. Quim. 20:32.

Yang S.-L., King R. A., Roberts M. F. (1990) The flavonoids of *Ageratina deltoidea*. Biochem. Syst. Ecol. 18:485–486.

Yang S.-L., Roberts M. F., O'Neill M. J., Bucar F., Phillipson J. D. (1995) Flavonoids and chromenes from *Artemisia annua*. Phytochemistry 38:255–257.

Yatsyuk Ya. K., Lyashenko S. S. (1969) Flavonoids of *Serratula inermis* Gilib. Chem. Nat. Cmpds. 5:46 [54].

Yatsyuk Ya. K., Lyashenko S. S., Segal G. M. (1970) Flavonoids of *Serratula bracteifolia*. Khim. Issled. Farm. 165.

Ybarra M. I., Catalan C. A. N., Diaz J. G., Herz W. (1992) A cyperane and trixanes from *Jungia polita*. Phytochemistry 31:3627–3629.

Ybarra M. I., Catalan C. A. N., Diaz J. G., Herz W. (1994) Two homofarnesane derivatives and other constituents from *Gochnatia glutinosa*. Phytochemistry 35:1585–1587.

Yeo P. F. (1971) × *Solidaster*, an intergeneric hybrid (Compositae). Baileya 18:27–32.

Ylstra B., Touraev A., Moreno R. M. B., Stöger E., van Tunen A. J., Vicente O., Mol J. N. M., Heberle-Bors E. (1992) Flavonols stimulate development, germination, and tube growth of tobacco pollen. Plant Physiol. 100:902–907.

Yoshitama K. (1981) Caffeic acid 4-ß-glucoside as the acyl moiety of the *Senecio cruentus* anthocyanin. Phytochemistry 20:186–187.

Yoshitama K., Abe K. (1977) Chromatographic and spectral characteristics of 3'-glycosylation in anthocyanidins. Phytochemistry 16:591–593.

Yoshitama K., Hayashi K. (1974) Concerning the structure of cinerarin, a blue anthocyanin from garden cineraria. Studies on anthocyanins LXVI. Bot. Mag. Tokyo. 87:33–40.

Yoshitama K., Hayashi K., Abe K., Kakisawa H. (1975) Further evidence for the glycoside structure of cinerarin. Studies on anthocyanins, LXVIII. Bot. Mag. Tokyo 88:213–217.

Yoshitama K., Kaneshige M., Ishikura N., Araki F., Yahara S., Abe K. (1994) A stable reddish purple anthocyanin in the leaf of *Gynura aurantiaca* cv. 'Purple Passion.' J. Plant Res. 107:209–214.

Youssef D., Frahm A. W. (1995) Constituents of the Egyptian *Centaurea scoparia.* III. Phenolic constituents of the aerial parts. Planta Medica 61:570–573.

Youssef S. A., Ibraheim Z. Z., Attia A. A. (1995) Flavonoids from *Asteriscus pygmaeus* (DC.) Cass. grown in Egypt. Bull. Pharm. Sci. Assiut Univ. 18:33–38.

Yu D., Xie F. (1987) Chemical constituents of *Chrysanthemum indicum* L. Yaoxue Xuebao 22:837.

Yu S., Fang N., Mabry T. J. (1986a) Flavonoids from *Ageratina saltillensis.* J. Nat. Prod. 49:1178–1179.

Yu S., Fang N., Mabry T. J. (1987a) Flavonoid aglycones from *Xanthocephalum gymnospermoides* var. *gymnospermoides.* Phytochemistry 26:2131–2133.

Yu S., Fang N., Mabry T. J. (1987b) Flavonoids from *Ageratina havanensis.* Rev. Latinoamer. Quim. 18:20–22.

Yu S., Fang N., Mabry T. J. (1988) Flavonoids from *Gymnosperma glutinosum.* Phytochemistry 27:171–177.

Yu S., Gage D. A., Fang N., Mabry T. J. (1986c) Flavonols and flavonol glycosides from *Eupatorium areolare* var. *leiocarpum.* J. Nat. Prod. 49:181–182.

Yu S., Norris J., Fang N., Mabry T. J. (1986b) Flavonoids of *Asanthus*, a segregate genus of *Brickellia.* J. Nat. Prod. 49:553–554.

Zaghloul A. M., El-Fattah H. A., Halim A. F. (1989) Chemical investigation of the aerial parts of *Anthemis melampodina* Dei. Mansoura J. Pharm. Sci. 5:23–33.

Zaghloul A. M., Salama O. M., Halim A. F. (1990) Chemical investigation of *Centaurea glomerata* Vahl. Mansoura J. Pharm. Sci. 6:61–68.

Zähringer U., Ebel J., Mulheim L. J., Lyne R. L., Grisebach H. (1979) Induction of phytoallexin synthesis in soybean. Dimethylallylpyrophosphate: trihydroxypterocarpan dimethylallyl transferase from elicitor-induced cotyledons. FEBS Lett. 101:90–92.

Zähringer U., Schaller E., Grisebach H. (1981) Induction of phytoallexin synthesis in soybean. Structure and reactions of naturally occurring and enzymatically prepared prenylated pterocarpans from elicitor-treated cotyledons and cell cultures of soybeans. Z. Naturforsch. 36c:234–241.

Zamorano R., Aguirre M. E., Muñoz de la Pena A., Cordano G., Medina J., Timmermann B. (1987) Flavonoids of *Baccharis concava* Pers. Bol. Soc. Chil. Quim. 32:101–103.

Zapesochnaya G. G., Ban'kovskii A. I. (1976) A flavonoid from *Achillea cartilaginea.* Chem. Nat. Cmpds. 12:733 [814].

Zapesochnaya G. G., Ban'kovskii A. I., Nakaidze Kh. A. (1972) A flavonoid and a phthalide from *Helichrysum polyphyllum.* Chem. Nat. Cmpds. 8:787 [804].

Zapesochnaya G. G., Evstratova R. I., Mukhametzhanov M. N. (1977) Methoxyflavones of some species of *Centaurea.* Chem. Nat. Cmpds. 13:590 [706].

Zavada M., Xu X. L., Edwards J. M. On the taxonomic status of *Lophiola aurea* Ker-Gawler. Rhodora 85:73–81.

Zdero C., Ahmed A. A., Bohlmann F., Mungai G. M. (1990a) Diterpenes and sesquiterpene xylosides from East African *Conyza* species. Phytochemistry 29:3167–3172.

Zdero C., Bohlmann F. (1987) Sesquiterpene lactones from the genus *Brachylaena*. Phytochemistry 26:2597–2601.

Zdero C., Bohlmann F. (1989a) Pseudoguaianolides and other constituents from *Anisopappus pinnatifidus* and *Antiphiona* species. Phytochemistry 28:1155–1161.

Zdero C., Bohlmann F. (1989b) Furoeudesmanes and other constituents from representatives of the *Pluchea* group. Phytochemistry 28:3097–3100.

Zdero C., Bohlmann F. (1990a) Sesquiterpene lactones from *Dicoma* species. Phytochemistry 29:183–187.

Zdero C., Bohlmann F. (1990b) Glaucolides, fulvenoguaianolides and other sesquiterpene lactones from *Pentzia* species. Phytochemistry 29:189–194.

Zdero C., Bohlmann F. (1990c) Glycerol derivatives and other constituents from *Metalasia* species. Phytochemistry 29:2179–2180.

Zdero C., Bohlmann F. (1990d) Systematics and evolution within the Compositae, seen with the eyes of a chemist. Plant Syst. Evol. 171:1–14.

Zdero C., Bohlmann F., Anderberg A., King R. M. (1991e) Eremophilane derivatives and other constituents from *Haeckeria* species and further Australian Inuleae. Phytochemistry 30:2643–2650.

Zdero C., Bohlmann F., Boldt P. E. (1989a) Pseudoguaianolides and other constituents from *Hymenoclea monogyra*. Rev. Latinoamer. Quim. 20:132–133.

Zdero C., Bohlmann F., King R. M. (1987) Chemistry of *Barnadesia* (Asteraceae). Phytologia 63:313–315.

Zdero C., Bohlmann F., King R. M. (1990b) Eudesmane derivatives and other constituents from *Apalochlamys spectabilis* and *Cassinia* species. Phytochemistry 29:3201–3206.

Zdero C., Bohlmann F., King R. M. (1992a) *ent*-Labdanes, manoyloxide and helipterol derivatives from *Chrysocephalum ambiguum*. Phytochemistry 31:1631–1638.

Zdero C., Bohlmann F., King R. M. (1992b) Clerodane and labdane derivatives from *Olearia teretifolia*. Phytochemistry 31:1703–1711.

Zdero C., Bohlmann F., King R. M., Robinson H. (1986a) Diterpene glycosides and other constituents from Argentinian *Baccharis* species. Phytochemistry 25:2841–2855.

Zdero C., Bohlmann F., King R. M., Robinson H. (1986b) a-Isocedrene derivatives, 5-methyl coumarins and other constituents from the subtribe Nassauviinae of the Compositae. Phytochemistry 25:2873–2882.

Zdero C., Bohlmann F., King R. M., Robinson H. (1987b) Pyrone derivatives from *Podolepis hieracioides* and sesquiterpene acids from *Cassinia longifolia*. Phytochemistry 26:187–190.

Zdero C., Bohlmann F., King R. M., Robinson H. (1989b) Sesquiterpene lactones and other constituents from *Helipterum* species. Phytochemistry 28:517–526.

Zdero C., Bohlmann F., Müller M. (1987a) Sesquiterpene lactones and other constituents from *Eriocephalus* species. Phytochemistry 26:2763–2775.

Zdero C., Bohlmann F., Mungai G. M. (1991a) Glaucolides and hirsutinolides from African *Vernonia* species. Phytochemistry 30:2653–2654.

Zdero C., Bohlmann F., Mungai G. M. (1991b) Carvoacetone derivatives and other constituents from representatives of the *Sphaeranthus* group. Phytochemistry 30:3297–3303.

Zdero C., Bohlmann F., Niemeyer H. M. (1988a) Diterpenes and 5-methyl coumarin derivatives from *Gypothamnium pinifolium* and *Plazia daphnoides*. Phytochemistry 27:2953–2959.

Zdero C., Bohlmann F., Niemeyer H. M. (1990c) A heliangolide, 3-hydroxyumbelliferone derivatives and diterpenes from *Bahia ambrosioides*. Phytochemistry 29:205–208.

Zdero C., Bohlmann F., Niemeyer H. M. (1990d) Sesquiterpene lactones from *Perityle emoryi*. Phytochemistry 29:891–894.

Zdero C., Bohlmann F., Niemeyer H. M. (1990e) *seco*-Labdanes and other constituents from *Ophryosporus floribundus*. Phytochemistry 29:3247–3253.

Zdero C., Bohlmann F., Niemeyer H. M. (1991c) An unusual dimeric sesquiterpene and other constituents from Chilean *Baccharis* species. Phytochemistry 30:1597–1601.

Zdero C., Bohlmann F., Niemeyer H. M. (1991d) Friedolabdanes and other constituents from Chilean *Haplopappus remyanus*. Phytochemistry 30:3669–3677.

Zdero C., Bohlmann F., Solomon J. (1988b) Further 5-methylcoumarin derivatives from *Mutisia orbignyana*. Phytochemistry 27:891–897.

Zdero C., Bohlmann F., Solomon J. C., King R. M., Robinson H. (1989c) *ent*-Clerodanes and other constituents from Bolivian *Baccharis* species. Phytochemistry 28:531–542.

Zdero C., Jakupovic J., Bohlmann F. (1990f) Diterpenes and other constituents from *Pteronia incana*. Phytochemistry 29:1231–1245.

Zdero C., Lehmann L., Bohlmann F. (1991f) Chemotaxonomy of *Athanasia* and related genera. Phytochemistry 30:1161–1163.

Zemtsova G. N., Molchanova L. P. (1978) Flavonoids and phenolic acids of *Pyrethrum corymbosum*. Chem. Nat. Cmpds. 14:337 [400].

Zemtsova G. N., Molchanova L. P. (1979a) Polyphenols of *Pyrethrum roseum*. Chem. Nat. Cmpds. 15:504 [577].

Zemtsova G. N., Molchanova L. P. (1979b) Flavonoids of *Xeranthemum annuum* L. Chem. Nat. Cmpds. 15:762 [856].

Zhang R., Zhang Y., Wang J., Lin Y., Shang B. (1988) Isolation and identification of flavonoids from shortscape fleabane *Erigeron breviscapus*. Zhongcaoyao 19:199–200.

Zhang X., Bremer K. (1993) A cladistic analysis of the tribe Astereae (Asteraceae) with notes on their evolution and subtribal classification. Plant Syst. Evol. 184:259–283.

Zheng G. Q. (1994) Cytotoxic terpenoids and flavonoids from *Artemisia annua*. Planta Medica 60:54–57.

Zheng S., Chen H., Shen X. (1990a) Chemical components of *Centaurea repens*. Gaodeng Xueziao Huaxue Xuebao 11:828.

Zheng S., Chen H., Shen X. (1990b) Study on flavonoids from *Saussurea gossypiphora* D. Don. Xibei Shifan Daxue Xuebao, Ziran Kexueban 4:42–45.

Zheng S., Yu J., Shen X. (1991) Chemical constituents of *Saussurrea gossypiphora*. Gaodeng Xueziao Huaxue Xuebao 12:1613–1616.

Zheng W. F., Tan R. X., Yang L., Liu Z. L. (1996) Two flavones from *Artemisia giraldii* and their antimicrobial activities. Planta Medica 62:160–162.

Zielinska M., Gill S. (1974) Flavonoid compounds of *Gaillardia aristata* Pursh. Rocz. Chem. 48:1111–1112.

Zielinska-Stasiek M., Gill S. (1977) Flavonoids in aboveground parts of *Gaillardia aristata* Pursh. Rocz. Chem. 51:921–927.
Zirkle C. (1935) *The Beginnings of Plant Hybridization*. Philadelphia: Univ. Pennsylvania Press.
Zohary M. (1982) *Plants of the Bible*. Cambridge: Cambridge Univ. Press.
Zurabishvili T. S. (1974) Quercimeritrin from *Achillea neilreichii*. Chem. Nat. Cmpds. 10:264 [253].

Common Names of Flavonoids Used in This Book and Their Equivalents

Common name	Derived common name	Formalized name
acacetin	apigenin 4'-methyl ether	5,7-dihydroxy-4'-methoxyflavone
acerosin		5,7,3'-trihydroxy-6,8,4'-trimethoxyflavone
alnusin, 3-methyl		5,7-dihydroxy-3,6-dimethoxyflavone
amentoflavone	3',8-biapigenin	
apigenin		5,7,4'-trihydroxyflavone
apigeninidin		5,7,4'-trihydroxyflavylium
apiin	apigenin 7-apiosylglucoside	5,7,4'-trihydroxyflavone 7-apiosylglucoside
arcapillin		5,2',4'-trihydroxy-6,7,5'-trimethoxyflavone
aromadendrin	dihydrokaempferol	3,5,7,4'-tetrahydroxyflavanone
artemetin	quercetagetin -3,6,7,3',4'-pentamethyl ether	5-hydroxy-3,6,7,3',4'-pentamethoxyflavone
aureusidin		4,6,3',4'-tetrahydroxyaurone
axillarin	quercetagetin 3,6-dimethyl ether	5,7,3',4'-tetrahydroxy-3,6-dimethoxyflavone
azaleatin	quercetin 5-methyl ether	3,7,3',4'-tetrahydroxy-5-methoxyflavone
baicalein	6-hydroxychrysin	5,6,7-trihydroxyflavone
betuletol	6-methoxykaempferol 4'-methyl ether	3,5,7-trihydroxy-6,4'-dimethoxyflavone
bonanzin	quercetagetin 3,6,3',4'-tetramethyl ether	5,7-dihydroxy-3,6,3',4'-tetramethoxyflavone
bracteatin		4,6,3',4',5'-pentahydroxyaurone
bractein	bracteatin 4-glucoside	4,6,3',4',5'-pentahydroxyaurone 4-glucoside
brickellin		5,2'-dihydroxy-3,6,7,4',5'-pentamethoxyflavone
butein		2',4',3,4-tetrahydroxychalcone
butin	5-deoxyeriodictyol	7,3',4'-trihydroxyflavanone
carlinoside	6-*C*-glucosyl-8-*C*-arabinosylluteolin	
carthamin		2',3',4',6',4-pentahydroxychalcone 2'-glucoside
casticin	quercetagetin 3,6,7,4'-tetramethyl ether	5,3'-dihydroxy-3,6,7,4'-tetramethoxyflavone
catechin		3,5,7,3',4'-pentahydroxyflavan
centaureidin	quercetagetin 3,6,4'-trimethyl ether	5,7,3'-trihydroxy-3,6,4'-trimethoxyflavone
cernuoside	aureusidin 4-glucoside	4,6,3',4'-tetrahydroxyaurone 4-glucoside

Common name	Derived common name	Formalized name
chalcone (per se)		*trans*-cinnamoylbenzene or phenylvinyl ketone
chalcononaringenin		2',4',6',4-tetrahydroxychalcone
chrysin		5,7-dihydroxyflavone
chrysoeriol	luteolin 3'-methyl ether	5,7,4'-trihydroxy-3'-methoxyflavone
chrysosplenetin	quercetagetin -3,6,7,3'-tetramethyl ether	5,4'-dihydroxy-3,6,7,3'-tetramethoxyflavone
chrysosplenol-C	quercetagetin 3,7,3'-trimethyl ether	5,6,4'-trihydroxy-3,7,3'-trimethoxyflavone
chrysosplenol-D	quercetagetin 3,6,7-trimethyl ether	5,3',4'-trihydroxy-3,6,7-trimethoxyflavone
cirsilineol		5,4'-dihydroxy-6,7,3'-trimethoxyflavone
cirsiliol		5,3',4'-trihydroxy-6,7-dimethoxyflavone
cirsimarin	cirsimaritin 4'-glucoside	5,4'-dihydroxy-6,7-dimethoxyflavone 4'-glucoside
cirsimaritin	scutellarein-6,7-dimethyl ether	5,4'-dihydroxy-6,7-dimethoxyflavone
coreopsin	butein 4'-glucoside	3',4',3,4-tetrahydroxychalcone 4'-glucoside
cytisoside	8-*C*-gucosylacacetin	5,7-dihydroxy-4'-methoxy-6-*C*-glucosylflavone
desmethylnobiletin		5-hydroxy-6,7,8,3',4'-pentamethoxyflavone
diosmetin	luteolin 4'-methyl ether	5,7,3'-trihydroxy-4'-methoxyflavone
echinacin	apigenin 7-glucoside 6''-*p*-coumarate	
embigenin	isovitexin 7,4'-dimethyl ether	5-hydroxy-7,4'-dimethoxy-6-*C*-glucosylflavone
eriodictyol		5,7,3',4'-tetrahydroxyflavone
ermanin	kaempferol 3,4'-dimethyl ether	5,7-dihydroxy-3,4'-dimethoxyflavone
eupafolin = nepetin		
eupalitin	6-methoxykaempferol 7-methyl ether	3,5,4'-trihydroxy-6,7-dimethoxyflavone
eupatilin		5,7-dihydroxy-6,3',4'-trimethoxyflavone
eupatin	quercetagetin 6,7,4'-trimethyl ether	3,5,3'-trihydroxy-6,7,4'-trimethoxyflavone
eupatolitin	quercetagetin 6,7-dimethyl ether	3,5,3',4'-tetrahydroxy-6,7-dimethoxyflavone
fastigenin		5,4'-dihydroxy-6,7,3'-trimethoxyflavone

Common name	Derived common name	Formalized name
fisetin	5-deoxyquercetin	3,7,3',4'-tetrahydroxyflavone
formononetin		7-hydroxy-4'-methoxyisoflavone
galangin		3,5,7-trihydroxyflavone
gardenin-B		5-hydroxy-6,7,8,4'-tetramethoxyflavone
gardenin-D		5,3'-dihydroxy-6,7,8,4'-tetramethoxyflavone
genkwanin	apigenin 7-methyl ether	5,4'-dihydroxy-7-methoxyflavone
gesnerin	apigeninidin glucoside	5,7,4'-trihydroxyflavylium glucoside
gnaphalin		5,7-dihydroxy-3,8-dimethoxyflavone
gossypetin	8-hydroxyquercetin	3,5,7,8,3',4'-hexahydroxyflavone
grantioidin		5-hydroxy-3,6,7,2',5'-pentamethoxyflavone
grantionin		7-hydroxy-6,3',5'-trimethoxyflavone
heliannone-A		2',4-dihydroxy-3',4'-dimethoxychalcone
heliannone-B		7,8-dimethoxy-4'-hydroxyflavanone
helmon		4,6,7,4'-tetrahydroxyaurone
herbacetin	8-hydroxykaempferol	3,5,7,8,4'-pentahydroxyflavone
hesperidin	hesperetin 7-glucoside	5,7,3'-trihydroxy-4'-methoxyflavanone 7-glucoside
hesperetin	eriodictyol 4'-methyl ether	5,7,3'-trihydroxy-4'-methoxyflavanone
hispidulin	6-methoxyapigenin	5,7,4'-trihydroxy-6-methoxyflavone
homoeriodictyol	eriodictyol 3'-methyl ether	5,7,4'-trihydroxy-3'-methoxyflavanone
hymenoxin		5,7-dihydroxy-6,8,3,'4'-tetramethoxyflavone
inucrithmin		3,7,3',4'-tetrahydroxy-6,5'-dimethoxyflavone
isoëtin	2'-hydroxyluteolin	5,7,2',4',5'-pentahydroxyflavone
isoliquiritigenin		2',4',4-trihydroxychalcone
isoökanin		6,7,3',4'-tetrahydroxyflavanone 7-glucoside
isoörientin	6-*C*-glucosylluteolin	
isoörientin	7,3'-dimethyl ether	
isorhamnetin	quercetin 3'-methyl ether	3,5,7,4'-tetrahydroxy-3'-methoxyflavone
isosakuranetin	naringenin 4'-methyl ether	5,7-dihydroxy-4'-methoxyflavanone

Common name	Derived common name	Formalized name
isosalipurposide	chalcononaringenin 2'-glucoside	2',4',6',4-tetrahydroxychalcone 2'-glucoside
isoscutellarein	8-hydroxyapigenin	5,7,8,4'-tetrahydroxyflavone
isoschaftoside	6-*C*-arabinosyl-8-*C*-glucosylapigenin	
isoscoparin	6-*C*-glucosylchrysoeriol	
isovitexin	6-*C*-glucosylapigenin	
jaceidin	quercetagetin 3,6,3'-trimethyl ether	5,7,4'-trihydroxy-3,6,3'-trimethoxyflavone
jacein	jaceosidin 7-glucoside	5,7,4'-trihydroxy-6,3'-dimethoxyflavone 7-glucoside
jaceosidin	6-methoxykaempferol 3'-methyl ether	5,7,4'-trihydroxy-6,3'-dimethoxyflavone
kaempferide	kaempferol 4'-methyl ether	3,5,7-trihydroxy-4'-methoxyflavone
kaempferol		3,5,7,4'-tetrahydroxyflavone
kulkulanine-B	3'-methoxyisoliquiritigenin	2',4',4-trihydroxy-3'-methoxychalcone
kumatakenin	kaempferol 3,7-dimethyl ether	5,4'-dihydroxy-3,7-dimethoxyflavone
laciniatin	quercetagetin 6,4'-dimethyl ether	3,5,7,3'-tetrahydroxy-6,4'-dimethoxyflavone
lanceolatin		2',4',3,4-tetrahydroxy-3'-methoxychalcone
lanceolin	lanceolatin 4'-glucoside	2',4',3,4-tetrahydroxy-3'-methoxychalcone 4'-glc
leptosidin	maritimetin 7-methyl ether	6,3',4'-trihydroxy-7-methoxyaurone
leptosin	leptosidin 6-glucoside	6,3',4'-trihydroxy-7-methoxyaurone 6-glucoside
leucocyanidin		3,4,5,7,3',4'-hexahydroxyflavan
leucopelargonidin		3,4,5,7,4'-pentahydroxyflavan
liquiritigenin		7,4'-dihydroxyflavanone
luteolin		5,7,3',4'-tetrahydroxyflavone
malvidin		3,5,7,4'-tetrahydroxy-3',5'-dimethoxyflavylium
marein	okanin 4'-glucoside	2',3',4',3,4-pentahydroxychalcone 4'-glucoside
maritimein	maritimetin 6-glucoside	6,7,3',4'-tetrahydroxyaurone 6-glucoside
maritimetin		6,7,3',4'-tetrahydroxyaurone
mikanin		3,5-dihydroxy-6,7,4'-trimethoxyflavone
morelosin-IV		3,5,6-trihydroxy-3',5'-dimethoxyflavone
morin		3,5,7,2',4'-pentahydroxyflavone

Common name	Derived common name	Formalized name
myricetin		3,5,7,3',4',5'-hexahydroxyflavone
naringenin		5,7,4'-trihydroxyflavanone
nepetin	6-methoxyluteolin	5,7,3',4'-tetrahydroxy-6-methoxyflavone
nevadensin		5,7-dihydroxy-6,8,4'-trimethoxyflavone
okanin		2',3',4',3,4'-pentahydroxychalcone
ombuin	quercetin 7,3'-dimethyl ether	3,5,4'-trihydroxy-7,3'-dimethoxyflavone
ononin		7-hydroxy-4'-methoxyisoflavone 7-glucoside
onopordin	8-methoxyluteolin	5,7,3',4'-tetrahydroxy-8-methoxyflavone
orientin	8-*C*-glucosylluteolin	
orobol		5,7,3',4'-tetrahydroxyisoflavone
oroxylin	6-methoxychrysin	5,7-dihydroxy-6-methoxyflavone
oxyayanin-B	quercetagetin 3,7,4'-trimethyl ether	5,6,3'-trihydroxy- 3,7,4'-trimethoxyflavone
pachypodol	quercetin 3,7,3'-trimethyl ether	5,4'-dihydroxy-3,7,3'-trimethoxyflavone
patuletin	6-methoxyquercetin	3,5,7,3',4'-pentahydroxy-6-methoxyflavone
pectolinarigenin	scutellarein 6,4'-dimethyl ether	5,7-dihydroxy-6,4'-dimethoxyflavone
penduletin		5,4'-dihydroxy-3,6,7-trimethoxyflavone
persicogenin	eriodictyol 7,4'-dimethyl ether	5,3'-dihydroxy-7,4'-dimethoxyflavanone
pilloin	luteolin 7,4'-dimethyl ether	5,3'-dihydroxy-7,4'-dimethoxyflavone
pinobanksin		3,5,7-trihydroxyflavanone
pinocembrin		5,7-dihydroxyflavanone
pinostrobin		5-hydroxy-7-methoxyflavanone
quercetagetin	6-hydroxyquercetin	3,5,6,7,3',4'-hexahydroxyflavone
quercetin		3,5,7,3',4'-pentahydroxyflavone
retusin	quercetin 3,7,3',4'-tetramethyl ether	5-hydroxy-3,7,3',4'-tetramethoxyflavone
rhamnazin	quercetin 7,3'-dimethyl ether	3,5,4'-trihydroxy-7,3'-dimethoxyflavone
rhamnetin	quercetin 7-methyl ether	3,5,3',4'-tetrahydroxy-7-methoxyflavone
rhamnocitrin	kaempferol 7-methyl ether	3,5,4'-trihydroxy-7-methoxyflavone
riparin		2',3',4',4-tetrahydroxychalcone
sakuranetin	naringenin 7-methyl ether	5,4'-dihydroxy-7-methoxyflavanone

Common name	Derived common name	Formalized name
salvigenin	scutellarein 6,7,4'-trimethyl ether	5-hydroxy-6,7,4'-trimethoxyflavone
santal	orobol 7-methyl ether	5,3',4'-trihydroxy-7-methoxyisoflavone
santin	6-methoxykaempferol 3,4'-dimethyl ether	5,7-dihydroxy-3,6,4'-trimethoxyflavone
saponarin	isovitexin 7-glucoside	5,7,4'-trihydroxy-6-*C*-glucosylflavone 7-glucoside
scaposin		5,7,3'-trihydroxy-6,8,4',5'-tetramethoxyflavone
schaftoside	6-*C*-glucosyl-8-*C*-arabinosylapigenin (α-L)	
schaftoside, iso	6-*C*-arabinosyl-8-*C*-glucosylapigenin (α-L)	
schaftoside, neo	6-*C*-glucosyl-8-*C*-arabinosylapigenin (β-L)	
schaftoside, neoiso	6-*C*-arabinosyl-8-*C*-glucosylapigenin (β-L)	
scutellarein	6-hydroxyapigenin	5,6,7,4'-tetrahydroxyflavone
sorbifolin	scutellarein 7-methyl ether	5,6,4'-trihydroxy-7-methoxyflavone
spinacetin	quercetagetin 6,3'-dimethyl ether	3,5,7,4'-dihydroxy-6,3'-dimethoxyflavone
squamatin		5,4'-dihydroxy-3,7,8,3',5'-pentamethoxyflavone
stillopsidin		2',4',5',3,4-pentahydroxychalcone
stillopsin	stillopsidin 4'-glucoside	2',4',5',3,4-pentahydroxychalcone 4'-glucoside
sudachitin		5,7,4'-trihydroxy-6,8,3'-trimethoxyflavone
sulfurein	sulfuretin 6-glucoside	6,3',4'-trihydroxyaurone 6-glucoside
sulfuretin		6,3',4'-trihydroxyaurone
swertijaponin	orientin 7-methyl ether	
tamarixetin	quercetin 4'-methyl ether	3,5,7,3'-tetrahydroxy-4'-methoxyflavone
tambulin	herbacetin 7,8,4'-trimethyl ether	3,5-dihydroxy-7,8,4'-trimethoxyflavone
taxifolin	dihydroquercetin	3,5,7,3',4'-pentahydroxyflavanone
tectochrysin	chrysin 7-methyl ether	5-hydroxy-7-methoxyflavone
ternatin	gossypetin 3,7,8,3'-tetramethyl ether	5,4'-dihydroxy-3,7,8,3'-tetramethoxyflavone
tomentin	quercetagetin 3,7-dimethyl ether	5,6,3',4'-tetrahydroxy-3,7-dimethoxyflavone
tricin		5,7,4'-trihydroxy-3',5'-dimethoxyflavone

Common name	Derived common name	Formalized name
velutin	luteolin 7,3'-dimethyl ether	5,4'-dihydroxy-7,3'-dimethoxyflavone
veronicafolin	quercetagetin 6,7,3'-trimethyl ether	3,5,4'-trihydroxy-6,7,3'-trimethoxyflavone
vicenin-2	6,8-di-*C*-glucosylapigenin	
vitexin	8-*C*-glucosylapigenin	
xanthomicrol		5,4'-dihydroxy-6,7,8-trimethoxyflavone

Chemical Index

Page numbers set in bold type identify pages with structures for the corresponding compound. The term "error" identifies compounds whose structures were originally reported incorrectly; see accompanying text for corrected structures. Detailed citations for the widespread compounds apigenin, luteolin, kaempferol, and quercetin (*per se*) are not included in the index, except for several of their glycosidic forms. Synonyms are provided for many flavonoids, e.g., 5-hydroxy-3,6,7,3',4'-pentamethoxyflavone = quercetagetin 3,6,7,3',4'-pentamethyl ether = artemetin. Compounds are listed in order of their substitution patterns (with primes ignored) in nearly all instances. Exceptions should be obvious.

Subject Index

Page numbers followed by a "t" indicate tabulated information. Numbers in bold type indicate pages where corresponding structures appear.

Taxon Index

Page numbers followed by a "t" indicate that the taxon is listed in a table.

Springer-Verlag and the Environment

WE AT SPRINGER-VERLAG FIRMLY BELIEVE THAT AN international science publisher has a special obligation to the environment, and our corporate policies consistently reflect this conviction.

WE ALSO EXPECT OUR BUSINESS PARTNERS – PRINTERS, paper mills, packaging manufacturers, etc. – to commit themselves to using environmentally friendly materials and production processes.

THE PAPER IN THIS BOOK IS MADE FROM NO-CHLORINE pulp and is acid free, in conformance with international standards for paper permanency.

GPSR Compliance

The European Union's (EU) General Product Safety Regulation (GPSR) is a set of rules that requires consumer products to be safe and our obligations to ensure this.

If you have any concerns about our products, you can contact us on ProductSafety@springernature.com

In case Publisher is established outside the EU, the EU authorized representative is:

Springer Nature Customer Service Center GmbH
Europaplatz 3
69115 Heidelberg, Germany

Batch number: 10732200

Printed by Printforce, the Netherlands